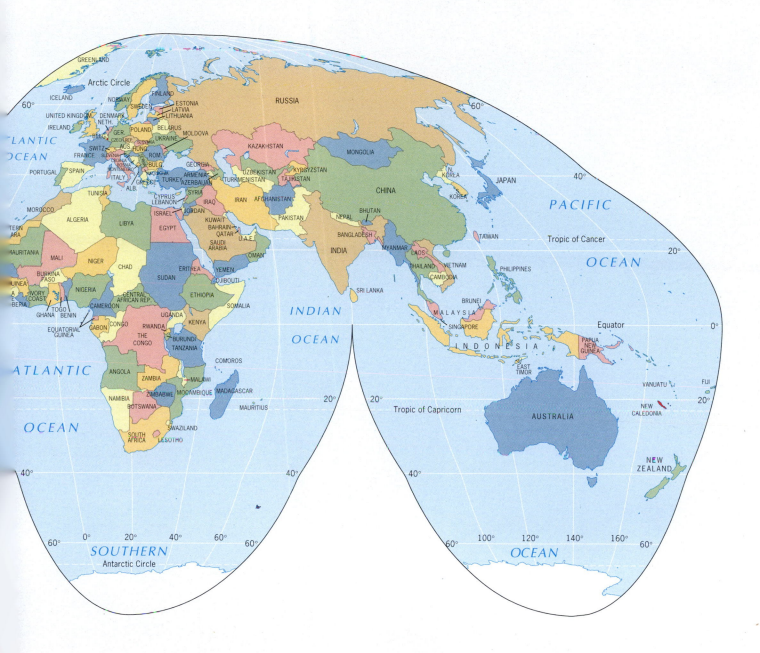

The Wiley Faculty Network

Where Faculty Connect

The Wiley Faculty Network is a faculty-to-faculty network promoting the effective use of technology to enrich the teaching experience. The Wiley Faculty Network facilitates the exchange of best practices, connects teachers with technology, and helps to enhance instructional efficiency and effectiveness. The network provides technology training and tutorials, including *Wiley PLUS* training, online seminars, peer-to-peer exchanges of experiences and ideas, personalized consulting, and sharing of resources.

Connect with a Colleague

Wiley Faculty Network mentors are faculty like you, from educational institutions around the country, who are passionate about enhancing instructional efficiency and effectiveness through best practices. You can engage a faculty mentor in an online conversation at **www.WhereFacultyConnect.com**

Participate in a Faculty-Led Online Seminar

The Wiley Faculty Network provides you with virtual seminars led by faculty using the latest teaching technologies. In these seminars, faculty share their knowledge and experiences on discipline-specific teaching and learning issues. All you need to participate in a virtual seminar is high-speed internet access and a phone line. To register for a seminar, go to **www.WhereFacultyConnect.com**

Connect with the Wiley Faculty Network

Web: **www.WhereFacultyConnect.com**
Phone: 1-866-4FACULTY

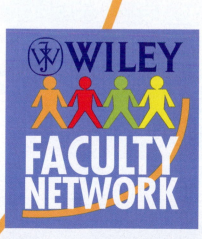

WILEY PLUS

www.wileyplus.com

Wiley is committed to making your entire *WileyPLUS* experience productive & enjoyable by providing the help, resources, and personal support you & your students need, when you need it. It's all here: www.wileyplus.com –

TECHNICAL SUPPORT:

- A fully searchable knowledge base of FAQs and help documentation, available 24/7
- Live chat with a trained member of our support staff during business hours
- A form to fill out and submit online to ask any question and get a quick response
- **Instructor-only** phone line during business hours: 1.877.586.0192

FACULTY-LED TRAINING THROUGH THE WILEY FACULTY NETWORK:

Register online: www.wherefacultyconnect.com

Connect with your colleagues in a complimentary virtual seminar, with a personal mentor in your field, or at a live workshop to share best practices for teaching with technology.

1ST DAY OF CLASS...AND BEYOND!

Resources You & Your Students Need to Get Started & Use *WileyPLUS* from the first day forward.

1st DAY OF CLASS AND BEYOND!

- 2-Minute Tutorials on how to set up & maintain your *WileyPLUS* course
- User guides, links to technical support & training options
- ***WileyPLUS for Dummies***: Instructors' quick reference guide to using *WileyPLUS*
- Student tutorials & instruction on how to register, buy, and use *WileyPLUS*

YOUR *WileyPLUS* ACCOUNT MANAGER:

Your personal *WileyPLUS* connection for any assistance you need!

WILEY PLUS QuickStart

SET UP YOUR *WileyPLUS* COURSE IN MINUTES!

Selected *WileyPLUS* courses with QuickStart contain pre-loaded assignments & presentations created by subject matter experts who are also experienced *WileyPLUS* users.

Interested? See and try WileyPLUS *in action!*
Details and Demo: www.wileyplus.com

THE WORLD TODAY

Concepts and Regions in Geography

To the memory of Theodorus J. M. van den Muysenberg,
career diplomat and observer of the world, 1923–2005.

FOURTH EDITION

THE WORLD TODAY

Concepts and Regions in Geography

H. J. de Blij

John A. Hannah Professor of Geography
Michigan State University

Peter O. Muller

Professor, Department of Geography and Regional Studies
University of Miami

Antoinette M. G. A. WinklerPrins

Associate Professor, Department of Geography
Michigan State University

John Wiley & Sons, Inc.

PUBLISHER	Jay O'Callaghan
EXECUTIVE EDITOR	Ryan Flahive
PROJECT EDITOR	Joan Kalkut
ASSISTANT EDITOR	Courtney Nelson
EDITORIAL ASSISTANT	Erin Grattan
SENIOR MEDIA EDITOR	Lynn Pearlman
MEDIA PROJECT MANAGER	Bridget O'Lavin
PRODUCTION EDITOR	Barbara Russiello
MARKETING MANAGER	Danielle Torio
SENIOR PHOTO EDITOR	Jennifer MacMillan
SENIOR DESIGNER	Kevin Murphy
INTERIOR DESIGN	Nancy Field
COVER DESIGN	David Levy
SENIOR ILUSTRATION EDITOR	Sigmund Malinowski
ELECTRONIC ILLUSTRATION	Mapping Specialists, Ltd.
FRONT COVER PHOTO	© Danny Lehman/© Corbis
BACK COVER PHOTO	© Jon Arnold Images/SUPERSTOCK

This book was set in 10.5/12.5 Times Roman by GGS Book Services PMG, Atlantic Highlands, NJ, and printed and bound by Quebecor World Dubuque. The cover was printed by Quebecor World Dubuque.

This book is printed on acid free paper. ∞

To order books or for customer service please, call 1-800-CALL WILEY (225-5945).

ISBN-13 978- 0-470-23713-7

Library of Congress Cataloging-in-Publication Data

de Blij, Harm J.
    The world today : concepts and regions in geography / H. J. de Blij, Peter O. Muller,
Antoinette M. G. A. WinklerPrins.—4th ed.
        p.   cm.
    Includes index.
    ISBN 978-0-470-23713-7 (pbk.)
    1. Geography--Textbooks.   I. Muller, Peter O.   II. WinklerPrins, Antoinette M. G. A.   III. Title.
G128.D394   2009
910—dc22

2008036133

Printed in the United States of America

10 9 8 7 6 5 4 3 2 1

PREFACE

READERS FAMILIAR WITH earlier editions of this book will readily note what is different about this one. This revised Fourth Edition of **The World Today** was prepared by the third author, Antoinette M. G. A. WinklerPrins, Associate Professor of Geography at Michigan State University, East Lansing.

We retained the new title introduced in the third edition, which signals that this book is about geographic ideas and concepts that have direct relevance to any understanding of what is happening in the world today. Our planet is in rapid transition—environmentally, economically, politically, and culturally—and many geographic factors influence the forces that are changing it. This book explains those factors and links them to the events you are witnessing.

Overall, the objective for this edition was to enhance the illustrations (including new time lines in Chapters 2 and 9), revise the maps to reflect current developments, and to add fresh photographs that make the book even more visually appealing. Many of the newly included photographs were taken by the new third author. Some maps from the previous edition were deleted but are still available on the book's website. Each individual chapter now features an attractive new map opener that focuses on the physical geography of the realm under discussion. The inviting chapter-opening layout continues to feature the concepts, ideas, and terms discussed in the text as well as the regions that are covered in each realm. The chapter openers highlight topics of current interest and relevance discussed in geographic context and display pertinent photographs.

A new boxed feature, *Environmentally Speaking*, appears in every chapter, highlighting a relevant environmental issue affecting each realm. Examples include air pollution in East Asia, hurricanes in Middle America, and deforestation in Southeast Asia. In addition, this new edition includes a popular feature from earlier editions: *From The Field Notes*, based on information recorded by the authors while in the regions that form the framework of this book. Many of the newly included photographs accompanied by *Field Notes* were taken by Antoinette WinklerPrins.

Numerous revisions update this edition of **The World Today**, involving such important topics as political change and economic development throughout the world; the growing importance of middle-income countries such as Brazil, Mexico, and South Africa; the rapidly increasing importance of China and India; and geopolitical competition in the Arctic.

One of the major assets of this book, the much-used table of global data by realm and country, has been moved to the back of the book for easier consultation and is now Appendix B. Although thoroughly updated, this book remains the same in one respect: its key objective is to link geography's perspectives to the problems and issues of today. Geographic knowledge by itself cannot solve the world's problems, but these problems cannot be effectively approached without it. In this respect, the importance of the geography course ahead is second to none.

Data Sources

Numerous print and Internet sources were consulted during the updating of the book. For all matters geographical, of course, we consult *The Annals of the Association of American Geographers, The Professional Geographer, The Geographical Review, The Journal of Geography,* and many other academic journals published regularly in North America—plus an array of similar periodicals published in English-speaking countries from Scotland to New Zealand.

All quantitative information was updated to the year of publication and checked rigorously. Hundreds of other modifications were made, many in response to readers' and reviewers' comments. New spellings of geographic names continue, and we pride ourselves in being a reliable source for current and correct usage.

The statistical data that constitute Appendix B are derived from numerous sources. As users of such data are aware, considerable inconsistency marks the reportage by various agencies, and it is often necessary to make informed decisions on contradictory information. For example, some sources still do not reflect the rapidly declining rates of population increase or life expectancies in AIDS-stricken African countries. Others list demographic averages without accounting for differences between males and females in this regard.

In formulating Appendix B we have used among our sources the United Nations, the Population

Reference Bureau, the World Bank, the Encyclopaedia Britannica *Books of the Year*, *The Economist* Intelligence Unit, the *Statesman's Year-Book*, and *The New York Times Almanac*. The urban population figures—which also entail major problems of reliability and comparability—are mainly drawn from the most recent database published by the United Nations' Population Division. For cities of less than 750,000, we developed our own estimates from a variety of other sources. At any rate, the urban population figures used here are estimates for 2009 and they represent *metropolitan-area totals* unless otherwise specified.

PEDAGOGY

We continue to devise ways to help students learn important geographic concepts and ideas, and to make sense of our complex and rapidly changing world. Continuing special features include the following:

Concepts, Ideas, and Terms.   Each chapter begins with a boxed sequential listing of the key geographic concepts, ideas, and terms that appear in the pages that follow. These are noted by numbers in the margins (e.g., **1**) that correspond to the introduction of each item in the text.

Two-Part Chapter Organization.   To help the reader to logically organize the material within chapters, we have broken the regional chapters into two distinct parts: first, *Defining the Realm* includes the general physiographic, historical, and human-geographic background common to the realm; the second section, *Regions of the Realm*, presents each of the distinctive regions within the realm.

List of Regions.   Also on the chapter-opening page, a list of the regions within the particular realm provides a preview and helps to organize the chapter.

Major Geographic Qualities.   Near the beginning of each chapter (except in the Introduction), we list, in boxed format, the major geographic qualities that best summarize that portion of the earth's surface.

What's Driving Geographic Change.   This second boxed feature appears near the break between the two sections of each chapter and focuses on the key forces shaping change in each part of the world.

Environmentally Speaking.   This stand-alone new box feature highlights an environmental issue relevant to each realm, with an accompanying photo.

From the Field Notes.   One of these appears in each chapter—a return of a popular feature from previous editions of the book. Most *Field Notes* are from our new author.

Table of Area and Demographic Data for the World's States.   This seven-page table (Appendix B; pp. A-2–A-7) lists the most important geographic and population data for each of the world's countries, updated to 2008. Definitions and brief discussions of each of the data categories reported is included with Appendix B.

Appendices and Glossary.   In addition to the data table that is Appendix B, we feature Appendix A that lists metric and customary units and their conversions. In the text we consistently use metric measurements with British units in parentheses. The Glossary and Index are available at the end of the book. Material that was included at the end of the book in previous editions is being made available on the book's website.

ANCILLARIES

A broad spectrum of print and electronic ancillaries are available to accompany *The World Today*. Additional information, including prices and ISBNs for ordering, can be obtained by contacting John Wiley & Sons, Inc. Go to: *www.wiley.com/college/deblij*

The Teaching and Learning Package

This Fourth Edition of *The World Today: Concepts and Regions in Geography* is supported by a comprehensive supplements package that includes an extensive selection of print, visual, and electronic materials.

Resources That Help Teachers Teach

People and Place **Lecture Launcher Videos.** Working closely with Wiley's Media Team, Erin Fouberg (Northern State University [SD]) has created a new collection of video lecture launchers that allow instructors to provide a visual context for key concepts, ideas, and terms that they plan to introduce during their lectures. These videos are available in a DVD format that is optimized for in-class presentation as well as in a streaming format that is available online. Streaming videos will also be made available to students in the context of *Wiley-PLUS* assignments that can be graded online and added to the *WileyPLUS* instructor gradebook.

WileyPLUS **Online Map Testing.**   We know that map tests are important to your course but also recognize that grading these map tests takes time out of your busy schedule. Therefore, we have built secure online tests to support every chapter of *The World Today* that can be customized to ask the questions that you want within a time frame that you set. With *WileyPLUS*, you are able to set all of the parameters for your tests. Moreover, your *WileyPLUS* grade book will automatically grade every test and provide you with the ability to measure and track each student's grade as well as uncover trends for the entire class. Your students will also have separate Map Quizzes for each chapter that allow them to study for your map tests at their own pace.

Online Geography On-Location Videos.   Because of their enduring popularity, we have digitized many of the videos from the original VHS series.

This rich collection of original and relevant footage was taken during H. J. de Blij's travels. These videos cover a wide range of themes and are integrated throughout each student's *WileyPLUS* eBook.

Google Earth Linking. The *WileyPlus eBook* links photos from the text to their actual location on the earth using Google Earth.

Geodiscoveries Interactions. A collection of activities, animations, and videos that allow students to explore key concepts in greater depth.

ConceptCaching Website *www.conceptcaching.com* ConceptCaching is a pedagogical technique that promotes student spatial awareness by relating specific concepts from their learning to their visual character and GPS coordinates. Photographs and GPS coordinates that demonstrate core concepts in geography are "cached" for viewing by core concept and by region. Additionally, all of the *From the Field Notes* in the text are linked to the ConceptCaching website.

Additional Videos and Podcasts. We have created a robust collection of streaming video resources and podcasts to support the Fourth Edition of *The World Today: Concepts and Regions in Geography*. Our streaming videos provide short lecture-launching video clips that can be used to introduce new topics, enhance your presentations, and stimulate classroom discussion.

***The World Today* 4e Instructor's Website** *www.wiley.com/college/deblij* This comprehensive website includes numerous resources to help you enhance your current presentations, create new presentations, and employ our pre-made PowerPoint presentations. These resources include:

- **Image Gallery.** We provide online electronic files for the line illustrations and maps in the text, which the instructor can customize for presenting in class (for example, in handouts, overhead transparencies, or PowerPoints).
- A complete collection of **PowerPoint presentations** is available in beautifully rendered, 4-color

format, which have been resized and edited for maximum effectiveness in large lecture halls.

- Prepared by Peter Muller, University of Miami, the *Test Bank* for the Fourth Edition of *The World Today* contains over 3000 test items including multiple-choice, true-false, and fill-in questions. It is distributed via the secure Instructor's website as electronic files, which can be saved into all major word processing programs.
- Also prepared by Peter Muller, **Clicker Questions** provide instructors with an opportunity for immediate feedback from their students. These clicker questions are available in a number of formats and also include system instructions and an instructor's manual.
- **Photo and Image Galleries** provide instructors with images from the text for use in their classrooms.

Wiley Faculty Network. This peer-to-peer network of faculty is ready to support your use of online course management tools and discipline-specific software-learning systems in the classroom. The WFN will help you apply innovative classroom techniques, implement software packages, tailor the technology experience to the needs of each individual class, and provide you with virtual training sessions led by faculty for faculty.

Course Management. Online course management assets are available to accompany the Fourth Edition of *The World Today*.

Resources That Help Students Learn

Student Companion Website *www.wiley.com/college/deblij* This easy-to-use and student-focused website helps reinforce and illustrate key concepts from the text. It also provides interactive media content that helps students prepare for exams and improve their grades. This website provides additional resources that complement the textbook and enhance students' understanding of geography:

- **Map Quizzes** help students master the place names that are crucial to their success in this

course. Three game-formatted place-name activities are provided for each chapter.

- **The Interactive Globe** allows the students to explore the earth in three dimensions. They are able to apply different textures that reveal spatial information about climate, soils, and demographics. Area and demographic data are available by clicking on each country or world realm.
- **Flashcards** offer an excellent way to drill and practice key concepts, ideas, and terms from the text.
- **Base Maps** are available for student study and practice.
- **Virtual Field Trips** and **Photo Gallery** allow students to "visit" many of the locations from their text.
- **Chapter Review Quizzes** provide immediate feedback to true-false, multiple-choice, and other short-answer questions.
- **Audio Pronunciation** is provided for over 2000 key words and place names from the text.
- **Annotated Web Links** allow students to further explore key topics on their own.
- **Area and Demographic Data** are provided for every country and world realm.

Student Study Guide. Text co-author Peter O. Muller and his geographer daughter, Elizabeth Muller Hames, have written a popular *Study Guide* to accompany the book that is packed with useful study and review tools. For each chapter in the textbook, the *Study Guide* gives students and faculty access to chapter objectives, content questions-and-answers, outline maps of each realm, sample tests, and more.

Wiley/National Geographic *College Atlas of the World*. Wiley is proud to offer the *College Atlas of the World* to your students through our exclusive partnership with the National Geographic Society. State-of-the-art cartographic technology plus award-winning design and content make this affordable, compact, yet comprehensive atlas the ultimate resource for every geography student. The past decade has seen an explosion of innovative and sophisticated digital mapping technologies—and the Society has been at the forefront of developing these powerful new tools to create the finest, most functional, and informative atlases available anywhere. This powerful resource can be purchased individually or at a significant discount when packaged with any Wiley textbook.

Social Geographies of the Modern World. This *free* 60-page supplement includes chapters that discuss:

- Global Disparities in Nutrition and Health;
- Geographies of Inequality: Race and Ethnicity;
- Gender Inequities in Geographic Perspective.

To arrange for your students to receive this *free* supplement with their copy of **The World Today**, please contact your local Wiley Sales Representative.

ACKNOWLEDGMENTS

Over the 38 years since the publication of the First Edition of *Geography: Realms, Regions, and Concepts* (and joined more recently by the appearance of this book), we have been fortunate to receive advice and assistance from literally hundreds of people. One of the rewards associated with the publication of a book of this kind is the steady stream of correspondence and other feedback it generates. Geographers, economists, political scientists, education specialists, and others have written us, often with fascinating enclosures. We make it a point to respond personally to every such letter, and our editors have communicated with many of our correspondents as well. Moreover, we have considered every suggestion made and many who wrote or transmitted their reactions through other channels will see their recommendations in print in this edition.

STUDENT RESPONSE

A major part of the correspondence we receive comes from student readers. We would like to take this opportunity to extend our deep appreciation to the several million students around the world who have studied from our books. In particular, we thank the students from more than 120 different colleges across the United States who took the time to send us their opinions. Students told us they found the maps and graphics attractive and functional. We have not only enhanced the map program with exhaustive updating but have added a number of new and updated maps to this Fourth Edition as well as making significant changes in many others. Generally, students have told us that they found the pedagogical devices quite useful. We have kept the study aids the students cited as effective: a boxed list of each chapter's key concepts, ideas, and terms (numbered for quick reference in both the box and text margins); a box summarizing each realm's major geographic qualities; and an extensive Glossary.

FACULTY FEEDBACK

In assembling the Fourth Edition, we are indebted to the following people for advising us on a number of matters:

CHRISTOPHER BADUREK, Appalachian State University
JILL (ALICE) BLACK, Missouri State University
BRYANT EVANS, Houston Community College
WILLIAM FLYNN, Oklahoma State University
HARI GARBHARRAN, Middle Tennessee State University
TRUMAN HARTSHORN, Georgia State University
UWE KACKSTAETTER, Front Range Community College, Westminster
CUB KAHN, Oregon State University
J. MIGUEL KANAI, University of Miami
CHRIS MAYDA, Eastern Michigan University
ROSANN POLTRONE, Arapahoe Community College
JOEL QUAM, College of DuPage
RHONDA REAGAN, Blinn College
PAUL ROLLINSON, Missouri State University
KATHLEEN SCHROEDER, Appalachian State University
DMITRII SIDOROV, California State University, Long Beach

In addition, several faculty colleagues from around the world assisted us with earlier editions, and their contributions continue to grace the pages of this book. Among them are:

JAMES P. ALLEN, California State University, Northridge
STEPHEN S. BIRDSALL, University of North Carolina
J. DOUGLAS EYRE, University of North Carolina
FANG YONG-MING, Shanghai, China
EDWARD J. FERNALD, Florida State University
RAY HENKEL, Arizona State University
RICHARD C. JONES, University of Texas at San Antonio
GIL LATZ, Portland State University (Oregon)
IAN MACLACHLAN, University of Lethbridge (Alberta)
MELINDA S. MEADE, University of North Carolina
HENRY N. MICHAEL, late of Temple University (Pennsylvania)
CLIFTON W. PANNELL, University of Georgia
J. R. VICTOR PRESCOTT, University of Melbourne (Australia)
JOHN D. STEPHENS, University of Washington
CANUTE VANDER MEER, University of Vermont

Faculty members from a large number of North American colleges and universities continue to supply us with vital feedback and much-appreciated advice. Our publishers arranged several feedback sessions, and we are most grateful to the following professors for showing us where the text could be strengthened and made more precise:

MARTIN ARFORD, Saginaw Valley State University (Michigan)
DONNA ARKOWSKI, Pikes Peak Community College (Colorado)
GREG ATKINSON, Tarleton State University (Texas)
DENIS BEKAERT, Middle Tennessee State University
THOMAS L. BELL, University of Tennessee
DONALD J. BERG, South Dakota State University
KATHLEEN BRADEN, Seattle Pacific University
DAVID COCHRAN, University of Southern Mississippi
DEBORAH CORCORAN, Southwest Missouri State University
MARCELO CRUZ, University of Wisconsin at Green Bay
WILLIAM V. DAVIDSON, Louisiana State University
LARRY SCOTT DEANER, Kansas State University
JASON DITTMER, Georgia Southern University
JAMES DOERNER, University of Northern Colorado

STEVEN DRIEVER, University of Missouri-Kansas City

ELIZABETH DUDLEY-MURPHY, University of Utah

DENNIS EHRHARDT, University of Louisiana-Lafayette

WILLIAM FORBES, Stephen F. Austin State University (Texas)

BILL FOREMAN, Oklahoma City Community College

ERIC FOURNIER, Samford University (Alabama)

GARY A. FULLER, University of Hawai'i

RANDY GABRYS ALEXSON, University of Wisconsin-Superior

WILLIAM GARBARINO, Community College of Allegheny County (Pennsylvania)

JON GOSS, University of Hawai'i

RICHARD J. GRANT, University of Miami

MARGARET M. GRIPSHOVER, University of Tennessee

SARA HARRIS, Neosho County Community College (Kansas)

SHIRLENA HUANG, National University of Singapore

INGRID JOHNSON, Towson State University (Maryland)

KRIS JONES, Saddleback College (California)

THOMAS KARWOSKI, Anne Arundel Community College (Maryland)

ROBERT KERR, University of Central Oklahoma

ERIC KEYS, University of Florida

JACK KINWORTHY, Concordia University-Nebraska

CHRISTOPHER LAINGEN, Kansas State University

UNNA LASSITER, Stephen F. Austin State University (Texas)

RICHARD LISICHENKO, Fort Hays State University (Kansas)

CATHERINE LOCKWOOD, Chadron State College (Nebraska)

GEORGE LONBERGER, Georgia Perimeter College

DALTON E. MILLER, Mississippi State University

ERNANDO F. MINGHINE, Wayne State University (Michigan)

VERONICA MORMINO, Harper College (Illinois)

TOM MUELLER, California University (Pennsylvania)

IRENE NAESSE, Orange Coast College (California)

JAN NIJMAN, University of Miami

VALIANT C. NORMAN, Lexington Community College (Kentucky)

RICHARD OLMO, University of Guam

PAI YUNG-FENG, New York City

J. L. PASZTOR, Delta College (Michigan)

IWONA PETRUCZYNIK, Mercyhurst College (Pennsylvania)

PAUL E. PHILLIPS, Fort Hays State University (Kansas)

ROSANN POLTRONE, Arapahoe Community College (Colorado)

RINKU ROY CHOWDHURY, Indiana University

JEFF POPKE, East Carolina University

DAVID PRIVETTE, Central Piedmont Community College (North Carolina)

RHONDA REAGAN, Blinn College (Texas)

A. L. RYDANT, Keene State College (New Hampshire)

NANCY SHIRLEY, Southern Connecticut State University

DEAN SINCLAIR, Northwestern State University (Louisiana)

RICHARD SLEASE, Oakland, North Carolina

SUSAN SLOWEY, Blinn College (Texas)

DEAN B. STONE, Scott Community College (Iowa)

JAMIE STRICKLAND, University of North Carolina at Charlotte

RUTHINE TIDWELL, Florida Community College at Jacksonville

IRINA VAKULENKO, Collin County Community College (Texas)

CATHY WEIDMAN, Austin, Texas

KIRK WHITE, York College of Pennsylvania

THOMAS WHITMORE, University of North Carolina, Chapel Hill

KEITH YEARMAN, College of DuPage (Illinois)

LAURA ZEEMAN, Red Rocks Community College (Colorado)

We also received input from a much wider circle of academic geographers. The list that follows is merely representative of a group of colleagues across North America to whom we are grateful for taking the time to share their thoughts and opinions with us:

MEL AAMODT, California State University-Stanislaus

WILLIAM V. ACKERMAN, Ohio State University

W. FRANK AINSLEY, University of North Carolina, Wilmington

SIAW AKWAWUA, University of Northern Colorado

DONALD P. ALBERT, Sam Houston State University

TONI ALEXANDER, Louisiana State University

R. GABRYS ALEXSON, University of Wisconsin-Superior

KHALED MD. ALI, Geologist, Dhaka, Bangladesh

NIGEL ALLAN, University of California-Davis

JOHN L. ALLEN, University of Wyoming

KRISTIN J. ALVAREZ, Keene State College (New Hampshire)

DAVID L. ANDERSON, Louisiana State University, Shreveport

KEAN ANDERSON, Freed-Hardeman University

JEFF ARNOLD, Southwestern Illinois College

JERRY R. ASCHERMANN, Missouri Western State College

JOSEPH M. ASHLEY, Montana State University

PATRICK ASHWOOD, Hawkeye Community College

THEODORE P. AUFDEMBERGE, Concordia College (Michigan)

JAIME M. AVILA, Sacramento City College

EDWARD BABIN, University of South Carolina-Spartanburg

ROBERT BAERENT, Randolph-Macon College

MARVIN W. BAKER, University of Oklahoma

GOURI BANERJEE, Boston University (Massachusetts)

MICHELE BARNABY, Pittsburg State University (Kansas)

J. HENRY BARTON, Thiel College (Pennsylvania)

CATHY BARTSCH, Temple University (Pennsylvania)

STEVEN BASS, Paradise Valley Community College (Arizona)

THOMAS F. BAUCOM, Jacksonville State University (Alabama)

KLAUS J. BAYR, Keene State College

DENIS A. BEKAERT, Middle Tennessee State University

JAMES BELL, Linn Benton Community College (Oregon)

KEITH M. BELL, Volunteer State Community College (Tennessee)

WILLIAM H. BERENTSEN, University of Connecticut

DONALD J. BERG, South Dakota State University

ROYAL BERGLEE, Morehead State University (Kentucky)

RIVA BERLEANT-SCHILLER, University of Connecticut

LEE LUCAS BERMAN, Southern Connecticut State University

RANDY BERTOLAS, Wayne State College

THOMAS BITNER, University of Wisconsin-Marshfield/Wood County

WARREN BLAND, California State University, Northridge

DAVIS BLEVINS, Huntington College (Alabama)

HUBERTUS BLOEMER, Ohio University

S. BO JUNG, Bellevue College (Nebraska)

R. DENISE BLANCHARD-BOEHM, Texas State University

MARTHA BONTE, Clinton Community College (Idaho)

GEORGE R. BOTJER, University of Tampa (Florida)

R. LYNN BRADLEY, Belleville Area College (Illinois)

KEN BREHOB, Elmhurst College, Illinois

JAMES A. BREY, University of Wisconsin-Fox Valley

ROBERT BRINSON, Santa Fe Community College (Florida)

REUBEN H. BROOKS, Tennessee State University

PHILLIP BROUGHTON, Arizona Western College

LARRY BROWN, Ohio State University

LAWRENCE A. BROWN, Troy State-Dothan (Alabama)

ROBERT N. BROWN, Delta State University (Mississippi)

STANLEY D. BRUNN, University of Kentucky

RANDALL L. BUCHMAN, Defiance College (Ohio)

MICHAELE ANN BUELL, Northwest Arkansas Community College

DANIEL BUYNE, South Plains College

MICHAEL BUSBY, Murray State University (Kentucky)

MARY CAMERON, Slippery Rock University (Pennsylvania)

MICHAEL CAMILLE, University of Louisiana at Monroe

MARY CARAVELIS, Barry University (Florida)

DIANA CASEY, Muskegon Community College (Michigan)

DIANN CASTEEL, Tusculum College (Tennessee)

BILL CHAPPELL, Keystone College

MICHAEL SEAN CHENOWETH, University of Wisconsin-Milwaukee

STANLEY CLARK, California State University, Bakersfield

JOHN E. COFFMAN, University of Houston (Texas)

DWAYNE COLE, Cornerstone University (Michigan)

JERRY COLEMAN, Mississippi Gulf Coast Community College

JONATHAN C. COMER, Oklahoma State University

BARBARA CONNELLY, Westchester Community College (New York)

FRED CONNINGTON, Southern Wesleyan University

WILLIS M. CONOVER, University of Scranton (Pennsylvania)

OMAR CONRAD, Maple Woods Community College (Missouri)

ALLAN D. COOPER, Otterbein College

WILLIAM COUCH, University of Alabama, Huntsville

BARBARA CRAGG, Aquinas College (Michigan)

GEORGES G. CRAVINS, University of Wisconsin-La Crosse

ELLEN K. CROMLEY, University of Connecticut

JOHN A. CROSS, University of Wisconsin-Osh Kosh

SHANNON CRUM, University of Texas at San Antonio

WILLIAM CURRAN, South Suburban College (Illinois)

KEVIN M. CURTIN, University of Texas at San Antonio

JOSE A. DA CRUZ, Ozarks Technical Community College

ARMANDO DA SILVA, Towson State University (Maryland)

MARK DAMICO, Green Mountain College (Vermont)

DAVID D. DANIELS, Central Missouri State University

RUDOLPH L. DANIELS, Morningside College (Iowa)

SATISH K. DAVGUN, Bemidji State University (Minnesota)

WILLIAM V. DAVIDSON, Louisiana State University

CHARLES DAVIS, Mississippi Gulf Coast Community College

JAMES DAVIS, Illinois College

JAMES L. DAVIS, Western Kentucky University

PEGGY E. DAVIS, Pikeville College

ANN DEAKIN, State University of New York, College at Fredonia

KEITH DEBBAGE, University of North Carolina at Greensboro

MOLLY DEBYSINGH, California State University, Long Beach

DENNIS K. DEDRICK, Georgetown College (Kentucky)

JOEL DEICHMANN, Bentley College

STANFORD DEMARS, Rhode Island College

TOM DESULIS, Spoon River College

THOMAS DIMICELLI, William Paterson College (New Jersey)

MARY DOBBS, Highland Community College, Wamego

SCOTT DOBLER, Western Kentucky University

D. F. DOEPPERS, University of Wisconsin-Madison

JAMES DOERNER, University of Northern Colorado

ANN DOOLEN, Lincoln College (Illinois)

STEVEN DRIEVER, University of Missouri-Kansas City

WILLIAM DRUEN, Western Kentucky University

ALASDAIR DRYSDALE, University of New Hampshire

KEITH A. DUCOTE, Cabrillo Community College (California)

ELIZABETH DUDLEY-MURPHY, University of Utah

WALTER N. DUFFET, University of Arizona

MIKE DUNNING, University of Alaska Southeast

CHRISTINA DUNPHY, Champlain College (Vermont)

ANTHONY DZIK, Shawnee State University (Kansas)

DENNIS EDGELL, Firelands BGSU (Ohio)

RUTH M. EDIGER, Seattle Pacific University

JAMES H. EDMONSON, Union University (Tennessee)

M. H. EDNEY, State University of New York-Binghamton

HAROLD M. ELLIOTT, Weber State University (Utah)

JAMES ELSNES, Western State College

ROBERT W. EVANS, Fresno City College (California)

EVERSON DICKEN, California State University, San Bernardino

DINO FIABANE, Community College of Philadelphia

G. A. FINCHUM, Milligan College (Tennessee)

IRA FOGEL, Foothill College (California)

RICHARD FOLEY, Cumberland College

ROBERT G. FOOTE, Wayne State College (Nebraska)

RONALD FORESTA, University of Tennessee

ELLEN J. FOSTER, Texas State University

G. S. FREEDOM, McNeese State University (Louisiana)

EDWARD T. FREELS, Carson-Newman College

PHILIP FRIEND, Inver Hills Community College

JAMES FRYMAN, University of Northern Iowa

OWEN FURUSETH, University of North Carolina at Charlotte

RICHARD FUSCH, Ohio Wesleyan University

GARY GAILE, University of Colorado-Boulder

EVELYN GALLEGOS, Eastern Michigan University & Schoolcraft College

GAIL GARBRANDT, Mount Union College, University of Akron

RICHARD GARRETT, Marymount Manhattan College

JERRY GERLACH, Winona State University (Minnesota)

MARK GISMONDI, Northwest Nazarene University

LORNE E. GLAIM, Pacific Union College (California)

SHARLEEN GONZALEZ, Baker College (Michigan)

DANIEL B. GOOD, Georgia Southern University

GARY C. GOODWIN, Suffolk Community College (New York)

S. GOPAL, Boston University (Massachusetts)

MARVIN GORDON, Lake Forest Graduate School of Business

ROBERT GOULD, Morehead State University (Kentucky)

MARY GRAHAM, York College of Pennsylvania

GORDON GRANT, Texas A&M University

PAUL GRAY, Arkansas Tech University

DONALD GREEN, Baylor University (Texas)

GARY M. GREEN, University of North Alabama

STANLEY C. GREEN, Laredo State University (Texas)

RAYMOND GREENE, Western Illinois University

MARK GREER, Laramie County Community College (Wyoming)

WALT GUTTINGER, Flagler College (Florida)

JOHN E. GYGAX, Wilkes Community College

LEE ANN HAGAN, College of Southern Idaho

RON HAGELMAN, University of New Orleans

W. GREGORY HAGER, Northwestern Connecticut Community College

RUTH F. HALE, University of Wisconsin-River Falls

JOHN W. HALL, Louisiana State University-Shreveport

PETER L. HALVORSON, University of Connecticut

DAVID HANSEN, Pennsylvania State University, Harrisburg & Schuylkill

MERVIN HANSON, Willmar Community College (Minnesota)

ROBERT C. HARDING, Lynchburg College (Virginia)

MICHELLE D. HARRIS, Bob Jones University (South Carolina)

SCOTT HARRIS, CGCS Drury University

ROBERT J. HARTIG, Fort Valley State College (Georgia)

SUZANNA HARTLEY, Shelton State Community College

TRUMAN A. HARTSHORN, Georgia State University

CAROL HAZARD, Meredith College

HARLOW Z. HEAD, Barton College

DOUG HEFFINGTON, Middle Tennessee State University

JAMES G. HEIDT, University of Wisconsin Center-Sheboygan

CATHERINE HELGELAND, University of Wisconsin-Manitowoc

NORMA HENDRIX, East Arkansas Community College

JAMES E. HERRELL, Otero Junior College (Colorado)

JAMES HERTZLER, Goshen College (Indiana)

JOHN HICKEY, Inver Hills Community College (Minnesota)

THOMAS HIGGINS, San Jacinto College (Texas)

EUGENE HILL, Westminster College (Missouri)

MIRIAM HELEN HILL, Indiana University Southeast

SUZY HILL, University of South Carolina-Spartanburg

ROBERT HILT, Pittsburg State University (Kansas)

SOPHIA HINSHALWOOD, Montclair State University (New Jersey)

PRISCILLA HOLLAND, University of North Alabama

MARK R. HOOPER, Freed-Hardeman University

R. HOSTETLER, Fresno City College (California)

ERICK HOWENSTINE, Northeastern Illinois University

LLOYD E. HUDMAN, Brigham Young University (Utah)

JANIS W. HUMBLE, University of Kentucky

JUANA IBANEZ, University of New Orleans

TONY IJOMAH, Harrisburg Area Community College (Pennsylvania)

WILLIAM IMPERATORE, Appalachian State University (North Carolina)

RICHARD JACKSON, Brigham Young University (Utah)

MARY JACOB, Mount Holyoke College (Massachusetts)

GREGORY JEANE, Samford University (Alabama)

SCOTT JEFFREY, Catonsville Community College (Maryland)

JERZY JEMIOLO, Ball State University (Indiana)

NILS I. JOHANSEN, University of Southern Indiana

DAVID JOHNSON, University of Southwestern Louisiana

INGRID JOHNSON, Towson State University (Maryland)

RICHARD JOHNSON, Oklahoma City University

SHARON JOHNSON, Marymount College (New York)

JEFFREY JONES, University of Kentucky

KRIS JONES, Saddleback College (California)

MARCUS E. JONES, Claflin College (South Carolina)

MOHAMMAD S. KAMIAR, Florida Community College, Jacksonville

MATTI E. KAUPS, University of Minnesota-Duluth

JO ANNE W. KAY, Brigham Young University, Idaho

PHILIP L. KEATING, Indiana University

DAVID KEELING, Western Kentucky University

COLLEEN KEEN, Gustavus Adolphus College (Minnesota)

ARTIMUS KEIFFER, Wittenberg University (Ohio)

GORDON F. KELLS, Mott Community College (Michigan)

KAREN J. KELLY, Palm Beach Community College, Boca Raton (Florida)

RYAN KELLY, Lexington Community College (Kentucky)

VIRGINIA KERKHEIDE, Cuyahoga Community College and Cleveland State University (Ohio)

TOM KESSENGER, Xavier University (Ohio)

MASOUD KHEIRABADI, Maryhurst University

SUSANNE KIBLER-HACKER, Unity College (Maine)

CHANGJOO KIM, Minnesota State University

JAMES W. KING, University of Utah

JOHN C. KINWORTHY, Concordia College (Nebraska)

ALBERT KITCHEN, Paine College (Georgia)

TED KLIMASEWSKI, Jacksonville State University (Alabama)

ROBERT D. KLINGENSMITH, Ohio State University-Newark

LAWRENCE M. KNOPP, JR., University of Minnesota-Duluth

LYNN KOEHNEMANN, Gulf Coast Community College

TERRILL J. KRAMER, University of Nevada

JOE KRAUSE, Community College of Indiana at Lafayette

BRENDER KREKELER, Northern Kentucky University; Miami University (Ohio)

ARTHUR J. KRIM, Cambridge, Massachusetts

MICHAEL A. KUKRAL, Rose-Hulman Institute of Technology

ELROY LANG, El Camino Community College (California)

RICHARD L. LANGILL, Saint Martin's University (Washington)

CHRISTOPHER LANT, Southern Illinois University

A. J. LARSON, University of Illinois-Chicago

PAUL R. LARSON, Southern Utah University

LARRY LEAGUE, Dickinson State University (North Dakota)

DAVID R. LEE, Florida Atlantic University

WOOK LEE, Texas State University

JOE LEEPER, Humboldt State University (California)

YECHIEL M. LEHAVY, Atlantic Community College (New Jersey)

SCOTT LEITH, Central Connecticut State University

JAMES LEONARD, Marshall University (West Virginia)

ELIZABETH J. LEPPMAN, St. Cloud State University (Minnesota)

JOHN C. LEWIS, Northeast Louisiana University

DAN LEWMAN JR., Southwest Mississippi Community College

CAEDMON S. LIBURD, University of Alaska-Anchorage

T. LIGIBEL, Eastern Michigan University

Z. L. LIPCHINSKY, Berea College (Kentucky)

ALLAN L. LIPPERT, Manatee Community College (Florida)

RICHARD LISICHENKO, Fort Hays State University (Kansas)

JOHN L. LITCHER, Wake Forest University (North Carolina)

LEE LIU, Central Montana State University

LI LIU, Stephen F. Austin State University (Texas)

WILLIAM R. LIVINGSTON, Baker College (Michigan)

CATHERINE M. LOCKWOOD, Chadron State College (Nebraska)

GEORGE E. LONGENECKER, Vermont Technical College

CYNTHIA LONGSTREET, Ohio State University

JACK LOONEY, University of Massachusetts, Boston

TOM LOVE, Linfield College (Oregon)

K. J. LOWREY, Miami University (Ohio)

JAMES LOWRY, Stephen F. Austin State University (Texas)

MAX LU, Kansas State University

ROBIN R. LYONS, University of Hawai'i-Leeward Community College

SUSAN M. MACEY, Texas State University

MICHAEL MADSEN, Brigham Young University, Idaho

RONALD MAGDEN, Tacoma Community College (Washington)

CHRISTIANE MAINZER, Oxnard College (California)

LAURA MAKEY, California State University, San Bernardino

HARLEY I. MANNER, University of Guam

ANTHONY PAUL MANNION, Kansas State University

GARY MANSON, Michigan State University

CHARLES MANYARA, Radford University (Virginia)

JAMES T. MARKLEY, Lord Fairfax Community College (Virginia)

SISTER MAY LENORE MARTIN, Saint Mary College (Kansas)

KENT MATHEWSON, Louisiana State University

PATRICK MAY, Plymouth State College (New Hampshire)

DICK MAYER, Maui Community College (Hawai'i)

SARA MAYFIELD, San Jacinto College, Central (California)

DEAN R. MAYHEW, Marine Maritime Academy

J. P. McFADDEN, Orange Coast College (California)

BERNARD McGONIGLE, Community College of Philadelphia

PAUL D. MEARTZ, Mayville State University (North Dakota)

DIANNE MEREDITH, California State University-Sacramento

GARY C. MEYER, University of Wisconsin, Stevens Point

JUDITH L. MEYER, Southwest Missouri State University

MARK MICOZZI, East Central University (Oklahoma)

JOHN MILBAUER, Northeastern State University

DALTON E. MILLER, Mississippi State University

RAOUL MILLER, University of Minnesota-Duluth

ROGER MILLER, Black Hills State University (South Dakota)

JAMES MILLS, State University of New York, College at Oneonta

INES MIYARES, Hunter College, CUNY (New York)

BOB MONAHAN, Western Carolina University

KEITH MONTGOMERY, University of Wisconsin-Madison

DAN MORGAN, University of South Carolina, Beaufort

DEBBIE MORIMOTO, Merced College (California)

JOHN MORTRON, Benedict College (South Carolina)

ANNE MOSHER, Syracuse University

BARRY MOWELL, Broward Community College (Florida)

TOM MUELLER, California University (Pennsylvania)

DONALD MYERS, Central Connecticut State University

ROBERT R. MYERS, West Georgia College

GARY NACHTIGALL, Fresno Pacific University (California)

YASER M. NAJJAR, Framingham State College (Massachusetts)

KATHERINE NASHLEANAS, University of Nebraska-Lincoln

JEFFREY W. NEFF, Western Carolina University

DAVID NEMETH, University of Toledo (Ohio)

ROBERT NEWCOMER, East Central University (Oklahoma)

WILLIAM NIETER, St. Johns University (New York)

WILLIAM N. NOLL, Highland Community College

VALIANT C. NORMAN, Lexington Community College (Kentucky)

RAYMOND O'BRIEN, Bucks County Community College (Pennsylvania)

PATRICK O'SULLIVAN, Florida State University

DIANE O'CONNELL, Schoolcraft College

JOHN ODLAND, Indiana University

DOUG OETTER, Georgia College and State University

ANNE O'HARA, Marygrove College (Michigan)

PATRICK OLSEN, University of Idaho

JOSEPH R. OPPONG, University of North Texas

LYNN ORLANDO, Holy Family University (Pennsylvania)

MARK A. OUMETTE, Hardin-Simmons University (Texas)

RICHARD OUTWATER, California State University, Long Beach

CISSIE OWEN, Lamar University (Texas)

MARY ANN OWOC, Mercyhurst College (Pennsylvania)

STEVE PALLADINO, Ventura College (California)

BIMAL K. PAUL, Kansas State University

SELINA PEARSON, Northwest-Shoals Community College (Alabama)

MAURI PELTO, Nichols College (Massachusetts)

JAMES PENN, Southeastern Louisiana University

LINDA PETT-CONKLIN, University of St. Thomas (Minnesota)

DIANE PHILEN, Lower Brule Community College (South Dakota)

PAUL PHILLIPS, Fort Hays State University (Kansas)

MICHAEL PHOENIX, ESRI, Redlands, California

JERRY PITZL, Macalester College (Minnesota)

BRIAN PLASTER, Texas State University

ARMAND POLICICCHIO, Slippery Rock University (Pennsylvania)

ROSANN POLTRONE, Arapahoe Community College (Colorado)

BILLIE E. POOL, Holmes Community College (Mississippi)

GREGORY POPE, Montclair State University (New Jersey)

JEFF POPKE, East Carolina University

VINTON M. PRINCE, Wilmington College (North Carolina)

GEORGE PUHRMANN, Drury University (Missouri)

DONALD N. RALLIS, Mary Washington College (Virginia)

RHONDA REAGAN, Blinn College (Texas)

DANNY I. REAMS, Southeast Community College (Nebraska)

JIM RECK, Golden West College (California)

ROGER REEDE, Southwest State University (Minnesota)

JOHN RESSLER, Central Washington University

JOHN B. RICHARDS, Southern Oregon State College

DAVID C. RICHARDSON, Evangel University (Missouri)

GRAY RINGLEY, Virginia Highlands Community College

SUSAN ROBERTS, University of Kentucky

CURT ROBINSON, California State University, Sacramento

WOLF RODER, University of Cincinnati

JAMES ROGERS, University of Central Oklahoma

PAUL A. ROLLINSON, AICP, Southwest Missouri State University

JAMES C. ROSE, Tompkins/Cortland Community College (New York)

THOMAS E. ROSS, Pembroke State University (North Carolina)

THOMAS A. RUMNEY, State University of New York, College at Plattsburgh

GEORGE H. RUSSELL, University of Connecticut

BILL RUTHERFORD, Martin Methodist College

RAJAGOPAL RYALI, Auburn University at Montgomery (Alabama)

PERRY RYAN, Mott Community College

JAMES SAKU, Frostburg State University (Maryland)

DAVID R. SALLEE, University of North Texas

RICHARD A. SAMBROOK, Eastern Kentucky University

EDUARDO SANCHEZ, Grand Valley State University (Michigan)

JOHN SANTOSUOSSO, Florida Southern College

GINGER SCHMID, Texas State University

BRENDA THOMPSON SCHOOLFIELD, Bob Jones University (South Carolina)

ADENA SCHUTZBERG, Middlesex Community College (Massachusetts)

ROGER M. SELYA, University of Cincinnati

RENEE SHAFFER, University of South Carolina

WENDY SHAW, Southern Illinois University, Edwardsville

SIDNEY R. SHERTER, Long Island University (New York)

HAROLD SHILK, Wharton County Jr. College (Texas)

NANDA SHRESTHA, Florida A&M University

WILLIAM R. SIDDALL, Kansas State University

DAVID SILVA, Bee County College (Texas)

JOSE ANTONIO SIMENTAL, Marshall University (West Virginia)

MORRIS SIMON, Stillman College (Alabama)

ROBERT MARK SIMPSON, University of Tennessee at Martin

KENN E. SINCLAIR, Holyoke Community College (Massachusetts)

ROBERT SINCLAIR, Wayne State University (Michigan)

JIM SKINNER, Southwest Missouri State University

BRUCE SMITH, Bowling Green State University (Ohio)

EVERETT G. SMITH, JR., University of Oregon

PEGGY SMITH, California State University, Fullerton

RICHARD V. SMITH, Miami University (Ohio)

JAMES SNADEN, Charter Oak State College (Connecticut)

DAVID SORENSON, Augustana College (Illinois)

SISTER CONSUELO SPARKS, Immaculata University (Pennsylvania)

CAROLYN D. SPATTA, California State University, Hayward

M. R. SPONBERG, Laredo Junior College (Texas)

DONALD L. STAHL, Towson State University (Maryland)

DAVID STEA, Texas State University

ELAINE STEINBERG, Central Florida Community College

D. J. STEPHENSON, Ohio University Eastern

HERSCHEL STERN, Mira Costa College (California)

REED F. STEWART, Bridgewater State College (Massachusetts)

NOEL L. STIRRAT, College of Lake County (Illinois)

JOSEPH P. STOLTMAN, Western Michigan University

DEAN B. STONE, Scott Community College

WILLIAM M. STONE, Saint Xavier University, Chicago

GEORGE STOOPS, Minnesota State University

Debra Straussfogel, University of New Hampshire

Jamie Strickland, University of North Carolina at Charlotte

Wayne Strickland, Roanoke College (Virginia)

Philip Sturm, Ohio Valley College, Vienna

Philip Suckling, Texas State University

Allen Sullivan, Central Washington University

Selima Sultana, University of North Carolina at Greensboro

Ray Sumner, Long Beach City College (California)

Christopher Sutton, Western Illinois University

T. L. Tarlos, Orange Coast College (California)

Wesley Teraoka, Leeward Community College

Michael Thede, North Iowa Area Community College

Derrick J. Thom, Utah State University

Curtis Thomson, University of Idaho

Ben Tillman, Texas Christian University

Cliff Todd, University of Nebraska, Omaha

Stanley Toops, Miami University (Ohio)

Richard J. Torz, St. Joseph's College (New York)

Harry Trendell, Kennesaw State University (Georgia)

Roger T. Trindell, Mansfield University (Pennsylvania)

Dan Turbeville, East Oregon State College

Norman Tyler, Eastern Michigan University

George Van Otten, Northern Arizona University

Gregory Veeck, Western Michigan University

C. S. Verma, Weber State College (Utah)

Kelly Ann Victor, Eastern Michigan University

Sean Wagner, Tri-State University (Indiana)

Graham T. Walker, Metropolitan State College of Denver

Montgomery Walker, Yakima Valley Community College (Washington)

Deborah Wallin, Skagit Valley College (Washington)

Mike Walters, Henderson Community College (Kentucky)

Linda Wang, University of South Carolina, Aiken

J. L. Watkins, Midwestern State University (Texas)

David Welk, Reedley College (California)

Kit W. Wesler, Murray State University (Kentucky)

Peter W. Whaley, Murray State University (Kentucky)

Macel Wheeler, Northern Kentucky University

P. Gary White, Western Carolina University (North Carolina)

W. R. White, Western Oregon University

Gary Whitton, Fairbanks, Alaska

Rebecca Wiechel, Wilmington College (North Carolina)

Mark Wiljanen, Eastern Kentucky University

Gene C. Wilken, Colorado State University

Forrest Wilkerson, Texas State University

P. Williams, Baldwin-Wallace College (Ohio)

Stephen A. Williams, Methodist College (North Carolina)

Deborah Wilson, Sandhills Community College (Nebraska)

Morton D. Winsberg, Florida State University

Roger Winsor, Appalachian State University (North Carolina)

William A. Withington, University of Kentucky

A. Wolf, Appalachian State University (North Carolina)

Joseph Wood, University of Southern Maine

Richard Wood, Seminole Junior College (Florida)

George I. Woodall, Winthrop College (North Carolina)

Stephen E. Wright, James Madison University (Virginia)

Leon Yacher, Southern Connecticut State University

Kyong Yup Chu, Bergen Community College (New Jersey)

Firooz E. Zadeh, Colorado Mountain College

Donald J. Zeigler, Old Dominion University (Virginia)

Robert C. Ziegenfus, Kutztown University (Pennsylvania)

PERSONAL APPRECIATION

For assistance with the map of North American indigenous people, we are greatly indebted to Jack Weatherford, Professor of Anthropology at Macalester College (Minnesota); Henry T. Wright, Professor and Curator of Anthropology at the University of Michigan; and George E. Stuart, President of the Center for Maya Research (North Carolina). The map of Russia's federal regions could not have been compiled without the invaluable help of David B. Miller, Senior Edit Cartographer at the National Geographic Society, and Leo Dillon of the Russia Desk of the U.S. Department of State. The map of Russian physiography was updated thanks to the suggestions of Mika Roinila of the State University of New York, College at New Paltz. Special thanks also go to Charles Pirtle, Professor of Geography at Georgetown University's School of Foreign Service for his advice on Chapter 4; to Bilal Butt (post-doctoral fellow, University of Wisconsin-Madison) for his significant help with Chapter 6; and to Charles Fahrer of Georgia College and State University for his suggestions on Chapter 6. Enormous gratitude goes to Beth Weisenborn, Virtual Course Coordinator of the Department of Geography at Michigan State University, and all the instructors of Geo 204 (World Regional Geography) for their invaluable suggestions to improve the book. Steve Aldrich, Joel Gruley, Bree Harrison, and Nathan Zukas, student assistants at Michigan State University, helped with updating the book. We also benefited from the insights offered by Narciso Barrera-Bassols of UNAM-Morelia, J. Christopher Brown of University of Kansas, William Doolittle of the University of Texas at Austin, Kyle Evered of Michigan State University, Brad Jokisch of Ohio University, Robert Kaiser of the University of Wisconsin-Madison, David Keeling, of Western Kentucky University, and Eric Perramond of Colorado College.

We also record our appreciation to those geographers who ensured the quality of this book's ancillary products: Elizabeth Muller Hames (M.A. in Geography, University of Miami) co-authored the *Study Guide*. At the University of Miami's Department of Geography and Regional Studies, Peter Muller is most grateful for the advice and support he continues to receive from faculty colleagues Douglas Fuller, Richard Grant, Miguel Kanai, Mazen Labban, Jan Nijman, Shouraseni Sen Roy, and Ira Sheskin as well as GIS Lab Manager Chris Hanson.

We are privileged to work with a team of professionals at John Wiley & Sons that is unsurpassed in the college textbook publishing industry. As authors we are acutely aware of these talents on a daily basis during the crucial production stage, especially the outstanding coordination and leadership skills of Production Editor Barbara Russiello, Illustration Editor Sigmund Malinowski, and Senior Photo Editor Jennifer MacMillan. Others who played a leading role in this process were Senior Designer Kevin Murphy with designer Nancy Field, indexer Barbara Holloway of WordCo, copy editor Betty Pessagno, Joyce Franzen of GGS Book Services PMG in Atlantic Highlands, New Jersey, and Don Larson of Mapping Specialists, Ltd., in Madison, Wisconsin. We much appreciated the leadership of Executive

Geosciences Editor Ryan Flahive, who was the prime mover in launching and guiding this book and was superbly assisted throughout the preparation of this latest edition by Courtney Nelson and Erin Grattan. Our College Marketing Manager, Danielle Torio, gave us considerable attention throughout the revision process, and her advice and enthusiasm were most welcome. We also thank Joan Kalkut, Lynn Pearlman, and Bridget O'Lavin, for their help and support. Beyond this immediate circle, we acknowledge the support and encouragement we have received over the years from many others at Wiley including Vice-President for Production Ann Berlin and Publisher Jay O'Callaghan.

Finally, and most of all, we express our gratitude to our spouses, Bonnie, Nancy, and Vince, for seeing us through the challenging schedule of this new edition of The World Today.

H.J. de Blij
Boca Grande, Florida

Peter O. Muller
Coral Gables, Florida

Antoinette M.G.A. WinklerPrins
East Lansing, Michigan

July 31, 2008

BRIEF CONTENTS

INTRODUCTION   1

CHAPTER 1   Europe   26

CHAPTER 2   Russia   74

CHAPTER 3   North America   104

CHAPTER 4   Middle America   132

CHAPTER 5   South America   162

CHAPTER 6   Subsaharan Africa   190

CHAPTER 7   North Africa/Southwest Asia   228

CHAPTER 8   South Asia   268

CHAPTER 9   East Asia   300

CHAPTER 10   Southeast Asia   342

CHAPTER 11   The Austral Realm   372

CHAPTER 12   The Pacific Realm   388

APPENDIX A   Metric Standard International (SI) and Customary Units and their Conversions   A-1

APPENDIX B   Area and Demographic Data for the World's States   A-2

Additional Appendices and other material available online:

Using the Maps   www.wiley.com/college/deblij

Opportunities in Geography   www.wiley.com/college/deblij

Pronunciation Guide   www.wiley.com/college/deblij

Concept Caching Website   www.ConceptCaching.com

Glossary   G-1

Index   I-1

List of Maps and Figures   M-1

CONTENTS

INTRODUCTION   1

A World on Maps   2
Maps in Our Minds   2
The Map Revolution   2

Geography's Perspective   3
Environment and Society   3

Location and Distribution   3
Scale and Scope   3

Geographic Realms   4

Realms and Regions   4
Regional Classification   5
Criteria for Geographic Realms   6
Criteria for Regions   8

The Physical Setting   8
Natural (Physical) Landscapes   9
Natural Hazards   9
Climate   10

Realms of Population   15
Major Population Clusters   15

Realms of Culture   16

Realms, Regions, and States   17
The State   18
States and Realms   18
States and Regions   18
Political Geography   18

Patterns of Development   18
Classification Schemes   18
Causes of Contrast   19
The Role of Corporate Power   20
The Specter of Debt   21
Globalization   21

FROM THE FIELD NOTES:
Globalization in China   23

**Regional Framework
and Geographic Perspective**   24

CHAPTER 1
EUROPE   26

Defining the Realm   28

Geographical Features   28
Europe's Eastern Border   28
Resources   29
Climates   29
Human Diversity   30
Locational Advantages   30

Landscapes and Opportunities   30
ENVIRONMENTALLY SPEAKING:
Heat Waves in Europe   31

Historical Geography   31
Ancient Greece and Imperial Rome   31
Triumph and Collapse   32
Rebirth and Royalty   32

The Revolutions of Modernizing Europe   33
The Agrarian Revolution   33
The Industrial Revolution   33
Political Revolutions   36

Contemporary Europe   36
FROM THE FIELD NOTES:
Amsterdam, The Netherlands   37
Language and Religion   37
Spatial Interaction   39
A Highly Urbanized Realm   39
A Changing Population   40

Europe's Modern Transformation   42
European Unification   42
New Money   44
Continued Expansion?   45
Centrifugal Forces   45
The EU's New Economic Geography   46

Regions of the Realm   47

Western Europe   48
Reunited Germany   49
France   51
Benelux   54
The Alpine States   54

The British Isles   55
States and Peoples   55
Roots of Devolution   55
A Discrete Region   56
End of Empire   56
The United Kingdom   57
Republic of Ireland   57

Northern (Nordic) Europe   58

Mediterranean Europe   60
Italy   61
The Iberian Peninsula   62
Greece   63
Cyprus and Malta   65

Eastern Europe   65
Cultural Legacies   65
The Geographic Framework   66
Countries Facing the Baltic Sea   67
The Landlocked Center   68
Countries Facing the Black Sea   68
Countries Facing the Adriatic Sea   70

CHAPTER 2

RUSSIA   74

A Troubled Realm   76
Russia's Changing Role   76

Defining the Realm   77
Transition Zones: Two Types   77
Russia's Physical Geography   78
Harsh Environments   78
Physiographic Regions   79
The Arctic   80
Russian Roots   83
A Vibrant Culture   83
Empire of the Czars   83
Communist Victory   83
Seven Fateful Decades   84
The Soviet Legacy   85
The Political Framework   86
The Soviet Economic Framework   88
Russia's Changing Political Geography   90
The Federal Framework of Russia   90
Changing Social Geographies   92
Russia's Demographic Disaster   93
Russia and the World   93
Russia and Europe   94
Unresolved Problems   94

Regions of the Realm   94
Russian Core and Peripheries   94
Central Industrial Region   95
Povolzhye : The Volga Region   95
The Urals Region   96
The Internal Southern Periphery   96
The External Southern Periphery   99
The Eastern Frontier   100
The Kuznetsk Basin (*Kuzbas*)   100
The Lake Baykal Area (*Baykaliya*)   101
Siberia   101
ENVIRONMENTALLY SPEAKING:
Forests in Russia   101
The Future   102

The Russian Far East   102
Mainland and Island   102
Post-Soviet Malaise   102

CHAPTER 3

NORTH AMERICA   104

Defining the Realm   106
Geographic and Social Contrasts   107
North America's Physical Geography   107
Physiography   107
Climate   109
Soils and Vegetation   109
Hydrography (Surface Water)   109
Indigenous North America   110
FROM THE FIELD NOTES:
Tribal Sovereignty in Arizona   111
The United States   111
Population in Time and Space   111
Cultural Geography   115
The Changing Geography
of Economic Activity   116
Canada   120
Population in Time and Space   121
Cultural/Political Geography   122
Economic Geography   124

Regions of the Realm   125
The North American Core   125
The Maritime Northeast   125
French Canada   126
The Continental Interior   126
The Ethanol Factor   126
Depopulation and Regional Change   127
The South   127
The Southwest   128
The Western Frontier   128
The Northern Frontier   129
The Pacific Hinge   129
ENVIRONMENTALLY SPEAKING:
California Wildfires   130

Challenges and Opportunities
in Maturing California   130
Growth and Change in the Pacific
Northwest   130

CHAPTER 4

MIDDLE AMERICA   132

Defining the Realm   135
Physiography   136
A Land Bridge   136
Island Chains   136
Legacy of Mesoamerica   136
The Lowland Maya   136
The Highland Civilizations   137
Collision of Cultures   137
Effects of the Conquest   137
Mainland and Rimland   138
Lingering Regional Contrasts   139
The Hacienda   140
The Plantation   140
Political Fragmentation   140
Independence   142

Regions of the Realm   142
Mexico   142
Physiography   142
Population Patterns   143
Revolution and Its Aftermath   144
FROM THE FIELD NOTES:
Days of the Dead in Mexico   145
Regions of Mexico   145
The Changing Geography
of Economic Activity   146
Mexico's Future   148
The Central American Republics   149
Altitudinal Zonation
of Environments   150
Population Patterns   150
Emergence from a Turbulent Era   151
Economic Dependence on the North   151
The Seven Republics   151

The Caribbean Basin   154
  Economic and Social Patterns   155
  The Greater Antilles   156
    ENVIRONMENTALLY SPEAKING:
      Hurricanes   158
  The Lesser Antilles   161

CHAPTER **5**

SOUTH AMERICA   162

Defining the Realm   164
  What Lies Ahead?   164
The Human Sequence   165
  Early South Americans   165
  The Iberian Invaders   167
  The Africans   167
  Longstanding Isolation   168
  Independence   168
Cultural Fragmentation   168
  Using the Land   169
    ENVIRONMENTALLY SPEAKING:
      Soybean Expansion in Brazil   169
  Cultural Landscapes   170
Economic Integration   170
Urbanization   170
  Regional Patterns   171
    FROM THE FIELD NOTES:
      Urban Agriculture
      in the Brazilian Amazon   172
  Causes and Challenges
    of Cityward Migration   172
  The "Latin" American City Model   172

Regions of the Realm   173
The North: Caribbean South America   174
  Colombia   174
  Venezuela   176
  The Guianas   177
The West: Andean South America   177
  Peru   177
  Ecuador   179
  Bolivia   180
  Paraguay   180

The South: Mid-Latitude South America   181
  Argentina   181
  Chile   182
  Uruguay   184
Brazil: Giant of South America   184
  Population Patterns   184
  African Heritage   185
  Inequality in Brazil   186
  Development Prospects   186
  Brazil's Subregions   188

CHAPTER **6**

SUBSAHARAN AFRICA   190

Cradle and Cauldron   192

Defining the Realm   193
Africa's Physiography   193
  Rifts and Rivers   194
  Continental Drift and Plate Tectonics   194
Natural Environments   195
  Wildlife   195
  Environmental Challenges to Farming   195
Land Issues   196
  Stolen Lands   196
    ENVIRONMENTALLY SPEAKING:
      Desertification in West Africa   197
Environment and Health   197
  Epidemics and Pandemics   197
  Africa's Latest Scourge   199
Africa's Historical Geography   199
  African Genesis   200
  Early Trade   200
  Early States   200
  Eastward Shift   201
  Beyond the West   201
  Bantu Migration   202
  The Colonial Transformation   202
Cultural Patterns   205
  African Languages   206
  Religion in Africa   207
Modern Map and Traditional Society   208
  Supranationalism   208
  Population and Urbanization   209

Regions of the Realm   209
Southern Africa   210
  Africa's Richest Region   210
  South Africa   212
  The Middle Tier   215
  The Northern Tier   216
East Africa   216
    FROM THE FIELD NOTES:
      Maasai Mara National Game Reserve,
      Kenya   216
  Madagascar   218
Equatorial Africa   219
  The Congo   219
  Across the River   220
West Africa   221
  Nigeria   222
  Coast and Interior   224
The African Transition Zone   225

CHAPTER **7**

NORTH AFRICA/
SOUTHWEST ASIA   228

Defining the Realm   230
  A 'Dry World'?   230
  Is This the 'Middle East'?   231
  An 'Arab World'?   231
  An 'Islamic World?'   231
Hearths of Culture   232
  Dimensions of Culture   233
  Rivers and Communities   233
  Decline and Decay   233
Stage for Islam   234
  The Faith   234
  The Arab-Islamic Empire   234
Islam Divided   236
  The Strength of Shi'ism   236
  Religious Revivalism
    in the Realm   237
  Islam and Other Religions   237
  The Ottoman Aftermath   238

The Power and Peril of Oil   240
Location and Dimensions
of Known Reserves   240
A Foreign Invasion   242

Regions of the Realm   243
Egypt and the Lower Nile Basin   245
Gift of the Nile   245
Valley and Delta   246
ENVIRONMENTALLY SPEAKING:
Aswan High Dam in Egypt   246
Subregions of Egypt   246
Divided Sudan   247

The Maghreb and Its Neighbors   248
Atlas Mountains   248
Colonial Impact   249
The Maghreb Countries   249
Adjoining Saharan Africa   249

The Middle East: Crucible of Conflict   250
War in Iraq   250
The Region's Other Countries   253

Arabian Peninsula   257
Saudi Arabia   257
On the Periphery   259

The Empire States   259
Turkey   260
Iran   261

Turkestan: The Six States of Central Asia   263
The States
of Former Soviet Central Asia   263
FROM THE FIELD NOTES:
Mahmud of Kashgar   263
Fractious Afghanistan   265

CHAPTER 8
SOUTH ASIA   268

Defining the Realm   270
Physiographic Regions of South Asia   271
Northern Mountains   272
River Lowlands   273

Southern Plateaus   273
The Annual Wet Monsoon   273
The Human Sequence   273
Hinduism and Buddhism   273
Aśoka's Mauryan Empire   274
The Power of Islam   275
The European Intrusion   275

South Asia's Population Dilemma   276
Population Change   278
Demographic Prospects for South Asia   278
Population Variability in the Realm   280

South Asia's Burden of Poverty   281
The Latest Invasions   281

Regions of the Realm   282
**Pakistan: South Asia's
Western Flank   282**
East and West, North and South   282
The Kashmir Issue   283
Forging Centripetal Forces   284
The Provinces   284
Pakistan's Prospects   285

India: South Asia's Giant   285
States and People   286
India's Changing Map   287
Centripetal Forces   289
Urbanization   290
FROM THE FIELD NOTES:
Squatters in Mumbai, India   290
Economic Geography   291
Globalization   291
Farming's Enduring Importance   292
The Energy Problem   293
Limitations on Manufacturing Growth   293
Improving Prospects   294
India West and East   294

**Vulnerable Bangladesh
on South Asia's Eastern Flank   295**
Hazard-Prone Territory   295
ENVIRONMENTALLY SPEAKING:
Flooding in Bangladesh   296
Limits to Opportunity   296

The Mountainous North   296
The Island South   298
Sri Lanka: South Asian Tragedy   298

CHAPTER 9
EAST ASIA   300

Defining the Realm   302
Natural Environments   302
Physiography and Population   303

Historical Geography   304
Early Cultural Geography   305
Early State Formation   305

Regions of the Realm   305
China Proper   306
Relative Location   307
Continental Size and Environment   307
Evolving China   308
A Century of Convulsion   309
The Opium Wars   310
The End of Dynastic Rule   312
Nationalists and Communists   313
Japan in China   313

The Rise of Communist China   313
China under Communist Rule   313
Political and Administrative Divisions   314

Population Issues   315
The One-Child Policy   315
The Minorities   316

People and Places of China Proper   317
The Northeast China Plain   317
The North China Plain   318
ENVIRONMENTALLY SPEAKING:
Air Pollution in China   319
Inner Mongolia   319
The Basins of the Chang/Yangzi   320
The Basins of the Xi (West)
and Pearl Rivers   321

Xizang (Tibet)   322
   Conquest by China   322

Xinjiang   323
   Physiography   324
   Energy and Technology   324

China's Pacific Rim   325
   FROM THE FIELD NOTES:
      Globalization in Xinjiang   325
   The Geography
      of Development   325
   Transforming
      the Economic Map   326

China's Pacific Rim Today   327
   Guangdong Province
      and the Pearl River Estuary   328
   Hong Kong   329
   The Burgeoning Pacific Rim   330
   Coast and Interior   330

China: Global Superpower?   331

Mongolia   332

The Jakota Triangle Region   332

Japan   332
   Colonial Wars and Recovery   332
   British Tutelage   333
   Spatial Contrasts
      and Constraints   334
   Modernization Japanese Style   335
   The Role of Relative Location   336
   Spatial Organization   336
   Coastal Development   337
   Japan's Pacific Rim Prospects   337

Korea   338
   War and Aftermath   338
   South Korea   339
   North Korea   339

Taiwan   340
   Peoples and Rulers   340
   A Strong Economy   340
   Geopolitical Risks   341

CHAPTER 10
SOUTHEAST ASIA   342

Defining the Realm   344
Colonialism's Heritage   345
Physical Geography   346
   Four Rivers   346
   Tropical Rainforests   346
   ENVIRONMENTALLY SPEAKING:
      Deforestation in Southeast Asia   347
Population Geography   347
   The Ethnic Mosaic   348
   Immigrants   349
How the Political Map Evolved   350
   The Colonial Imprint   350
   Southeast Asia's
      Political Geography   353
   Boundaries   353
   State Territorial Morphology   355

Regions of the Realm   356
Mainland Southeast Asia   356
Indochina   356
   Vietnam   357
   Cambodia   358
   Laos   359
   The Mighty Mekong   359
Thailand   359
   The Restive Peninsular South   360
   Bangkok on the Chao Phraya   361
Myanmar   361
   FROM THE FIELD NOTES:
      Yangon, Myanmar   362
   Morphology and Structure   362
Insular Southeast Asia   364
Mainland-Island Malaysia   364
   Ethnic Components   364
   The Dominant Peninsula   364
   Malaysian Borneo   365

Brunei   365
Singapore   366
Indonesia   367
   The Major Islands   368
   Diversity in Unity   369
   Transmigration
      and the Outer Islands   369
East Timor   370
The Philippines   370
   Muslim Insurgency   370
   People and Culture   370
   Prospects   371

CHAPTER 11
THE AUSTRAL REALM   372

Defining the Realm   374
Land and Environment   374
   Climates   375
   The Southern Ocean   375
   Biogeography   376
   The Human Impact   376

Regions of the Realm   377
Australia   377
   Distance   377
   Core and Periphery   378
   A Federal State   379
   An Urban Culture   379
   Economic Geography   380
   Australia's Future   382
   ENVIRONMENTALLY SPEAKING:
      Drought in Australia   383
New Zealand   384
   Human Spatial Organization   385
   FROM THE FIELD NOTES:
      Tourism in New Zealand   386
   The Maori Factor
      and New Zealand's Future   386

CHAPTER 12
THE PACIFIC REALM   388

Defining the Realm   390
Colonization and Independence   390
The Pacific Realm
   and Its Marine Geography   391
   The State at Sea   391
   UNCLOS Intervention   391

Regions of the Realm   394
Melanesia   395
   ENVIRONMENTALLY SPEAKING:
   Sea-Level Rise in the Pacific   396

Micronesia   396
Polynesia   396
   FROM THE FIELD NOTES:
   Pre-European Hawai'i   397
A Final Caveat: Pacific and Antarctic   398

APPENDIX A
METRIC (STANDARD
INTERNATIONAL [SI] AND
CUSTOMARY UNITS AND
THEIR CONVERSIONS   A-1

APPENDIX B
AREA AND DEMOGRAPHIC DATA
FOR THE WORLD'S STATES   A-2

Additional Appendices and other material
available online:

Using the Maps   www.wiley.com/college/deblij
Opportunities
   in Geography   www.wiley.com/college/deblij
Pronunciation
   Guide   www.wiley.com/college/deblij
Concept Caching
   Website   www.ConceptCaching.com
Glossary   G-1
Index   I-1
List of Maps and Figures   M-1

FOURTH EDITION

THE WORLD TODAY

Concepts and Regions in Geography

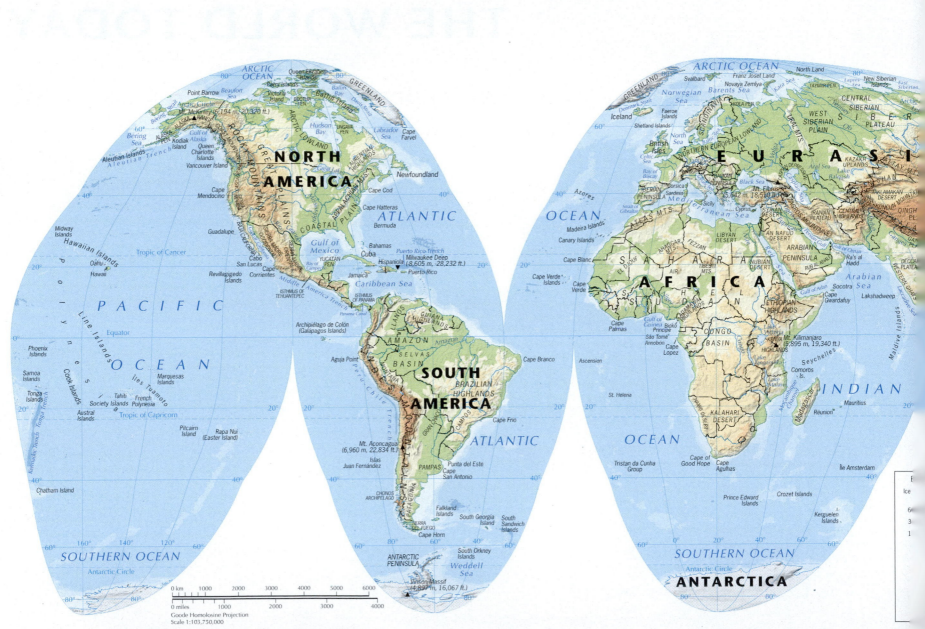

FIGURE G-1   *Map:* © H. J. de Blij, P. O. Muller, and John Wiley & Sons, Inc.

Introduction

In This Chapter

- Maps in our minds
- A geographic approach to understanding the world today
- Why earthquakes happen and how they get started
- Global climate change
- How corporations are bigger than countries
- Is globalization everywhere?

CONCEPTS, IDEAS, AND TERMS

1	Mental maps
2	Spatial perspective
3	Scale
4	Geographic realm
5	Transition zone
6	Regional concept
7	Absolute location
8	Relative location
9	Formal region
10	Spatial system
11	Hinterland
12	Functional region
13	Natural landscape
14	Continental drift
15	Tectonic plate
16	Pacific Ring of Fire
17	Ice age
18	Glaciation
19	Interglacial
20	Desertification
21	Global climate change
22	Population distribution
23	Urbanization
24	Cultural landscape
25	State
26	European state model
27	Development
28	Core-periphery relations
29	Core areas
30	Regional disparities
31	Globalization

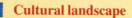

Photos: *(upper left)* Mali, © Bruno Morandi/Robert Harding World Imagery/Getty Images, *(right)* Tokyo, © A. WinklerPrins.

WHAT ARE YOUR expectations as you open this book? You have signed up for a course that will take you around the world to try to understand how it functions today. Hopefully, you will also discover how interesting and unexpectedly challenging the discipline of geography is. We hope that this course, and this book, will open new vistas and bring new perspectives, and help you navigate this increasingly complex and often daunting world.

And what a time this is to be studying geography! The world is undergoing dramatic transformations on many fronts. The United States remains the world's superpower, but challengers such as China and India are emerging. Long-term climate change has the potential to alter the prospects of countries and peoples everywhere. Instant natural disasters have lasting geopolitical as well as economic consequences. Energy crises, enmeshed with cultural conflict of which extremist Islamic terrorism is only one manifestation, threaten stability and security. When the Soviet Union collapsed in the early 1990s, the Cold War ended and the map changed as new countries took their place on the global stage; there was much talk of a 'new world order' that would mark a major step forward for the planet. Has this really happened? A new world order would require an alleviation of poverty, the global adoption of human rights standards, and the universal acceptance of representative government. We have a long way to go, and our task is to understand the changes and to make sense of where the world is headed. Geography, as you will discover, is a powerful ally in this mission.

A WORLD ON MAPS

Just a casual glance at the pages that follow reveals a difference between this and other textbooks: there are almost as many maps as there are pages. Geography is more closely identified with maps than any other discipline, and we urge you to give as much (or more!) attention to the maps in this book as you do to the text. It is often said that a picture is worth a thousand words, but a map can be worth a million. When we write 'see Figure XX,' we really mean it . . . and we hope that you will get into the habit. We humans are territorial creatures, and the boundaries that fence off our 190 or so countries reflect our divisive ways. But other, less visible borders—between religions, languages, rich and poor—partition our planet as well. When political and cultural boundaries are at odds, there is nothing like a map to summarize the circumstances. Just look at Fig. 7-13. How clear were the implications of this map to those who made the decision to send American troops into war?

Maps in Our Minds

All of us carry in our minds maps of what psychologists call our *activity space*: the apartment building or house we live in, the streets nearby, the way to school or business, the general layout of our hometown or city. You will know what lane to use when you turn into a shopping mall, or where to park at the movie theater. You can probably draw from memory a pretty good map of your hometown. These **1 mental maps** allow you to navigate your activity space with efficiency, predictability, and safety. When you arrived as a first-year student on a college or university campus, a new mental map will have started forming. At first you needed a hard-copy map to find your way around, but soon you dispensed with that because your mental map was sufficient. And it will continue to improve as your activity space expands.

If a well-formed mental map is useful for decisions in daily life, then an adequate mental map is surely indispensable when it comes to decision making in the wider world. You can give yourself an interesting test.

Choose some part of the world, beyond North America, in which you have an interest or about which you have a strong opinion—for example, Israel, Taiwan, Afghanistan, North Korea, or Venezuela. On a blank piece of paper, draw a map that reflects your impression of the regional layout there: the country, its internal divisions, major cities, neighbors, seas (if any), and so forth. That is your mental map of the place. Put it away for future reference, and try it again at the end of this course. You will have proof of your improved mental-map inventory.

The Map Revolution

The maps in this book show larger and smaller parts of the world in various contexts, some depicting political frameworks, others displaying ethnic, cultural, economic, environmental, and other features unevenly distributed across our world. But *cartography* (the making of maps) has undergone a dramatic technological revolution—a revolution that continues. Earth-orbiting satellites with special on-board scanners and television cameras transmit remotely sensed information to computers on the surface, recording the expansion of deserts, the shrinking of glaciers, the depletion of forests, the growth of cities, and myriad other geographic phenomena. Earthbound computers possess ever-expanding capabilities not only to sort this information but also to display it graphically. This allows geographers to develop *geographic information systems (GISs)*, presenting on-screen information within seconds that would have taken months to assemble just a few decades ago.

Nevertheless, satellites—even spy satellites—cannot record everything that occurs on the earth's surface. Sometimes the transition zones between ethnic groups or cultural sectors can be discerned by satellites, for example, in changing types of houses or religious shrines, but this kind of information tends to require on-the-ground verification through field research and reporting. No satellite view of Iraq could show you the distribution of Sunni and Shia Muslim adherents. Many of the boundaries you see on the maps in this book cannot be seen from space because long stretches are not even marked on the ground. So the maps you are about to 'read' have their continued uses. They summarize complex situations and allow us to begin forming lasting mental maps of the areas they represent.

GEOGRAPHY'S PERSPECTIVE

Geography has been described as the most interdisciplinary of disciplines. That is a testimonial to geography's historic linkages to many other fields, ranging from geology to economics and from sociology to political science. And, as has been the case so often in the past, geography is in the lead on this point. Today, *interdisciplinary* studies and research are more prevalent than ever. The old barriers between disciplines are breaking down.

This should not suggest that college and university departments are no longer relevant; they are just not as exclusive as they used to be. These days, you can learn some useful geography in economics departments and some good economics in geography departments. But each discipline still has its own particular way of looking at the world.

In a very general way, we can visualize three key perspectives when we try to figure out how the world works. One is the historic or chronological (you have heard the expression 'if you do not learn the lessons of history, you will be doomed to repeat them'). History's key question is *when*. A second perspective centers on the systems people have invented to stabilize their interactions, from the economic to the political. Here the question is *how*. The third perspective is the geographic—the spatial—and the key questions are *where* and *why there*. We seek description and explanation of the patterns of human activity on this

earth. This approach is called the **2** **spatial perspective** and has defined geography from its beginnings.

Environment and Society

There is another glue that binds geography and has done so for a very long time: an interest in the relationships between human societies and the natural (physical) environment. Geography sits at the intersection of the social and natural sciences and integrates perspectives from both, being the only discipline to do so explicitly. This perspective comes into play frequently: environmental change is in the news on a daily basis in the form of global climate change, but this current surge of global warming is only the latest phase of endless climatic and ecological fluctuation. Geographers are involved in understanding current environmental issues not only by considering climate change in the context of the past, but also by looking carefully at the implications of global climate change for human societies.

LOCATION AND DISTRIBUTION

Geographers, therefore, need to be conversant with the location and distribution of salient features on the earth's surface. This includes the natural (physical) world simplified in Figure G-1 (chapter opener map) as well as the human world, and our inquiry will view these in temporal (historical) as well as spatial perspective. We take a penetrating look at the overall geographic framework of the contemporary world, the still-changing outcome of thousands of years of human achievement and failure, movement and stagnation, stability and revolution, interaction and isolation. The spatial structure of cities, the layout of farms and fields, the networks of transportation, the configurations of rivers, the patterns of climate—all these form part of our investigation. As you will find, geography employs a comprehensive spatial vocabulary

with meaningful terms such as area, distance, direction, clustering, proximity, accessibility, and many others we will encounter in the pages that follow. For geographers, some of these terms have more specific definitions than is generally assumed. There is a difference, for example, between *area* and *region*, and between *boundary* and *frontier*. Other terms, such as *location* and *pattern*, can have multiple meanings. The vocabulary of geography holds some surprises.

Scale and Scope

One prominent item in this vocabulary is the term **3** **scale**. Whenever a map is created, it represents all or part of the earth's surface at a certain level of detail. Obviously, Figure G-1 has a very low level of detail; it is little more than a general impression of the distribution of land and water as well as lower and higher elevations on the earth's surface. A few prominent features such as the Himalayas and the Sahara are named, but not the Appalachians or the Gobi Desert. At the bottom of the map you can see that one inch at this scale must represent about 1650 miles of the real world, leaving the cartographer little scope to insert information.

A map such as Figure G-1 is called a *small-scale* map because the ratio between map distance and real-world distance, expressed as a fraction, is very small at 1:103,750,000. Increase that fraction, and you can represent less territory—but also enhance the amount of detail the map can represent. In Figure G-2, note how the fraction increases from the smallest (1:103,000,000) to the largest (1:1,000,000). Montreal, Canada, is just a dot on Map A but an urban region on Map D.

Does this mean that world maps like G-1 are less useful than larger-scale maps? It all depends on the *purpose of the map*. In this chapter, we often use world maps to show global distributions as we set the stage for the more detailed discussions to follow. In later chapters, the scale tends to become larger as we focus on smaller areas, even on individual countries

EFFECT OF SCALE

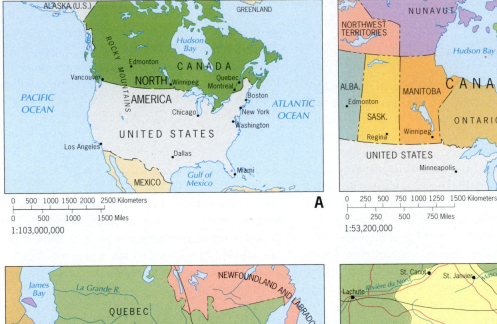

FIGURE G-2

© H. J. de Blij, P. O. Muller, and John Wiley & Sons, Inc.

hoods **4** **geographic realms**. Each of these realms possesses a particular combination of environmental, cultural, and organizational properties. These characteristic qualities are imprinted on the landscape, giving each realm its own traditional attributes and social settings. As we come to understand the human and environmental makeup of these geographic realms, we learn not only *where* they are located (as we noted, a key question in geography), but also *why they are located where they are*, how they are constituted, and what their future is likely to be in our fast-changing world. Figure G-3, therefore, forms the framework for our investigation.

REALMS AND REGIONS

Geographers, like other scholars, seek to establish order from the countless data that confront them. Biologists have established a system of classification to categorize the many millions of plants and animals into a hierarchical system of seven ranks. In descending order, we humans belong to the animal *kingdom*, the *phylum* (division) named chordata, the *class* of mammals, the *order* of primates, the *family* of hominids, the *genus* designated *Homo*, and the *species* known as *Homo sapiens*. Geologists classify the earth's rocks into three major (and many subsidiary) categories and then fit these categories into a complicated geologic time scale that spans hundreds of millions of years. Historians define eras, ages, and periods to conceptualize the sequence of the events they study.

Geography, too, employs systems of classification. When geographers deal with urban problems, for instance, they use a classification scheme based

and cities. But whenever you read a map, be aware of the scale, because the scale is a guide to its utility.

GEOGRAPHIC REALMS

Ours may be a globalized, interconnected world, a world of international trade and travel, migration and movement, tourism and television, financial flows

and Internet traffic, a world that, in some contexts, is taking on the properties of a 'global village'—but that village still has neighborhoods. Their names are Europe, South America, Southeast Asia, and others familiar to us all. Like the neighborhoods of a city or town, these global neighborhoods may not have sharply defined borders, but their persistence, after tens of thousands of years of human dispersal, is beyond doubt. Geographers call such global neighbor-

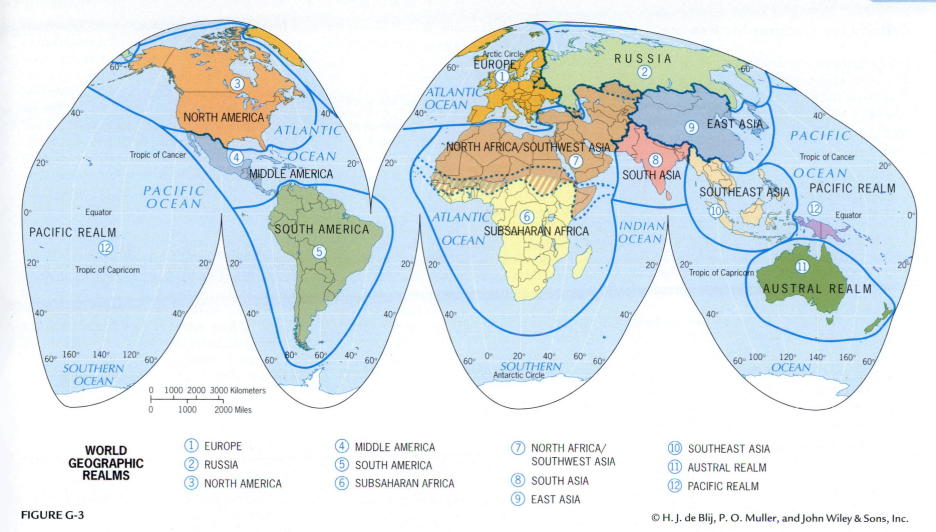

WORLD GEOGRAPHIC REALMS

① EUROPE
② RUSSIA
③ NORTH AMERICA
④ MIDDLE AMERICA
⑤ SOUTH AMERICA
⑥ SUBSAHARAN AFRICA
⑦ NORTH AFRICA/ SOUTHWEST ASIA
⑧ SOUTH ASIA
⑨ EAST ASIA
⑩ SOUTHEAST ASIA
⑪ AUSTRAL REALM
⑫ PACIFIC REALM

FIGURE G-3

© H. J. de Blij, P. O. Muller, and John Wiley & Sons, Inc.

on the sizes and functions of the places involved. Some of the terms in this classification are part of our everyday language: megalopolis, metropolis, city, town, village, hamlet.

Regional Classification

In regional geography, the focus of this book, our challenge is different. We, too, need a hierarchical framework for the areas of the world we study, from the largest to the smallest. But our classifica-tion system is horizontal, not vertical. It is *spatial*. There are four levels:

1. *Landmasses and Oceans* (Fig. G-1). This is geography's equivalent to the biologists' over-arching kingdoms (of plants and animals), al-though the issue arises as to whether the icy sur-face of the Antarctic 'continent' constitutes a landmass. Since no permanent human popula-tion has become established on these glaciers, no regional geography has as yet evolved.

2. *Geographic Realms* (Fig. G-3). Based on a com-bination of physical and human factors, these are the most comprehensive divisions of the inhab-ited world but are not standard.

3. *Geographic Regions* More specific criteria di-vide the great geographic realms into smaller re-gions, as will be demonstrated at the conclusion of this chapter.

4. *Subregions, Domains, Districts* Subdivisions of regions, sometimes based on single factors, are mapped for specific purposes.

Criteria for Geographic Realms

In any classification system, criteria are the key. Not all animals are mammals; the criteria for inclusion in that biological class are more specific and restrictive. A dolphin may look and act like a fish, but both anatomically and functionally dolphins belong to the class of mammals.

- *Physical and Human* Geographic realms are based on sets of spatial criteria. They are the largest units into which the inhabited world can be divided. The criteria on which such a broad regionalization is based include both physical (that is, natural) and human (or social) yardsticks. On the one hand, South America is a geographic realm because physically it is a continent and culturally it is dominated by a set of social norms. The realm called South Asia, on the other hand, lies on a Eurasian landmass shared by several other geographic realms; high mountains, wide deserts, and dense forests combine with a distinctive social fabric to create this well-defined realm centered on India.

- *Functional* Geographic realms are the result of the interaction of human societies and natural environments, a *functional* interaction revealed by farms, mines, fishing ports, transport routes, dams, bridges, villages, and countless other features that mark the landscape. According to this criterion, Antarctica is a continent but not a geographic realm.

- *Historical* Geographic realms must represent the most comprehensive and encompassing definition of the great clusters of humankind in the world today. China lies at the heart of such a cluster, as does India. Africa constitutes a geographic realm from the southern margin of the Sahara (an Arabic word for desert) to the Cape of Good Hope and from its Atlantic to its Indian Ocean shores.

    Figure G-3 displays the 12 world geographic realms based on these criteria. As we will show in more detail later, waters, deserts, and mountains as well as cultural and political shifts mark

the borders of these realms. We will discuss the position of these boundaries as we examine each realm.

Geographic Realms: Margins and Mergers

Where geographic realms meet, **5** **transition zones**, not sharp boundaries, mark their contacts. We need only remind ourselves of the border zone between the geographic realm in which most of us live, North America, and the adjacent realm of Middle America. The line in Figure G-3 coincides with the boundary between Mexico and the United States, crosses the Gulf of Mexico, and then separates Florida from Cuba and the Bahamas. But Hispanic influences are strong in North America north of this boundary, and the U.S. economic influence is strong south of it. The line, therefore, represents an ever-changing zone of regional interaction. Again, there are many ties between South Florida and the Bahamas, but the Bahamas resemble a Caribbean more than a North American society.

In Africa, the transition zone from Subsaharan to North Africa is so wide and well defined that we have put it on the world map; elsewhere, transition zones tend to be narrower and less easily represented. In these early years of the twenty-first century, such countries as Belarus (between Europe and Russia) and Kazakhstan (between Russia and Muslim Southwest Asia) lie in inter-realm transition zones. Remember, over much (though not all) of their length, borders between realms are zones of regional change. As you will see, transition zones are often places of tension and/or conflict.

Geographic Realms: Changing Times

Had we drawn Figure G-3 before Columbus made his voyages (1492–1504), the map would have looked different (see Fig. G-4): Amerindian states and peoples would have determined the boundaries in the Americas; Australia and New Guinea would have constituted one realm, and New Zealand would have been part of the Pacific Realm. The colonization,

---

WHAT'S DRIVING GEOGRAPHIC CHANGE IN THE WORLD

- Environmental extremes and weather swings signal a growing likelihood of **abrupt climate change** that will threaten coastlines and affect food production.

- Remember the population explosion of the twentieth century? We are heading now for a twenty-first century **population implosion** as some world regions see their numbers shrink and others will follow suit. But high growth rates will persist longest in regions afflicted by poverty.

- In a world of regions and countries, look for **subnational entities** such as provinces, States, communities, and even islands to demand more rights and greater autonomy, countering a push by national governments to cooperate in their common interest.

- The bipolar world of the twentieth century, when the USSR and the West were locked in a Cold War, has given way to a world dominated by a **single superpower**, the United States of America. But competitors are on the rise.

- Issues involving **culture**, ranging from religious to economic and from historical to political, are driving local and global tension and conflict. In an era of weapons of mass destruction, this makes our understanding of others' viewpoints ever more important.

lude to a multipolar world? We will address such questions in many chapters.

Criteria for Regions

The spatial division of the world into geographic realms establishes a broad global framework, but for our purposes a more refined level of spatial classification is needed. This brings us to an important organizing concept in geography: the **6** **regional concept**. To continue the analogy with biological taxonomy, we now go from phylum to order. To establish regions within geographic realms, we need more specific criteria.

Let us use the North American realm to demonstrate the regional idea. When we refer to a part of the United States or Canada (e.g., the South, the Midwest, or the Prairie Provinces), we employ a regional concept—not scientifically but as part of everyday communication. We reveal our *perception* of local or distant space as well as our mental image of the region we are describing.

But what exactly is the Midwest? How would you draw this region on the North American map? Regions are easy to imagine and describe, but they can be difficult to outline on a map. One way to define the Midwest is to use the borders of States[1]: certain States are part of this region, others are not. You could use agriculture as the chief criterion: the Midwest is where corn and/or soybeans occupy a certain percentage of the farmland. Look ahead to Fig. 3.15 where you will notice that a different name for this region is being used, the heartland, because of the differing criteria (agriculture) defining it. Each method results in a different delimitation; a Midwest based on States is different from a Midwest based on

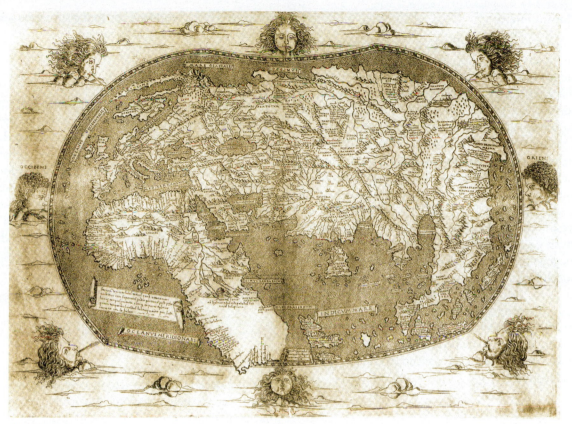

FIGURE G-4   This world map circa 1492 shows only the Eastern Hemisphere because the Western Hemisphere had not yet been 'discovered' and so was not cartographically displayed by European cartographers. It is one of the first maps to depict the Cape of Good Hope, Madagascar, and Zanzibar. The distortions in this map are meant to display a seemingly safer and easier western sailing route to reach the East Indies than by sailing around Africa. Attributed to Francesco Rosselli, circa 1492.   (© The Granger Collection)

Europeanization, and Westernization of the world changed that map dramatically. During the four decades after World War II relatively little change took place, but since 1985 far-reaching realignments have again been occurring.

Geographic Realms: Two Categories

The world's geographic realms can be divided into two categories: (1) those dominated by one major political entity, in terms of territory or population or both (North America/United States, Middle America/Mexico, South America/Brazil, South Asia/India, East Asia/China, Southeast Asia/Indonesia as well as Russia and Australia), and (2) those that contain many countries but no dominant state (Europe, North Africa/Southwest Asia, Subsaharan Africa, and the Pacific Realm). For several decades in the twentieth century two major powers, the United States and the former Soviet Union, dominated the world and competed for global influence. Today, the United States is dominant, but its influence is dwindling. What lies ahead? Will China and/or some other power challenge U.S. supremacy? Is our map of realms a pre-

---

[1]Throughout this book, we will capitalize *State* when this term refers to an administrative subdivision of a country, for example, the U.S. *State* of Nebraska or the Australian *State* of Victoria. Since this term is also synonymous with country (for example, the state of Mexico), we use the lower case when referring to such a national *state*.

farm production or on industrial location. Therein lies an important principle: regions are devices that allow us to make spatial generalizations, and they are based on artificial criteria we establish to help us construct them.

Area

Given these different dimensions of the same region, we can identify properties that all regions have in common. To begin with, all regions have *area*. This observation would seem obvious, but there is more to this idea than meets the eye. Regions may be intellectual constructs, but they are not abstractions: they exist in the real world, and they occupy space on the earth's surface.

Boundaries

It follows that regions have *boundaries*. Occasionally, nature itself draws sharp dividing lines, for example, along the crest of a mountain range or the margin of a forest. More often, regional boundaries are not self-evident, and we must determine them using criteria that we establish for that purpose. For example, to define a citrus-growing agricultural region, we may decide that only areas where more than 50 percent of all farmland stands under citrus trees qualify to be part of that region.

Location

All regions also have *location*. Often the name of a region contains a locational clue, as in Amazon Basin or Indochina (a region of Southeast Asia lying between India and China). Geographers refer to the **7** **absolute location** of a place or region by providing the latitudinal and longitudinal extent of the region with respect to the earth's grid coordinates. A far more practical measure is a region's **8** **relative location**, that is, its location with reference to other regions. Again, the names of some regions reveal aspects of their relative locations, as in *Eastern* Europe and *Equatorial* Africa.

Homogeneity

Many regions are marked by a certain *homogeneity*, or sameness. Homogeneity may lie in a region's human (cultural) properties, its physical (natural) characteristics, or both. Siberia, a vast region of northeastern Russia, is marked by a sparse human population that resides in widely scattered, small settlements of similar form, frigid climates, extensive areas of permafrost (permanently frozen subsoil), and cold-adapted vegetation. This dominant uniformity makes it one of Russia's natural and cultural regions, extending from the Ural Mountains in the west to the Pacific Ocean in the east. When regions display a measurable and often visible internal homogeneity, they are called **9** **formal regions**. But not all formal regions are visibly uniform. For example, a region may be delimited by the area in which, say, 90 percent of the people or more speak a particular language. This cannot be seen in the landscape, but the region is a reality, and we can use this criterion to draw its boundaries accurately. It, too, is a formal region.

Regions as Systems

Other regions are marked *not* by their internal sameness but by their functional integration—that is, the way they work. These regions are defined as **10** **spatial systems** and are formed by the areal extent of the activities that define them. Take the case of a large city with its surrounding zone of suburbs, urban-fringe countryside, satellite towns, and farms. The city supplies goods and services to this encircling zone, and it buys farm products and other commodities from it. The city is the heart, the *core* of this region, and we call the surrounding zone of interaction the city's **11** **hinterland**. But the city's influence wanes on the outer periphery of that hinterland, and there lies the boundary of the functional region of which the city is the focus. A **12** **functional region**, therefore, is usually forged by a structured, urban-centered system of interaction. It has a core and a periphery. As we shall see, core-periphery contrasts in

some parts of the world are becoming strong enough to endanger the stability of countries.

Interconnections

All human-geographic regions are *interconnected*, being linked to other regions. We know that the borders of geographic realms sometimes take on the character of transition zones, and so do neighboring regions. Trade, migration, education, television, computer linkages, and other interactions blur regional boundaries. These are just some of the links in the fast-growing interdependence among the world's peoples, and they reduce the differences that still divide us. Understanding these differences will lessen them further.

THE PHYSICAL SETTING

This book focuses on the geographic realms and regions produced by human activity over thousands of years. But we should not forget the natural environments in which all this activity took place because we can still recognize the role of these environments in how people make their living. Certain areas of the world, for example, presented opportunities for plant and animal domestication that other areas did not. The people who happened to live in those favored areas learned to grow wheat, rice, or root crops, and to domesticate oxen, goats, or llamas. We can still discern those early *patterns of opportunity* on the map in the twenty-first century. From such opportunities came adaptation and invention, and thus arose villages, towns, cities, and states. But people living in different environments found it much harder to achieve this organization. The Americas, for instance, had no large animals that could be domesticated except the llamas. This meant that societies created agricultural systems that did not involve ploughing as there were no draught animals. When Europeans introduced cows, horses, and other livestock, this change completely revolutionized the environments and cultural systems of the Western Hemisphere. The modern map carries many imprints of the past.

Natural (Physical) Landscapes

The landmasses of Planet Earth present a jumble of **13** **natural landscapes** ranging from rugged mountain chains to smooth coastal plains (Fig. G-1). Certain continents are readily linked with a dominant physical feature—for example, North America and its Rocky Mountains, South America with its Andes and Amazon River Basin, Europe with its Alps and Rhine and Danube River basins, Asia with its Himalaya Mountains and numerous river basins, and Africa with its Sahara and Congo River Basin. Physical features have long influenced human activity and movement—even today. Mountain ranges formed barriers to movement but also channeled the spread of agricultural and technical innovations. Today the Taliban, al-Qaeda, and Chechnyan fugitives still use rugged mountains to hide out. Large deserts similarly formed barriers, as did rivers, although rivers do permit accessibility and connectivity between people. As we study each of the world's geographic realms, we will find that physical landscapes continue to play significant roles in this modern world. That is one reason why the study of world regional geography is so important: it puts the human map in environmental as well as regional perspective.

Natural Hazards

Our planet may be 4.6 billion years old, but it is far from placid. As you read this chapter, earth tremors are shaking the still-thin crust on which we live, volcanoes are erupting, storms are raging. Even the very continents are moving measurably, pulling apart in some areas, colliding in others. Hundreds of thousands of human lives are lost to natural calamities in almost every decade, and such calamities have at times altered the course of history.

About a century ago a geographer named Alfred Wegener, a German scientist, used spatial analysis to explain something that is obvious even from a small-scale map like Figure G-1: the apparent, jigsaw-like fit of the landmasses, especially across the South Atlantic Ocean. He concluded that the landmasses on the map are actually pieces of a supercontinent that existed hundreds of millions of years ago (he called it *Pangaea*) that drifted away when, for some reason, that supercontinent broke up. His hypothesis of **14** **continental drift** set the stage for scientists in other disciplines to search for a mechanism that might make this possible, and much of the answer to that search proved to lie in the crust beneath the ocean surface. Today we know that the continents are 'rafts' of relatively light rock that rest on slabs of heavier rock called **15** **tectonic plates** (Fig. G-5) whose movement is propelled by giant circulation cells in the red-hot magma below (when this molten magma reaches the surface through volcanic vents, it is called lava).

Inevitably, moving tectonic plates collide. When they do, earthquakes and volcanic eruptions result, and the physical landscape is thrown into spectacular relief. Compare Figures G-5 and G-6, and you can see the outlines of the tectonic plates in the distribution of these hazards to human life. The May 12, 2008, earthquake in Sichuan province, China, measured 7.9 on the Richter scale. Although a shallow quake, it was in a densely populated region. At least 70,000 people died and millions were made homeless. One of the earth's oceans is almost completely encircled

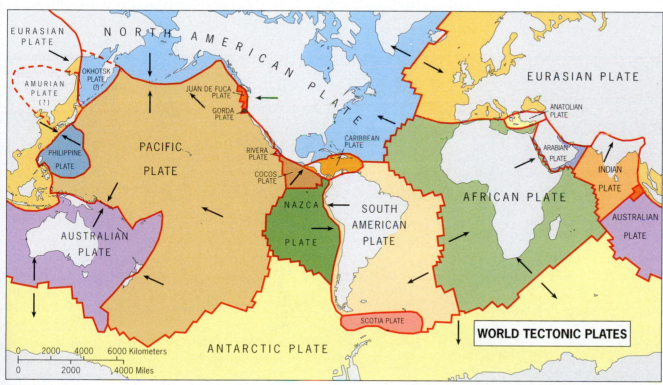

FIGURE G-5

© H. J. de Blij, P. O. Muller, and John Wiley & Sons, Inc.

Earthquakes occur frequently around the world. In May 2008, a massive earthquake registering 7.9 on the Richter scale devastated large areas of Sichuan province in southeast China and was felt throughout the region. The quake killed over 70,000 people, including many thousands of children, and left millions homeless. (© Liu Jin/Agence France-Presse/Getty Images)

Climate

The prevailing climate constitutes a key factor in the geography of realms and regions (in fact, some regions are essentially *defined* by climate). But we should always remember that climates change and that the climate predominating in a certain region today may not be the climate prevailing there several thousand years ago. Any map of climate, including the maps in this chapter, is but a still-picture of our always-changing world.

Ice Ages and Climate Change

Climatic conditions have swung back and forth for as long as the earth has had an atmosphere. Periodically, an **17** **ice age** lasting tens of millions of years chills the planet and causes massive ecological change. One such ice age occurred while Pangaea was still in one piece, between 250 and 300 million

years ago. Another started about 35 million years ago, and we are still experiencing it. The current epoch of this ice age, on average the coldest yet, is called the *Pleistocene* and has been going on for nearly two million years.

In our time of global warming this may come as a surprise, but we should remember that an ice age is not a period of unbroken, bitter cold. Rather, an ice age consists of surges of cold, during which glaciers expand and living space shrinks, separated by warmer phases when the ice recedes and life spreads poleward again. The cold phases are called **18** **glaciations**, and they tend to last longest, although milder spells create some temporary relief. The truly warm phases, when the ice recedes poleward and mountain glaciers melt away, are known as **19** **interglacials**. We are living in one of these interglacials today. It even has a geologic name: the *Holocene*.

Imagine this: just 18,000 years ago, great ice-sheets had spread all the way to the Ohio River Val-

by active volcanoes and earthquake epicenters. Appropriately, this is called the **16** **Pacific Ring of Fire**.

It is useful to compare Figure G-6 to Figure G-3 to see which of the world's geographic realms are most susceptible to the hazards inherent in crustal instability. Russia, Europe, Africa, and Australia are relatively safe; in other realms the risks are far greater in one sector than in others (western as opposed to eastern North and South America, for example). As we will find, for certain parts of the world the details on Figure G-5 present a clear and present danger. Some of the world's largest cities (e.g., Tokyo, Mexico City) lie in areas most vulnerable to sudden disaster.

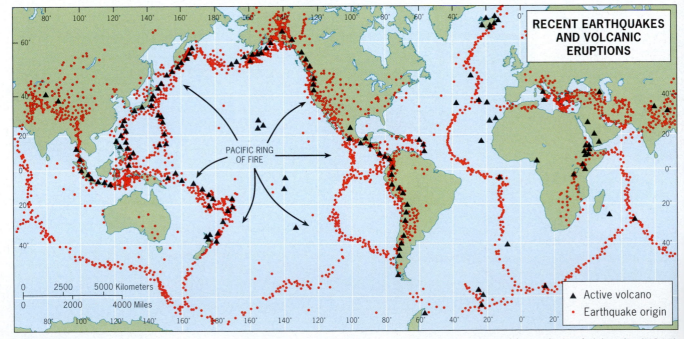

FIGURE G-6

RECENT EARTHQUAKES AND VOLCANIC ERUPTIONS

PACIFIC RING OF FIRE

▲ Active volcano
● Earthquake origin

Data courtesy U.S. National Oceanic and Atmospheric Administration (NOAA)

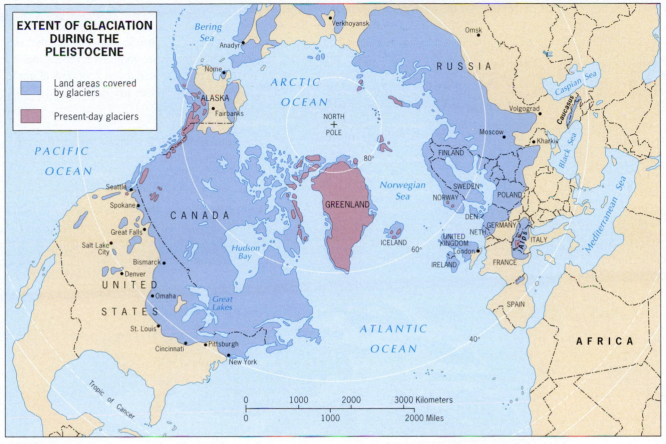

EXTENT OF GLACIATION DURING THE PLEISTOCENE

Land areas covered by glaciers

Present-day glaciers

FIGURE G-7

© H. J. de Blij, P. O. Muller, and John Wiley & Sons, Inc.

ley, covering most of the Midwest; this was the zenith of a glaciation that had lasted about 100,000 years, the *Wisconsinan Glaciation* (Fig. G-7). The Antarctic Icesheet was bigger than ever, and even in the tropics, great mountain glaciers pushed down valleys and onto plateaus. But then Holocene warming began, the continental and mountain glaciers receded, and ecological zones that had been squeezed between the advancing glaciers now spread north and south. In Europe, particularly, where humans had arrived from Africa via Southwest Asia during one of the milder spells of the Wisconsinan Glaciation, living space expanded and human numbers grew.

Global Climate Change

Consider the transformation that our planet—and our ancestors—experienced as the ice receded. Huge slabs of ice thousands of miles across slid into the oceans from Canada and Antarctica. Rivers raged with silt-clogged meltwater and formed giant deltas. Sea level rose rapidly, submerging vast coastal plains. Animals and plants migrated into newly opened lands at higher elevations and latitudes. As climatic zones shifted, moist areas fell dry, and arid areas turned wet. In Southwest Asia, where humans had begun to cultivate crops and towns signaled the rise of early states, these changes, such as increasing aridity, often destroyed what had been achieved. In

northern tropical Africa, where rivers flowed and grasses fed wildlife, a region extending from Atlantic shores to the Red Sea fell dry in a very short time around 5000 years ago. We know it today as the Sahara. We call the process **20** **desertification**.

Temperatures reached their present-day levels about 7000 years ago, but the effects of this Holocene interglacial warmth took more time to take hold, as the Sahara's later desiccation shows. Soils formed on newly exposed rock, but generally faster in warm and moist areas than in dry and still-cooler zones. Pine forests that had migrated southward during the Wisconsinan Glaciation now moved north, and equatorial rainforests expanded in all directions. Some of the world's early empires formed in areas that had been inhospitable during the Wisconsinan: the Roman Empire at the western end of Eurasia, the Han Empire at the eastern end. Europeans and Chinese traded with each other along the Silk Route, now open for business.

A great deal of this environmental history remains imprinted today (see Fig. G-3). The great river basins of East Asia, where humans exploited agricultural opportunities thousands of years ago, still anchor populous China today. The great Ganges Basin, where one of early humankind's great population explosions may have taken place, remains the core of a huge modern state, India. The Roman Empire's legacies infuse much of European culture.

The past 1000 years have witnessed some troubling environmental developments. Around 1300, the first warning signals of a return to cooler conditions caused crop failures and social dislocation, first in Europe and later in China. This episode, which worsened during the 1600s and has come to be known as the *Little Ice Age*, had major impacts on the human geography of Eurasia and on peoples elsewhere in the world as well, accompanied as it was by long-term changes in climate. A return to pre-Little Ice Age conditions commenced in the early nineteenth century, but then the exploding human population began to have a stronger impact on the atmosphere, and humanity itself became a factor in global climate change.

Today, we are living in an era of **21 global climate change**, particularly natural global warming that has been accelerated by anthropogenic (human-source) causes. Since the Industrial Revolution, we have been emitting gases that have enhanced nature's **greenhouse effect** wherein the sun's radiation becomes trapped in the earth's atmosphere. This is leading to a series of climate changes, especially the overall warm-

ing of the globe. In 2007 the Intergovernmental Panel on Climate Change (IPCC) released an important series of documents that present indisputable scientific evidence on global climate change and the possible implications of these changes. The experts predict an increase of 2–3°C (3.6–5.4°F) overall for the globe, but with significant regional variability (e.g., more at higher latitudes, less at lower latitudes). Precipitation patterns are predicted to become more variable, particularly in regions where they are already seasonal. This change in temperature may seem small but will have significant impacts on global climate patterns, agricultural zones, and the quality of human lives. The full ramifications are not known, but scenarios are being modeled so that societies can confront the changes that are coming. Leaders of some countries are more skep-

tical than others, and some have already made greater adjustments than others. One of the most significant ramifications of global climate change is that the Arctic icecap is melting faster than even recent models predicted, with environmental and geopolitical implications. We pick up this issue in Chapter 2.

Climatic Regions

We have just learned how variable climate can be, but in a human lifetime we see little evidence of this variability. We talk about the **weather** (the immediate state of the atmosphere) in a certain place at a given time, but as a technical term **climate** defines the aggregate, total record of weather conditions at a place, or in a region, over the entire period during which records have been kept.

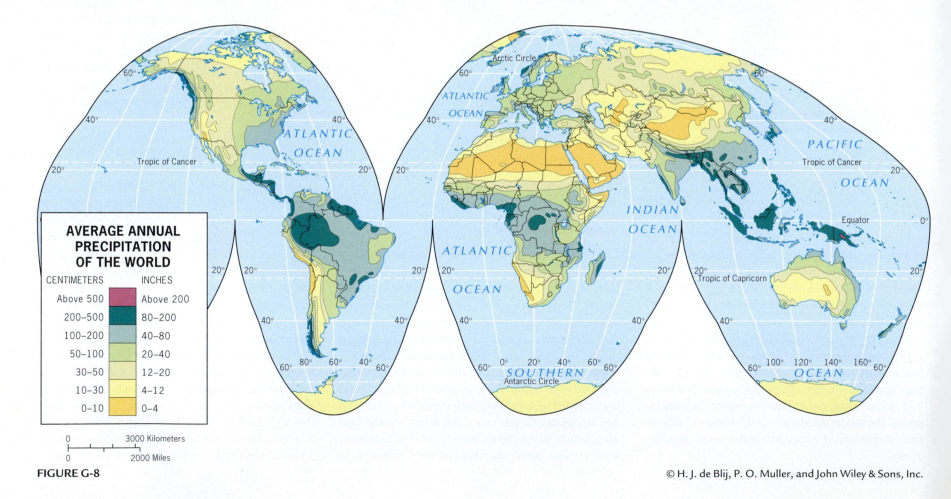

FIGURE G-8   © H. J. de Blij, P. O. Muller, and John Wiley & Sons, Inc.

WORLD CLIMATES
After Köppen–Geiger

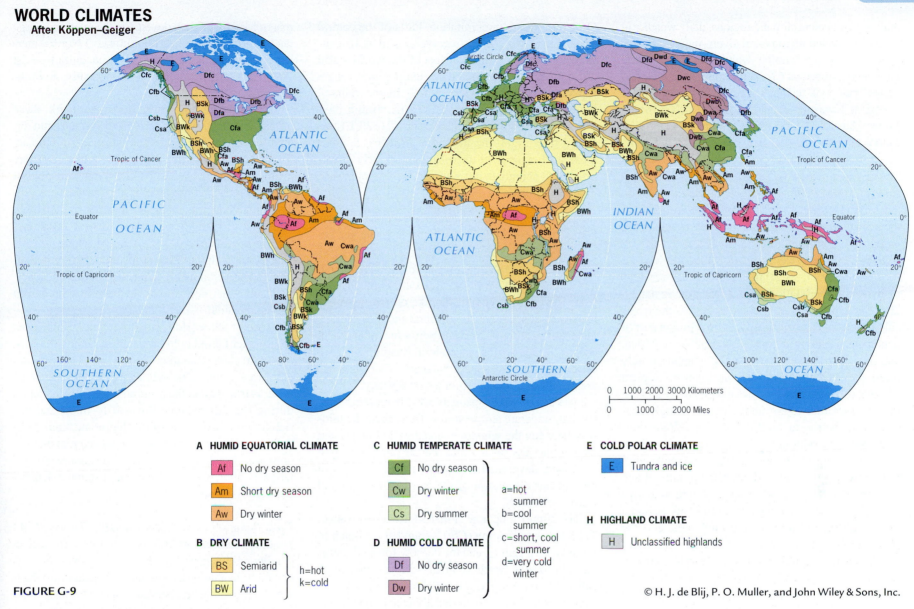

FIGURE G-9

© H. J. de Blij, P. O. Muller, and John Wiley & Sons, Inc.

A HUMID EQUATORIAL CLIMATE
- Af — No dry season
- Am — Short dry season
- Aw — Dry winter

B DRY CLIMATE
- BS — Semiarid
- BW — Arid

  h=hot
  k=cold

C HUMID TEMPERATE CLIMATE
- Cf — No dry season
- Cw — Dry winter
- Cs — Dry summer

D HUMID COLD CLIMATE
- Df — No dry season
- Dw — Dry winter

a=hot summer
b=cool summer
c=short, cool summer
d=very cold winter

E COLD POLAR CLIMATE
- E — Tundra and ice

H HIGHLAND CLIMATE
- H — Unclassified highlands

The key records are those of temperature and precipitation, their seasonal variations and extremes, but other data (humidity, wind direction, and the like) also come into play at the larger scale. Precipitation across the planet, for example, is quite variable (Fig. G-8), but this map cannot show *when*, seasonally, this precipitation arrives. Average temperatures, too, mean very little: what is needed at minimum is information on daily and monthly ranges.

For our discussion of global realms and regions, a comprehensive map is enormously important and useful. Efforts to create such a map have gone on for about a century, and Figure G-9, based on a system devised by Wladimir Köppen and modified by Rudolf Geiger, is one of the most useful because of its comparative simplicity.

The Köppen-Geiger map represents climatic regions through a system of letter symbols. The first

(capital) letter is the critical one: the **A** climates are humid and tropical, the **B** climates are arid, the **C** climates are mild and humid, the **D** climates show increasing extremes of seasonal heat and cold, and the **E** climates reflect the frigid conditions at and near the poles.

Figure G-9 merits your attention because familiarity with it will help you understand much of what follows in this book. The map has practical utility, too.

Although it depicts climatic regions, daily weather in each color-coded region is relatively standard. If, for example, you are familiar with the weather in the large area mapped as *Cfa* in the southeastern United States, you will feel at home in Uruguay (South America), Kwazulu-Natal (South Africa), New South Wales (Australia), and Fujian Province (China). Let us look at the world's climatic regions in some detail.

Humid Equatorial (A) Climates The humid equatorial, or tropical, climates are characterized by high temperatures all year and by heavy precipitation. In the *Af* subtype, the rainfall arrives in substantial amounts every month; but in the *Am* areas, the arrival of the annual wet *monsoon* (the Arabic word for season [see Fig. 8-2]) marks a sudden enormous increase in precipitation. The *Af* subtype is named after the vegetation that develops there—the tropical rainforest. The *Am* subtype, prevailing in part of peninsular India, in a coastal area of West Africa, and in sections of Southeast Asia, is appropriately referred to as the monsoon climate. A third tropical climate, the savanna (*Aw*), has a wider daily and annual temperature range and a more strongly seasonal distribution of rainfall. As Figure G-8 indicates, savanna rainfall totals tend to be lower than those in the rainforest zone, and savanna seasonality is often expressed in a 'double maximum.' Each year produces two periods of increased rainfall separated by pronounced dry spells. In many savanna zones, inhabitants refer to the long rains and the short rains to identify those seasons; a persistent problem is the unpredictability of the rain's arrival, which global warming will exacerbate. Savanna soils are not among the most fertile, and when the rains fail hunger looms. Savanna regions are far more densely peopled than rainforest areas, and millions of residents of the savanna subsist on what they cultivate. Rainfall variability is their principal environmental challenge.

Dry (B) Climates Dry climates occur in both lower and higher latitudes. The difference between the *BW* (true desert) and the moister *BS* (semiarid steppe) varies but may be taken to lie at about 25 centimeters

(10 in) of annual precipitation. Parts of the central Sahara in North Africa receive less than 10 centimeters (4 in) of rainfall. Most of the world's arid areas have an enormous daily temperature range, especially in subtropical deserts. In the Sahara, there are recorded instances of a maximum daytime shade temperature of over 49°C (120°F) followed by a nighttime low of 9°C (48°F). Soils in these arid areas tend to be thin and poorly developed; soil scientists have an appropriate name for them—Aridisols.

Humid Temperate (C) Climates As the map shows, almost all these midlatitude climate areas lie just beyond the Tropics of Cancer and Capricorn (23½° North and South latitude, respectively). This is the prevailing climate in the southeastern United States from Kentucky to central Florida, on North America's west coast, in most of Europe and the Mediterranean, in southern Brazil and northern Argentina, in coastal South Africa, in eastern Australia, and in eastern China and southern Japan. None of these areas suffers climatic extremes or severity, but the winters can be cold, especially away from water bodies that moderate temperatures. These areas lie midway between the winterless equatorial climates and the summerless polar zones. Fertile and productive soils have developed under this regime, as we will note in our discussion of the North American and European realms.

The humid temperate climates have a wide range. Their moist expression (*Cfb*) is found along much of northwestern Europe and the densely forested northwestern coast of North America, a climate zone often called Marine West Coast for short. Its dry expression (*Csa* and *Csb*), is frequently called a Mediterranean climate with characteristic dry summers. This is a small climate zone on a global scale, with pockets along coastal Southern Europe and northwestern Africa, southwestern tips of Australia and Africa, central Chile, and Central and Southern California, yet a very important climate type in terms of specialized agriculture such as grapes, olives, and other fruit.

Humid Cold (D) Climates The humid cold (or 'snow') climates may be called the continental cli-

mates, for they seem to develop in the interior of large landmasses, as in the heart of Eurasia or North America. No equivalent land areas at similar latitudes exist in the Southern Hemisphere; consequently, no *D* climates occur there.

Great annual temperature ranges mark these humid continental climates, and cold winters and relatively cool summers are the rule. In a *Dfa* climate, for instance, the warmest summer month (July) may average as high as 21°C (70°F), but the coldest month (January) might average only −11°C (12°F). Total precipitation, much of it snow, is not high, ranging from over 75 centimeters (30 in) to a steppe-like 25 centimeters (10 in). Compensating for this paucity of precipitation are cool temperatures that inhibit the loss of moisture from evaporation and evapotranspiration (moisture loss to the atmosphere from soils and plants).

Some of the world's most productive soils lie in areas under humid cold climates, including the U.S. Midwest, parts of southern Russia and Ukraine, and Northeast China. The winter dormancy (when all water is frozen) and the accumulation of plant debris during the fall balance the soil-forming and enriching processes. The soil differentiates into well-defined, nutrient-rich layers, and substantial organic humus accumulates. Even where the annual precipitation is light, this environment sustains extensive coniferous forests.

Cold Polar (E) and Highland (H) Climates Cold polar (*E*) climates are differentiated into true icecap conditions, where permanent ice and snow keep vegetation from gaining a foothold, and the tundra, which may have average temperatures above freezing up to four months of the year. Like rainforest, savanna, and steppe, the term *tundra* is vegetative as well as climatic, and the boundary between the *D* and *E* climates in Figure G-9 corresponds closely to that between the northern coniferous forests and the tundra.

Finally, the *H* climates—unclassified highlands mapped in gray (Fig. G-9)—resemble the *E* climates. High elevations and the complex topography of major mountain systems produce complex climates, including near-Arctic conditions above the tree line,

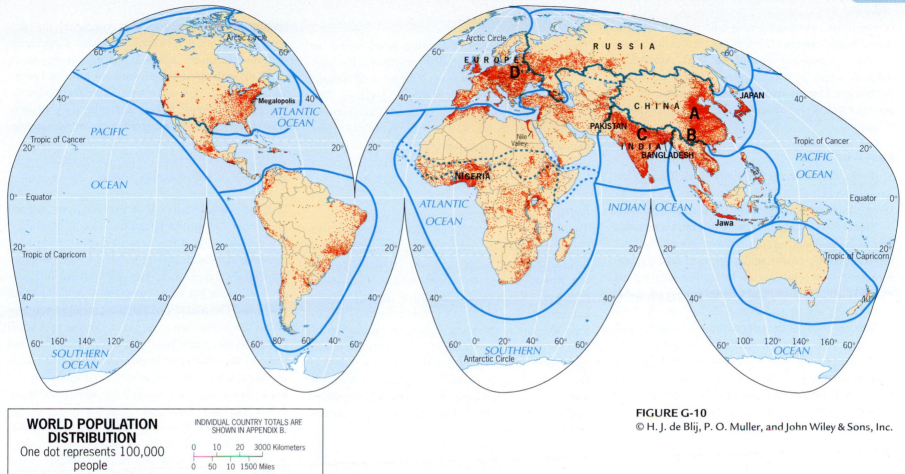

FIGURE G-10
© H. J. de Blij, P. O. Muller, and John Wiley & Sons, Inc.

even in the lowest latitudes such as the equatorial section of the high Andes of South America.

Let us not forget an important qualification concerning Figure G-9: this is a still-picture of a changing scene, a single frame from an ongoing film. Climates are changing, and less than a century from now climatologists are likely to be modifying the climate maps to reflect new data. Who knows: we may have to redraw even those familiar coastlines. Environmental change is a never-ending challenge.

REALMS OF POPULATION

Earlier we noted that population numbers by themselves do not define geographic realms or regions. Population distributions, and the functioning society that gives them common ground, are more significant

criteria. That is why we can identify one geographic realm (the Austral) with less than 25 million people and another (East Asia) with more than 1.5 billion inhabitants. Neither population numbers nor territorial size alone can delimit a geographic realm. Nevertheless, the map of world population distribution (Fig. G-10) suggests the relative location of several of the world's geographic realms, based on the strong clustering of population in certain areas. Before we examine these clusters in some detail, remember that the world's human population now totals just over 6.7 billion—six thousand seven hundred million people confined to the landmasses that constitute less than 30 percent of our planet's surface, much of which is arid desert, rugged mountain terrain, or frigid tundra. (Remember that Fig. G-10 is another still-picture of an ever-changing scene: the rapid

growth of humankind continues.) After thousands of years of slow growth, world population during the nineteenth and twentieth centuries grew at an increasing rate. That rate has recently been slowing down, even imploding in some parts of the world, but consider this: it took about 17 centuries after the birth of Christ for the world to add 250 million people to its numbers; now we are adding 250 million about every three and a half years.

Major Population Clusters

One way to present an overview of the location of people on the planet is to create a map of **22** **population distribution** (Fig. G-10). As you see on the map's legend, every dot represents 100,000 people, and the clustering of large numbers of people in certain areas as well as the near-emptiness of others is immediately evident. There is a technical difference between population distribution and *population density*, which is another way of showing where people are. Density maps reveal the number of persons per unit area, requiring a different cartographic technique.

East Asia

Still the world's greatest population cluster, *East Asia* lies centered on China and includes the Pacific-facing Asian coastal zone from the Korean Peninsula to Vietnam. Not long ago, we would have reported this as a dominantly rural, farming population, but rapid economic growth and associated urbanization are changing this picture. In the interior river basins of the Huang (Yellow) and Chang/Yangzi (*A* and *B* on the map), and in the Sichuan Basin between these two letters, most of the people remain farmers, and, as Appendix B shows, farmers still outnumber city-dwellers in China as a whole. But the great cities of coastal and near-coastal China are attracting millions of new inhabitants, and interior cities are growing rapidly as well. By 2020, the East Asia cluster will be more urban than rural.

South Asia

The *South Asia* population cluster lies centered on India and includes its populous neighbors, Pakistan and Bangladesh. This huge agglomeration of humanity focuses on the wide plain of the Ganges River (*C* in Fig. G-10). It is nearly as large as that of East Asia and at present growth rates will overtake East Asia in less than 20 years. A larger percentage of the people remain farmers, although pressure on the land is greater, while farming is less efficient than in East Asia.

Europe

The third-ranking population cluster, *Europe*, also lies on the Eurasian landmass but at the opposite end from China. The European cluster, including western Russia, counts over 700 million inhabitants, which puts it in a class with the two larger Eurasian concentrations—but there the similarity ends. In Europe, the key to the linear, east-west orientation of the axis of population (*D* in Fig. G-10) is not a fertile river basin but a zone of raw materials for industry. Europe is among the world's most highly urbanized and industrialized realms, its human agglomeration sustained by factories and offices rather than paddies and pastures.

The three world population concentrations just discussed (East Asia, South Asia, and Europe) account for more than 3.7 billion of the world's 6.7 billion people. No other cluster comes close to these numbers. The next-ranking cluster, *Eastern North America*, is only about one quarter the size of the smallest Eurasian concentrations. As in Europe, the population in this area is concentrated in major metropolitan complexes; the rural areas are now relatively sparsely settled. Geographic realms and regions, therefore, display varying levels of **23** **urbanization**, the percentage of the total population living in cities and towns. Some regions are urbanizing much more rapidly than others, a phenomenon we will explain as we examine the world's realms.

REALMS OF CULTURE

Imagine yourself in a boat on the Nile River, headed upstream (south) from Khartoum, Sudan. The desert sky is blue, the heat is searing. You pass villages on the shore that look much the same: low, square or rectangular dwellings, some recently whitewashed, others gray, with flat roofs, wooden doors, and small windows. The minaret of a modest mosque may rise above the houses, and you get a glimpse of a small central square. There is very little vegetation; here and there a hardy palm tree stands in a courtyard. People on the paths wear long white robes and headgear, also white, that looks like a baseball cap without the visor. A few goats lie in the shade. Along the river's edge lie dusty farm fields that yield to the desert in the distance. At the foot of the river's bluff lie some canoes.

All this is part of the central Sudan's rural **24** **cultural landscape**, the distinctive attributes of a society imprinted on its portion of the world's physical stage. The cultural landscape concept was first articulated in the 1920s by a University of California geographer named Carl Sauer, who stated that "a cultural landscape is fashioned from a natural landscape by a culture group" and that "culture is the agent; the natural environment the medium." What this means is that people, starting with their physical environment and using their culture as their agency, fashion a landscape that is layered with forms such as buildings, gardens, and roads, and also modes of dress, aromas of food, and sounds of music.

Continue your journey southward on the Nile, and you will witness a remarkable transition. Quite suddenly, the square, solid-walled, flat-roofed houses of central Sudan give way to round, wattle-and-thatch, conical-roofed dwellings of the south. You may note

The population explosion of the mid-to-late twentieth century may be over, with overall global population stabilizing and parts of the world losing population. Yet population growth is still a reality in many parts of the world. Through globalization and migration, an increasing interaction among many different peoples expands.
(© AP/Wide World Photos)

that clouds have appeared in the sky: it rains more here, and flat roofs will not do. The desert has given way to green. Vegetation, natural as well as planted, grows between houses, flanking even the narrow paths. The villages seem less orderly, more varied. People ashore wear a variety of clothes, the women often in colorful dresses, the adult men in shirts and slacks, but shorts when they work the fields, although you see more women wielding hoes than men. You have traveled from one cultural landscape into another, from Arabized, Islamic Africa to animist or Christian Africa. You have crossed the boundary between two geographic realms.

No geographic realm, not even the Austral Realm, has just one single cultural landscape, but cultural landscapes help define realms as well as regions. The cultural landscape of the high-rise North American city with its sprawling suburbs differs from that of Brazil and South America; the organized terraced paddies of Southeast Asia are unlike anything to be found in the rural cultural landscape of neighboring Australia. Variations of cultural landscapes *within* geographic realms, such as between highly urbanized and dominantly rural (and more traditional) areas, help us define the world's regions.

REALMS, REGIONS, AND STATES

Our analysis of the world's regional geography requires data, and it is crucial to know the origin of these data. Unfortunately, we do not have a uniformly sized grid that we can superimpose over the globe: we must depend on the world's 190-plus countries to report vital information. Irregular as the boundary framework shown on the world map (see front endpaper or Fig. G-11) may be, it is all we have. Fortunately, all large and populous countries are subdivided into provinces, States, or other

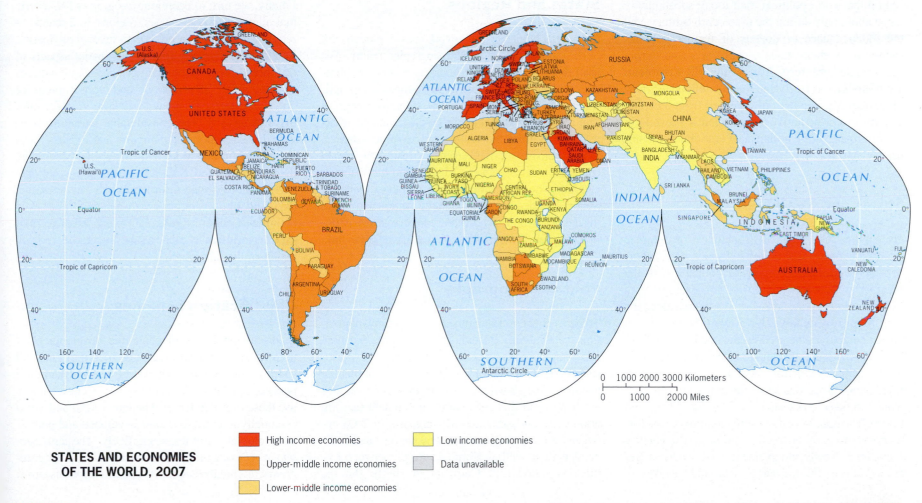

**STATES AND ECONOMIES
OF THE WORLD, 2007**

- 🟥 High income economies
- 🟧 Upper-middle income economies
- 🟨 Lower-middle income economies
- 🟨 Low income economies
- ⬜ Data unavailable

FIGURE G-11

Adapted with permission from *World Bank Atlas*, pp. 7, 64. The World Bank, 2007.

internal entities, and their governments provide information on each of these subdivisions when they conduct their census.

The State

The 25 **state** as a political, social, and economic entity has been developing for thousands of years, ever since agricultural surpluses made possible the growth of large and powerful towns that could command hinterlands and control peoples far beyond their walls. But the modern state is a relatively recent phenomenon. The boundary framework we see on the world political map today substantially came about only during the nineteenth century, and the independence of dozens of former colonies (which made them states as well) occurred during the twentieth century. Today, the 26 **European state model**—a clearly and legally defined territory inhabited by a population governed from a capital city by a representative government—prevails in the aftermath of the collapse of colonial and communist empires.

States and Realms

As Figures G-3 and G-10 suggest, geographic realms are mostly assemblages of states, and the borders between realms frequently coincide with the boundaries between countries—for example, between North America and Middle America along the U.S.-Mexico boundary. But a realm boundary can also cut *across* a state, as does the one between Subsaharan Africa and the Muslim-dominated realm of North Africa/Southwest Asia. Here the boundary takes on the properties of a wide transition zone, but it still divides states such as Nigeria, Chad, and Sudan. The transformation of the margins of the former Soviet Union is creating similar cross-country transitions. Newly independent states such as Belarus (between Europe and Russia) and Kazakhstan

(between Russia and Muslim Southwest Asia) lie in transition zones of regional change.

Most often, however, geographic realms consist of groups of states whose boundaries also mark the limits of the realms. Look at Southeast Asia, for instance. Its northern border coincides with the political boundary that separates China (a realm practically unto itself) from Vietnam, Laos, and Myanmar (Burma). The boundary between Myanmar and Bangladesh (which is part of the South Asian realm) defines its western border. Here, the state boundary framework helps delimit geographic realms.

States and Regions

The global boundary framework is even more useful in delimiting regions *within* geographic realms. We shall discuss regional divisions every time we introduce a geographic realm, but an example is appropriate here. In the Middle American realm, we recognize four regions. Two of these lie on the mainland: Mexico, the giant of the realm, and Central America, which consists of the seven comparatively small countries located between Mexico and the Panama-Colombia border (which is the boundary with the South American realm). Central America often is misdefined in news reports; the correct regional definition is based on the politico-geographical framework.

Political Geography

To our earlier criteria of physical geography, population distribution, and cultural geography, we now add political geography as a determinant of world-scale geographic regions. In doing so, we should be aware that the global boundary framework continues to change and that boundaries are created (e.g., in 2006 between Serbia and Montenegro; in 2008 between Serbia and Kosovo as Kosovo declared independence) as well as eliminated (e.g., between former West and East Germany in 1990). But the

overall system, much of it resulting from colonial and imperial expansionism, has endured, despite the predictions of some geographers that the 'boundaries of imperialism' would be replaced by newly negotiated ones in the postcolonial period.

In Chapters 2 and 12, we will discuss a recent development in boundary-making: the extension of boundaries onto and into the oceans and seas. This process has been dividing up the last of the earth's open frontiers, with uncertain consequences.

PATTERNS OF DEVELOPMENT

Finally, we turn to economic geography for criteria that allow us to place states, regions, and realms in regional groupings based on their level of prosperity. Economic geography focuses on spatial aspects of how people make a living; thus it deals with patterns of production, distribution, and consumption of goods and services. As with all else in this world, these patterns exhibit considerable variation. Individual states report their imports and exports, farm and factory production, and many other economic data to the United Nations and other international agencies. From such information we can determine the comparative economic well-being of the world's countries. Economic geographers use the concept of 27 **development** to gauge a state's economic, social, and institutional growth.

Classification Schemes

In discussing levels of economic development, it used to be that the world was divided into First and Third World countries. This classification system still persists in many conversations, but we choose not to use these outdated terms. The terms *First* and *Third* essentially referred to rich or developed and poor or undeveloped countries, respectively. The term *Second World* was rarely used, but referred to the communist (command economy) countries, for this terminology was a creation of the Cold War when the First World and the Second World fought over (literally and figuratively) peoples of the developing or Third World. Some observers even added a Fourth World for very poor countries. Given the fall of command economics, this classification has lost its relevance today. Not only that, but there is a much larger continuum of income levels for countries around the world.

We feel that the classification scheme offered by the World Bank (one of the agencies that monitor economic conditions around the world) is much more suitable and relevant today. It sorts countries into four groups: (1) high-income, (2) upper-middle-income, (3) lower-middle-income, and (4) low-income countries. These groupings display regional clustering when mapped, as you can see in Figure G-11. Certain geographic-realm boundaries stand out on this economic map, including those between North America and Middle America as well as between Australia and Southeast Asia. Within several geographic realms, you can also see some distinct regional boundaries: between Western and Eastern Europe, between the oil-rich Arabian Peninsula and the countries of the Middle East to its northwest, and between some countries in Southern Africa and its numerous much poorer neighbors to the north.

Causes of Contrast

Figure G-11 reveals the level of economic development among the world's countries—based, we should remind ourselves, on the political framework as the reporting grid. But the political entities as pictured were created and may not reflect the entire story of economic development. The spatial contrasts in well-being have been attributed to many factors, including climate (as a conditioner of human capacity), cultural heritage (e.g., receptivity or resistance to change and innovation), colonial exploitation, neocolonialism, and globalization. Obviously, the distribution of accessible natural resources and the relative location of countries play their parts.

Certain areas and peoples have had more opportunities for interaction and exchange than others. Also, richer countries perpetuate their advantages by erecting high tariffs (tolls) against the products of poorer countries and by subsidizing their own producers (such as farmers) when they compete on world markets. Europe, the United States, and Japan do this even as they advocate free trade and open markets, giving more than $300 billion to their farmers and thus depriving poor-country farmers of the opportunity to compete on the global 'level playing field' they claim to envision. Such unfair policies have much to do with the pattern you see in Figure G-11.

One way to explain these contrasts today is through the use of the concept of **28 core-periphery relations** developed by the sociologist Immanuel Wallerstein in his highly influential book, *World-Systems Theory*. This theory postulates that **29 core areas**—the most dominant and productive—need peripheries in order to remain central and dominant. In fact, the relationship between core and periphery is often exploitative—the core 'exploiting' the periphery. The periphery is where resources are and where the cheap labor is—in sum, where the costs of production can be kept lower. This concept can be applied globally, regionally, nationally, and even locally. For example, most people accept the fact that the global economic engines are still centered on North America, Western Europe, and Japan. With respect to this core, the rest of the world is the periphery and is exploited by the core for its resources and cheap labor (e.g., China). But taking North America by itself we see that it, too, has a core—the east and west coasts for the United States and 'Main Street' for Canada (Chapter 3). The flyover zone in between is its periphery, exploited for resources and with lower costs of doing business. This works at an individual State level as well. Take the State of Michigan; it has a core in its southeast centered on Detroit, and a

periphery in the north, especially its Upper Peninsula, which it exploits for its resources. All realms and regions have cores and peripheries, and many states do as well.

Core-periphery relations have existed for many centuries. During the eras of exploration and colonialism, these relations became both visible and

Countries have cores and peripheries just as the world as a whole does. Brazil, the fifth largest country in terms of area, has an extensive periphery deep in the interior and away from its coastal core–the Amazon Basin (see Fig. 5-1). This region has long been exploited for its resources. A century ago, wild rubber was harvested from the forest; more recently gold, iron ore, and various other minerals as well as lumber have been extracted. Today the region faces unprecedented pressure both within and outside of Brazil between those who wish to preserve its majestic forests and its biodiversity and those who want to benefit from its resources and economically develop the country. Here bauxite (the raw material to make aluminum) is being mined near the company town of Porto Trombetas, in the western part of Pará State in the Brazilian Amazon region. After clearing the forest, the mining is done mechanically by Mineração Rio do Norte (a multinational conglomerate), partially processed on-site, and then shipped by barge elsewhere for processing into aluminum. One of the largest bauxite mines in the world, it produces about 16.3 million tons of bauxite per year. After extraction, the ground is restored and the forest replanted.   (© A. WinklerPrins)

Transnational corporations—many but not all are based in the United States—dominate commerce around the world. Some have larger economies than countries have. Given their economic strength, they can exert enormous power, sometimes convincing countries to change policies. This General Motors plant is located in Toluca, Mexico, a city near Mexico City and part of Mexico's industrial core. GM is one of the largest companies in the world.   (© AP/Wide World Photos)

strong as European empires exploited their colonies for resources and labor. With independence many new states had hoped to throw off these relationships, but modern times have, in many cases, worsened, not improved, this core-periphery dichotomy. Infrastructure development and power relations remain dominated by the core, and this continuing relationship is referred to as *neocolonialism*.

Centers of dominance remain in Western Europe, North America, and Japan where power, money, influence, and decision-making (and enforcing) organizations are concentrated. It is here that the terms of an economic system that perpetuate the core's primacy are dictated, and countries of the periphery remain 'stuck' in an exploitative relationship whose terms of engagement they have no role in setting.

The core's impact on the periphery takes many forms. As core countries ensure that the prices of commodities from periphery sources remain low and that competition from countries on the margin is suppressed, their wealth and opportunity attract skilled and professional people from afar. How many thousands of doctors, engineers, and teachers from lower income countries such as India are working in England and the United States? This export of human talent is often referred to as the 'brain drain' and can perpetuate core-periphery relations as the periphery suffers from a lack of skilled human resources.

The countries of the periphery, therefore, face severe problems. They are pawns in a global economic game whose rules they cannot change. Their internal problems are intensified by the aggressive involve-

ment of core-area interests. The way they use resources (such as whether to produce food for local consumption or to produce export crops) is strongly affected by foreign interference. They suffer far more from environmental degradation, overpopulation, and mismanagement than core-area countries do. Their inherited disadvantages have grown, not lessened, over time. The widening gaps that result between enriching cores and persistently impoverished peripheries threaten the future of the world. Better than 'national' economic statistics, an index representing **30** **regional disparity** would give us a truer picture of the economic and social conditions in geographic realms, their component regions, and individual countries (see, e.g., Fig. 5-12).

Today we are witnessing a push back from the periphery. Slowly, new alliances are forming, especially between periphery countries such as India, Brazil, and South Africa (e.g., IBSA). They have become allies in their joint effort to turn the tide against organizations such as the World Trade Organization (see pp. 21–22). Specifically, they seek to change the terms of trade that will more fairly treat countries disadvantaged in global competition by the trade barriers and tariffs imposed by the core. Global trade talks are at a virtual standstill today because of the rising periphery's demand that the terms of engagement be altered.

The Role of Corporate Power

We have alerted you to be cautious about learning about the world solely along political lines. Certainly, the world is divided in this way, but the reality is that above the level of many states are corporations, usually multi- or transnational corporations, that wield more economic and hence political clout than many countries. They have offices and operations in many countries and are influential in terms of their economic strength that they can impose their political will on the states in which they operate. The following table lists the top ten companies in the world in terms of their earned income for 2007.

These corporations all have incomes that rank them among the fifty largest economies in the world. Given that there are around 190 countries in the world, one can envision the kind of economic power these corporations hold.

Rank	Company	2007 Revenue (in millions of $)
1	Wal-Mart	351,139
2	Exxon Mobil	347,254
3	Royal Dutch Shell	318,845
4	British Petroleum (BP)	274,316
5	General Motors	207,349
6	Toyota Motor	204,746
7	Chevron	200,567
8	DaimlerAG	190,191
9	ConocoPhillips	172,451
10	Total Group	168,357

The Specter of Debt

The core continues to control the periphery in part by ensuring that the periphery remains in debt. All of us are aware of the dangers of going too deeply into debt. Borrow too much money, and payments of interest and principal may leave too little for routine needs such as food and clothing, not to mention repairs and replacement of equipment. Individuals, families, villages, and cities must devise budgets and balance incomes against expenditures.

So it is with countries. National governments, like individuals and families, must sometimes borrow to make ends meet. If a drought curtails domestic food production, a government may borrow against future oil or ore exports to pay for urgently needed staples. Should domestic harvests fail in the following year as well, further borrowing will be necessary. This means that the government's future budgets will be burdened by ever-higher debt retirement payments, limiting its ability to spend on schools, roads, and clinics. Should the decision be made to raise taxes to help make the payments, the result may be a slowdown in the economy. The cycle of debt is hard to escape.

Underlying Causes

All too often it is not just a food emergency or some other unavoidable problem, but a government's mismanagement, that leads to excessive debt. In the postcolonial period, many governments of former colonies sought shortcuts to boost their economies, building dams, factories, and ports that did not justify their huge investments. Of course, the governments and corporations of the former colonial powers were all too willing to do the building, making large profits and even lending the new governments money to build still more.

Soon, many of the civilian governments that took over from the colonial powers, weakened by such errors and losing the confidence of their citizens, fell prey to military takeovers. And the military dictatorships that replaced them needed weapons to maintain their hold on power—weapons readily sold for credit by the arms manufacturers in the wealthy core countries. During the Cold War, dictatorial regimes in the periphery became close allies of both superpowers, and proxy wars were fought in Asia, Africa, and the Americas. By the time the Cold War ended and representative government began to return to some (though not all) of these countries, they were so deeply in debt to the rich core countries that they had no prospect of recovery.

Debt Relief

External debt is not a problem just for countries in the periphery. Even rich countries (these days prominently including the United States) have national debts. What matters is a country's ability to service (repay) its debt and avoid a financial crisis arising from such payments. In the case of a rich country, a growing national debt might lead to inflation and higher taxes; in a poor country, it might entail financial collapse. Periodically, institutions based in the core will offer debt relief for countries most unable to pay off what they owe. However, the underlying reasons for the debt are rarely addressed, and so poor countries easily fall back into debt, enabling the core to continue exploiting the periphery.

Globalization

If a geographic concept can arouse strong passions, **31** **globalization** is it. Many definitions exist; we consider globalization to be the increase in economic and cultural integration around the world. Cultural dimensions (such as the growth of world music) are part of the process but it is most frequently thought of in economic terms. To leading economists, politicians, and businesspeople, this is the best of all worlds: the march of international capitalism, open markets, and free trade. To millions of poor farmers and powerless citizens in debt-mired African and Asian countries, it represents a system that will keep them in poverty and subservience forever. Economic geographers can prove that global economic integration allows the overall economies of poorer countries to grow faster: the ***Gross National Income (GNI)***[2] of those economies that engage in more international trade (and thus are more globalized) rises, while the GNI of those with less international trade actually declines. But the problem is that those poorest countries contain more than 2 billion people, and their prospects in this globalizing world are worsening, not rising, especially as the terms of trade remain set by the rich core countries. In places as diverse as the Philippines, Kenya, and Nicaragua, income per person has shrunk, and globalization is seen as a culprit, not a cure.

One-Way Street?

Globalization in the economic sphere is proceeding under the auspices of the World Trade Organization

---

[2]**GNI** or **gross national income** is the total income earned from all goods and services produced by the citizens of a country, within or outside of its borders, during a calendar year, usually expressed on a per capita basis. **GDP** or **gross domestic** (national) **product** is a term that is frequently used but refers only to the total value of all the goods and services produced in a given political entity, expressed on a per capita basis.

Globalization, or the implications of it, is not desired by many in the world. Here, in Kunda, a coastal town in Kenya, demonstrators held placards to protest a meeting of trade ministers whose goal was to give fresh impetus to World Trade Organization (WTO) talks on global commerce. In the developing world, anti-globalization sentiments usually are about the loss of control over livelihoods and resources.   (© AP/Wide World Photos)

(WTO), of which the United States is the leading architect. To join, countries must agree to open their economies to foreign trade and investment and to adhere to a set of economic rules discussed annually at ministerial meetings. In mid 2008, the WTO had 151 member-states, all expecting benefits from their participation. But the driving forces of globalization, mostly high-income countries, did not always do their part in giving the poorer countries the chances the WTO seemed to promise. The case of the Philippines is often cited: joining the WTO, its government assumed, would open world markets to Filipino farm products, which can be brought to those markets cheaply because farm workers' wages in the Philippines are very low. Economic

planners forecast a huge increase in farm jobs, a slow rise in wages, and a major expansion in produce exports. Instead, the Filipino farmers found themselves competing against U.S. and European farmers, who receive subsidies for domestic production as well as the export of their products—and losing out. Meanwhile, low-priced, subsidized American maize (corn) appeared on Filipino markets. As a result, the Philippine economy lost several hundred thousand farm jobs, wages went down, and WTO membership severely damaged its agricultural sector. Not surprisingly, the notion of globalization is not popular among rural Filipinos.

This story has parallels involving banana and sugar growers in the Caribbean, cotton and sugar

producers in various African countries, and maize (corn), fruit, and vegetable farmers in Mexico. Economic globalization has not been the two-way street most WTO members, especially the poorer countries, had hoped. Nor is rich-country protectionism confined to agriculture. When cheap foreign steel began to threaten what remains of the U.S. steel industry in early 2002, the American government erected tariffs to guard domestic producers against unwanted competition. These tariffs were removed in late 2003, however, and the U.S. steel industry was forced to compete on the global market.

An Open World

When meetings of economic ministers take place in major cities, they attract thousands of vocal and sometimes violent opponents who demonstrate and occasionally riot to express their opposition. This happens because globalization is seen as more than an economic strategy to perpetuate the advantages of rich countries (the *core*): it is also viewed as a cultural threat. Globalization constitutes Americanization, these protesters say, eroding local traditions, endangering moral standards, and menacing the social fabric. Such views are not confined to the disadvantaged, poorer countries still losing out in our globalizing world. In France the leader of a so-called Peasant Federation became a national hero after his sympathizers destroyed a McDonald's restaurant in what he described as an act of cultural self-defense. McDonald's restaurants, he argued, promote the consumption of junk food over better, more traditional fare. Elsewhere the target is violent American movies or the lyrics of certain American popular music.

Nevertheless, the process of globalization—in cultural as well as economic spheres—continues to change our world, for better and for worse. In the United States, opponents of globalization bemoan the loss of American jobs to foreign countries where labor is cheaper. In theory, globalization breaks down barriers to international trade, stimulates commerce, brings jobs to remote places, and promotes social, cultural, political, and other kinds of exchanges. High-tech workers in India are employed by computer firms based in California. Japanese cars are assembled in Thailand. American shoes are made in China. McDonald's restaurants spread standards of service and hygiene as well as familiar (and standard) menus from Tokyo to Tel Aviv. If wages and standards of employment are lower in newly involved countries than in the global core, the gap is shrinking, employment conditions are more open to scrutiny, and global openness is a byproduct of globalization.

A New Revolution

We should be aware that the current globalization process is a revolution—but also that it is not the first of its kind. It can be argued that the first 'globalization revolution' occurred during the fifteenth century, extending into the nineteenth and early twentieth centuries, when Europe's exploration and then colonial expansion spread ideas, inventions, products, and habits around the world. Colonialism transformed the world as the European powers built cities, transport networks, dams, irrigation systems, power plants, and other facilities, often with devastating impact on local traditions, cultures, and economies. From goods to games (soap to soccer), people in much of the world started doing similar things. The largest of all colonial empires, that of Britain, made English a worldwide language, a key element in the current, second globalization process.

The current globalization is even more revolutionary than the precolonial and colonial phases because it is driven by more modern, higher-speed communications. When the British colonists planned the construction of their ornate Victorian government and public buildings in (then) Bombay, the architectural drawings had to be prepared in London and sent by boat to India. When the Chinese government in the 1980s decided to create a Manhattan-like commercial district on the riverfront in Shanghai (Pudong), the plans were drawn in the United States, Japan, and

Western Europe and transmitted to Shanghai via the Internet. One container ship carrying products from China to the American market hauls more cargo than a hundred colonial-era boats.

Also, as the chapters that follow will show frequently, even as the world's national political boundaries are being tightened for fear of illegal immigration and terrorism, economic boundaries are becoming increasingly porous. Economic alliances enable manufacturers to send raw materials and finished products across borders that once inhibited such exchanges. Groups of countries forge economic unions whose acronyms (EU, NAFTA, Mercosur/l) stand for freer trade. The ultimate goal of the World Trade Organization is to lower the remaining trade barriers the world over, boosting not just regional commerce but also global trade.

The Future

As with all revolutions, the overall consequences of the ongoing globalization process are uncertain.

FROM THE FIELD NOTES

Globalization in China.   "I had just visited the top of the Oriental Pearl Tower in the Pudong district of Shanghai when I encountered this McDonald's 'hut' selling only soft-serve ice-cream. The hut was located along the walkway that follows the Huangpu River, a tributary of the Chiang Jiang, capitalizing on the many pedestrians visiting the district. One of the aspects of globalization is its perceived homogenization of many elements of culture. McDonald's has often been the 'face' of all that is wrong with globalization. This despite its attempts in most parts of the world to adapt its menu to local demand and taste. Here the localization effort was to limit the menu to only ice-cream.

The Pudong district had once been one of the poorest districts in Shanghai until it was declared a special economic zone and razed to become a new commercial district. Today it is a veritable forest of skyscrapers, and the Oriental Pearl (TV) Tower offers visitors astounding views when the air pollution is not too bad." (© A. WinklerPrins)

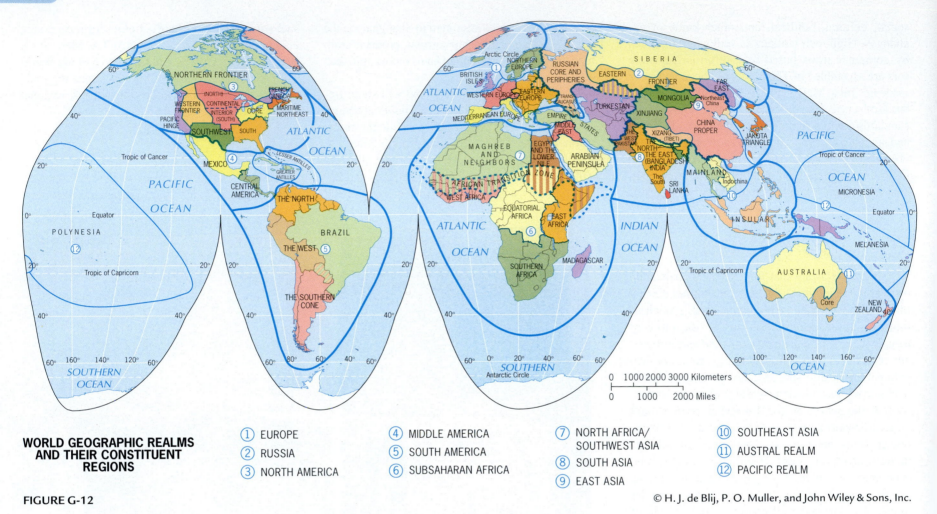

**WORLD GEOGRAPHIC REALMS
AND THEIR CONSTITUENT
REGIONS**

① EUROPE
② RUSSIA
③ NORTH AMERICA

④ MIDDLE AMERICA
⑤ SOUTH AMERICA
⑥ SUBSAHARAN AFRICA

⑦ NORTH AFRICA/
   SOUTHWEST ASIA
⑧ SOUTH ASIA
⑨ EAST ASIA

⑩ SOUTHEAST ASIA
⑪ AUSTRAL REALM
⑫ PACIFIC REALM

FIGURE G-12

© H. J. de Blij, P. O. Muller, and John Wiley & Sons, Inc.

Critics emphasize that one of its outcomes is a growing gap between rich and poor, a polarization of wealth that will destabilize the world. Core-periphery contrasts are intensified, not lessened, by globalization as the poor in peripheral societies are exploited by core-based corporations. Proponents argue that, as with the Industrial Revolution, it will take time for the benefits to spread—but that globalization's ultimate effects will be advantageous to all.

The world is functionally shrinking; we will find evidence for this in these chapters. But the 'global village' still retains its distinctive neighborhoods; globalization has not erased identity. Some argue that the rise of globalization and homogenization is simultaneously giving rise to *localization*, a re-engagement with local traditions, languages, and cuisines. Locally grown food and native varieties of crops, a resurgence of local language dialects, and the revival of local crafts and music flourish. In reality, globalization is an infusion of new with a resurgence of old ways. In the chapters that follow, we use the vehicle of geography to visit and investigate them.

REGIONAL FRAMEWORK AND GEOGRAPHIC PERSPECTIVE

At the beginning of this chapter, we discussed how geographers like to classify geographical information and outlined a map of the great geographic realms of the world (Fig. G-3). We then addressed the task of dividing those realms into regions, using criteria ranging from physical geography to cultural geography to economic geography. The result is Figure G-12, which includes both realms and regions.

On this map, note that we display not only the great geographic realms but also the regions into which they subdivide. The numbers in the legend reveal the order in which the realms and regions are discussed, starting with Europe (1) and ending with the Pacific Realm (12).

As this introductory chapter demonstrates, our world regional survey is no mere description of places and areas. We have combined the study of realms and regions with a look at geography's ideas and concepts—the notions, generalizations, and basic theories that make the discipline what it is. We continue this method in the chapters ahead so that we will become better acquainted with the world and with geography. By now you are aware that geography is a wide-ranging, multifaceted discipline. It is often described as a social science, but that is only half the story. In fact, geography straddles the divide between the social and the physical (natural) sciences. Many of the ideas and concepts you will encounter have to do with the multiple interactions between human societies and natural environments.

Regional geography allows us to view the world in an all-encompassing way. As we have seen, regional geography borrows information from many sources to create an overall image of our divided world. Those sources are not random. They represent topical or **systematic geography**. Research in the systematic fields of geography makes our world-scale generalizations possible. As Figure G-13 shows, these systematic fields relate closely to those of other disciplines. Cultural geography, for example, is allied with anthropology; it is the spatial perspective that distinguishes cultural geography. Economic geography focuses on the spatial dimensions of economic activity; political geography concentrates on the spatial imprints of political behavior.

THE RELATIONSHIP BETWEEN REGIONAL AND SYSTEMATIC GEOGRAPHY

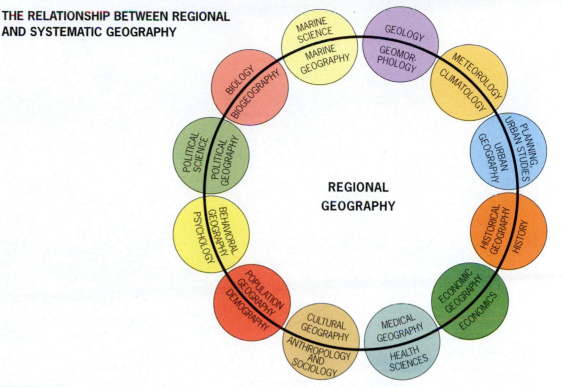

FIGURE G-13

© H. J. de Blij, P. O. Muller, and John Wiley & Sons, Inc.

Other systematic fields include historical, medical, behavioral, environmental, and urban geography. We will also draw on information from biogeography, marine geography, population geography, and climatology (as we did earlier in this chapter).

These systematic fields of geography are so named because their approach is global, not regional. Take the geographic study of cities, urban geography. Urbanization is a worldwide process, and urban geographers can identify certain human activities that all cities in the world exhibit in one form or another. But cities also display regional properties. The typical Japanese city is quite distinct from, say, the Brazilian city. Regional geography, therefore, borrows from the systematic field of urban geography, but it injects this regional perspective.

In the following chapters we call upon these systematic fields to give us a better understanding of the world's realms and regions. As a result, you will gain insights into the discipline of geography as well as the regions we investigate. This will prove that geography is a relevant and practical discipline as we attempt to comprehend and cope with our fast-changing world.

1

EUROPE

In This Chapter

- Why Europe is becoming the world's home for the aged
- Europe's unifying express hits several roadblocks
- Provinces behaving like countries rile national governments
- How Islam is changing Europe's cultural geography
- Why the British are different
- Problems unresolved in fractious Eastern Europe

CONCEPTS, IDEAS, AND TERMS

1. Land hemisphere
2. Physiography
3. Infrastructure
4. Local functional specialization
5. *The Isolated State*
6. Model
7. Industrial Revolution
8. Nation-state
9. Nation
10. Centrifugal forces
11. Centripetal forces
12. Indo-European language family
13. Complementarity
14. Transferability
15. Intervening opportunity
16. Primate city
17. Supranationalism
18. Devolution
19. Four Motors of Europe
20. Regional state
21. Site
22. Situation
23. Conurbation
24. Landlocked location
25. Break-of-bulk
26. *Entrepôt*
27. Shatter belt
28. Balkanization
29. Exclave
30. Irredentism

REGIONS

WESTERN EUROPE
THE BRITISH ISLES
NORTHERN (NORDIC) EUROPE
MEDITERRANEAN EUROPE
EASTERN EUROPE

Photos: (*upper left*) Paris, France, © A. WinklerPrins; (*above*) Porto, Portugal, © V. WinklerPrins.

FIGURE 1-1   *Map:* © H. J. de Blij, P. O. Muller, and John Wiley & Sons, Inc.

IT IS APPROPRIATE to begin our investigation of the world's geographic realms in Europe because over the past five centuries Europe and Europeans have influenced and changed the rest of the world more than any other realm or people has done. European empires spanned the globe and transformed societies far and near. European colonialism propelled the first wave of globalization. Millions of Europeans migrated from their homelands to other areas of the Old World as well as to the New, changing (and sometimes nearly obliterating) traditional cultures and creating new societies from Australia to North America. Colonial power and economic incentive combined to impel the movement of millions of imperial subjects from their ancestral homes to distant lands: Africans to the Americas, Indians to Africa,

Chinese to Southeast Asia, Malays to South Africa's Cape, Native Americans from east to west. In agriculture, industry, politics, and other spheres, Europe generated revolutions—and then exported those revolutions across the world, thereby consolidating the European advantage.

But throughout much of that 500-year period of European hegemony, Europe also was a cauldron of conflict. It was a fragmented and fractious realm with religious, territorial, and political disputes that precipitated bitter wars that even spilled over into the colonies. And during the twentieth century, Europe twice plunged the world into war. The terrible, unprecedented toll of World War I (1914–1918) was not enough to stave off World War II (1939–1945), which ended with the first-ever use of nuclear

weapons in Japan. In the aftermath of that war, Europe's weakened powers lost most of their colonial possessions, and a new rivalry emerged: an ideological Cold War between the communist Soviet Union and the capitalist United States. This Cold War lowered an Iron Curtain across the heart of Europe, leaving most of the east under Soviet control and most of the west in the American camp. Western Europe proved resilient, overcoming the destruction of war and the loss of colonial power to regain economic strength. Meanwhile, the Soviet communist experiment failed at home and abroad, and in 1990 the last vestiges of the Iron Curtain were lifted. Since then, a massive effort has been underway to reintegrate and reunify Europe from the Atlantic coast to the Russian border, the key geographic story of this chapter.

Defining the Realm

GEOGRAPHICAL FEATURES

As Figures 1-1 (see chapter opener map) and 1-2 show, Europe is a realm of peninsulas and islands on the western margin of the world's largest landmass, Eurasia. It is a realm of 590 million people and 40 countries, but is territorially quite small. Yet despite its modest proportions it has had—and continues to have—a major impact on world affairs. For many centuries Europe has been a hearth of achievement, innovation, and invention.

Europe's Eastern Border

The European realm is bounded on the west, north, and south by Atlantic, Arctic, and Mediterranean waters, respectively. But where is Europe's eastern limit? Some scholars place it at the Ural Mountains, deep inside Russia, thereby recognizing a European

MAJOR GEOGRAPHIC QUALITIES OF Europe

1. The European realm lies on the western extremity of the Eurasian landmass, a locale of maximum efficiency for contact with the rest of the world.

2. Europe's lingering and resurgent world influence results largely from advantages accrued over centuries of global political and economic domination.

3. The European natural environment displays a wide range of topographic, climatic, and soil conditions and is endowed with many industrial resources.

4. Europe is marked by strong internal regional differentiation (cultural as well as physical), exhibits a high degree of functional specialization, and provides multiple exchange opportunities.

5. European economies are dominated by manufacturing, and the level of productivity has been high;

levels of development generally decline from west to east.

6. Europe's nation-states emerged from durable power cores that formed the headquarters of world colonial empires. A number of those states are now plagued by internal separatist movements.

7. Europe's rapidly aging population is generally well off, highly urbanized, well educated, and enjoys long life expectancies.

8. A growing number of European countries are experiencing population declines; in many of these countries, the natural decrease is partially offset by immigration.

9. Europe has made significant progress toward international economic integration and, to a lesser extent, political coordination.

EUROPE

- Western Europe
- British Isles
- Northern (Nordic) Europe
- Mediterranean Europe
- Eastern Europe
- Railroads
- Roads

0   250   500 Kilometers
0 100 200 300 Miles

National capitals are underlined

FIGURE 1-2

© H. J. de Blij, P. O. Muller, and John Wiley & Sons, Inc.

Russia and, presumably, an Asian one as well. Our realm definition places Europe's eastern boundary between Russia and its numerous European neighbors to the west. This definition is based on several geographic factors, including European-Russian contrasts in territorial dimensions, population size, cultural properties, and historic aspects, all discernible on the maps in Chapters 1 and 2.

Resources

Europe's peoples have benefited from a large and varied store of raw materials. Whenever the opportunity or need arose, the realm proved to contain what was required. Early on, these requirements included cultivable soils, rich fishing waters, and wild animals that could be domesticated; in addition, extensive forests provided wood for houses and boats. Later, coal and mineral ores propelled industrialization. More recently, Europe proved to contain substantial deposits of oil and natural gas.

Climates

From the balmy shores of the Mediterranean Sea to the icy peaks of the Alps, and from the moist woodlands and moors of the Atlantic fringe to the semiarid prairies north of the Black Sea, Europe presents an almost infinite range of natural environments (Fig. 1-3). Compare Western Europe and eastern North America on Figure G-9 and you will see the moderating influence of the warm

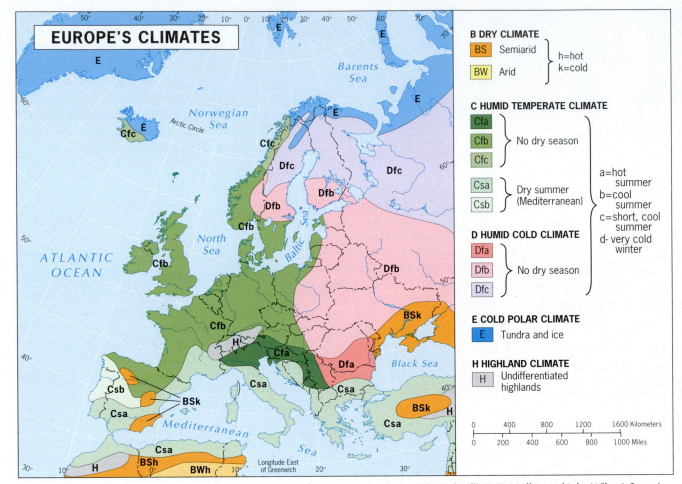

EUROPE'S CLIMATES

B DRY CLIMATE

BS	Semiarid	h=hot
BW	Arid	k=cold

C HUMID TEMPERATE CLIMATE

Cfa
Cfb — No dry season
Cfc

Csa — Dry summer
Csb — (Mediterranean)

a=hot summer
b=cool summer
c=short, cool summer
d- very cold winter

D HUMID COLD CLIMATE

Dfa
Dfb — No dry season
Dfc

E COLD POLAR CLIMATE

E Tundra and ice

H HIGHLAND CLIMATE

H Undifferentiated highlands

0 400 800 1200 1600 Kilometers
0 200 400 600 800 1000 Miles

FIGURE 1-3

© H. J. de Blij, P. O. Muller, and John Wiley & Sons, Inc.

avenues of seaborne trade and conquest. Hundreds of miles of navigable rivers, augmented by an unmatched system of canals, open the interior of Europe to its neighboring seas and to the shipping lanes of the world.

Also consider the scale of the maps of Europe in this chapter. Europe is a realm of moderate distances and close proximities. Short distances and large cultural differences make for intense interaction, leading to the constant circulation of goods and ideas. That has been the hallmark of Europe's geography for over a millennium.

LANDSCAPES AND OPPORTUNITIES

In the Introduction we noted the importance of physical geography in the definition of geographic realms. The natural landscape with its array of landforms (such as mountains and plateaus) is a key element in the total physical geography—or **2** **physiography**—of any part of the terrestrial world. Other physiographic components include climate and the physical features that mark the natural landscape, such as vegetation, soils, and water bodies.

Europe's area may be small, but its landscapes are varied and complex. Regionally, we identify four broad units: the Central Uplands, the southern Alpine Mountains, the Western Uplands, and the North European Lowland (Fig. 1-4).

The *Central Uplands* form the heart of Europe. It is a region of hills and low plateaus well endowed with raw materials whose farm villages grew into towns and cities when the Industrial Revolution transformed this realm.

ocean current known as the North Atlantic Current and its onshore windflow.

Human Diversity

The European realm is home to peoples of numerous cultural-linguistic stocks, including not only Latins, Germanics, and Slavs but also minorities such as Finns, Magyars (Hungarians), Basques, and Celts. This diversity of ancestries continues to be an asset as

well as a liability. Although it has generated interaction and exchange, it has also resulted in conflict and war.

Locational Advantages

Europe also has outstanding locational advantages. Its *relative location*, at the heart of the **1** **land hemisphere**, creates maximum efficiency for contact with much of the rest of the world. A 'peninsula of peninsulas,' Europe is nowhere far from the ocean and its

Heat Waves in Europe. Europe has long enjoyed relatively mild climates for its latitude. Winters in western, southern, and even parts of northern Europe have been mild; summers rarely hot, except in parts of the Mediterranean. This pattern is predicted to change as global climate change will likely increase the likelihood of more frequent and severe heat waves in Europe. In 2003 a severe and extended heat wave contributed to the death of an estimated 35,000 people, mostly in France, but also in the Netherlands and elsewhere on the continent. Europeans do not typically use air-conditioning in buildings (not even in offices or commercial spaces) or their cars; they have not felt the need to. Refrigerating food has also been limited. But as temperatures are predicted to become hotter by several degrees Celsius, Europeans may need to rethink these habits. More heat waves, though with far fewer deaths, hit in 2006. Shown here is a heat wave in Paris during the summer of 2006. Wildfires may also become a greater issue than heretofore. During the summer of 2007, Greece endured numerous wildfires that ran out of control and cost more than 60 people their lives, as well as considerable property damage.   (© Francois Guillot/ AFP/Getty Images)

The *Alpine Mountains*, a highland region named after the Alps, extend from the Pyrenees on the French-Spanish border to the Balkan Mountains near the Black Sea, and include Italy's Appennines and Eastern Europe's Carpathians.

The *Western Uplands*, geologically older, lower, and more stable than the Alpine Mountains, extend from Scandinavia through western Britain and Ireland to the heart of the Iberian Peninsula in Spain.

The *North European Lowland* extends in a lengthy arc from southeast Britain and central France across Germany and Denmark into Poland and Ukraine, from where it continues well into Russia. Also known as the Great European Plain, this has been an avenue for human migration and invasions time after time, so

that complex cultural and economic mosaics developed here together with a jigsaw-like political map. As Figure 1-4 shows, many of Europe's major rivers and connecting waterways serve this populous region, where many of Europe's leading cities (Paris, Amsterdam, Copenhagen, Berlin, Warsaw) are located.

HISTORICAL GEOGRAPHY

Modern Europe was peopled in the wake of the Pleistocene's most recent glacial retreat and global warming—a gradual warming that caused tundra to give way to deciduous forest and ice-filled valleys to

turn into grassy vales. On Mediterranean shores, Europe witnessed the rise of its first great civilizations on the islands and peninsulas of Greece and later in what is today Italy.

Ancient Greece and Imperial Rome

Ancient Greece lay exposed to influences radiating from the advanced civilizations of Mesopotamia and the Nile Valley, and in their fragmented habitat the Greeks laid the foundations of European civilization. Their achievements in political science, philosophy, the arts, and other spheres have endured

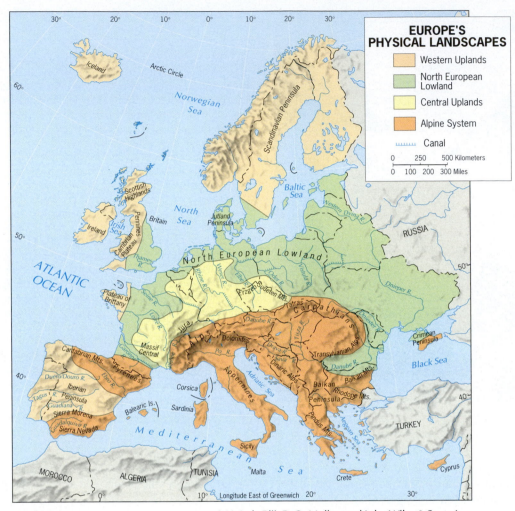

FIGURE 1-4   © H. J. de Blij, P. O. Muller, and John Wiley & Sons, Inc.

Triumph and Collapse

Roman rule brought disparate, isolated peoples into the imperial political and economic sphere. By guiding (and often forcing) these groups to produce particular goods or materials, the Romans launched Europe down a road for which it would become famous: **4** **local functional specialization**. The workers on Elba, a Mediterranean island, mined iron ore. Those near Cartagena in Spain produced silver and lead. Certain farmers were taught irrigation to produce specialty crops. Others raised livestock for meat or wool. The *production of particular goods by particular people in particular places* became and remained a hallmark of the realm.

The Romans also spread their language across the empire, setting the stage for the emergence of the *Romance languages*; they disseminated Christianity; and they established durable systems of education, administration, and commerce. But when their empire collapsed in the fifth century, disorder ensued, and massive migrations soon brought Germanic and Slavic peoples to their present positions on the European stage. Capitalizing on Europe's weakness, the Arab-Berber Moors from North Africa, energized by Islam, conquered most of Iberia and penetrated France. Later, the Ottoman Turks invaded Eastern Europe and reached the outskirts of Vienna.

Rebirth and Royalty

Europe's revival—its *Renaissance*—did not begin until the fifteenth century. After a thousand years of feudal turmoil marking the Dark and Middle Ages, powerful monarchies began to lay the foundations of modern states. The discovery of continents and riches across the oceans opened a new era of ***mercantilism***, the competitive accumulation of wealth chiefly in the form of gold and silver. Best placed for this competition were the kingdoms of Western Europe. Europe was on its way to colonial expansion and world domination.

for 25 centuries. But the ancient Greeks never managed to unify their domain, and their persistent conflicts proved fatal when the Romans challenged them from the west. By 147 BC, the last of the sovereign Greek intercity leagues (alliances) had fallen to the Roman conquerors.

The center of civilization and power shifted to Rome in present-day Italy. Borrowing from Greek culture, the Romans created an empire that stretched from Britain to the Persian Gulf and from the Black Sea to Egypt; they made the Mediterranean Sea a Roman lake carrying armies to distant shores and goods to imperial Rome. With an urban population that probably exceeded 1 million, Rome was the first metropolitan-scale urban center in Europe.

The Romans founded numerous other cities throughout their empire and linked them to the capital through a vast system of highway and water routes, facilitating political control and enabling economic growth in their provinces. It was an unparalleled **3** **infrastructure**, much of which long outlasted the empire itself.

THE REVOLUTIONS OF MODERNIZING EUROPE

Even as Europe's rising powers reached for world domination overseas, they fought with each other in Europe itself. Powerful monarchies and land-owning ('landed') aristocracies had their status and privilege challenged by ever-wealthier merchants and businesspeople. Demands for political recognition grew; cities mushroomed with the development of industries; the markets for farm products burgeoned; and Europe's population, more or less stable at about 100 million since the sixteenth century, began to increase.

The Agrarian Revolution

We know Europe as the focus of the Industrial Revolution, but before this momentous development occurred another revolution was already in progress: the *agrarian revolution*. Port cities and capital cities thrived and expanded, and their growing markets created economic opportunities for farmers. This led to revolutionary changes in land ownership and agricultural methods. Improved farm practices, better equipment, superior storage facilities, and more efficient transport to the urban markets marked a revolution in the countryside. The colonial merchants brought back new crops (such as the potato from South America, which quickly became a European staple), and market prices rose, drawing more and more farmers into the economy.

Agricultural Market Model

The transformation of Europe's farmlands reshaped its economic geography, producing new patterns of land use and market links. The economic geographer Johann Heinrich von Thünen (1783–1850) published a pioneering work entitled **5** *The Isolated State* in 1826, chronicling the geography of Europe's agricultural transformation.

Von Thünen used information from his own farmstead to build what today we call a **6** model (an idealized representation of reality that demonstrates its most important properties) of the location of productive activities in Europe's farmlands. Since a model is an abstraction that must always involve assumptions, von Thünen postulated an ideal self-contained area (hence the *isolation*) with a single market center, flat and uninterrupted land without impediments to cultivation or transportation. In such a situation, transport costs would be directly proportional to distance.

Von Thünen's model revealed four zones or rings of land use encircling the market center (see Fig. 1-5). Innermost and directly adjacent to the market would lie a zone of intensive farming and dairying, yielding the most perishable and highest priced products. Immediately beyond lay a zone of forest used for timber and firewood (still a priority in von Thünen's time). Next there would be a ring of field crops, for example, grains or potatoes. A fourth zone would contain pastures and livestock. Beyond lay wilderness, from where the costs of transport to market would become prohibitive.

In many ways, von Thünen's model was the first analysis in a field that would eventually become known as **location theory**. Von Thünen knew, of course, that the real Europe did not present the conditions postulated in his model. It did demonstrate, however, the economic-geographic forces that shaped the new Europe, which is why the model is still being discussed today. More than a century after the publication of *The Isolated State*, geographers Samuel van Valkenburg and Colbert Held produced a map of twentieth-century European agricultural intensity, revealing a striking, ring-like concentricity focused on the vast urbanized area lining the North Sea—now the dominant market for a realmwide, macroscale 'Thünian' agricultural system (Fig. 1-5).

The Industrial Revolution

The term **7** Industrial Revolution suggests that an agrarian Europe was suddenly swept up in a wholesale industrialization that changed the realm in a few decades. In reality, seventeenth- and eighteenth-century Europe had been industrializing in many spheres, long before the chain of events known as the Industrial Revolution began. From the textiles of England and Flanders to the iron farm implements of Saxony (in present-day Germany), from Scandinavian furniture to French linens, Europe had already entered a new era of local functional specialization. It would therefore be more appropriate to call what happened next the period of Europe's *industrial intensification*.

British Primacy

In the 1780s, the Scotsman James Watt and others devised a steam-driven engine, which was soon adopted for numerous industrial uses. At about the same time, coal (converted into carbon-rich coke) was recognized as a vastly superior substitute for charcoal in smelting iron. These momentous innovations had a rapid effect. The power loom revolutionized the weaving industry. Iron smelters, long dependent on Europe's dwindling forests for fuel, could now be concentrated near coalfields. Engines could move locomotives as well as power looms. Ocean shipping entered a new age.

Britain had an enormous advantage, for the Industrial Revolution occurred when British influence reigned worldwide and the significant innovations were achieved in Britain itself. The British controlled the flow of raw materials, they held a monopoly over products that were in global demand, and they alone possessed the skills necessary to make the machines that manufactured the products. Soon the fruits of the Industrial Revolution were being exported, and the modern industrial spatial organization of Europe began to take shape (see insert, Fig. 1-6). In Britain, manufacturing regions, densely populated and heavily urbanized, developed near coalfields in the English Midlands, at Newcastle to the northeast, in southern Wales, and along Scotland's Clyde River around Glasgow.

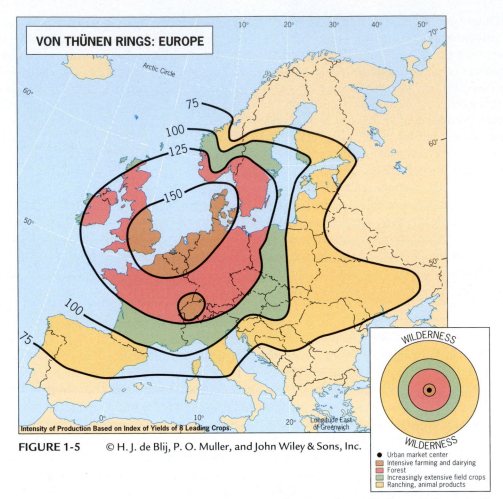

FIGURE 1-5   © H. J. de Blij, P. O. Muller, and John Wiley & Sons, Inc.

Inset: Adapted with permission from Samuel van Valkenburg & Colbert C. Held,
Europe, 2 rev. ed. (New York: John Wiley & Sons, Inc., 1952).

of potential buyers. Although the Industrial Revolution thrust other places into prominence, London did not lose its primacy: industries in and around the British capital multiplied.

Consequences of Clustering

The industrial transformation of Europe, like the agrarian revolution, became the focus of geographic research. One of the leaders in this area was the economic geographer Alfred Weber (1868–1958), who published a spatial analysis of the process titled *Concerning the Location of Industries* (1909). Unlike von Thünen, Weber focused on activities that take place at particular points rather than over large areas. His model, therefore, represented the factors of industrial location—the clustering or dispersal of places of intense manufacturing activity.

One of Weber's most interesting conclusions has to do with what he called *agglomerative* (concentrating) and *deglomerative* (dispersive) forces. It is often advantageous for certain industries to cluster together, sharing equipment, transport facilities, labor skills, and other assets of urban areas. This is what made London (as well as Paris and other cities that were not situated on rich deposits of industrial raw materials) attractive to many manufacturing plants that could benefit from agglomeration and from the large markets that these cities anchored. As Weber found, however, excessive agglomeration may lead to disadvantages such as competition for increasingly expensive space, congestion, overburdening of infrastructure, and environmental pollution. Manufacturers may then move away, and deglomerative forces will increase.

The Industrial Revolution spread eastward from Britain onto the European mainland throughout the middle and late nineteenth century (see inset, Fig. 1-6). Population skyrocketed, emigration mushroomed, and industrializing cities burst at the seams. European states already had acquired colonial empires before this revolution started; now colonialism gave Europe an unprecedented advantage in its dominance over the rest of the world.

Resources and Industrial Development

In mainland Europe, a belt of major coalfields extends from west to east, roughly along the southern margins of the North European Lowland, due eastward from southern England across northern France and Belgium, Germany (the Ruhr), western Bohemia in the Czech Republic, Silesia in southern Poland, and the Donets Basin (Donbas) in eastern Ukraine. Iron ore is found in a broadly similar belt, and the industrial map of Europe reflects the resulting concentrations of economic activity (Fig. 1-6). Another set of manufacturing zones emerged in and near the growing urban centers of Europe, as the same map demonstrates. London—already Europe's leading urban focus and Britain's richest domestic market—was typical of these developments. Many local industries were established here, taking advantage of the large supply of labor, the ready availability of capital, and the proximity of so great a number

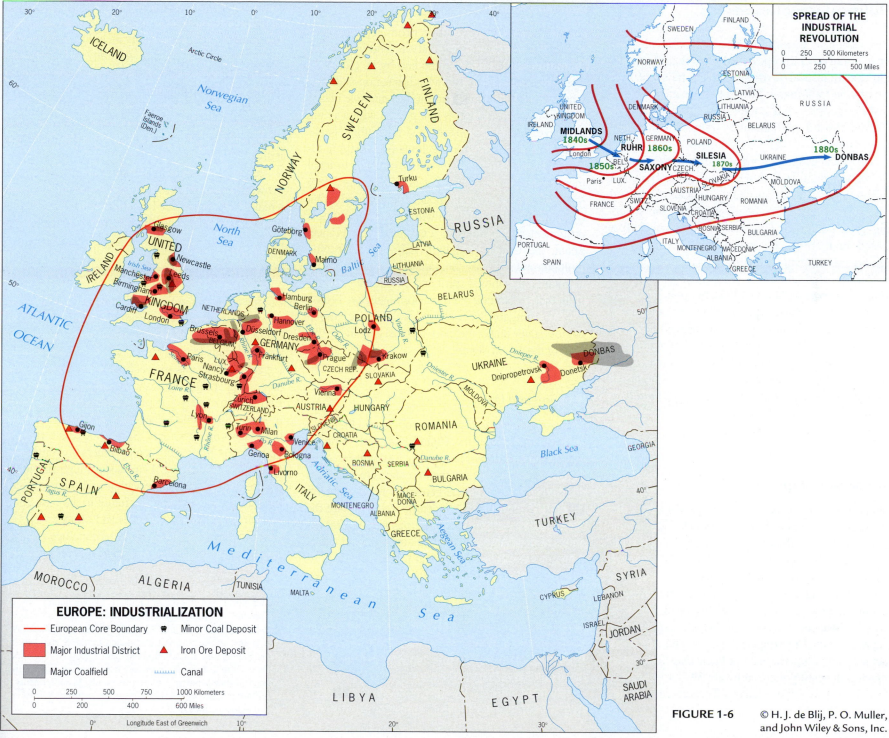

SPREAD OF THE INDUSTRIAL REVOLUTION

MIDLANDS 1840s
RUHR 1860s
SAXONY 1850s
SILESIA 1870s
1880s DONBAS

EUROPE: INDUSTRIALIZATION

— European Core Boundary
⚒ Minor Coal Deposit
🟥 Major Industrial District
▲ Iron Ore Deposit
⬜ Major Coalfield
⬝⬝⬝ Canal

FIGURE 1-6   © H. J. de Blij, P. O. Muller, and John Wiley & Sons, Inc.

Political Revolutions

Revolution in a third sphere—the political—had been going on in Europe even longer than the agrarian and industrial revolutions. Europe's *political revolution* took many different forms and affected diverse peoples and countries, but in general it headed toward parliamentary representation and democracy.

Historical geographers point to the Peace (Treaty) of Westphalia in 1648 as a key step in the evolution of Europe's state system, ending decades of war and recognizing territories, boundaries, and the sovereignty of countries. This treaty's stabilizing effect lasted until 1806, by which time revolutionary changes were again sweeping across Europe.

The most dramatic of these changes was the French Revolution (1789–1795), but political transformation had come much earlier to the Netherlands, Britain, and the Scandinavian countries. Other parts of Europe remained under the control of authoritarian (dictatorial) regimes headed by monarchs or despots. Europe's patchy political revolution lasted into the twentieth century, and by then **nationalism** (national spirit, pride, patriotism) had become a powerful force in European politics.

A Fractured Map

When you look at the political map of Europe, the question that arises is, how did so small a geographic realm come to be divided into so many political entities? Europe's map is a legacy of its feudal and royal periods, when powerful kings, barons, dukes, and other rulers, rich enough to fund armies and powerful enough to exact taxes and tribute from their domains, created bounded territories in which they reigned supreme. Royal marriages, alliances, and conquests actually simplified Europe's political map. In the early nineteenth century there still were 39 German states; Germany as we know it today did not emerge until the 1870s.

State and Nation

Europe's political revolution produced a form of political-territorial organization known as the **8 nation-state**, a state embodied by its culturally distinctive population. But what is a nation-state and what is not? The term **9 nation** has multiple meanings. In one sense it refers to a people with a single language, a common history, a similar ethnic background. It also is used for feelings of belonging together. In the sense of *nationality* it relates to legal membership in the state, that is, citizenship. Very few states today are so homogeneous culturally that the culture is conterminous with the state. Europe's prominent nation-states of a century ago—France, Spain, the United Kingdom, Italy—have become multicultural societies, their nations defined more by an intangible national spirit and emotional commitment than by cultural or ethnic homogeneity. Today, Poland, Hungary, Iceland, and Sweden are among the few states that come close to satisfying the definition of the nation-state.

Division and Unity

Mercantilism and colonialism empowered the states of Western Europe; the United Kingdom (Britain) was the superpower of its day. But all countries, even Europe's nation-states in their heyday, are subject to divisive stresses. Political geographers use the term **10 centrifugal forces** to identify and measure the strength of such division, which may result from religious, racial, linguistic, political, or regional factors. Language is often a centrifugal force, as it is shorthand for cultural group. In Belgium today, the division between the Flemish (Dutch-speaking) North and the Walloon (French-speaking) South is so strong that the country did not have a permanent government for months in 2007 and 2008 as various factions could not agree on an agenda for the country. It could break up in the future.

Centrifugal forces are measured against **11 centripetal forces**, the binding, unifying glue of the state. General satisfaction with the system of government and administration, legal institutions, and other functions of the state (notably including its treatment of minorities) can ensure stability and continuity when centrifugal forces threaten. The former state of Yugoslavia[1] is an example because its dictator, Tito, forcibly held the country together. When he was no longer in power, centrifugal forces exceeded the weak centripetal forces that had bound the state, and it disintegrated.

Europe's political revolution continues. Today a growing group of European states is trying to create a realmwide union (the EU) that might some day become a European superstate.

CONTEMPORARY EUROPE

Europe today is modern and efficient. Its airports are generally spacious, with immigration and security operating smoothly. Public transport, from high-speed intercity trains to local buses, is excellent and well integrated and seems to reach even the most remote places. Superhighways connect major cities, although Europe's relatively compact, old, and densely-built-up urban areas tend to become easily congested and traffic jams are legendary. Some cities have virtually banned cars from their center; others charge money for driving downtown. Ultramodern high-rise architecture and historic preservation give Europe a cultural landscape welding the future to the past. From nuclear power plants (France) and offshore oil platforms (Norway) to flood-control technology (Netherlands, see photo p. 54) and mega-ship construction

---

[1]Between 1919 and 1991, the name *Yugoslavia* referred to the multicultural state shown in Figure 1-23. When civil war destroyed the old order, parts of Yugoslavia became independent states, including Slovenia, Croatia, Bosnia, and Macedonia. The largest of these states, dominated by the Serbs, kept the name Yugoslavia. But in 2002 the then-Yugoslav government announced that it would agree to a name change, calling their country Serbia-Montenegro after its two main components. In 2006 Montenegro voted to secede from the state; Kosovo declared independence in 2008.

Curtain' embarked on a massive effort not only to recover, but also to put old divisions permanently behind them. Assisted by the United States, they moved far along the road to unification—especially in the economic arena. But, as we will see later, Europe's old fractiousness has not yet been overcome.

Language and Religion

Europe's territory is just over 60 percent the size of the United States, but the population of Europe's 41 countries is about twice as large as America's. This population of more than 590 million speaks numerous languages, almost all of which belong to the **12 Indo-European language family** (Fig. 1-7). But most of those languages are not mutually understandable. When the unification effort began, one major problem was to determine which languages to recognize as official. That problem still prevails, although English has become the realm's unofficial *lingua franca*. Europe's multilingualism remains a major barrier to integration.

Nevertheless, the unification effort has gone so far that in recent years Europeans have been debating the terms of a proposed constitution. In this debate, another divisive element of the regional culture has come to the fore: religion. Europe's cultural heritage is steeped in Christian traditions, but Christianity has not unified Europeans. Sectarian strife between Catholics and Protestants that previously plunged Europe into bitter conflict still divides communities and, as in Northern Ireland, can still arouse violence. Religion and politics remain closely connected; religious-based political parties such as the Christian Democrats in Germany exist in many countries. Eastern Europe has a history of strife between Christian Orthodox and Islamic faiths. Today, a new factor has entered the religious landscape: the rise of Islam in Western Europe, resulting mainly from the recent immigration of millions of Muslims from North Africa and Southwest Asia. When some of the framers of the European Constitution wanted

Amsterdam, The Netherlands. "It was early in the morning in Amsterdam, and I witnessed the city waking up. I was on a long layover at Amsterdam's Schiphol Airport and took the train in to the city (only 15 minutes!) to spend a few hours. Amsterdam is a typical old European city, very compact in the center, with very narrow streets. Many of these have been converted to pedestrian and bicycle-only throughways. Although cars have not been officially banned from the city, they functionally have been as parking barely exists and the traffic pattern steers cars away from the center. I call it the anti-car city. People walk, take public transportation (tram, bus, or subway), or bicycle to work. In the Netherlands today 25 percent of the population uses a bicycle to get to work."
(© A. WinklerPrins)

www.conceptcaching.com

(Finland), as well as in countless other enterprises, Europe is in the vanguard of the world. Nine of Europe's national economies usually rank among the top 20 in the world. Collectively, Europe's economies outrank even the U.S. economy. Collectively—but that is the problem. Europe does not have a long-term history of collective action. It has long been fractious and has a long history of disagreement and warfare between its many cultures. At the end of World War II (1939–1945), Europeans saw their ravaged realm divided in the Cold War between Western and Soviet spheres. The countries on the western side of the 'Iron

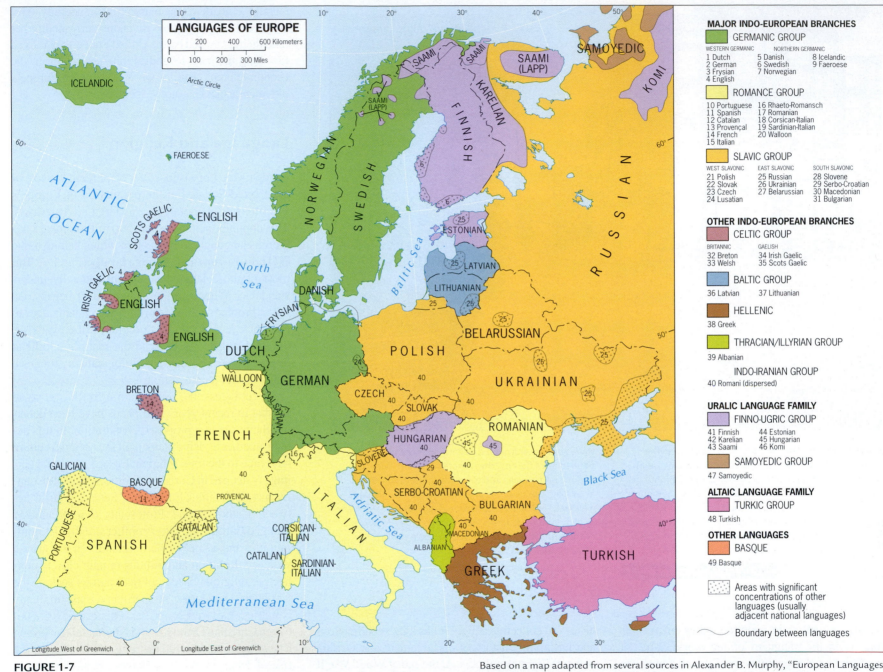

LANGUAGES OF EUROPE

MAJOR INDO-EUROPEAN BRANCHES

GERMANIC GROUP

WESTERN GERMANIC	NORTHERN GERMANIC
1 Dutch	5 Danish	8 Icelandic
2 German	6 Swedish	9 Faeroese
3 Frysian	7 Norwegian	
4 English		

ROMANCE GROUP

10 Portuguese	16 Rhaeto-Romansch
11 Spanish	17 Romanian
12 Catalan	18 Corsican-Italian
13 Provençal	19 Sardinian-Italian
14 French	20 Walloon
15 Italian	

SLAVIC GROUP

WEST SLAVONIC	EAST SLAVONIC	SOUTH SLAVONIC
21 Polish	25 Russian	28 Slovene
22 Slovak	26 Ukrainian	29 Serbo-Croatian
23 Czech	27 Belarussian	30 Macedonian
24 Lusatian		31 Bulgarian

OTHER INDO-EUROPEAN BRANCHES

CELTIC GROUP

BRITANNIC	GAELISH
32 Breton	34 Irish Gaelic
33 Welsh	35 Scots Gaelic

BALTIC GROUP

36 Latvian   37 Lithuanian

HELLENIC

38 Greek

THRACIAN/ILLYRIAN GROUP

39 Albanian

INDO-IRANIAN GROUP

40 Romani (dispersed)

URALIC LANGUAGE FAMILY

FINNO-UGRIC GROUP

41 Finnish	44 Estonian
42 Karelian	45 Hungarian
43 Saami	46 Komi

SAMOYEDIC GROUP

47 Samoyedic

ALTAIC LANGUAGE FAMILY

TURKIC GROUP

48 Turkish

OTHER LANGUAGES

BASQUE

49 Basque

⋯⋯ Areas with significant concentrations of other languages (usually adjacent national languages)

﹏ Boundary between languages

FIGURE 1-7

Based on a map adapted from several sources in Alexander B. Murphy, "European Languages," in Tim Unwin, ed., *A European Geography* (Harlow, UK: Addison Wesley Longman, 1998), p. 38.

to state that the realm's cultural heritage was Judeo-Christian-Islamic, they were outvoted by those who preferred no reference to religion at all. Meanwhile, as mosques overflow with the immigrant faithful, churches stand nearly empty as secularism is on the rise among nonimmigrant Europeans. Some scholars have gone so far as to describe Europe as being in a 'post-Christian' era.

For so small a realm, Europe's cultural geography is sharply varied. The popular image of Europe tends to be formed by British pageantry, French countrysides, German cities—but go beyond this core, and you will find isolated Slavic communities in the mountains facing the Adriatic, Muslim towns in poor and isolated Albania, Roma (Gypsy) villages in the interior of Romania, farmers using methods unchanged for centuries in rural Poland. That map of 40 countries does not begin to reflect the diversity of European cultures.

Spatial Interaction

If not culture, what does unify Europe? The answer lies in this realm's outstanding opportunities for productive interaction. The ancient Romans realized it, but they developed a system of production that focused on Rome, not primarily on regional exchange. Modern Europeans seized on the same opportunities to create a regionwide network of spatial interaction whose structure intensified as time went on, linking places, communities, and countries in countless ways. The geographer Edward Ullman envisaged this process as operating on three principles:

13 Complementarity When one area produces a surplus of a commodity required by another area, regional complementarity prevails. The mere existence of a particular resource or product is no guarantee of trade: it must be needed elsewhere. When two areas each require the other's products, double complementarity occurs. Europe has countless examples of complementarity, from local communities to entire countries. To cite one:

industrial Italy needs coal from Western Europe; Western Europe needs Italy's farm products.

14 Transferability The ease with which a commodity can be transported by producer to consumer defines its transferability. Distance and physical obstacles can raise the cost of a product to the point of unprofitability. But Europe is small, distances are short, and Europeans have built the world's most efficient transport system of roads, railroads, and canals linking navigable rivers.

15 Intervening Opportunity Potential trade between two places, even if they are in a position of strong complementarity and high transferability, will develop only in the absence of a closer, intervening source of supply. To use our earlier example, if Switzerland proved to contain large coal reserves close to the Italian border, Italy would avail itself of that opportunity, reducing or eliminating its imports from more distant Western Europe.

A Highly Urbanized Realm

Figure 1-8 illustrates today's population distribution in Europe. With its early industrialization and accompanying urbanization and mercantilism, Europe is now a highly urbanized realm. Overall, about three of every four Europeans live in towns and cities, an average that is far exceeded in the west (see Appendix B) but not yet attained in much of the east. Large cities are production centers as well as marketplaces, and they also form the crucibles of their nations' cultures.

Flower market on the Grand Place in Brussels, Belgium. Europe faces an aging population that, given life expectancies, will have many years to pass the time after retirement. This has implications for the provision of social and health services for the elderly, much of which needs to be paid for by a shrinking work force.   (© Louis Grandadam/Stone/Getty Images)

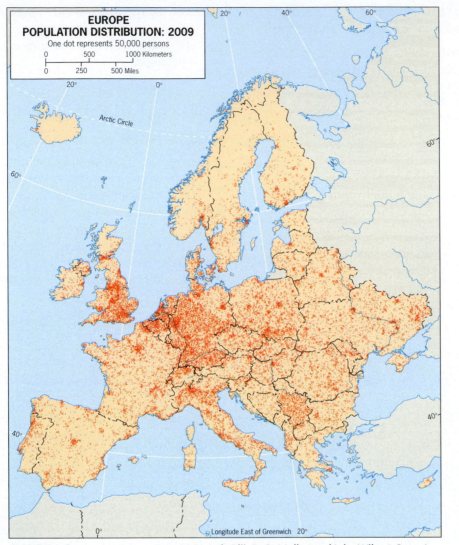

**EUROPE
POPULATION DISTRIBUTION: 2009**
One dot represents 50,000 persons

0   500   1000 Kilometers

0   250   500 Miles

FIGURE 1-8   © H. J. de Blij, P. O. Muller, and John Wiley & Sons, Inc.

During the 1930s, the geographer Mark Jefferson studied the pivotal role of great cities in the development of national cultures. He postulated a 'law' of the **16 primate city**, which stated that "a country's leading city is always disproportionately large and exceptionally expressive of national culture." This obviously debatable 'law' still has relevance: Paris personifies France in countless ways; nothing in the United Kingdom rivals London; Athens anchors and dominates Greece. Europe's primate cities have also grown disproportionately since the end of World War II, gaining rather than losing primacy in this still-urbanizing realm.

A Changing Population

When a population urbanizes, average family size declines, and so does the overall rate of natural increase. There was a time when Europe's population was 'exploding,' sending millions to the New World and the colonies and still growing at home. But today Europe's indigenous population is shrinking. To keep a population from declining, the (statistically) average woman must bear 2.1 children and birthrates need to be at par with or exceed the deathrate. This is the case for only a handful of European countries (Appendix B). Several large countries are in fact recording negative rates of natural increase, including Germany and Italy. Eastern European countries are recording even more dire percent changes such as Ukraine's −0.8 rate of natural increase—the lowest in the world today.

Such *negative population growth* poses serious challenges for any country, but particularly those where social welfare structures are in place. When the population pyramid becomes top-heavy, the number of workers whose taxes pay for the social services of the aged goes down, leading to reduced pensions and dwindling funds for health care. Governments that impose ever-increasing taxes to overcome the population deficit endanger the business climate. Europeans are already heavily taxed, so options are limited. Europe, and especially Western Europe, is experiencing a ***population implosion*** that will be a formidable challenge for decades to come.

Immigrant Infusions

Meanwhile, ***immigration*** is partially offsetting the losses European countries face. Millions of Turks, Algerians, Moroccans, West Africans, West Indians, Indonesians, and even South Americans are changing the social fabric of what once were unicultural nation-states. One key dimension of this change is the spread of Islam in Europe (Fig. 1-9). Muslim populations in Eastern Europe (such as those in Albania, Kosovo, and Bosnia) are local, Slavic communities converted during the period of Ottoman rule. The Muslim sectors of Western European countries, on the other hand, represent more recent immigrations.

The vast majority of the Muslim immigrants are intensely devout, politically aware, and culturally insular. They continue to arrive in a Europe where native populations are stagnant or declining, where Christian religious institutions are weakening, where secularism is rising rapidly, where political positions

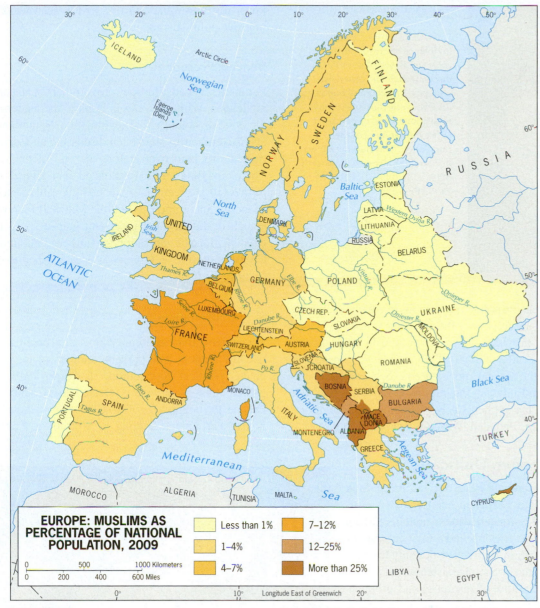

FIGURE 1-9   © H. J. de Blij, P. O. Muller, and John Wiley & Sons, Inc.

EUROPE: MUSLIMS AS PERCENTAGE OF NATIONAL POPULATION, 2009

- Less than 1%
- 1–4%
- 4–7%
- 7–12%
- 12–25%
- More than 25%

The Growing Multicultural Challenge

In truth, European governments have not done enough to foster the very integration they see Muslims rejecting. The French assumed that their North African immigrants would aspire to assimilation as Muslim children went to public schools from their urban public-housing projects; instead, they got into disputes over the dress codes of Muslim girls. The Germans for decades would not award German citizenship to children of immigrant parents born on

A Turkish market in the Kreuzberg district of Berlin, capital city of Germany. Western Europe's ethnic composition has changed significantly in the last several decades, and this has been challenging for a part of the world not used to being an immigrant society. Turks and North Africans bring their Islamic faith with them as well as cultural traditions that are very different from native ways of life. (© Petforsberg/Alamy)

often appear to be anti-Islamic, and where cultural norms are incompatible with Muslim traditions. Many young men, unskilled and uncompetitive in a European Union (EU) where unemployment is already high, get involved in petty crime or the drug trade, are harassed by the police, and turn to their faith for solace and reassurance. Unlike most other immigrant groups, Muslim communities also tend to resist assimilation, making Islam the essence of their identity. In Britain alone there are more than 1500 mosques, in themselves constituting a transformation of local cultural landscapes.

Segregated immigrant communities in the suburbs (*banlieus*) of Paris often erupt in violence because youth are unemployed, culturally disconnected, and otherwise unhappy. Vehicles and buildings are frequently burned—often in retaliation to real or perceived police mishandling of the population. This burning car was in Pierrefitle, a community north of Paris, in 2005.   (© AP/Wide World Photos)

German soil. The British found their modest efforts at integration stymied by the Muslim practice of importing brides from Islamic countries, which had the effect of perpetuating segregation. In many European cities, the housing projects that form the living space of Muslim residents (by now many of them born and retired there) are dreadful, barren, impoverished environments unseen by Europeans whose images of Paris, Manchester, or Frankfurt are far different.

The social and political implications of Europe's demographic transformation are numerous and far-reaching. European societies are now attempting to restrict immigration in various ways, and political parties with anti-immigrant platforms are gaining ground rapidly. Multiculturalism poses a growing challenge in a changing Europe.

EUROPE'S MODERN TRANSFORMATION

As we already pointed out, much of Europe lay shattered at the end of World War II, its cities and towns devastated, its infrastructure wrecked, its economies devastated. In 1947, U.S. Secretary of State George C. Marshall proposed a European Recovery Program designed to help counter all this dislocation and to create stable political conditions in which democracy would survive. Over the next four years, the United States provided about $13 billion in assistance to Europe (almost $100 billion in today's money). Because the Soviet Union refused U.S. aid and forced its Eastern European satellites to do the same, the Marshall Plan applied solely to 16 European countries, including defeated (West) Germany and Turkey.

European Unification

The Marshall Plan did far more than stimulate European economies. It confirmed European leaders' conclusion that their countries needed a joint economic-administrative structure not only to coordinate the financial assistance, but also to ease the flow of resources and products across Europe's mosaic of boundaries, to lower restrictive trade tariffs, and to seek ways to improve political cooperation.

Benelux Precedent

For all these needs Europe's governments had some guidelines. While in exile in Britain, the leaders of three small countries—Belgium, the Netherlands, and Luxembourg—had been discussing an association of this kind even before the end of the war. There, in 1944, they formulated and signed the *Be-Ne-Lux* Agreement, intended to achieve total economic integration. When the Marshall Plan was launched, the Benelux precedent helped speed the creation of the Organization for European Economic Cooperation (OEEC), which was established to coordinate the investment of America's aid.

Supranationalism

Recognizing that economic cooperation could stabilize the region and prevent another major war, greater political cooperation ensued. In 1949, the participating governments created the Council of Europe, the beginnings of what was to become a European Parliament meeting in Strasbourg, France.

Immediately after midnight January 1, 2007, Romania joined the European Union, but this crowd in Bucharest could not wait. This photo was taken on the evening of December 31, 2006, as the EU flag was hoisted while Romania was still in the final countdown to a momentous event in its long and troubled history.   (© Landov LLC)

Europe was embarked on still another political revolution, the formation of a multinational union involving a growing number of European states. Geographers define 17 **supranationalism** as the voluntary association in economic, political, or cultural spheres of three or more independent states willing to yield some measure of sovereignty for their mutual benefit. In later chapters we will encounter other supranational organizations, including the North American Free Trade Agreement (NAFTA), but none has reached the plateau achieved by the European Union (EU).

Policies and Priorities

In Europe, the key initiatives arose from the Marshall Plan and lay in the economic arena; political integration came more haltingly. Under the Treaty of Rome, six countries joined to become the European Economic Community (EEC) in 1958, also called the Common Market. In 1973 the United Kingdom, Ireland, and Denmark joined, and the renamed European Community (EC) had nine members. As Figure 1-10 shows, membership reached 15 countries in 1995, after the organization had been renamed yet one more time to become the European Union (EU). In 2004 momentous expansion occurred as ten new countries were admitted. Geographically, these ten came in three groups: the three Baltic states (Estonia, Latvia, and Lithuania); five contiguous states in Eastern Europe extending from Poland and the Czech Republic through Slovakia, Hungary, and Slovenia; and two Mediterranean island-states, the ministate of Malta and the still-divided state of Cyprus (Fig. 1-10). In 2007 two more formerly communist-controlled countries, Romania and Bulgaria, were admitted.

The European Union is not just a paper organization for bankers and manufacturers. It is becoming more visible in the daily lives of its member countries' citizens. Taxes tend to be high in Europe, and those collected in the richer member-states are used to subsidize growth and development in the less prosperous ones. This is one of the burdens of

Supranationalism in Europe

1944	Benelux Agreement signed.
1945	World War II ends
1947	Marshall Plan created (effective 1948–1952).
1948	Organization for European Economic Co-operation (OEEC) established.
1949	Council of Europe created.
1952	European Coal and Steel Community (ECSC) Agreement goes into effect.
1958	Treaty of Rome goes into effect. It established the European Economic Community (EEC), also known as the Common Market and 'The Six.'
1958	European Atomic Energy Community (EURATOM) Treaty goes into effect.
1960	European Free Trade Association (EFTA) Treaty goes into effect.
1961	United Kingdom, Ireland, Denmark, and Norway apply for EEC membership.
1963	France vetoes United Kingdom EEC membership; Ireland, Denmark, and Norway withdraw applications.
1967	EEC–ECSC–EURATOM Merger Treaty goes into effect.
1967	European Community (EC) inaugurated.
1968	All customs duties removed for intra-EC trade; common external tariff established.
1973	United Kingdom, Denmark, and Ireland admitted as members of EC, creating 'The Nine.' Norway rejects membership in the EC by referendum.
1979	First general elections for a European Parliament held; new 410-member legislature meets in Strasbourg. European Monetary System established.
1981	Greece admitted as member of EC, creating 'The Ten.'
1985	Greenland, acting independently of Denmark, withdraws from EC.
1986	Spain and Portugal admitted as members of EC, creating 'The Twelve.' Single European Act ratified, targeting a functioning European Union in the 1990s.
1987	Turkey and Morocco make first application to join EC. Morocco is rejected; Turkey is told that discussions will continue.
1990	Charter of Paris signed by 34 members of the Conference on Security and Cooperation in Europe (CSCE). Former East Germany, as part of newly reunified Germany, incorporated into EC.
1991	Maastricht meeting charts European Union (EU) course for the 1990s.
1993	Single European Market goes into effect. Modified European Union Treaty ratified, transforming EC into EU.
1995	Austria, Finland, and Sweden admitted into EU, creating 'The Fifteen.'
1990	Schengen Agreement goes into effect. This permits free passage from one country to the next without border checks once admitted into one of the 'Schengen' countries.
1999	European Monetary Union (EMU) goes into effect. Helsinki summit discusses fast-track negotiations with six prospective members and applications from six others; prospects for Turkey considered in longer term.
2001	Denmark's voters reject EMU participation by 53 to 47 percent.
2002	The euro is introduced as historic national currencies disappear in 12 countries.
2003	First draft of a European Constitution is published to mixed reviews from member-states.
2004	Historic expansion of EU from 15 to 25 countries with the admission of Cyprus, the Czech Republic, Estonia, Hungary, Latvia, Lithuania, Malta, Poland, Slovakia, and Slovenia.
2005	The proposed European Constitution is rejected by voters in France and the Netherlands.
2007	Romania and Bulgaria are admitted, bringing total EU membership to 27 countries. Slovenia and Malta adopt the euro currency.

EUROPEAN SUPRANATIONALISM

- Original EEC members (joined 1958)
- Later EC/EU members (joined 1973–present)
- Discussions in progress
- No early prospects
- Countries voting against membership
- € Euro adopters as of 2008

| 0 | 200 | 400 | 600 Kilometers |
| 0 | 100 | 200 | 300 Miles |

FIGURE 1-10

© H. J. de Blij, P. O. Muller, and John Wiley & Sons, Inc.

membership that is not universally popular in the EU, to say the least. But it has strengthened the economies of Portugal, Greece, and other national and regional economies to the betterment of the entire organization. Some countries also object to the terms and rules of the Common Agricultural Policy (CAP), which, according to critics, supports farmers far too much and, according to others, far too little.

New Money

Ever since the first steps toward European unification were taken, EU planners dreamed of a time when the EU would have a single currency not only to symbolize its strengthening unity but also to establish a joint counterweight to the American dollar. A European Monetary Union (EMU) became a powerful objective of EU planners, but many observers thought it unlikely that member-states would in the end be willing to give up their historic marks, francs, guilders, liras, and escudos. Yet it happened. In 2002, twelve of the (then) fifteen EU countries withdrew their currencies and began using the euro, with the United Kingdom (Britain), Denmark, and Sweden opting to stay out (Fig. 1-10). Others, such as Slovenia and Malta, have joined. Many of the newly admitted Eastern European countries are not ready yet to adopt the euro but will likely do so in the future. The European Monetary Union marches onward, and the euro has become a strong world currency, challenging the dollar in global financial markets and global acceptance.

Continued Expansion?

Expansion has always been an EU objective, and the subject has always aroused passionate debate. Will the incorporation of weaker economies undermine the strength of the whole? Despite such misgivings, the EU has expanded significantly, changing the organization in the process.

Numerous structural implications arise from expansion, affecting all EU countries. Governance is an issue, with 27 different voices all wanting to be heard. Disputes over representation at EU's Brussels headquarters arose quickly. Later, even before the new members' 2004 and 2007 accessions, Poland was demanding that the representative system favor medium-sized members (such as Poland and Spain) over larger ones such as Germany and France. A common agricultural policy is even more difficult to achieve, given the poor condition of farming in most of the new member countries. Also, some of the former EU's poorer states (e.g., Greece and Portugal), which were on the receiving end of the subsidy program that aided their development, now will have to pay up to support the much poorer new eastern members.

The expansions of 2004 and 2007 are having major geographic consequences for all of Europe, not just the EU. But will there be more expansions? While there are countries that have chosen to stay out of the EU to retain their independence (Norway, Iceland, and Switzerland), there are more countries knocking on the EU's door. These latter countries can be divided into three groups:

1. *States emerging from former Yugoslavia.* In the western Balkan Peninsula cluster of countries, only Slovenia has achieved EU membership. The remaining countries are among Europe's poorest and most ethnically divided states, but they contain 21 million people whose circumstances could worsen if they are kept outside 'the club.' Better, EU leaders reason, to move them toward membership, although the process is likely to take many years. Discussions with Croatia began in 2005, and negotiations toward formal discussions with the others were authorized in the same year. If all the former Yugoslav republics become EU members, only Albania will remain an outsider.

2. *States emerging from the former Soviet Union.* Among the remaining former Soviet republics that are part of this realm (Ukraine, Moldova, and Belarus), only Ukraine has shown signs of interest in better relations with Europe, but even here the prospect of membership is dim. A strongly divided electorate and Russia's objections will delay any formal discussions. No prospects exist for Belarus or Moldova, although they both now border on the EU.

3. *Turkey.* This NATO member first applied for membership in 1987 but has been told repeatedly that it needs to progress in the areas of social standards, human rights, and economic policies before its accession can be contemplated. Some EU leaders would like to include a mainly Muslim country in what Islamic states sometimes call the 'Christian Club,' but the possible accession of Turkey leaves the current EU bitterly divided. Admitting a large Muslim country could redefine what it means to be European, challenging deeply ingrained feelings and ideas of 'Europeanness.'

Still, the fact that the EU now reaches deeply into Eastern Europe, encompasses 27 members, has a common currency used in many of its member-states, has a parliament, is developing a constitution, and is even considering expansion beyond the realm's borders constitutes a tremendous and historic achievement in this fragmented, fractious part of the world. The EU now has a combined population of close to 500 million representing one of the world's richest markets.

It is remarkable that all this has been accomplished in little more than half a century. Some of the EU's leaders want more than economic union: they envisage a United States of Europe, a political as well as an economic competitor for the United States. To others, such a 'federalist' notion is an idea not even to be mentioned (the British are especially wary of such an idea). Whatever happens, Europe is going through still another of its revolutionary changes, and when you study its evolving map you are looking at history in the making.

Centrifugal Forces

For all its dramatic progress toward unification, Europe remains a realm of geographic contradictions. Europeans are well aware of their history of conflict, division, and repeated self-destruction. Will supranationalism finally overcome the centrifugal forces that have so long and so frequently afflicted this part of the world?

Devolutionary Pressures

It is ironic that even as Europe's states have been working to join forces in the EU, many of those same states are confronting severe centrifugal stresses. For example, the former Eastern European country called Czechoslovakia underwent a breakup before both new countries joined the EU. The term **18** **devolution** has come into use to describe the powerful centrifugal forces whereby regions or peoples within a state, through negotiation or active rebellion, demand and gain political strength and sometimes autonomy at the expense of the center. Most states exhibit some level of internal regionalism, but the process of devolution is set into motion when a key centripetal binding force—the nationally accepted idea of what a country stands for—erodes to the point that a regional drive for autonomy, or for outright secession, is launched.

The Case of the United Kingdom

As Figure 1-11 shows, devolution is affecting numerous European countries. States large and small, young and old, EU members and non-EU members must deal with the problem. Even the long-stable United Kingdom is affected. England, the historic core area of the British Isles, dominates the UK in

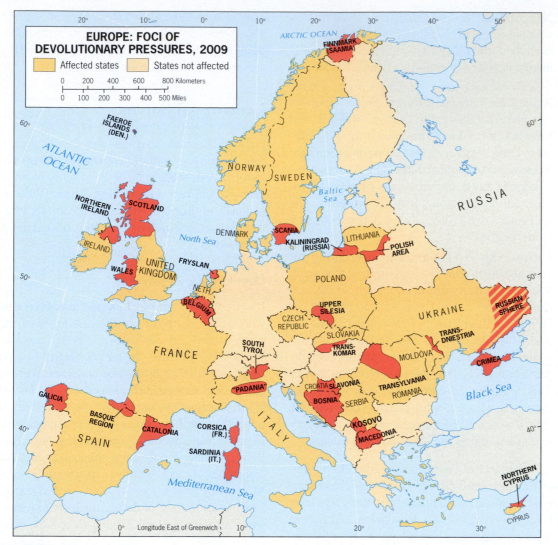

FIGURE 1-11   © H. J. de Blij, P. O. Muller, and John Wiley & Sons, Inc.

Devolution Elsewhere

Even while this devolutionary process was ongoing, the British government still joined the European Union on behalf not only of the English, but also of the Scots, Welsh, and Northern Irelanders seeking a new relationship with London. As the map shows, the UK is not alone in such a contradiction. Spain faces severe devolutionary forces in its Basque area and lesser ones in Catalonia and Galicia; France contends with a secessionist movement on its island of Corsica; Belgium is severely affected by Flemish-Walloon separatism; Italy confronts devolutionary pressures in South Tyrol and Lombardy. In recent decades Eastern Europe has been a cauldron of devolution as Yugoslavia and Czechoslovakia collapsed, Moldova fragmented, Montenegro seceded from Serbia-Montenegro in 2006, and Kosovo declared independence from Serbia in 2008.

The EU's New Economic Geography

At the outset of this chapter, we used as our frame of reference the old designations of Western and Eastern, Northern and Southern Europe. Certainly, those regional names still have some validity: Western Europe remains the core of the EU; the cultural landscapes of Northern and Southern Europe, though converging, still differ markedly; Eastern Europe still is, on average, the poorer, lagging part of the realm. But the birth and growth of the European Union have generated a new economic landscape that transcends the old.

Engines of Growth

By investing heavily in regional infrastructure and by smoothing the flows of money, labor, and products, European planners have dramatically reduced the divisive effects of their national boundaries. And by acknowledging demands for greater freedom of action by their provinces, States, departments, and other administrative units of their countries, Euro-

terms of population as well as political and economic power. The country's three other entities—Scotland, Wales, and Northern Ireland—were acquired over several centuries and attached to England (hence the 'United' Kingdom). But neither time nor representative democratic government was enough to eliminate all latent regionalism in these three components of the UK. During the 1960s and 1970s, the British government confronted a virtual civil war in North-

ern Ireland and rising tides of nationalism in Scotland and Wales. In 1997, the government in London gave the Scots and Welsh the opportunity to vote for greater autonomy in new regional parliaments that would have limited but significant powers over local affairs. The Scots voted overwhelmingly in favor and the Welsh by a slim majority—and thus a major devolutionary step was taken in one of Europe's oldest, most durable, and most unified states.

pean leaders unleashed a wave of economic energy that transformed some of these units into powerful engines of growth. In the next section of this chapter, when we look at regions and states, we will have to pay attention to these emerging powerhouses, which tend to be centered on major cities.

Four of these growth centers are especially noteworthy, to the point that geographers refer to them as the **19** **Four Motors of Europe**: (1) France's *Rhône-Alpes Region*, centered on the country's second-largest city, Lyon; (2) *Lombardy* in northern Italy, focused on the industrial city of Milan; (3) *Catalonia* in northeastern Spain, anchored by the cultural and manufacturing center of Barcelona; and (4) *Baden-Württemberg* in Germany, headquartered at the high-tech city of Stuttgart. Often, the local governments in these subregions simply bypass the governments in their national capitals, dealing not only with each other but even with foreign governments as their business networks span the globe. In this they are imitated by other provinces, all seeking to foster their local economies and, in the process, strengthen their political position relative to the state.

Regional States

The lowering of political barriers to trade and the resulting increase in regional transferability are having a further effect on Europe's economic map: States or provinces on opposite sides of political boundaries can now cooperate to achieve common economic goals. Already, the hinterlands of the cities anchoring the Four Motors extend into neighboring countries. The Japanese economist Kenichi Ohmae termed these

WHAT'S DRIVING GEOGRAPHIC CHANGE IN THE REALM

- The big issue in Europe today: should Islamic **Turkey** be invited to join the 'Christian Club'? The bureaucrats want it, European voters are showing signs of doubt, and many Turks are getting fed up with European demands and conditions. It's a recipe for trouble.

- The **economic integration** of significantly poorer countries into the now 27-strong European Union is a challenge that will stretch the unification experiment going on in Europe today. Can Europe remain economically vibrant while trying to bring 'up' its eastern brethren?

- **Catalonia**, a Spanish Autonomous Community, is demanding that it be formally recognized as a 'nation,' transferring a set of powers ranging from tax collection to security from Madrid to Barcelona. The Spanish parliament has approved in principle, and if Catalonia succeeds, get ready for others to follow suit.

- Terrorism, disruptive violence, and escalating asylum costs are among factors causing European countries to reassess their **immigration** policies. Europe never was a melting pot, and public opinion, especially in Western Europe, is turning against a growing Muslim population whose integration will nevertheless be key to Europe's future.

- The serious **population implosion** being experienced in various parts of the realm poses the challenge for Western Europe to continue to maintain a high level of social welfare while Eastern countries try to improve their economies with a declining population.

cross-border growth engines **20** **regional states**, new phenomena on the old map of Europe. Elsewhere, notably along the borders between the fifteen pre-2004 EU member-states and the newly admitted twelve, cross-border investment is creating start-up regional states that will further alter the economic landscape.

All these developments underscore how far-reaching and irreversible Europe's current transformation is. European leaders understood the need for it; the Marshall Plan jump-started it; good economic times sustained it; and the end of the Cold War galvanized it. But Europe's ultimate supranational goals have not yet been attained. The problems of the old Eastern Europe remain evident in the data presented in Appendix B. The sense of have and have-not, core and periphery, still pervades EU negotiations. Many Europeans, perhaps a majority, feel that the EU project is a bureaucratic edifice constructed by leaders who are out of touch with grass-roots Europe. Europe's quest for enduring unity continues.

Regions of the Realm

Our objective in this and later regional discussions is to become familiar with the regional and national frameworks of the realms under investigation, and to discover how individual states fit into and function within the larger mosaic. Especially in this era

of globalization, and even in an integrating Europe, the state still plays a key role in the process. Where they are located, how they interact, and what their prospects may be in this changing world are among the key questions we will address.

Europe presents us with a particular challenge because of its numerous countries and territories plus a set of *microstates* such as Monaco, San Marino, and Andorra that do not qualify as full-fledged countries. So it is not surprising that the old

FIGURE 1-12

© H. J. de Blij, P. O. Muller, and John Wiley & Sons, Inc.

geographic division into five regions remains helpful. In Figure 1-12: the colors represent (1) Western Europe, (2) the British Isles, (3) Northern (Nordic) Europe, (4) Mediterranean Europe, and (5) Eastern Europe. This regionalization scheme employs the formal-region concept, but we can also approach Europe in other ways. Focusing on Europe's supranational and subnational development in a functional-region context, we recognize a European core (delimited by the red line in Fig. 1-12) and a highly varied periphery beyond it. We employ a combination of perspectives in the discussion that follows.

WESTERN EUROPE

Is it still appropriate to view 'Western' Europe as a regional concept? Consider this: the three Benelux countries started the ball rolling toward postwar European integration; France and Germany have been the main proponents and remain the most influential forces in the European Union; and the ideals of the EU are nowhere closer to achievement than in Western Europe. For example, the Schengen Agreement, which makes border-crossings even easier than EU regulations stipulate, involves primarily Western European countries. As Figure 1-12 suggests, Western Europe is the very heart of Europe's coreland.

Counting the microstate of Liechtenstein, there are eight states in Western Europe: dominant Germany and France, the three Benelux countries, and three landlocked mountain states

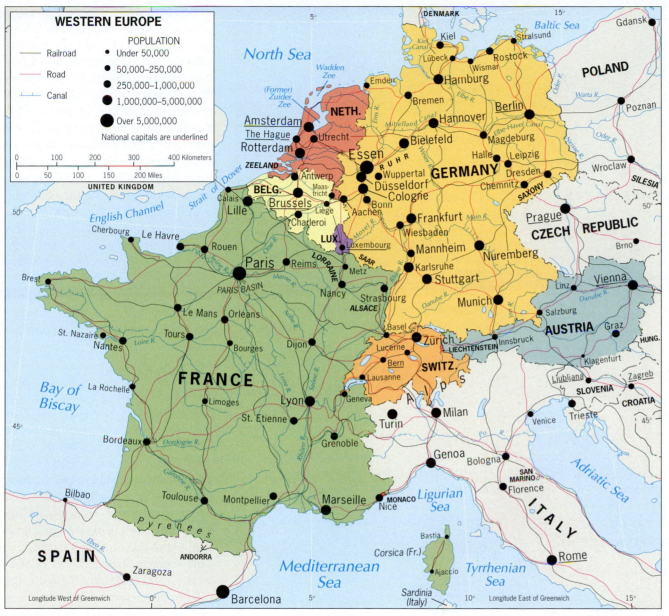

WESTERN EUROPE

POPULATION

- —— Railroad
- —— Road
- ⌐⌐ Canal

- • Under 50,000
- ● 50,000–250,000
- ● 250,000–1,000,000
- ● 1,000,000–5,000,000
- ● Over 5,000,000

National capitals are underlined

0   100   200   300   400 Kilometers
0   50   100   150   200 Miles

FIGURE 1-13

© H. J. de Blij, P. O. Muller, and John Wiley & Sons, Inc.

(Fig. 1-13). France is territorially the region's largest state, but Germany is Europe's most populous. And no European country has more neighbors than Germany. So let us focus first on this, Europe's largest and most productive economy.

Reunited Germany

Twice during the twentieth century Germany plunged Europe and the world into war, until, in 1945, the defeated and devastated German state was divided into two parts, West and East (see the delimitation in red in Fig. 1-14). Its eastern boundaries were also changed, leaving the industrial district of Silesia in newly defined Poland, that of Saxony in communist-ruled and Soviet-controlled East Germany, and the Ruhr in West Germany (see Fig. 1-14). Aware that these were the industrial centers that had enabled Nazi Germany to seek world domination through war, the victorious allies laid out this new boundary framework to make sure this would not happen again.

In the aftermath of World War II, Soviet and Allied administration of East and West Germany differed. Soviet rule in East Germany was established on the Russian communist model and, given the extreme hardships the USSR had suffered at German hands during the war, was harshly punitive. The American-led authority in West Germany was less strict and aimed more at rehabilitation. When the Marshall Plan was instituted, West Germany was included, and its economy recovered rapidly. Meanwhile, West Germany was reorganized politically into a modern federal state on democratic foundations.

Revival

West Germany's economy thrived. Between 1949 and 1964, its GNI tripled while industrial output rose 60 percent. It absorbed millions of German-speaking refugees from Eastern Europe (and many escapes from communist East Germany as well). Since unemployment was virtually nonexistent, hundreds of thousands of Turks and other foreign guest workers arrived to take jobs Germans could not fill or did not want.

STATES (LÄNDER) OF REUNIFIED GERMANY

GDP PER CAPITA,
IN EUROS, 2002

- Over 26,000
- 20,000–26,000
- Below 20,000
- Railroads
- Roads

City population
- · Under 50,000
- • 50,000–250,000
- ● 250,000–1,000,000
- ● 1,000,000–5,000,000
- ● Over 5,000,000

0   50   100 Kilometers
0   25   50 Miles
National capitals are underlined

FIGURE 1-14

© H. J. de Blij, P. O. Muller, and John Wiley & Sons, Inc.

Simultaneously, West Germany's political leaders participated enthusiastically in the OEEC and in the negotiations that led to the six-member Common Market. Geography worked in West Germany's favor: it had common borders with all but one of the EEC member-states. Its transport infrastructure, rapidly rebuilt, was second to none in the realm. More than compensating for its loss of Saxony and Silesia were the expanding Ruhr (in the hinterland of the Dutch port of Rotterdam) and the newly emerging industrial complexes centered on Hamburg in the north, Frankfurt (the leading financial hub as well) in the center, and Stuttgart in the south. West Germany exported huge quantities of iron, steel, motor vehicles, machinery, textiles, and farm products.

Setback and Shock

No economy grows without setbacks and slowdowns, however, and Germany experienced such problems in the 1970s and 1980s, when energy shortages, declining competitiveness on world markets, lagging modernization (notably in the aging Ruhr), and social dilemmas involving rising unemployment, an aging population, high taxation, and a backlash against foreign resident workers roiled West German society. And then, quite suddenly, the collapse of the communist Soviet Union opened the door to reunification with East Germany.

Germany Restored

In 1990, West Germany had a population of about 62 million and East Germany 17 million. Communist misrule in the East had yielded outdated factories, crumbling infrastructures, polluted environments, drab cities, inefficient farming, and inadequate legal and other institutions. Reunification was more a rescue than a merger, and the cost to West Germany was enormous. When the West German government imposed sales-tax increases and an income-tax surcharge on its citizens, many Westerners doubted the wisdom of reunification. It was projected that it would take decades to reconstruct Virginia-sized East Germany: more than ten years later, exports from the former East still contributed only about 7 percent of the national total. Regional disparity would afflict Germany for a very long time to come.

The Federal Republic

Before reunification, West Germany functioned as a federal state consisting of ten States or Länder (Fig. 1-14). East Germany had been divided under communist rule into fifteen districts including East Berlin. Upon reunification, East Germany was reorganized into six new States based on traditional provinces within its borders. Figure 1-14 makes a key point: regional disparity in terms of income per person remains a serious problem between the former East and West. Note that five of former East Germany's six States (Berlin being the sole exception) are in the lowest income category, while most of the ten former West German States rank in the two highest income categories. In the first decade of this century, Germany's economy was stagnant, raising unemployment and slowing former East Germany's recovery. Nonetheless, the gap continues to narrow, and with 82.1 million inhabitants Germany is again exerting its dominance over a mainland Europe in which it has no peer.

This image of Paris shows clearly the work of Baron Haussmann's urban planning of the mid-nineteenth century. Before this time, the city was crowded and unsanitary; his new designs cleared the congestion and created the wide radial street pattern for which Paris is now so famous. He also improved the city's water supply and sanitation system.   (© iStockphoto)

France

German dominance in the European Union is a constant concern in the other leading Western European country. The French and the Germans have been rivals in Europe for centuries. France (population: 61.7 million) is an old state, by most measures the oldest in Western Europe. Germany is a young country, created in 1871 after a loose association of German-speaking states had fought a successful war against . . . the French.

Territorially, France is much larger than Germany, and the map suggests that France has a superior relative location, with coastlines on the Mediterranean Sea, the Atlantic Ocean, and, at Calais, even a window on the North Sea. But France does not have any good natural harbors, and oceangoing ships cannot navigate its rivers and other waterways far inland. France has no equivalent to Rotterdam either internally or externally.

Figures 1-8 and 1-13 reveal a significant demographic contrast between France and Germany. France has one dominant city, Paris, at the heart of the Paris Basin, France's core area. No other city in France comes close to Paris in terms of population or centrality: Paris has 9.8 million residents, whereas its closest rival, Lyon, has only 1.4 million. Germany has no city to match Paris, but it does have a number of cities with populations between 1 and 5 million. And as Appendix B shows, Germany is much more highly urbanized overall than France.

Paris: Site and Situation

Why should Paris, without major raw materials nearby, have grown so large? Whenever geographers investigate the evolution of a city, they focus on two important locational qualities: its **21** site (the physical attributes of the place it occupies) and its **22** situation (its location relative to surrounding areas of productive capacity, other cities and towns, barriers to access and movement, and other aspects of the greater regional framework in which it lies).

The site of the original settlement at Paris lay on an island in the Seine River, a defensible place where the river was often crossed. This island, the *Île de la Cité*, was a Roman outpost 2000 years ago; for centuries its security ensured continuity. Eventually the island became overcrowded, and the city expanded along the banks of the river (Fig. 1-15A).

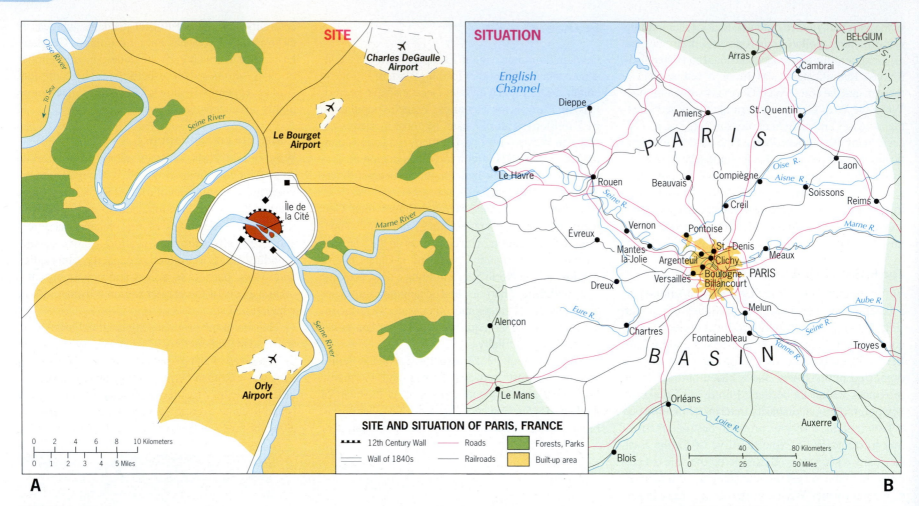

FIGURE 1-15

© H. J. de Blij, P. O. Muller, and John Wiley & Sons, Inc.

Soon the settlement's advantageous situation stimulated its growth and prosperity. Its fertile agricultural hinterland thrived, and, as an enlarging market, Paris's focality increased steadily. The Seine River is joined near Paris by several navigable tributaries (the Oise, Marne, and Yonne). When canals extended these waterways even farther, Paris was linked to the Loire Valley, the Rhône-Saône Basin, Lorraine (an industrial area in the northeast), and the northern border with Belgium. When Napoleon reorganized France and built a radial system of roads—followed later by railroads—that focused on Paris from all parts of the country, the city's primacy was assured

(Fig. 1-15B). The only disadvantage in Paris's situation lies in its seaward access: oceangoing ships can sail up the Seine River only as far as Rouen.

Modern France

Paris, in accordance with Weber's agglomeration principle, grew into one of Europe's greatest cities. French industrial development was less spectacular, but northern French agriculture remained Europe's most productive and varied, exploiting the country's wide range of soils and climates and enjoying state subsidies and protections. Today France's economic geography is marked by new high-tech industries. It

is a leading producer of high-speed trains, aircraft, fiber-optic communications systems, and space-related technologies. It also is the world leader in nuclear power, which currently supplies more than 75 percent of its electricity and thereby reduces its dependence on foreign oil imports.

Napoleon's Legacy

When Napoleon reorganized France in the early 1800s, he broke up the country's large traditional subregions and established more than 80 small *départements* (additions and subdivisions later increased this number to 96). Each *département* had

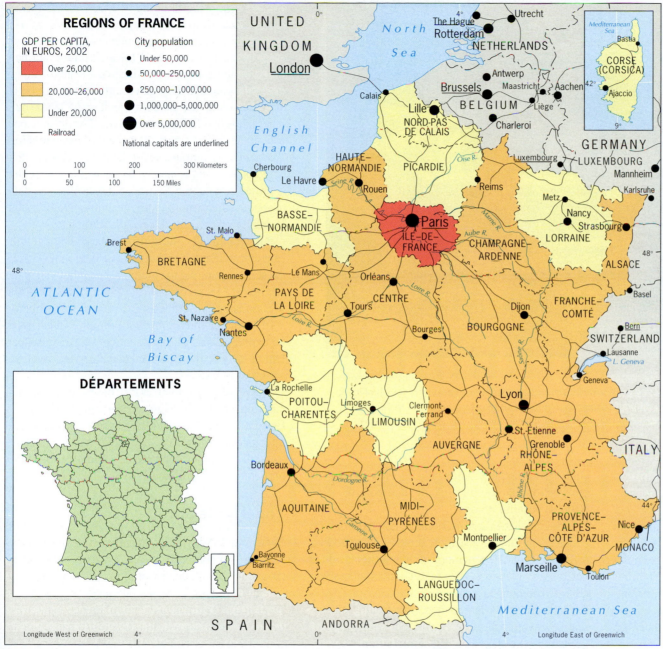

REGIONS OF FRANCE

GDP PER CAPITA,
IN EUROS, 2002

- Over 26,000
- 20,000–26,000
- Under 20,000

City population
- Under 50,000
- 50,000–250,000
- 250,000–1,000,000
- 1,000,000–5,000,000
- Over 5,000,000
- Railroad

National capitals are underlined

100   200   300 Kilometers
50   100   150 Miles

DÉPARTEMENTS

FIGURE 1-16   © H. J. de Blij, P. O. Muller, and John Wiley & Sons, Inc.

representation in Paris, but the power was concentrated in the capital, not in the individual *départements*. France became a highly centralized state and remained so for nearly two centuries (see inset map, Fig. 1-16). Only the island *département* of Corsica produced a rebel movement, whose violent opposition to French rule continued for decades and even touched the mainland. In 2003 the voters in Corsica rejected an offer of special status for their island, including limited autonomy. They wanted more, and trouble lies ahead.

Decentralizing the State

Today, France is decentralizing. A new subnational framework of 22 historically significant provinces, groupings of *départements* called *regions* (Fig. 1-16), has been established to accommodate the devolutionary forces felt throughout Europe and throughout the world. These regions, though still represented in Paris, have substantial autonomy in such areas as taxation, borrowing, and development spending. The cities that anchor them benefit because they are the seats of governing regional councils that can attract investment, not only within France but also from abroad.

Lyon, France's second city and headquarters of the region named Rhône-Alpes, has become a focus for growth industries and multinational firms. This region is evolving into a self-standing economic powerhouse that is becoming a driving force in the European economy; indeed, it is one of the Four Motors of Europe with its own international business connections to countries as far away as China and Chile.

After terrible floods in 1953, the Dutch constructed large storm surge barriers that can be closed during heavy storms to prevent disasters inland. This, along with a complex system of dikes and pumps, keeps the country (much of it lies below sea level) dry and protected from the sea. Although threatened by sea level rise, the country has developed significant technologies to counter the forces of nature. The storm surge barrier above is part of the Neeltje Jans complex in Zeeland province, The Netherlands.   (© AP/Wide World Photos)

project so far, the draining of much of the former Zuider Zee (Southern Sea), now known as IJsselmeer, began in 1932 and ended in 1990. What was once seafloor is now used for agriculture and new cities. In the southern province of Zeeland, islands are being connected by dikes and storm surge barriers, protecting the **polders** (reclaimed lands) and other lands that lie below sea level. Twenty-five percent of the country, with 60 percent of its population, lies below current sea level. This is a country that could be severely impacted by the sea-level rise that is projected to come from global warming. To date, however, the wealthy and technologically astute Dutch have been able to keep the sea at bay.

Geographically, **Belgium** is marked by a significant cultural fault line that extends diagonally across the state, separating the Dutch-related Flemish in the northwest (58 percent of the population) from the Walloons in the southeast (31 percent). Devolution is an imminent threat here, but Belgium's great asset, its capital, Brussels, headquarters of the European Union, may be the glue needed to keep this fractious country together.

Luxembourg, small as it is, has the distinction of recording by far the highest per-capita GNI in all of Europe (Appendix B). Financial, service, tourist, and information technology industries make this country, still run by a Grand Duke, the most prosperous **ministate** in the realm.

The Alpine States

Switzerland, Austria, and the **microstate** of Liechtenstein on their border share an absence of coasts and the mountainous topography of the Alps—and little else (Fig. 1-13). Austria speaks one language; the Swiss speak four—German in the north, French in the west, Italian in the southeast, and Rhaeto-Romansch in the central and eastern highlands (Fig. 1-7). Austria has a large primate city; multicultural Switzerland does not. Austria has a substantial range of domestic raw materials; Switzerland does not. Austria is twice the size of Switzerland and has a larger population, but far more trade crosses the Swiss Alps between Western and Mediterranean Europe than crosses Austria.

France has one of the world's most productive and diversified economies, based in one of humanity's richest cultures and vigorously promoted and protected (notably its heavily subsidized agricultural sector). Although France and Germany agree on many aspects of EU integration, they tend to differ on important issues. Old, historically centralized France is less eager than young, federal Germany to push political integration in supranational Europe.

Benelux

Three countries are crowded into the northwest corner of Western Europe: Belgium, the Netherlands, and tiny Luxembourg, collectively referred to by their first syllables (*Be-Ne-Lux*). The major contrasts that evolved between the agriculturally productive

Netherlands and the industrially developed Belgium yielded a double complementarity that led to the 1944 Benelux customs union.

The Benelux countries, home to about 28 million people, rank among the most densely populated on earth. The regional geography of the **Netherlands** is noted for its *Randstad* (edge-city), a triangular urban core area dominated by the constitutional capital, Amsterdam, Europe's largest port, Rotterdam, and the seat of government, The Hague, but with a 'green' (agricultural) center. This **23** **conurbation**, as geographers call large multimetropolitan complexes formed by the coalescence of two or more urban areas, forms a ring-shaped complex that encircles a still-rural center.

For centuries the Dutch have been expanding their living space—not by warring with their neighbors but by wresting land from the sea. The greatest

Switzerland, not Austria, is in most ways the leading state in Alpine Western Europe (Appendix B). Mountainous terrain and **24** landlocked location can constitute crucial barriers to economic development, tending to inhibit the dissemination of ideas and innovations, obstruct circulation, constrain farming, and divide cultures. That is why Switzerland is such an important lesson in human geography. Through the skillful maximization of their opportunities (including the transfer needs of their neighbors), the Swiss have transformed their seemingly restrictive environment into a prosperous state. They used the waters cascading from their mountains to generate hydroelectric power to develop highly specialized industries. Swiss farmers perfected ways to optimize the productivity of mountain pastures and valley soils. Swiss leaders converted their country's isolation into stability, security, and neutrality, making it a world banking giant, a global magnet for money. Zürich, in the German sector, is the financial center; Geneva, in the French sector, is one of the world's most international cities. The Swiss do not feel that they need to join the EU; they have not done so, and they increasingly oppose immigration.

Austria, which joined the EU in 1995, is a remnant of the Austro-Hungarian Empire and has a historical geography that is far more reminiscent of unstable Eastern Europe than that of Switzerland. Even

Austria's physical geography seems to demand that the country look eastward: it is at its widest, lowest, and most productive in the east, where the Danube links it to Hungary, its old ally in the anti-Muslim wars of the past.

Vienna, by far the Alpine subregion's largest city, also lies on the country's eastern perimeter. One of the world's most expressive primate cities with magnificent architecture and monumental art, Vienna today is Western Europe's easternmost city, situated on the doorstep of fast-changing Eastern Europe. But Austria's cultural landscapes and economic and political standards are those of Western Europe, a nation-state of the European core, its relative location enormously enhanced by the EU's eastward expansions of 2004 and 2007.

THE BRITISH ISLES

Off the coast of mainland Western Europe lie two major islands, surrounded by a constellation of tiny ones, that constitute the British Isles, a discrete region of the European realm (Fig. 1-17). The larger of the two major islands, which also lies nearest to the mainland (a mere 34 kilometers or 21 mi, at the closest point), is the island called *Britain*; its smaller neighbor to the west is *Ireland*.

Transportation in a mountainous country such as Switzerland can be very difficult, but the Swiss have built a train system that moves people with ease through difficult terrain. This train connects the famed ski town of St. Moritz with villages in the Engadine Valley in the southeast part of the country near the Italian border. (© V. WinklerPrins)

States and Peoples

The names attached to these islands and the countries they encompass are the source of some confusion. They are still called the British Isles, even though British dominance over most of Ireland ended in 1921. The nation-state that occupies Britain and the northeastern corner of Ireland is officially called the United Kingdom of Great Britain and Northern Ireland—United Kingdom for short and UK by abbreviation. But this country often is referred to simply as Britain, and its people are known as the British. The nation-state of Ireland officially is the Republic of Ireland (*Eire* in Irish Gaelic), but it does not include the whole island of Ireland.

How convenient it would be if physical and political geography coincided! Unfortunately, the two do not. During the long British occupation of Ireland, which is overwhelmingly Catholic, many Protestants from northern Britain settled in northeastern Ireland. In 1921, when British domination ended, the Irish were set free—except in that corner in the north, where London kept control to protect the area's Protestant settlers. That is why the country to this day is officially known as the United Kingdom of Great Britain and Northern Ireland.

Roots of Devolution

Northern Ireland (Fig. 1-17) was home not only to Protestants from Britain, but also to a substantial population of Irish Catholics who found themselves on the wrong side of the border when Ireland was liberated. Ever since, conflict has intermittently engulfed Northern Ireland and spilled over into Britain and even, in the form of terrorism, into Western Europe. It has been one of Europe's most costly struggles.

Although all of Britain lies in the United Kingdom, political divisions exist here as well. England is the largest of these units, the center of power from which the rest of the region was originally brought under unified control. The English conquered Wales in the Middle Ages, and Scotland's link to England, cemented when a Scottish king ascended the English throne in 1603, was ratified by the Act of Union of 1707. Thus England, Wales, Scotland, and Northern Ireland became the United Kingdom.

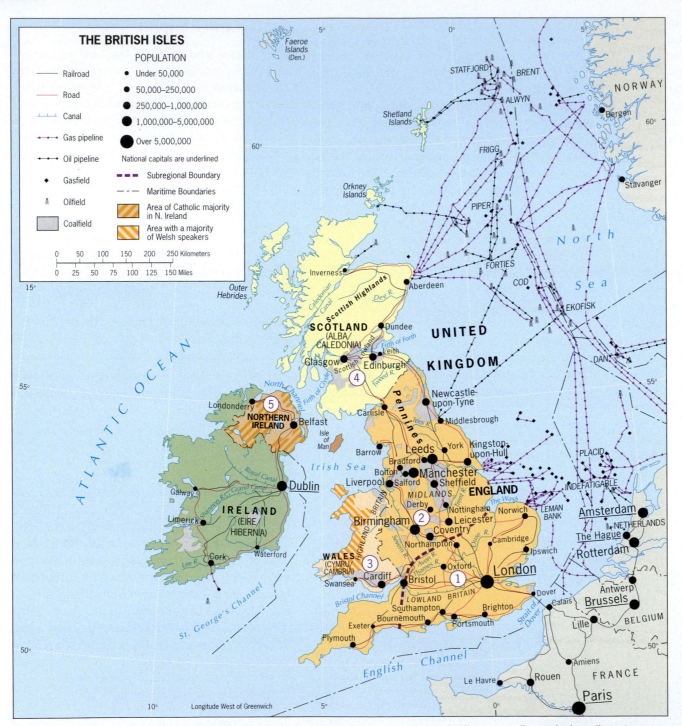

THE BRITISH ISLES

POPULATION

— Railroad
— Road
—┬— Canal
—●— Gas pipeline
—●— Oil pipeline
◆ Gasfield
⚒ Oilfield
▨ Coalfield

● Under 50,000
● 50,000–250,000
● 250,000–1,000,000
● 1,000,000–5,000,000
● Over 5,000,000

National capitals are underlined
▪▪▪ Subregional Boundary
—·— Maritime Boundaries
▨ Area of Catholic majority in N. Ireland
▨ Area with a majority of Welsh speakers

0   50   100   150   200   250 Kilometers
0   25   50   75   100   125   150 Miles

FIGURE 1-17

© H. J. de Blij, P. O. Muller, and John Wiley & Sons, Inc.

A Discrete Region

The British Isles form a distinct region of Europe for several reasons. Britain's insularity provided centuries of security from turbulent Europe, protecting the evolving British nation as it achieved a system of parliamentary government that had no peer in the Western world. Having united the Welsh, Scots, and Irish, the British set out to forge what would become the world's largest colonial empire. An era of mercantilism and domestic manufacturing (the latter based on water power from streams flowing off the Pennines, Britain's mountain backbone) foreshadowed the momentous Industrial Revolution, which transformed Britain—and much of the world. British cities became synonyms for specialized products as the smokestacks of factories rose like forests over the urban scene. London on the Thames River anchored an English core area that mushroomed into the headquarters of a global political, financial, and cultural empire. As recently as World War II, the narrow English Channel ensured the United Kingdom's impregnability against German invasion, giving the British time to organize their war machine. When the United Kingdom emerged from that conflict as a leading power among the victorious allies, it seemed that its superpower role in the postwar era was assured.

End of Empire

Two unanticipated developments changed that prospect: the worldwide collapse of colonial empires and the rapid resurgence of mainland Europe. Always ambivalent about the EC and EU, and with its first membership application vetoed by the French in 1963, Britain (admitted in 1973) has worked to restrain moves toward tighter integration. When most member-states adopted the new euro in favor of their national currencies, the British kept their pound sterling and delayed their participation in the EMU. As for a federalized Europe, to Britain this prospect is out of the question. In this as in other respects, Britain's historic, insular standoffishness continues.

The United Kingdom

As we noted earlier, the British Isles as a region consists of two political entities: the United Kingdom and Ireland. The UK, with an area about the size of Oregon and a population of over 60 million, is by European standards quite a large country. Based on a combination of physiographic, historical, cultural, economic, and political criteria, the United Kingdom can be divided into five subregions:

1. *Southern England.* Centered on the gigantic London metropolitan area, this is the UK's most affluent subregion, with one-third of the country's population. Financial, communications, engineering, and energy-related industries cluster in this economically and politically dominant area. London exemplifies the momentum of long-term agglomeration; today this subregion benefits anew not only from its superior links to the European mainland but also as one of the three leading 'world cities' on the global scene.
2. *Northern England.* As the map suggests, this subregion should really be called Northern, Central, and Western England. Its center, appropriately called the Midlands, was the focus of the Industrial Revolution and the spectacular rise of Manchester, the source; Liverpool, the great port; Birmingham, an unmatched industrial city; and many other manufacturing cities and towns. But here obsolescence has overtaken what was once ultramodern, and the North now suffers from rustbelt conditions and high unemployment among immigrants from South Asia, Africa, and the Caribbean.
3. *Wales.* This nearly rectangular, rugged territory was a refuge for ancient Celtic peoples, and in its western counties more than half the inhabitants still speak Welsh. Because of the high-quality coal reserves in its southern tier, Wales too was engulfed by the Industrial Revolution, and Cardiff, the capital, was once the world's leading coal exporter. But the fortunes of Wales also declined, and many Welsh emigrated. Among the 3 million who remained, however, the flame of Welsh nationalism survived, and in 1997 the voters approved the establishment of a Welsh Assembly to administer public services in Wales, a first devolutionary step.
4. *Scotland.* Nearly twice as large as the Netherlands and with a population about the size of Denmark's, Scotland is a major component of the United Kingdom. As Figure 1-17 suggests, most of Scotland's more than 5 million people live in the Scottish Lowlands anchored by Edinburgh, the capital, in the east and Glasgow in the west. Attracted there by the labor demands of the Industrial Revolution (coal and iron reserves lay in the area), the Scots developed a world-class shipbuilding industry. Decline and obsolescence were followed by high-tech development, notably in the hinterland of Glasgow, and Scottish participation in the exploitation of oil and gas reserves under the North Sea (Fig. 1-17), which transformed the eastern ports of Aberdeen and Leith (Edinburgh). But many Scots feel that they are disadvantaged within the UK and should play a major role in the EU. Therefore, when the British government put the option of a Scottish parliament before the voters in 1997, 74 percent approved. Many Scots still hope that total independence lies in their future and in 2007, local elections gave the independence-minded Scottish National Party more seats in parliament than any of the other parties.
5. *Northern Ireland.* Prospects of devolution in Scotland pale before the devastation caused by political and sectarian conflict in Northern Ireland. With a population of 1.8 million occupying the northeastern one sixth of the island of Ireland, this area represents the troubled legacy of British colonial rule. A declining majority, now about 54 per-

cent of the people in Northern Ireland, trace their ancestry to Scotland or England and are Protestants; a growing minority, currently around 45 percent, are Roman Catholics, who share their Catholicism with virtually the entire population of the Irish Republic on the other side of the border. Although Figure 1-17 suggests that there are majority areas of Protestants and Catholics in Northern Ireland, no clear separation exists; they live mostly in clusters throughout the territory, including walled-off neighborhoods in the major cities of Belfast and Londonderry. Partition is no solution to a conflict that has raged for over three decades at a cost of thousands of lives; Catholics accuse London as well as the local Protestant-dominated administration of discrimination, whereas Protestants accuse Catholics of seeking union with the Republic of Ireland. The territory remains precarious.

Republic of Ireland

What Northern Ireland is missing through its conflicts is shown by the Irish Republic itself: a booming service-based economy, the fastest-growing in all of Europe for a time around the turn of this century, when Ireland was dubbed the *Celtic Tiger*. Burgeoning cities and towns (and rising real-estate prices), mushrooming industrial parks, bustling traffic, and construction everywhere reflect this new era. For the first time ever, workers of Irish descent returned from foreign places to take jobs at home; non-Irish immigrants also arrived, posing some new social problems for a closely knit, long-isolated society.

The Republic of Ireland fought itself free from British colonial rule just three generations ago. Its cool, moist climate had earlier led to the adoption of the potato from South America as a staple crop, but excessive rain and reliance on only one variety of potato resulted in a blight in the late 1840s, which caused famine and cost over 1 million lives. Another two million Irish emigrated.

Independence in 1921 did not bring economic prosperity, and Ireland stagnated until the 1990s. Then, European telecommunications service industries saw labor and locational opportunities in Ireland,

and soon Ireland was Europe's leading call center for the realm's rapidly expanding toll-free telephone market. Other service industries followed suit, benefiting from Ireland's well-educated but not highly paid labor pool. Changing economic conditions worldwide and in Europe slowed Ireland's economy in the early 2000s, but growth has now resumed and the country continues to be an EU success story.

NORTHERN (NORDIC) EUROPE

North of Europe's Western European core area lies a disconnected group of six countries that exemplify core-periphery contrasts. Northern Europe is a region of difficult environments: generally cold climates, poorly developed soils, limited mineral resources, and long distances. Together, the six countries in this region—Sweden, Norway, Denmark, Finland, Estonia, and Iceland—contain just over 26 million inhabitants, which is a lower total population than that of Benelux and only one seventh that of Western Europe. The overall land area, on the other hand, is almost the size of the entire European core. Here in the peripheral north, national core areas lie in the south: note the location of the capitals of Helsinki, Stockholm, and Oslo at approximately the same latitude (Fig. 1-18).

Northern Europe's remoteness, isolation, and environmental severity also have had positive effects for this region. The countries of the Scandinavian Peninsula lay removed from the wars of mainland Europe, although Norway was overrun by Nazi Germany during World War II. The three major

FIGURE 1-18

© H. J. de Blij, P. O. Muller, and John Wiley & Sons, Inc.

languages—Danish, Swedish, and Norwegian—are mutually intelligible, which creates one of the criteria delimiting this region. Another regional criterion is the overwhelming adherence to the same Lutheran church in each of the Scandinavian countries (Norway, Sweden, and Denmark), Iceland, and Finland. Furthermore, democratic and representative governments emerged early, and individual rights and social welfare have long been carefully protected. Women participate more fully in government and politics here than in any other region of the world.

Sweden is the largest Nordic country in terms of both population and territory. Most Swedes live south of 60° North latitude (which passes through Uppsala), in what is climatically the most moderate part of the country (Fig. 1-18). Here lie the capital, core area, and, as Figure 1-6 shows, the main industrial districts; here, too, are the main agricultural areas that benefit from the lower relief, better soils, and milder climate.

Sweden long exported raw or semifinished materials to industrial countries, but today the Swedes are making finished products themselves, including automobiles, electronics, stainless steel, furniture (IKEA is a Swedish company), and glassware. Much of this production is based on local resources, including a major iron ore reserve at Kiruna in the far north (there is a steel mill at Luleå). Swedish manufacturing, in contrast to that of several Western European countries, is based in dozens of small and medium-sized towns specializing in particular products. Energy-poor Sweden was a pioneer in the development of nuclear power, but a national debate over the risks involved has reversed that course.

Norway does not need a nuclear power industry to supply its energy needs. It has found its economic opportunities on, in, and beneath the sea. Norway's fishing industry, now augmented by highly efficient fish farms, long has been a cornerstone of the economy, and its merchant marine spans the world. But since the 1970s, Norway's economic life has been transformed by the bounty of oil and natural gas discovered in its sector of the North Sea.

With its limited patches of cultivable soil, high relief, extensive forests, frigid north, and spectacularly fjorded coastline, Norway has nothing to compare to Sweden's agricultural or industrial development. Its cities, from the capital Oslo and the North Sea port of Bergen to the historic national focus of Trondheim as well as Arctic Hammerfest, lie on the coast and have difficult overland connections. The isolated northern province of Finnmark has even become the scene of a movement for autonomy among the reindeer-herding indigenous Saami (Fig. 1-11). Norway has been described as a necklace, its beads linked by the thinnest of strands. Nor has this configuration not constrained national development. Norway typically has one of the lowest rates of unemployment and is one of the richest countries in the world.

Norwegians have a strong national consciousness and a spirit of independence. In 1994, when Sweden and Finland voted to join the European Union, the Norwegians again said no. They did not want to trade their economic independence for the regulations of a larger, even possibly safer, Europe.

Denmark, territorially small by Scandinavian standards, has a population of 5.4 million, second largest in the Nordic region after Sweden. It consists of the Jutland Peninsula and several islands to the east at the gateway to the Baltic Sea; it is on one of these islands, Sjaelland, that the capital of Copenhagen is located. Copenhagen, the 'Singapore of the Baltic,' has long been a port that collects, stores, and transships large quantities of goods. This **25** **break-of-bulk** function exists because many oceangoing vessels cannot enter the shallow Baltic Sea, making the city an **26** **entrepôt** where transfer facilities and activities prevail. The completion of the Øresund bridge-tunnel link to southern Sweden enhanced Copenhagen's situation in 2000 (Fig. 1-18).

Denmark remains a kingdom, and in centuries past Danish influence spread far beyond its present confines. Remnants of that period now challenge Denmark's governance. Greenland came under Danish rule after union with Norway (1380) and remained a Danish domain when that union ended (1814). In 1953, Greenland's status changed from colony to province, and in 1979 the 60,000 inhabitants were given home rule with an Inuit name: *Kalaallit Nunaat*. They promptly exercised their rights by withdrawing from the European Union, of which they had become a part when Denmark joined. Another restive dependency is the Faroe Islands, located between Scotland and Iceland. These 17 small islands and their 45,000 inhabitants were awarded self-government in 1948, complete with their own flag and currency, but even this was not enough to defuse demands for total independence. A referendum in mid-2001 confirmed that not even Denmark is immune from Europe's devolutionary forces (Fig. 1-11).

Finland, territorially almost as large as Germany, has only 5.3 million residents, most of them concentrated in the triangle formed by the capital, Helsinki, the textile-producing center, Tampere, and the shipbuilding center, Turku (Fig. 1-18). A land of evergreen forests and glacial lakes, Finland has an economy that has long been sustained by wood and wood product exports. But the Finns, being a skillful and productive people, have developed a diversified economy in which the manufacture of precision machinery and telecommunications equipment (prominently including cell phones) as well as the growing of staple crops are key.

As in Norway and Sweden, environmental challenges and relative location have created Nordic cultural landscapes in Finland, but the Finns are not a Scandinavian people; their linguistic and historic links are instead with the Estonians across the Gulf of Finland. **Estonia**, northernmost of the three Baltic states, is part of Nordic Europe by virtue of its ethnic and linguistic ties to Finland. But during the period of Soviet control from 1940 to 1991, Estonia's demographic structure changed drastically: today about 25 percent of its 1.3 million inhabitants are Russians, most of whom came there as colonizers.

After a difficult period of adjustment, Estonia today is forging ahead of its Baltic neighbors (see Appendix B) and catching up with its Nordic counterparts. Busy traffic links Tallinn, the capital, with Helsinki, and a new free-trade zone at Muuga Harbor facilitates commerce with Russia. But more important for Estonia's future was its entry into the European Union in 2004.

Iceland is a volcanic, glacier-studded island in the frigid waters of the North Atlantic just south of the Arctic Circle. Inhabited by people with Scandinavian ancestries (population: 288,000), Iceland and its small neighboring archipelago, the Westermann Islands, are of special scientific interest because they lie on the Mid-Atlantic Ridge, where the Eurasian and North American tectonic plates of the earth's crust are diverging and new land can be seen forming (see Fig. G-5).

Iceland's population is almost totally urban, and the capital, Reykjavik, contains about half the country's inhabitants. The nation's economic geography is almost entirely oriented toward the surrounding waters, whose seafood harvests give Iceland one of the world's highest standards of living—but at the risk of overfishing. Disputes over fishing grounds and fish quotas have intensified in recent decades; the Icelanders argue that, unlike the Norwegians or the British, they have little or no alternative economic opportunity.

MEDITERRANEAN EUROPE

South of Europe's core lie the six countries that constitute the Southern or Mediterranean region: Italy, Spain, Portugal, Greece, and the island countries of Cyprus and Malta (Fig. 1-19). Like Northern Europe, this is a region of peninsulas, and again it is discontinuous. It lies separated from the European core, and core-periphery contrasts in some areas are quite sharp. A degree of continuity, dating from Greco-Roman times, marks the region's languages and religion, lifeways, and cultural landscapes. As Figure 1-3 shows, natural environments in this region are dominated by a climatic regime that bears its very name—Mediterranean. Dry, hot summers are the norm, so that moisture is often in short supply during the growing season, and specially adapted plants mark local agriculture.

In terms of raw materials, Southern (Mediterranean) Europe is not as well endowed as the European core, as reflected by the map of industrial complexes (Fig. 1-6). Only northern Italy and northern Spain (the former through massive imports of coal

FIGURE 1-19

© H. J. de Blij, P. O. Muller, and John Wiley & Sons, Inc.

and iron ore) have become part of the core area. Furthermore, the region has been largely deforested, and the limited and highly seasonal water supply constrains Southern Europe's hydroelectric opportunities.

A key contrast between Northern and Southern Europe lies in their populations. Although territorially smaller than Northern Europe, Mediterranean Europe has nearly five times as many people (almost 128 million in 2008). Population distribution continues to reflect the agricultural bases of the preindustrial era, with large concentrations in coastal lowlands and fertile river basins, although the growth of major industrial centers, notably in northern Italy and northern Spain, has superimposed a new mosaic. Still, urbanization in Southern Europe is far below that in Western or Northern Europe, or in the British Isles (Portugal ranks among the lowest levels in the realm). Other data indicate that living standards in Mediterranean Europe also lag behind (Appendix B). All this is changing, of course, with the EU's expansion into Eastern Europe. Because of the recent expansions into Eastern Europe, Southern Europe is no longer the lowest-ranking EU region in terms of GNI, living standards, or urbanization.

Italy

Centrally located in the Mediterranean region, most populous of the Mediterranean states, best connected to the European core, and economically most advanced is Italy (59 million), a charter member of Europe's Common Market.

Administratively, Italy is organized into 20 regions, many with historic roots dating back centuries (Fig. 1-20). Several of these regions have become powerful economic entities centered on major cities, such as Lombardy (Milan) and Piedmont (Turin); others are historic hearths of Italian culture, including Tuscany (Florence) and Veneto (Venice). These regions in the northern half of Italy stand in strong social, economic, and political contrast to such southern regions as Calabria (the 'toe' of the Italian 'boot') and Italy's two

FIGURE 1-20 © H. J. de Blij, P. O. Muller, and John Wiley & Sons, Inc.

major Mediterranean islands, Sicily and Sardinia. Not surprisingly, Italy is often described as two countries—a progressive north and a stagnant south or *Mezzogiorno*.

North of the Ancona Line

North and south are bound by the ancient headquarters, Rome, which lies astride the narrow transition zone between Italy's contrasting halves. This zone is referred to as the *Ancona Line*, after the city on the Adriatic coast where the zone reaches the other side of the peninsula (Fig. 1-20). Whereas Rome remains Italy's capital and cultural focus, the functional core area of Italy has shifted northward into Lombardy in the Po River Basin. Here lies Southern Europe's leading manufacturing complex, in which a large, skilled labor force and ample hydroelectric power from Alpine and Appennine slopes combine with a host of imported raw materials to produce a wide range of machinery and precision equipment. The Milan-Turin-Genoa triangle exports appliances, instruments, automobiles, ships, and many specialized products. Meanwhile, the Po Basin, lying on the margins of the region's dominant Mediterranean climatic regime, enjoys a more even pattern of rainfall distribution throughout the year, making it a productive agricultural zone as well.

Metropolitan Milan embodies the new, modern Italy. Not only is Milan (at 4.1 million) Italy's largest city and leading manufacturing center—making Lombardy one of Europe's Four Motors—but it also is the country's financial and service-industry headquarters. Today the Milan area, a cornerstone of the European core, has just 9 percent of Italy's population but accounts for one third of the entire country's national income.

The Mezzogiorno

As in Germany, the lowest-income regions of Italy lie concentrated in one part of the country: the Mezzogiorno in the south. But Italy's north-south disparity continues to grow, which has led to taxpayers' revolts over the subsidies the state pays to the poorer southern regions. In truth, the south receives the bulk of Italy's illegal immigrants, whether from Africa across the Mediterranean or from the states of former Yugoslavia and Albania across the Adriatic, and it is these workers, willing to work for low wages, who move north to take jobs in the factories of Milan and Turin. Italian southerners tend to stay where they were born. Sicily and the Mezzogiorno underscore the problems of a periphery.

The Iberian Peninsula

At the western end of Southern Europe lies the Iberian Peninsula, separated from France and Western Europe by the rugged Pyrenees and from North Africa by the narrow Strait of Gibraltar (Fig. 1-21). Spain (population: 45.7 million) occupies most of this compact Mediterranean landmass. Portugal lies in its southwestern corner.

Imperial Romans, Muslim Moors, and Catholic kings left their imprints on Iberia, notably the boundary between Spain and Portugal dating from the twelfth century. The golden age of discovery and colonialism was followed by dictatorial rule and economic stagnation.

Today both Spain and Portugal are democracies, and their economies are doing well, helped by EU subsidies. As Figure 1-21 shows, Spain followed the leads of Germany and France and decentralized its administrative structure, creating 17 regions called Autonomous Communities (ACs). Every AC has its own parliament and administration that control planning, public works, cultural affairs, education, environmental matters, and even, to some extent, international commerce. Each AC can negotiate its own degree of autonomy with the central government in Madrid. This new system, however, has not been enough to defuse Spain's most problematic devolutionary issue, which involves the Basques in the AC mapped as Basque Country (Fig. 1-21, especially the top inset map).

Of particular note is Catalonia, the triangular AC in Spain's northeastern corner, adjacent to France. Centered on prosperous, productive Barcelona, Catalonia is Spain's leading industrial area, an AC with a population of 7.2 million. It is imbued with a fierce nationalism, endowed with its own language and culture, and economically it is one of the Four Motors of Europe. Note that while most of Spain's industrial raw materials lie in the northwest, its major industrial development has taken place in the northeast, where innovations and skills propel a high-technology-driven regional economy. In recent years, Catalonia—with 6 percent of Spain's territory and 15 percent of its population—has annually produced 25 percent of all Spanish exports and nearly 40 percent of its industrial exports. Such economic strength translates into political power, and in Spain the issue of Catalonian separatism is never far from the surface.

Here are two items that keep Catalans angry: first, the fact that Spain's first high-speed rail line, built with EU funds, runs south from Madrid to poor, agricultural Andalusia and not to rich, industrial Barcelona; and second, that the Madrid government is contemplating the southward diversion of river water from the Rio Ebro through dams and pipelines.

The Monument to the Discoveries sits on the waterfront of the Tagus River in Lisbon's Belém district. Built in 1960, standing 52 m (170 ft) tall, it commemorates the 500 year anniversary of the death of Henry the Navigator, who laid the foundation for Portugal's rise as a maritime power in the fifteenth and sixteenth centuries. The figures depicted all played prominent roles in Portugal's golden age of exploration. (© V. WinklerPrins)

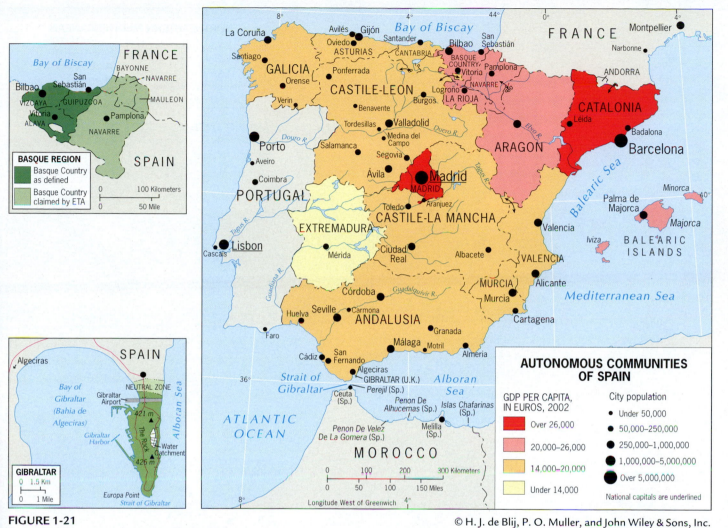

AUTONOMOUS COMMUNITIES OF SPAIN

GDP PER CAPITA, IN EUROS, 2002
- Over 26,000
- 20,000–26,000
- 14,000–20,000
- Under 14,000

City population
- Under 50,000
- 50,000–250,000
- 250,000–1,000,000
- 1,000,000–5,000,000
- Over 5,000,000

National capitals are underlined

FIGURE 1-21

© H. J. de Blij, P. O. Muller, and John Wiley & Sons, Inc.

the southwestern corner of the Iberian Peninsula. Since one rule of EU membership is that the richer members assist the poorer ones, Portugal has benefited with a massive renovation project in the capital, Lisbon, as well as in the modernization of surface transport routes.

Unlike Spain, which has major population clusters on its interior plateau as well as its coastal lowlands, the Portuguese are concentrated along and near the Atlantic coast. Lisbon and the second city, Porto, are coastal cities; the best farmlands lie in the moister western and northern zones of the country. But the farms are small and inefficient, and although Portugal remains dominantly rural, it must import as much as half of its foodstuffs. Exporting textiles, wines, corks, and fish, and running up an annual deficit, the indebted Portuguese economy remains a far cry from those of other European countries of similar dimensions.

As Figure 1-21 indicates, Spain's capital and largest city, Madrid, lies near the geographic center of the state. It also lies along an economic-geographic divide. Catalonia and Madrid are Spain's most prosperous ACs; the contiguous group of five ACs between the Basque Country and Aragon rank next. Tourism (notably along the Mediterranean coast) and winegrowing (especially in La Rioja) contribute importantly here. Ranking below this cluster of ACs are the four ACs of the northwest: Galicia, Castile-Leon, and industrialized Asturias and Cantabria. Incomes in these regions are below the national average because industrial obsolescence, dwindling raw-material sources, and emigration have plagued their economies. Worst off, however, are the three large ACs to the south of Madrid: Extremadura, Castile-La Mancha, and Andalusia. Drought, inadequate land reform, scarce resources, and remoteness from Spain's fast-growing northeast are among the factors that inhibit development here. Even as EU members seek to reduce the differences among themselves, individual countries face the geographic consequences of focused growth that deepens internal divisions.

The state of **Portugal** (population: 10.6 million), a comparatively poor country that has benefited enormously from its admission to the EU, occupies

Greece

The eastern segment of the region that we define as Southern Europe is dominated by Greece, an outlier of both Mediterranean Europe and the European Union. Greece's land boundaries are with Turkey, Bulgaria, Macedonia, and Albania; as Figure 1-22 reveals, it also owns islands just offshore from mainland Turkey. Altogether, the Greek archipelago numbers some 2000 islands, ranging in size from Crete (8335 square kilometers [3218 sq mi]) to small specks of land in the Cyclades. In addition, Greeks represent the great majority on the now-divided island of Cyprus.

FIGURE 1-22

© H. J. de Blij, P. O. Muller, and John Wiley & Sons, Inc.

Ancient Greece was a cradle of Western civilization, and later it was absorbed by the expanding Roman Empire. For some 350 years beginning in the mid-fifteenth century, Greece was under the sway of the Ottoman Turks. Greece regained independence in 1827, but not until nearly a century later, through a series of Balkan wars, did it acquire its present boundaries. During World War II, Nazi Germany occupied and ravaged the country, and in the postwar period the Greeks have quarreled with the Turks, the Albanians, and the newly independent Macedonians. Today, Greece finds itself between the Muslim world of Southwest Asia and the Muslim communities of Eastern Europe; it is still the only noncontiguous mainland EU member even after the recent accession of twelve new states.

Implications of EU Expansion

Volatile as Greece's surroundings are, Greece itself is a country on the move, an EU success story to rival Ireland's. Political upheavals in the 1970s and economic stagnation in the 1980s are all but forgotten now that Greece's economy is booming. The locomotive for the Balkans, Greece is a beacon for the EU in a crucial part of the world. Infrastructure improvements prompted by the 2004 Olympics focused on Athens, where more than one third of the population is concentrated, include new subways, a new beltway, and a new airport.

With the accession of several Eastern European members to the EU, Greece faces a great challenge: it will lose much of its EU subsidy to these poorer countries. Greece's democratic institutions still need strengthening, and corruption remains a serious problem. Educational institutions need modernization. And while Greece is on better terms today with its fractious neighbors and invests in development projects in Macedonia and Bulgaria (for example, an oil pipeline from Thessaloniki to a Greek-owned refinery in Skopje, Macedonia), it will take skillful diplomacy to navigate the shoals of historic discord.

Urban and Rural Greece

Modern Greece is a nation of 11.1 million centered on historic Athens, one of the realm's great cities. With its port of Piraeus, metropolitan Athens contains about 40 percent of the Greek population, making it one of Europe's most congested and polluted urban areas. Athens is the quintessential primate city; the monumental architecture of ancient Greece still dominates its cultural landscape. The Acropolis and other prominent landmarks attract a steady stream of visitors; tourism is one of Greece's leading sources of foreign revenues, and Athens is only the beginning of what the country has to offer.

Deforestation, increasing wildfires, soil erosion, and variable rainfall make farming difficult in much of Greece, but the country remains strongly agrarian. It is self-sufficient in staple foods, and farm products continue to figure strongly among exports. But other sectors of the economy, including manufacturing (textiles) and the service industries, are growing rapidly. The challenge for Greece will be to maintain its growth after the momentous EU expansions of this decade.

Cyprus and Malta

Cyprus lies in the far northeastern corner of the Mediterranean Sea, much closer to Turkey than to Greece (Fig. 1-22), but is peopled dominantly by Greeks rather than Turks. In 1571, the Turks conquered Cyprus, then ruled by Venice, and controlled it until 1878 when the British took over. Most of the island's Turks arrived during the Ottoman period; the Greeks have been there longest.

When the British were ready to give Cyprus independence after World War II, the 80 percent Greek majority mostly preferred union with Greece. Ethnic conflict followed, but in 1960 the British granted Cyprus independence under a constitution that prescribed majority rule but guaranteed minority rights.

This fragile order broke down in 1974, and civil war engulfed the island. Turkey sent in troops and massive dislocation followed, resulting in the partition of Cyprus into northern Turkish and southern Greek sectors (Fig. 1-19, inset map). In 1983, the 40 percent of Cyprus under Turkish control, with about 100,000 inhabitants (and some 30,000 Turkish soldiers), declared itself the independent Turkish Republic of Northern Cyprus. Only Turkey recognizes this ministate (which now contains a population of almost 200,000); the international community recognizes the government on the Greek side as legitimate. With about 900,000 residents, a relatively prosperous economy based on agriculture and tourism, and strong links with Europe, the Greek-side government qualified the south for 2004 membership in the European Union.

The potential for serious conflict over Cyprus has not disappeared. In effect, the Green Line that separates the Turkish and Greek communities constitutes not just a regional border but a boundary between geographic realms.

The Mediterranean region contains one other ministate, **Malta**, located south of Sicily. Malta is a small archipelago of three inhabited and two uninhabited islands with a population under 400,000 (Fig. 1-22, inset). An ancient crossroads and culturally rich with Arab, Phoenician, Italian, and British infusions, Malta became a British dependency and served British shipping and its military. It suffered terribly during World War II bombings, but despite limited natural resources recovered strongly during the postwar period. Today Malta has a booming tourist industry and a relatively high standard of living, is a relatively new member of the European Union, and subsequently adopted the euro.

EASTERN EUROPE

As Figure 1-12 shows, Eastern Europe is not only territorially the largest region in the European realm: it also contains more countries (18) than any other European region. Almost all of Eastern Europe lies outside the core, and the problems of the periphery affect many of its countries. From the North European Lowland in Poland to the rugged highlands of the south, this is a region of physiographic, cultural, and political fragmentation. Open plains, major rivers, strategic mountains, isolated valleys, and crucial corridors all have influenced Eastern Europe's tumultuous migrations, epic battles, foreign invasions, and imperial episodes. Illyrians, Slavs, Turks, Hungarians, and other peoples converged on this region from near and far. Ethnic and cultural differences have kept them in chronic conflict.

Geographers call this region a **27** **shatter belt**, a zone of persistent splintering and fracturing. Geographic terminology uses several expressions to describe the breakup of established order, and these tend to have their roots in this part of the world. One of them is **28** **balkanization**. The southern half of Eastern Europe is referred to as the Balkans or Balkan Peninsula, after the name of a mountain range in Bulgaria. Balkanization denotes the recurrent division and fragmentation of this part of Eastern Europe, and it is now applied to any place where such processes take place. *Ethnic cleansing* refers to the forcible removal of entire populations through forced migration from their homelands or extermination by a stronger power bent on taking their territories. Due to recurring tensions over minorities in the newly created territories and countries, ethnic cleansing has taken place in this part of Europe following the breakup of Yugoslavia.

Cultural Legacies

Each episode in the historical geography of Eastern Europe has left its legacy in the cultural landscape. Twenty centuries ago the Roman Empire ruled much of it (Romania is a cartographic reminder of this period); during the past half-century, the Soviet Empire controlled almost all of it. In the intervening two millennia, Christian Orthodox church doctrines spread from the southeast, and Roman Catholicism advanced from the northwest. Turkish (Ottoman) Muslims invaded and created an empire that reached the environs of Vienna. By the time the Austro-Hungarian Empire ousted the Turks, millions of Eastern Europeans had been converted to Islam. Albania today remains a dominantly Muslim country. In the twentieth century, Eastern Europe became a battleground between superpowers, and the complicated map reflects the results through 1991 (Fig. 1-23).

The subsequent collapse of the Soviet Union freed several of Russia's neighbors, and with only one exception—Belarus—these countries turned their gaze from Moscow to the west, specifically to the European Union and its economic promise. This changed the map of Eastern Europe, shifting the realm boundary eastward and adding five countries to the region (Fig. 1-24): Latvia, Lithuania, Belarus, Moldova, and Ukraine. Meanwhile, two of Eastern Europe's established states fell apart: Czechoslovakia peacefully, Yugoslavia violently.

The Geographic Framework

Eastern Europe is changing so rapidly that it is sometimes difficult to keep up. The number of countries in this region is not stable: in 2006 Montenegro seceded from Serbia, leaving it a landlocked state, and Kosovo declared independence in 2008, although it is not yet a fully recognized sovereign country. Further devolution in the future is possible.

FIGURE 1-23

© H. J. de Blij, P. O. Muller, and John Wiley & Sons, Inc.

FIGURE 1-24

© H. J. de Blij, P. O. Muller, and John Wiley & Sons, Inc.

This region could be subdivided in many ways, but we have chosen to group Eastern Europe's 18 countries into four subregions based on their relative location:

1. Countries Facing the Baltic Sea
2. The Landlocked Center
3. Countries Facing the Black Sea
4. Countries Facing the Adriatic Sea

Countries Facing the Baltic Sea

Poland dominates this subregion, which also includes Lithuania and Latvia and, by virtue of relative location, Belarus (although Belarus does not possess a Baltic Sea coast). Wedged between coastal Poland and Lithuania is the Russian **29** exclave of Kaliningrad, which is not functionally a part of Eastern Europe (see Chapter 2).

Figure 1-23 displays Poland's most recent boundary shifts: in the aftermath of World War II, the whole country shifted westward, losing land to the (then) USSR in the east and gaining it at the expense of Germany to the west. Situated on the North European Lowland, Poland traditionally was an agrarian country, but during the Soviet-communist period Silesia (once part of Germany) became its industrial heartland, and Katowice, Wroclaw, and Krakow grew into major industrial cities amid some of the world's worst environmental pollution. The Soviets invested far less in agriculture, collectivizing farms without modernizing the technology and leaving post-Soviet farming in abysmal condition.

By many measures, Poland is the most important state that joined the EU in 2004: with 38 million people, it has half the total population of the ten 2004 entrants; territorially, it is larger than all the others combined; and its economy is the biggest by far.

Lithuania (3.4 million) and Latvia (2.3 million) face the Baltic Sea north of Poland. **Lithuania**, remnant of the Grand Duchy of Lithuania that once dominated Eastern Europe from the Baltic to the Black Sea, is centered on the interior capital of Vilnius. When Kaliningrad became a Russian exclave as a World War II prize, Lithuania was left with only about 80 kilometers (50 mi) of Baltic coastline and a small port (Klaipeda) that was not even connected to Vilnius by rail. But since Lithuania (like Latvia and Estonia) fell under Soviet domination, that seemed to matter little as its trade was with the Soviet Union, not the outside world. In 2003 Lithuania recorded the highest growth rate among European states, which demonstrated its readiness for the EU and enabled it to join in 2004.

Latvia is centered on the Baltic port of Riga and experienced far more Russian immigration than Lithuania did: today Latvians make up only about 59 percent of the population. Its economy was geared almost totally to Moscow, and after the half-century of Soviet domination over the Baltics (1940–1991) ended, Latvia was left with huge economic problems and few options. In recent years, however, Latvia's economy has reoriented and improved, enabling it also to join the EU in 2004.

Belarus has no Baltic coast, but it borders all three of the coastal states in this subregion. It is a transitional country, and some would argue it belongs in the Russian realm. About 80 percent of its nearly 10 million people are Belarussians ('White' Russians), a West Slavic people. Only some 13 percent are (East Slavic) Russians. A small Polish minority complains of mistreatment and discrimination. Devastated during World War II, Belarus became one of Moscow's most loyal satellites, and the Soviets made Mensk (Minsk), the capital, into a large industrial center. But in the post-Soviet era, Belarus has lagged badly, and its government has retained powers reminiscent of the communist era. Un-

like its neighbors, Belarus has no interest in joining the EU. On the contrary, its overtures have been toward Moscow: it seeks to rejoin Russia in some formal way.

The Landlocked Center

Three states comprise this subregion: the Czech Republic, Slovakia, and Hungary. Until 1993, the first two formed the state of Czechoslovakia, but their 'velvet divorce' broke it up (Fig. 1-24). The Czechs got the better of the deal; Slovakia is this subregion's poorest country by far. Both joined the EU in 2004, as did Hungary. The **Czech Republic** (10.3 million) centers on Bohemia, the mountain-enclosed core area that contains the historic capital, Prague. This province always has been cosmopolitan and Western in its exposure, outlook, development, and linkages. Prague lies in the basin of the Elbe River, Bohemia's traditional outlet through northern Germany to the North Sea. It is a classic primate city, its cultural landscape faithful to Czech traditions; but it also is an industrial center. The encircling mountains contain many valleys with small towns that specialize, Swiss-style, in fabricating high-quality goods. In Eastern Europe, the Czechs always were the leaders in technology and engineering; even during the communist period, their products found markets in foreign countries near and far.

The separation of the Czech and Slovak societies created a virtual nation-state in the Czech Republic, with only a small minority of several hundred thousand Roma, or Gypsies, among the Czechs. The treatment of this minority became a human rights issue in the discussions leading to Czech membership in the EU in 2004.

As Figure 1-25 shows, **Slovakia** inherited Czechoslovakia's largest minority, the ethnic Hungarian community, which now constitutes about 10 percent of the country's 5.4 million inhabitants and is concentrated in the southern zone along the Danube River. The Slovaks and Hungarians came to terms, but initially the country remained mired in inefficient, corrupt, Soviet-style government. But

then a reform-minded administration was elected, and Slovakia became an Eastern European success story, capped by admission to the EU in 2004.

Hungary almost approaches the status of nation-state, but Hungarians (Magyars) live not only in Slovakia but also in Serbia, Austria, and Romania (Fig. 1-25), all remnants of a time when Hungary ruled much of this region. Hungarian governments have repeatedly expressed support for these ethnic cohorts in neighboring countries, a practice referred to as **30 irredentism**. The term derives from a nineteenth-century campaign by Italy to incorporate an Italian-speaking area of Austria, calling it *Italia Irredenta* (Unredeemed Italy). In the case of Hungary, its recent manifestation was the so-called status law giving persons of Hungarian ancestry living in neighboring countries certain rights (including work permits) in the 'motherland.' The governments of those neighboring countries did not like this 'law' applying to its citizens, and neither did the EU leadership. It has been modified, but public opinion in Hungary still supports the notion.

The Hungarians moved into the middle Danube River Basin more than a thousand years ago from an Asian source; they have neither Slavic nor Germanic roots. They converted their fertile lowland into a thriving state while retaining their cultural and linguistic identity, eventually forging an imperial power in the region. The twin-cities capital astride the Danube, Buda and Pest (better known as Budapest), is a primate city nearly ten times the size of the next largest town in Hungary—reflecting the continuing rural character of the country.

With a population of 10 million and a considerable and varied resource base, Hungary is stable economically. It joined the European Union in 2004.

Countries Facing the Black Sea

Four countries form Eastern Europe's Black Sea quadrant: Ukraine, Moldova, Romania, and Bulgaria (Fig. 1-24). All except Moldova have coastlines on the Black Sea, but none of their core areas or capital cities lies on the coast. This reflects an inward orien-

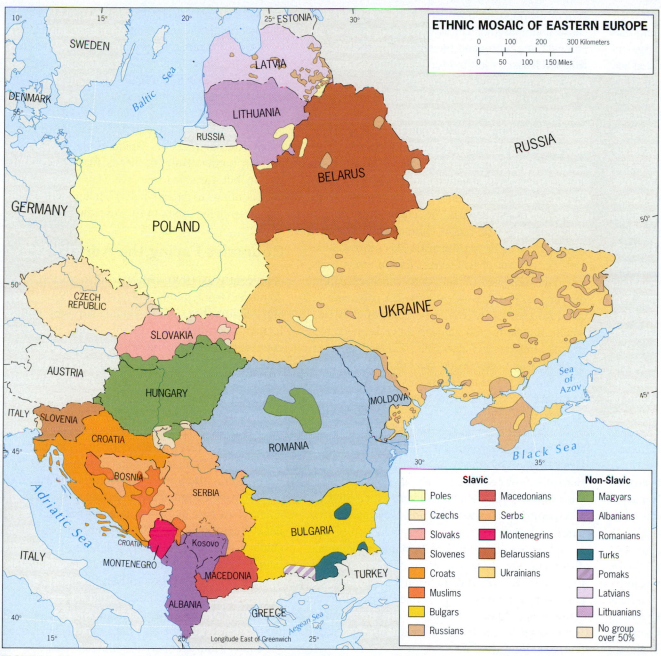

ETHNIC MOSAIC OF EASTERN EUROPE

0   100   200   300 Kilometers

0   50   100   150 Miles

Slavic

- Poles
- Czechs
- Slovaks
- Slovenes
- Croats
- Muslims
- Bulgars
- Russians
- Macedonians
- Serbs
- Montenegrins
- Belarussians
- Ukrainians

Non-Slavic

- Magyars
- Albanians
- Romanians
- Turks
- Pomaks
- Latvians
- Lithuanians
- No group over 50%

FIGURE 1-25   Adapted (in part) with permission from George Hoffman, ed., *Europe in the 1990s: A Geographic Analysis*, 6 rev. ed., p. 551.

tation that characterizes the subregion as a whole. Two qualified to join the EU in 2007.

Ukraine is Eastern Europe's most populous country (46.1 million); territorially, it is the largest state in the entire realm. Its capital, Kiev (Kyyiv), is a major historic, cultural, and political focus. Briefly independent before the communist takeover in Russia, Ukraine regained its sovereignty as a much-changed country in 1991. Once a land of farmers tilling its famously fertile soils, Ukraine emerged from the Soviet period with a huge industrial complex in its east—and with a large (20 percent-plus) Russian minority. Ukraine's boundaries also changed during the Soviet era. In 1954, a Soviet dictator capriciously transferred the entire Crimea Peninsula, including its Russian inhabitants, to Ukraine as a reward for its productivity.

The Dnieper River forms a useful geographic reference to comprehend Ukraine's spatial division (Fig. 1-24). To its west lies agrarian, rural, mainly Roman Catholic Ukraine; in its great southern bend and eastward lies industrial, urban, Russified (and Russian Orthodox) Ukraine. Soviet planners built a key industrial complex in the Donets Basin (*Donbas* for short) based on local coal and iron ore deposits. Meanwhile, the Russian Soviet Republic supplied Ukraine with oil and gas.

As Figure 1-25 shows, Ukraine, with the exception of its eastern and urban-concentrated Russian minority, has an ethnically homogeneous population by East European standards. Ukraine is a critically important country for Europe's future, but it suffers from numerous problems ranging from political mismanagement and corruption to a faltering economy and rising crime. Yet this country has access to international shipping lanes, a large resource base, massive farm production, educated and skilled labor, and a large domestic market. In 2004, its problems were evident when Ukraine became a political battleground with geostrategic overtones. In a key election, the pro-Russian East and the pro-European West each had a presidential candidate. Moscow overtly supported the eastern candidate; Europe clearly preferred his opponent. When the votes were tallied, the map showed the political 'fault line' dividing the country, with the pro-Russian candidate victorious. This, however, provoked an 'Orange Revolution' when electoral fraud swiftly came to light. That peaceful protest led to a second election a few weeks later in which the the pro-European candidate won. But the dividing line remains unchanged. The new president immediately sought to reassure Russia, even as his government made it clear that its aspirations are to join the EU at some future time. Russia's leader, however, retaliated in the winter of 2006 by drastically raising the price of natural gas it supplies to Ukraine, and then, when Ukraine sought to negotiate, by curtailing the supply—affecting not only Ukraine, but also European consumers depending on the same pipelines. Once again, Ukraine's weakness and vulnerability cost it dearly.

Moldova, Ukraine's small and impoverished neighbor, is Europe's poorest country. A Romanian province seized by the Soviets in 1940, it was made into a landlocked 'Soviet Socialist Republic.' A half-century later, along with other such 'republics,' Moldova gained independence when the Soviet Union collapsed. Romanians remain in the majority among its 4 million people, but most of the Russians and Ukrainians (each about 13 percent) have moved across the Dniester River to a strip of land between that river and the Ukrainian border, proclaiming there a 'Republic of Transdniestria' (see Fig. 1-11). But such separatist efforts form only one of Moldova's many problems. Its economy, dominated by farming, is in decline; an estimated 40 per-

cent of the population works outside its borders because unemployment in Moldova is as high as 30 percent; smuggling and illegal arms trafficking are rife; and Russia's support for Transdniestria's separatists keeps the country in turmoil.

Romania (population 21.5 million) entered the European Union in 2007 and demonstrates the EU leadership's ability to ease requirements in order to expand the organization. Romania's circumstances are only slightly better than Moldova's, yet it was able to improve its situation enough to be admitted. Romania was once an energy exporter, but its reserves are nearly depleted. Its communist dictatorship ended in 1989, but its current political system still is far less democratic than the EU used to require. Only 55 percent of Romanians live in urban areas, and in the rural parts of the country subsistence farming persists. The state is deeply in debt, and many talented Romanians are leaving in search of work elsewhere. Relations with Hungary have been strained over the treatment of Romania's Hungarian minority; European observers complain about Romania's treatment of its Roma (Gypsy) residents. The subsidies that come with EU membership should help this country considerably.

Romania's drab and decaying capital, Bucharest, anchors an interior core area linked by rail to the Black Sea port of Constanta, Once known as Eastern Europe's most civilized society, its primate city the Paris of the region, Romania today exemplifies the core-periphery contrasts that make further EU expansion challenging.

Across the Danube lies Romania's neighbor, **Bulgaria**, southernmost of this subregion's countries. The rugged Balkan Mountains form Bulgaria's physiographic backbone, separating the Danube and Maritsa basins. As the map shows, Bulgaria has five neighbors, several of which are in political turmoil.

The Bulgarian state appeared in 1878, when the Russian czar's armies drove the Turks out of this area. The Slavic Bulgars, who form 84 percent of the population of 7.6 million, were loyal allies of Moscow dur-

ing the Soviet period. But they did not treat their Turkish minority, about 9 percent of the population, very kindly, closing mosques, prohibiting use of the Turkish language, and forcing Turkish families to adopt Slavic names. Conditions for the remaining Turks improved somewhat after the end of the Soviet period.

Bulgaria has a Black Sea coast and an outlet, the port of Varna, but the country does not generate much external trade; the capital, Sofia, lies near the Serbian border. Bulgaria needs further economic reform, but was able to achieve what was necessary to be admitted to the EU in 2007. It, too, will benefit greatly from the influx of EU subsidies.

Countries Facing the Adriatic Sea

As recently as 1990, only two Eastern European countries fronted the Adriatic Sea: Yugoslavia and Albania (Fig. 1-23). Albania survives, but the former Yugoslavia has splintered into at least six countries—and its disintegration may not be over.

We turn first to the former Yugoslavia (Land of the South Slavs), a country extending from Austria to Greece, thrown together in 1918 after World War I, containing 7 major and 17 smaller ethnic and cultural groups (Fig. 1-23). The Slovenes and Croats in the north were Roman Catholics; the Serbs in the south adhered to the Serbian Orthodox church. Several million Muslims also formed part of the cultural mosaic. Two alphabets were in use in separate regions of the country. At first the Royal House of Serbia dominated Yugoslavia; after 1945, a communist dictatorship personified by World War II hero Marshal Tito held Yugoslavia together. But when the communist system collapsed in Eastern Europe and the Soviet Union disintegrated, so did the Yugoslav state.

Communist social planners laid the groundwork for the disaster that befell Yugoslavia. They divided the country into six internal 'republics' based on the Soviet model, each dominated (except Bosnia) by one major group. Since all these 'republics' inevitably incorporated minorities, the state guaranteed their rights, albeit through autocratic means.

When the communist system collapsed, individual 'republics' proclaimed their independence—that is, the majorities in these entities did so. What remained of the machine of state, still dominated by the Serbs, tried to halt this disintegration. That effort soon failed, and new countries named Slovenia, Croatia, Bosnia, Macedonia, and, by subtraction, Serbia appeared on the map. Minorities in these new states, however, objected and, in Croatia and Bosnia, rose against those who were advocating statehood. The result was a catastrophic conflict just when grandiose Euro-unification schemes were under way. Europe, which twice during the twentieth century had plunged the world into war and had vowed that genocide would never again cloud its horizons, failed this first test of its declarations. EU members, led by Germany, quickly accorded official recognition to the post-Yugoslavian states, even before the concerns of the frightened minorities could be considered. Then, European powers stood by as more than 250,000 people were killed, perhaps a million more were injured, ethnic cleansing and mass executions depopulated entire areas, refugees streamed into neighboring countries, and historic treasures were demolished.

Devolutionary processes continue in the twenty-first century. Today, the still-evolving map emerging from the former Yugoslavia consists of six countries: Slovenia, Croatia, Bosnia, Macedonia, Serbia, and Montenegro[2] (Fig. 1-24). **Slovenia** was the first 'republic' to secede and proclaim independence. Ethnically the most homogeneous but territorially small with only 2 million of the former Yugoslavia's 23 million citizens, Slovenia lay farthest from the Serbian power core and seized its opportunity. This Alpine-mountain state, with the most productive economy among the republics even before independence, today has this subregion's highest per-capita GNI by far. It joined the EU in 2004 and adopted the euro in 2007.

Croatia's crescent-shaped territory, with prongs along the Hungarian border and along the Adriatic coast, only partially reflects the distribution of Croats in the former Yugoslavia (Fig. 1-25). About 90 percent of Croatia's 4.4 million people are Croats, but another 800,000 live in a broad zone within southern Bosnia. Croatia's Serb minority—about 12 percent when independence came—has dwindled under Croatian pressure to less than 5 percent. Recent democratic reforms have improved Croatia's prospects.

Bosnia was the cauldron of calamity. Multicultural, effectively landlocked, and situated between the Croatian state to the west and the Serbian stronghold to the east, Bosnia fell victim to disastrous conflict among Serbs, Croats, and Bosniaks (now the official name for Bosnia's Muslims, who constituted about 50 percent of the population). In 1995, a U.S. diplomatic initiative resulted in a truce that partitioned Bosnia as shown in Figure 1-26, making the country a fragile federation.

Macedonia was the southernmost 'republic' of the former Yugoslavia and became a state with 2 million inhabitants. More than 60 percent of these inhabitants are Macedonian Slavs, about 25 percent

SERBIA AND ITS NEIGHBORS

Dayton Accords Partition Line

0   50   100 Kilometers
0   25   50 Miles

National capitals are underlined

Longitude East of Greenwich

FIGURE 1-26

© H. J. de Blij, P. O. Muller, and John Wiley & Sons, Inc.

[2]As of mid 2008, Kosovo's political future remained unresolved.

are Muslim Albanians, and the remainder are Turks, Serbs, and Roma (Gypsies). Small, landlocked, and poor, Macedonia first faced the ire of Greece, which argued that the name *Macedonia* was Greek property, then confronted a massive flow of refugees entering from war-torn Kosovo, and most recently coped with an Albanian autonomy movement in its northwestern corner, where this minority is concentrated. Accommodation of some of the Muslims' demands, coupled with political concessions, has reduced the tensions.

Serbia is the name of what is left of the larger domain once ruled by the Serbs, who were dominant in the former Yugoslavia. Serb minorities still remain in Bosnia, Croatia, and other former Yugoslav 'republics,' but the Serbian sphere of influence continues to shrink. Centered on the historic capital of Belgrade on the Danube River, Serbia still is the largest of the post-Yugoslav states, but the country potentially faces still more territorial losses. While its official population is about 8 million, Serbia incorporates a Hungarian minority of about 400,000 in its northern province of *Vojvodina*.

Meanwhile, Serbia's hopes to join the EU are being thwarted by the lingering effects of its recent history of misdeeds during the collapse of the former Yugoslavia. A record of mass murder and ethnic cleansing sent its former ruler, Slobodan Milosevic, to The Hague for trial, where he died during the legal proceedings, but the current government has been unable to arrest and extradite other offenders, assumed to be hiding among Serb minorities outside Serbia. The EU will not begin admission procedures until those culprits have been arrested. Meanwhile, Serbia remains in limbo, its economy and social conditions beset by uncertainty.

Montenegro, with a mere 630,000 inhabitants (about one third of them Serbs, mostly clustered in the tiny territory's northeast near the Serbian border), voted in 2006 to separate from the country called Serbia-Montenegro. Once an independent kingdom, Montenegro was forced into the Serb-dominated sphere after World War I and stayed with Serbia when Yugoslavia disintegrated. As a result, it suffered from the sanctions imposed on Serbia-Montenegro, its economy collapsing and a black market taking over. Montenegro has spectacular Adriatic shorelines and scenic mountains, and tourists are expected to return now that the political issue is settled.

Kosovo declared independence in 2008, wanting to become Europe's forty-first and the world's newest state (Fig. 1-26). Full international recognition has not yet occurred. Therefore, as of mid 2008, Kosovo is not a sovereign nation, yet it is operating separately from Serbia. This small territory was administered by NATO after Serb-inflicted atrocities of the late 1990s and is now under EU administration. It is an overwhelmingly Albanian-Muslim territory, with a population of nearly 2 million living in rugged terrain and with an economy in shambles.

It is appropriate to save our discussion of **Albania** for last because even in this turbulent region Albania is unusual. It is the only dominantly Muslim state in Europe; some 70 percent of its 3.3 million people adhere to Islam. Albania also rivals Moldova as the poorest country in Europe, ranking lowest on several indices of well-being. It has by far the fastest rate of population growth in the realm.

In February 2008, Kosovo, a province of Serbia, declared independence from that country. Many countries around the world recognized the new country. A majority have not, however, and the territory remains under European Union protection. Here a Kosovo youth group, Self Determination, tries to break through a police cordon. (© AP/Wide World Photos)

Most Albanians subsist on livestock herding and farming on the one fifth of this mountainous, earthquake-prone country that can be used for agriculture. Thousands of Albanians, seeking a better life, have tried to reach Italy across the Adriatic Sea.

Kosovar and Macedonian Muslims look to Albania for support, but the government headquartered in Tirane cannot embark on irredentist campaigns. Albania has its own cultural divide between the poverty-stricken Gegs in the north and the somewhat-better-off Tosks in the south. National unity is its primary goal.

With 590 million inhabitants in 40 countries, including some of the world's highest-income economies, a politically stable and economically integrated Europe would be a superpower in the twenty-first century. But Europe's political geography is anything but stable, as devolutionary forces and cultural conflict continue to trouble the realm. Moreover, economic integration, despite the momentous EU expansions of 2004 and 2007, still involve less than two thirds of the realm's countries, a process that will become more difficult as the European Union confronts applications from marginally qualifying states in the east. Europe always has been a realm of revolutionary change, and it remains so today.

2

RUSSIA

CONCEPTS, IDEAS, AND TERMS

1. Climatology
2. Greenhouse effect
3. Global climate change
4. Continentality
5. Weather
6. Tundra
7. Taiga
8. Permafrost
9. Colonialism
10. Imperialism
11. Forward capital
12. Russification
13. Federation
14. Command economy
15. Unitary state system
16. Distance decay
17. Population decline
18. Core area

In This Chapter

- The geography of Russian nationalism.
- Russia's colonies and the Soviet Empire
- How Russia copes with distance and isolation
- The geopolitical implications of a melting Arctic ice cap
- Continued unrest in Transcaucasia
- Where Russia meets China

REGIONS

RUSSIAN CORE AND PERIPHERIES
EASTERN FRONTIER
SIBERIA
RUSSIAN FAR EAST

Photos: (*upper left*) Moscow, Russia © C. Bowman/Robert Harding World Imagery/Getty Images; (*above*) Yerevan, Armenia © Stephane Victor/Lonely Planet Images/Getty Images)

FIGURE 2-1 *Map:* © H. J. de Blij, P. O. Muller, and John Wiley & Sons, Inc.

OES THE RUSSIAN state constitute a geographic realm? Look at Figures G-3, and 2-1 (see chapter opener map), and you will see that it clearly does. Russia is the planet's giant. Not only is the Russian state the largest in the world in terms of territorial size, but it is nearly four times as large as all of South Asia including India, three times as large as Europe, and one-third larger than East Asia including China. Along with its large territorial size, Russia has a substantial population by world standards and a strong cultural identity. It also has more neighbors than any other country in the world. With its northerly location and legendary harsh climates, the Russian realm is vivid and clear-cut.

Three small countries also form part of this Russian realm, not because they display strong regional similarities with the Russian state but because they have been strongly influenced by Russia in the past—and because they are more closely linked with Russia than with any other realm. Armenia, Georgia, and Azerbaijan lie in the land corridor between the Black and the Caspian seas, where Russian power has historically collided with Muslim culture.

A Troubled Realm

Russia is vast and formidable, but it also is a realm in trouble. As we will see, Russia's population is rapidly declining and will, if present demographic trends continue, shrink from 140.6 million today to only about 110 million by 2050. Russia's social fabric is fraying, the gap between rich and poor is widening continuously, and relationships between Russians and minorities are eroding. Russia's government is an uneasy mix of democracy and autocracy, its legal and financial institutions weaker than they should be. The Russian economy continues to depend mainly on the sale of oil and natural gas; needed are diversification and foreign investment. Crime is rife and corruption is endemic. Russia's once-fearsome armed forces have experienced a period of disarray and failure. All this while Russia still possesses huge arsenals of Soviet-era weapons of mass destruction, and governance remains a challenge in the Caucasus.

Russia's Changing Role

Modern Russia was the cornerstone of the communist Soviet Union, whose collapse in 1991 ended the Cold War and ushered in a new era. In the years that have followed, world attention has been focused elsewhere: on the rise of China, on terrorism, on European unification, on global warming. But while Russia no longer occupies center stage, it remains a force to be reckoned with. Its leaders are determined to resurrect Russia's stature as a global power. Increasingly, Russia is asserting itself in international diplomacy, sometimes obstructing Western initiatives. And Russia has some of the world's largest fossil-fuel reserves; Europe already depends critically on Russian natural gas supplies. Russia will play a crucial role in the twenty-first century world.

Here is Tula, about 160 kilometers (100 miles) south of Moscow. Tula was Russia's second city during the Middle Ages, endowed with an impressive kremlin (fortress) and thriving as a commercial center. Tula's twenty-first century townscape is a mixture of historic buildings (built between 1514 and 1521), drab Soviet-era tenements (unchanged since 1964), and a few more modern apartment buildings. Old habits persist: where there is some fertile soil, you are likely to see vegetable plots, not flower gardens.
(© Mauro Galligani/Contrasto/Redux Pictures)

FIGURE 2-2   © H. J. de Blij, P. O. Muller, and John Wiley & Sons, Inc.

Defining the Realm

As noted, the Russian geographic realm is virtually conterminous with the Russian state, with only the three small countries of the Transcaucasus (a term meaning across the Caucasus Mountains) breaking this monopoly. The geographic effect of their inclusion is that, in this area, the realm borders Turkey and Iran (Figs. G-3 and 2-2).

TRANSITION ZONES: TWO TYPES

The Russian realm displays yet another geographic feature, evident even at the small scale of Figure G-3: a set of *transition zones* along its periphery. The two most obvious of these transition zones lie in Belarus and Kazakhstan. Belarus is not part of the Russian state, but, as we noted in Chapter 1, Belarus is a Slavic country in which many vestiges of the former Soviet Union survive. Today, Belarus has a more autocratic regime than Russia itself, a throwback to communist times. The country's rulers have even appealed to Russia to consider formal amalgamation, but the Russians have declined. Hence Belarus lies between two realms, Russia on its eastern doorstep and the European Union—in the form of Poland—on its western.

The transition zone involving Kazakhstan has a different origin. As Figure G-3 shows, this zone does not affect the entire country, but only its northern sector bordering Russia. During the period of Soviet communism, when Russia ruled 14 entities beyond its borders, millions of Russians left Russia and moved to what were, in effect, Russia's Eurasian colonies. One of the largest emigrations of this kind made northern Kazakhstan a virtual extension of Russia, integrating this area (where Russia's space port was built) tightly into the Soviet framework. You can still see the effects of this in the transport networks in Figure 2-10, although many Russians have migrated back to Russia since 1991.

In addition, substantial Russian minorities now find themselves on the wrong side of the border in the ring of countries once ruled from Moscow. After a difficult period of adjustment for both sides, most of the new governments have come to terms with their Russian minorities (Moldova, as we noted in Chapter 1, remains the exception).

The population shifts still visible on the map were part of an immense social experiment that Russia launched in the early part of the twentieth century. Its goal was an egalitarian communist state that would be a model for the rest of the world, an alternative to rapacious capitalism. The experiment failed, but not before Russia and its Eurasian empire had been trans-formed at gigantic cultural, economic, political, and environmental cost. The geographic story of this chapter centers on Russia's postcommunist reorganization and its changing role in the world.

RUSSIA'S PHYSICAL GEOGRAPHY

Russia's physiography is dominated by vast plains and plateaus rimmed by rugged mountains (Fig. 2-1). Only the Ural Mountains break an expanse of low re-lief that extends from the Polish border to eastern Siberia. In the western part of this huge plain, where the northern pine forests meet the midlatitude grass-lands, the Slavs established their domain.

Harsh Environments

The historical geography of Russia is the story of Slavic expansion from its populous western heartland across interior Eurasia to the east, and into the mountains and deserts of the south. This eastward march was hampered not only by vast distances but also by harsh natural conditions. As the northernmost populous country on earth, Russia has virtually no natural barriers against the onslaught of Arctic air. Moscow lies farther north than Edmonton, Canada, and St. Petersburg lies at latitude 60° North—the latitude of the southern tip of Greenland. Winters are long, dark, and bitterly cold in most of Russia; summers are short and growing seasons limited. Many a Siberian frontier outpost was doomed by cold, snow, and hunger.

It is therefore useful to view Russia's past, present, and future in the context of its **1 climatology**. This field of geography investigates not only the distribution of climatic conditions all over the globe but also the processes that generate this spatial arrangement. The earth's atmosphere traps heat received as radiation from the sun, but this **2 greenhouse effect** varies over planetary space—and over time. As we noted in the introductory chapter, much of what is today Russia was in the grip of a glaciation until the onset of the warmer Holocene. But even today, with a natural warming cycle in progress augmented by human activity, Russia still suffers from severe cold and associated drought. If the enhanced **3 global climate change** continues, Russia may benefit. But that may take generations.

Currently, precipitation totals, even in western Russia, range from modest to minimal because the warm, moist air carried across Europe from the North Atlantic Ocean loses much of its warmth and moisture by the time it reaches Russia. Figures G-8, G-9, and 2-3 reveal the consequences. Russia's cli-

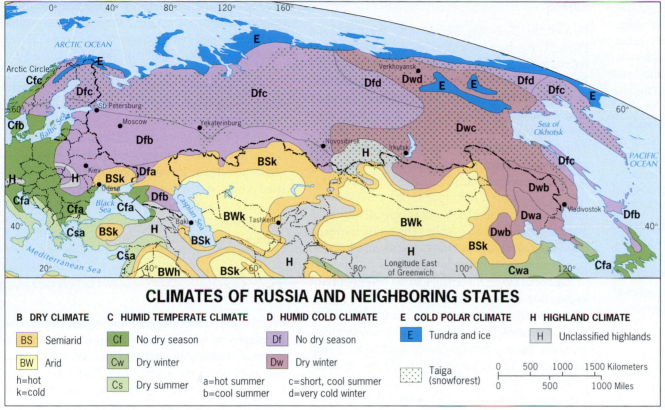

CLIMATES OF RUSSIA AND NEIGHBORING STATES

B DRY CLIMATE	C HUMID TEMPERATE CLIMATE	D HUMID COLD CLIMATE	E COLD POLAR CLIMATE	H HIGHLAND CLIMATE
BS Semiarid	Cf No dry season	Df No dry season	E Tundra and ice	H Unclassified highlands
BW Arid	Cw Dry winter	Dw Dry winter		
h=hot	Cs Dry summer	a=hot summer c=short, cool summer	Taiga (snowforest)	
k=cold		b=cool summer d=very cold winter		

0   500   1000   1500 Kilometers
0   500   1000 Miles

FIGURE 2-3   © H. J. de Blij, P. O. Muller, and John Wiley & Sons, Inc.

---

matic **4** **continentality** (inland climatic environment remote from moderating and moistening maritime influences) is expressed by its prevailing ***Dfb*** and ***Dfc*** conditions. Compare the Russian map to that of North America (Fig. G-9), and you will note that, except for a small corner off the Black Sea, Russia's climatic conditions resemble those of the U.S. Upper Midwest and southern Canada. Along its entire north, Russia has a zone of ***E*** climates, the most frigid on the planet. In these Arctic latitudes originate the polar air masses that dominate its environments.

Climates and Peoples

By studying Russia's climates, we can begin to understand what the map of population distribution (see Fig. 2-12) shows. The overwhelming majority of the country's people are concentrated in the west and southwest, where climatic conditions were least difficult at a time when farming was the mainstay for most of the people. The map is a legacy that will mark Russia's living space for generations to come.

Environment and Politics

Climate and weather (there *is* a distinction: ***climate*** is a long-term average, whereas **5** **weather** refers to existing atmospheric conditions at a given place and time) have always challenged Russia's farmers. Conditions are most favorable in the west, but even there temperature extremes, variable and undependable rainfall, and short growing seasons make farming difficult. During the Soviet period, fertile and productive Ukraine supplied much of Russia's food needs, but even then Russia often had to import grain. Soviet rulers wanted to reduce their country's dependence on imported food, and their communist planners built major irrigation projects to increase crop yields in the colonized republics of Central Asia. As we will see in Chapter 7, some of these attempts to overcome nature's limitations spelled disaster for the local people.

Physiographic Regions

To assess the physiography of this vast country, refer again to Figure 2-2. Note how mountains and deserts encircle Russia: the Caucasus in the southwest ⑧; the Central Asian Ranges in the center ⑦; the Eastern Highlands facing the Pacific from the

Bering Sea to the East Sea (Sea of Japan) ⑥. The Kamchatka Peninsula has a string of active volcanoes in one of the world's most earthquake-prone zones (Fig. G-5). Warm subtropical air thus has little opportunity to penetrate Russia, while cold Arctic air sweeps southward without impediment. Russia's Arctic north is a gently sloping lowland broken only by the Urals and Eastern Highlands.

Russia's vast and complex physical stage can be divided into eight physiographic regions, each of which, at a larger scale, can be subdivided into smaller units. In the Siberian region, one criterion for such subdivision is the vegetation. The Russian language has given us two terms to describe this vegetation: **6 tundra**, the treeless plain along the Arctic shore where mosses, lichens, and some grasses survive, and **7 taiga**, the mostly coniferous forests that begin to the south of where the tundra ends, and extend over vast reaches of Siberia, which means the 'sleeping land' in Russian.

The Russian Plain

The Russian Plain ① is the eastward continuation of the North European Lowland, and here the Russian state formed its first *core area*. Travel north from Moscow at its heart, and the countryside soon is covered by coniferous (needleleaf) forests like those of Canada; to the south lie the grain fields of southern Russia and, beyond, those of the Ukraine. Note the Kola Peninsula and Barents Sea in the far north: warm water from the North Atlantic Drift flows around northern Norway and keeps the port of Murmansk ice free most of the year. Russia's only ice-free port on the European side of the country's landmass, Murmansk, is critical to its navy. The melting of the Arctic ice sheet, discussed below, may change this. The Russian Plain is bounded on the east by the Ural Mountains ②; though not a high range, it is topographically prominent because it separates two extensive plains. The range of the Urals is more than 3200 kilometers (2000 mi) long and reaches from the shores of the Kara Sea to the border with Kazakhstan. It is not a barrier to east-west transportation, and its southern end is densely populated. Here the Urals yield minerals and fossil fuels.

Siberia

East of the Urals lies Siberia. The West Siberian Plain ③ has been described as the world's largest unbroken lowland; this is the vast basin of the Ob and Irtysh rivers. Over the last 1600 kilometers (1000 mi) of its course to the Arctic Ocean, the Ob falls less than 90 meters (300 ft). In Figure 2-2, note the dashed line that extends from the northern Urals to the East Siberian Sea and includes the Arctic Lowland. North of the line, water in the ground is permanently frozen; this **8 permafrost** creates another obstacle to permanent settlement. Looking again at the West Siberian Plain, we see that the north is permafrost-ridden and the central zone is marshy. The south, however, has such major cities as Omsk and Novosibirsk, beyond the reach of permafrost conditions.

East of the West Siberian Plain the country begins to rise, first into the Central Siberian Plateau ④, another sparsely settled, remote, permafrost-affected region. Here winters are long and cold, and summers are short; the area remains barely touched by human activity. Beyond the Yakutsk Basin ⑤, the terrain becomes mountainous and the relief high. The Eastern Highlands ⑥ are a jumbled mass of ranges and ridges, precipitous valleys, and volcanic mountains. Lake Baykal lies in a trough that is over 1500 meters (5000 ft) deep—the deepest rift lake in the world. On the Kamchatka Peninsula, volcanic Mount Klyuchevskaya reaches nearly 4750 meters (15,600 ft).

Encircling Mountains

The northern part of region ⑥ is Russia's most inhospitable zone, but southward along the Pacific coast the climate is less severe. Nonetheless, this is a true frontier region. The forests provide opportunities for lumbering, a fur trade exists, and there are gold and diamond deposits.

Mountains also mark the southern margins of much of Russia. The Central Asian Ranges ⑦, from the Kazakh border in the west to Lake Baykal in the east, contain many glaciers whose annual meltwaters send alluvium-laden streams to enrich farmlands at lower elevations. The Caucasus ⑧, in the land corridor between the Black and Caspian seas, form an extension of Europe's Alpine Mountains and exhibit a similarly high *relief* (range of elevations). Here Russia's southern border is sharply defined by *topography* (surface configuration).

As our physiographic map suggests, the more habitable terrain in Russia becomes latitudinally narrower from west to east. Beyond the southern Urals, the zone of settlement in Russia becomes discontinuous (in Soviet times it extended into northern Kazakhstan, where a large Russian population still lives). Isolated towns did develop in Russia's vast eastern reaches, even in Siberia, but the tenuous ribbon of settlement does not widen again until it reaches the country's Far Eastern Pacific Rim.

The Arctic

Frigid Arctic environments dominate the physical geography of northern Russia. All of the country's northern coastline along the Arctic Ocean lies on the poleward side of the Arctic Circle. The Arctic Ocean is frozen for much of the year, although some warmth from the North Atlantic Drift and from upwelling waters in the far west keep the ports of Murmansk and Arkhangelsk open longer. Ports like this (and even St. Petersburg) would never have developed the way they did if Russia had better access to the world's oceans, but this has been one of its historic impediments. One of the chief goals of Russia's colonial policies was to gain such access. They never succeeded.

Now it looks as though nature will give Russia a helping hand. If global warming permanently melts significant parts of the Arctic ice, the Arctic Ocean will come to play a very different role in Russia's future.

Arctic Melting

When global climate is relatively stable, polar ice expands over land and sea during winters and recedes during summers. In recent years, the scientific literature as well as newspapers and television have been reporting that Arctic ice, including the Greenland ice

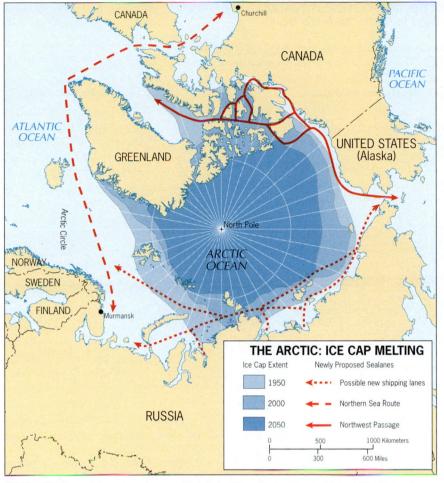

FIGURE 2-4   © H. J. de Blij, P. O. Muller, and John Wiley & Sons, Inc.

THE ARCTIC: ICE CAP MELTING

Ice Cap Extent   Newly Proposed Sealanes

1950   ◄┈┈┈ Possible new shipping lanes

2000   ◄┈ ┈ Northern Sea Route

2050   ◄━━ Northwest Passage

0   500   1000 Kilometers
0   300   600 Miles

cap, is melting at a rapid rate. Winters are not cold enough to bring all the ice back (Fig. 2-4). According to some reports, the year 2007 recorded shrinkage of the ice at a rate ten times faster than the known long-term average. At this rate, some scientists predict, the whole Arctic could be ice-free during summers by the end of this century.

Although polar regions are known to display variable climate, and episodes of shrinking of permanent ice are known to have occurred in the past, scientists attribute the current evidence of rapid melting to human-enhanced global warming. An Intergovernmental Panel on Climate Change (IPCC) has repeatedly confirmed the role of human activities in intensifying, and thus worsening, the greenhouse effect to which we made reference in the Introduction (see p. 12). It is obvious that international action is needed to address this global threat, but United States leadership has been lacking in this crucial arena.

Consequences of Arctic Opening

Sustained Arctic warming is already having significant impact on species of wildlife concentrated in these ecologically sensitive environments. The intricate web of relationships among species and their environments on the one hand, and among species themselves on the other, can be critically affected by temperature changes. The polar bear is perhaps the most prominent animal at risk: it depends on ample sea ice to hunt and rear cubs, and diminished ice and more open water forces it to swim greater distances and to rely on ever-fewer places to rear its young. If ice-free Arctic summers are indeed in the offing, the polar bear could become extinct by 2100. Rapid ecosystem change will also endanger seal, bird, fish, and other populations.

In turn, such developments will also affect human populations such as the Inuit, still living in the Arctic region and likewise adapted to the harsh environments prevailing there. Their traditions, already under pressure from political and economic forces resulting from their incorporation into modern states, will be further affected by environmental change that is likely to alter, or even destroy, the ways of life they developed over thousands of years.

Still another consequence of sustained Arctic warming has to do with the region's resources. As technologies of exploration and exploitation become more efficient, more and more oil and natural gas reserves are not only identified but can be reached by drills and pumps—even under the sea and beneath deeper waters. As the ice disappears, the Arctic Ocean's floor may become the next oil frontier, and some reports estimate that the region may contain as much as one-quarter of the world's remaining supply. If so, this will entail the building of oil platforms, the risk of oil spills where they would do incalculable damage to fragile ecosystems and where cleanup operations would be hampered, the presence of polluting oil tankers, and probably the construction of pipelines and terminals in sensitive coastal areas. Such activities will activate additional shipping from freighters to cruise ships, and other manifestations of development in a part of the world long protected from it by distance and nature

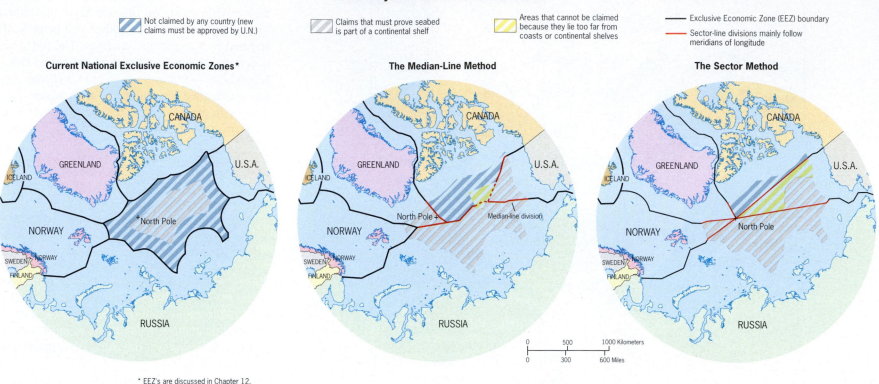

THE ARCTIC: GEOPOLITICAL SITUATION
Three Ways to Divide the Arctic Ocean

Not claimed by any country (new claims must be approved by U.N.)

Claims that must prove seabed is part of a continental shelf

Areas that cannot be claimed because they lie too far from coasts or continental shelves

—— Exclusive Economic Zone (EEZ) boundary

—— Sector-line divisions mainly follow meridians of longitude

Current National Exclusive Economic Zones*

The Median-Line Method

The Sector Method

* EEZ's are discussed in Chapter 12.

Boundaries based on a map by Dave Monahan of the Center for Coastal and Ocean Mapping/ Joint Hydrographic Center, University of New Hampshire, Durham, New Hampshire

FIGURE 2-5

© H. J. de Blij, P. O. Muller, and John Wiley & Sons, Inc.

Trade and Geopolitics

The permanent melting of large segments of Arctic ice will alter the sea lanes as well as the resource picture in the region. The Northwest Passage, the fabled northerly route between the Atlantic and Pacific Oceans through the Arctic islands of Canada, has been the object of hope and despair for centuries— and caused the loss of much human life. Not until 1944 did a boat make it through in a single season, and that was long before the effect of global warming began to be felt. But now that possibility opens up all kinds of prospects: the passage from Tokyo to New York would be shortened from 18,200 kilometers to 14,000 kilometers (11,300 to 8700 mi), not

only with none of the delays at the Panama Canal but also without the tonnage restrictions imposed on vessels by the dimensions of that canal. Other plans include a maritime route between Murmansk (Russia) and Churchill (Canada), perhaps as early as 2040 based on current temperature projections. Regular commercial traffic through the Northwest Passage would have worldwide economic impact.

This is only one reason why the melting of Arctic ice has serious geopolitical implications. Who will own the waters and waterways vacated by the melting ice? Although many countries have an interest in the changing situation, Russia is among those most concerned because of its lengthy frontage on the

Arctic Ocean and its historic links with Arctic environs (Fig. 2-5). Canada has a major stake for obvious geographic reasons, as do the United States through Alaska and Denmark because it governs Greenland. Norway faces the Arctic Ocean and Iceland touches the Arctic Circle, but Sweden can only claim geographic proximity.

Although there are international rules to govern the drawing of maps in remote, icebound and maritime areas, those rules tend not to be enforced until a reason arises to implement them. When that happens, the negotiations can become quite contentious. As we will see in more detail in Chapter 12, countries with coastlines can claim sovereignty over waters and even submerged 'land' up to 230 nautical miles from their shores, and when that submerged land contains valuable resources, get ready for a struggle. So it is in the long-neglected Arctic, where what some are calling a frenzy of exploration and claiming has begun. The Russians sent a mini-submarine to plant a metal flag on the Arctic Ocean floor at the North Pole (see photo); the United States is opening a new Coast Guard station to patrol the narrow Bering Strait; Norway is expanding its natural gas industry by exploiting previously inaccessible reserves beneath Arctic waters, and Canada is building a new army training base at Resolute and a new deep-water port at Nanisvik on Baffin Island to protect the Northwest Passage. This passage is considered an interior waterway by the Canadians, but other Arctic countries claim it as an international waterway.

On Friday, August 3, 2007, an operator of a Russian mini-submarine planted a titanium capsule with the Russian flag at the North Pole, under shrinking Arctic sea ice. This highly symbolic moment, claiming much of the Arctic Ocean floor for Moscow, was broadcast live on Russian television. Changing environmental conditions in the Arctic are prompting a flurry of scientific exploration that will result in needed diplomacy to sort out which country can claim which part of the Arctic for itself.   (Visar Kryeziu/© AP/ Wide World Photos)

As Figure 2-5 shows, the Arctic can be divided up among claimant states in several different ways by lines marking Exclusive Economic Zones (EEZs). These are zones in which claimant countries declare exclusive right to the resources, such as minerals and fish. Such zones are well established in less remote areas of the world but still open to negotiation here in the Arctic; *median lines*, which divide waters by drawing a line exactly down the middle; and *sectors*, pie-shaped entities that converge on the North Pole. Just by looking at these three maps you can see how the remaining 'open' area shrinks in each option. The Arctic is an international quarrel in the making, and some of the participants are global heavyweights. Stay tuned!

RUSSIAN ROOTS

The name *Russia* evokes cultural-geographic images of a stormy past: terrifying czars, conquering Cossacks, rousing revolutionaries, clashing cultures. Russians repulsed the Tatar (Mongol) hordes, forged a powerful state, colonized a vast contiguous empire, defeated Napoleon, adopted communism, and, when the communist system failed, lost most of their imperial domain.

Russia as we know it today started out as small unified states (*Russes*) in the region now known as the Ukraine. After challenges from the Mongols, the Russian empire that emerged was as much a colonial power as various European countries were. Figure 2-6 presents a time line of Russian and Soviet history highlighting the events that are key to understanding the geography of this realm.

A Vibrant Culture

Precommunist Russia may be described as a culture of extremes. It was a culture of strong nationalism, resistance to change, and despotic rule. Enormous wealth was concentrated in a small elite. Powerful rulers and bejeweled aristocrats perpetuated their privileges at the expense of millions of peasants and serfs who lived in dreadful poverty. The Industrial Revolution arrived late in Russia, and a middle class was slow to develop. Yet the Russian nation gained the loyalty of many of its citizens, and its writers and artists were among the world's greatest. Authors such as Tolstoy and Dostoyevsky chronicled the plight of the poor; composers celebrated the indomitable Russian people, and, as Tchaikovsky did in his *1812 Overture*, commemorated their victories over foreign foes.

Empire of the Czars

Under the czars, Russia grew from nation into empire through the process of **9 colonialism**. Russia was a hearth of **10 imperialism**. The czar's insatiable demands for wealth, territory, and power sent Russian armies across the plains of Siberia, through the deserts of interior Asia, and into the mountains along Russia's rim (Fig. 2-7). One of the czars, Peter the Great (ruled from 1682 to 1725), built St. Petersburg as a **11 forward capital** on the doorstep of Swedish-held Finland as a way of penetrating into Europe. Russian pioneers ventured farther in the other direction as well, entering Alaska, traveling down the Pacific coast of North America, and planting the Russian flag near San Francisco in 1812. At its height, the Russian Empire extended across 22 million square kilometers (8.5 million sq mi). As Russia's empire expanded, however, its internal weaknesses gnawed at the power of the czars. Peasants rebelled. Unpaid (and poorly fed) armies mutinied. When the czars tried to initiate reforms, the aristocracy objected. The empire at the beginning of the twentieth century was ripe for revolution, which began in 1905.

Communist Victory

The last czar, Nicholas II, was overthrown in 1917, and civil war followed. The victorious communists led by V. I. Lenin soon swept away much of the Russia of the past. The Russian flag disappeared, and the

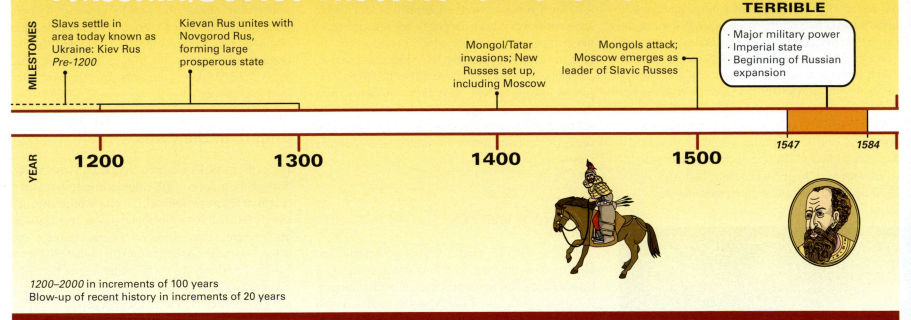

Russian/Soviet Historic Time Line

IVAN THE TERRIBLE

MILESTONES

Slavs settle in area today known as Ukraine: Kiev Rus *Pre-1200*

Kievan Rus unites with Novgorod Rus, forming large prosperous state

Mongol/Tatar invasions; New Russes set up, including Moscow

Mongols attack; Moscow emerges as leader of Slavic Russes

- Major military power
- Imperial state
- Beginning of Russian expansion

YEAR

1200   1300   1400   1500

1547   1584

1200–2000 in increments of 100 years
Blow-up of recent history in increments of 20 years

FIGURE 2-6

czar and his family were executed. The old capital of Russia, St. Petersburg, was renamed Leningrad in honor of the revolutionary leader. Moscow, in the interior of the country, was chosen as the capital for a country with a new name, the *Soviet Union*. Eventually, this Union consisted of 15 political entities, each a Soviet Socialist Republic. Russia was just one of these republics, and the name *Russia* disappeared from the international map.

But on the Soviet map, Russia was the giant, the dominant republic, the Slavic center. Not for nothing was the communist revolution known as the Russian Revolution. The other republics of the Soviet Union were for minorities the czars had colonized or for countries that fell under Soviet sway later, but none could begin to match Russia. The Soviet Empire was the legacy of czarist expansionism, and the new

communist rulers were Russian first and foremost (with the exception of Josef Stalin, who was a Georgian). Russians moved by the millions to the non-Russian republics, where the **12 Russification** of the empire proceeded, just as the British and French and Dutch and Portuguese were also moving to their colonies in large numbers. The Soviet Union was a Russian colonial empire, and like all colonial empires it was doomed to failure.

Seven Fateful Decades

Officially, the Union of Soviet Socialist Republics (USSR) endured from 1924 to 1991. The year 1924 was a fateful one in Russian and Soviet history, marking the death of Lenin, the ideological organizer, and

the beginning of nearly three decades of rule by Stalin, the ruthless tyrant. During Stalin's time, many of the peoples under Russian control suffered unimaginably. In pursuit of communist reconstruction, Stalin and his henchmen starved millions of Ukrainian peasants to death, forcibly relocated entire ethnic groups (including a people known as the Chechens), exterminated 'uncooperative' or 'disloyal' peoples, and purged the Communist Party time and again. Death camps proliferated in Siberia; many of the country's most creative people were eliminated. The full extent of these horrors may never be known.

On December 25, 1991, the inevitable occurred: the Soviet Union ceased to exist, its economy a shambles, its political system shattered, the communist experiment a failure. The Soviet hammer-and-sickle flag flying atop the Kremlin was lowered for the last

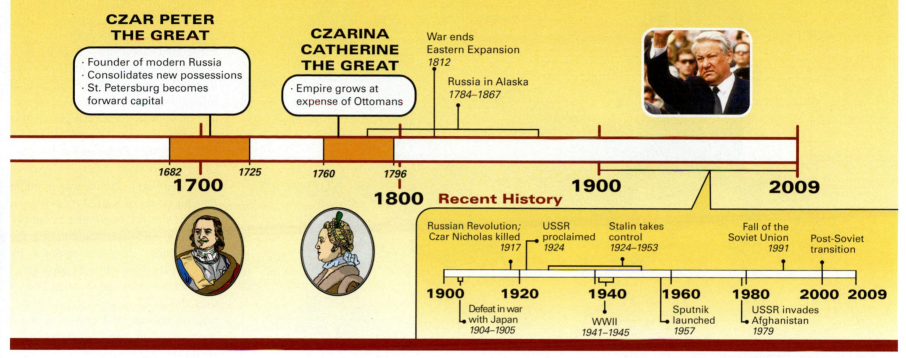

CZAR PETER THE GREAT
- Founder of modern Russia
- Consolidates new possessions
- St. Petersburg becomes forward capital

CZARINA CATHERINE THE GREAT
- Empire grows at expense of Ottomans

War ends Eastern Expansion *1812*

Russia in Alaska *1784–1867*

1682   1725   1760   1796

1700   **1800**   **1900**   **2009**

Recent History

Russian Revolution; Czar Nicholas killed *1917*

USSR proclaimed *1924*

Stalin takes control *1924–1953*

Fall of the Soviet Union *1991*

Post-Soviet transition

1900   **1920**   **1940**   **1960**   **1980**   **2000 2009**

Defeat in war with Japan *1904–1905*

WWII *1941–1945*

Sputnik launched *1957*

USSR invades Afghanistan *1979*

(Photo of Boris Yeltsin © Andre Durand/AFP/Getty Images)
© H. J. de Blij, P. O. Muller, and John Wiley & Sons, Inc.

time and replaced by the white, red, and blue Russian tricolor. The republics were set free, and world maps had to be redrawn. Now Russia and its former colonies had to adjust to the realities of a changing world order. It proved to be a difficult, sometimes desperate journey.

THE SOVIET LEGACY

When the Soviet system failed and the Soviet Socialist Republics became independent states, Russia was left without the empire that had taken centuries to build and consolidate—and that contained crucial agricultural and mineral resources. No longer did Moscow control the farms of Ukraine

and the oil and natural gas reserves of Central Asia. But look again at Figure 2-7 and you will see that, even without its European and Central Asian colonies, Russia remains an empire. Russia lost the republics on its periphery, but Moscow still rules over a domain that extends from the borders of Finland to North Korea. Inside that domain Russians are in the overwhelming majority, but many subjugated nationalities, from Tatars to Yakuts, still inhabit ancestral homelands. Accommodating these many indigenous peoples is one of the challenges facing the Russian Federation today.

In the 1990s, Russia began to reorganize in the aftermath of the collapse of the Soviet Union. This reorganization cannot be understood without reference to the seven decades of Soviet communist rule that went before.

The era of communism may have ended in the Soviet Empire, but its effects on Russia's political and economic geography will long remain. Seventy years of centralized planning and implementation cannot be erased overnight; regional reorganization toward a market economy cannot be accomplished in a day.

While the world of capitalism celebrates the failure of the communist system in the former Soviet realm, it should be remembered why communism found such fertile ground in the Russia of the 1910s and 1920s. Before the Revolution, Russia was infamous for the wretched serfdom of its peasants, the cruel exploitation of its workers, the excesses of its nobility, and the ostentatious palaces and riches of the czars. Ripples from the Western European Industrial Revolution introduced a new age of misery for

GROWTH OF THE RUSSIAN EMPIRE

Grand Duchy of Moscow 1462, Russia 1533

Territory Gained   **Western Border**
1533–1598   1725–1801   ——— 1864
1598–1689   1801–1945   ——— 1920
1689–1725

ALASKA Permanent settlements were established in 1784. Territory sold to United States in 1867. Fort Ross, California built by Russians in 1812. Relinquished in 1840.

GRAND DUCHY OF WARSAW Gained by Russia under the Vienna Settlement (1815). Lost in 1918 on the formation of an independent Poland.

FINLAND gained by Russia from Sweden 1809. Independent since 1918.

ESTONIA incorporated into Russia 1721. Independent 1918–1940; 1991–.

PECHENGA (PETSAMO) Area ceded by Finland to Russia in 1940.

KALININGRAD OBLAST under Soviet administration 1945–1991. Now Russian.

BELORUSSIA (BELARUS) Russian 1795–1920 Reincorporated into the Soviet Union in 1939. Independence in 1991.

SAKHALIN under joint Russo-Japanese control 1854–1875. Became Russian in 1875. Southern part ceded to Japan in 1905. Reincorporated 1945.

Incorporated into Ukraine from Czechoslovakia in 1945.

MOLDAVIA (BESSARABIA) Russian 1812–1918. Incorporated into Romania from 1918 to 1940. The territory then passed back to the USSR. Independence as Moldova in 1991.

KURILE ISLANDS Divided between Russia and Japan 1854. Passed to Japan in 1875. Incorporated into USSR 1945.

ARDAHAN AND KARS Changed hands between Russia and Turkey several times in the 19th century. Annexed by Russia in 1878 and returned to Turkey in 1921.

TUVA made protectorate in 1911. Joined the USSR in 1944.

MANCHURIA occupation 1901–1905

THE KHANATE OF BUKHARA Became a Russian vassal in 1868 and then THE KHANATE OF KHIVA 1873. They were merged into the Soviet system in 1920.

Russian 1871–1881

THE KWANYUNG TERRITORY Leased to Russia 1898–1905 and 1945–1955.

Tributary to the Tsar 1731–1824

Tributary to the Tsar 1734–1822

FIGURE 2-7

© H. J. de Blij, P. O. Muller, and John Wiley & Sons, Inc.

those laboring in factories. There were workers' strikes and ugly retributions, but when the czars finally tried to better the lot of the poor, it was too little too late. There was no democracy, and the people had no way to express or channel their grievances. Europe's democratic revolution passed Russia by, and its economic revolution touched the czars' domain only slightly. Most Russians, and tens of millions of non-Russians under the czars' control, faced exploitation, corruption, starvation, and harsh subjugation. When the people began to rebel in 1905, there was no hint of what lay in store; even after the full-scale Revolution of 1917, Russia's political future hung in the balance.

The Political Framework

Russia's great expansion had brought many nationalities under czarist control; now the revolutionary government sought to organize this heterogeneous ethnic mosaic into a smoothly functioning state. The czars had conquered, but they had done little to bring Russian culture to the peoples they ruled. The Geor-

gians, Armenians, Tatars, and residents of the Muslim states of Central Asia were among dozens of individual cultural, linguistic, and religious groups that had not been Russified. In 1917, however, the Russians themselves constituted only about one-half of the population of the entire empire. Thus it was impossible to establish a Russian state instantly over this vast political region, and these diverse national groups had to be accommodated.

The question of the nationalities became a major issue in the young Soviet state after 1917. Lenin, who brought the philosophy of Karl Marx to Russia, talked from the beginning about the "right of self-determination for the nationalities." The first response by many of Russia's subject peoples was to proclaim independent republics, as they did in Ukraine, Georgia, Armenia, Azerbaijan, and even in Central Asia. But Lenin had no intention of permitting the Soviet state to break up. In 1923, when his blueprint for the new Soviet Union went into effect, the last of these briefly independent units was fully absorbed into the sphere of the Moscow regime. Ukraine, for example, declared itself independent in 1917 and managed to sustain this initiative until 1919. But in that year the Bolsheviks set up a provisional government in Kiev, the Ukrainian capital, thereby ensuring the incorporation of the country into Lenin's Soviet framework.

The Communist System

The political framework of the Soviet Union was based on the ethnic identities of its many incorporated peoples. Given the size and cultural complexity of the empire, it was impossible to allocate territory of equal political standing to all the nationalities; the communists controlled the destinies of well over 100 peoples, both large nations and small isolated groups. It was decided to divide the vast realm into Soviet Socialist Republics (SSRs), each of which was delimited to correspond broadly to one of the major nationalities. At the time, Russians constituted about half of the developing Soviet Union's popula-

tion, and, as Figure 2-8 shows, they also were (and still are) the most widely dispersed ethnic group in the realm. The Russian Republic, therefore, was by far the largest designated SSR, comprising just under 77 percent of total Soviet territory.

Within the SSRs, smaller minorities were assigned political units of lesser rank. These were called

Autonomous Soviet Socialist Republics (ASSRs), which in effect were republics within republics; other areas were designated Autonomous Regions or other nationality-based units. It was a complicated, cumbersome, often poorly designed framework, but in 1924 it was launched officially under the banner of the Union of Soviet Socialist Republics (USSR).

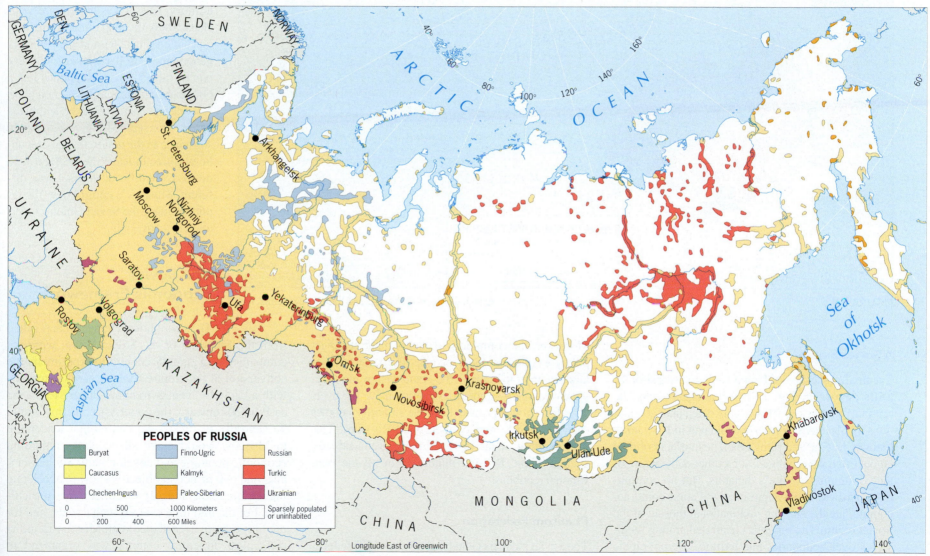

FIGURE 2-8

© H. J. de Blij, P. O. Muller, and John Wiley & Sons, Inc.

FIGURE 2-9

© H. J. de Blij, P. O. Muller, and John Wiley & Sons, Inc.

involves the sharing of power between a country's central government and its political subdivisions (provinces, States, or, in the Soviet case, Socialist Republics). The centerpiece of the tightly controlled Soviet federation was the Russian Republic. With half the vast state's population, the capital city, the realm's *core area*, and over three-quarters of the Soviet Union's territory, Russia was the empire's nucleus. Study the map of the former Soviet Union (Fig. 2-9), and an interesting geographic corollary emerges: every one of the 15 Soviet Republics had a boundary with a non-Soviet neighbor. Not one was spatially locked within the others. This seemed to give geographic substance to the notion that any republic was free to leave the USSR if it so desired. Reality, of course, was different. Moscow's control over the republics made the Soviet Union a federation in theory only.

The Soviet Economic Framework

A Soviet Empire

Eventually, the Soviet Union came to consist of 15 SSRs (shown in Fig. 2-9), including not only the original republics of 1924 but also such later acquisitions as Moldova (formerly Moldavia in Romania), Estonia, Latvia, and Lithuania. The internal political layout often was changed, sometimes at the whim of the communist empire's dictators. But no communist apartheid-like system of segregation could accommodate the shifting multinational mosaic of the Soviet realm. The republics quarreled among themselves over boundaries and territory. Demographic changes, migrations, war, and economic factors soon made much of the layout of the 1920s obsolete.

Moreover, the communist planners made it Soviet policy to relocate entire peoples from their homelands in order to better fit the grand design and to reward or punish—sometimes capriciously. The overall effect, however, was to move minority peoples eastward and to replace them with Russians. This *Russification* of the Soviet Empire produced substantial ethnic Russian minorities in all the non-Russian republics.

Phantom Federation

The Soviet planners called their system a **13** **federation**. We focus in more detail on this geographic concept in Chapter 11, but we note here that federalism

The geopolitical changes that resulted from the founding of the Soviet Union were accompanied by a gigantic economic experiment: the conversion of the empire from a czarist autocracy with a capitalist veneer to communism. From the early 1920s onward, the country's economy would be centrally planned—the communist leadership in Moscow would make all decisions regarding economic planning and development. Soviet planners had two principal objectives: (1) to accelerate industrialization and (2) to collectivize agriculture. For the first time ever on such a scale, and for the first time in accordance with Marxist-Leninist principles, an entire country was organized to work toward national goals prescribed by a central government.

A Command Economy

This centralized economic planning as practiced in the USSR is called a **14** **command economy**. State planners assigned the production of particular manufactures to particular places, often disregarding the rules of economic geography. For example, the manufacture of railroad cars might be assigned (as indeed it was) to a factory in Latvia. No other factory anywhere else would be permitted to produce this equipment—even if supplies of raw materials would make it cheaper to build them near, say, Volgograd 1931 kilometers (1200 mi) away. Despite an expanded and improved transport network (Fig. 2-10), such practices made manufacturing in the USSR extremely expensive, and the absence of competition

FIGURE 2-10

© H. J. de Blij, P. O. Muller, and John Wiley & Sons, Inc.

made managers complacent and workers less productive than they could be.

Soviet planners never imagined that their experiment would fail and that a market-driven economy would replace their command economy. When that happened, the transition was predictably difficult; indeed, it is far from over, and it is putting severe strains on the now more democratic state.

RUSSIA'S CHANGING POLITICAL GEOGRAPHY

When the USSR dissolved in 1991, Russia's former empire devolved into 14 independent countries, and Russia itself was a changed nation. Russians now made up about 83 percent of the population of just under 150 million, a far higher proportion than in the days of the Soviet Union. But numerous minority peoples remained under Moscow's new flag, and millions of Russians found themselves under new governments in the former republics.

Soviet planners had created a complicated administrative structure for their Russian Soviet Federative Socialist Republic, and Russia's postcommunist leaders had to use this framework to make their country function. In 1992, most of Russia's internal republics, autonomous regions, and other components of the administrative hierarchy signed a document known as the Russian Federation Treaty, committing them to cooperate in the new federal system. At first a few units refused to sign, including Tatarstan, scene of Ivan the Terrible's brutal conquest more than four centuries ago, and a republic in the Caucasus periphery, then known as Chechenya-Ingushetia, where Muslim rebels waged a campaign for independence. As the map shows, Chechnya-Ingushetiya split into two separate republics, whose names are now spelled Chechenya and Ingushetia (see Fig. 2-16). Eventually, only Chechnya refused to sign the Russian Federation Treaty, and subsequent Russian military intervention led to a prolonged and violent conflict, with disastrous consequences for Chechnya's people and infrastruc-

Chechnya has been plagued by fighting with separatist rebels for much of the past dozen years, and its capital city, Groznyy, still bears scars of battle with bombed-out apartment buildings lining the streets. Things are now stable, and reconstruction has begun. Here, conditions are safe enough for Chechen children to ride bicycles near destroyed houses in Groznyy, Chechnya. (© AP/Wide World Photos)

ture (the capital, Groznyy, was completely destroyed). The Chechnya war continues today (see photo) and is a disaster for Russia's government as well.

The Federal Framework of Russia

The spatial framework of the still-evolving Russian Federation is as complex as that of the Russian Federative Socialist Republic of communist times. As the twenty-first century opened, the Federation consisted of 89 entities: 2 Autonomous Federal Cities, Moscow and St. Petersburg (Leningrad), 21 Republics, 11 Autonomous Regions (Okrugs), 49 Provinces (Oblasts), and 6 Territories (Krays). The 21 Republics, recognized to accommodate substantial ethnic minorities in the population, lie in several clusters.

When Russia's post–1991 government took over, it faced complicated administrative problems. Not only were some entities reluctant to sign the Russian Federation Treaty, but despite the ranking implied by the list above, some regions were more equal than others—that is, some regions were used to privilege under the old system, and their local leaders expected that to continue. Meanwhile, a multinational, multicultural state that had been accustomed to authoritarian rule and government control over virtually everything—from factory production to everyday life—now had to be governed in a new way. Democratization of the political system, transition to a market economy, the sale of state-owned industries (privatization), and other far-reaching changes had to come quickly or the country risked chaos.

Unitary and Federal Options

Russia's leaders knew that their options were limited. They could continue to hold as much power as possible at the center, making decisions in Moscow that would apply to all the Republics, Regions, and other subdivisions of the state. Such a **15 unitary state system**, with its centralized government and administration, marked authoritarian kingdoms of the past and serves totalitarian dictatorships of the present. Or they could share power with the Republics and Regions, allowing elected regional leaders to come to Moscow to represent the interests of their people. This is the federal system Russia chose as the only way to accommodate the country's economic and cultural diversity.

In a *federal system*, the national government usually is responsible for matters such as defense, foreign policy, and foreign trade. The Regions (or provinces, States, or other subdivisions) retain authority over affairs ranging from education to transportation. A federal system does not create unity out of diversity, but it does allow diverse components of the state to co-exist, their common interests represented by the national government and their regional interests by their local administrations. Some countries

owe their survival as coherent states to their federal frameworks. India and Australia are cases in point.

But to maintain a generally acceptable balance of power between the center and Regions (or States) is difficult. Disputes over States' rights continue to roil the American political scene more than two centuries after the Constitution was adopted. Early on, the Russian government decided to end, for all practical purposes, the hierarchical regional system the Soviets had established. The Republics retained their special status, but all the others—Okrugs, Krays, and so on—were designated as Regions. This was intended to address the favoritism of Soviet times and to streamline the system generally.

Federal Administrative Districts

In 2000, the Putin administration moved to diminish the influence of the Regions by creating a new spatial framework that combines the 89 Regions, Republics, and other entities into seven new administrative units—not to enhance their influence in Moscow, but to increase Moscow's authority over them (Fig. 2-11). As the map shows, each of these new Federal Districts has a capital, elevating such cities as Rostov and Novosibirsk to a status secondary to Moscow. In a related move, the Russian president proposed, and the parliament approved, that Regional governors be appointed rather than elected—thereby concentrating

FIGURE 2-11

© H. J. de Blij, P. O. Muller, and John Wiley & Sons, Inc.

RUSSIAN REALM POPULATION DISTRIBUTION: 2009
One dot represents 50,000 persons

0   600   1200 Kilometers
0   300   600 Miles

Longitude East of Greenwich

FIGURE 2-12   © H. J. de Blij, P. O. Muller, and John Wiley & Sons, Inc.

still more power in Moscow. Russia's federal system appears to be moving in a unitary direction.

Problems of Size and Distance

The new Russian government also faced an old Soviet problem: the sheer size of the country, its vast distances, and the remoteness of many of its Regions. Geographers refer to the principle of **16 distance decay** to explain how increasing distances between places tend to reduce interactions among them. Because Russia is the world's largest country, distance is a significant factor in the relationships between the capital and thinly populated outlying areas (Fig. 2-12). Furthermore, Moscow lies in the far west of the giant country, half a world away from the shores of the Pacific. Not surprisingly, one of the most obstreperous Regions has been remote Primorskiy, the Region of Vladivostok.

CHANGING SOCIAL GEOGRAPHIES

None of this political maneuvering reflects in any way what is happening in the daily lives of most of Russia's citizens. Since the end of communist rule, Russia's social geography has changed quite drastically, for the better for some but for the worse for many. These changes include:

1. *Revival of religion.* The communist regime was avowedly atheist, and the end of Soviet rule revived the long-suppressed Russian Orthodox Church. In the cultural landscape, this is reflected by the reconstruction of cathedrals and churches as well as the reappearance of public religious events and ceremonies.

2. *Failure of the pension system.* During the Soviet era, millions of Russians depended on state pensions for their retirement. Political and financial turmoil during the post-Soviet years caused the system to falter, impoverishing many Russian families and widening the gap between rich and poor. Street people and the destitute are now seen in numbers unknown during the communist era, a significant change in Russia's social landscape.

3. *Rise of the oligarchs.* During the chaotic 1990s, the new Russian government sold state industries to private entrepreneurs (often friends of the new leadership) at low prices in return for political and other favors. The newly rich oligarchs went on a binge of conspicuous consumption, further widening the gap between rich and poor. Some took their money and went abroad; others stayed and took to political meddling. After 2000, the government began a campaign to crack down on these newly rich, and some were imprisoned.

4. *Crime and corruption.* Organized crime exploded after the fall of Soviet rule, in part because of the end of communism's iron-fisted rule and also because law enforcement in the new Russia could not cope with the new epidemic. Crime syndicates thrived and extended their networks abroad, aided by endemic corruption in Russia itself. Street crime continues to threaten Russians' quality of life.

5. *New freedoms.* Russian citizens enjoy freedoms unimaginable under communist rule, and surveys confirm that a large majority prefer the new order, despite the problems arising from the transition. Greater (not complete) freedom of the press and other media brought a new era of candor. Street protests, unheard of during Soviet times, now occur frequently, as older citizens complain about pension-plan changes, victims' families denounce

official failures in terrorist-attack investigations, and voters march to condemn the decision to appoint rather than elect Regional governors. A new era that some in Russia call a liberal revolution has arrived.

RUSSIA'S DEMOGRAPHIC DISASTER

All this pales, however, against what some geographers refer to as Russia's demographic disaster: its rapid decline in population coupled with the deterio-

Since the break-up of the Soviet Union, Russian society has suffered from social disruption that has resulted in an epidemic of self-destructive behavior. Alcoholism and drug dependence are pervasive, accident and suicide rates high. Here a man drinks vodka from a plastic glass at a snack bar. Partly as a consequence of this social collapse, Russia is suffering from severe population decline.   (© NewsCom)

rating health of its citizens. When the Soviet Union collapsed in 1991, Russia's population totaled about 149 million. By 2008, this number was down to 140.6 million—despite the immigration of several million ethnic Russians from the former Soviet Republics beyond Russia's borders. Since the end of communist rule, Russia has seen about 10 million more deaths than births. Uncertainty and disillusionment in the future as well as severe dislocation are contributing factors to this population decline.

Undoubtedly, the transition from Soviet rule is one cause: uncertainty tends to cause families to have fewer children, and abortion is widespread in Russia. But the birth rate has stabilized at around 10 per thousand; it is Russia's death rate that has skyrocketed, now recording more than 16 per thousand. This produces an annual population loss of over 0.5 percent, or more than three-quarters of a million per year.

Russian males are the ones most affected. Male life expectancy dropped from 71 in 1991 to 59 in 2006 (female life expectancy has also declined, but markedly less, to 72). Males are more likely to be afflicted by alcoholism and related diseases, by AIDS (which is severely underreported in Russia, according to international agencies), by heavy smoking, and by suicide, accidents, and murder. On average, a Russian male is nine times more likely to die a violent or accidental death than his EU counterpart. Fewer than half of today's Russian male teenagers will survive to age 60.

If this **17** **population decline** continues, as noted earlier Russia will have only about 100 million citizens by 2050, possibly even fewer, raising doubts about the future of the state itself. Look again at that map of the Districts (Fig. 2-11) and consider this: since 1991, the Far East District has lost 17 percent of its population, Siberia 5 percent, the Northwest 9 percent, and the South 12 percent. Only the Central District, the one around Moscow, has a loss on a par with that of European countries (0.2 percent).

What is the answer? Russia's leaders hope that improvements in the social circumstances of the average Russian will reduce the rate of population loss.

Campaigns against alcoholism and careless lifestyles are under way. Immigration is another option, and should Russia relax its rules, hundreds of thousands of Koreans and Chinese would move into the Russian Far East and counter the outflow of Russians there. But Moscow is not eager to see its eastern frontier transformed into an extension of East Asia. So Russia's population dilemma continues, clouding its future as a viable state.

RUSSIA AND THE WORLD

The expansions of the European Union in 2004 and 2007 have reached Russia's borders. Former components of the Soviet Empire, such as Latvia and Lithuania, now are members of NATO. How will Russia's relationship with Europe evolve, and what are the prospects for Russia in the wider world?

To Europeans, Russia has always been the enigmatic colossus to the east, alternately bungling and threatening but never in tandem with the West. When the Berlin Wall came down and the Soviet banner folded, there were hopeful predictions of a 'European Russia' finally joining its erstwhile ideological adversaries in a regional partnership that would extend from the Atlantic to the Pacific. Such forecasts were based on real as well as idealized evidence: Russians' ethnic ties with Eastern Europe; the revival of Christian churches after decades of communist atheism; the sprouting of Russian democracy; the budding of a market economy; the long-term dependence of Europe on Russian energy supplies.

Reality was rather different, however. Russian democracy remained tinged by an authoritarian streak that scared investors. The media continued to face government interference. The Russian Orthodox Church suppressed efforts by other churches to become reestablished. Russian leaders objected vigorously, then acquiesced as NATO expanded toward its borders. Russia was unhelpful during efforts to mitigate the collapse of Yugoslavia. And Russia's treatment of minorities and migrants seemed to violate EU standards. All this occurred against the background of a military nuclear arsenal still capable of destroying the world in an afternoon.

Russia and Europe

Nevertheless, Russia has more in common with Europe than it has with any other realm of the world, and the Europeanization of Russia, whatever form it takes, is likely to be a hallmark of the twenty-first century. The process was given an inadvertent boost by al-Qaeda's 9/11 attack on New York and Washington, D.C., when Russian, European, and American leaders realized that they faced a common enemy far more capable than they had estimated. One result was a softening of Western criticism of Russian efforts to suppress Islamic dissidents in Chechnya and elsewhere in the Federation. Over the longer term, however, Russia's well-being and its ability to reverse its social deterioration will depend in large measure on its integration into Europe and its adoption of European norms.

Unresolved Problems

Russia's enormous territory endows it with numerous neighbors, and its imperial history leaves it with many border problems. In Central Asia, millions of

Russians still reside in the former Soviet Republics including neighboring Kazakhstan (see Chapter 7). In East Asia, China and Russia have settled a number of boundary disputes (see Fig. 2-17 inset), but for the future the larger question is likely to involve China's 'lost' territories east of the Amur and Ussuri rivers, taken by Russia more than a century ago.

Other significant external problems Russia confronts include: (1) the ownership of oil and gas reserves in the Caspian Basin, where maritime boundaries in the Caspian Sea have not yet been satisfactorily delimited; (2) settlement of the Kurile Islands issue with Japan, a legacy of World War II still not settled; (3) relations with its Transcaucasian neighbor, Georgia; and (4) efforts by the authoritarian regime of Belarus to reunite in some formal way with the Russian state. But perhaps Russia's greatest challenge of all, in the international arena, is to sustain its residual position as a credible force in world affairs.

WHAT'S DRIVING GEOGRAPHIC CHANGE IN THE REALM

- Russia's government increasingly is taking on the hallmarks of an **autocratic regime**, and while many Russians do not like it, most seem to feel that Russia needs strong guidance more than it needs freedoms taken for granted in democracies.

- Russian citizens are angry at Caucasus-based **Islamic terrorists**, at their government's inability to stabilize that border area, and at the incompetence of police and security forces when acts of terror occur.

- In mid-2006, then Russian president Vladimir Putin proposed a ten-year plan to reverse the decline in Russia's **birth rate**, calling for subsidies and financial incentives for mothers to have more children, paid maternity leave, support for adoptive parents, and paid prenatal care. In 2007 he subsequently launched a program of incentives to lure back people from the vast Russian diaspora, the results of which are not yet known.

- Russia relentlessly plays its **energy card**, using its state-controlled oil and gas industries to play needy neighbors off against each other and to threaten others.

- **Global climate change** could be positive for Russia as Arctic sealanes and resources become accessible and forests grow larger.

Regions of the Realm

Russia is highly varied in its physiography and so diverse in its cultural landscape that we need to regionalize it. Figure 2-13 outlines a four-region framework: the Russian Core and its Peripheries west of the Urals, the Eastern Frontier, Siberia, and the Far East. As we will see, each of these massive regions contains major subregions.

RUSSIAN CORE AND PERIPHERIES

The heartland of a state is its **18** **core area**. Here much of the population is concentrated, and here lie its leading cities, major industries, densest transport networks, most intensively cultivated lands, and other key components of the country. Core areas of long standing strongly reflect the imprints of culture and history. The Russian core area, broadly defined, extends from the western border of the Russian realm to the Ural Mountains in the east (Fig. 2-13). This is the Russia of Moscow and St. Petersburg, of the Volga River and its industrial cities.

FIGURE 2-13

© H. J. de Blij, P. O. Muller, and John Wiley & Sons, Inc.

Under the czars, St. Petersburg was the focus of Russian political and cultural life, and Moscow was a distant second city. Today, however, St. Petersburg has none of Moscow's locational advantages, at least not with respect to the domestic market. It lies well outside the Central Industrial Region near the northwestern corner of the country, 650 kilometers (400 mi) from Moscow. Neither is it better off than Moscow in terms of resources: fuels, metals, and foodstuffs must all be brought in, mostly from far away. The former Soviet emphasis on self-sufficiency even reduced St. Petersburg's asset of being on the Baltic coast because some raw materials could have been imported much more cheaply across the Baltic Sea from foreign sources than from domestic sites in distant Central Asia (only bauxite deposits lie nearby, at Tikhvin).

Yet St. Petersburg was at the vanguard of the Industrial Revolution in Russia, and its specialization and skills have remained important. Today, the city and its immediate environs contribute about 10 percent of the country's manufacturing, much of it through fabricating high-quality machinery.

Central Industrial Region

At the heart of the Russian Core lies the Central Industrial Region (Fig. 2-14). The precise definition of this subregion varies, for all regional definitions are subject to debate. Some geographers prefer to call this the Moscow Region, thereby emphasizing that for over 400 kilometers (250 mi) in all directions from the capital, everything is oriented toward this historic focus of the state. As Figure 2-10 shows, Moscow has maintained its decisive centrality: roads and railroads converge in all directions from Ukraine in the south; from Mensk (Belarus) and the rest of Eastern Europe in the west; from St. Petersburg and the Baltic coast in the northwest; from Nizhniy Novgorod (formerly Gorkiy) and the Urals in the east; from the cities and waterways of the Volga Basin in the southeast (a canal links Moscow to the Volga,

Russia's most important navigable river); and even to the subarctic northern periphery that faces the Barents Sea, where the strategic naval port of Murmansk and lumber-exporting Arkhangelsk lie.

Two Great Cities

Moscow (population: 10.9 million) is the megacity focus of an area that includes some 50 million inhabitants (more than one-third of the country's total population), many of them concentrated in such major cities as Nizhniy Novgorod, the automobile-producing Soviet Detroit; Yaroslavl, the tire-producing center; Ivanovo, the heart of the textile industry; and Tula, the mining and metallurgical center where lignite (brown coal) deposits are worked.

St. Petersburg (the former Leningrad) remains Russia's second city, with a population of 5.3 million.

Povolzhye: The Volga Region

A second region within the Russian Core is the *Povolzhye*, the Russian name for an area that extends along the middle and lower valley of the Volga River. It would be appropriate to call this the Volga Region, for this greatest of Russia's rivers is its lifeline and most of the Povolzhye's cities lie on its banks (Fig. 2-14). In the 1950s, a canal was completed to link the lower Volga with the lower Don River (and thereby the Black Sea).

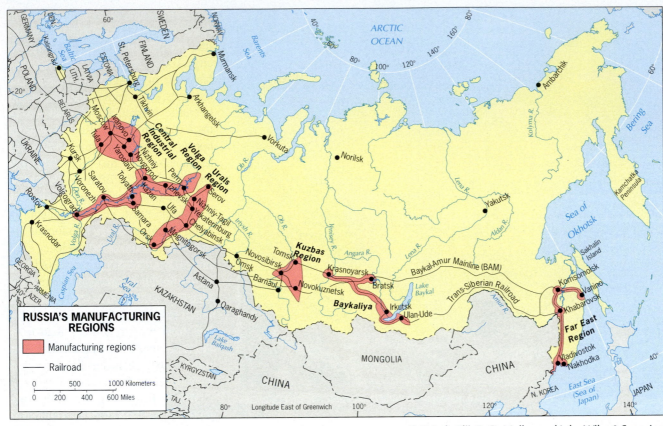

FIGURE 2-14

© H. J. de Blij, P. O. Muller, and John Wiley & Sons, Inc.

exceeds 25 million, and the cities of Samara (formerly Kuybyshev), Volgograd, Kazan, and Saratov all have populations between 1.0 and 1.3 million. Manufacturing has also expanded into the middle Volga Basin, emphasizing more specialized engineering industries. The huge Fiat-built auto assembly plant in Tolyatti, for example, is one of the world's largest of its kind.

The Urals Region

The Ural Mountains form the eastern limit of the Russian Core. They are not particularly high; in the north they consist of a single range, but southward they broaden into a hilly zone. Nowhere are they an obstacle to east-west transportation. An enormous storehouse of metallic mineral resources located in and near the Urals has made this area a natural place for industrial development. Today, the Urals Region, well connected to the Volga and Central Industrial Regions, extends from Serov in the north to Orsk in the south (Fig. 2-14).

The Central Industrial, Volga, and Urals regions form the anchors of the Russian core area. For decades they have been spatially expanding toward one another, their interactions ever more intensive. These regions of the Russian Core stand in sharp contrast to the comparatively less developed, forested, Arctic north and the remote upland to the south between the Black and Caspian seas.

The Volga River was an important historic route in old Russia, but for a long time neighboring regions overshadowed it. The Moscow area and Ukraine were far ahead in industry and agriculture. The Industrial Revolution that came late in the nineteenth century to the Moscow Region did not have much effect in the Povolzhye. Its major function remained the transit of foodstuffs and raw materials to and from other regions.

Changing Times

This transport function is still important, but the Povolzhye has changed. First, World War II brought furious development because the Volga River, located east of Ukraine, was far from the German armies that invaded from the west. Second, in the postwar era the Volga-Urals Region for some time was the largest known source of petroleum and natural gas in the entire Soviet Union. From near Volgograd (formerly Stalingrad) in the southwest to Perm on the Urals' flank in the northeast lies a belt of major oilfields (Fig. 2-15).

Third, the transport system has been greatly expanded. The Volga-Don Canal directly connects the Volga waterway to the Black Sea; the Moscow Canal extends the northern navigability of this river system into the heart of the Central Industrial Region; and the Mariinsk canals provide a link to the Baltic Sea. Today, the Volga Region's population

The Internal Southern Periphery

Nothing in Figure 2-13 suggests the magnitude of the problems Russia faces in its highly fractured mountainous southern periphery, between the Black

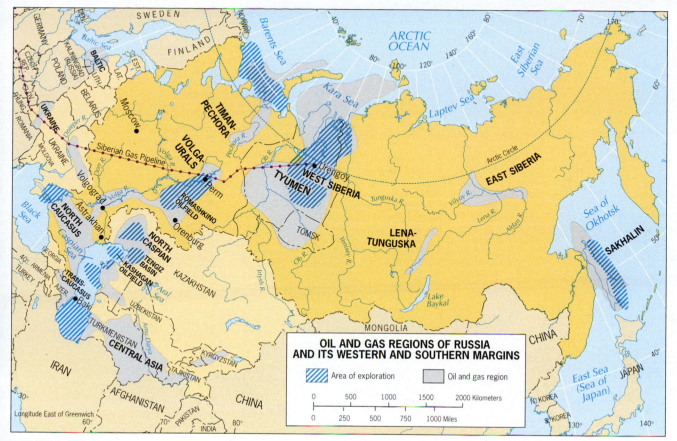

FIGURE 2-15

© H. J. de Blij, P. O. Muller, and John Wiley & Sons, Inc.

OIL AND GAS REGIONS OF RUSSIA
AND ITS WESTERN AND SOUTHERN MARGINS

Area of exploration   Oil and gas region

and mainly Christian Georgia and Armenia. That would have been challenging enough, but then came the next complicating factor: major oil reserves were discovered in coastal Azerbaijan, and the pipelines would lead from there to Russia through those troublesome, non-Russian minority republics. In Figure 2-16, just follow the pipeline from Baki (Baku) in Azerbaijan through Dagestan and Chechnya toward the Black Sea coast!

Disaster in Chechnya

When the Soviet Union collapsed in 1991 and the new Russian government took over, there was little to indicate that Chechnya would become Russia's flashpoint. True, during World War II Stalin had exiled the entire Chechen population to Central Asia, inflicting a terrible death toll. But other minorities had also been treated badly, and Stalin's successor, Nikita Khrushchev, had allowed the survivors to return during the 1950s. The Chechens, like some of the other minorities in this Internal Southern Periphery, were Muslims—but so were the Ingush and others elsewhere. And Groznyy, Chechnya's capital, had become a crucial oil-industry center, pipeline junction, and service hub. Furthermore, one-quarter of Chechnya's population of 1.2 million was Russian by the time the Soviet Union collapsed.

When representatives of Chechnya's Muslim majority refused to sign the Russian Federation Treaty and demanded independence for their republic, they seemed at first to pose a political, not a strategic challenge. But within months the Russian army was trying to secure control over the Chechens, who responded with hit-and-run attacks within their territory and in neighboring republics as well, inflicting heavy casualties on the Russians. In the intermittent war that

Sea to the west and the Caspian Sea to the east. Take a look at Figure 2-16, and you will see a crowded mosaic of small minority republics along Russia's border, capable of generating large problems. One of these republics, Chechnya, has become a symbol of Moscow's failure to stabilize its southern flank.

Islam and Oil

In this much-contested region, the czars' armies drove south toward the sea, but European colonial competitors and local peoples stopped their advance. When energized Islamic forces drove northward, the Russians met the challenge and stalled their thrust, creating a jumbled cultural landscape of Muslim and non-Muslim clusters. The Soviet planners who inherited this chaos tried to bring order to it by creating a tier of ethnic republics extending from Dagestan in the east to Adygeya in the west (Fig. 2-16). Dagestan, on the Caspian Sea, is a good example of the futility of this scheme: this republic never had a majority, its 2 million people comprising some 30 ethnic groups using about 80 languages. To go from one valley to the next in Dagestan is to go from one cultural world to another.

What the Soviets could not foresee was how important this restive region was to become economically. The Soviet Union extended beyond Russia's borders, and thus beyond this frontier, into three neighboring territories: mainly Muslim Azerbaijan

SOUTHERN RUSSIA:
INTERIOR AND EXTERIOR PERIPHERIES

POPULATION

- Under 50,000
- 50,000–250,000
- 250,000–1,000,000
- 1,000,000–5,000,000
- Over 5,000,000

National capitals are underlined

| | | | | |
| Railroad |
| Road |
| Canal |
| Oil pipeline |
| Proposed oil pipeline |

0 50 100 150 200 250 300 350 Kilometers

0 50 100 150 200 Miles

Longitude East of Greenwich

FIGURE 2-16   © H. J. de Blij, P. O. Muller, and John Wiley & Sons, Inc.

98

Inset map:

CHECHNYA
Kargalinskaja
Russian North (Plains)
Iscerskaja
Selkovskaja
Cervlënnaja
Terek R.
INGUSHETIYA
Groznyy
Gudermes
Sunzha R.
Argun
Shali
Urban-Industrial Middle Zone
NORTH OSSETIA
Vedeno
Chechen South (Mountains)
CAUCASUS MTS.
Shatoy
DAGESTAN
Argun R.

0 20 40 Kilometers
0 10 20 30 Miles

Main map labels:

KAZAKHSTAN
UKRAINE
Sloviansk
Luhansk
Volgograd
Horlivka
Makiyivka
Donetsk
Shakhty
Don River
Mariupol
Rostov
Tsimlyansk Res.
KALMYKIYA
Sea of Azov
Volga River
Astrakhan
Elista
RUSSIA
Tikhoretsk
Kropotkin
Krasnodar
Armavir
Stavropol
Novorossiysk
Budennovsk
Maykop
ADYGEYA
Mineral'nyye Vody
Cherkessk
KARACHAYEVO-CHERKESSIYA
Pyatigorsk
Sochi
KABARDINO-BALKARIYA
Nalchik
Groznyy
ABKHAZIA
Caspian Sea
CAUCASUS
Vladikavkaz
INGUSHETIYA
CHECHNYA
Sokhumi
NORTH OSSETIA
Nazran
DAGESTAN
Makhachkala
Black Sea
SOUTH OSSETIA
Pankisi Gorge
MOUNTAINS
Samtredia
Kutaisi
Tskhinval
Gori
Supsa
Rioni R.
Batumi
AJARIA
Tbilisi
GEORGIA
Rustavi
Borcka
Lake Mingacevir
Ceyhan
AZERBAIJAN
TURKEY
Sumgait
Trabzon
Kura R.
Torul
Gyumri
Gänja
Baki (Baku)
ARMENIA
NAGORNO-KARABAKH
Kura R.
Yerevan
Xankändi (Stepanakert)
L. Sevan
Agri
Aras R.
Armenian controlled
NAXCIVAN
AZER.
AZERBAIJAN PROVINCE
Van
Marand
Ardabil
Tabriz
IRAN
Lake Urmia

followed, Groznyy was reduced to rubble, Chechen attacks and Russian retaliation killed tens of thousands of civilians, and Chechen terrorists brought their campaign to Moscow itself, bombing apartment buildings and subway trains, taking an entire theater audience hostage, and blowing up airliners departing from the capital. Nearer to their Caucasus Mountain hideouts, they seized a hospital, killing patients and doctors, and in 2004 at a school in Beslan, in the Republic of North Ossetia, caused the deaths of more than 300 students, teachers, and parents. It is all part of a campaign that defies Russia's efforts to bring all of its minorities into the new framework, and Moscow is determined not to let it succeed. Chechnya's secession, the Russian government fears, would open the floodgates of separatism in this multicultural country.

The External Southern Periphery

Beyond Russia's tier of internal republics lies still another periphery, legally outside the Russian state but closely tied to it nevertheless. This is *Transcaucasia*, historically a battleground for Christians and Muslims, Armenians and Turks, Russians and Persians. Today it is a subregion containing three former Soviet Socialist Republics: landlocked Armenia, coastal Georgia on the Black Sea, and Azerbaijan on the land-encircled Caspian Sea. Although these three countries are independent states today, they are so closely bound up with Russia that they remain, functionally, part of the Russian realm (Fig. 2-16).

Armenia

As Figure 2-2 shows, landlocked Armenia (population: 3.0 million) occupies some of the most rugged and mountainous terrain in the earthquake-prone Transcaucasus. The Armenians are an embattled people who adopted Christianity 17 centuries ago and for more than a millennium sought to secure their ancient homeland here on the margins of the Muslim world. During World War I, the Ottoman Turks massacred much of the Christian Armenian minority and drove the survivors from eastern Anatolia and what is now

Iraq into the Transcaucasus. This event remains a highly sensitive political issue. At the end of that war in 1918, an independent Armenia arose, but its autonomy lasted only two years. In 1920, Armenia was taken over by the Soviets; in 1936, it became one of the 15 constituent republics of the Soviet Union. The collapse of the Soviet Empire gave Armenia what it had lost three generations earlier: independence.

Or so it seemed. Soon afterward, the Armenians found themselves at war with neighboring Azerbaijan over the fate of some 150,000 Armenians living in Nagorno-Karabakh, a pocket of territory surrounded by Azerbaijan. Such a separated territory is called an **exclave**, and this one had been created by Soviet sociopolitical planners who, while acknowledging the cultural (Christian) distinctiveness of this cluster of Armenians, nevertheless gave (Muslim) Azerbaijan jurisdiction over it.

That was a recipe for trouble. In the ensuing conflict, Armenian troops entered Azerbaijan and gained control over the exclave, even ousting Azerbaijanis from the zone between the main body of Armenia and Nagorno-Karabakh (Fig. 2-16). The international community, however, has not recognized Armenia's occupation, and officially the territory remains a part of Azerbaijan. After nearly a decade into the twenty-first century, the matter still remained unresolved.

Georgia

Of the three former Soviet Republics in Transcaucasia, only Georgia has a Black Sea coast and thus an outlet to the wider world. Smaller than South Carolina, Georgia is a country of high mountains and fertile valleys. Its social and political geographies are complicated. The population of 4.4 million is more than 70 percent Georgian but also includes Armenians (8 percent), Russians (6 percent), Ossetians (3 percent), and Abkhazians (2 percent). The Georgian Orthodox Church dominates the religious community, but about 10 percent of the people are Muslims, most of them concentrated in Ajaria in the southwest.

Unlike Armenia and Azerbaijan, Georgia has no exclaves, but its political geography is problematic nonetheless (Fig. 2-16). Within Georgia's borders

lie three minority-based autonomous entities: the Abkhazian and Ajarian Autonomous Republics, and the South Ossetian Autonomous Region.

Sakartvelos, as the Georgians call their country, has a long and turbulent history. Tbilisi, the capital for 15 centuries, lay at the core of an empire around the turn of the thirteenth century, but the Mongol invasion ended that era. Next, the Christian Georgians found themselves in the path of wars between Islamic Turks and Persians. Turning northward for protection, the Georgians were annexed in 1800 by the Russians, who were looking for warm-water ports. Like other peoples overpowered by the czars, the Georgians took advantage of the Russian Revolution to reassert their independence; but the Soviets reincorporated Georgia in 1921 and proclaimed a Georgian Soviet Socialist Republic in 1936. Josef Stalin, the communist dictator who succeeded Lenin, was a Georgian.

Georgia is renowned for its scenic beauty, warm and favorable climates, agriculture (especially tea), timber, manganese, and other products. Georgian wines, tobacco, and citrus fruits are much in demand. The diversified economy could support a viable state, but has yet to prosper. Tensions with Russia regarding sovereignty over some of Georgia's minority areas contribute to its continued political instability.

Azerbaijan

Azerbaijan is the name of an independent state *and* of a province in neighboring Iran. The Azeris (short for Azerbaijanis) on both sides of the border have the same ancestry: they are a Turkish people divided between the (then) Russian and Persian empires by a treaty signed in 1828. By that time, the Azeris had become Shi'ite Muslims, and when the Soviet communists laid out their grand design for the USSR, they awarded the Azeris their own republic. On the Persian side, the Azeris were assimilated into the Persian Empire, and their domain became a province. Today, this former Soviet Socialist Republic is the independent state of Azerbaijan (population: 8.7 million), and the 10 million Azeris to the south live in the Iranian province.

During the brief transition to independence and at the height of their war with the Armenians, the dominantly Muslim Azeris tended to look southward, toward Iran. But geographic realities dictate a more practical orientation. Azerbaijan possesses huge reserves of oil and natural gas; under the Soviets it was one of Moscow's chief regional sources of fuels. The center of the oil industry is Baki (Baku), the capital on the shore of the Caspian Sea—but the Caspian Sea is a lake. To export its oil, Azerbaijan needs pipelines, but those of Soviet vintage link Baki to Russia's Black Sea terminal of Novorossiysk.

In the post-Soviet period, Azerbaijan has played its energy card to national advantage. Its authoritarian government is sometimes compared to that of Belarus, but unlike Belarus it has tried to establish some independence from Moscow. Oil and gas have made this easier: Washington, which supposedly promotes democracy, does not press the issue when it negotiates for Azerbaijan's oil, and the United States has become its most important customer. During Soviet and immediate post-Soviet times, oil from Baki continued to flow exclusively through Novorossiysk (Fig. 2-16), but the United States negotiated the construction of an alternate route across Georgia and Turkey to the Mediterranean oil terminal at Ceyhan, thus avoiding Russian transit altogether. Oil began to flow through this outlet in 2006. Soviet state companies once extracted Azerbaijan's oil, but now American, French, British, and Japanese oil companies are developing the country's Caspian reserves.

Azerbaijan's nearly 9 million inhabitants have not benefited greatly from their country's energy riches. Given its Muslim roots and relative location, the time may come when this country turns toward the Islamic world. For the present, it is caught in the worldwide web of energy demand and supply.

THE EASTERN FRONTIER

From the eastern flanks of the Ural Mountains to the headwaters of the Amur River, and from the latitude of Tyumen to the northern zone of neighboring Kazakhstan, lies Russia's vast Eastern Frontier region, product of a gigantic experiment in the eastward extension of the Russian Core (Fig. 2-13). As the maps of cities and surface communications suggest, this Eastern Frontier is more densely peopled and more fully developed in the west than in the east; at the longitude of Lake Baykal, settlement has become linear, marked by ribbons and clusters along the east-west railroads. Two subregions dominate the geography: the Kuznetsk Basin in the west and the Lake Baykal area in the east.

The Kuznetsk Basin (*Kuzbas*)

Some 1450 kilometers (900 mi) east of the Urals lies another of Russia's primary regions of heavy manufacturing resulting from the communist period's centralized planning: the Kuznetsk Basin, or *Kuzbas* (Fig. 2-14). In the 1930s, it was opened up as a supplier of raw materials (especially coal) to the Urals, but that function became less important as local industrialization accelerated. The original plan was to move coal from the Kuzbas west to the Urals and allow the returning trains to carry iron ore east to the coalfields. However, good-quality iron ore deposits were subsequently discovered near the Kuznetsk Basin itself. As the new resource-based Kuzbas industries grew, so did its urban centers. The leading city, located just outside the region, is Novosibirsk, which stands at the intersection of the Trans-Siberian Railroad and the Ob River as the symbol of Russian enterprise in the vast eastern interior. To the northeast lies Tomsk, one of the oldest Russian towns in all of Siberia, founded in the seventeenth century and now caught up in the modern development of the Kuzbas Region. Southeast of Novosibirsk lies Novokuznetsk, a city that produces steel for the region's machine and metal-working plants and aluminum products from Urals bauxite.

In the fall of 2004, Russia experienced its worst terrorist attack as rebel suicide bombers held children and parents hostage in a school in Beslan, North Ossetia (see Fig. 2-16). A botched rescue effort on the part of the Russian government set off the suicide bombers, resulting in the death of 333 staff, students, and other adults. Here a woman cries as she walks past photos of victims of the school siege on the second anniversary of the event.   (© AP/Wide World Photos)

The Lake Baykal Area (*Baykaliya*)

East of the Kuzbas, development becomes more insular, and distance becomes a stronger adversary. North of the Tyva Republic and eastward around Lake Baykal, larger and smaller settlements cluster along the two railroads to the Pacific coast (Fig. 2-10). West of the lake, these rail corridors lie in the headwater zone of the Yenisey River and its tributaries. A number of dams and hydroelectric projects serve the valley of the Angara River, particularly the city of Bratsk. Mining, lumbering, and some farming sustain life here, but isolation dominates it. The city of Irkutsk, near the southern end of Lake Baykal, is the principal service center for a vast Siberian region to the north and for a lengthy east-west stretch of southeastern Russia.

Beyond Lake Baykal, the Eastern Frontier really lives up to its name: this is southern Russia's most rugged, remote, forbidding country. Settlements are rare, many being mere camps. The Buryat Republic is part of this zone; the territory bordering it to the east was taken from China by the czars and may become an issue in the future. Where the Russian-Chinese boundary turns southward, along the Amur River, the region called the Eastern Frontier ends and Russia's Far East begins.

SIBERIA

Before we assess the potential of Russia's Pacific Rim, we should remember that the ribbons of settlement just discussed hug the southern perimeter of this giant country, avoiding the vast Siberian region to the north (Fig. 2-13). Siberia extends from the Ural Mountains to the Kamchatka Peninsula—a vast, bleak, frigid, forbidding land. Larger than the conterminous United States but inhabited by only an estimated 15 million people, Siberia quintessentially symbolizes the Russian environmental plight: vast distances, cold temperatures worsened by strong Arctic winds, difficult terrain, poor soils, and limited options for survival.

ENVIRONMENTALLY SPEAKING . . .

Forests in Russia. The boreal forests of northern Russia, including Siberia, are one of the last remaining stands of substantial forest in the world. Vast expanses of slow-growing coniferous needleleaf forests (such as fir trees) cover much of this region south of the **tree line**. How has this forest, which looms so large in our imagination, been affected by environmental change? Analysis of satellite images taken over time indicates significant degradation of these forests by the lumber industry. On the other hand, wetter and warmer conditions due to global climate change are permitting trees to colonize and expand into areas that had been too cold and dry for survival. This has more than offset the degradation caused by the lumber industry, resulting in an overall expansion of the forest. (© John Cancalosi/ DRK photo)

But Siberia also has resources. From the days of the first Russian explorers and Cossack adventurers, Siberia's riches have beckoned. Gold, diamonds, and other precious minerals were found. Later, metallic ores including iron and bauxite were discovered. Still more recently, the Siberian interior proved to contain massive quantities of oil and natural gas (Fig. 2-15) that now contribute significantly to Russia's energy supply and exports.

As the physiographic map (Fig. 2-2) shows, major rivers—the Ob, Yenisey, and Lena—flow gently northward across Siberia and the Arctic Lowland into the Arctic Ocean. Hydroelectric power development in the basins of these rivers has generated electricity used to extract and refine local ores, and run the lumber mills that have been set up to exploit the vast Siberian forests.

The Future

The human geography of Siberia is fragmented, and most of the region is virtually uninhabited (Fig. 2-8). Ribbons of Russian settlement have developed; the Yenisey River, for instance, can be traced on this map of Russian peoples (a series of small settlements north of Krasnoyarsk), and the upper Lena Valley is similarly fringed by ethnic Russian settlement. Yet hundreds of miles of empty territory separate these ribbons and other islands of habitation.

The political geography of eastern Siberia is marked by the growing identity of Sakha (the Yakut Republic). As additional resources are discovered here (including oil and natural gas), this republic, centered on the capital, Yakutsk, will become more important.

Siberia, Russia's freezer, is stocked with goods that may become mainstays of future national development. Already, precious metals and mineral fuels are bolstering the Russian economy. In time, we may expect Siberian resources to play a growing role in the economic development of the Eastern Frontier and the Russian Far East as well. One step in that process was already taken during Soviet times: the completion of the BAM (Baykal-Amur Mainline) Railroad in the 1980s. This route, lying north of and parallel to the old Trans-Siberian Railroad, extends 3540 kilometers (2200 mi) eastward from Tayshet (near the important center of Krasnoyarsk) directly to the Far East city of Komsomolsk (Fig. 2-10). In the post-Soviet era, the BAM Railroad has been beset by equipment breakdowns, workers' strikes, and melting permafrost that threatens infrastructure stability. Nonetheless, it is a key element of the infrastructure that will serve the Eastern Frontier's economic growth in the twenty-first century.

THE RUSSIAN FAR EAST

Imagine this: a country with 8000 kilometers (5000 mi) of Pacific coastline, two major ports, large interior cities nearby, huge reserves of resources ranging from minerals to fuels to timber, directly across from one of the world's largest economies—all this at a time when the Asian Pacific Rim was the world's fastest-growing economic region. Would not that country have burgeoning cities, busy harbors, growing industries, and expanding trade?

In the Russian Far East (Fig. 2-13), the answer is—no. Activity in the port of Vladivostok is a shadow of what it was during the Soviet era, when it was the communists' key naval base. The nearby container terminal at Nakhodka suffers from break-

The port of Vladivostok, in Russia's Far East, could be part of the dynamic Pacific Rim. Instead it is a forgotten outpost of the stressed Russian state.   (© ITAR-TASS/Sovfoto/eastfoto)

downs and inefficiencies. The railroad to western Russia carries just a fraction of the trade it did during the 1970s and 1980s. Cross-border trade with China is minimal. Trade with Japan is inconsequential. The region's cities are grimy, drab, moribund. Utilities are shut off for hours at a time because of fuel shortages and system breakdowns. Outdated factories are shut, their workers dismissed. Political relations with Moscow are poor. The population is dwindling. There is potential here, but little of it has been realized.

Mainland and Island

As a region, the Russian Far East consists of two parts: the mainland area extending from Vladivostok to the Eastern Highlands and the large island of Sakhalin (Figs. 2-2; 2-17). This is cold country: icebreakers have to keep the ports of Vladivostok and Nakhodka open throughout the winter. Winters here are long and bitterly cold; summers are brief and cool. Although the population is small (about 7 million), food must be imported because not much can be grown. Most of the region is rugged, forested, and remote. Vladivostok, Khabarovsk, and Komsomolsk are the only cities of any size. Nakhodka and the newer BAM Railroad terminal at Vanino are smaller towns; the entire population of Sakhalin is about 700,000 (on an island the size of Caribbean Hispaniola [18.2 million]). Offshore lie productive fishing grounds. Most important of all are the huge oil and gas reserves recently discovered on and around Sakhalin, whose exploitation may soon launch a new energy-based era.

Post-Soviet Malaise

This part of Russia is suffering from being a far periphery of a state with a core on the other side of the world. The Soviet regime rewarded people willing to move to this region with housing and subsidies. The communists, like the czars before them, realized the importance of this frontier (Vladivostok means "We Own the East"), and they used every possible incen-

products westward by train, and they received food and other consumer goods from the Russian heartland.

But these Soviet-era 'rewards' have not held in the post-Soviet transition. The new economic order has canceled the region's communist-era advantages: the Trans-Siberian Railroad now must charge the real cost of transporting products from the Eastern Frontier to the Russian Core. State-subsidized industries must compete on market principles, their subsidies having ended. The decline of Russia's armed forces has hit Vladivostok hard. The fleet lies rusting in port; service industries have lost their military markets; the shipbuilding industry has no government contracts. Coal miners in the Bureya River Valley (a tributary of the Amur) go unpaid for months and go out on strike; coal-fired power plants do not receive fuel shipments, and cities and towns go dark.

Locals put much of the blame for their region's failure on Moscow, and with reason. The Far East contains only five administrative regions with little political clout in Moscow, which is the way Moscow appears to want it.

But, this is the Pacific Rim, which is booming elsewhere. It is likely that despite its stagnation, Russia's Far East will figure prominently in Russian (and probably world) affairs. Here, Russia meets Japan at sea and China on land. China is booming economically and is putting demographic pressure on the shrinking Russian popula-

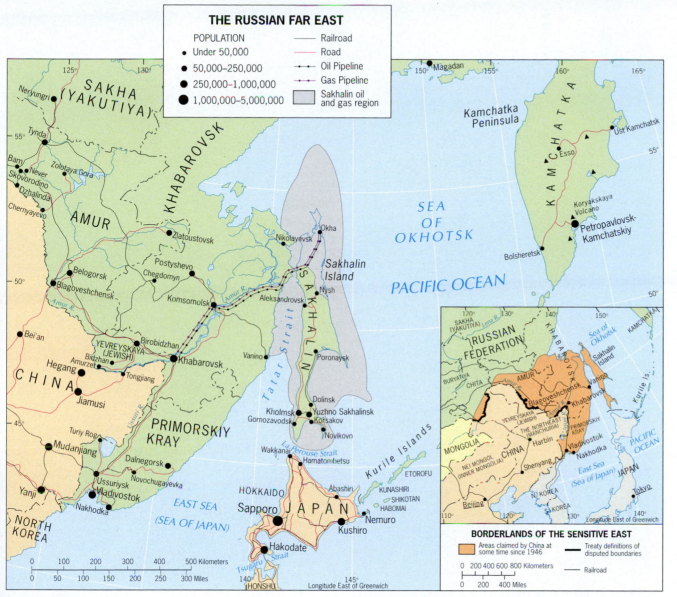

FIGURE 2-17

© H. J. de Blij, P. O. Muller, and John Wiley & Sons, Inc.

tive to develop it and link it more strongly to Russia's distant western core. Freight rates on the Trans-Siberian Railroad, for example, were about 10 percent of their real costs; the trains were always fully loaded in both directions. Vladivostok was a military base and a city closed to foreigners, and Moscow in-

vested heavily in its infrastructure. Komsomolsk in the north and Khabarovsk near the region's center were endowed with state-owned industries using local resources: iron ore from Komsomolsk, oil from Sakhalin, timber from the ubiquitous forests. The steel, chemical, and furniture industries sent their

tion (Fig. 2-11). Despite its seeming remoteness, vast resources ranging from Sakhalin's fuels to Siberia's lumber, Sakha's gold to Khabarovsk's metals, may help enlarge Russia's foothold on the Pacific Rim, a window on the ocean on whose shores the world is being transformed.

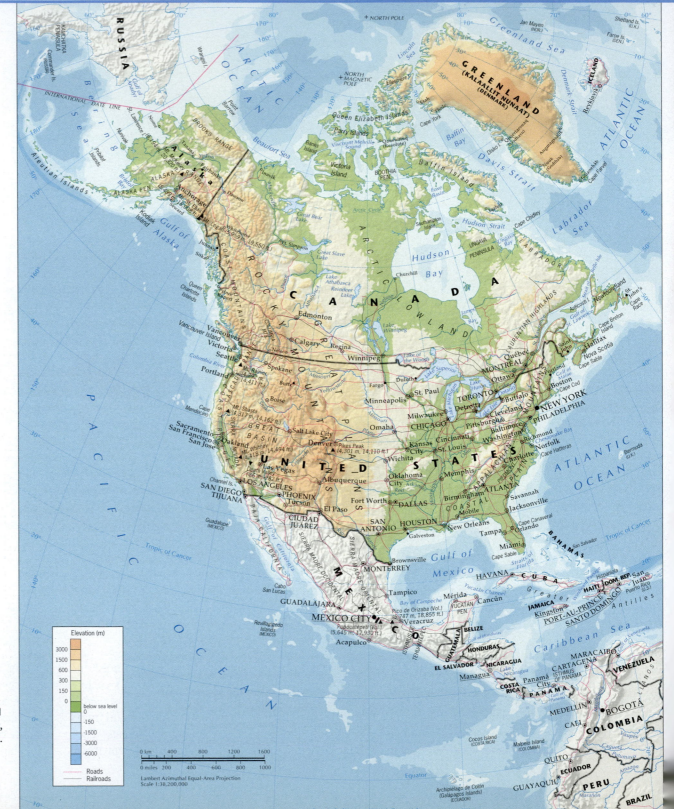

FIGURE 3-1

Map: © H. J. de Blij, P. O. Muller,
and John Wiley & Sons, Inc.

3

NORTH AMERICA

In This Chapter

- Spectacular scenery, violent weather
- North America before the European invasion
- What makes Americans and Canadians move so often
- Immigration and the changing map of the United States
- Canada and Quebec: Still together, but forever?
- China's impact on the Pacific Hinge

CONCEPTS, IDEAS, AND TERMS

1. Cultural pluralism
2. Physiographic province
3. Rain shadow effect
4. Migration
5. Push/pull factors
6. Sunbelt
7. American Manufacturing Belt
8. Megalopolis
9. Central Business District (CBD)
10. Inner City
11. Ghetto
12. Edge City
13. Mosaic culture
14. Productive activities
15. Economies of scale
16. Rustbelt
17. Postindustrial economy
18. Technopole
19. Ecumene
20. Regional state
21. Pacific Rim

REGIONS

NORTH AMERICAN CORE
MARITIME NORTHEAST
FRENCH CANADA
CONTINENTAL INTERIOR
SOUTH
SOUTHWEST
WESTERN FRONTIER
NORTHERN FRONTIER
PACIFIC HINGE

Photos: (*upper left*) New York City © A. WinklerPrins; (*above*) Southern Utah © A. WinklerPrins.

105

NORTH AMERICA AS a geographic name means different things to different people. To physical geographers, North America refers to the landmass that extends from Alaska to Panama. As Figure G-5 shows, the North American continent is the dominant feature of the North American tectonic plate, which reaches almost—but not quite—to Panama. To human geographers, North America (sometimes inappropriately called Anglo America) signifies a geographic realm consisting of two large countries with much in common, Canada and the United States. We therefore recognize three realms in the Americas: (1) *North America*, propelled by the United States; (2) *Middle America*, anchored by Mexico; and (3) *South America*, dominated by Brazil.

Defining the Realm

Defined in the context of human geography, North America is constituted by two of the world's most highly advanced countries by virtually every measure of human development. Blessed by an almost endless range of natural resources and bonded by trade as well as culture, Canada and the United States are locked in a mutually productive embrace that is reflected by the statistics. In an average year of the recent past, more than 80 percent of Canadian exports went to the United States and about two-thirds of Canada's imports came from its southern neighbor. For the United States, Canada is its leading export market and its number one source of imports.

Both countries also rank among the world's most highly urbanized, each with 80 percent of their popu-

MAJOR GEOGRAPHIC QUALITIES OF North America

1. North America encompasses two of the world's biggest states territorially. Canada is the second largest in size; the United States is third.

2. Both Canada and the United States are federal states, but their political systems differ. Canada's is adapted from the British parliamentary system and is divided into ten provinces and three territories. The United States separates its executive and legislative branches of government, and it consists of 50 States, the Commonwealth of Puerto Rico, and a number of island territories under U.S. jurisdiction in the Caribbean Sea and the Pacific Ocean.

3. Both Canada and the United States are plural societies. Although ethnicity is increasingly important, Canada's pluralism is most strongly expressed in regional bilingualism. In the United States, major divisions occur along racial and ethnic lines.

4. Many of Quebec's French-speaking citizens support a movement that seeks independence for the province. In 1995 there was a referendum in which separation of Quebec from Canada was only narrowly averted. In 2006 Quebec was recognized as a distinct nation within a united Canada, settling Canada's devolution for now.

5. North America's population, not large by international standards, is highly urbanized and the world's most mobile. Largely propelled by a continuing wave of immigration, the realm's population total is expected to grow by more than 40 percent over the next half-century.

6. By world standards, this is a rich realm where high incomes and high rates of consumption prevail. North America possesses a highly diversified resource base, but nonrenewable fuel and mineral deposits are consumed prodigiously.

7. North America is home to one of the world's great manufacturing complexes. The realm's industrialization generated its unparalleled urban growth, but a new postindustrial society and economy are rapidly maturing in both countries.

8. The two countries heavily depend on each other for supplies of critical raw materials (e.g., Canada is the leading source of U.S. energy imports) and have long been each other's chief trading partners. Today, the North American Free Trade Agreement (NAFTA), which also includes Mexico, is linking all three economies ever more tightly as the last barriers to international trade and investment flow are dismantled.

9. North Americans are the world's most mobile people. Although plagued by increasing congestion problems, the realm's networks of highways, commercial air routes, and cutting-edge telecommunications are the most efficient in the world.

lations concentrated in cities that have evolved into metropolitan agglomerations. The old core cities are now encircled by rings of still-expanding suburban development, which draw an ever greater proportion of the metropolitan population and activity base into their newly urbanized environments. Nothing symbolizes the North American city as strongly as the skyscrapered panoramas of New York, Chicago, and Toronto—or the vast, beltway-connected suburban expanses of Los Angeles, Washington, and Houston.

North Americans also are the most mobile people in the world. Commuters stream into and out of suburban business centers and central-city downtowns by the millions each working day; most of them drive cars, whose numbers have multiplied more than six times faster than the human population since 1970. Moreover, each year nearly one out of every six individuals changes his or her residence.

GEOGRAPHIC AND SOCIAL CONTRASTS

Although Canada and the United States share many historical, cultural, and economic qualities, they also differ in significant ways, as can be seen on the map. The United States, somewhat smaller territorially than Canada, occupies the heart of the North American continent and, as a result, encompasses a greater environmental range. The U.S. population is dispersed across most of the country, forming major concentrations along both the (north-south-trending) Atlantic and Pacific coasts. The overwhelming majority of Canadians, however, live in an interrupted east-west corridor that lies across southern Canada, mainly within 320 kilometers (200 mi) of the U.S. border. The United States also encompasses North America's northwestern extension, Alaska. Offshore Hawai'i, however, belongs in the Pacific Realm (Chapter 12).

Demographic contrasts also exist between the two countries because the U.S. population is nearly ten times larger than Canada's. The United States is the world's third most populous country with just over 300 million while Canada ranks 36th with 33 million. Comparatively small as its population may be, however, Canada has varied cultures and traditions ranging from multilingualism (French has equal status with English) to ethnic diversity that, as we will note later, have strong spatial expression. Thus Canada shares with the United States a characteristic feature of this realm: the prevalence of **1** **cultural pluralism**. In the United States, this is exhibited by a range of ethnic backgrounds from European and African to Asian and Native American. And while the civil rights movement and other initiatives have eliminated *de jure* (legal) segregation in the United States, ethnic clustering and *de facto* (real-world) self-segregation continue to mark the social landscape. In part, this is due to a troubling aspect of this wealthiest of all realms: poverty and deprivation still afflict a substantial minority as income gaps between the successful rich and the struggling poor continue to widen.

NORTH AMERICA'S PHYSICAL GEOGRAPHY

Before we examine the human geography of the United States and Canada more closely, we need to consider the physical setting in which they are rooted. The North American continent extends from the Arctic Ocean to Panama (see chapter opener map, Fig. 3-1), but we will confine ourselves here to the territory north of Mexico—a geographic realm that still stretches from the near-tropical latitudes of southern Florida and Texas to subpolar Alaska and Canada's far-flung northern periphery. The remainder of the North American continent comprises a separate realm, *Middle America*, which is covered in Chapter 4.

Physiography

North America's physiography is characterized by its clear, well-defined division into physically homogeneous regions called **2** **physiographic provinces**. Each region is marked by considerable uniformity in relief, climate, vegetation, soils, and other environmental conditions, resulting in a scenic sameness that comes readily to mind. For example, we identify such regions when we refer to the Rocky Mountains, the Great Plains, and the Appalachian Highlands. However, not all the physiographic provinces of North America are so easily delineated.

Figure 3-2 maps the complete layout of the continent's physiography and includes a cross-sectional terrain profile along the 40th parallel. The most obvious aspect of this map of North America's physiographic provinces is the north-south alignment of the continent's great mountain backbone, the Rocky Mountains, whose rugged topography dominates the western segment of the continent from Alaska to New Mexico. The major feature of eastern North America is another, much lower chain of mountain ranges called the Appalachian Highlands, which also trend approximately north-south and extend from Canada's Atlantic Provinces to Alabama. The orientation of the Rockies and Appalachians is important because, unlike Europe's Alps, they do not form a topographic barrier to polar or tropical air masses flowing southward or northward, respectively, across the continent's interior.

Between the Rocky Mountains and the Appalachians lie North America's vast interior plains, which extend from the Mackenzie Delta on the Arctic Ocean to the Gulf of Mexico. We can subdivide these into several provinces: (1) the great Canadian Shield, which is the geologic core area containing North America's oldest rocks; (2) the Interior Lowlands, covered largely by glacial debris laid down by ice, meltwater, and wind during the Pleistocene glaciation; and (3) the Great Plains, the extensive sedimentary surface that rises gently westward toward the Rocky Mountains. Along the southern margin, these interior plainlands merge into the Gulf-Atlantic Coastal Plain, which stretches from southern Texas along the seaward margin of the Appalachian Highlands and the neighboring Piedmont until it ends at Long Island just to the east of New York City.

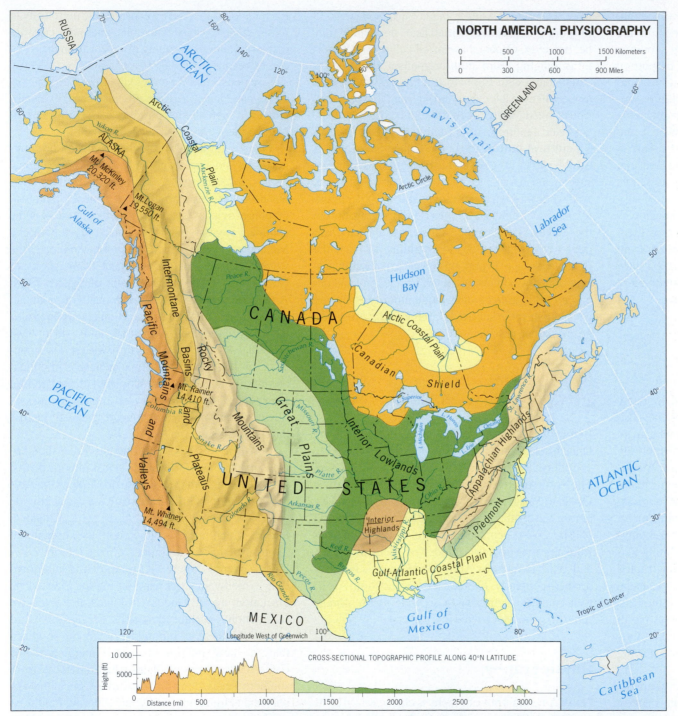

NORTH AMERICA: PHYSIOGRAPHY

| 0 | 500 | 1000 | 1500 Kilometers |

| 0 | 300 | 600 | 900 Miles |

CROSS-SECTIONAL TOPOGRAPHIC PROFILE ALONG 40°N LATITUDE

FIGURE 3-2

© H. J. de Blij, P. O. Muller, and John Wiley & Sons, Inc.

On the western side of the Rocky Mountains lies the zone of Intermontane Basins and Plateaus. Within the conterminous United States, this physiographic province includes: (1) the Colorado Plateau in the south, with its thick sediments and spectacular Grand Canyon; (2) the lava-covered Columbia Plateau in the north, which forms the watershed of the Columbia River; and (3) the central Basin-and-Range country (Great Basin) of Nevada and Utah, which contains several extinct lakes from the glacial period as well as the surviving Great Salt Lake. This province is called *intermontane* because of its position between the Rocky Mountains to the east and the Pacific coast mountain system to the west.

From the Alaskan Peninsula to Southern California, the west coast of North America is dominated by an almost unbroken corridor of high mountain ranges that originated from the contact between the North American and Pacific Plates (Fig. G-5). The major components of this coastal mountain belt include California's Sierra Nevada, the Cascades of Oregon and Washington, and the long chain of highland massifs that line the British Columbia and southern Alaska coasts. Three broad valleys—which contain dense populations—are the only noteworthy interruptions: California's Central (San Joaquin-Sacramento) Valley; the Cowlitz-Puget Sound lowland of Washington State, which extends southward into western Oregon's Willamette Valley; and the lower Fraser Valley, which slices through southern British Columbia's coast range.

Climate

The world climate map (Fig. G-9) clearly depicts the various climatic regimes and regions of North America. In general, temperature varies latitudinally—the farther north one goes, the cooler it gets. Regional land-and-water-heating differentials, however, distort this broad pattern. Because land surfaces heat and cool far more rapidly than water bodies, yearly temperature ranges are much larger where *continentality* (interior remoteness from the sea) is greatest.

Western Rain Shadow

Precipitation generally tends to decline toward the west (except for the Pacific coastal strip itself) as a result of the **3 rain shadow effect**. This occurs because Pacific air masses, driven by prevailing winds, carry their moisture onshore but soon collide with the Sierra Nevada-Cascades wall, forcing them to rise—and cool—in order to crest these mountain ranges. Such cooling is accompanied by major condensation and precipitation, so that by the time these air masses descend and warm along the eastern slopes to begin their journey across the continent's interior, they have already deposited much of their moisture. Thus the mountains produce a downwind shadow of dryness, which is reinforced whenever eastward-moving air must surmount other ranges farther inland, especially the massive Rockies.

Arid and Humid America

This semiarid (and in places truly arid) environment extends so deeply into the central United States that a broad division can be made between Arid (western) and Humid (eastern) America, which face each other along a fuzzy boundary that is best viewed as a wide transitional zone. Although the separating criterion of 50 cm (20 in) of annual precipitation is easily mapped (see Fig. G-8), generally the north-south *isohyet*[1] can

[1]An isohyet is a line connecting all places receiving the same precipitation per year, in this case, 50 centimeters.

and does swing widely across the drought-prone Great Plains from year to year because highly variable warm-season rains from the Gulf of Mexico come and go in unpredictable fashion. Indeed, drought is a constant environmental hazard throughout Arid America, and the past decade has been one of the driest on record, possibly linked to global climate change.

On the other hand, precipitation in Humid America is far more regular. The prevailing westerly winds (blowing from west to east—winds are always named for the direction *from* which they come), which normally come up dry for the large zone west of the 100th meridian, pick up considerable moisture over the Interior Lowlands and distribute it throughout eastern North America. A large number of storms develop here on the highly active weather front between tropical Gulf air to the south and polar air to the north. Even if major storms do not material-ize, local weather disturbances created by sharply contrasting temperature differences are always a danger (as recent disastrous weather and floods have proven). There are more tornadoes (nature's most vi-olent weather) in the central United States each year than anywhere else in the world. And in winter the northern half of this region receives large amounts of snow, particularly around the Great Lakes.

Figure G-9 shows the limited extent of humid tem-perate (C) climates in Canada, which occur only along the narrow Pacific coastal zone, and the prevalence of cold in Canadian environments. East of the Rocky Mountains, Canada's most *moderate* climates corre-spond to the *coldest* of the United States. Nonetheless, southern Canada shares the environmental conditions that mark the Upper Midwest and Great Lakes areas of the United States, so that agricultural productivity in the Prairie Provinces and in Ontario is substantial. Canada is a leading food exporter (chiefly wheat), de-spite its comparatively short growing season.

Soils and Vegetation

The broad environmental partitioning into Humid and Arid America is also reflected in the distribution of the realm's soils and vegetation. For farming pur-poses there is usually sufficient soil moisture to sup-port crops where annual precipitation exceeds the critical 50 centimeters; where the yearly total is less, soils may still be fertile (especially in the Great Plains), but irrigation is often necessary to achieve their full agricultural potential. As for vegetation, the Humid/Arid America dichotomy is again a valid generalization: the natural vegetation of areas re-ceiving more than 50 centimeters of water annually is mostly forest, whereas the drier climates give rise to mostly grassland cover.

Hydrography (Surface Water)

Surface water patterns in North America are domi-nated by the two major drainage systems that lie be-tween the Rockies and the Appalachians: (1) the five Great Lakes (Superior, Michigan, Huron, Erie, and Ontario) that drain into the St. Lawrence River, and (2) the mighty Mississippi-Missouri river net-work, fed by such major tributaries as the Ohio, Tennessee, and Arkansas rivers. Both are products of the last episode of Pleistocene glaciation, and to-gether they amount to nothing less than the best nat-ural inland waterway system in the world. Human intervention has further enhanced this network of navigability, mainly through the building of canals that link the two systems as well as the St. Lawrence Seaway.

Elsewhere, the northern east coast of the continent is well served by a number of short rivers leading in-land from the Atlantic. In fact, a number of northeast-ern seaboard cities of the United States—such as Washington, D.C., Baltimore, and Philadelphia—are located at the waterfalls that marked the limit to tidewater navigation (hence their designation as *Fall Line cities*). Rivers in the Southeast and west of the Rockies at first offered little practical value because of their orientation and the difficulty of navigating them. In the Far West, however, the Colorado and Columbia rivers have become important as suppliers of drinking and irrigation water as well as hydroelec-tric power.

INDIGENOUS NORTH AMERICA

When the first Europeans set foot on North American soil, the continent was occupied by millions of people whose early ancestors had reached the Americas from Asia, via Alaska and probably also across the Pacific, more than 13,000 years before (and possibly as long as 30,000 years ago). In search of Asia, the Europeans misnamed them Indians, but the historic affinities of these earliest Americans were with the peoples of eastern and northeastern Asia, not India. In North America these *Native Americans* or *First Nations*—as they are now called in the United States and Canada, respectively—had organized themselves into hundreds of nations with a rich mo-saic of languages and a great diversity of cultures (Fig. 3-3). Farmers grew crops the Europeans had never seen; other nations depended chiefly on fishing, herding, hunting, or some combination of these. Elaborate houses, efficient watercraft, effective weaponry, decorative clothing, and wide-ranging art forms distinguished the aboriginal nations. Certain nations had formulated sophisticated health and medical practices; ceremonial life was complex and highly developed; and political institutions were mature and elaborate.

The eastern nations were the first to bear the brunt of the European invasion. By the end of the eighteenth century, ruthless, land-hungry settlers had driven most of the Native American peoples living along the Atlantic and Gulf coasts from their homes and lands, beginning a westward push that was to devastate indigenous society. The U.S. Congress in 1789 proclaimed that "Indian . . . land and property shall never be taken from them without their consent," but in fact this is just what happened. One of the sorriest episodes in American history involved the removal of the eastern Cherokee, Chickasaw, Choctaw, Creek, and Seminole from their homelands in the Southeast through forced marches a thousand miles westward to reservations in Oklahoma. One-fourth of the entire Cherokee population died along the way from exposure, starvation, and disease, and the others fared little better. Congress approved treaties that would at least protect the native peoples of the Plains (Fig. 3-3) and those farther to the west, but after the mid-nineteenth century the white settlers ignored those guarantees as well. A half-century of war left what remained of North America's nations with about 4 percent of U.S. territory in the form of mostly resource-poor reservations. Only in the last few decades have several Native American communities been able to improve their economic situations by building and operating casinos on their lands.

In what is today Canada, too, the comparatively small First Nations population was overwhelmed by the numbers and power of European settlers, and decimated by the diseases they introduced. Efforts at restitution and recognition of First Nations rights, however, have gone farther in Canada than in the United States.

As we noted earlier, two large countries, with geographic similarities as well as differences, constitute the North American realm. Canada and the United States share physiographic provinces as well as border-straddling economic and cultural subregions (for example, large-scale grain cultivation in the Great Plains). It would be possible to base the regionalization of this realm on contrasts between the two states; but as we will see in the second part of this chapter, quite a different set of regions emerges from our geographic analysis. To fully appreciate this regional framework, we must examine in some detail the changing human geography of the United States and Canada individually.

NORTH AMERICA: INDIGENOUS DOMAINS

ARCTIC
CALIFORNIA
GREAT BASIN
NORTHEAST WOODLAND
NORTHWEST COAST
PLAINS
PLATEAU
SOUTHEAST WOODLAND
SOUTHWEST
SUBARCTIC

0   500   1000 Kilometers
0   300   600 Miles
Longitude West of Greenwich

FIGURE 3-3   © H. J. de Blij, P. O. Muller, and John Wiley & Sons, Inc.

Tribal Sovereignty in Arizona. "Long known as the Papago Indians, the Tohono O'odham nation straddles the Arizona/Mexico border. I had an opportunity to visit their reservation in 2002 as part of a group of geographers on a field trip to the borderlands. The sign in the photo indicates the switch in political jurisdiction that takes place upon entering the reservation. Within the reservation, tribal law supersedes federal law, and tribal police are the enforcers. This has recently become a heightened issue as more illegal immigrants are crossing into the United States via the Tohono reservation. The Tohono do not have the personnel to deal with the flood of illegals, nor do they have the police force needed to combat the violence and crime that accompany such an exposed crossing point. Given their own transnational rights (they are permitted to cross back and forth between the United States and Mexico because their native lands preceded that boundary), the Tohono are opposed to a wall being built on their land. But with their reservation being squeezed on both sides by walls and heightened vigilance (by U.S. Border Patrol and citizen vigilantes), more illegals are crossing onto the Tohono O'odham reservation. Some cooperation between tribal and federal authorities now exists, but the situation is a complex one." (© A. WinklerPrins)

www.conceptcaching.com

THE UNITED STATES

As we note throughout this book, the administrative components of states are assuming ever-greater importance in the scheme of things. In Europe, all the major states have reorganized their political-economic systems, forged regions from smaller units, and devolved power to the provinces. As a result, it is increasingly important to know not just Spain but also Catalonia, not just Italy but also Lombardy. So it is in the United States. We should have a clear mental map of the functional layout of this federation, which is why it may be helpful to take a few moments to review Figure 3-4.

Population in Time and Space

The current population distribution of the United States is shown in Figure 3-5, a map that is the latest still in a motion picture that has been unreeling for four centuries since the founding of the first permanent European settlement on the northeastern coast. Slowly at first and then with accelerating speed after 1800, Americans and Canadians took charge of their remarkable continent and pushed the settlement frontier westward to the Pacific.

To understand the current population map, we need to review the major forces that have shaped, and continue to shape, the distribution of Americans and their activities. Since its earliest days, the United States has attracted a steady influx of immigrants who were rapidly assimilated into the societal mainstream. Within the country, people have sorted themselves out to maximize their proximity to existing economic opportunities, and they have shown little resistance to relocating as the nation's evolving economic geography has successively favored different sets of places over time. These movements continue to redistribute millions of Americans, but before we trace the evolution of the population map we need to acquaint ourselves with the process of migration.

The Role of Migration

4 **Migration** refers to a change in residential location intended to be permanent, and it has long played a key role in the human geography of this realm. After thousands of years of native settlement, European explorers reached North American shores beginning with Columbus in 1492 (and probably centuries before that). The first permanent colonies were established in the early 1600s, and from them evolved the modern United States of America. Between 1835 and 1935, perhaps as many as 75 million Europeans departed for distant shores—most of them bound for the Americas. Some sought religious freedom; others escaped poverty and famine; still others simply hoped for a better life.

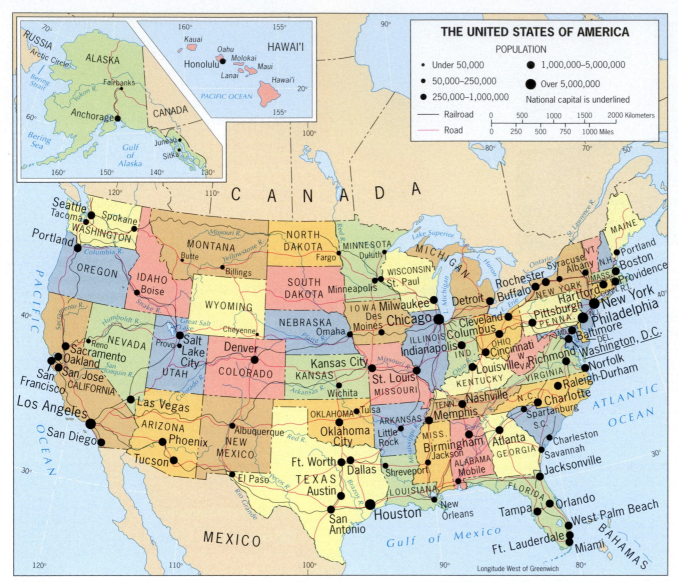

FIGURE 3-4

© H. J. de Blij, P. O. Muller, and John Wiley & Sons, Inc.

counter-stream of returning migrants who cannot adjust.)

Push and Pull Factors. Studies of migration also conclude that several factors are at work in the migration process. **5 Push factors** motivate people to move away; **pull factors** attract them to new destinations. To those early Europeans the United States was a new frontier, a place where one might acquire a piece of land and where the opportunities were reported to be unlimited. That perception has never changed, although the opportunities may now lie in the wage economy, and international immigration continues to significantly shape the demographic complexion of the United States in the twenty-first century. Immigration is at an all-time high, with over 38 million foreign-born residents (12.6 percent of the population) in 2007. What is different today is that Middle America, and East and Southeast Asia have replaced Europe as the dominant source.

Ongoing Migrations. Within the conterminous United States today, the leading internal migration flow that continues to transform the population map is the persistent drift of people and livelihoods toward the Southwest and Southeast—the so-called **6 Sunbelt**. In addition, a pair of lesser but long-standing migratory flows still play a major role: (1) the growth of metropolitan areas, first triggered by the late-nineteenth-century Industrial Revolution, which since the 1960s has been largely rechanneled from the central cities to the suburban ring; and (2) the movement of African-Americans from the rural South to the urban North, which since the 1970s has become a stronger return flow.

Studies of the *migration decision* indicate that migration flows vary in size with: (1) the perceived degree of difference between one's home, or source, and the destination; (2) the effectiveness of flows of information about the destination which migrants sent to those who stayed behind waiting to decide; and (3) the distance between the source and the destination. More than a century ago, the British social scientist Ernst Georg Ravenstein studied the migration process, and many of his conclusions remain valid today. (For example, he posited that every migration stream from source to destination produces a

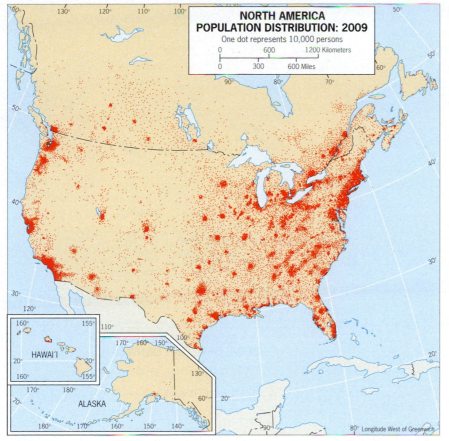

NORTH AMERICA
POPULATION DISTRIBUTION: 2009
One dot represents 10,000 persons

FIGURE 3-5   © H. J. de Blij, P. O. Muller, and John Wiley & Sons, Inc.

ica was surpassing Europe as the world's mightiest industrial power. The impact of industrial urbanization occurred simultaneously at two levels of generalization. At the national or *macroscale*, a system of new cities swiftly emerged, specializing in the collection, processing, and distribution of raw materials and manufactured goods, linked together by an efficient web of railroad lines. Within that urban network, at the local or *microscale*, individual cities prospered in their new roles as manufacturing centers, generating an internal structure that still forms the framework of most large central cities.

The cultural mix in many large cities in North America is enormous as can be seen in this image of rush hour in the center of New York's midtown Manhattan. (© Stock Collection Blue/Alamy)

Pre-Twentieth-Century Population Patterns

The current distribution of the U.S. population is rooted in the colonial era of the seventeenth and eighteenth centuries that was dominated by England and France. The French sought mainly to organize a lucrative fur-trading network, while the English established settlements (beginning in 1607) along what is now the northeastern U.S. seaboard. These British colonies quickly became differentiated in their local economies, a diversity that later shaped American cultural geography. Three separate seaboard colonies became *culture hearths*: the northern colony of New England specialized in commerce; the southern

Chesapeake Bay colony emphasized the plantation farming of tobacco; and the Middle Atlantic area lying in between was home to a number of smaller, independent-farmer colonies. These three culture hearths became the source areas of subsequent westward migration movements.

Post–1900 Industrial Urbanization

The Industrial Revolution occurred almost a century later in the United States than in Europe, but when it did cross the Atlantic in the 1870s, it took hold and advanced so robustly that only 50 years later Amer-

The American Manufacturing Belt. Industrialization and the accompanying growth of the urban system reconfigured the realm's economic landscape. The most notable regional transformation was the emergence of the North American Core, or **7** **American Manufacturing Belt**, which contained the lion's share of industrial activity in both the United States and Canada. As Figure 3-6 shows, the geographic form of the Core Region—which includes southern Ontario—was a near rectangle whose four corners were Boston, Milwaukee, St. Louis, and Baltimore. Within this region, manufacturing is heavily concentrated into a dozen districts centered on the cities mapped in Figure 3-6.

Megalopolitan Growth. Transportation breakthroughs at the subregional scale permitted progressive urban decentralization; the expanding peripheries of major cities coalesced to form a number of conurbations. Also known as a **8** **megalopolis**, the most important by far is the Atlantic Seaboard Megalopolis (Fig. 3-7), the 1000-kilometer (600-mi) urbanized northeastern coastal strip extending from southern Maine to Virginia that contains metropolitan Boston, New York, Philadelphia, Baltimore, and Washington. This was and continues to be the economic heartland of the Core; the seat of U.S. government, business and culture; and the trans-Atlantic trading interface between much of North America and Europe. Six other primary conurbations have also emerged: *Lower Great Lakes* (Chicago-Detroit-Cleveland-Pittsburgh), *Piedmont* (Atlanta-Charlotte-Raleigh/Durham), *Florida* (Jacksonville-Tampa-Orlando-Miami), *Texas* (Houston-Dallas/Fort Worth-Austin-San Antonio), *California*

FIGURE 3-6

© H. J. de Blij, P. O. Muller, and John Wiley & Sons, Inc.

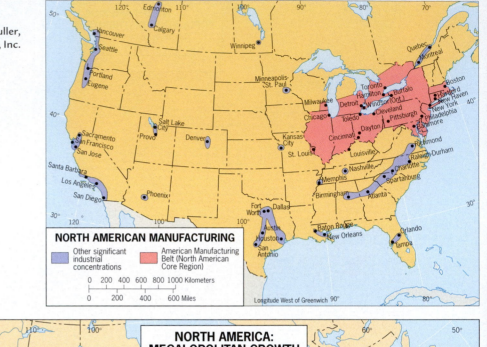

FIGURE 3-7 © H. J. de Blij, P. O. Muller, and John Wiley & Sons, Inc.

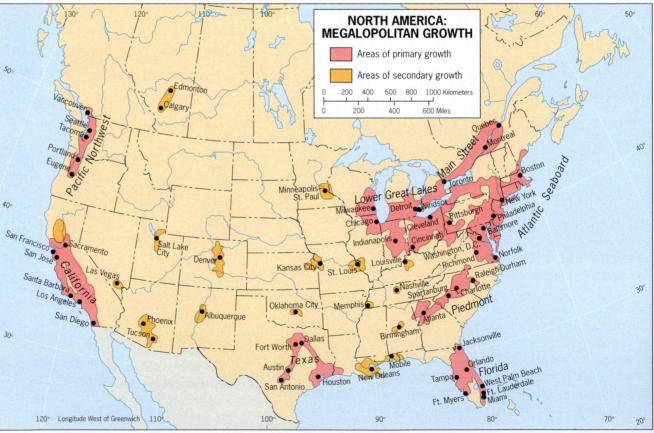

(San Diego-Los Angeles-San Francisco), and the *Pacific Northwest* (Portland-Seattle-Vancouver). Note that the last spills across the border into Canada, which has also spawned its own nationally predominant conurbation—***Main Street*** (Windsor-Toronto-Montreal-Quebec City).

Changing Structure of the American Metropolis

The spatial organization of American cities reflects the growth of transportation corridors (Fig. 3-8 is an idealized American metropolis). A metropolis typically radiated away from the original **9 Central Business District (CBD)**, or 'downtown,' of the old city, toward the early suburbs on the periphery. First horse-drawn trolleys traveled these corridors, later electric streetcars, which in turn were replaced with highways. Sometimes highways and light rail would follow the same corridor, as they still do in parts of

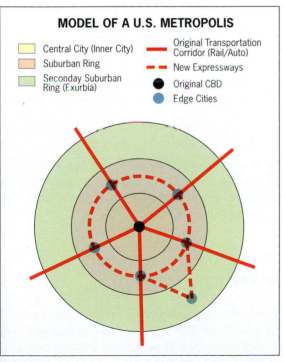

MODEL OF A U.S. METROPOLIS

Central City (Inner City)
Suburban Ring
Seconday Suburban Ring (Exurbia)
Original Transportation Corridor (Rail/Auto)
New Expressways
Original CBD
Edge Cities

FIGURE 3-8 © H. J. de Blij, P. O. Muller, and John Wiley & Sons, Inc.

Chicago. The increasing use of the automobile, especially after World War II, accelerated the development of highways and ring roads in and around major cities, and America increasingly turned from building compact cities to widely dispersed metropolises. By 1970, the new intraurban expressway network had set the stage for suburban rings to transform themselves from residential preserves into complete outer cities. Newly urbanized suburbs started capturing major economic activities, and their rise came at the expense of the central city, which saw its status diminish to that of coequal.

Social Fragmentation. History of the social geography of the American metropolis was marked by the development of a residential mosaic of ever more specialized groups. A century ago the electric streetcar, which introduced affordable transit for all, allowed the immigrant-dominated city population to sort itself into ethnically uniform neighborhoods. When the United States sharply curtailed immigration in the 1920s, industrial managers recruited African-American workers by the thousands for the factories of Manufacturing Belt cities from the rural South. This influx had an immediate impact on the social geography of the industrial city: whites were unwilling to share their living space, and the result was the segregation of these newest migrants into geographically separate, all-black areas. By the 1950s, these mostly **10 inner city** areas became large expanding **11 ghettos**, speeding the departure of many white central-city communities to the suburbs and reinforcing the trend toward a racially divided urban society.

Suburbs and Intraurban Restructuring. As ties to the central city loosened, the suburban ring was transformed into a full-fledged outer city. Its independence was accelerated by the rise of major new suburban nuclei, multipurpose activity nodes that grew up around large shopping centers with prestigious images that attracted industrial parks, office campuses and high-rises, entertainment facilities, and even major league sports stadiums

and arenas. This burgeoning new suburban downtown or **12 Edge City** is nothing less than an automobile-age version of the CBD. The newest spatial elements of the contemporary urban complex are assembled in the model shown in Figure 3-8. The outer city today anchors a multicentered metropolis consisting of the traditional CBD and a constellation of coequal suburban downtowns.

The position of the central city within the new multinodal metropolis of realms is eroding. The CBD increasingly serves the less affluent residents of the inner city and those working downtown. As manufacturing employment declined, many large cities adapted by shifting to service industries, including recreation such as large sports facilities, which were meant to embody downtown commercial revitalization. Residential reinvestment has also occurred in many downtown-area neighborhoods, but beyond the CBD the vast inner city remains the problem-ridden domain of low- and moderate-income people, with most forced to reside in ghettos.

Cultural Geography

In the United States over the past two centuries, the contributions of a wide spectrum of immigrant groups have shaped—and continue to shape—a rich and varied cultural complex. Great numbers of these newcomers were willing to set aside their original cultural baggage in favor of assimilation into the emerging culture of their adopted homeland, which itself was a hybrid nurtured by constant infusions of new influences.

Language and Religion

Although linguistic variations play a far more important role in Canada, more than one-fifth of the U.S. population speaks a primary language other than English. Differences in English usage are also evident at the subnational level in the United States, where regional variations (***dialects***) can still be noted despite the recent trend toward a truly national society.

North America's kaleidoscope of religious faiths, dominated by Christianity, contains important spatial variations (the details are mapped in Fig. 7-2). Many major Protestant denominations are clustered in particular regions, with Southern Baptists localized in the southeastern quadrant of the United States, Lutherans in the Upper Midwest, and Mormons focused on Utah. Roman Catholics are most visibly concentrated in Manufacturing Belt metropolises and the Mexican borderland zone. Jews are most heavily clustered in the suburbs of Megalopolis, Southern California, South Florida, and the Midwest.

Ethnic Patterns and Immigration

Ethnicity (national ancestry) has always played a key role in American cultural geography. Today, whites of European background no longer dominate the increasingly diverse U.S. ethnic tapestry, with ethnics of color and non-European origin comprising a steadily expanding proportion that will surpass 50 percent by 2050. In the late 1990s, Hispanic-Americans surpassed African-Americans to become the nation's largest minority, and today they account for just under 15 percent of the U.S. total. The spatial distribution of the four largest ethnic minorities is mapped in Figure 3-9.

Immigration has long influenced the ethnic complexion of the United States, and during the first half of this decade about 957,000 legal immigrants annually entered the country (plus another estimated 650,000 illegal immigrants). The source areas, however, have changed dramatically over the past half-century. During the 1950s, just over 50 percent came from Europe, 25 percent from Middle and South America, 15 percent from Canada, and only about 5 percent from Asia; today, over 41 percent come from Middle and South America, 30 percent from Asia, and only about 18 percent from Europe and Canada combined.

The Emerging Mosaic Culture

American cultural geography continues to evolve. What is now taking place is a new fragmentation into the emerging nationwide **13** **mosaic culture**, an increasingly heterogeneous complex of separate, uniform tiles that cater to more specialized groups than ever before. No longer based solely on such broad divisions as income, race, and ethnicity, today's residential communities of interest are also forming along the dimensions of age, occupational status, and especially lifestyle.

The Changing Geography of Economic Activity

The economic geography of the United States today is the product of all of its history and its endowment of natural resources, with people and activities overcoming the tyranny of distance to organize a continent-

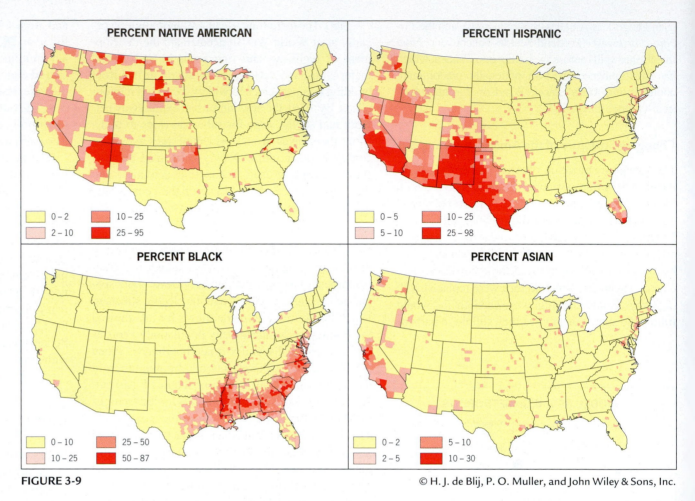

FIGURE 3-9

© H. J. de Blij, P. O. Muller, and John Wiley & Sons, Inc.

wide spatial economy that has taken full advantage of agricultural, industrial, and urban development opportunities. Yet, despite these past achievements, American economic geography today is again in the throes of restructuring as the transition is completed from industrial to postindustrial society.

Major Components of the Spatial Economy

Economic geography is heavily concerned with the locational analysis of **14 productive activities**. Four major sets may be identified:

- *Primary Activity* The extractive sector of the economy in which workers and the environment come into direct contact, especially in mining and agriculture.

- *Secondary Activity* The manufacturing sector, in which raw materials are transformed into finished industrial products.

- *Tertiary Activity* The services sector, including a wide range of activities from retailing to finance to education to routine office-based jobs.

- *Quaternary Activity* Today's dominant sector, involving the collection, processing, and manipulation of information; a subset, sometimes referred to as *Quinary Activity*, is managerial decision making in large organizations.

Historically, each of these activities has successively dominated the U.S. labor force for a period over the past 200 years, with the quaternary sector now dominant. The approximate current breakdown by major sector of employment is agriculture, 2 percent; manufacturing, 16 percent; services, 17 percent; and quaternary, 65 percent (with about 10 percent in the quinary sector). We now review these major productive components of the spatial economy.

Mineral and Energy Resources

The North American realm is blessed with abundant deposits of mineral and energy resources. North America's mineral resources are localized in three zones: the Canadian Shield north of the Great Lakes, mostly in Canada; the Appalachian Highlands; and scattered areas across the mountain ranges of the West. The Shield's most noteworthy minerals are iron ore, nickel, uranium, and copper. Besides vast deposits of coal, the Appalachian region also contains iron ore in central Alabama. The western mountain zone contains significant deposits of coal, copper, lead, zinc, and gold. The United States has been able to take much advantage of these resources in its economic development.

The realm's most vital energy resources are its fossil fuel deposits, petroleum (oil), natural gas, and coal (Fig. 3-10). The leading oil-production areas of the United States are located along and offshore from the Texas-Louisiana Gulf Coast; in the Midcontinent district, extending through western Texas, Oklahoma, and eastern Kansas; and along Alaska's North Slope facing the Arctic Ocean. Canada's leading oilfields lie in a crescent curving southeastward from northern Alberta to southern Manitoba. The distribution of natural gas supplies resembles the geography of oilfields because both fuels are usually found in the floors of ancient shallow seas. The realm's coal-production zones, which rank among the largest in

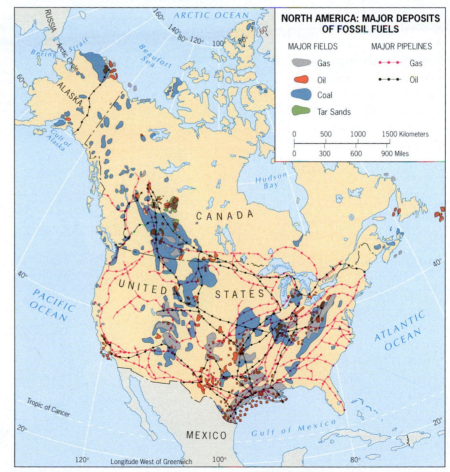

FIGURE 3-10 © H. J. de Blij, P. O. Muller, and John Wiley & Sons, Inc.

the world, are found in Appalachia, the northern U.S. Great Plains/southern Alberta, and southern Illinois/western Kentucky.

Agriculture

Despite the post–1900 emphasis on developing the nonprimary sectors of the spatial economy, agriculture remains an important element in America's economy. It produces significant proportions of the world's output of maize (corn), soy, and wheat, although today only about 1 percent of the U.S. population still resides on farms (2 percent of its work force). Vast expanses of the U.S. landscape are clothed with fields of grain or support great herds of livestock that are sustained by pastures and fodder crops. High-technology mechanization has revolutionized farming; this industrialization has been accompanied by a sharp reduction in the agricultural workforce. Competition from other global agricultural producers such as Brazil and China is also changing the agricultural heartland.

The farm resource regions of U.S. agricultural production are shown in Figure 3-11. The spatial organization of U.S. agriculture developed within the framework of the *von Thünen model*. As in Europe (see pp. 33–34), the early-nineteenth-century, original-scale model of town and hinterland expanded outward (driven by constantly improving transportation technology) from a local *isolated state* to encompass the entire continent by 1900. The U.S. version of the supercity anchoring this macro-Thünian regional system is the northeastern Megalopolis, already the dominant food market and transport focus of the entire country. Although at a local scale this model does not hold—the circular rings of the model are not apparent in Figure 3-11—on a

As the price of a barrel of oil (and a gallon of gasoline) skyrockets, it becomes profitable to derive oil from sources other than liquid reserves. The Canadian province of Alberta contains vast deposits of oil sands in which the oil is mixed with sand, requiring a relatively expensive and complicated process to separate it. The quantity of oil locked in these sands is estimated to rival the remaining reserves of Saudi Arabia, and a huge project is under way around the town of Fort McMurray to recover it. Americans who assume that Canada will sell this oil to the United States should be aware that the Chinese are offering to fund construction of a pipeline to Canada's Pacific coast; the rising cost of a barrel of oil has political as well as economic consequences. (© AP/Wide World Photos)

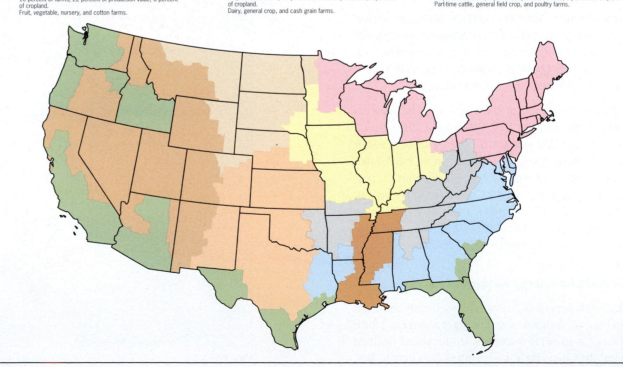

FARM RESOURCE REGIONS

BASIN AND RANGE
Largest share of nonfamily farms, smallest share of U.S. cropland.
4 percent of farms, 4 percent of value of production, 4 percent of cropland.
Cattle, wheat, and sorghum farms.

EASTERN UPLANDS
Most small farms of any region.
15 percent of farms, 5 percent of production value, and 6 percent of cropland.
Part-time cattle, tobacco, and poultry farms.

FRUITFUL RIM
Largest share of large and very large family and nonfamily farms.
10 percent of farms, 22 percent of production value, 8 percent of cropland.
Fruit, vegetable, nursery, and cotton farms.

HEARTLAND
Most farms (22 percent), highest value of production (23 percent), and most cropland (27 percent).
Cash grain and cattle farms.

MISSISSIPPI PORTAL
Higher proportions of both small and larger farms than elsewhere.
5 percent of farms, 4 percent of value, 5 percent of cropland.
Cotton, rice, poultry, and hog farms.

NORTHERN CRESCENT
Most populous region.
15 percent of farms, 15 percent of value of production, 9 percent of cropland.
Dairy, general crop, and cash grain farms.

NORTHERN GREAT PLAINS
5 percent of farms, 6 percent of production value, 17 percent of cropland.
Wheat, cattle, and sheep farms.

PRAIRIE GATEWAY
Second in wheat, oat, barley, rice, and cotton production.
13 percent of farms, 12 percent of production value, 17 percent of cropland.
Cattle, wheat, sorghum, cotton, and rice farms.

SOUTHERN SEABOARD
Mix of small amd larger farms.
11 percent of farms, 9 percent of production value, 6 percent of cropland.
Part-time cattle, general field crop, and poultry farms.

FIGURE 3-11

© H. J. de Blij, P. O. Muller, and John Wiley & Sons, Inc.

continental scale many spatial regularities can be observed in this real-world application. This is a map of *farm resource regions*, a nine-region framework recently developed by the U.S. Department of Agriculture (USDA) for planning purposes. To date, it is the most comprehensive attempt to assemble the components of the changing regional geography of agricultural activity in the United States.

Manufacturing

The geography of North America's industrial production has long been dominated by the Manufacturing Belt in the core of the continent (Fig. 3-6). The emergence of this region was propelled by (1) superior access to the Megalopolis national market that formed its eastern edge and (2) proximity to industrial resources, particularly iron ore and coal for the pivotal steel industry that arose in its western half. We noted earlier that manufacturers had a strong locational affinity for cities, and the internal structure of the Belt became organized around a dozen urban-industrial districts interconnected by a dense transportation network. As these industrial centers expanded, they swiftly achieved **15** **economies of scale**, savings accruing from large-scale production in which the cost of manufacturing a single item was further reduced as factories mechanized assembly lines, specialized their workforces, and purchased raw materials in massive quantities.

Changing Industrial Geography. This production pattern served the nation well throughout the remainder of the industrial age. Today, however, much of the Manufacturing Belt is aging, and the distribution of American industry is changing, giving rise to the term **16** **Rustbelt** for this core area. As transportation costs equalize among U.S. regions, as energy costs now favor the south-central oil- and gas-producing States, as high-technology manufacturing advances steadily reduce the need for lesser skilled labor, and as locational decision making intensifies its attachment to noneconomic factors, industrial management has increasingly demonstrated its willingness to relocate to regions it perceives as more desirable in the South and West as well as foreign locations where conditions may be even more economically favorable.

Nonetheless, parts of the Manufacturing Belt have been resisting this trend. This is particularly true of the industrial Midwest, whose prospects have improved since the early Rustbelt days of the 1970s. Shedding that label in recent years, Midwestern manufacturers are reinventing their operations by rooting out inefficiencies, investing in cutting-edge factories and technologies, and maintaining their competitiveness in the global marketplace.

The Postindustrial Revolution

High-technology, white-collar, office-based activities are the leading growth industries of the **17** **post-industrial economy**. Most are relatively footloose and therefore responsive to such noneconomic

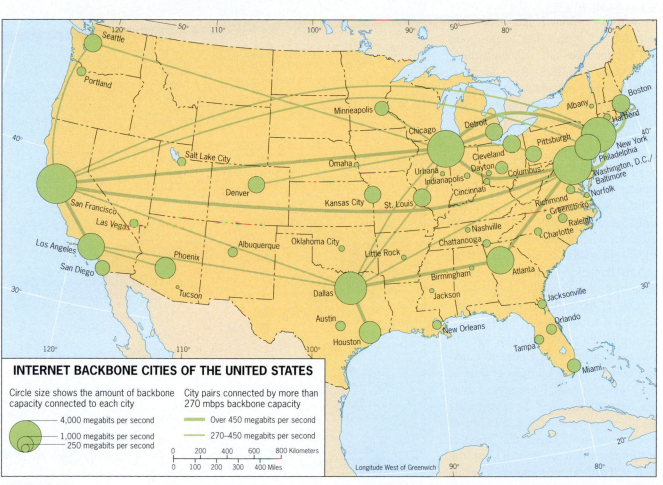

INTERNET BACKBONE CITIES OF THE UNITED STATES

Circle size shows the amount of backbone capacity connected to each city

- 4,000 megabits per second
- 1,000 megabits per second
- 250 megabits per second

City pairs connected by more than 270 mbps backbone capacity

- Over 450 megabits per second
- 270–450 megabits per second

0   200   400   600   800 Kilometers
0 100 200 300 400 Miles

Longitude West of Greenwich

FIGURE 3-12   Adapted with permission from H. J. de Blij, ed., *Oxford Atlas of North America*, p. 41. Oxford University Press, 2005.

locational forces as geographic prestige, local amenities, and proximity to recreational opportunities. Northern California's *Silicon Valley*—the world's leading center for computer research and development and the headquarters of the U.S. microprocessor industry—epitomizes the blend of locational qualities that attract a critical mass of high-tech companies to a given locality. These include: (1) a world-class research university (Stanford); (2) a large pool of highly skilled labor; (3) proximity to a cosmopolitan urban center (San Francisco); (4) abundant venture capital; and (5) a locally based network of global business linkages.

The development of Silicon Valley is so significant to the new post-industrial era that it has been conceptualized as the first technopole. **18 Technopoles** are planned techno-industrial complexes that innovate, promote, and manufacture the hardware and software products of the new informational economy. On the landscape of an edge city, where almost all technopoles are located, their signature is a low-density cluster of ultramodern buildings laid out as a campus. From Silicon Valley, technopoles have not only spread to many parts of North America but are now also becoming a global phenomenon (e.g., around Bengaluru [Bangalore] in India), as we note in several other chapters.

One of the hallmarks of the post-industrial revolution is the Internet, whose recent and continuing growth has been driven by the need to transmit ever greater quantities of data at ever higher speeds. Fiber-optic cables are best able to meet these demands and have given rise to a network of backbones or linkages that connect information-rich centers across the planet. The leading component of this still-emerging global infrastructure is

its U.S. network, which handles up to 75 percent of the world's Internet traffic and is dominated by a set of major backbone cities and their interconnections (Fig. 3-12). Note that the hierarchy of backbone capacity does not match the urban population hierarchy. The primate city, New York, ranks in the second tier of backbone cities, while the operations of Silicon Valley and the federal government, respectively, catapult metropolitan San Francisco and Washington, D.C., to the topmost level.

CANADA

Like the United States, Canada is a federal state, but it is organized differently. Canada is divided into ten provinces and three territories (Fig. 3-13).

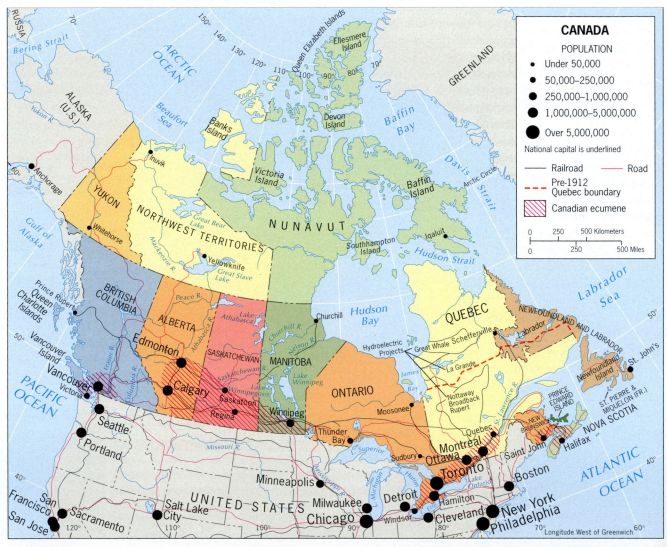

FIGURE 3-13   © H. J. de Blij, P. O. Muller, and John Wiley & Sons, Inc.

The ten provinces—where almost all Canadians live—range in territorial size from tiny, Delaware-sized Prince Edward Island to sprawling Quebec, more than twice the area of Texas. Beginning in the east, the four Atlantic Provinces are Nova Scotia, New Brunswick, Prince Edward Island, and New-foundland and Labrador. To their west lie Quebec and Ontario, Canada's two biggest provinces. Most of western Canada consists of three Prairie Provinces—Manitoba, Saskatchewan, and Alberta. In the far west, facing the Pacific, lies British Columbia.

The three territories—Yukon, the Northwest Ter-ritories, and Nunavut—together occupy a massive area half the size of Australia but are inhabited by only about 100,000 people. Nunavut is the newest addition to Canada's political map and deserves spe-cial mention. Created in 1999, this new territory is the outcome of a major aboriginal land claim agree-ment between the Inuit people (formerly called Eski-mos) and the federal government, and encompasses all of Canada's eastern Arctic.

In population size, Ontario (12.8 million) and Quebec (7.8 million) are again the leaders; British Columbia ranks third with 4.4 million; next come the three Prairie Provinces with a combined total of 5.5 million; the Atlantic Provinces are the smallest, together containing 2.3 million. Canada's total pop-ulation of 32.9 million is only slightly larger than one-tenth the size of the U.S. population.

Though comparatively small, Canada's popula-tion is divided by culture and tradition, and this divi-sion has a pronounced regional expression. Just under 60 percent of Canada's citizens speak English as their mother tongue, 23 percent speak French, and the remaining 17 percent other languages. The spatial clustering of more than 85 percent of the country's *Francophones* (French-speakers) in Quebec accen-tuates this Canadian social division along ethnic and linguistic lines (Fig. 3-14). Quebec is the historic and emotional focus of French (or Québécois) culture in Canada, and over the past few decades a strong na-tionalist movement emerged in this province to de-mand outright separation from the rest of the country. The independence movement crested in the mid-1990s, but this ethnolinguistic division continues to

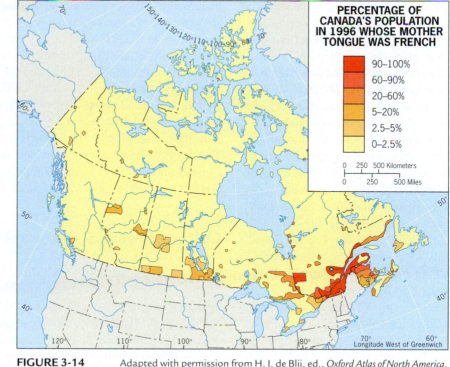

PERCENTAGE OF CANADA'S POPULATION IN 1996 WHOSE MOTHER TONGUE WAS FRENCH

- 90–100%
- 60–90%
- 20–60%
- 5–20%
- 2.5–5%
- 0–2.5%

FIGURE 3-14   Adapted with permission from H. J. de Blij, ed., *Oxford Atlas of North America*, p. 44. Oxford University Press, 2005.

be the cornerstone of Canada's cultural pluralism and may again pose a threat to national unity in the future.

Population in Time and Space

The map showing the distribution of Canada's pop-ulation (Fig. 3-5) reveals that only about one-eighth of this enormous country can be classified as its **19** ecumene—the inhabitable zone of permanent settlement. As Figure 3-13 indicates, the Canadian ecumene is dominated by a discontinuous strip of population clusters that lines the U.S. border. We can identify four such clusters on the map, the largest by far being Main Street (home to more than six out of every ten Canadians). As noted earlier (Fig. 3-7), *Main Street* is the megalopolis that stretches across southernmost Quebec and Ontario, from Quebec City on the lower St. Lawrence River southwest through Montreal and Toronto to Windsor on the De-

troit River. The three lesser clusters are: (1) the Saint John-Halifax crescent in central New Brunswick and Nova Scotia; (2) the prairies of southern Alberta, Saskatchewan, and Manitoba; and (3) the southwest-ern corner of British Columbia, focused on Canada's third-largest metropolis, Vancouver.

Pre-Twentieth-Century Canada

Compared to European expansion in the United States, penetration of the Canadian interior lagged well behind. In terms of political geography, Canada did not unify before the last third of the nineteenth century—and then mainly because of fears the United States was about to expand in a northerly direction.

The evolution of modern Canada is deeply rooted in a longstanding bicultural division. Its origin lies in the fact that it was the French, not the British, who were the first European colonizers of present-day

Canada, with *New France* growing to encompass the St. Lawrence Basin, the Great Lakes region, and the Mississippi Valley. A series of wars between the English and French subsequently ended in France's defeat and the cession of New France to Britain in 1763. By the time London took control of its new possession, the French had made considerable progress in their North American domain. The British, anxious to avoid a war of suppression and preoccupied with problems in their other American colonies, gave former French Quebec—the territory extending from the Great Lakes to the mouth of the St. Lawrence—the right to retain its legal and land-tenure systems as well as freedom of religion.

After the American War for Independence, London was left with a region it called British North America but whose cultural imprint still was decidedly French. The Revolutionary War drove many thousands of English refugees northward, and soon difficulties arose between them and the French. In 1791, heeding appeals by these new settlers, the British Parliament divided Quebec into two provinces: Upper Canada, the region upstream from Montreal centered on the north shore of Lake Ontario, and Lower Canada, the valley of the St. Lawrence. Upper and Lower Canada became, respectively, the provinces of (English-speaking) Ontario and (Francophone) Quebec (Fig. 3-13). This earliest cultural division did not work well and in 1867 finally led to the British North America Act, which established the Canadian federation (consisting initially of Upper and Lower Canada, New Brunswick, and Nova Scotia, later to be joined by the other provinces and territories). Under this Act, Ontario and Quebec were again separated, but this time Quebec was given important guarantees: the French civil code was left unchanged, and the French language was protected in Parliament and in the courts.

Canada Since 1900

By 1900, the Canadian federation was making major strides toward regional development and the spatial integration of a continentwide economy. The

transcontinental Canadian Pacific Railway had been completed to Vancouver, along the way spawning the settlement of the fertile Prairie Provinces whose wheat-raising economy expanded steadily as immigrants arrived from the east and abroad. Industrialization also began to stir, and by 1920, Canadian manufacturing had surpassed agriculture as the leading source of national income. As noted earlier (Fig. 3-6), the dominant zone of industrial activity is the Toronto-Hamilton-Windsor corridor of southern Ontario, the crucible of Canada's Industrial Revolution that took hold during World War I (1914–1918).

Cultural/Political Geography

The historic cleavage between Canada's French- and English-speakers has resurfaced over the past four decades to dominate the country's cultural and political geography. By the time the Canadian federation observed its centennial in 1967, it had become evident that Quebecers regarded themselves as second-class citizens; they believed that bilingualism meant that French-speakers had to learn English but not vice versa; and they perceived that Quebec was not getting its fair share of the country's wealth. Since the 1960s, the intensity of ethnic feelings in Quebec has risen in surges despite the federal government's efforts to satisfy the province's demands. During the 1970s, while a separatist political party came to power in Quebec, a new federal constitution was drawn up in Ottawa. In 1980, Quebec's voters solidly rejected independence when given that choice in a referendum. But the new constitution did *not* satisfy the Quebecers, and throughout the 1980s and early 1990s the Ottawa government struggled unsuccessfully to devise a plan, acceptable to all the provinces, that would keep Quebec in the Canadian federation.

Quebec's 1995 Referendum and Its Aftermath

By 1995, with Canada's interest in constitutional reform exhausted, a second referendum on Quebec's sovereignty could no longer be put off. With the

reenergized separatist party again leading the way, the Francophone-dominated electorate very nearly approved independence (which attracted an astonishing 49.4 percent of the vote), an outcome now regarded by Canadians as their country's near-death experience. Subsequently, despite calls by many separatists for a follow-up vote that might turn narrow defeat into victory, plans for a third referendum were shelved because opinion polls in Quebec have shown erosion in public support for secession. Several reasons lie behind that shift in attitude: (1) the implementation of provincial laws that firmly established the use of French and the primacy of Québécois culture; (2) the increased bilinguality of Quebec's Anglophones (English-speakers), which has calmed Francophone fears concerning assimilation into Canada's English-speaking mainstream; (3) the arrival of a new wave of immigrants (some Francophone, some Anglophone) from Asia, Africa, Eastern Europe, and the Caribbean, who have accelerated the weakening of language barriers by settling in both French-speaking and English-speaking neighborhoods; and (4) the landslide defeat of the ruling separatist Parti Québécois in the provincial election of 2003.

Although the issue of Quebec's separation from the Canadian federation has receded in this decade, it has not been buried and could resurface. Anticipating that possibility, the federal government has armed itself by adopting the Clarity Act, which mandates that any future referendum on independence must be based on a clearly worded question that passes with the support of a clear majority (i.e., well over 50 percent) before Ottawa would open negotiations on separation.

The most insightful analyses of public opinion in the province today suggest that what Quebecers really want is a better relationship with the rest of Canada, but not outright independence. There is also considerable evidence that the Québécois regard themselves as a *distinct nation* and wish to be treated accordingly. That is precisely the recognition they received under a new parliamentary resolution passed at the end of 2006—but with the important

An unmistakably French cultural landscape in North America: the Rue Saint-Louis in Quebec City, the capital of Quebec Province. There is not an English sign in sight in this center of French/Québécois culture. The green spires in the background belong to Le Chateau Frontenac, a grand hotel built on the site of the old Fort St. Louis. French-speaking, Roman Catholic Quebec lies at the core of French Canada.   (© SUPERSTOCK)

breakthroughs have been achieved with the creation of Nunavut and a treaty for limited tribal self-government in northern British Columbia. Nonetheless, this fastest-growing segment of the Canadian population is plagued by serious social and economic problems, and after seven years of self-rule Nunavut's development has been minimal. Elsewhere, confrontations over aboriginal land claims in both populated and frontier areas continue to proliferate, but governments and private landowners are notoriously slow to resolve such issues.

A foremost concern among indigenous peoples is that the federal government protect their aboriginal rights against the provinces. This is especially true for the First Nations of Quebec's northern frontier, the Cree, whose historic domain covers more than half of the province of Quebec as it appears on current maps. Administration of the Cree was assigned to Quebec's government in 1912, a responsibility that a move toward independence might well invalidate. In any case, the Cree would also likely be empowered to seek independence. This would leave the French-speaking remnant of Quebec with only about 45 percent of the province's present territory. As Figure 3-13 shows, the territory of the Cree is not an unproductive wilderness: it contains vital facilities of the James Bay Hydroelectric Project, a massive scheme of dikes, dams, and artificial lakes that has transformed much of northwestern Quebec and generates electrical power for a huge market within and outside the province. In 2002, after the federal courts supported its attempts to block construction of the Project, the Cree negotiated a groundbreaking treaty with the Quebec government whereby this First Nation dropped its opposition in return for a portion of the income earned from electricity sales as well as the granting of power to control its own economic and community development.

added caveat . . . *within a united Canada*. This was followed by a resounding defeat of the Parti Québécois in the provincial election of March 2007, and the emergence of a new political movement whose guiding vision replaced independence with autonomy—a Quebec that would remain part of Canada but with as-yet-undefined increased powers. Even if the supporters of separatism never again come close to attaining their goal, they can nonetheless claim several successes: (1) the achievement of Francophone cultural and economic dominance in Quebec; (2) the strengthening position of the French language throughout the Canadian federation; and

(3) an upsurge of federal spending in the province that now surpasses the collection of Ottawa-bound tax revenues, a monetary inflow that is likely to increase as the country's energy windfall continues to propel rising national income.

The Rise of the First Nations

On the wider Canadian scene, the culture-based Quebec crisis has stirred the ethnic feelings of the country's 1.4 million native (First Nations and Inuit) peoples. Their assertions have also received a sympathetic hearing in Ottawa, and in recent years

Persistent Regionalism

Finally, the events of the past three decades have profoundly impacted Canada's national political landscape, and not only in Quebec. Regionalism has also intensified in the increasingly affluent west, whose leaders opposed federal concessions to Quebec and insisted that the equal treatment of all ten provinces is a basic principle that precludes the designation of special status for anyone. Recent federal elections have clearly revealed the emerging fault lines that surround Quebec and set the western provinces (British Columbia, Alberta, Saskatchewan, and Manitoba) off from the rest of the country. Even the remaining Ontario-led center and the eastern bloc of Atlantic Provinces vote divergently. Thus, with its national unity under pressure, Canada today confronts the coalescing forces of *devolution* that threaten to transform the new fault lines into permanent fractures.

Economic Geography

As in the United States, a diversified resource base has supported the growth of Canada's spatial economy. We noted earlier that the Canadian Shield is endowed with major mineral deposits and that oil and natural gas are extracted in sizeable quantities in Alberta. Canada has long been a leading agricultural producer and exporter, especially of wheat and other grains from its breadbasket in the Prairie Provinces. Postindustrialization has caused substantial employment decline in the manufacturing sector, with Southern Ontario's industrial heartland being the most adversely affected. On the other hand, Canada's robust *tertiary and quaternary sectors* (which today employ more than 70 percent of the total workforce) are creating a host of new economic opportunities.

Canada-U.S. Trade

Canada's economic future is also going to be strongly affected by the continuing development of its trading relationships. These include the 1989 United States-Canada Free Trade Agreement (today

more than four-fifths of Canada's exports go to the United States, from which it also derives about two-thirds of its imports) and the 1994 North American Free Trade Agreement (NAFTA), which added Mexico to the trading partnership. Because these free-trade agreements increasingly impact the Canadian spatial economy, they are likely to weaken domestic east-west linkages and to strengthen international north-south ties. Since many local cross-border linkages built on geographical and historical commonalities are already well developed, they can be expected to intensify in the future: the Atlantic Provinces with neighboring New England; Quebec with New York State; Ontario with Michigan; the Prairie Provinces with the Upper Midwest; and British Columbia with the (U.S.) Pacific Northwest. Such functional reorientations, of course, constitute yet another set of powerful devolutionary forces confronting the Ottawa government.

Emerging Regional States?

The rising importance of this framework of transnational regions straddling the U.S.-Canadian border recalls Kenichi Ohmae's regional state concept introduced in Chapter 1. A [20] **regional state** is a 'natural economic zone' that defies old borders and is shaped by the global economy of which it is a part; its leaders deal directly with foreign partners and negotiate the best terms they can with the national governments under which they operate. Writing about Canada, Ohmae identified a Pacific Northwest (the Seattle-Vancouver axis) and a Great Lakes regional state (the intertwined Ontario-Michigan industrial complex), and warned that the manner in which Ottawa's leaders dealt with these new economic entities would be critical to the survival of the Canadian state. Indeed, these growing transnational interactions are increasingly evident in the regional configuration of North America, to which we now turn.

WHAT'S DRIVING GEOGRAPHIC CHANGE IN THE REALM

● Despite their mutual membership (with Mexico) in the North American Free Trade Agreement, **relations** between the United States and Canada are rebuilding from a low ebb as a result of disagreements over issues ranging from trade (especially lumber) to security (Canada did not support the Iraq invasion).

● The United States has a much smaller percentage of foreign-born residents than Canada, but **immigration** is a far more contentious issue in the United States than it is in now immigrant-friendly Canada.

● **Regionalism**, political as well as economic, is troubling Canada. The separatist movement in French-speaking Quebec continues to simmer, and wealth disparities among Canada's provinces are straining the federal fabric in other ways.

● **Global warming** will likely create an ice-free passage around the northern perimeter of this realm, and a possible geopolitical struggle may ensue in the Arctic Ocean (see Chapter 2), although new trade routes may become viable.

● A variety of **natural events** that often result in disasters will continue to affect this realm, especially the United States. The frequency and/or severity of hurricanes, tornadoes, droughts, floods, and wildfires are likely related to global environmental change and may increase in the future.

Regions of the Realm

This realm is a bit different from many others inasmuch as it encompasses only two large countries. Yet, if we were to erase the political boundary between the United States and Canada and apply our regional framework to this realm, we would see a realm divided a different way. Here we present our interpretation of the regions of North America (Fig. 3-15).

THE NORTH AMERICAN CORE

The Core Region (Fig. 3-6)—synonymous with the American Manufacturing Belt—was introduced earlier. This region was key to America's industrialization and propelled it to global economic dominance. Much of the manufacturing of this region has moved elsewhere as the U.S. spatial economy reconfigures. But make no mistake: this is still the geographic heart of North America. Here, one finds over one third of the population and the largest cities—as well as the leading business complexes, cultural centers, and busiest transportation facilities.

Postindustrial development has also spawned new growth centers in the Core, especially in the northeastern Megalopolis. The Boston area, richly endowed with research facilities, continues to attract innovative high-tech businesses to the Route 128 freeway corridor that girdles the central city; and Greater New York is a world city and remains the national leader in finance, advertising, and corporate decision-making activity. The Core Region metropolis that has gained the most from postindustrialization is Washington, D.C. As quaternary activities have blossomed, and as the U.S. federal government extended its ties to the private sector, Washington and its surrounding outer city of Maryland and Virginia suburbs (interconnected by the 105-kilometer [66-mi] Beltway encircling the capital city) have

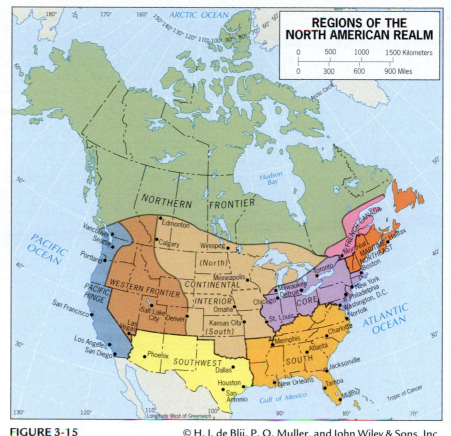

FIGURE 3-15 © H. J. de Blij, P. O. Muller, and John Wiley & Sons, Inc.

amassed an enormous complex of office, research, high-tech, and consulting firms. The most important development has occurred along the expressway corridor between Virginia's Tyson's Corner on the Beltway and Dulles International Airport. Since 1995, many telecommunications and Internet companies have been attracted to this part of northern Virginia, a leading fiber-optic-cable hub thanks to its local cluster of federal intelligence agencies (see Fig. 3-12). This major, still-developing technopole is already being called the technological capital of the

United States, because more technical workers are now employed here than in Silicon Valley.

THE MARITIME NORTHEAST

The Maritime Northeast consists of upper New England (Vermont, New Hampshire, Maine) and the neighboring Atlantic Provinces of easternmost Canada (Fig. 3-15). New England, one of the realm's culture hearths, has retained a strong regional identity for almost 400 years. Even with the urbanized southern half (Massachusetts, Connecticut, Rhode Island) lying in the Core, the six New England States still share many characteristics. The Maritime Northeast region also extends northeastward to encompass most of Canada's four Atlantic Provinces of New Brunswick, Nova Scotia, Prince Edward Island, and Newfoundland and Labrador.

A long association based on economic and cultural similarities has tied northern New England to Atlantic Canada. Both have a strong maritime orientation, are rural in character, possess difficult environments with limited land resources and relatively poor soils, and were historically bypassed in favor of more fertile inland areas. Economic growth in upper New England has always lagged behind the rest of the realm, with development centered on fishing (once) rich offshore waters, forestry, and farming in the few fertile lowlands available. Recreation and tourism have boosted the regional economy in recent times, with scenic coasts and mountains attracting millions from the neighboring Core Region.

The Atlantic Provinces have also experienced hard economic times since the 1980s. Most adversely affected was the groundfish industry as offshore stocks of flounder, haddock, and cod became severely depleted through overfishing. New opportunities, never easy to come by here, are most promising in remote Newfoundland Island, where major offshore oil deposits (see Fig. 3-10) were discovered in the 1990s. The construction and opening of seabed drilling platforms and coastal support facilities soon followed, and the long-term expectation is that oil can transform the economy of Newfoundland and Labrador along the lines of Norway's over the past three decades.

FRENCH CANADA

Francophone Canada constitutes the effectively settled, southern portion of Quebec, which straddles the central and lower St. Lawrence Valley from where that river crosses the Ontario-Quebec border to its mouth in the Gulf of St. Lawrence. Also included is a sizeable concentration of French speakers, known as the Acadians, who reside just beyond Quebec's provincial boundary in neighboring New Brunswick and some in New Hampshire (Fig. 3-15). The Old World charm of Quebec's cities is matched by an equally unique rural settlement landscape introduced by the French: narrow rectangular farms, known as long lots, are laid out in sequence perpendicular to the St. Lawrence, other rivers, and the roads that parallel them, thereby allowing each farm access to an adjacent transportation corridor.

The economy of French Canada, however, is no longer rural and exhibits urbanization rates similar to those of the rest of the country. Industrialization is widespread, supported by cheap hydroelectric energy generated at huge dams in northern Quebec, but relatively little of the region's manufacturing could be classified as high-tech. Tertiary and postindustrial commercial activities are concentrated around Montreal, and tourism and recreation are also important to the regional economy. But the long-term health of these sectors is tied to the resolution of Quebec's political status within Canada.

THE CONTINENTAL INTERIOR

The Continental Interior extends across the center of both the conterminous United States and the southern tier of Canada (Fig. 3-15). Throughout the Continental Interior, economic activity is oriented toward farming. Its leading metropolises are major processing and marketing points for pork and beef packing, flour milling, and soybean, sunflower, and canola oil production, all of which are increasingly exported. With few exceptions—most notably in the region's northeastern corner where the barren but mineral-rich Canadian Shield meets the Great Lakes—agriculture is the predominant feature of the rural landscape. Because of its proximity to the national food market on the northeastern seaboard and its situation in Humid America (Fig. 3-11), mixed crop-and-livestock and corn farming dominate the eastern portion of the Continental Interior. In the western portion grain farming, especially wheat and soybeans, dominate the fertile but semiarid environment of the central and western Great Plains on the dry side of the 100th meridian (line of longitude). The latter area also contains Canada's agricultural heartland north of the 49th-parallel (line of latitude) border. Although these fertile Prairie Provinces are less subject to drought than the U.S. high plains to the south, they are situated at a more northerly latitude and thus have a shorter growing season.

The Ethanol Factor

Anyone who has followed the headlines of the past three years is aware of the rising importance of ethanol in the farmlands of the central United States. The United States is now the world's foremost producer of this gasoline substitute, and as corn has become the leading source of ethanol and other biofuels, the agricultural landscape of the Continental Interior is being increasingly impacted. Because the more than 130 modest-sized ethanol refineries that have been built are widely dispersed across the region, the production of these alternative fuels is not dominated by large, centralized, agribusiness corporations. Thus the perception is taking hold that local control of this booming industry will prevail and that ethanol can become the cornerstone of a new prosperity that will revitalize much of rural America.

How realistic is this rosy scenario? Despite substantial government subsidies for corn (maize) farmers, currently strong retail demand, and handsome profits, biofuel production is a decidedly risky longer-term proposition. The projected annual output of ethanol in 2008 would only be sufficient to replace about 4 percent of U.S. gasoline consumption. Even if the *entire* corn crop could be converted to ethanol, it would only account for 12 percent of total gasoline consumption (and thus would minimally enhance national energy independence). Far more seriously, so much of the corn crop (about 50 percent

As of mid-2008, 139 ethanol refineries were operating in the United States, with another 62 under construction or expansion. This plant is located in Liddonia, Missouri (population: approximately 600), at the southern edge of the largest cluster of refineries that is focused on Iowa, southern Minnesota, and the eastern Dakotas. Ethanol is basically alcohol distilled from corn mash and was first utilized as an additive to gasoline to reduce vehicle emissions. But with the recent surge in oil prices, it is increasingly being used to run cars and trucks as a fuel blend that contains up to 85 percent ethanol/15 percent gasoline. Although many potential problems are involved, they are not yet constraining the development of this new energy industry. Ethanol refining is particularly attractive to small towns on railroad lines in less accessible farming areas, where converting local corn to more valuable biofuel is also a more cost-effective way of shipping their product to distant markets.   (© Peter Newcomb/Redux Pictures)

in 2007) is already being used to produce biofuels that the United States and other countries are experiencing food shortages in this commodity—signified by rising corn prices since 2005, especially for cattle feed. Moreover, the economics of these biofuels are proving to be problematic, because corn is a much less efficient source of ethanol than the sugarcane that has been so successful in reducing Brazil's gasoline consumption. In the United States, not only is more energy required to produce a gallon of ethanol versus a gallon of gasoline, but that gallon of ethanol yields one-third less mileage than the same quantity of gasoline. Finally, the benefits of ethanol-based development may be overstated: in many of the small towns in which biorefineries have operated for the past few years, employment gains have been modest and have not reversed the outflow of population.

Depopulation and Regional Change

The dream of ethanol-based salvation notwithstanding, individual agricultural opportunities are declining in the Continental Interior. This is particularly evident in the less intensively farmed Great Plains in the region's western half, where the exodus of younger people is accelerating. Left behind is an ever more elderly and poorer population, thinly dispersed across a constellation of stressed rural communities, struggling to survive. The Great Plains depopulation trend, in fact, is rapidly plummeting toward demographic collapse: fewer than 12 million residents remained on the Plains in 2005, a population total equal to only about half that of the New York metropolitan area. As Great Plains society unravels, a policy debate intensifies as to whether or not the best solution is planned abandonment and the return of at least a portion of these grasslands to their natural state (the Buffalo Commons).

Despite the key role that agriculture plays in the regional economy, the nonfarming sectors of the Continental Interior continue to develop and diversify, especially in and around major metropolitan centers. Minnesota has been the most successful in promoting such growth, capitalizing on the reputations of that State's stable economy, highly educated workforce, track record as a product innovator, and recreational opportunities. And in the urban areas of the eastern Great Plains, led by Omaha, Nebraska, telemarketing has become a major pursuit.

THE SOUTH

The American South occupies the realm's southeastern corner, extending from the lush Bluegrass Basin of northern Kentucky to the swampy bayous of Louisiana's Gulf Coast, and from the knobby hills of West Virginia to the flat sandy islands off southernmost Florida (Fig. 3-15). Of the realm's nine regions, none has undergone more overall change during the past half-century. For more than 100 years after the ruinous Civil War, the South languished in economic stagnation, but the 1970s finally witnessed a reassessment of the nation's perception of the region that launched a still-ongoing wave of growth and change.

Driven by the forces that created the Sunbelt phenomenon, people and activities streamed into the urban South. Cities such as Atlanta, Charlotte, and Tampa became booming metropolises practically overnight, and conurbations swiftly formed in such places as southeastern and central Florida, the Carolina Piedmont, and the Texas-Louisiana Gulf Coast. Yet for all the growth that has taken place, the South remains a region beset by many economic problems because the geography of its development has been decidedly uneven. Although several metropolitan and a few favored farming areas have benefited, many others containing sizeable populations have not, and the juxtaposition of progress and backwardness is encountered all across the Southern landscape. Not surprisingly, the gap between rich and poor is wider here than in any other U.S. region as the aftermath of Hurricane Katrina made abundantly clear to the world.

The transformation of the South is also changing its demographic complexion. Racially and economically, the region's population is being reshaped by the immigration of affluent whites and blacks from the North. Ethnically, the steady influx of Hispanics and Asians is intensifying the region's cultural heterogeneity. The growth of the South's Hispanic population is especially strong: Mexican and Central American migrants continue to stream in to fill agricultural and lower-skilled industrial jobs that others do not take.

Three decades of rapid change have filled the tiles of the Southern population mosaic with an ever more diverse array of social and lifestyle groups, a clear signal that this once outcast region is completing its convergence with the rest of the country. Indeed, this 30-year transition period has ended with the South not only fully absorbed into the national economy and society, but also becoming a player on the international stage. During the 1990s, European and Japanese manufacturers were active investors in the region, but industrial jobs are proving hard to hang onto as factories now relocate outside the United States to take advantage of cheaper foreign labor. In terms of relative location, the most promising postindustrial opportunities in this era of NAFTA lie to the south in Middle and South America, with potential gateway cities such as Miami, Atlanta, and New Orleans competing to develop the new trade connections.

THE SOUTHWEST

Not long ago, most geographers did not identify a distinct southwestern region, but this booming area—constituted by the States of Texas, New Mexico, and Arizona—is now solidly on the regional map of North America (Fig. 3-15). The Southwest is unique in the United States because it is a bicultural regional complex where immigrating Anglo-Americans join a rapidly expanding Mexican-American population, which traces its roots to the Spanish colonial era. With its large Native American population added in, the Southwest can be regarded as a *tricultural* region.

The Southwest has developed through the ability to run air conditioners in places that would otherwise be unbearable due to the heat. Engineering has also made it possible to bring water to an otherwise arid environment, an issue that threatens the future of the region because there is not enough water available in this rapidly growing region. But the postindustrial economy is doing well here and will continue to at-

tract migrants from the declining Core. In Texas, for example, a booming growth triangle cornered by Dallas-Fort Worth, Houston, and San Antonio specializes in health-care services, high-tech manufacturing, and computer research and fabrication.

THE WESTERN FRONTIER

To the north of the western half of the Southwest lies the burgeoning Western Frontier region. In its east-west extent, this mountain-studded plateau region stretches between (and includes) the Rocky Mountains and the Sierra Nevada-Cascades chain; its longer north-south axis reaches from northern Arizona's Grand Canyon country to the edge of the subarctic snowforest west of the Canadian Rockies (Fig. 3-15). The region's heartland encompasses Utah, Nevada, Idaho, western Wyoming, and western Colorado, and it is here that the old intermountain West is most actively reinventing itself. Until recently, such an opportunity would have been unthinkable because of this region's remoteness, dryness, and sparse population. But the Western Frontier's geographic prospects are changing significantly as the postindustrial national economy matures, advances in communications and transportation erode interregional differences in the costs of doing business, and amenities become the leading locational preference of employers.

The label *Western Frontier* further emphasizes that the transformation under way is being driven by the influx of people and activities from outside the region. Indeed, the populations of the four States of its heartland all rank among the fastest-growing since 1990, including two million disenchanted Californians who have migrated into this region since 1995. In the economic-geographic sphere, the region's identity is being reforged as traditional resource-based industries (mining, timber, livestock grazing) decline, while immigrating and new home-grown companies contribute to a postindustrial economy with thousands of high-tech manufacturing and specialized services jobs in rapidly growing urban areas such as Las Vegas, Salt Lake City, and Denver. Workers are attracted to these new jobs and to the nearby high-amenity and recreation areas.

Lying near the Sangre de Cristo Mountains, Santa Fe, New Mexico is a tricultural region. Native Americans have been here for thousands of years, and continue to be a significant demographic and cultural presence. The Spanish arrived in the mid-sixteenth century; some of the earliest Roman Catholic missions in the United States were founded at that time. This church is St. Francis Cathedral and was started in the mid-eighteenth century. Many of the area's Hispanics trace their ancestry directly to original Spanish settlers. Americans of European background have been in New Mexico since their westward migration in the nineteenth century. Today Santa Fe attracts many artists and tourists who come to enjoy the various cultural amenities and pleasant mountain environment. (© A. WinklerPrins)

Although most of the Western Frontier's development is confined to urban areas, the effects of this rapid growth are penetrating deeply into the region's vast rural zones. Causing some of the problems of the region are the spillover effects of suburbanization—especially environmental blight—on adjacent areas. But many other problems result from a social collision of lifestyles as recreation-seeking city-dwellers with different ideas about nature and resource use invade ever deeper into a countryside whose longtime, nonaffluent residents increasingly struggle to preserve their way of life.

THE NORTHERN FRONTIER

The northern half of the realm, lying poleward of roughly 52° North latitude, constitutes the Northern Frontier region (Fig. 3-15). This is by far North America's biggest regional subdivision, covering almost 90 percent of Canada and all of the U.S. State of Alaska. Indeed, if this huge territory were a separate country, it would rank as the world's fifth largest in size.

The latitudinal extent of this region leaves no doubt as to why it is designated northern, and most of its people and activities are understandably concentrated in its southern half. Unquestionably, this region is also a classic *frontier* that represents a thrust into undeveloped country, one marked by the long-term advance of a line across Canada's northern periphery that absorbs the areas it crosses into the national spatial economy. Boom and bust cycles over the past century have controlled the rate of movement of this line, and the still sparsely populated Northern Frontier continues to be mostly inhabited by Native Americans and First Nations, with only a limited intrusion of Anglo-European settlement based on the extraction of newly discovered resources.

As noted earlier, the Canadian Shield, which covers the eastern half of the Northern Frontier (Fig. 3-2), is one of the world's richest storehouses of mineral resources. Most developing areas are focused on mining complexes along the Shield's southern margins that extract such metallic ores as nickel, copper, and zinc. The Northern Frontier is also endowed with an abundance of other resources (fossil fuels, hydroelectric power opportunities, timber, fisheries), which are dominated by substantial oil and gas reserves lying east of the Canadian Rockies between the Yukon and the conterminous U.S. border (Fig. 3-10; also see photo, p. 118).

The productive activities of the Northern Frontier are linked together within a far-flung network of mines, oil- and gasfields, pulp mills, and hydropower stations that have spawned hundreds of small settlements and thousands of miles of transport and communications facilities. Though dominant, this commercial sector is part of a wider, dual regional economy that also includes a native sector. This native economy, which adheres to the traditional values of the indigenous peoples (First Nations), has long been tied to the commercial economy as a source of wages. It has also been heavily land-based, and in recent years First Nations throughout the region have become more assertive and filed legal actions to recover lands that Europeans took in the past without negotiating treaties. Both the courts and the federal government have been sympathetic, and the creation of Nunavut in 1999 (Fig. 3-13) has been the most dramatic breakthrough to date. It now seems clear that First Nations will increasingly be consulted on future resource development in their traditional homelands. However, because vast tracts of northern lands are likely to be involved, this represents a potential threat to national unity should related issues of self-government arise.

The Northern Frontier also contains Alaska, whose geographic opportunities and challenges differ somewhat from those of the Canadian North. Besides being the largest U.S. State, Alaska earns a sizeable income from oil production on its northeastern Arctic slope (Fig. 3-10). Fossil fuels are likely to continue driving Alaska's development, even though supplies are dwindling in the key Prudhoe Bay oilfield. Besides the existence of significant additional oil reserves on the North Slope—most notably in the controversial Arctic National Wildlife Refuge—enormous supplies of natural gas are potentially available for development.

THE PACIFIC HINGE

The Pacific coastlands of the contiguous United States and southwesternmost Canada, which comprise the realm's ninth region (Fig. 3-15), have been a powerful lure to migrants since the Oregon Trail was pioneered more than 160 years ago. Unlike the remainder of western North America south of 50° North latitude, the strip of land between the Sierra Nevada-Cascade mountain wall and the sea receives adequate moisture. It also possesses a far more hospitable environment, with generally delightful weather south of San Francisco, highly productive farmlands in California's Central Valley, and such scenic glories as the spectacular waters surrounding San Francisco, Seattle, and Vancouver. Most of the major development here took place during the post–World War II era, accommodating enormous population and economic growth along this outermost edge of the conterminous United States.

Today, in terms of its economic geography, the west coast no longer represents an end but rather a beginning—a gateway to an abundance of growing opportunities that in recent times have blossomed on many of the distant shores encircling the Pacific Basin. That is why we use the term *hinge*, for how this region increasingly forms an interface between North America and the booming, still-emerging **21** **Pacific Rim**. This regional term has come into use to describe the interconnections of a string of economies that have redrawn the map of countries facing the Pacific. This a superb example of a *functional region*: economic activity in the form of capital flows, raw-material movements, and trade linkages are generating urbanization, industrialization, and labor migration between countries across a vast ocean. Its leading participants are the United States, Canada, Japan, coastal China, South Korea, Taiwan, Thailand, Malaysia, and Singapore, and such Southern Hemisphere locales as Australia and South America's Chile. (The Pacific Rim concept will be discussed in later chapters as well.)

California Wildfires. In the fall of 2007, it seemed that all of Southern California was on fire with Santa Ana wind-induced wildfires running rampant. Over 2000 square kilometers (732 sq mi or 202,350 ha/500,000 ac) burned, and over 1500 homes were destroyed in the zone from Santa Barbara County south to the Mexican border. Luckily, there was no loss of life.

Wildfires have long been a problem in the State, but they seemed to have reached an unprecedented scale in 2007. Why do wildfires ignite in California, and why do they seem to be getting worse? The natural vegetation of Southern California is chaparral, a vegetation type that needs to burn periodically to remain a healthy ecosystem. California's population has tripled in the last 50 years and as a result, people are living in areas where formerly there was nothing in the way of these fires. Of the new housing in the region, 50 percent has been built in 'severe-fire' zones, and not necessarily using fire-retarding technologies. Long exposure to Smokey the Bear campaigns, which stressed the need to put out fires at all cost, has resulted in the suppression of natural burning, leading to a build-up of dry, dead vegetation. Adding to this condition are California's long years of drought and the semiarid environment's vulnerabilities to the hot and dry Santa Ana winds (these are winds that flow from the *interior* across the mountains *toward* the coast).

In the photo, these residents of Santa Clarita, in Los Angeles County, are watching a threatening fire very close by. Fires can easily start, sometimes caused naturally by lightning but mostly by people. A stray spark from a tossed cigarette butt, a forgotten barbeque or campfire quickly results in a conflagration. Global climate change models predict that more intense fires are likely in California. The best course is to learn to live with them by limiting urban sprawl in severe fire zones and by constructing houses using fire-retardant technologies when populations do expand into these zones. (Krista Kennell/© Corbis)

Environmental hazards bedevil this entire corridor, including inland droughts, coastal-zone flooding, mudslides, wildfires, and particularly earthquakes—with the ominous San Andreas Fault practically the axis of megalopolitan coalescence. To all these hazards, humans have added their own abuses of California's fragile habitat, from overuse of water supplies (requiring vast aqueduct systems to import water from hundreds of kilometers away) to the chronic air pollution of most of its metropolitan areas.

Despite serious challenges, Californians have built one of the realm's most productive economic machines. Indeed, if the Golden State were an independent country, today its economy would rank as the sixth largest in the world. Southern California constitutes one of the State's two leading subregions. Its cornerstone Los Angeles-area economy practically reinvented itself during the 1990s after the sudden end to the Cold War devastated its aerospace and defense industries. By the early 2000s, a turnaround began as Southern California successfully diversified its economy and developed new growth sectors such as the entertainment industry and foreign trade. Most importantly, it built the nation's largest single metropolitan manufacturing complex, with products ranging from high-tech medical equipment to clothing produced by its burgeoning, immigrant-staffed garment industry. Northern California, the State's other major subregion focused on the San Francisco Bay Area, has experienced a smoother path since the 1990s, driven by the steady expansion of the world's leading technopole, Silicon Valley.

Growth and Change in the Pacific Northwest

Continuing up the coast, the northern portion of the region is centered on the Pacific Northwest, which extends northward from Oregon's Willamette Valley through the Cowlitz-Puget Sound lowland of western Washington State into the adjoining coastal zone of British Columbia beyond the Canadian border. This subregion of the Pacific Hinge is also deeply involved in the pioneering and production of

Challenges and Opportunities in Maturing California

More than 60 years of unrelenting growth have taken their toll, and whereas the west coast States and British Columbia are now savoring the prospects of their Pacific Rim location, they must also confront the less pleasant consequences of regional maturity. This is especially true in California, where the massive development of America's most populated and multiethnic State (almost 40 million in 2008, 20 percent larger than all of Canada) has been overwhelmingly concentrated in the teeming conurbation extending south from San Francisco through San Jose, the San Joaquin Valley, the Los Angeles Basin, and the southwestern coast into San Diego at the Mexican border.

computer technology, especially in the technopole centered on suburban Seattle's Redmond. Originally developed by timber and fishing, the Pacific Northwest found its impetus for industrialization in the Columbia River dams built in the 1930s and 1950s, which generated cheap hydroelectricity that attracted aluminum and aircraft manufacturers. North of the Canadian border, British Columbia's development clusters around its economic core, Vancouver. This city is the most Asianized metropolis in North America: it not only lies closer to East Asia on the air and sea routes than the cities to its south, but it also got a head start in forging trade and investment linkages to the western Pacific Rim thanks to its large and growing Asian community. Today, ethnic Chinese residents make up more than 20 percent of the metropolitan population, whose total Asian component approaches a remarkable 40 percent.

With North America ever more tightly enmeshed in the global economy, nowhere are the realm's international linkages more apparent today than in the Pacific Hinge. From the northern end of this region to the Mexican border, its geographic advantages as a gateway to the Pacific Rim are providing opportunities undreamed of barely a decade ago. Asianizing British Columbia now exports close to 50 percent of its goods to the countries of the western Pacific Rim; China alone has sparked a resurgence of the province's mining industries, importing ever increasing quantities of coal and metallic ores. The rest

Vancouver, British Columbia, is a city that has benefited greatly from the development of the Pacific Rim. Before the return of Hong Kong to China in 1997, many thousands of wealthy Hong Kong Chinese immigrated to Vancouver, contributing to today's very large Asian population in its metropolitan region. These immigrants played a significant role in the economic growth of the city, and its skyline has been transformed over the past two decades.   (© A. WinklerPrins)

of the Vancouver–Seattle–Portland corridor is following suit, led by its high-technology industries. In California, the foreign trade sector of the State's gigantic economy has tripled since 1990, and today more than one-sixth of its jobs are export-dependent. Both Northern and Southern California are expected to thrive as these trends continue, with metropolitan Los Angeles becoming the financial, manufacturing, and trading capital of the eastern Pacific Rim as well as its transport hub (Los Angeles overtook New York in the 1990s as the leading U.S. seaport in the value of goods passing through it). Thus the midlatitude West Coast no longer serves as North America's back door, but forms the hinges of a new front door to the Pacific arena thrown wide open to foster international interactions of every kind.

4

MIDDLE AMERICA

In This Chapter

- How NAFTA changed the economic geography of Mexico
- Indigenous peoples demand recognition and rights
- Changes on the horizon for the Panama Canal
- Cuba: Future star of the Caribbean?
- The debate over Puerto Rico's status continues

CONCEPTS, IDEAS, AND TERMS

1	Land bridge
2	Archipelago
3	Culture hearth
4	Mainland-Rimland framework
5	Mestizo
6	Hacienda
7	Plantation
8	Acculturation
9	Transculturation
10	Ejido
11	Maquiladora
12	Dry canal
13	Altitudinal zones
14	Remittances
15	Tropical deforestation
16	Mulatto

REGIONS

MEXICO
CENTRAL AMERICA
CARIBBEAN BASIN

Photos: (*upper left*) Morelia, Mexico (© A. WinklerPrins);
(*above*) Antigua, Guatemala (© A. WinklerPrins).

FIGURE 4-1 *Map:* © H. J. de Blij, P. O. Muller, and John Wiley & Sons, Inc.

MIDDLE AMERICA IS a realm of vivid contrasts, turbulent history, continuing political turmoil, and an uncertain future. The realm encompasses all the lands and islands between the United States to the north and South America to the south (see chapter opener map, Fig. 4-1). Its regions are mapped in Figure 4-2 and include: (1) the substantial landmass of *Mexico*; (2) the narrowing strip of land to its southeast that constitutes *Central America*; and (3) the many large and small islands—respectively known as the *Greater Antilles* and *Lesser Antilles*—of the Caribbean Sea to the east.

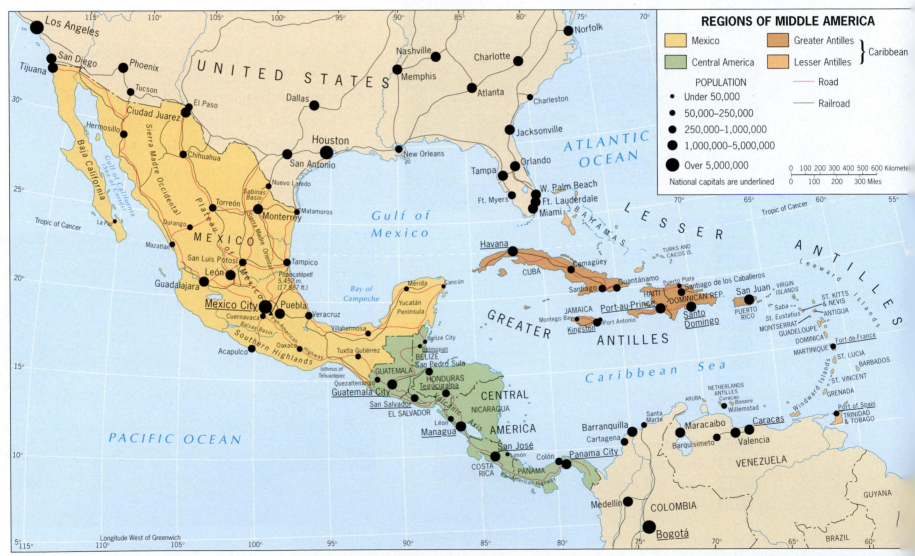

FIGURE 4-2

© H. J. de Blij, P. O. Muller, and John Wiley & Sons

Middle America is a realm of soaring volcanoes and forested plains, of mountainous islands and flat coral cays. Moist tropical winds sweep in from the east, watering windward (wind-facing) coasts while leaving leeward (wind-protected) areas dry. Soils vary from fertile volcanic to desert barren. Spectacular scenery abounds, and tourism is one of the realm's leading industries.

Population distribution patterns reflect both the realm's physical and cultural geographies (Fig. 4-3). Mexico contains 57 percent of Middle America's population, its largest cluster centered on Mexico City (already the leading cultural focus when the Europeans arrived nearly 500 years ago). The rest of the realm's population is about evenly divided between Central America and the Caribbean Basin, but it too exhibits considerable internal geographic variation.

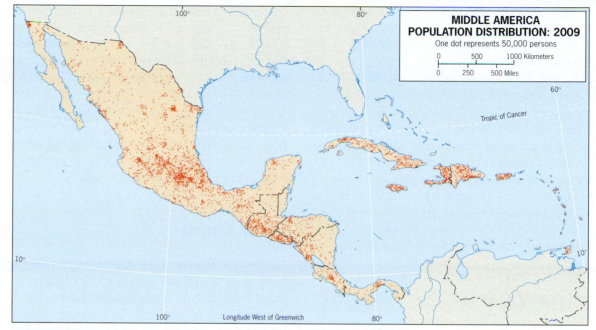

FIGURE 4-3   © H. J. de Blij, P. O. Muller, and John Wiley & Sons, Inc.

Defining the Realm

Is Middle America a discrete geographic realm? Some geographers combine Middle and South America into a realm they call Latin America, citing the dominant Iberian (Spanish and Portuguese) heritage and the prevalence of Roman Catholicism. But these are ethnic labels, and we have chosen to avoid such labels in naming our realms. In Middle America, large populations exhibit African and Asian as well as European ancestries. And nowhere in the Americas has native Amerindian culture contributed to modern civilization as strongly as it has in Mexico. The Caribbean Basin is a patchwork of independent states, territories in political transition, and residual colonial dependencies. The Dominican Republic speaks Spanish, but adjacent Haiti uses French; Dutch is spoken in Aruba, while English is spoken in Jamaica and even in parts of coastal Nicaragua. Middle America thus gives vivid definition to concepts of cultural-geographical pluralism.

MAJOR GEOGRAPHIC QUALITIES OF Middle America

1. Middle America is a fragmented realm that consists of all the mainland countries from Mexico to Panama and all the islands of the Caribbean Basin to the east.

2. Middle America's mainland constitutes a crucial barrier between Atlantic and Pacific waters. In physiographic terms, this is a land bridge that connects the continental landmasses of North and South America.

3. Middle America is a realm of intense cultural and political fragmentation. The political geography defies unification efforts, but countries and regions are beginning to work together for economic purposes and to solve mutual problems.

4. Middle America's cultural geography is complex. African influences dominate the Caribbean and coastal areas, whereas Spanish and Amerindian traditions survive on the mainland.

5. The realm contains the Americas' least developed territories. New economic opportunities may help alleviate Middle America's endemic poverty.

6. In terms of area, population, and economic potential, Mexico dominates the realm.

7. Mexico is reforming its economy and has experienced major industrial growth. Its hopes for continuing this development are tied to overcoming its remaining economic problems and to expanding trade with the United States and Canada under the North American Free Trade Agreement (NAFTA).

Middle America is occasionally called *Central America*, but that name actually refers to a region within the Middle American realm (Fig. 4-2). Central America comprises the republics that occupy the strip of mainland between Mexico and Panama: Guatemala, Belize, Honduras, El Salvador, Nicaragua, and Costa Rica. Although Panama itself is regarded here as belonging to Central America, it should be noted that many Central Americans do not consider Panama to be part of their region because for most of its history that country was a part of South America's Colombia. But, as we define it, *Middle America* includes all the mainland and island countries and territories that lie between the United States and the continent of South America.

PHYSIOGRAPHY

As Figure 4-1 shows, fragmented Middle America is a realm of high relief, studded with active volcanoes. Figure G-5 reminds us why: the Caribbean, North American, South American, and Cocos tectonic plates converge here, creating dangerous landscapes subject to earthquakes and landslides. Add to this the realm's exposure to Atlantic hurricanes, and it amounts to some of the highest-risk real estate on earth.

A Land Bridge

The funnel-shaped mainland, a 6000-kilometer (3800-mi) connection between North and South America, is wide enough in the north to contain two major mountain chains and a vast interior plateau, but narrows to a slim 65-kilometer (40-mi) ribbon of land in Panama. Here this strip of land, or *isthmus*, bends eastward so that Panama's orientation is east-west. Thus mainland Middle America is what physical geographers call a **1 land bridge**, an isthmian link between continents.

If you examine a globe, you can see other present and former land bridges: Egypt's Sinai Peninsula between Asia and Africa, the (now-broken) Bering land bridge between northeasternmost Asia and Alaska, and the shallow waters between New Guinea and Australia. Such land bridges, though temporary features in geologic time, have played crucial roles in the dispersal of animals and humans across the planet. But even though mainland Middle America forms a land bridge, its internal fragmentation has always inhibited movement. Mountain ranges, swampy coastlands, and dense rainforests make contact and interaction difficult.

Island Chains

As shown in Figure 4-1, the approximately 7000 islands of the Caribbean Sea stretch in a lengthy arc from Cuba and the Bahamas east and south to Trinidad, with numerous outliers outside (such as Barbados) and inside (e.g., the Cayman Islands) the main chain. The four large islands—Cuba, Hispaniola (containing Haiti and the Dominican Republic), Puerto Rico, and Jamaica—are called the *Greater Antilles*. All the remaining smaller islands are called the *Lesser Antilles*.

The entire Antillean **2 archipelago** (island chain) consists of the crests and tops of mountain chains that rise from the floor of the Caribbean, the result of collisions between the Caribbean Plate and its neighbors (Fig. G-5). Some of these crests are relatively stable, but elsewhere they contain active volcanoes, and almost everywhere in this realm earthquakes are an ever-present danger—in the islands as well as on the mainland.

LEGACY OF MESOAMERICA

Mainland Middle America was the scene of the emergence of a major ancient civilization. Here lay one of the world's true **3 culture hearths**, a source area from which new ideas radiated and whose population could expand and make significant material and intellectual progress. Agricultural specialization, urbanization, and transport networks developed, and writing, science, art, and other spheres of achievement saw major advances. Anthropologists refer to the Middle American culture hearth as *Mesoamerica*, which extended southeast from the vicinity of present-day Mexico City to central Nicaragua. Its development is especially remarkable because it occurred in very different geographic environments, each presenting obstacles that had to be overcome in order to unify and integrate large areas. First, in the low-lying tropical plains of what is now northern Guatemala, Belize, and Mexico's Yucatán Peninsula, and overlapping chronologically in Guatemala's highlands to the south, the Maya civilization arose more than 3000 years ago. Later, far to the northwest on the high plateau in central Mexico, a number of sequential and simultaneous civilizations established themselves, culminating in the Aztecs whose civilization was centered on the largest city ever to exist in pre-Columbian times.

The Lowland Maya

The Maya civilization is the only major culture hearth in the world that arose in the lowland tropics. Its great cities, with their stone pyramids and massive temples, still yield new archaeological information today. Maya culture reached its zenith from the third to the tenth centuries AD.

The Maya civilization, anchored by a series of city-states, unified an area larger than any of the modern Middle American countries except Mexico. Its population probably totaled between 2 and 3 million at its peak; certain Maya languages are still actively used in the area today, reflecting the continued existence of Mayan peoples today. The Maya city-states were marked by dynastic rule that functioned alongside a powerful religious hierarchy, and the great cities that now lie in ruins were primarily ceremonial centers. We also know that Maya culture produced skilled artists and scientists, and that these people achieved a great deal in agriculture and trade. They grew cotton, created a rudimentary textile industry, and exported cotton cloth by seagoing canoes to other parts of Middle America in return for valuable raw materials. They domesticated the turkey and grew cacao, developed writing systems, and studied astronomy.

The Highland Civilizations

In what is today the intermontane highland zone of Mexico, significant cultural developments were also taking place. Here, just north of present-day Mexico City, lay Teotihuacán, the first true urban center in the Western Hemisphere. It prospered for nearly seven centuries after its founding a few centuries before the Christian era, even before the Maya were active farther south.

A Once and Future Core Area

The Teotihuacán civilization died out, but was followed by a sequence of other peoples, culminating with the Aztec state, the pinnacle of organization and power in pre-Columbian Middle America. It is thought to have originated in the early fourteenth century with the founding of a settlement on an island in a lake that lay in the *Valley of Mexico*, the area surrounding what is today Mexico City. This urban complex, named Tenochtitlán, was a functioning city with an active market as well as a ceremonial center, and would become the greatest city in the Americas and the capital of a large powerful state. The Aztecs soon gained control over the entire Valley of Mexico, a pivotal 50-by-65 kilometer (30-by-40-mi) mountain-encircled basin that is still the heart of the modern state of Mexico. Both elevation and interior location moderate its climate; for a tropical area, it is quite dry and very cool. The basin's lakes formed a valuable means of internal communication, and the Aztecs built canals to connect several of them. This fostered busy canoe traffic, bringing agricultural and other products to the cities. Many subjects paid tribute to the headquarters of the ruling nobility.

A Regional Empire

The Aztecs' expansion of their empire was driven by their desire to subjugate peoples in order to extract taxes and tribute. With its heartland consolidated, the new Aztec state soon began to conquer territories to the east and south. To the west they were never able to conquer the Purepecha (Tarascan) peoples, and neither were their Spanish successors. As Aztec influence spread throughout Middle America, the state grew ever richer, its population mushroomed, and its cities thrived and expanded.

The Aztecs produced a wide range of impressive accomplishments, although they were better borrowers and refiners than they were innovators. They developed intensive agriculture with complex irrigation systems, and they built elaborate walls to terrace slopes where soil erosion threatened. Indeed, all of Mesoamerica's Amerindians contributed immensely to the world's agricultural staples by domesticating some of the world's most important crops: maize (corn), the sweet potato, various varieties of beans, the tomato, squash, cacao beans (the raw material of chocolate), and tobacco. As a result, Mesoamerica remains a critical hearth for agricultural genetic material, and there is great concern today about the infiltration of genetically modified varieties into the world's stock of maize (corn) diversity.

COLLISION OF CULTURES

The Western world all too often believes that history began only when Europeans, with their superior power, arrived in an area, rendering whatever had existed there previously of little significance. Middle America confirms this misperception: the great, feared Aztec state apparently fell before a relatively small band of Spanish invaders in an incredibly short period of time (1519–1521). But let us not lose sight of a few realities. At first, the Aztecs believed the Spaniards were 'White Gods' whose arrival had been predicted by Aztec prophecy. Hernán Cortés, for all his 508 soldiers, did not singlehandedly overthrow this powerful empire: he ignited a rebellion by Amerindian peoples who had fallen under Aztec domination and had seen their relatives carried off for human sacrifice to Aztec gods. Preceded by the lethal diseases the Europeans brought with them, Cortés with his horses and guns was joined by peoples who rose against their Aztec oppressors as the Spaniards headed toward Tenochtitlán and defeated the Aztecs.

Effects of the Conquest

Spain's defeat of Middle America's dominant indigenous state opened the door to Spanish penetration and supremacy. Throughout the realm, the confrontation between Hispanic and native cultures spelled cultural disaster for the Amerindians: a catastrophic decline in population (perhaps as high as 90 percent) due to introduced diseases and enslavement, rapid deforestation, pressure on vegetation from newly introduced grazing animals, substitution of Spanish wheat for maize (corn) on cropland, the shift from communal land management to private ownership, and the concentration of Amerindians into newly built towns.

The Rural Impact

Middle America's cultural landscape—its great cities, its terraced fields, its dispersed indigenous villages—was thus drastically modified. Unlike the Amerindians, who had used stone as their main building material, the Spaniards employed great quantities of wood and used charcoal for heating, cooking, and smelting metal. The onslaught on the forests was immediate, and rings of deforestation swiftly expanded around the colonizers' towns. The Spaniards also introduced large numbers of cattle and sheep into places where there had been no livestock, and people and animals now had to compete for available food (requiring the opening of vast areas of marginal land that further disrupted the region's food-production balance). Moreover, the Spaniards introduced their own grain crops and farming equipment, and soon

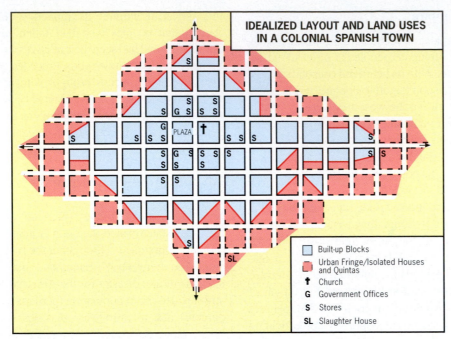

FIGURE 4-4   Adapted with permission from Charles S. Sargent, Jr., "The Latin American City," in Brian & Olwyn Blouet, eds., *Latin America and the Caribbean: A Systematic and Regional Geography*, 5 rev. ed., p. 161. John Wiley & Sons, Inc. 2006.

FIGURE 4-5   © H. J. de Blij, P. O. Muller, and John Wiley & Sons, Inc.

large fields of wheat began to encroach upon the small plots of maize that the natives cultivated.

The New Urban Settlements

The Spaniards' most far-reaching cultural changes derived from their traditions as town dwellers. To facilitate domination, the Amerindians were moved off their land into nucleated villages and towns that the Spaniards established and laid out. In these settlements, the Spaniards could exercise the kind of rule and administration to which they were accustomed (Fig. 4-4). The internal focus of each Spanish town was the central *plaza* or market square, around which both the local church and government buildings were located. The surrounding street pattern was deliberately laid out in *gridiron* form, so that any insurrections by the resettled Amerindians could be contained by having a small military force seal off the affected blocks and then root out the troublemakers.

Each town was located near what was thought to be good agricultural land (which was often not so good), so that the Amerindians could go out each day and work in the fields. Packed tightly into these towns and villages, they came face to face with Spanish culture. They were converted to Catholicism and forced to speak Spanish and pay taxes and tribute to a new master. Nonetheless, the nucleated indigenous village survived under colonial (and later postcolonial) administration and is still a key feature of Amerindian areas in southeastern Mexico and inner Guatemala, where to this day native languages prevail over Spanish (Fig. 4-5).

Church and State

Once the indigenous population was conquered and resettled, the Spaniards were able to pursue another primary goal in their New World territory: the exploitation of its wealth (especially gold and silver)

for their own benefit. Lucrative trade, commercial agriculture, livestock ranching, and especially mining, were avenues to affluence. And wherever the Spaniards ruled—in towns, farms, mines, or indigenous villages—the Roman Catholic Church was the supreme cultural force transforming Amerindian society. Although church and state did not always agree on the process, Jesuits and soldiers did work together to advance the frontiers of New Spain.

MAINLAND AND RIMLAND

In Middle America outside Mexico, only Panama, with its twin attractions of interoceanic transit and gold deposits, became an early focus of Spanish activity. From there, following the Pacific side of the isthmus, Spanish influence radiated northwestward through Central America and into Mexico. The major arena of international competition in Middle America, however, lay not on the Pacific side but on the islands and coasts of the Caribbean Sea. The Spanish were less interested in lowland areas when highlands were available: here the British gained a

FIGURE 4-6

© H. J. de Blij, P. O. Muller, and John Wiley & Sons, Inc.

CARIBBEAN REGION:
COLONIAL SPHERES ca. 1800

| Br. | British | Sp. | Spanish |
| Du. | Dutch | Fr. | French |

Lingering Regional Contrasts

These contrasts between the Middle American highlands on the one hand and the coastal areas and Caribbean islands on the other were conceptualized by John Augelli into the **4** **Mainland-Rimland framework** (Fig. 4-7). Augelli recognized (1) a Euro-Amerindian highland **Mainland**, which consisted of continental Middle America from Mexico to Panama, excluding the Caribbean coastal belt from mid-Yucatán southeastward; and (2) a Euro-African **Rimland**, which included this coastal zone (including all of Belize) as well as the islands of the Caribbean. The terms *Euro-Amerindian* and *Euro-African* underscore the cultural heritage of each region. On the Mainland, European (Spanish) and Amerindian influences are paramount and also include **5** **mestizo** sectors where the two ancestries mixed. In the Rimland, the heritage is African and European. Some newer interpretations of the framework also include three South American entities in the Rimland: Guyana, Suriname, and French Guiana (see p. 177 in Chapter 5) because of their similar colonial histories.

As Figure 4-7 shows, the Mainland is subdivided into several areas based on the strength of the Amerindian legacy. The Rimland is also subdivided, with the most obvious division the one between the mainland-coastal plantation zone and the islands. Note, too, that the islands themselves are classified according to their cultural heritage (Figs. 4-6, 4-7).

Supplementing these contrasts are regional differences in outlook and orientation. The Rimland was an area of sugar and banana plantations, of high accessibility, of seaward exposure, and of maximum cultural

foothold on the mainland, controlling a narrow coastal strip that extended southeast from Yucatán to what is now Costa Rica. As the colonial-era map (Fig. 4-6) shows, in the Caribbean the Spaniards faced not only the British but also the French and Dutch, all interested in the lucrative sugar trade, all searching for instant wealth, and all seeking to expand their empires.

Much later, after centuries of European colonial rivalry in the Caribbean Basin, the United States entered the picture and made its influence felt in the coastal areas of the mainland, not through conquest

but through the introduction of widespread, large-scale, banana plantation agriculture. The effects of these plantations were as far-reaching as the impact of colonialism on the Caribbean islands. Because the diseases the Europeans had introduced were most rampant in these hot, humid lowlands (as well as the Caribbean islands to the east), the Amerindian population that survived was too small to provide a sufficient workforce. This labor shortage was remedied through the trans-Atlantic slave trade from Africa that transformed the Caribbean Basin's demography (see Fig. 6-8).

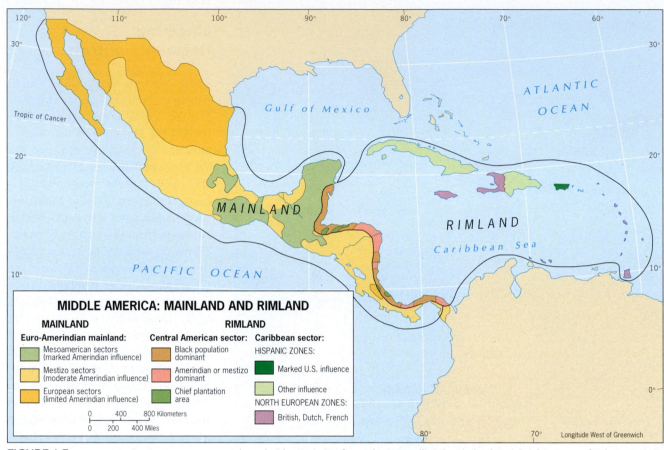

FIGURE 4-7   Adapted with permission from John P. Augelli, "The Rimland-Mainland Concept of Culture Areas in Middle America," *Annals of the AAG*, 52 (1962): 119–129. © Association of American Geographers, 1962.

The Plantation

The **7 plantation** was conceived as something entirely different from the hacienda. Robert West and John Augelli list five characteristics of Middle American plantations that illustrate the differences between hacienda and plantation: (1) plantations are located in the humid tropical coastal lowlands; (2) plantations produce for export almost exclusively—usually a single crop; (3) capital and skills are often imported so that foreign ownership and an outflow of profits occur; (4) labor is seasonal—needed in large numbers mainly during the harvest period—and such labor has been imported because of the scarcity of Amerindian workers; and (5) with its 'factory-in-the-field' operation, the plantation is more efficient in its use of land and labor than the hacienda. The objective was not self-sufficiency but profit, and wealth rather than social prestige is a dominant motive for the plantation's establishment and operation. It was central in the colonial enterprises, especially in the Rimland.

contact and mixture. The Mainland, being farther removed from these contacts, was an area of greater isolation. The Rimland was the region of the great *plantation*, and its commercial economy was therefore susceptible to fluctuating world markets and tied to overseas investment capital. The Mainland was the region of the *hacienda*, which was more self-sufficient and less dependent on external markets.

The Hacienda

This contrast between plantation and hacienda land tenure in itself constitutes strong evidence for the Rimland-Mainland division. The hacienda was a Spanish institution, but the modern plantation, Augelli argued, was the concept of Europeans of more northerly origin. In the **6 hacienda**, Spanish landowners possessed a domain whose productivity they might never push to its limits: the very possession of such a vast estate brought with it social prestige and a comfortable lifestyle. Native workers lived on the land—which may once have been *their* land—and had plots where they could grow their own subsistence crops. All this is written as though it is mostly in the past, but the legacy of the hacienda system, with its inefficient use of land and labor, still exists throughout mainland Middle America and into South America (Chapter 5).

POLITICAL FRAGMENTATION

Today continental Middle America is fragmented into eight countries, all but one of which has Hispanic origins (Belize is the exception). The largest of them all is Mexico—the giant of Middle America—whose 1,960,000 square kilometers (756,000 sq mi) constitute more than 70 percent of the realm's entire land area (the Caribbean region included) and whose 112 million people outnumber those of all the other countries and islands of Middle America combined.

Contrasting land uses in the Middle American Rimland and Mainland give rise to some very different rural cultural landscapes. Huge stretches of the realm's best land continue to be controlled by (often absentee) landowners whose haciendas yield export or luxury crops, or foreign corporations that raise fruits for transport and sale on their home markets. The banana plantation shown here (*left*) lies near the Caribbean coast of Honduras and is owned by United Brands Company. The vast fields of banana plants stand in strong contrast to the lone peasant who ekes out a bare subsistence from small cultivable plots of land, often in high-relief countryside where grazing some goats or other livestock is the only way to use most of the land. (*Left:* © J. P. Courau/D. Donne Bryant Stock Photography; *Right:*© Jean-Gerard Sidaner/Photo Researchers)

The cultural variety in Caribbean Middle America is much greater. Here Hispanic-influenced Cuba dominates: its area is almost as large as that of all the other islands put together, and its population of 11.4 million is well ahead of the next-ranking country—the Dominican Republic (9.3 million), also of Spanish heritage. As we noted, however, the Caribbean is hardly an arena of exclusive Hispanic cultural heritage: for example, Cuba's southern neighbor, Jamaica (population 2.8 million), has a strong African and British legacy, while to the east in Haiti (8.9 million) the strongest imprints have been African and French. The Lesser Antilles also exhibit great cultural diversity. There are the (once Danish) U.S. Virgin Islands; French Guadeloupe and Martinique; a group of British-influenced islands, including Barbados, St. Lucia, and Trinidad and Tobago; and the Dutch St. Maarten (shared with the French) and

WHAT'S DRIVING GEOGRAPHIC CHANGE IN THE REALM

● The expansion of the **Panama Canal** will proceed with Chinese involvement. Will Nicaragua proceed with a competing canal?

● The **Netherlands Antilles**, long on the Middle American map, dissolved in 2008. Curaçao and St. Maarten became self-governing parts of the Kingdom of the Netherlands (Aruba already had this status). Bonaire, together with St. Eustatius and Saba, became Dutch municipalities.

● **Mexico**'s economy continues to depend heavily on oil exports, but the country's reserves are declining. Mexico needs to diversify its economy to avoid long-term economic problems and to keep more of its workforce at home.

● **Natural hazards** (hurricanes and tectonic activity) will continue to plague the region and hinder the economic development of its many small economies.

● **NAFTA** (North American Free Trade Agreement) and **CAFTA-DR** (Central American Free Trade Agreement–Dominican Republic) will continue to shape economic change in all participating countries.

the A-B-C islands of the Netherlands Antilles—Aruba, Bonaire, and Curaçao—off the northwestern Venezuelan coast.

Independence

Independence movements stirred Middle America at an early stage. On the mainland, revolts against Spanish authority (beginning in 1810) achieved independence for Mexico by 1821 and for the Central American republics by the end of the 1820s. The United States, concerned over European designs in the realm, proclaimed the Monroe Doctrine in 1823 to deter any European power from reasserting its authority in the newly independent republics or from further expanding its existing domains. By the end of the nineteenth century, the United States itself had become a major force in Middle America. The Spanish-American War of 1898 made Cuba independent and put Puerto Rico under the U.S. flag; soon afterward, the Americans were in Panama constructing the Panama Canal. Meanwhile, with U.S. corporations driving a boom based on huge banana plantations, the Central American republics had become colonies of the United States in all but name. It is from this era that the term *banana republic* comes.

Independence came to the Caribbean Basin in fits and starts. Afro-Caribbean Jamaica as well as Trinidad and Tobago, where the British had brought a large South Asian population, attained full sovereignty from the United Kingdom in 1962; other British islands (among them Barbados, St. Vincent, and Dominica) became independent later. France, however, retains Martinique and Guadeloupe as overseas *départements* of the French Republic, and the Dutch islands are at various stages of autonomy.

Regions of the Realm

Middle America consists of four geographic regions: (1) Mexico, the giant of the realm in every respect; (2) Central America, the string of seven small republics occupying the land bridge to South America; (3) the four large islands that constitute the Greater Antilles of the Caribbean; and (4) the numerous islands of the Caribbean's Lesser Antilles (Fig. 4-2).

MEXICO

By virtue of its physical size, large population, cultural identity, resource base, economy, and relative location, the state of Mexico by itself constitutes a geographic region in this multifaceted realm. The U.S.-Mexican boundary on the map crosses the continent from the Pacific (see photo at right) to the Gulf, but Mexican cultural influences penetrate deeply into the southwestern States and North American impacts reach far into Mexico. To Mexicans the border is a reminder of territory lost to the United States in historic conflicts; to North Americans it is a symbol of economic contrasts and illegal immigration.

Physiography

The physiography of Mexico is reminiscent of that of the western United States, although its environments are more tropical. Figure 4-8 shows several prominent features: the elongated Baja (Lower) California Peninsula in the northwest, the far eastern Yucatán Peninsula, and the Isthmus of Tehuantepec in the southeast where the Mexican landmass tapers to its narrowest extent. Here in the southeast, Mexico most resembles Central America physiographically; a mountain backbone forms the isthmus, curves southeast into Guatemala, and extends northwest toward Mexico City. Shortly before reaching the capital, this mountain range divides into two chains, the Sierra Madre Occidental in the west and the Sierra Madre

El Norte (the United States) remains a strong lure despite anti-immigrant sentiments and physical impediments, such as this wall. Here potential illegal immigrants are by the wall that marks the border between the United States and Mexico in Tijuana, Mexico. When such great differences in wealth and opportunity exist side by side, as they do on the border between the North American and Middle American realms, it is difficult to stop the flow.   (© AP/Wide World Photos)

Population Patterns

Mexico's population grew rapidly during the last three decades of the twentieth century, doubling in just 28 years; but demographers have recently noted a sharp drop in fertility, and they are predicting that Mexico's population (currently 112 million) will stop growing altogether by about 2050. This will have enormous implications for the country's economy and for its internal and external population movements.

Where the People Live

The distribution of population across Mexico's 31 internal States is shown in Figures 4-9 and 4-3. The largest concentration, containing more than half the Mexican people, extends across the densely populated 'waist' of the country from Veracruz State on the eastern Gulf Coast to Jalisco State on the Pacific. The center of this corridor is dominated by the most populous State, Mexico (**3** on the map), at whose heart lies the Federal District of Mexico City (**9**). In the dry and rugged terrain to the north of this central corridor lie Mexico's least-populated States. Southern Mexico also exhibits a sparsely peopled periphery in the hot and humid lowlands of the Yucatán Peninsula.

Pull and Push

Another major feature of Mexico's population map is urbanization, driven by the **pull** of the cities (with their perceived opportunities for upward mobility) in tandem with the **push** of the economically stagnant countryside. Today, 75 percent of the Mexicans reside in towns and cities. Mexico City, one of the world's megacities, now totals approximately 28.2 million

FIGURE 4-8

© H. J. de Blij, P. O. Muller, and John Wiley & Sons, Inc.

Oriental in the east (Figs. 4-1, 4-8). These diverging ranges frame the funnel-shaped Mexican heartland, the center of which consists of the rugged, extensive Plateau of Mexico (the Valley of Mexico lies near its southeastern end). As Figure G-9 reveals,

Mexico's climates are marked by dryness, particularly in the broad, mountain-flanked north. Most of the better-watered areas lie in the southern half of the country where the major population concentrations have developed.

people and is home to fully 25 percent of the national population. Among the other leading cities are Guadalajara, Puebla, and León in the central population corridor, and Monterrey, Ciudad Juarez, and Tijuana in the northern U.S. border zone (Fig. 4-9). Urbanization rates at the other (southern) end of Mexico, however, are at their lowest in remote uplands where Amerindian society has been least touched by modernization. NAFTA has contributed to the pull and push as its manufacturing plants have pulled workers to towns and cities while agricultural change has pushed people out of the countryside. This same push also impels people to migrate, often illegally, to the United States.

A Mix of Cultures

Nationally, the Amerindian imprint on Mexican culture remains quite strong. Today, 60 percent of all Mexicans are *mestizos*, 22 percent are predominantly Amerindian, and about 8 percent are full-blooded Amerindians; only 10 percent are principally of European descent. Certainly the Mexican Amerindian has been Europeanized, but the Amerindianization of modern Mexican society is so powerful that it would be inappropriate here to speak of one-way, European-dominated 8 **acculturation**. Instead, what took place in Mexico is 9 **transculturation**—the two-way exchange of culture traits between societies in close contact. In the southeastern periphery (Fig. 4-5), several hundred thousand Mexicans still speak only an Amerindian language, and millions more still use these languages in everyday conversation even though they also speak Mexican Spanish. The latter has been strongly shaped by Amerindian influences, as have Mexican modes of dress, foods and

FIGURE 4-9   © H. J. de Blij, P. O. Muller, and John Wiley & Sons, Inc.

cuisine, sculpture and painting, architectural styles, and folkways. This fusion of heritages, which makes Mexico unique, is the product of an upheaval that began to reshape the country a century ago.

Revolution and Its Aftermath

Modern Mexico was forged in a revolution that began in 1910 and set into motion events that are still unfolding today. At its heart, this revolution was about the redistribution of land, an issue that had not been resolved after Mexico freed itself from Spanish colonial control in the early nineteenth century. As late as 1900, more than 8000 haciendas had blanketed virtually all of Mexico's good farmland, and about 95 percent of all rural families owned no land whatsoever and toiled as *peones* (landless, constantly indebted serfs) on the haciendas. The triumphant revolution produced a new constitution in 1917 that launched a program of expropriation and parceling out of the haciendas to rural communities.

Days of the Dead in Mexico. "In October 2005, I was in the State of Michoacán, Mexico, in the days leading up to Halloween. In most of Mexico, the Days of the Dead (October 31–November 2) are a major celebration. Much less ghoulish than the American Halloween, the Days of the Dead are a celebration of the lives of all those who have passed. Special pastries and candies are made, as are crafts such as *papier mâché* skulls and skeletons (see above). These skeletons, known as 'Katrinas' are dressed up and displayed, not in deathly positions, but in postures that demonstrate an enjoyment of life. Entire families spend the night of November 1 in cemeteries sitting by cleaned and decorated graves of departed family members. Altars of favorite foods of the deceased are set out, and there is a joyful reconnection with the memories of those who have passed. This celebration demonstrates the fusion of indigenous belief systems, which included ancestor worship, and the holy days imposed by the Catholic Church as these Days of the Dead are celebrated by the Church's All Saints' (November 1) and All Souls' (November 2) Days."   (© A. WinklerPrins.)

www.conceptcaching.com

Land Reform

Mexico, alone among Middle America's countries with large Amerindian populations, has made significant strides toward undertaking land reform and has done so without major dislocation. Since 1917, more than half the cultivated land of Mexico has been redistributed, mostly to peasant communities consisting of 20 families or more. On such farmlands, known as **10 ejidos**, the government holds title to the land, but the use rights are parceled out communally to villages and then to individuals for cultivation by those communities. This system of land management is an Amerindian legacy, and not surprisingly most *ejidos* lie in central and southern Mexico, where Amerindian agricultural traditions are strongest. Approximately 50 percent of Mexico's land continues to be held in such 'social landholdings' (*ejidos* plus communal indigenous communities). These landholdings are fragmented, however, providing low agricultural yields; therefore considerable poverty persists in the countryside. In the early 1990s the Mexican government set out to privatize *ejidos* in an effort to improve agricultural production. This has been a failure as less than 10 percent of *ejidos* have been privatized.

Chiapas and Oaxaca

In the far southern periphery of Mexico lie the poorest of the 31 States in the country, Chiapas and Oaxaca (shown in lavender in Fig. 4-11). These two are mostly indigenous States that have more in common with Central America than with Mexico. Their rapidly growing population is heavily Amerindian and dominated by families of peasant farmers who eke out a precarious existence cultivating tiny mountainous plots. For centuries, the better valley soils had been incorporated into the estates of the large landholders, a system that endures in these two States virtually unaffected by the land redistribution that reshaped so much of rural Mexico. In response to this lack of development, during the 1980s the Mexican government pledged to introduce new services and programs in Chiapas, but its main efforts were half-hearted: an ineffective coffee-raising scheme and a feeble attempt to privatize Amerindian lands. This only intensified the longstanding bitterness of the Chiapans, and a radical group of Mayan peasant farmers began to organize to resume the historic struggle of Amerindians to gain land and fair treatment.

On January 1, 1994, this organization, now calling itself the Zapatista National Liberation Army (ZNLA), launched a guerrilla war with coordinated attacks on several Chiapan towns. By taking the name of Zapata (a legendary leader of the 1910 revolution) and by timing its insurgency to coincide with the birth of NAFTA, the ZNLA achieved maximum impact and publicity. The Mexican army's response was heavy-handed, and suddenly Mexico confronted a major domestic challenge.

As the map shows, Chiapas lies on Mexico's periphery, far from the burgeoning core area and even farther from NAFTA developments. The Zapatistas demanded greater political autonomy (on the European devolutionary model) and control over local affairs. And they brought to the fore an uncomfortable reality of Mexican culture: the low and disadvantaged status of Mexico's 9 million ethnic Amerindians, who choose to preserve their pre-Hispanic cultural traditions and whose ancestral homes are widely distributed throughout the country.

In Oaxaca State, a different yet parallel civil uprising emerged in 2006. During the late spring, public school teachers in this poor State went on strike when their grievances went unheeded. Chief among their complaints were poor pay and inadequate services for poor children. But they also called for the resignation of the State's governor, who was deemed corrupt and excessively repressive. As the strike escalated, the protestors occupied the central square of Oaxaca City, and several demonstrations ensued. Supporting the teachers' strike were the Zapatistas from Chiapas and several other groups. Eventually, the federal police were called in to disperse the protestors, but not without violence. Teachers have now gone back to work, but discontent and instability continue.

Regions of Mexico

Physiographic, demographic, economic, historical, and cultural criteria combine to reveal a regionally diverse Mexico extending from the lengthy ridge of

FIGURE 4-10   © Robert C. West & John P. Augelli, *Middle America: Its Lands and Peoples*, 3rd edition, p. 340, © 1989. Adapted by permission
of Pearson Education, Inc., Upper Saddle River, N.J.

tive and discontinuous but is changing northern Mexico significantly. This is true even in Yucatán, where Mérida and environs are strongly affected by NAFTA development. To go from comparatively well-off northern Yucatán to much poorer southern Chiapas is to see the whole range of Mexico's regional geography.

The Changing Geography of Economic Activity

During the last two decades of the twentieth century and in the first years of the twenty-first, Mexico's economic geography has changed, and in some respects progressed—though not without setbacks. During the early 1990s, the implementation of NAFTA led to an economic boom as Mexico became part of a free-trade zone and market comprising over 400 million people. This boom transformed urban landscapes along the 3115-kilometer (1936-mi) border between Mexico and the United States, but it could not, of course, close the economic gap between the two sides.

Today, Mexico is in progressive transition in many spheres: its democratic institutions are strengthening; its economy is more robust than it was at the time of NAFTA's inception; its once-rapid population growth is declining; its social fabric (notably relations with Amerindian minorities) is improving. The familiar problems of a country in transition, such as unchecked urbanization, inadequate infrastructure, corruption, and violent crime (especially involving the drug trade), will continue to afflict Mexico for decades to come. But, as the following discussion confirms, Mexico is a far stronger economy and society today than it was just one generation ago.

Baja California to the tropical lowlands of the Yucatán Peninsula, and from the economic frenzy of the NAFTA North to the Amerindian traditionalism of the Chiapan southeast (Fig. 4-10). In the Core Area, anchored by Mexico City, and in the West, centered on Guadalajara, lies the transition zone from the more Hispanic-mestizo north to the more Amerindian-infused mestizo south. East of the Core Area lies the Gulf Coast, once dominated by major irrigation projects and huge livestock-raising schemes but now the mainland center of Mexico's petroleum industry. The dry, scrub-vegetated Balsas Lowland separates

the Core Area from the rugged, Pacific-fronting Southern Highlands, where Acapulco's luxury hotels stand in stark contrast to the Amerindian villages and *ejidos* of the interior, scene of major land reform in the wake of the 1910 revolution.

The dry, vast north stands in sharp contrast to these southern regions: only in the Northwest was there significant sedentary Amerindian settlement when the Spanish arrived. Huge haciendas, major irrigation projects, and some large cities separated by great distances mark this area, now galvanized by the impact of NAFTA. The NAFTA North region is still forma-

Agriculture

Although traditional subsistence agriculture and the output of the inefficient *ejidos* have not changed a great deal in the poorer areas of rural Mexico, larger-scale commercial agriculture has diversified during the past three decades and made major gains with respect to both domestic and export markets. The country's arid northern tier has led the way as major irrigation projects have been built on streams flowing down from the interior highlands. Along the booming northwest coast of the mainland, which lies within a day's drive of Southern California, mechanized large-scale cotton production now supplies an increasingly profitable export trade. Here, too, wheat and winter vegetables are grown, with fruit and vegetable cultivation attracting foreign investors.

But many Mexican small farmers are having a tough time of it, for two reasons. First, cheap maize (corn) from the United States, grown by American farmers heavily subsidized by the government, floods Mexico's markets, so that local farmers cannot make a profit. Second, the United States tried to create barriers against the import of low-priced Mexican produce such as avocados and tomatoes—violating the very rules NAFTA is supposed to stand for. Ironically, the average Mexican consumer is now faced with increased food prices, especially for staples such as the white maize (corn) used to make tortillas.

Energy and Industry

Until 1990, Mexican industry, as that in any developing economy, was subject to state control over income-generating natural resources, especially oil and natural gas. The government set up industries such as the country's first steel plant (opened in Monterrey in 1903) without allowing competition. Employment in agriculture far exceeded that in manufacturing or mining, despite Mexico's substantial reserves of silver and other metals, high-quality coal (especially in Coahuila State), and additional inventory ranging from antimony to zinc. Inefficiency prevailed, wages were low, inflation was rife, and Mexico's public debt was one of the highest in the world. The Mexican economy suffered from ups and downs directly related to world oil prices; when the price of oil dropped, federal expenditures had to be cut and deficits soared. In 2007, oil still accounted for more than one third of total government revenues, a dangerous situation in a volatile world (the distribution of Mexican oil and gas reserves is mapped in Fig. 4-8).

During the 1980s, successive Mexican governments began trying to address these problems as part of the preparations for NAFTA. Some of the inefficient mining and manufacturing operations were sold to private investors, and Mexico's financial systems (banking, tax collection) were strengthened. But with the inception of NAFTA in 1994, Mexico's industrial geography changed dramatically. Although the Border Industrialization Program had begun decades earlier, it exploded once NAFTA started. This program encouraged assembly plants in Mexico to assemble imported, duty-free raw materials and components into finished products, which were then exported back tariff-free into the U.S. market. These assembly plants, called 11 **maquiladoras**, were initially located as close to the U.S. border as possible to take advantage of proximity to the U.S. market and the much lower wages on the Mexican side. As a result, manufacturing employment in the cities and towns along that border, from Tijuana in the west to Matamoros in the east, expanded rapidly. After only seven years of NAFTA's existence, there were 4000 factories with more than 1.2 million workers in the border zone and in northern Yucatán accounting for nearly one third of Mexico's industrial jobs and 45 percent of its total exports.

NAFTA's impact on Mexico has been far-reaching. Mexico has benefited from foreign investment, job creation, tax receipts, and technology transfer. But Mexican employees work long hours for low wages with few benefits and live in the most basic shacks and slum dwellings encircling the burgeoning towns of NAFTA North (Fig. 4-10). And there is no job security.

Ten years after NAFTA's start, hundreds of American and other foreign corporations that had moved their factories to northern Mexico decided to relocate once again—to East and Southeast Asia where wages were even lower than the U.S.$2.00 per hour average paid by the maquiladoras. While factories assembling heavy and bulky items such as vehicles and refrigerators were still better off right across the U.S. border, others producing lighter and smaller goods such as electronic equipment and cameras moved to China, Vietnam, and other countries where wages were less than half of Mexico's. As a result, thousands of Mexican workers found themselves unemployed—and many crossed the border into the United States.

How can Mexico counter this trend? Here is one indicator: while maquiladora jobs were lost to Asia in textiles, jobs were being added in electronics, and Mexican company managers were in short supply. Education in high-tech and management fields is therefore part of the answer. In the Northeast, the city of Monterrey in the high-income State of Nuevo Léon has a substantial international business community and modern industrial facilities that have attracted major multinational companies. The *Technológico* campus near Monterrey's airport, a degree-granting college supported by both government and private enterprise, lies at the heart of a network of more than 30 campuses throughout Mexico offering degrees in information technology, engineering, and administration. In the West, near Guadalajara, in the less prosperous State of Jalisco, one can discern the beginnings of a Silicon Valley of Mexico. In such places, the outlines of the landscape of the global economy are evident.

Geography of Inequality

The statistics of economic geography often hide their implications for the people involved. Without doubt NAFTA has enriched Mexico in general and northern Mexico in particular, and almost all the northern States bordering the United States have per-capita incomes above the national average. Northern Mexico has historically been better off than the south—and NAFTA has widened the gap. NAFTA has impacted not just the industries it represents, but also Mexico's farmers.

One stipulation of the agreement was that Mexico would lower and then drop its tariff barriers against U.S. and Canadian farm produce, including Mexico's staple, maize. But the United States generously subsidizes its maize (corn) farmers, who as a result can market maize in Mexico for prices even lower than Mexican farmers have to charge in order to make a small profit. In the process, hundreds of thousands of Mexican farmers are being put out of business. Not surprisingly, the U.S. government finds itself having to build walls to keep illegal immigrants out.

Mexico's north-south divide is starkly evident from the economic data (Fig. 4-11). Generally, the annual per-capita income in the northern States exceeds U.S.$10,000, but in the southern States it falls below $5000. Economic growth in the northern States has averaged more than 4 percent in recent years; in southern States it is less than 2 percent. Far more people are poor in the south than in the north. Mexico's infrastructure, already inadequate, serves the south far less well than the north. Whatever the index—literacy, electricity use, water availability—the south lags.

These still-widening contrasts were thrown into sharp relief in 2006, when Mexico's presidential election was contested by three candidates, of whom the two leading ones were a conservative and a populist. When the ballots were counted, there was a near-tie, a result so close that the populist at first refused to concede. The north was won by the conservative candidate and the south by the populist. Mexico's electoral geography thus reflected the serious social consequences of the country's regional economic disparities.

Mexico's Future

Since the mid-1990s, Mexico has strengthened its democratic institutions, brought greater stability to the economy, absorbed the impacts of NAFTA, and improved relations with its Amerindian minorities. Mexico's rate of population growth continues to decline, and while millions of Mexicans have migrated to the United States since NAFTA's inception, fraying the social fabrics of its communities, even this disruption has a positive side: remittances from these workers in recent years are estimated to amount to between U.S. $25 and $30 billion, which is much greater than official foreign investment.

But as we have noted, globalization has fractured Mexico's economic and political geography, reflecting success in the north and failure in the south and deepening historic regional disparities. Northern Mexico is becoming more like North America, and southern Mexico more like Central America. The future of the country depends on the government's success in closing this gap and on its ability to spread the positive effects of NAFTA from north to south. One such government effort, a federal antipoverty program that requires families to keep their children in school, is having good

MEXICO: ECONOMIC DISPARITIES

GDP PER CAPITA
National average=100

	Above 149
	125–149
	100–124
	75–99
	50–74
	Below 50

BORDER-ZONE MAQUILADORA EMPLOYMENT, 2005
- 100,000–300,000
- 50,000–100,000
- 15,000–50,000
- 5,000–15,000

— Railroad
— Road
— NAFTA Highway

0 100 200 300 400 500 600 Kilometers
0 100 200 300 400 Miles

1 QUERÉTARO
2 HIDALGO
3 MEXICO
4 MORELOS
5 TLAXCALA
6 AGUASCALIENTES
7 GUANAJUATO
8 MEXICO CITY (DISTRITO FEDERAL)

FIGURE 4-11

© H. J. de Blij, P. O. Muller, and John Wiley & Sons, Inc.

effect. Another, the improvement of infrastructure to reduce the isolation of southern communities, has been less successful. This cannot be done by government alone, but Mexico's history of reliance on government, inefficient and corrupt as it has long been, is not easily overcome. Take a look at Figure 4-8: the dominance of Mexico City, the seat of power, is reflected by the national transport system. Most roads go to and through Mexico City, creating a costly bottleneck that disadvantages the movement of people and products between the south and the north.

Mexico's future is now inextricably bound up with the United States: when the American economy stalls, Mexico's suffers, and vice versa. Mexico's proximity to the American market has other negative consequences. The United States is the world's richest single market for illegal narcotics, and drug cartels that once operated from Colombia are now based in some Mexican cities, contributing to Mexico's rapidly rising crime wave.

Mexico is a country in tumultuous transition, a land of many opportunities in need of regional integration and the expansion of modernization. Its tourist industry—ranging from the great Maya cities of the Yucatán Peninsula to the beaches of the Mexican Riviera on the Pacific coast—could contribute far more to the economy than it presently does. This failure is yet another consequence of inadequate infrastructure. Mexico's energy industries are burdened by insufficient investment in research and renovation. In the maquiladora industries, the drive should be toward higher-value manufacturing, thereby reducing vulnerability to Asian competition. At times the Mexican government reverts to its old penchant for megaprojects, such as building the so-called **12 dry canal** across the narrow-

est part of the country from the Pacific port of Salina Cruz in Oaxaca State to the Mexican Gulf port of Coatzacoalcos in Veracruz State (Fig. 4-8). This overland railroad would compete with the Panama Canal. However, there are less spectacular and more effective ways to spend money—for example, on improving access roads, electricity supply, and irrigation systems for small farmers in the impoverished south. Mexico has made significant progress in the past two decades, but old problems continue to slow the pace.

THE CENTRAL AMERICAN REPUBLICS

Crowded onto the narrow segment of the Middle American land bridge between Mexico and the South American continent are the seven countries of Central America (Fig. 4-12). Territorially, they are all quite small; their population sizes range from Guatemala's 13.7 million down to Belize's 315,000.

FIGURE 4-12

© H. J. de Blij, P. O. Muller, and John Wiley & Sons, Inc.

Physiographically, the land bridge here consists of a highland belt flanked by coastal lowlands on both the Caribbean and Pacific sides (Fig. 4-1). These highlands are studded with volcanoes, and local areas of fertile volcanic soils are scattered throughout them. From earliest times, the region's inhabitants have been concentrated in this upland zone, where tropical temperatures are moderated by elevation and rainfall is sufficient to support a variety of crops.

Altitudinal Zonation of Environments

Continental Middle America and the western margin of South America are areas of high relief and strong environmental contrasts. Even though settlers have always favored temperate intermontane basins and valleys, people also inhabit the tropical lowlands. In each of these zones, distinct local climates, soils, vegetation, crops, domestic animals, and modes of life prevail. Such **13** **altitudinal zones** (diagrammed in Fig. 4-13) are known by specific names.

The lowest of these vertical zones, from sea level to about 750 meters (2500 ft), is known as the *tierra caliente*, the 'hot land' of the coastal plains and low-lying interior basins where tropical (plantation) agriculture predominates. Above this zone lie the tropical highlands containing Middle and South America's largest population clusters, the *tierra templada* of temperate land reaching up to about 1800 meters (6000 ft). Temperatures here are cooler; prominent among the commercial crops is coffee, while maize (corn) and wheat are the staple grains. Still higher, from about 1800 meters to nearly 3600 meters (12,000 ft), is the *tierra fría*, the cold country of the higher elevations (found mostly in South America's Andes Mountains;

see Chapter 5) where hardy crops such as potatoes and barley are mainstays, as well as livestock grazing. Above the tree line, which marks the upper limit of the *tierra fría*, lies the *tierra helada*; this fourth altitudinal zone, extending from about 3600 to 4500 meters (15,000 ft), is so cold and barren that it can support only the grazing of hardy livestock. The highest zone of all is the *tierra nevada*, a zone of permanent snow and ice associated with the loftiest Andean peaks. The varied human geography of Middle America and western South America is closely related to these diverse environments.

Population Patterns

Figure 4-3 indicates that Central America's population tends to concentrate in the uplands of the *tierra templada* and that population densities are generally greater toward the Pacific than toward the Caribbean side. The most significant exception is El Salvador, whose political boundaries confine its people mostly

to its tropical *tierra caliente*, tempered here by the somewhat cooler Pacific offshore. On the opposite side of the isthmus, Belize has the typically sparse population of the hot, wet Caribbean coastlands and their infertile soils. Panama is the only other exception to the rule, but for different reasons. Economic development has focused on the Panama Canal and the coasts, although new settlement is now moving onto the mountain slopes of the Pacific side.

Central America, we noted earlier, actually begins within Mexico (in Chiapas and also in the Yucatán), and the region's republics face many of the same problems as Mexico's periphery. Population pressure is one of these. The environment is mountainous and arable land limited. Much of the best cultivable land has been appropriated by plantations that produce coffee (in the highlands) and bananas (in the lowlands), leaving many people to eke out a subsistence living on marginal land. Also, unlike Mexico, Central America's population growth is not yet slowing down significantly except in Costa Rica and Panama.

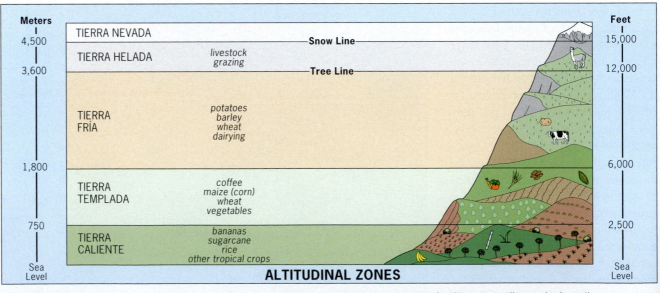

FIGURE 4-13

© H. J. de Blij, P. O. Muller, and John Wiley & Sons, Inc.

Emergence from a Turbulent Era

Devastating inequities, repressive governments, external interference, and the frequent unleashing of armed forces have destabilized Central America for much of its modern history. The roots of these upheavals are old and deep, and today the region continues its struggle to emerge from a period of turmoil that lasted through the 1980s into the mid-1990s.

Central America is not a large region, but because of its physiography it contains many isolated, comparatively inaccessible locales. Conflicts between Amerindian population clusters and mestizo groups are endemic to the region, and contrasts between the privileged and the poor are especially harsh. Dictatorial rule by local elites followed authoritarian rule by Spanish colonizers.

Economic Dependence on the North

Political leaders and economic planners in the United States have long advocated a free-trade agreement to boost the economies of Central American countries. In 2005, this campaign resulted in the approval by the U.S. Congress of the Central American Free Trade Agreement (CAFTA-DR) linking the powerful U.S. economy to five small Central American economies (Guatemala, El Salvador, Honduras, Nicaragua, and Costa Rica) and one Caribbean economy (the Dominican Republic). This is not the first time such an integrative effort has been tried in Central America: the Central American Common Market was created in 1960 but fell apart within a decade.

Today, however, the region is more stable, the United States has stopped supporting dictators, democratic governments are holding their own, and trade between the republics and the United States, though still modest, is expanding. It is true, as critics of CAFTA-DR argue, that many of its free-trade provisions were already in place, the result of earlier bilateral agreements, and that other tariffs will

be reduced gradually over a period of two decades. But CAFTA-DR will do more than enhance trade: it will signal to the world that the region has turned a corner. It will encourage investment, force governments to be more open and less corrupt, expand trade among the five Central American republics themselves, promote economic diversification, and thereby reduce excessive dependence on farm products.

As a whole, Central America is highly dependent on **14 remittances** sent home by its citizens who have emigrated. Remittances usually consist of small amounts of money (on average, $300) from legal and illegal migrants. Many Central Americans leave their homes, work in the United States, and send home much of what they earn. For some countries these payments represent their most important source of income. For example, 10 percent of Guatemala's GNI comes from remittances, totaling U.S.$2.6 billion. Income from remittances usually exceeds foreign aid significantly, but for some countries creates an awkward dependency on former citizens who have emigrated.

The Seven Republics

Guatemala, the westernmost of Central America's republics, has more land neighbors than any other. Straight-line boundaries across the tropical forest mark much of the border with Mexico, creating the box-like region of Petén between Chiapas State on the west and Belize on the east; also to the east lie Honduras and El Salvador (Fig. 4-12). This heart of the ancient Maya Empire, which remains strongly infused by Amerindian culture and tradition, has just a small window on the Caribbean but a longer Pacific coastline. Guatemala was still part of Mexico when the Mexicans threw off the Spanish yoke, and although independent from Spain after 1821, it did not become a separate republic until 1838. Mestizos, not the Amerindian majority, secured the country's independence.

Most populous of the seven republics with 13.7 million inhabitants (mestizos are in the majority with

55 percent, Amerindians 43 percent), Guatemala has seen much conflict. Repressive regimes made deals with U.S. and other foreign economic interests that stimulated development, but at a high social cost. Over the past half-century, military regimes have dominated political life. The deepening split between the deep poverty of the Amerindians and the better-off mestizos, who here call themselves *ladinos*, generated a civil war that started in 1960 and has since claimed more than 200,000 lives as well as 50,000 'disappearances.' An overwhelming number of the victims were of Mayan descent; the mestizos control the government, army, and land-tenure system. Recent years have seen reconciliation and slow improvement for Guatemala's majority with demographic governments that have taken a greater interest in improving the lives of the marginalized. Guatemala's economic geography has considerable potential but has long been shackled by the unending internal conflicts that have kept the income of 80 percent of the population below the poverty line. The country's mineral wealth includes nickel in the highlands and oil in the lower-lying north. Agriculturally, volcanic soils are fertile and moisture is ample over highland areas large enough to produce a wide range of crops including excellent coffee. If violence and instability can be kept at bay, tourism has considerable potential in this beautiful country.

Belize, strictly speaking, is not a Central American republic in the same tradition as the other six and is best included in the Rimland (Fig. 4-7). Until 1981, this country, a wedge of land between northern Guatemala, Mexico's Yucatán Peninsula, and the Caribbean, was an African-infused dependency of the United Kingdom known as British Honduras. Slightly larger than Massachusetts and with a minuscule population of only about 315,000, Belize has been more reminiscent of a Caribbean island than of a continental Middle American state. Guatemala has long held that Belize is part of its territory, although this border dispute is not in conflict. Today, all that is changing as the demographic complexion of Belize is being reshaped. Thousands of residents have recently emigrated (many to the United States) and are being replaced by tens of thousands of Spanish- and Maya-speaking immigrants from neighboring Guatemala and other Central American countries. Their proportion of the Belizean population has risen from 33 to nearly 50 percent since 1980. Within the next few years the newcomers will likely be a majority, Spanish will become the *lingua franca*, and Belize's cultural geography may exhibit an expansion of the Mainland at the expense of the Rimland.

The Belizean transformation extends to the economic sphere as well. No longer just an exporter of sugar, bananas, and citrus, Belize is producing new commercial crops, and its seafood-processing and clothing industries have become major revenue earners. Also important is tourism, which annually lures more than 150,000 vacationers to the country's Mayan ruins, resorts, and newly legalized casinos; a growing speciality is ecotourism, based on the natural attractions of the country's rainforest and offshore reefs. Belize is also known as a center for *offshore banking*—a financial haven for foreign companies and individuals who want to avoid paying taxes in their home countries.

Honduras is a country on hold as it continues to struggle and rebuild its battered infrastructure and economy. In 1998 Hurricane Mitch struck Honduras, and the consequences were catastrophic as massive floods and mudslides were unleashed across the country, killing 9200 people, demolishing more than 150,000 homes, destroying 21,000 miles of roadway and 335 bridges, and rendering 2 million homeless. Also devastated was the critical agricultural sector that employed two-thirds of Honduras's labor force, accounted for nearly a third of its GDP, and earned more than 70 percent of its foreign revenues.

With 7.8 million inhabitants, about 90 percent mestizo, Honduras still has years to go to overcome the complete devastation of its infrastructure and economy after Hurricane Mitch. Agriculture, livestock, forestry, and limited mining formed the mainstays of the pre-1998 economy, with the familiar Central American products—bananas, coffee, shellfish, and apparel—earning most of the external income.

Honduras, in direct contrast to Guatemala, has a lengthy Caribbean coastline and a small window on the Pacific (Fig. 4-12). The country also occupies a critical place in the political geography of Central America, flanked as it is by Nicaragua, El Salvador, and Guatemala—all continuing to grapple with the aftermath of years of internal conflict and, most recently, natural disaster. The road back to economic viability is an arduous one, but once traversed will still leave four out of five Hondurans in poverty and the country with little overall improvement in its development prospects. This country continues to rely greatly on remittances from its emigrés, which now account for more than one sixth of the GDP.

El Salvador is Central America's smallest country territorially—smaller even than Belize—but with a population about 25 times as large (7.3 million), it is the most densely peopled. With Belize, it is one of only two continental republics that lack coastlines on both the Caribbean and Pacific sides (Fig. 4-12). El Salvador adjoins the Pacific in a narrow coastal plain backed by a chain of volcanic mountains, behind which lies the country's heartland. Unlike neighboring Guatemala, El Salvador has a quite homogeneous population (90 percent mestizo and just 1 percent Amerindian). Yet ethnic homogeneity has not translated into social or economic equality, or even opportunity. Coffee cultivation has been particularly important to El Salvador; it is mostly produced on the large landholdings of a few wealthy landowners and on the backs of a subjugated peasant labor force. The military supported this system and over the years repeatedly suppressed violent and desperate peasant uprisings.

From 1980 to 1992, El Salvador was torn by a devastating civil war exacerbated by outside arms supplies from the United States (supporting the government) and Nicaragua (aiding the Marxist rebel forces). But ever since the negotiated end to that war, efforts have been under way to prevent a recurrence because El Salvador is having difficulty overcoming its legacy of severe inequality. The civil war did have one positive result: affluent citizens who left the country and did well in the United States and elsewhere remit substantial funds back home, which now provide the largest single source of foreign revenues. This has helped stimulate such industries as apparel and footwear manufacturing, as well as food processing. But a major stumbling block to revitalization of the agricultural sector has again been land reform.

Nicaragua (population 5.9 million) is best approached by reexamining the map (Fig. 4-12), which underscores the country's pivotal position in the heart of Central America. The Pacific coast follows a southeasterly direction, but the Caribbean coast is oriented north-south so that Nicaragua forms a triangle of land with its lakeside capital, Managua, located in a valley on the mountainous, earthquake-prone, Pacific side. The country's core area has always been located on this side. The Caribbean side, where the uplands yield to a coastal plain of rainforest, savanna, and swampland, has for centuries been home to Amerindian peoples such as the Miskito, who have been remote from the focus of national life.

Until the end of the 1970s, Nicaragua was a typical Central American republic, ruled by a dictatorial government and exploited by a wealthy land-owning minority, its export agriculture dominated by huge plantations owned by foreign corporations. It was a situation ripe for insurgency, and in 1979 leftist rebels overthrew the government. But the new regime quickly produced its own excesses, resulting in civil war through most of the 1980s, and a conflict in which the United States got involved by arming the rebel *Contras*, who were trying to undermine the left-wing Sandinista regime. This strife ended in 1990, and since then more democratic governments have been voted into office. In 2006, Nicaraguans elected a former Sandinista leader, Daniel Ortega, back into office, to the consternation of the U.S. government.

Nicaragua's economy has been a leading casualty of this turmoil, and for the past two decades it has ranked as continental Middle America's poorest country. Hurricane Mitch struck here too, devastating the country's farms and driving tens of thousands into the impoverished towns. This occurred at a time when the agricultural sector was recovering and land reform promised a better life for some 200,000 peasant families. Nicaragua's options are limited, however. For years there has been talk of a trans-isthmus canal that would compete with the Panama Canal further south. Figure 4-12 indicates the proposed route of this canal: it would be mostly a water route, with a relatively short overland or dry canal on the Pacific side. This canal would prove a considerable boon to the country's economy.

Costa Rica differs significantly from its neighbors. Although the country's Hispanic imprint is similar to that found elsewhere on the Mainland, it always had a very low percentage of Amerindians and lay remote from regional strife, won independence early, and perhaps most importantly, internal political stability has prevailed there over much of the past 175 years. It is bordered by two volatile countries, Nicaragua to the north and Panama to the east, yet Costa Rica is a nation with an old democratic tradition, and, in this cauldron, it has had no standing army for the past sixty years! Instead it has invested money in its social infrastructure.

From a physiographic perspective, Costa Rica is similar to its neighbors, with environmental zone divisions that parallel the coasts. The most densely settled area is the central highland zone, lying in the cooler *tierra templada*, whose heartland is the *Valle Central* (Central Valley). This fertile basin contains the country's main coffee-growing area and the leading population cluster focused on San José (Fig. 4-12)—the most cosmopolitan urban center between Mexico City and the primate cities of northern South America. To the east of the highlands are the hot and rainy Caribbean lowlands, a sparsely populated segment of Rimland where many plantations have been abandoned and replaced by subsistence farming. Between 1930 and 1960, the U.S.-based United Fruit Company shifted

most of the country's banana plantations from this crop-disease-ridden coastal plain to Costa Rica's third zone—the plains and gentle slopes of the Pacific coastlands. This move gave the Pacific zone a major boost in economic growth, and it is now an area of diversifying and expanding commercial agriculture.

The long-term development of Costa Rica's economy has given it the region's highest standard of living, literacy rate, and life expectancy (see Appendix B). Agriculture continues to dominate (with bananas, coffee, tropical fruits, and seafood the leading exports), although tourism, especially ecotourism, is expanding steadily. Costa Rica is widely known for its superb scenery and for its significant efforts to protect what is left of its diverse tropical flora and fauna. **15 Tropical deforestation** is a regionwide outcome of the colonial enterprise and continued pressures on shrinking forest resources. Even in Costa Rica more than 80 percent of the original forest has vanished—but enough remains to attract more than a million visitors annually.

Still, the country's veneer of development and high standard of living cannot mask serious problems. In terms of social structure, about one-quarter of its population of 4.4 million remains trapped in a cycle of poverty, and the huge gap between the poor and the affluent is constantly widening. With volatile neighbors and an economy that remains insufficiently diversified against risk, Costa Rica is only one step ahead of its regional partners.

Panama owes its existence to the idea of a canal connecting the Atlantic and Pacific oceans to avoid the lengthy circumnavigation of South America. In the 1880s, when Panama was still part of neighboring Colombia, a French company tried and failed to build such a waterway here. By the turn of the twentieth century, U.S. interest in a Panama canal rose sharply, and in 1903 the United States proposed a treaty that would permit a renewed effort at construction across Colombia's Panamanian isthmus. When the Colombian Senate refused to go along, Panamanians rebelled and

the United States supported this uprising by preventing Colombian forces from intervening. The Panamanians, at the behest of the United States, declared their independence from Colombia, and the new republic immediately granted the United States rights to the Canal Zone, averaging about 16 kilometers (10 mi) in width and just over 80 kilometers (50 mi) in length.

Soon canal construction commenced, and this time the project succeeded as American technology and medical advances triumphed over a formidable set of obstacles. The Panama Canal (see inset map, Fig. 4-12) was opened in 1914, a symbol of U.S. power and influence in Middle America. The Canal Zone was held by the United States under a treaty that granted it "all the rights, powers, and authority" in the area "as if it were the sovereign of the territory." Such language might suggest that the United States held rights over the Canal Zone in perpetuity, but the treaty nowhere stated specifically that Panama permanently yielded its own sovereignty in that transit corridor. In the 1970s, as the canal was transferring more than 14,000 ships per year (that number is now only slightly lower, but the cargo tonnage is up significantly) and generating hundreds of millions of dollars in tolls, Panama sought to terminate U.S. control in the Canal Zone. Delicate negotiations began. In 1977, an agreement was reached on a staged withdrawal by the United States from the territory, first from the Canal Zone and then from the Panama Canal itself (a process completed on December 31, 1999).

Panama today reflects some of the usual geographic features of the Central American republics. Its population of 3.4 million is 70 percent mestizo and also contains substantial Amerindian, white, and black minorities. Spanish is the official language, but English is also widely used. Ribbon-like and oriented east-west, Panama's topography is mountainous and hilly. Eastern Panama, especially Darien Province adjoining Colombia, is densely forested, and here is the only remaining gap in the intercontinental Pan American Highway. Most of the rural population lives in the

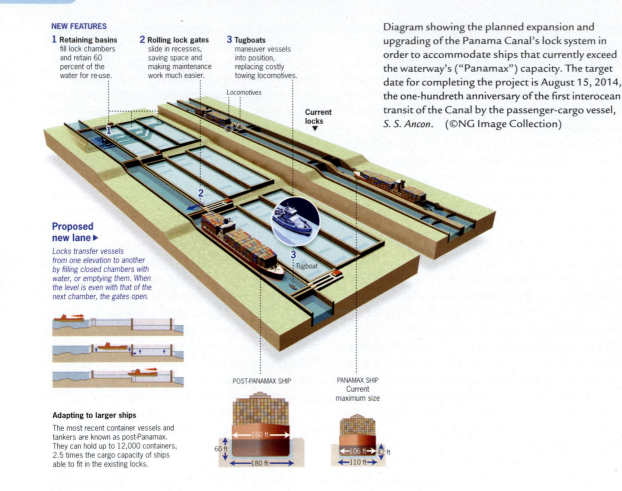

NEW FEATURES

1 Retaining basins fill lock chambers and retain 60 percent of the water for re-use.

2 Rolling lock gates slide in recesses, saving space and making maintenance work much easier.

3 Tugboats maneuver vessels into position, replacing costly towing locomotives.

Locomotives

Current locks ▼

Proposed new lane ▶

Locks transfer vessels from one elevation to another by filling closed chambers with water, or emptying them. When the level is even with that of the next chamber, the gates open.

Adapting to larger ships

The most recent container vessels and tankers are known as post-Panamax. They can hold up to 12,000 containers, 2.5 times the cargo capacity of ships able to fit in the existing locks.

POST-PANAMAX SHIP

160 ft
60 ft
180 ft

PANAMAX SHIP
Current maximum size

106 ft — 42 ft
110 ft

Diagram showing the planned expansion and upgrading of the Panama Canal's lock system in order to accommodate ships that currently exceed the waterway's ("Panamax") capacity. The target date for completing the project is August 15, 2014, the one-hundreth anniversary of the first interocean transit of the Canal by the passenger-cargo vessel, *S. S. Ancon*.   (©NG Image Collection)

uplands west of the canal; there, Panama produces bananas, shrimps and other seafood, sugarcane, coffee, and rice. Much of the urban population is concentrated in the vicinity of the waterway, anchored by the cities at each end of the canal.

Near the northern end of the Panama Canal lies the city of Colón, site of the Colón Free Zone, a huge trading entrepôt designed to transfer and distribute goods bound for South America. It is augmented by the Manzanillo International Terminal, an ultramodern port facility capable of transshipping more than 1000 containers a day. By 2002 China had become the third-largest user of the Canal (after the United States and Japan) and accounted for more than 20 percent of the cargo entering the Colón Free Zone.

Near the southern end lies Panama City, the 'Miami' of the Caribbean because of its waterfront location and skyscrapered skyline. The capital is the financial center that handles the funds generated by the Canal, but its high-rise profile also reflects the proximity of Colombia's illicit drug industry and associated money-laundering and corruption. The Panamanians' pervasive poverty presents a stark contrast to the city's modern image.

The Panama Canal has political as well as economic implications for Panama. For many years, Panama has recognized Taiwan as an independent entity and has sponsored resolutions to get Taiwan readmitted to the United Nations (where it was ousted in favor of China in 1971). In return, Taiwan has

spent hundreds of millions of dollars in Panama in the form of investment and direct aid. But now, China's growing presence in the Canal and the Colón Free Zone is causing a conundrum. Panama has awarded a contract to a Hong Kong-based (and thus Chinese) firm to operate and modernize the ports at both ends of the Canal, resulting in investments approaching $500 million. Now, China presses Panama for a switch in recognition, using this commitment as leverage.

Meanwhile, many huge ships now sailing the oceans are too large to transit the Canal. The Panama Canal Authority has therefore embarked on a plan to modernize the waterway by building new locks that will be able to handle vessels twice as large as the current maximum. Not only will this reverse the decline in the share of world cargo passing through the Canal, but it will also increase the Canal's efficiency. That will make potential competitors (such as Mexico with its dry-canal plan and Nicaragua with a scheme that would exploit Lake Managua in a mixed land-water venture) less eager to make huge investments in such projects.

In 2006 Panamanian voters overwhelmingly approved the modernization plan, which will involve building a new set of locks that will be 40 percent longer and 60 percent wider than the existing ones (see diagram at left). The cost will be financed by charging ships more for passage and by international loans (China is sure to play its role), and the project is due to be completed in time for the Canal's centenary in 2014. To what extent it will help ordinary Panamanians escape poverty (which afflicts 40 percent of the population) remains uncertain.

THE CARIBBEAN BASIN

As Figures 4-1 and 4-2 reveal, the Caribbean Basin, Middle America's island region, consists of a broad arc of numerous islands extending from the western tip of Cuba to the southern coast of Trinidad. The four larger islands or *Greater Antilles* (Cuba, Hispaniola, Jamaica, and Puerto Rico) are clustered in the western half of this arc. The smaller islands or

Lesser Antilles extend to the east in a crescent-shaped zone from the Virgin Islands to Trinidad and Tobago. (Breaking this tectonic-plate-related regularity are the Bahamas and the Turks and Caicos, north of the Greater Antilles, and numerous other islands too small to appear on a map at the scale of Fig. 4-1.)

On these islands, whose combined land area constitutes only 9 percent of Middle America, lie 33 states and several other entities. Europe's colonial flags have not totally disappeared from this region, and the U.S. flag flies over Puerto Rico. The populations of these states and territories, however, comprise 21 percent of the entire geographic realm, making this the most densely peopled part of the Americas.

Economic and Social Patterns

Untold wealth flowed to the European colonists who invaded this region, enslaved and eventually obliterated the indigenous Amerindian (Carib and Arawak) population, and brought to these islands in bondage the Africans to work on the sugar plantations that made the colonists rich. But competition from elsewhere ended the near-monopoly Caribbean sugar had enjoyed on European markets, and the region is today one of the poorest of the world (see Fig. 4-3). Environmental, demographic, political, and economic circumstances combine to impede development in the region at almost every turn. The Caribbean Basin is scenically beautiful, but it is a difficult place to make a living, and poverty and emigration are the norm.

After the sugar trade collapsed, millions were pushed into a life of subsistence, malnutrition, and even hunger. Population growth created ever-greater pressure on the limited land. The American market for sugar allowed some island countries to revive their sugar exports, including the Dominican Republic and Jamaica. After 1959 Cuba's exports went to the Soviet Union, but when the USSR disintegrated in 1991 Cuba's sugar industry collapsed as well. Some agricultural diversification occurred in the Lesser Antilles, where bananas, spices, and sea-island cotton replaced sugar, but farming is a risky

profession in the Caribbean with frequent hurricanes and volcanic hazards. Foreign markets are not dependable, and producers are trapped in a disadvantageous international economic system they cannot change.

Not surprisingly, many farmers and their families simply abandon their land and leave for towns and cities, both within their countries as well as outside. Although the Caribbean region is less urbanized, overall, than other parts of the Americas, it is far more urbanized than Subsaharan Africa or the Pacific Realm. Nearly two-thirds of the Caribbean's population now live in cities such as Santo Domingo in the Dominican Republic, Havana in Cuba, Port-au-Prince in Haiti, and San Juan in Puerto Rico. Many of the region's cities reflect the poverty of the citizens who were driven to seek refuge there. Port-au-Prince has some of the world's worst slums, and desolate squatter settlements surround cities such as Kingston, Jamaica, and others.

Ethnicity and Advantage

The human geography of the Caribbean region carries strong imprints of the cultures of Subsaharan

The Caribbean idyll—a couple walk along a beach in Grenada. North American and European tourists enjoy this part of the world as a playground and a place to thaw out during long winters. Nonetheless, this beautiful façade masks the reality of the Caribbean societies who do not benefit fully from the northern dollars and euros but who do suffer from the consequent environmental degradation.   (© Barry Tessman/National Geographic/Getty Images)

Africa. The legacy of European domination also lingers. The historical geography of Cuba, Hispaniola, and Puerto Rico is suffused with Hispanic culture; Haiti and Jamaica carry stronger African legacies. But the reality of this ethnic diversity is that European lineages still hold the advantage. Hispanics tend to be in the best positions in the Greater Antilles; people who have mixed European-African ancestries, and who are described as **16 mulatto**, rank next. The largest lower part of this social pyramid is also the least advantaged: the Afro-Caribbean majority. In virtually all societies of the Caribbean, the minorities hold disproportionate power and exert overriding influence. In Haiti, the mulatto minority accounts for barely 5 percent of the population but has long held most of the power. In the adjacent Dominican Republic, the pyramid of power puts Hispanics (16 percent) at the top, the mixed sector (73 percent) in the middle, and the Afro-Caribbean minority (11 percent) at the bottom. Historic advantage has a way of perpetuating itself.

The composition of the population of the islands is further complicated by the presence of Asians from both China and India. During the nineteenth century, the emancipation of slaves and ensuing local labor shortages brought some far-reaching solutions. Some 100,000 Chinese immigrated to Cuba as indentured laborers, and Jamaica, Guadeloupe, and especially Trinidad saw nearly 250,000 South Asians arrive for similar purposes. To the African-modified forms of English and French heard in the Caribbean, therefore, can be added several Asian languages. The ethnic and cultural variety of the plural societies of Caribbean America is indeed endless.

Tourism: Promising Alternative?

Given the Caribbean region's limited economic options, does the tourist industry offer better opportunities? Opinions on this question are divided. The resort areas, scenic treasures, and historic locales of Caribbean America attract well over 20 million visitors annually, with about half of these tourists traveling on Florida-based cruise ships. Certainly, Caribbean tourism is a prospective money-maker for many islands. In Jamaica alone, this industry now accounts for about one-sixth of the gross domestic product and employs more than one-third of the labor force.

But Caribbean tourism also has serious drawbacks. The industry can be cyclical and tied to the health of the economy in source countries. The invasion of poor communities by affluent tourists contributes to rising local resentment, which is further fueled by the glaring contrasts of shiny new hotels towering over substandard housing and luxury liners gliding past poverty-stricken villages. Tourism can debase the local culture, which often is adapted to suit the visitors' tastes at hotel-staged 'culture' shows. From an environmental perspective tourism is also problematic. Cruise ships can harm coral reefs; the emphasis on white sandy beaches impacts coastal deposition processes; and too many hotels can overwhelm sewage and water infrastructure. While tourism does generate income in the Caribbean, the intervention of island governments and multinational corporations removes opportunities from local entrepreneurs in favor of large operators and major resorts.

The Greater Antilles

The four islands of the Greater Antilles contain five political entities: Cuba, Jamaica, Haiti, the Dominican Republic, and Puerto Rico (Fig. 4-2). Haiti and the Dominican Republic share the island of Hispaniola.

FIGURE 4-14

© H. J. de Blij, P. O. Muller, and John Wiley & Sons, Inc.

Havana, Cuba. Cubans are well taken care of by their government in terms of health care and education. Housing and other infrastructure are dilapidated, however. Individual freedoms are curtailed, and the economy is in tatters after subsidies from the Soviet Union dried up when that state fell apart in 1991. With Fidel turning over control to his brother Raúl, it remains to be seen what kind of change the island will see.   (© Vince J. WinklerPrins)

Cuba, the largest Caribbean island-state in terms of both territory (111,000 square kilometers/43,000 sq mi) and population (11.4 million), lies only 145 kilometers (90 mi) from the southern tip of Florida (Fig. 4-14). Havana, the now-dilapidated capital, lies almost directly across from the Florida Keys on the northwest coast of the elongated island. Cuba was a Spanish possession until the late 1890s when, with American help in the Spanish-American War, it attained independence. Fifty years later, a U.S.-backed dictator was in control, and by the 1950s Havana had become an American playground. The island was ripe for revolution, and in 1959 Fidel Castro's insurgents gained control, thereby making Cuba a communist dictatorship and a Soviet client. Castro's rule survived the collapse of the Soviet Empire despite the loss of subsidies and sugar markets on which it had long relied.

Sugar was Cuba's economic mainstay for many years; the plantations, once the property of rich landowners, extend all across the territory. Today sugarcane is losing its position as the leading Cuban foreign exchange earner. Mills are being closed down, and the canefields are being cleared for other crops and for pastures. Cuba has other economic opportunities, however, especially in its highlands. There are three mountainous areas, of which the southeastern chain, the Sierra Maestra, is the highest and most extensive. These highlands create considerable environmental diversity as reflected by extensive, timber-producing tropical forests and varied soils on which crops ranging from tobacco to subtropical and tropical fruits are grown. Rice and beans are the staples, but Cuba cannot meet its needs and so must import food. The savannas of the center and west support livestock. Although Cuba has only limited mineral reserves. its nickel deposits are extensive and have been mined for a century.

Early in the twenty-first century, Cuba found a crucial new supporter in Venezuela's leader, Hugo Chávez. Cuba has no domestic petroleum reserves, but Venezuela is oil-rich and, since 2003, has been providing all of the fuel that Cuba needs. In return, Castro sent 30,000 health workers and other professionals to Venezuela, where they aid the poor.

Poverty, crumbling infrastructure, crowded slums, and unemployment mark the Cuban cultural landscape, but Cuba's regime still has support among the general population. During the Castro period, much was done to bring the Afro-Cuban population into the mainstream through education and health provisions. Take a look at Appendix B and you will notice that Cuba has indices of health care and literacy that rival those in developed countries. Cubans point to Guatemala, Nicaragua, and El Salvador and ask whether those countries are better off than they are under Castro. But in the United States, the view is different: Cuba, exiles and locals agree, could be the shining star of the Caribbean, its people free, its tourist economy booming, its products flowing to

Hurricanes. This man is surveying storm-surge damage caused by Hurricane Dean, which devastated Jamaica in August 2007. Hurricanes are a way of life in Middle America with frequent hits on the islands as well as the mainland. In addition, mainland Middle America is affected by both the Atlantic and Pacific hurricanes. Hurricanes form in these latitudes, a combination of wind patterns and warm water, and are rotating tropical cyclones. In the United States, we are well aware of the devastation of hurricanes, but with our vast economic resources we have the ability to recover relatively quickly from disaster. In contrast, in many of the small, lower-income countries in Middle America, entire economies can be wiped out by a single storm, with little to no buffer to help with recovery. Many countries have improved their warning systems for affected areas, reducing death tolls due to hurricanes in recent years, but damage to crops and infrastructure remain severe and may take years to overcome. Natural events such as hurricanes can wipe out any significant gains in development, increasing vulnerabilities to the next event.

Since 1995, both the number and intensity of hurricanes have increased. Is this trend due to a natural cycle, or is it related to global warming? There appears to be some link, though no clear association has been established. At this point, many experts agree on at least an indirect link between global warming and hurricane frequency and intensity. Global warming is leading to an increase in sea-surface temperatures and atmospheric water vapor, and these two conditions may result in more frequent and more intense hurricanes. Climatologists maintain that a decadal cycle called the Atlantic Multi-Decadal Oscillation also influences the frequency and intensity of hurricanes. (© AP/Wide World Photos)

American markets. It could be "the future Ireland of the Caribbean," a Cuban geographer suggested recently. Much will have to change for that prediction to come true.

Jamaica lies across the deep Cayman Trench from southern Cuba, and a cultural gulf separates these two countries as well. Jamaica, a former British dependency, has an almost entirely Afro-Caribbean population. As a member of the British Commonwealth, Jamaica still recognizes the British monarch as the chief of state, represented by a governor-general. The effective head of government in this democratic country, however, is the prime minister. English remains the official language here, and British customs still linger.

Smaller than Connecticut and with 2.8 million people, Jamaica has experienced a steadily declining GNI over the past few decades despite its relatively slow population growth. Tourism has become the largest source of income, but the markets for bauxite (aluminum ore), of which Jamaica is a major exporter, have dwindled. Like other Caribbean countries, Jamaica has trouble making money from its sugar exports. Jamaican farmers also produce crops ranging from bananas to tobacco, but the country faces the disadvantages on world markets common to those in the periphery. Meanwhile, Jamaica must import all of its oil and much of its food because the densely populated coastal flatlands suffer from overuse and shrinking harvests.

The capital, Kingston, on the south coast, reflects Jamaica's economic struggle. Almost none of the hundreds of thousands of tourists who visit the country's beaches, stop off for a few hours at cruise ship ports, or explore its ship ports, or explore its limestone towers and caverns, even get a glimpse of what life is like for the ordinary Jamaican.

Haiti, the poorest state in the Western Hemisphere by virtually every measure, occupies the western part of the island of Hispaniola, directly across the Windward Passage from eastern Cuba (Fig. 4-15). Arawaks, not Caribs, formed the dominant indigenous population here, but the Spanish

economic and social collapse. By 2003, Haiti's GNI per capita had fallen to less than half that of Jamaica, a level lower than that of many poor Subsaharan African countries; foreign aid makes possible most of the country's limited public expenditures. Health conditions are dreadful: malnutrition is common, AIDS is rampant, diseases ranging from malaria to tuberculosis are rife, but hospital beds and doctors are in short supply. Conditions in and around the capital, Port-au-Prince, are among the worst in the world—less than 1000 kilometers (625 mi) from the United States.

The **Dominican Republic** has a larger share of the island of Hispaniola than Haiti (Fig. 4-15) in terms of both territory and population. Fly along the north-south border between the two countries, and you see a crucial difference: to the west, Haiti's hills and plains are treeless and gulleyed, its soils eroded, and its streams silt-laden. To the east, forests drape the countryside and streams run clear.

Indigenous Caribs inhabited this eastern part of Hispaniola, and when the European colonists arrived on the island they were in the process of driving the Arawaks westward. Spanish colonists made this a prosperous colony, but then Mexico and Peru diverted Spanish attention from Hispaniola and the territory was ceded to France. But Hispanic culture lingered, and after the 1804 revolution in Haiti the French gave eastern Hispaniola back to Spain. After the Dominican Republic declared its independence in 1821, Haitian forces invaded it and occupied the republic until 1844, creating an historic animosity that persists today.

The mountainous Dominican Republic has a wide range of natural environments and a far stronger

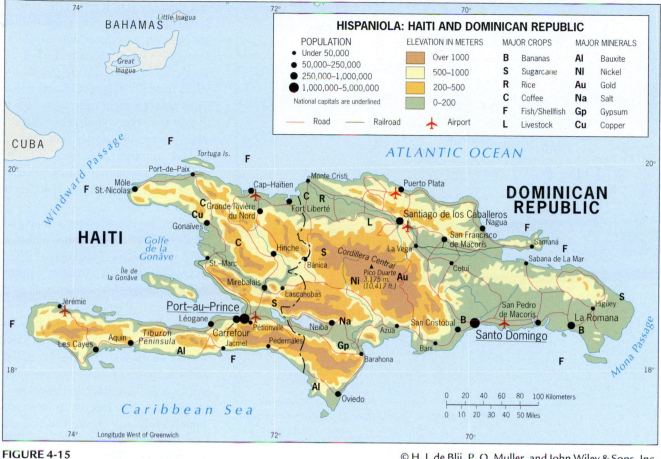

HISPANIOLA: HAITI AND DOMINICAN REPUBLIC

POPULATION
- Under 50,000
- 50,000–250,000
- 250,000–1,000,000
- 1,000,000–5,000,000

National capitals are underlined

—— Road   —— Railroad   ✈ Airport

ELEVATION IN METERS
- Over 1000
- 500–1000
- 200–500
- 0–200

MAJOR CROPS
- B Bananas
- S Sugarcane
- R Rice
- C Coffee
- F Fish/Shellfish
- L Livestock

MAJOR MINERALS
- Al Bauxite
- Ni Nickel
- Au Gold
- Na Salt
- Gp Gypsum
- Cu Copper

FIGURE 4-15

© H. J. de Blij, P. O. Muller, and John Wiley & Sons, Inc.

colonists who first took Hispaniola killed many, worked most of the survivors to death on their plantations, and left the others to die of the diseases they brought with them. French pirates established themselves in coastal coves along Hispaniola's west coast even as Spanish activity focused on the east, and just before the eighteenth century opened France formalized the colonial status of Saint Domingue. During the following century, French colonists established vast plantations in the valleys of the northern mountains and in the central plain, laid out elaborate irrigation systems, built dams,

and brought in a large number of Africans in bondage to work the fields. Prosperity made the colonists rich, but the African workers suffered terribly. They rebelled and, in a momentous victory over their European oppressors, established the independent republic of Haiti (resurrecting the original Arawak name for it) in 1804.

Strife among Haitian groups, American intervention, mismanagement, and dictatorship doomed the fortunes of the republic. In the 1990s, the United States attempted to help move Haiti toward more representative government, but the country was in

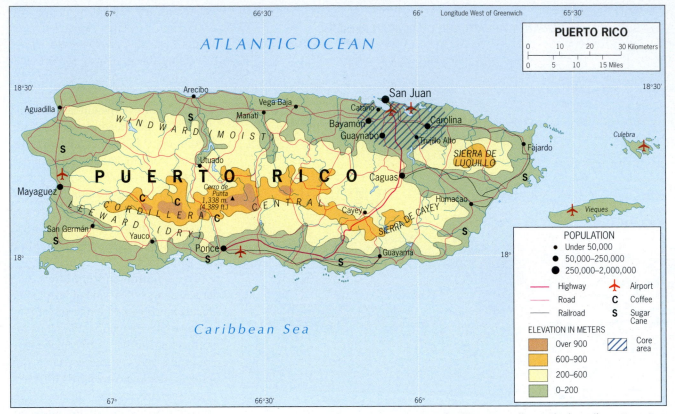

FIGURE 4-16

© H. J. de Blij, P. O. Muller, and John Wiley & Sons, Inc.

lous than Oregon. It fell to the United States more than a century ago during the Spanish-American War of 1898. Since the Puerto Ricans had been struggling for some time to free themselves from Spanish control, this transfer of power was, in their view, only a change from one colonial power to another. As a result, the first half-century of U.S. administration was difficult, and it was not until 1948 that Puerto Ricans were permitted to elect their own governor.

When the island's voters approved the creation of a Commonwealth in a 1952 referendum, Washington, D.C., and San Juan, the two seats of government, entered into a complicated arrangement. Puerto Ricans are U.S. citizens but pay no federal taxes on local incomes. The Puerto Rican Federal Relations Act governs the island under the terms of its own constitution and awards it considerable autonomy. Puerto Rico also receives a large annual subsidy and other financial aid from Washington, totaling over U.S. $4.2 billion in 2005.

resource base than Haiti. Nickel, gold, and silver have long been exported along with sugar, tobacco, coffee, and cacao, but tourism (the great opportunity lost to Haiti) is the leading industry. A long period of dictatorial rule punctuated by revolutions and U.S. military intervention ended in 1978 with the first peaceful transfer of power following a democratic election.

Political stability brought the Dominican Republic rich rewards, and during the late 1990s the economy, based on manufacturing, high-tech industries, and remittances from Dominicans abroad as well as tourism, grew at an average 7 percent per year. But in the early 2000s the economy imploded, not only be-

cause of the downturn in the world economy but also because of bank fraud and corruption in government. Suddenly the Dominican peso collapsed, inflation skyrocketed, jobs were lost, and blackouts prevailed. As the people protested, lives were lost and the self-enriched elite blamed foreign financial institutions that were unwilling to lend the government more money. Yet the economy is recovering and the country was able to join CAFTA-DR in 2007.

Puerto Rico is the largest U.S. domain in Middle America, the easternmost and smallest island of the Greater Antilles (Fig. 4-16). This 9000-square-kilometer (3500-sq-mi) island, with a population of 3.9 million, is larger than Delaware and more popu-

Despite these apparent advantages in the Caribbean, Puerto Rico has not thrived under U.S. administration. Long dependent on a single-crop economy (sugar), the island based its industrialization during the 1950s and 1960s on its comparatively cheap labor, tax breaks for corporations, political stability, and special access to the U.S. market. As a result, pharmaceuticals, electronic equipment, and apparel top today's list of exports, not sugar or bananas. But this industrialization failed to stem a tide of emigration that carried more than 1 million Puerto Ricans to New York City alone. The same wages that favored corporations kept many Puerto Ricans poor or unemployed. By some estimates, unemployment on the island stands at about 45 percent today. Most receive federal support. Another 30 percent work in the public sector, that is, in government. Puerto Rico's welfare-state condition discourages initiative, so that many people who could be working do not because their federal subsidy would decrease accordingly. Economists report that workers from the Dominican Republic even come to Puerto Rico to take low-wage jobs Puerto Ricans do not want. Although Puerto Ricans are vocal in demanding political change, it is perhaps not surprising that successive referendums during the 1990s resulted in retention of the status quo—continuation of Commonwealth status rather than either Statehood or independence. The issue will no doubt continue to pose a formidable challenge to American statecraft in the years ahead.

The Lesser Antilles

As Figure 4-2 shows, the Greater Antilles are flanked by two clusters of islands: the extensive Bahamas-Turks/Caicos archipelago to the north and the Lesser Antilles to the east and south. The Lesser Antilles are grouped geographically into the Leeward Islands and the Windward Islands, a climatologically incorrect reference to the prevailing airflows in this tropical area. The Leeward Islands extend from the U.S. Virgin Islands to the French dependencies of Guadeloupe and Martinique, and the Windward Islands from St. Lucia to the Netherlands Antilles off the Venezuelan coast (Fig. 4-2). It should be noted that these countries and territories share the environmental risks of this region: earthquakes, volcanic eruptions, and hurricanes; that they face to varying degrees similar economic challenges in the form of limited domestic resources, high population densities, soil deterioration, land fragmentation, and market limitations; that tourism and offshore banking has become the leading industry for many; that polit-

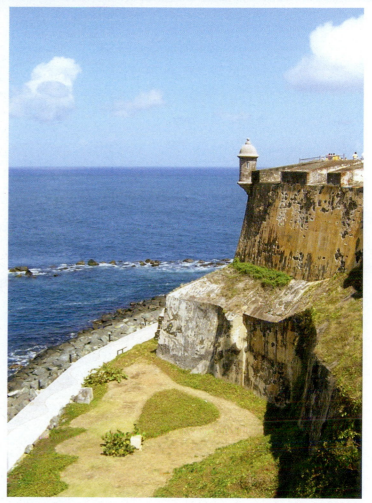

Puerto Rico was claimed for Spain by Columbus early in the colonization process. Its natural harbor at San Juan had strategic importance in the colonial enterprise as ships sailing together from Spain used the harbor as a main destination, resupply center, and transshipment point. At the entrance to the harbor, the Spanish built a fort to protect it from foreign attack; it is the oldest Spanish fort in the New World. Today, 'El Morro,' as the fort is known, is a tourist attraction for the many visitors to the island. (© A. WinklerPrins)

ical status ranges from complete sovereignty to continuing dependency; and that cultural diversity is strong not only between but also within islands. In economic terms, the GNI figures given in Appendix B may look encouraging, but these figures conceal the social reality in virtually every entity: the gap between the fortunate few who are well off and the great majority who are poor remains enormous.

Middle America is a physically, culturally, and economically fragmented and diverse geographic realm that defies generalization. Given its strong Amerindian presence, its North American infusions, and its lingering Western European traditions, this is a realm of hybridization, of mixed cultures and traditions that offer resilience to its people as they confront challenging economic conditions.

FIGURE 5-1

Map: © H. J. de Blij, P. O. Muller, and John Wiley & Sons, Inc.

5

SOUTH AMERICA

In This Chapter

- The growing power of indigenous peoples
- U.S. initiatives in economics and politics less welcome today
- Tentative efforts toward economic integration: China lends a hand
- Is Brazil a superpower in the making?
- The poor performance of rich Argentina
- Chile: Star of the realm?

CONCEPTS, IDEAS, AND TERMS

1 Amerindian
2 *Altiplano*
3 Land alienation
4 Plural society
5 Commercial agriculture
6 Subsistence agriculture
7 *Cerrado*
8 Free Trade Area of the Americas (FTAA)
9 Urbanization
10 Rural-to-urban migration
11 Megacity
12 'Latin' American city model
13 Informal sector
14 *Barrio (favela)*
15 Insurgent state
16 Failed state
17 El Niño
18 Forward capital
19 Growth pole

REGIONS

THE NORTH
THE WEST
THE SOUTH
BRAZIL

Photos: *(upper left)* São Paulo, Brazil © A. WinklerPrins; *(above)* Belém, Brazil © A. WinklerPrins.

163

OF ALL THE continents, South America has the most familiar shape—a giant triangle connected to Middle America's tenuous land bridge to North America. South America also lies not only south but mostly east of North America. Lima, the capital of Peru—one of the continent's westernmost cities—lies farther east than Miami, Florida. Thus South America juts out much more prominently into the Atlantic Ocean toward Southern Europe and Africa than does North America. Lying so far eastward means that South America's western flank faces a much wider Pacific Ocean, with the distance from Peru to Australia nearly twice that from California to Japan. But it also means that the eastern flank is closer to Africa.

As if to reaffirm South America's eastward orientation, the western margins of the continent are rimmed by one of the world's longest and highest mountain ranges, the Andes, a gigantic wall that extends unbroken from Tierra del Fuego near the continent's southern tip in Chile to northeastern Venezuela in the far north (Fig. 5-1, chapter opener map). The other major physiographic feature of South America dominates its central north—the Amazon Basin; this vast humid-tropical amphitheater is drained by the mighty Amazon River, which is fed by several major tributaries. Much of the remainder of the continent can be classified as plateau, with the most important components being the Brazilian Highlands that cover most of Brazil southeast of the Amazon Basin, the Guiana Highlands located north of the lower Amazon Basin, and the cold Patagonian Plateau that blankets the southern third of Argentina. Figure 5-1 also reveals two other noteworthy river basins beyond Amazonia: the Paraná-Paraguay Basin of south-central South America, and the Orinoco Basin in the far north that drains interior Colombia and Venezuela.

Defining the Realm

South America is a realm in dramatic transition, and it is not clear where this transition will lead. During much of the twentieth century, South American countries were in frequent political turmoil; dictatorial regimes ruled from one end of the realm to the other; unstable governments fell with damaging frequency. Widespread poverty, harsh regional disparities, poor internal surface connections, limited international contact, and economic stagnation prevailed.

Today, at the end of the first decade of the twenty-first century, there is significant change. Democratic governments have replaced authoritarian regimes. South America's countries are becoming more interconnected and cooperative. New transport routes cross international borders, and new settlement frontiers have opened. High energy and commodity prices have fueled economic growth in several countries.

What Lies Ahead?

But achieving economic growth and democratic statecraft rarely come without setbacks, and over the past decade South America has seen reversals as well as gains. The realm's giant, Brazil, has launched a major campaign against poverty and an effort to maintain financial rigor that have run up against endemic corruption in government. The economy of Argentina is recovering after an implosion that shook the country to its core late in the twentieth century. Whereas Chile is the realm's economic success story, emerging from the horrors of the Pinochet era as a stable and vibrant democracy with a thriving economy, neighboring Bolivia is teetering on becoming a failed state despite

MAJOR GEOGRAPHIC QUALITIES OF South America

1. South America's physiography is dominated by the Andes Mountains in the west and the Amazon Basin in the central north. Much of the remainder is plateau country.

2. Half of the realm's area and half of its population are concentrated in one country—Brazil.

3. South America's population remains concentrated along the continent's coasts. Most of the interior is sparsely peopled, but sections of it are now undergoing significant development and environmental change.

4. Interconnections among the states of the realm have been poor but are improving rapidly. Economic integration has become a major force, particularly in southern South America.

5. Regional economic contrasts and disparities, both in the realm as a whole and within individual countries, are strong.

6. Cultural pluralism prevails in almost all of the realm's countries.

7. Rapid urban growth continues to mark much of the South American realm, and the overall level of urbanization today is on a par with or higher than North America and Europe.

a realization of its energy riches. And on the north coast lies Venezuela, its oil reserves by far the largest in the hemisphere and among the largest in the world, and its political life dominated by a one-time coup leader whose closest ideological ally is Cuba's communist ruler and whose major adversary is the U.S. government.

Today, the United States hopes to foster democracy and encourage regional economic integration, but many South Americans remember past U.S. toleration of, and even support for, the realm's former dictators. Venezuela's populist dictator champions the poor and uses high oil prices to counter American influence, campaigning vigorously against the notion of a **Free Trade Area of the Americas (FTAA)** (discussed further below) and warning South American governments against capitalist plots. Such advice finds a ready market because the great majority of South Americans remain poor. By some measures, the disparity between rich and poor is wider in this realm than in any other, and wealth is disproportionately concentrated in the hands of a small minority (the richest 20 percent of the realm's inhabitants control 70 percent, while the poorest 20 percent own only 2 percent). The question of the day is whether South America can sustain its recovery against political, ideological, and economic odds.

THE HUMAN SEQUENCE

Although modern South America's largest populations are situated along coastal areas (Fig. 5-2), during pre-Columbian times the Andes Mountains and some interior lowland areas contained the most densely peopled and best organized states on the continent. Even though the origins of these civilizations continue to be debated, it is still generally accepted that the Incas were descendants of ancient peoples who came to South America via the Middle American land bridge. Thus, for thousands of years before the Europeans arrived in the sixteenth century, indigenous **1** **Amerindian** societies had been developing in South America.

FIGURE 5-2   © H. J. de Blij, P. O. Muller, and John Wiley & Sons, Inc.

Early South Americans

About one thousand years ago, a number of regional cultures thrived in Andean valleys and basins and at places along the Pacific coast. By AD 1300, the Incas had established themselves in an elongated, high-altitude basin called an **2** **altiplano** at Cuzco in the high Andes (Fig. 5-3). As the term implies, these *altiplanos* are high-elevation valleys between parallel Andean ranges, filled and floored by sediments

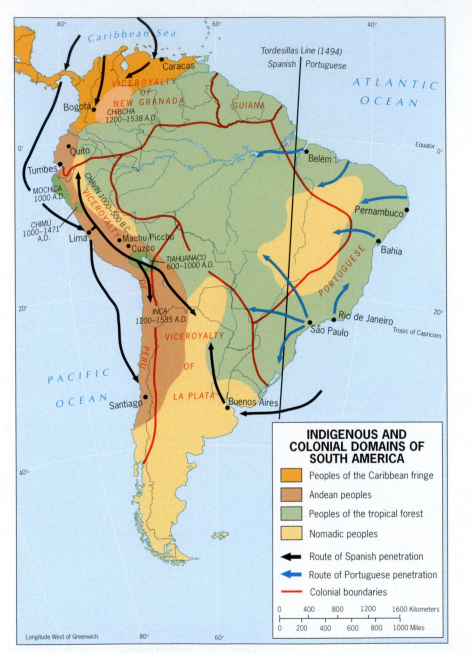

FIGURE 5-3 © H. J. de Blij, P. O. Muller, and John Wiley & Sons, Inc.

The Inca State

The Incas were great military strategists, but even more impressive was their ability to integrate vanquished peoples into a stable and efficiently functioning state. They were expert road and bridge builders, colonizers, and administrators, and in a short period they unified an empire that extended from present-day Colombia to Chile (Fig. 5-3). In fact, the Incas were in a minority in their own far-flung state, becoming a ruling elite in a rigidly class-structured society. So centralized and controlled was this state that a takeover at the top was enough to gain power over the whole empire—as a small army of Spanish invaders realized in the 1530s. Already weakened through the silent ally of epidemic disease, the Inca system broke down, with little resistance, and the state collapsed, leaving behind a cultural landscape studded with magnificent structures of which Machu Picchu, in present-day Peru, is the most spectacular. It also left a legacy of contrasting social values between Hispanic and Amerindian citizens that remains a part of life in Andean South America to this day.

Other Amerindian Civilizations

In addition to the mighty Inca, South America was home to a number of other Amerindian civilizations. Most of these are much less well known despite leaving behind important legacies. Recent research in the Amazon demonstrates that there were once settled populations there as well. Along the banks of the Xingu River, a major tributary of the Amazon River, for example, communities on the order of several thousand lived a sedentary life, utilizing the resources of the river as well as the forests. Farming systems were complex and included the cultivation of forests to become food-producing components of the landscape. These civilizations peaked in the early 1400s, just before the time that the Americas were being discovered by Europeans and were decimated by diseases. Current estimates are that approximately 90 percent of native Amazonians died from diseases introduced by Europeans, even before Europeans were able to describe native livelihoods. Since they built their structures with

from the adjacent mountains. By building terraces on steep hillsides, the Incas exploited fertile soils and mountain meltwaters to achieve specialized agriculture that sustained an ever-growing population, and from their mountainous abode they extended their authority over the peoples of other *altiplanos* as well as others living along the Pacific coast.

forest materials, these decomposed quickly; as a result, their settlements and cultivated forests were soon overgrown, and the landscape was reclaimed by the fast-growing tropical rainforest. Unlike the Incas in the Andes and the Maya and Aztecs in Middle America, these native Amazonians did not leave us stone temples; instead they left us a wild rainforest filled with patches of species that were once cultivated as well as a hint of lifeways now past.

Modern Heirs

South America's densest Amerindian population clusters extend geographically in a wide arc from present-day Chile through the highlands of Bolivia, Peru, and Ecuador to Colombia (Fig. 5-2). This detail matters because South America's long-marginalized indigenous peoples are experiencing a social, political, and economic awakening. They are not alone in this—we have already noted the Zapatista movement based in southern Mexico's Chiapas State, and we will see that Amerindians in Brazil are resisting encroachment by farmers—but they have had greater impact in such countries as Ecuador and Bolivia, where they form a larger proportion of the population. Amerindian leaders are emerging to exercise political influence and to bring the plight of the realm's indigenous peoples to regional as well as international attention. They were conquered, decimated by foreign diseases, robbed of their best lands, subjected to forced labor, denied the right to grow traditional crops, socially discriminated against, and swindled out of their fair share of the revenues from resources in their (modern) countries. Today they are the poorest of the poor, but they are asserting themselves. As we will discover, there are even calls for secession and independence for Amerindian-dominated areas of Andean states. It may not come to this, but the Bolivarian Revolution that gave the poor of Venezuela their champion is more popular in Andean South America than the preferences of their own governments—such as FTAA and globalization.

The Iberian Invaders

In South America as in Middle America, the location of major settlements of indigenous peoples largely determined the direction of the thrusts of European invasion. The Incas, like Mexico's Maya and Aztec peoples, had accumulated gold and silver, occupied highland areas, possessed productive farmlands, and constituted a ready labor force. Not long after the defeat of the Aztecs in 1521, Francisco Pizarro sailed southward along the continent's northwestern coast, learned of the existence of the Inca Empire, and withdrew to Spain to organize its overthrow. He returned to the Peruvian coast in 1531 with 183 men and two dozen horses, and easily toppled a civilization already crippled by European diseases and internal rivalries. In 1533, his party rode victorious into Cuzco.

At first, the Spaniards kept the Incan imperial structure intact by permitting the crowning of an emperor who was under their control. But soon the breakdown of the old order began. The new order that gradually emerged in western South America placed the indigenous peoples in serfdom to the Spaniards. Great haciendas were formed by **3 land alienation** (the takeover of indigenously held land by foreigners), taxes were instituted, and a forced-labor system was introduced to maximize the profits of exploitation.

Lima, the west coast headquarters of the Spanish conquerors, soon became one of the richest cities in the world, its wealth based on the exploitation of vast Andean silver deposits. The city also served as the capital of the Viceroyalty of Peru, as the Spanish authorities quickly integrated the new possession into their colonial empire (Fig. 5-3). Subsequently, when Colombia and Venezuela came under Spanish control and, later, when Spanish settlement expanded in what is now Argentina and Uruguay, two additional viceroyalties were added to the map: New Granada and La Plata.

Portuguese Brazil

Meanwhile, another vanguard of the Iberian invasion was penetrating the east-central part of the continent, the coastlands of present-day Brazil. This area had become a Portuguese sphere of influence because Spain and Portugal had signed a treaty in 1494 to recognize a north-south line 370 leagues west of the Cape Verde Islands as the boundary between their New World spheres of influence. This border ran approximately along the meridian of 46° 37' W longitude, thereby cutting off a sizeable triangle of eastern South America for Portugal's exploitation (Fig. 5-3). But a brief look at the political map of South America (Fig. 5-4) shows that this treaty did not limit Portuguese colonial territory to the east of the Tordesillas Treaty dividing line (see Fig. 5-3). Instead, Brazil's boundaries were bent far inland to include almost the entire Amazon Basin, and the country came to be only slightly smaller in territorial size than all the other South American countries combined. This westward thrust was the result of Spanish neglect of their lowland territories and Portuguese and Brazilian concurrent penetration, particularly by the *Paulistas*, the settlers of São Paulo who needed Amerindian slave labor to run their plantations.

The Africans

As Figure 5-3 shows, the Spaniards initially got very much the better of the territorial partitioning of South America—not just in land quality but also in the size of the aboriginal labor force. When the Portuguese began to develop their territory, they turned to the same lucrative activity that their Spanish rivals had pursued in the Caribbean—the plantation cultivation of sugar for the European market. As the indigenous labor force was shrinking, the Portuguese, too, found their labor force in the same source region, as millions of Africans were brought in bondage to the tropical Brazilian coast north of Rio de Janeiro. In fact, the overwhelming majority of enslaved Africans came to Brazil (just under 50 percent of the total). Not surprisingly, Brazil now has South America's largest black population, which is still heavily concentrated in the country's northeastern States. With Afro-Brazilians today accounting for 45 percent of Brazil's total population, the Africans decidedly constitute the

FIGURE 5-4   © H. J. de Blij, P. O. Muller, and John Wiley & Sons, Inc.

there was little interest in developing the American lands for their own sake. Only after those who had made Spanish and Portuguese America their home and who had a stake there rebelled against Iberian authority did things begin to change, and then very slowly. Thus South America was saddled with the values, economic outlook, and social attitudes of eighteenth-century Iberia—not the best tradition from which to begin the task of forging modern nation-states.

Independence

Certain isolating factors had their effect even during the wars for independence. Spanish military strength was always concentrated in Lima, and those territories that lay farthest from this center of power—Argentina and Chile—were the first to establish their independence from Spain (in 1816 and 1818, respectively). In the north Simón Bolívar led the burgeoning independence movement, and in 1824 two decisive military defeats there spelled the end of Spanish power in South America.

This joint struggle, however, did not produce unity because no fewer than nine countries emerged from the three former viceroyalties. It is not difficult to understand why this fragmentation took place. With the Andes intervening between Argentina and Chile and the Atacama Desert between Chile and Peru, overland distances seemed even greater than they really were, and these obstacles to contact proved quite effective. Hence, from their outset the new countries of South America began to grow apart amid friction and wars. Only within the past few decades have the countries of this realm finally begun to recognize the mutual advantages of increasing cooperation and to make lasting efforts to steer their relationships in this direction.

CULTURAL FRAGMENTATION

When we speak of the interaction of South American countries, it is important to keep in mind just who does the interacting. The fragmentation of colonial South America into ten individual republics, and the subse-

third major immigration of foreign peoples into South America.

Longstanding Isolation

Despite their adjacent location on the same continent, their common language and cultural heritage, and their shared national problems, the countries that

arose out of South America's Spanish viceroyalties have until quite recently existed in a considerable degree of isolation from one another. Distance and physiographic barriers reinforced this separation, and the realm's major population agglomerations adjoined the coast, mainly the eastern and northern coasts (Fig. 5-2). The viceroyalties existed primarily to extract riches and fill Spanish coffers. In Iberia

quent postures of each of these states, was the work of a small minority that constituted the landholding, upper-class elite. Thus in every country a vast majority—be they Amerindians in Peru or people of African descent in Brazil—could only watch as their European masters struggled with each other for supremacy.

Using the Land

South America, then, is a continent of **4 plural societies**, where Amerindians of different cultures, Europeans from Iberia and elsewhere, Africans mainly from western tropical Africa, and Asians from India, Japan, and Indonesia cluster in adjacent areas. The result is a cultural kaleidoscope of almost endless variety, whose internal divisions are also reflected in the realm's economic landscape. This is readily visible in the map of South America's dominant livelihood, agriculture (Fig. 5-5). Here **5 commercial** or market (for-profit) and **6 subsistence** (primarily for household use) farming exist side by side to a greater degree than anywhere else in the world. The geography of commercial agricultural systems (as shown by areas of soybean and non-soy grain production) was initially tied to the distribution of landholders of European background, while subsistence farming (such as highland mixed subsistence–market, agroforestry, and shifting cultivation) is associated with the spatial patterns

ENVIRONMENTALLY SPEAKING . . .

Soybean Expansion in Brazil. As Figure 5-5 shows, soybean cultivation covers a large part of the agricultural landscape in South America. In the past several decades the amount of land in soy has increased enormously. According to the USDA, acreage has increased 65 percent in the last five years, and the *soy frontier* is marching northward and eastward from its beginnings in the south of Brazil and adjacent areas of Argentina and Paraguay into the savanna-like **7 cerrado** and the Amazon rainforest. In the last decade Brazil, in particular, has accelerated its production of soybeans, and the country is now a leading producer of this global commodity. Most of it is exported to Europe and China where it is used as livestock feed. The environmental impact of this crop has been great and has taken on two forms: (1) the direct conversion of savanna and rainforest vegetation into soybean fields, and (2) the indirect clearing of the Amazon rainforest as land speculation associated with its possible cultivation escalates. Both of these biomes are hotbeds of biodiversity. Deforestation is also accompanying the road building and paving initiated to improve the infrastructure needed to export soy from Brazil. (© D. Donne Bryant/Stock Photography)

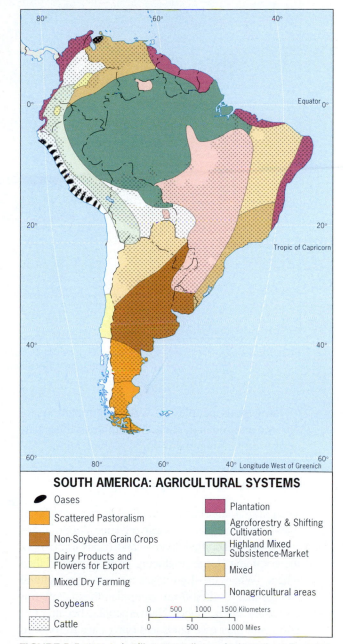

SOUTH AMERICA: AGRICULTURAL SYSTEMS

- Oases
- Scattered Pastoralism
- Non-Soybean Grain Crops
- Dairy Products and Flowers for Export
- Mixed Dry Farming
- Soybeans
- Cattle
- Plantation
- Agroforestry & Shifting Cultivation
- Highland Mixed Subsistence-Market
- Mixed
- Nonagricultural areas

0   500   1000   1500 Kilometers
0   500   1000 Miles

FIGURE 5-5 © H. J. de Blij, P. O. Muller, and John Wiley & Sons, Inc.

FIGURE 5-6

© H. J. de Blij, P. O. Muller, and John Wiley & Sons, Inc.

of indigenous peoples as well as populations of African and Asian descent. Nonetheless, these patterns are changing in this era of globalization.

Cultural Landscapes

The cultural landscape of South America, similar to that of Middle America, is a layered one. Amerindians cultivated and crafted diverse landscapes throughout the continent, some producing greater impacts than others. When the Europeans arrived, the cultural change that resulted from depopulation severely impacted the environment. Native peoples became minorities in their own lands and Europeans introduced

crops, animals, and ideas about land ownership and land use that changed South America irreversibly. They also brought in Africans from various locations of Subsaharan Africa. Europeans from non-Iberian Europe also started immigrating to South America, especially during the first half of the twentieth century. Japanese settlers arrived in Brazil and Peru during the same era. All of these elements have contributed to the present-day ethnic composition in this realm. Figure 5.6 shows the distinct clusters of Amerindian and African cultural dominance, as well as areas where these groups are hardly present and people of European descent dominate.

ECONOMIC INTEGRATION

As noted above, the separatism that has so long characterized international relations in this realm is giving way as South American countries discover the benefits of forging new partnerships with one another. With mutually advantageous trade the catalyst, a new continentwide spirit of cooperation is blossoming at every level. Periodic flareups of boundary disputes now rarely escalate into open conflict. Cross-border rail, road, and pipeline projects, stalled for years, are multiplying steadily. In southern South America, five formerly contentious nations are developing the *hidrovia* (water highway), a system of river locks that is opening most of the Paraná-Paraguay Basin to barge transport. Investments today flow freely from one country to another, particularly in the agricultural sector. Similar ideas have been proposed to connect the Paraná-Paraguay rivers to the Amazon River system.

Recognizing that free trade could solve many of the realm's economic-geographic problems, governments are pursuing several avenues of economic supranationalism. In 2008, South America's republics were affiliating with the following major trading blocs:

Mercosur/l (Mercosur in Spanish; Mercosul in Portuguese). Launched in 1995 by countries of the Southern Cone and Brazil, this Common Market established a free-trade zone and customs union linking Brazil, Argentina, Uruguay, Paraguay, and now Venezuela. Bolivia, Chile, Colombia, Ecuador, and Peru are associate members. This organization is becoming the dominant free-trade organization for South America.

Andean Community. Formed as the Andean Pact in 1969 but restarted in 1995 as a customs union with common tariffs for imports, this bloc is made up of Colombia, Peru, Ecuador, and Bolivia. Venezuela was a member until it withdrew in 2006.

Union of South American Nations (UNASUR). Founded in 2008 in Brasília, Brazil, the twelve independent countries of South America signed a treaty to create a union envisioned as similar to the European Union (see Chapter 1) with the goal of a continental parliament, a coordinated defense effort, one passport for all its citizens, and greater cooperation on infrastructure development. However, significant disagreement between member-states about details still puts these efforts years into the future. UNASUR had been preceded by the South American Community of Nations.

8 **Free Trade Area of the Americas (FTAA)** The United States and other NAFTA proponents have tried to move this hemispheric free-trade idea forward, but it has been resisted by peasants and workers in South America, and formally opposed by Mercosur/l. So long as the terms of trade remain set by the North, the Southern partners will be reluctant toward this initiative.

URBANIZATION

As in most other realms, South Americans are leaving the land and moving to the cities. South America started relatively early in this **9** **urbanization** process, which intensified throughout the twentieth century. With South America's urban population now at 80 percent, the realm ranks with those of Europe and the United States. The urban population of South America has grown annually by about 5 percent since 1950, while the increase in rural areas was less than 2 percent. These numbers underscore not only the dimensions but also the durability of the **10** **rural-to-urban migration** from the countryside to the cities.

Regional Patterns

The generalized spatial pattern of South America's urban transformation is displayed in Figure 5-7, which shows a *cartogram* of the continent's population.[1] Here we see not only the realm's countries in population-space relative to each other, but also the proportionate sizes of individual large cities within their total national populations.

Regionally, southern South America is the most highly urbanized. Today in Argentina, Chile, and Uruguay, almost all of the population resides in cities. Ranking next in urbanization is Brazil. The next highest group of countries borders the Caribbean in the north. Not surprisingly, the Andean countries constitute the realm's least urbanized zone. Figure 5-7 tells us a great deal about the relative positions

[1]A cartogram is a specially transformed map in which countries and cities are represented in proportion to their populations. Those containing large numbers are blown up in population-space, while those containing lesser numbers are shrunk in size accordingly.

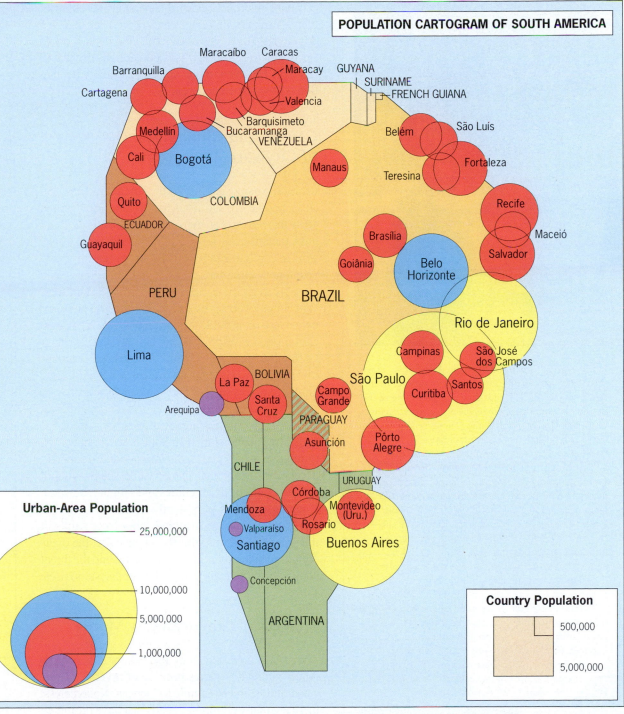

FIGURE 5-7

Adapted and updated with permission from Richard L. Wilkie, *Latin American Population and Urbanization Analysis: Maps and Statistics, 1950–1982.* UCLA Latin American Center, 1984.

Urban Agriculture in the Brazilian Amazon. "How do recently arrived migrants to the city survive? Many come with few skills except their own labor and spend their days scraping together a living from several often very menial jobs such as shining shoes, selling individual candies or cigarettes, or hawking newpapers. For many migrants the move to the city also comes with a shock as they enter into a fully waged economy. Many rural agriculturalists lived in a system of subsistence and barter, wherein the exchange of money was not necessarily the main system of exchange. In rural areas, food is obtained by growing it or in exchange for work or other nonmonetary goods and services. Research has shown that in rapidly urbanizing places such as the Brazilian Amazon, people continue to grow some of their own food as a survival mechanism. This is called urban agriculture. As shown here, people grow fruits and vegetables in backyards; they also cultivate grains such as corn in empty lots and unused public spaces. Families and friends share food among themselves in elaborate networks of exchange that are informal yet critical to urban survival." (© A. WinklerPrins)

Concept Caching
www.conceptcaching.com

of major metropolises in their countries. Three of them—Brazil's São Paulo and Rio de Janeiro, and Argentina's Buenos Aires—rank among the world's **11** **megacities** (cities whose populations exceed 10 million). But even in the Amazon Basin the population is now 70 percent urban.

Causes and Challenges of Cityward Migration

In South America, as in Middle America, Africa, and Asia, people are attracted to the cities and driven from the poverty of the rural areas. Both *pull* and *push* factors are at work. Rural land reform has been slow in coming, and for this and other reasons every year tens of thousands of farmers simply give up and leave, seeing little or no possibility for economic advancement. The cities lure them because they are perceived to provide opportunity—the chance to earn a regular wage. Visions of education for their children, better medical care, upward social mobility, and the excitement of life in a big city draw hordes to places such as São Paulo and Caracas.

But the actual move can be traumatic. Cities in developing countries are surrounded and often invaded by squalid slums, and this is where the urban immigrant most often finds a first—and frequently permanent—abode in a makeshift shack without even the most basic amenities and sanitary facilities. Unemployment is persistently high, often exceeding 25 percent of the available labor force. But still the people come, hopeful for a better life, the overcrowding in the shantytowns worsens, and the threat of epidemic-scale disease (and other disasters) rises.

The 'Latin' American City Model

The urban experience in the South and Middle American realms varies because of diverse historical, cultural, and economic influences. Nonetheless, there are many common threads that have prompted

geographers to search for useful generalizations. One is the model of the intraurban spatial structure of the **12** **'Latin' American city** proposed by Ernst Griffin and Larry Ford (Fig. 5-8).

Model and Reality

As noted in Chapter 1, the idea behind a *model* is to create an idealized representation of reality, displaying as many key real-world elements as possible. In the case of South America's cities, the basic spatial framework of city structure, which blends traditional elements of South and Middle American culture with modernization forces now reshaping the urban scene, is a composite of radial sectors and

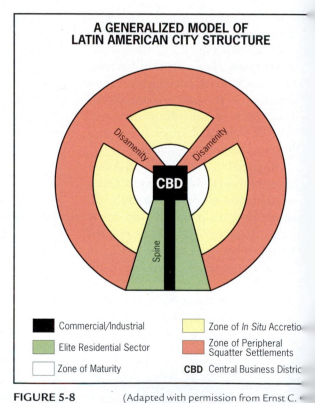

A GENERALIZED MODEL OF LATIN AMERICAN CITY STRUCTURE

Legend:
- Commercial/Industrial
- Elite Residential Sector
- Zone of Maturity
- Zone of *In Situ* Accretio
- Zone of Peripheral Squatter Settlements
- **CBD** Central Business Distric

FIGURE 5-8   (Adapted with permission from Ernst C. and Larry R. Ford, "A Model of Latin Americ Structure," *Geographical Review*, Vol. 70 (1980 © American Geographical Society,

concentric zones. Anchoring the model is the **CBD**, which is the primary business, employment, and entertainment focus of the surrounding metropolis. The **CBD** contains many modern high-rise buildings but also mirrors its colonial beginnings. As was shown in Figure 4-4, by colonial law, the Spanish colonizers laid out their cities around a central square, or plaza, dominated by a church and government buildings. Santiago's *Plaza de Armas*, Bogotá's *Plaza Bolívar*, and Buenos Aires' *Plaza de Mayo* are classic examples. The plaza was the hub of the city, which later outgrew its old center as new commercial districts formed nearby; but to this day the plaza remains an important link with the past.

Radiating outward from the urban core along the city's most prestigious axis is the commercial **spine**, which is adjoined by the **elite residential sector** (shown in green in Fig. 5-8). This widening corridor is essentially an extension of the **CBD**, featuring offices, shopping, and housing for the upper and upper-middle classes.

The three remaining concentric zones are home to the less fortunate residents of the city, with income level and housing quality decreasing as distance from the **CBD** increases. The **zone of maturity** in the inner city contains housing for the middle class, who invest sufficiently to keep their aging dwellings from deteriorating. The adjacent **zone of *in situ* accretion** is one of much more modest housing interspersed with unkempt areas, representing a transition from inner-ring affluence to outer-ring poverty and taking on slum characteristics.

WHAT'S DRIVING GEOGRAPHIC CHANGE IN THE REALM?

- **Amerindian** South Americans are flexing their political muscle. In Bolivia, an ethnic Aymara is president and opposes foreign exploitation of the country's energy resources. In Ecuador, a new president was elected with major support from Amerindians and is trying to balance their needs with the overall needs of the country.

- **Brazil** is on the move, already an industrialized country and now an agricultural commodity superpower. The leader in ethanol and flex-fuel automobiles, Brazil is increasingly becoming a political leader of the Global South, allied formally with India and South Africa and challenging U.S. hegemony on trade issues.

- **China** in the picture: from Argentina to Brazil to Peru, Chinese business executives are buying commodities and offering to finance trans-Andean routes to South America's **west coast** to integrate the realm more effectively with Asia's Pacific Rim.

- The **Free Trade Agreement of the Americas (FTAA)** has stalled with resistance by Brazil, and also with opposition from Venezuela, Ecuador, and Bolivia, who oppose U.S. policies in the region and ideologically align themselves more with Cuba.

The outermost **zone of peripheral squatter settlements** is home to the millions of relatively poor and unskilled workers who have recently migrated to the city. Here many newcomers earn their first cash income by becoming part of the **13** **informal sector**, where workers are undocumented and money transactions are beyond the control of government. The settlements consist mostly of self-help housing, vast shantytowns known as **14** *barrios* in Spanish-speaking South America and *favelas* in Brazil. Some of their entrepreneurial inhabitants succeed more than others, transforming parts of these shantytowns into beehives of activity that can propel resourceful workers toward a middle-class existence.

A final structural element of many South American cities forms an inward, narrowing sectoral extension of the zone of peripheral squatter settlements and is known as the **zone of disamenity**. It consists of undesirable land along highways, railroad corridors, riverbanks, and other low-lying areas; people are so poor that they are forced to live in the open. Thus this realm's cities present enormous contrasts between poverty and wealth, squalor and comfort—harsh contrasts all too frequently seen in the cityscape.

Regions of the Realm

South America divides geographically into four rather clearly defined regions (Fig. 5-4):

1. *The North* consists of five entities that display a combination of Caribbean and South American features: Colombia, Venezuela, and the three small Guianas that represent historic colonial footholds of Britain (Guyana), the Netherlands (Suriname formerly Dutch Guiana), and France (French Guiana).

2. *The West* is formed by four republics that share a strong Amerindian cultural heritage as well as powerful influences resulting from their Andean physiography: Ecuador, Peru, Bolivia, and, transitionally, Paraguay.

3. *The South*, often called the Southern Cone, consists of three countries that all exhibit very strong European influences: Argentina, Chile, and Uruguay, plus aspects of Paraguay.

4. *Brazil* by itself accounts for just about half the continent in terms of territory as well as population. Here the dominant Iberian influence is Portuguese, not Spanish; and here Africans, not Amerindians, form a significant element in the population and cultural landscape.

THE NORTH: CARIBBEAN SOUTH AMERICA

The countries of South America's northern tier have something in common as each has a coastal plantation zone similar to the Caribbean colonial model. Especially in the three Guianas, early European plantation development entailed the forced immigration of African laborers and eventually the absorption of this element into the population matrix. In addition to the Africans, Europeans brought over tens of thousands of South and Southeast Asians as contract laborers who stayed as settlers, creating a unique cultural mix in South America. Today, the three Guianas still display the coastal orientation and plantation dependency left over from the colonial period, although the logging of their tropical forests is penetrating and ravaging the interior. In Venezuela and Colombia, however, colonial efforts in farming, ranching, and mining drew the population inland, overtaking the coastal-plantation economy and creating diversified economies.

Figure 5-9 shows that not only Venezuela but also Colombia is Caribbean in its orientation: in Venezuela, coastal oil and natural gas

reserves have replaced plantations as the economic mainstay, and Colombia's Andean valleys open toward the north where roads, railways, and pipelines lead toward Caribbean ports such as Cartagena and Barranquilla. Colombia's Pacific coast is hardly a factor in the national economy despite the outlet at Buenaventura. But, as we will see, the North's locational advantages are countered by physical, economic, and political obstacles that continue to prevent the realization of their full potential.

Colombia

Imagine a country more than twice the size of France, with a physical geography so varied that it can produce crops ranging from the temperate to the tropical, possessing world-class reserves of oil, plus 3200 kilometers (2000 mi) of coastline on Atlantic and Pacific waters. Situated in the northwestern corner of South America, closer than any of its neighbors to the markets of the north, with a single language and one dominant religion, would such a country not assuredly thrive amidst the burgeoning economic geography of its hemisphere? The answer is no. Colombia today is a country as troubled as any in the world, handicapped by long-term civil unrest and violence, its economy imperiled, its politics destabilized, its future a gigantic question mark. Colombia's cultural uniformity has not produced social cohesion. The same physical geography that di-

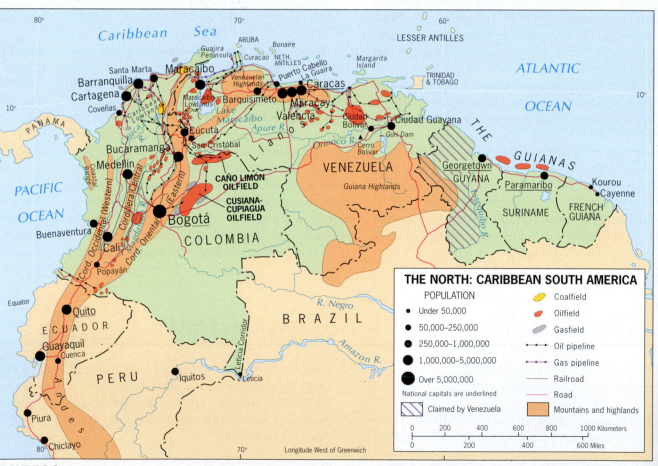

FIGURE 5-9

© H. J. de Blij, P. O. Muller, and John Wiley & Sons, Inc.

versifies its environments also divides its population clusters, which are so tenuously interconnected that the country totals less than 400 kilometers (250 mi) of paved four-lane highway. Its proximity to U.S. markets is a curse as well as a blessing: at the root of Colombia's current problems lies its role in the international drug trade, especially cocaine.

People and Resources

As Figure 5-9 shows, western Colombia is mountainous while the east lies in the Orinoco and Amazon basins. Most of the country's 48.2 million people live in the western and northern parts of Colombia. Physiography has contributed to the concentration of this population into more than a dozen separate clusters. Some of them are located in the Caribbean Lowlands; others lie in the Magdalena and Cauca valleys between the parallel ranges of the Andes; and still others concentrate in basins within the mountains themselves, the most important being the cluster around the capital city of Bogotá. The country's renowned coffee crop comes from the *tierra templada* zones on the Andean slopes. Colombia's belt of oil and natural gas reserves flanks the border with Venezuela from Lake Maracaibo south to the Caño Limón oilfield, and then continues southwest along the Andean front through the country's largest deposits, the Cusiana-Cupiagua oilfield (Fig. 5-9). Meanwhile, the sparsely populated east and south have become the production centers for the coca crop that sustains the country's illicit narcotics industry.

Cocaine's Cost

With its fuel, mineral, and agricultural productivity, Colombia might have been on the fast track toward prosperity, but the rise of the narcotics industry, fueled by outside (especially U.S.) demand, coupled with the legacy of violence, has crippled the country. Today, Colombia remains in combat. In the cities, narcoterrorists repeatedly commit appalling acts of violence; in the countryside, rebel forces and drug-financed armies of the political left fight equally vicious paramilitary forces of the right. Civilians by the thousands die in the crossfire of all these belligerent groups, and countless more are dislocated. Meanwhile, U.S. involvement in the war against drug cultivation and production has complicated the situation further.

Insurgent State(s)

Parts of Colombia were/are beyond the control of its government and its national armed forces and were essentially ruled by rebels who 'protected' the local citizenry against Bogotá's efforts to reestablish authority there. In effect, insurgents in several parts of Colombia created their own domains, and some of their rulers even negotiated their status with the legitimate government in the capital.

What happened here is a process long studied by political scientists. It involves three stages: (1) *contention*, the period of initial rebellion; (2) *equilibrium*, when the rebels gain effective control over some of the national territory, as has happened in parts of Colombia; and (3) *counteroffensive*, in which the government and the rebels, usually following the breakdown of negotiations, engage in armed conflict that will decide the future of the country. The geographer Robert McColl translated this sequence into spatial terms, suggesting that the equilibrium stage represents the emergence of an **15 insurgent state**. The formation of an insurgent state is the key stage in the process because it creates an informally bounded territory, usually with a distinct core area centered on a captured town, establishes a parallel government, and provides schools and other social services that substitute for those the formal state may have provided.

Colombia in the 1990s contained several enclaves that were taking on the properties of insurgent states. This was because Colombia confronted not just one but several different guerrilla forces, each operating in a particular area of the country. Thus the M-19 Movement focused on the zone near the Panamanian border. The Popular Liberation Army was active in the Cauca Valley. The National Liberation Army (ELN) operated in the Arauca area on the Venezuelan border and in the middle Magdalena Valley. And the Revolutionary Armed Forces of Colombia (FARC), by far the most powerful insurgent group, concentrated on the central-south and east.

In 1999, as an incentive to negotiate, FARC compelled the government to demilitarize a large zone south of Bogotá, in effect leaving the insurgents there in control of an area the size of Switzerland. Also important was FARC's simultaneous initiative to take over the far southeastern town of Mitú and its environs, an area whose remoteness made it particularly difficult for the Colombian army to regain control. And although the insurgents claimed they were suppressing drug production in the zones under their control, the evidence indicates that coca cultivation skyrocketed in the FARC-held insurgent state.

A Nation in Crisis

Colombia's twin scourges, narcotics and insurgencies, have set the country back immeasurably. Not only has the state been unable to provide security for its citizens; perhaps as many as a million farmers have been driven off their lands. In 2002, then newly elected President Alvaro Uribe (he was reelected in 2006 to a second term) began a twin campaign to defeat the insurgents militarily and to persuade them through legal means to give up their arms. His military action drove the militias out of most of the areas they had taken over, and his legal lure, which involved reduced sentences for those willing to give up their weapons while confessing to their misdeeds, attracted thousands of takers. None of this pacified Colombia entirely, but it reduced the level of violence and offered hope to a weary population. The counteroffensive stage appeared to be tilting to the state, not the insurgents.

Nonetheless, Colombia remains in a turbulent condition not all of its own making. About 70 percent of the cocaine consumed in the United States comes from Colombia, and of the rest, most goes to Brazil or on to Europe. American support for Colombia's antidrug campaign (Plan Colombia), logistical as well as financial, raises political issues in Bogotá, where extraditions of Colombian drug kingpins to the United States are not universally popular.

What keeps the cocaine flowing, many Colombians argue, are external, especially North American, markets, so stopping the consumption there will benefit Colombia.

Flowers as an Alternative to Violence

Cocaine is a lucrative industry and a dangerous one. Interestingly, however, Colombia does have other important economic activities, especially the production of cut flowers. In fact, Colombia is the second largest cut flower producer in the world (after The Netherlands), accounting for 14 percent of global production. The great savanna east of Bogotá is dotted with innumerable hothouses where roses, carnations, and many other flowers (about 50 varieties) are grown under highly controlled conditions primarily for the U.S. market. This is a growth industry, and in 2005 it produced $900 million worth of flowers. The industry employs more people than traditional agriculture, but with a feminized workforce (65 percent of the workforce is female) that is poorly unionized. Significant environmental and health consequences are associated with flowers inasmuch as they are cultivated using large amounts of agrotoxins that poison not only the immediate physical environment but also the people who work in the industry. Birth defects and cancer rates have been found to be significantly higher for workers in this industry. There is a dark underside to this beautiful industry.

Venezuela

Much of what is important in Venezuela is concentrated in the northern and western parts of the country, where the Venezuelan Highlands form the eastern spur of the north end of the Andes system. Most of Venezuela's 27.9 million people are concentrated in these uplands, which include the capital of Caracas. The Venezuelan Highlands are flanked by the Maracaibo Lowlands to the west and by a vast savanna country, known as the **Llanos**, in the Orinoco Basin to the south and east (Fig. 5-9).

Taking advantage of its temperate climate in the highlands as well as its relatively low-wage labor pool and lax environmental laws, Colombia has seen its cut-flower industry grow steadily in recent years. Colombia is now the world's second-largest producer of cut flowers, exporting about 50 varieties of flowers, including most of the roses bought in the United States. Here a woman cuts gerbera daisies in a hothouse near Bogotá to be shipped to the United States ahead of Valentine's Day, which is the biggest day of the year for fresh-cut flower sales in the United States.   (© AP/Wide World Photos)

The Maracaibo Lowlands constitute one of the world's leading oil-producing areas. Much of the oil is drawn from reserves that lie beneath the shallow lake at its center. Actually, Lake Maracaibo is a misnomer, for the 'lake' is open to the ocean and is in fact a gulf with a very narrow entry. The **Llanos** on the southern side of the Venezuelan Highlands and the Guiana Highlands in the country's southeast are less economically developed. The **Llanos** slope gently from the base of the Andean spur to the Orinoco River. Its mixture of savanna grasses and scrub woodland support cattle grazing on higher ground, but wet-season flooding of the more fertile lower-lying areas has inhibited the **Llanos'** commercial agricultural potential. Economic integration of this interior zone with the rest of Venezuela has so far been spearheaded by the exploitation of rich iron ores on the northern flanks of the Guiana Highlands.

Oil and Politics

Despite these assets, since 1998 Venezuela has been in upheaval as longstanding economic and social problems finally intensified to push the country into a new era of radical political change. A major reason was that oil had not bettered the lives of most

Venezuelans because the government had acquired the habit of living off oil profits, forcing the country to suffer the consequences of the long global oil depression that began in the early 1980s. With more and more Venezuelans enraged at the way their oil-rich country was approaching bankruptcy without making progress toward the more equitable distribution of the national wealth, voters turned in an extreme direction in the 1998 presidential election. They elected Hugo Chávez, a former colonel who had led a failed military coup, and they gave him a mandate to be a strongman type of leader.

Chávez has indeed pursued this course since entering office in 1999, sweeping aside Congress and the Supreme Court, supervising the rewriting of the Venezuelan Constitution in his own image, and proclaiming himself the leader of a 'peaceful leftist (Bolivarian) revolution' that would transform the country. Although he professes that social equality ranks highest on his agenda, Chávez has stirred up racial divisions by actively promoting mestizos (64 percent of the population) over those of European background (20 percent). And in the international arena, Chávez sparks controversy at every turn: angering the government of neighboring Colombia by expressing his 'neutrality' in its confrontation with cocaine-producing insurgent forces; unsettling another neighbor, Guyana, by aggressively reviving a century-old territorial claim to the western zone of that country (the striped area in Fig. 5-9); embracing Fidel Castro and providing communist Cuba with oil in exchange for the assignment to Venezuela of thousands of doctors and technicians; and challenging the globalization plans of the United States throughout South America, using the windfall income from high-priced oil to reward allies in this effort with substantial subsidies. Chávez easily won reelection in late 2006. In late 2007, he attempted to rewrite the country's constitution to give him unprecedented power and unlimited terms as president. The referendum failed, however, and Chávez had to admit defeat. Yet, Chávez continues to position himself as the champion of Venezuela's (and South America's) poor, trumpeting his Bolivarian Revolu-

tion as a local alternative to what he describes as the insidious encroachment of U.S. imperialism. He has taken Venezuela to center stage in the ideological contest in this realm, casting doubt on the virtues of democracy and free enterprise, castigating elites for not doing enough to reduce the gap between rich and poor, and urging the downtrodden to assert themselves. His message resonates across the continent.

The Guianas

The three small entities that form the remainder of the region remind us why the term *Latin* America constitutes an excessive generalization for South as well as Middle America. Formerly known as British, Dutch, and French Guiana, two are now independent (Guyana since 1966, Suriname since 1975) and the third, French Guiana, has representation in the French Assembly. None has a population exceeding 1 million (see Appendix B), and per-capita incomes resemble Caribbean rather than South American norms.

British-influenced Guyana's population of under 800,000 is divided about equally between people of African and Asian descent, which has caused social problems since before independence and political crises ever after. For the people living in the small coastal villages along the coast, economic options are few; for the government, Venezuela's claim to a large part of the national territory (Fig. 5-9) is a constant worry. Guyana, however, has at times claimed parts of its neighbor Suriname, where political instability and joblessness drove more than 100,000 residents—more than one-fifth of the population—to migrate to the Netherlands. Suriname is somewhat better off economically, with a large bauxite mine, self-sufficiency in rice, and some banana and fish exports. But ethnic problems recur here too, with people of African descent (including those who escaped slavery by living deep in the interior forests) slightly in the majority over those of Asian descent. French Guiana remains officially a *département* of the French Republic, with the space center at Kourou its main

asset but the lowest per-capita income among ordinary inhabitants.

As a South American region, the North continues to be volatile and rife with actual and potential conflict on the doorstep of the Caribbean. Legal (oil) and illicit (cocaine) exports move continuously to North American markets. Indigenous peoples, confined to the mountains of southern Colombia and to the Amazonian interiors of Colombia and Venezuela, are not part of the national fabric.

THE WEST: ANDEAN SOUTH AMERICA

The second regional grouping of South American states—the Andean West (Fig. 5-10)—encompasses Peru, Ecuador, Bolivia, and transitional Paraguay. All these countries have high numbers of people who are ethnically Amerindian (Bolivia, about 55 percent; Peru, about 45 percent; and Ecuador, about 25 percent), as Figure 5-6 shows. Paraguay, not an Andean country but strongly Amerindian-infused, is transitional to the next region, the South. These countries also exhibit other similarities: their incomes are low, they are comparatively economically unproductive, and they exemplify relentless poverty. Today, the population no longer passively tolerates these desperate conditions, and uprisings continue to occur throughout the Andean West.

Peru

Peru straddles the Andean spine for more than 1600 kilometers (1000 mi) and is the largest of the region's four republics in both territory and population (29.1 million). Physiographically and culturally, Peru divides into three subregions: (1) the desert coast, the European-mestizo region; (2) the Andean highlands or *Sierra*, the Amerindian region; and (3) the eastern slopes, the sparsely populated Amerindian-mestizo interior (Fig. 5-10).

FIGURE 5-10

© H. J. de Blij, P. O. Muller, and John Wiley & Sons, Inc.

Three Subregions

Lima and its port, Callao, lie at the center of the desert *coastal strip*, and it is symptomatic of the cultural division prevailing in Peru that for nearly 500 years the capital city has been positioned on the periphery, not in a central location in a basin of the Andes. From an economic point of view, however, the Spaniards' choice of a headquarters on the Pacific coast proved to be sound, for the coastal subregion has become commercially the most productive part of the country. A thriving fishing industry contributes significantly to the export trade; so do the products of irrigated agriculture in some 40 oases distributed all along the arid coast, which include fruits such as citrus, olives, and avocadoes, and vegetables such as asparagus (a big money-maker) and lettuce (Fig. 5-5).

The *Sierra* (Andean) subregion occupies about one-third of the country and is the ancestral home of the largest component in the total population, the Quechua-speakers who had been subjugated by the Inca rulers when the Spanish conquerors arrived. Their survival during the harsh colonial regime was made possible by their adaptation to the high-altitude environments they inhabited, but their social fabric was ripped apart by communalization, forced cropping, religious persecution, and outward migration to towns and haciendas where many became serfs. Another Amerindian people, the Aymara, live in the area of Lake Titicaca, making up about 6 percent of Peru's population but larger numbers in neighboring Bolivia.

Although these Amerindian people make up nearly half of the population of Peru, their political influence remains slight—even during the term of former President Alejandro Tolédo, who was of Amerindian descent and had strong support in the Andean districts. Nor is the Andean subregion a major factor in Peru's commercial economy—except for its mineral storehouse, which yields copper, zinc, and lead from mining centers, the largest of which is Cerro de Pasco. In the high valleys and intermontane basins, the Amerindian population is clustered either in isolated villages, around which people practice a precarious subsistence agriculture, or in the more favorably located and fertile areas where they are tenants, peons on white- or mestizo-owned haciendas. Most of these people never receive an adequate daily caloric intake or balanced diet of any sort. The wheat produced around Huancayo, for example, is mainly exported and would in any case be too expensive for the Amerindians themselves to buy. Potatoes, barley, and corn are among the subsistence crops grown here in the *tierra fría* zone, and in the *tierra helada* of the higher basins the Amerindians graze their llamas, alpacas, cattle, and sheep.

Of Peru's three subregions, the *Oriente*, or East—the inland slopes of the Andes and the Amazon-drained, rainforest-covered *montaña*—is the most isolated. The focus of the eastern subregion, in fact, is Iquitos, a city that looks east rather than west and can be reached by oceangoing vessels sailing 3700 kilometers (2300 mi) up the Amazon River across northern Brazil. Iquitos grew rapidly during the Amazon wild-rubber boom of a century ago and then declined; now it is finally growing again and reflects Peruvian plans to open up the eastern interior. Petroleum was discovered west of Iquitos in the 1970s, and since 1977, oil has flowed through a pipeline built across the Andes to the Pacific port of Bayóvar. But what the Oriente needs most is a highway to the coast. Prospects for the development of the region could improve with better highway connectivity to the Pacific Ocean. Construction of such a highway, known as the *Carreterra Transoceanica* (Transoceanic Highway), began in 2005 with significant funding from Brazil, which sees this new road as an outlet for its products to the Pacific Rim. However, this highway project concerns environmentalists who would rather not see the *Oriente* developed.

An Energy Age

Peruvian planners see their country on the verge of an energy age as interior exploration is proving large reserves of gas and oil east of the Andes. Gas will flow from the Camisea reserve (north of Cuzco) across the Andes to a conversion plant on the Paracas Peninsula south of Lima, from where a 3-kilometer (2-mi) ocean-floor pipeline will carry it to an off-shore loading platform for tankers that will transport it to the U.S. market (Fig. 5-10).

Insurgency to Stability

By the turn of this century, Peru appeared to have put behind it the lengthy period of instability during which its government was threatened by well-organized guerrilla movements. The most serious threat, which for a time during the 1980s seemed likely to achieve an insurgent state in its Andean base, was the so-called Maoist-inspired *Sendero Luminoso* (Shining Path) movement. But forceful action under the later-disgraced president of Japanese ancestry, Alberto Fujimori, defeated the rebel movement and returned Peru to political stability and economic growth.

The social cost of the Fujimori regime, however, was high. Democratic principles were violated and corruption was endemic. The succeeding Tolédo administration soon ran into trouble on economic and political fronts, and Tolédo's successor, Alan García, elected in 2006, is trying to unify and move forward a culturally divided country.

Ecuador

On the map, Ecuador, smallest of the Andean West republics (population 13.9 million), appears to be just a corner of Peru. But that would be a misrepresentation because Ecuador possesses a full range of regional contrasts (Fig. 5-10). It has a coastal belt; an Andean zone that may be narrow (under 250 kilometers or 150 mi) but by no means of lower elevation

Due to its location along the collision zone of two tectonic plates, the western side of South America is always threatened with tectonic activity—earthquakes or volcanic eruptions. In mid-August 2007, a powerful magnitude-eight earthquake devastated the southern coast of Peru centered on the town of Pisco. The earthquake killed at least 510 people, injured more than 1500, and displaced thousands. (© AP/Wide World Photos)

than in other Andean countries; and an *Oriente*—an eastern lowland subregion that is sparsely populated and was long economically marginalized, yet is now the center of Ecuador's oil production.

Ecuador's Pacific coastal zone consists of a belt of hills interrupted by lowlands, of which the most important lies in the south between the hills and the Andes, drained by the Guayas River. Guayaquil—the country's largest city, main port, and leading commercial center—forms the focus of this subregion. Unlike Peru's coastal strip, Ecuador's is not a desert: it consists of fertile tropical plains, mostly *tierra caliente*. These lowlands support a thriving commercial agricultural economy built around shrimp, bananas, cacao, cattle-raising, and coffee on the hillsides. The coastal zone is home to most of the 46 percent of Ecuadorians who are mestizos, and from time to time they express their displeasure with the central government based in the Andean city of Quito by proclaiming a desire for autonomy or even secession.

The great majority of the Amerindians (who constitute 25 percent of Ecuador's population) and most of those of European ancestry live in the central Andean zone, where land tenure reform remains a contentious issue and rural-urban migration is high. The differing interests of the Guayaquil-dominated coastal lowland and the Andean-highland subregion which focuses on the capital (Quito) have long fostered a deep regional cleavage between the two. In 2006, a left-leaning and anti-American president, Rafael Correa, was elected with strong support from the large Amerindian population. His charge is to bring stability to an unstable and discontented populace.

In the early 2000s, Ecuador, like Peru, began rapidly expanding its energy industry based on a growing volume of proven reserves not only in the *Oriente* but also in the coastal zone. So large are these reserves that experts compare Ecuador's energy resources to those of Mexico and Nigeria, which would rank this country among South America's leaders in this respect. Oil flows from beneath the rainforests of the *Oriente* via a trans-Andean pipeline to the port of Esmeraldas; another 500-kilometer (300-mi) pipeline was put into operation in 2003. Oil and gas yields about 50 percent of Ecuador's export revenues, and the government expects this proportion to grow rapidly in the years ahead.

Bolivia

From Ecuador southward through Peru, the Andes broaden until in Bolivia they reach a width of some 720 kilometers (450 mi). Here, between the eastern and western Andean cordilleras (ranges), lies the *Altiplano* proper (Fig. 5-10; as noted earlier, an *altiplano* is an elongated, high-altitude basin). On the boundary between Peru and Bolivia, freshwater Lake Titicaca lies at 3700 meters (12,507 ft) above sea level (the highest lake in the world) and helps make the adjacent *Altiplano* livable by ameliorating the coldness in its vicinity, where the snow line lies just above the plateau surface. On the surrounding cultivable land, potatoes and grains have been raised for centuries dating back to pre-Inca times, and the Titicaca Basin still supports a major cluster of Aymara farmers. This portion of the *Altiplano* is the heart of modern Bolivia and also contains the capital city, La Paz.

The landlocked state of Bolivia is the product of the Hispanic impact, and the country's Amerindians (who make up 55 percent of the national population of 9.5 million) no more escaped the loss of their land than did their Peruvian or Ecuadorian counterparts. What made the richest Europeans in Bolivia wealthy, however, was not land but minerals. The town of Potosí in the eastern cordillera became a legend for the immense deposits of silver in its vicinity; tin, zinc, copper, and several ferroalloys were also discovered there.

Today oil and natural gas, exported to Argentina and Brazil, are major sources of foreign revenues, and zinc has replaced tin as the leading metal export. But Bolivia's economic prospects will always be impeded by the loss of its outlet to the Pacific Ocean during its war with Chile in the 1880s, despite its transit rights and dedicated port facilities at Antofagasta.

More critical than its economic limitations or its landlocked situation is Bolivia's social predicament. The government's history of mistreatment of indigenous people and the harsh exploitation of its labor force hangs heavily over a society where about 70 percent of the people, largely Amerindian, live in poverty. In recent years, however, this underrepresented majority has been making an impact on national affairs. Earlier in this decade, violent opposition to a government plan to export natural gas to the United States via a new pipeline to the Chilean coast led to chaos and the government's resignation. In 2005, Bolivian voters elected their first president of Amerindian ancestry, Evo Morales, who proceeded to nationalize the country's natural gas resources and force out multinational corporations.

All this reflects not only the social cleavages in Bolivian society but also the country's sharp regional contrasts. In the lowland east, centered on Santa Cruz Province and its capital, the hacienda system persists almost unchanged from colonial times, its profitable agriculture supporting a wealthy aristocracy. Now this eastern region proves to contain Bolivia's rich energy resources as well. While the government in La Paz struggles to meet growing Amerindian demands for a share in the action, the prosperous east sees its objectives (among them that pipeline to export gas via the Chilean coast) thwarted. A strong separatist movement in the lowland east is growing among landowners. In 2008, referendums were held that sought greater autonomy for the region, thus undermining land reform in this hacienda-rich region. Neither the haciendas nor the energy industry, however, could do without the Amerindian labor force working here. Toward the end of this decade, Bolivia's future remained clouded, potentially a **16 failed state** whose centrifugal forces threaten its institutions.

Paraguay

Earlier we noted the transitional situation of Paraguay (see Figs. 5-4 and 5-10). In terms of economic development, Paraguay is more typical of the Andean West than of the richer South. Ethnically, too, the pattern is more characteristic of the West: about 95 percent of Paraguay's 6.5 million people are mestizo, but with so pervasive an Amerindian influence that any European ancestry is almost totally submerged. As for languages, Amerindian Guaraní is so widely spoken alongside Spanish that the country is surely one of the world's most completely bilingual. The physiography, however, is decidedly non-Andean because all of Paraguay lies to the east of the mountains in the center of the Paraná-Paraguay Basin.

Also note that two major rivers link Paraguay to the South: the Paraguay River, on which the capital of Asunción is located, and the Paraná River, along Paraguay's eastern border, where three countries—Paraguay, Brazil, and Argentina—meet in what is known as the *Triple Frontier*.

Persistent Stagnation

Apart from the Guianas, Paraguay long has been South America's least-developed country by most measures, the result of a combination of factors including not only its landlocked situation but also land expropriation, subjugation of the poor, power struggles, mismanagement, and corruption. In 2007, nearly two thirds of the population lived at or below the poverty level. Meanwhile, about 300,000 better-off Brazilians have acquired many of eastern Paraguay's most productive haciendas and farms.

Remote and dictatorially run countries ruled by corrupt oligarchies make attractive targets for terrorist organizations. The Triple Frontier area, notably the lawless town of Ciudad del Este, became a place of concern during the current global War on Terror, when a map of it was found in an al-Qaeda hideout in Kabol, Afghanistan. An investigation exposed financial ties involving millions of dollars between certain local businesspeople and terrorist-linked organizations in the Middle East. Bringing Paraguay into the mainstream of South American development is more than an economic objective.

THE SOUTH: MID-LATITUDE SOUTH AMERICA

The three southern countries of South America—Argentina, Chile, and Uruguay—constitute a region often referred to as the Southern Cone because of its pointed, ice-cream-cone shape (Fig. 5-11). As noted earlier, Paraguay has strong links to this region and in some ways forms part of it, but social contrasts between Amerindian-infused Paraguay and the more culturally European Southern Cone remain sharp. This is the region where Mercosur/l was initiated in 1995, the hemisphere's second-largest trading bloc after NAFTA.

Argentina

The largest Southern Cone country by far is Argentina, whose territorial size ranks second only to Brazil in this geographic realm; its population of 39.8 million ranks third after Brazil and Colombia. Argentina exhibits a great deal of physical-environmental variety within its boundaries, and the vast majority of the Argentines are concentrated in the physiographic subregion known as the *Pampa* (a Spanish word meaning plain). Figures G-10 and 5-2

FIGURE 5-11   © H. J. de Blij, P. O. Muller, and John Wiley & Sons, Inc.

underscore the degree of clustering of Argentina's inhabitants on the land and in the cities of the Pampa. It also shows the relative emptiness of the other six subregions: the scrub-forest *Chaco* in the northwest; the mountainous *Andes* in the west, along whose crestline lies the boundary with Chile; the arid plateaus of *Patagonia* south of the Rio Colorado; and the undulating transitional terrain of intermediate *Cuyo, Entre Rios* (also known as Mesopotamia because it lies between the Paraná and Uruguay rivers), and the *North*.

The Argentine Pampa is the product of the past 150 years. During the second half of the nineteenth century, when the great grasslands of the world were being opened up (including those of the interior United States, Russia, and Australia), the economy of the Pampa began to emerge. The food needs of industrializing Europe grew by leaps and bounds, and the advances of the Industrial Revolution—railroads, more efficient ocean transport, refrigerated ships, and agricultural machinery—helped to make large-scale commercial meat and grain production in the Pampa not only feasible but also highly profitable. Large haciendas were laid out and farmed by tenant workers; railroads radiated ever farther outward from the booming capital of Buenos Aires and brought the entire Pampa into production.

A Culture Urban and Urbane

Argentina, as we noted in our introductory chapter, once was one of the richest countries in the world. Its historic affluence is still reflected in its architecturally splendid cities whose plazas and avenues are flanked by ornate public buildings and private mansions. This is true not only of the capital, Buenos Aires, at the head of the Rio de la Plata estuary—it also applies to interior cities such as Mendoza and Córdoba. The cultural imprint is dominantly Spanish, but the cultural landscape was diversified by a massive influx of Italians and smaller but influential numbers of British, French, and German immigrants. A sizeable immigration from Lebanon resulted in the diffusion of Arab ancestry to more than 12 percent of the Argentinian population.

Argentina has long been one of the realm's most urbanized countries: 90 percent of its population is concentrated in cities and towns, a higher percentage than Western Europe or the United States. About one third of all Argentinians live in metropolitan Buenos Aires, also by far the leading industrial complex where processing Pampa products dominates. Córdoba has become the second-ranking industrial center and was chosen by foreign automobile manufacturers as the car-assembly center for the expanding Mercosur/l market. One in three Argentinian wage earners is engaged in manufacturing, another indication of the country's economic progress. But what concentrates the urban population is the processing of products from the vast, sparsely peopled interior: Tucumán (sugar), Mendoza (wines), Santa Fe (forest products), and Salta (livestock). Argentina's product range is enormous. There is even an oil reserve near Comodoro Rivadávia on the coast of Patagonia. Yerba maté, now a niche coffee shop drink, comes mostly from the Entre Rios subregion and adjacent areas of Brazil and Paraguay.

Economic Boom and Bust

Despite all these riches, Argentina's economic history is one of boom and bust. With just under 40 million inhabitants, a vast territory with diverse natural resources, adequate infrastructure, and good international linkages, Argentina could still be one of the world's wealthiest countries, as it once was. But political infighting and economic mismanagement have combined to ruin a vibrant and varied economy. What began as a severe recession toward the end of the 1990s became an economic collapse in the first few years of the twenty-first century.

The never-ending problem for Argentina has been corrupt politics and associated mismanagement. Following an army coup in 1946, Juan Perón got himself elected president, bankrupted the country, and was succeeded by a military junta that plunged the country into its darkest days, culminating in the Dirty War of 1976–1983 which saw more than ten thousand Argentinians disappear without a trace. In 1982, this ruthless military clique launched an invasion of the British-held Malvinas (Falkland Islands), resulting in a major defeat for Argentina. By the time civilian government replaced the discredited junta, inflation was soaring and the national debt had become staggering. Economic revival during the 1990s was followed by another severe downturn that exposed the flaws in Argentina's fiscal system, including scandalously inefficient tax collection and unconditional federal handouts to the politically powerful provinces.

In 2003, a new administration took office led by President Néstor Kirchner, who hailed from a small Patagonian province, not from one of the country's traditional power centers. He vowed to revive Argentina's fortunes and deal with corruption, military immunity from prosecution, and foreign economic intervention. In 2007, he decided not to run for re-election and instead selected his wife, Sen. Cristina Kirchner, to succeed him. She won the election and is now president of Argentina. She started with high hopes to continue the country's strong economic performance, easing its social tensions and rebuilding foreign engagement, but has thus far fallen short.

Chile

For 4000 kilometers (2500 mi) between the crestline of the Andes and the coastline of the Pacific lies the narrow strip of land that is the Republic of Chile (Fig. 5-11). On average just 150 kilometers (90 mi) wide (and only rarely over 250 kilometers or 150 mi in width), Chile is the world's quintessential example of what *elongation* means to the functioning of a state. Accentuated by its north-south orientation, this severe territorial attenuation results in Chile extending across numerous environmental zones, and it has contributed to the country's external political, internal administrative, and general economic stresses. Nonetheless, throughout most of their modern history, the Chileans have made the best of this potentially disastrous centrifugal force and have instead taken advantage of its varied geography.

Three Subregions

Chile is a three-subregion country with about 90 percent of its 16.8 million people concentrated in what is called Middle Chile, where Santiago, the capital and largest city, and Valparaíso, the chief port, are located. North of Middle Chile lies the Atacama Desert, which is wider, drier (in fact it is the driest place on earth), and colder than the coastal desert of Peru. South of Middle Chile, the coast is broken by a plethora of fjords and islands, the topography is mountainous, and the climate—wet and cool near the Pacific—soon turns drier and colder against the Andean interior. South of the latitude of Chiloé Island, there are no permanent overland routes and hardly any settlements. These three subregions are also apparent on the map of agricultural systems (Fig. 5-5). Some intraregional differences exist between northern and southern Middle Chile, the country's core area. Northern Middle Chile, the land of the hacienda and of Mediterranean climate (*Cs*) with its dry summer season, is an area of (usually irrigated) crops that include wheat, corn, grapes, apples, and peaches. Livestock raising and fodder crops also take up much of the productive land, but continue to give way to the more efficient and profitable cultivation of fruits for export, especially to North America (see photo). Southern Middle Chile, the destination of many immigrants from both the north and Europe (especially Germany) is a better-watered area where raising cattle has predominated. But here, too, more lucrative fruits, vegetables, and grains are changing the area's agricultural specializations.

Prior to the 1990s, the arid Atacama region in the north accounted for more than half of Chile's foreign revenues. The Atacama Desert contains the world's largest exploitable deposits of nitrates, which was the country's economic mainstay before the discovery of methods of synthetic nitrate production a century ago. Subsequently, copper became the chief export (Chile again possesses the world's largest reserves, which accounts for about half of export revenues). It is found in several places, but the main concentration lies on the eastern margin of the Atacama near Chuquicamata, not far from the port of Antofagasta.

Political and Economic Transformation

Chile today is emerging from a development boom that transformed its economic geography during the 1990s, a growth spurt that established its reputation as South America's greatest success story. Following the withdrawal of its brutal military dictatorship in 1990, Chile embarked on a program of free-market economic reform that brought stable growth, lowered inflation and unemployment, reduced poverty, and attracted massive foreign investment. The last is of particular significance because these new international connections enabled the export-led Chilean economy to utilize its varied geography and diversify and develop in some badly needed new directions. Copper remains the single leading export, but many other mining ventures have been launched. In the agricultural sphere, fruit and vegetable production for export has soared because Chile's harvests coincide with the winter farming lull in the affluent countries of the Northern Hemisphere. Industrial expansion is occurring as well, though at a more leisurely pace; new manufactures include a modest array of goods that range from basic chemicals to computer software. Forestry has also taken off in the south of the country.

Chile's newly internationalized economy has propelled the country to forge a prominent role for itself on the global trading scene. The Japanese expressed interest in building ties, and for a time during the 1990s Japan was its leading trading partner. This foothold in the Pacific Rim arena notwithstanding, Chile's trade linkages are now more heavily oriented toward its own hemisphere. The Chileans quickly realized that regional economic integration offered the best opportunities for sustained progress. Chile has affiliated with Mercosur/l as an associate member, though in 2000 it decided against seeking full membership. Chile does have a free-trade agreement with the United States, however, which it negotiated in 2002.

The port of Valparaíso, Chile. Ever wonder where your fruit comes from in January? Check out the packaging—chances are that it has come from Chile, where it is summer during the North American winter. Grapes, apples, peaches, and nectarines may all have been grown in Chile's productive Central Valley. Here fruit is being packed into special ships that keep it from ripening while en route to North America—a voyage of two weeks.   (Both: © A. WinklerPrins)

A Chilean Model?

Chile's economic and democratic transformation is widely touted as a Chilean model to be followed by other Middle and South American countries. But Chile benefits from an unusual combination of environmental opportunities, and some economic geographers argue that Chile has in fact not taken good advantage of all of them. Exports should be more diversified, and Chile's dependence on high copper prices entails a major risk. Despite reform efforts by Chile's president, Michelle Bachelet, the education system remains inadequate. Income inequality—that realm-wide affliction—has improved, but nearly 20 percent of Chileans still live in poverty. The situation of Chile's small number of indigenous people has been brought to national attention by the Mapuche, an Amerindian group of the forested south. They are in a stand-off with the government regarding use and management of forest resources. Nevertheless, compare Chile to virtually any other country in this realm: the Chilean model may not always apply, but the Chilean example certainly provides hope for the future.

Uruguay

Uruguay, unlike Argentina or Chile, is compact, small, and rather densely populated. This former buffer state has become a fairly prosperous agricultural country. Figures 5-1 and G-9 show the similarity of physical conditions on the two sides of the Plata estuary. Montevideo, the coastal capital, contains more than 40 percent of the country's population of 3.3 million; from here, railroads and roads radiate outward into the productive agricultural interior (Fig. 5-11). In the immediate vicinity of Montevideo lies Uruguay's major farming area, which produces vegetables and fruits for the metropolis as well as wheat and fodder crops. Most of the rest of the country is used for grazing cattle and sheep, with beef products, wool and textile manufactures, and hides dominating the export trade. Tourism is another major economic activity as Argentines, Brazilians, and other visitors increasingly flock to the Atlantic beaches at Punta del Este and other thriving resort towns.

BRAZIL: GIANT OF SOUTH AMERICA

By any measure, Brazil is South America's giant. It is so large that it has common boundaries with all the realm's other countries except Ecuador and Chile (Fig. 5-4). Its tropical and subtropical environments range from the equatorial rainforest of the Amazon Basin to the humid temperate climate of the far south. Territorially, Brazil occupies just under 50 percent of South America and is exceeded on the global stage only by Russia, Canada, the United States, and China. In population size, it accounts for just over half the realm's total and is again the world's fifth-largest country (surpassed only by China, India, the United States, and Indonesia). The Brazilian economy, with its modern industrial base, is now the eleventh largest on earth, and it is likely to become a world force as the twenty-first century unfolds.

Population Patterns

Brazil's population of 192.4 million is at least as diverse as that of the United States. In a pattern quite familiar in the Americas, the indigenous inhabitants of the country were decimated following the European invasion (fewer than 200,000 Amerindians now survive mainly in the North–Amazonian–part of the country). Africans came in great numbers, too, and today there are about 12 million Afro-Brazilians in Brazil. Significantly, however, there was also much racial mixing, and 82 million Brazilians have mixed European, African, and minor Amerindian ancestries. The remaining 98 million—now barely in the majority at 50.5 percent—are mainly of European origin, the descendants of immigrants from Portugal, Italy, Germany, and Eastern Europe. The complexion of the population was further diversified by the arrival of Lebanese and Syrian Muslims following the collapse of the Ottoman Empire nearly a century ago. Estimates of the Muslim component in the Brazilian population suggest that about 10 million can claim to be of Arab descent, of whom about 1 million are nominally Muslims. But practicing Muslims are far fewer in number: in 2005 Brazil had only about 50 functioning mosques.

Another significant, though small, minority began arriving in Brazil in 1908: the Japanese, who today are concentrated in São Paulo State. The more than 1 million Japanese-Brazilians form the largest ethnic Japanese community outside Japan, and in their multicultural environment they have risen to the top ranks of Brazilian society as business leaders, urban professionals, farmers—and even as politicians in the city of São Paulo. Committed to their Brazilian homeland as they are, the Japanese community also retains its contacts with Japan, resulting in many trade connections. Recently, too, there has been a limited amount of return migration to Japan as the Japanese struggle with an aging population and the need for immigrants in their economy (see Chapter 9).

A National Culture

Brazilian society, to a greater degree than is true elsewhere in the Americas, has made progress in dealing with its racial divisions. To be sure, blacks are still the least advantaged among the country's major population groups, and community leaders continue to complain about discrimination. But ethnic mixing in Brazil is so pervasive that hardly any group is unaffected, and official census statistics about blacks and Europeans are almost meaningless. What the Brazilians do have is a true national culture, expressed in an historic adherence to the Catholic faith (now eroding under Protestant-evangelical and secular pressures), in the universal use of Portuguese as the common language ('Brazilian'), and in a set of lifestyles in which soccer, beach culture, distinctive music and dance, and a growing national consciousness and pride are fundamental ingredients.

Changing Demography

Brazil's population grew rapidly during the world's twentieth-century population explosion. But over the past three decades, the rate of natural increase has slowed from nearly 3.0 percent to 1.4 percent, and the average number of children born to a Brazilian woman has been nearly halved from 4.4 in 1980 to 2.3 in 2006. These dramatic reductions occurred in the absence of any active family-planning policies by Brazil's federal or State governments and in direct contradiction to the teachings of the Catholic Church, reflecting religion's changing role in this former bastion of Roman Catholicism. Brazil's rapid urbanization, its prevalent economic uncertainties, relatively open discussion of sexuality, and the widespread use of contraceptives are among the factors influencing this significant change.

African Heritage

As noted, Afro-Brazilians still suffer disproportionately from the social ills of this relatively integrated nation. And yet Brazil's culture is infused with African themes, a quality that has marked it from the very beginning. Three centuries ago, the Afro-Brazilian sculptor and architect affectionately called Aleijadinho was Brazil's most famous artist. The world-famous composer Heitor Villa-Lobos used numerous Afro-Brazilian folk themes in his music, and African rhythms are the basis of so much of Brazilian popular music. So many Africans were brought in bondage to the city and hinterland of Salvador (Bahia State) from what is today Benin (formerly Dahomey) in West Africa that Bahia has become a veritable outpost of African culture. Candomblé, a religion arising from the arrival of African cultures in Brazil, is concentrated in Salvador and Bahia (Fig. 5-12). Its presence there has led to a reconnection with modern Benin as Afro-Brazilians travel to West Africa

FIGURE 5-12

© H. J. de Blij, P. O. Muller, and John Wiley & Sons, Inc.

to search for their roots and West Africans come to Bahia to experience the cultural landscape to which their ancestors gave rise.

In some parts of Bahia, the rural Northeast, and urban areas in Brazil's core you can almost imagine being in Africa. As a nation, Brazil is taking an increased interest in African linkages, not only for cultural connections but because of links with other former Portuguese colonies such as Angola and Moçambique.

Inequality in Brazil

For all its accomplishments in multiculturalism, Brazil remains a country of stark, appalling social inequalities (Fig. 5-13). Although such inequality is hard to measure precisely, South America is often cited as the geographic realm exhibiting the world's sharpest division between affluence and poverty. In South America, Brazil has the widest gap of all.

The Scourge of Poverty

Data that underscore this situation include the following: today the richest 10 percent of the population own two-thirds of all the land and control more than half of Brazil's wealth. The poorest 20 percent of the people live in the most squalid conditions prevailing anywhere on the planet, even including the megacities of Africa and Asia. According to UN estimates, in this age of adequate available (but not everywhere affordable) food, about half the population of Brazil suffers from some form of malnutrition and, in the northeastern States, even hunger. Packs of young orphans and abandoned children roam the cities, sleeping where they can, stealing or robbing when they must. Some of the world's most magnificent central cities are ringed and sectored by the most wretched *favelas* (see photo on page 187) where poverty, misery, and crime converge.

Most Brazilian voters had long believed that these conditions were worsening, that what economic development was taking place benefited those already well off, and that powerful corporations and foreign investors were to blame for the spiral of rising and growing unemployment. When the latest official reports announced that the number of Brazilians afflicted by poverty had increased by 50 percent since 1980, the people demanded change and, as we will note, they achieved that in 2002.

Development Prospects

Brazil is richly endowed with mineral resources, including enormous iron and aluminum ore reserves, extensive tin and manganese deposits, and sizeable oil- and gasfields, especially the recently discovered deepwater Tupi field off the coast of the Atlantic Ocean (Fig. 5-12). Other significant energy developments involve massive new hydroelectric facilities and the successful substitution of sugarcane-based alcohol (gasohol) for gasoline—allowing well over half of Brazil's cars to use this fuel instead of costly imported petroleum, making Brazil one of the few energy self-sufficient industrialized countries. Besides these natural endowments, most Brazilian soils sustain a bountiful agricultural output that makes the country a global leader in the production and export of coffee, soybeans, and orange juice concentrate. Industrial-scale commercial agriculture, in fact, is now the fastest growing economic sector, driven by mechanization and the opening of a major new farming frontier in the fertile grasslands of southwestern Brazil (see Environmentally Speaking box, p. 169).

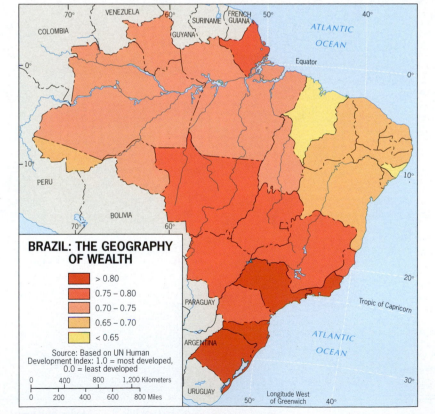

BRAZIL: THE GEOGRAPHY OF WEALTH

- > 0.80
- 0.75 – 0.80
- 0.70 – 0.75
- 0.65 – 0.70
- < 0.65

Source: Based on UN Human Development Index: 1.0 = most developed, 0.0 = least developed

FIGURE 5-13   © H. J. de Blij, P. O. Muller, and John Wiley & Sons, Inc.

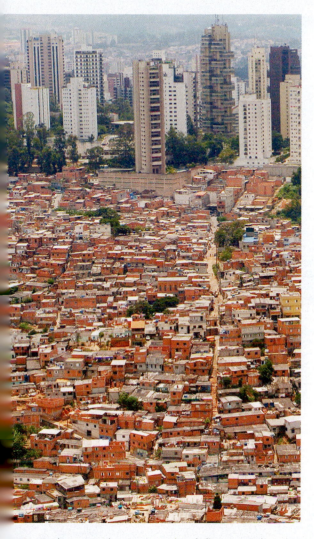

São Paulo may not have an imposing skyline to match New York or a skyscraper to challenge the Sears Tower of Chicago, but what this Brazilian megacity does have is mass. São Paulo is more than just another city—it is a vast conurbation of numerous cities and towns containing more than 25 million inhabitants who constitute the third-largest human agglomeration on the earth. This view shows part of the razor-sharp edge of the CBD, a concrete jungle with little architectural distinction but vibrant urban culture. In the foreground, juxtaposed against the affluence of downtown, the teeming inner city reflects the opposite end of the social spectrum—a chaotic jumble of *favelas* that mainly house the newest urban migrants, especially from Brazil's hard-pressed Northeast. São Paulo, demographers say, gives us a glimpse of the urban world of the future.   (© AP/Wide World Photos)

Industrialization has propelled Brazil's rise as a global economic power. Much of the momentum for this continuing development was unleashed in the early 1990s after the government opened the country's long-protected industries to international competition and foreign investment. These new policies are proving to be effective because productivity has risen by over a third since 1990, as Brazilian manufacturers attain world-class quality in particular sectors. You may very well fly in a Brazil-made passenger jet the next time you travel by air within the United States. During the mid-1990s, the revenues from industrial exports surpassed those from agriculture. Commerce with Argentina made that member of Mercosur/l one of Brazil's leading trade partners. On the global stage, Brazil became a formidable presence in other ways. The country's enormous and easily accessible iron ore deposits, the relatively low wages of its workers, and the mechanized efficiency of its steelmakers enable Brazil to produce that commodity at half the cost of steel made in the United States. This causes American steel producers to demand government protection through tariffs, which goes against purported U.S. principles of free trade.

A New Era

Even though Brazil has become a major player on the world's economic stage, it has not escaped the cycles of boom and bust that affect transforming economies. In the mid- and late twentieth century, Brazil needed loans from international lending agencies to weather the downturns, and the government had to agree to terms that included the privatization of public companies ranging from telephones to utilities. This often meant layoffs, rising prices to consumers, and even civil unrest. Brazil's currency, the *real* (plural *reais*), was devalued in 1999, resulting in the outflow of billions of dollars withdrawn from Brazil's banks by panicked depositors. Strikes and marches against increased gas prices, highway tolls, and utility charges paralyzed the country for days. Newspapers reported scandals and corruption involving leading public figures and hundreds of millions of *reais*. In the political campaign that led up to the presidential election of 2002, these issues guaranteed the defeat of the right-of-center government held responsible for it all. In that election a leftist candidate, Luiz Inácio Lula da Silva, won in a landslide victory. He faced the daunting task of adhering to Brazil's existing financial commitments while steering a new course toward more fairness, openness, and honesty in government. Although there were a few scandals that forced a run-off, he was reelected in 2006.

Antipoverty Programs

The left-of-center government inherited Brazil's massive social problems without adequate resources to address them. The new administration focused on two key areas: land reform and poverty reduction. At the turn of this century, 1 percent of Brazilians owned 50 percent of all productive land. But the Lula da Silva government was able to settle only about 60,000 landless families per year, far below what was needed. This resulted in protests and disruptions, which in turn scared off investors, affecting the jobless rate. By 2006, the government had found the resources to increase the rate of settlement, and voters reelected President Lula da Silva. In the poverty arena, the administration began a *Fome Zero* (Zero Hunger) program, but in the process created a huge, corruption-riddled bureaucracy. Officially, Brazil counts some 40 million people who live on half the minimum wage of about U.S. $80 per month or less, and the question was how to aid the poorest of the poor with modest government handouts. By 2006, the government's well-intended but muddled plan had been replaced by a far more efficient subsidy program that was having a significant impact in the poorest States. Based on a similar program in Mexico, this *Bolsa Familiar* (family fund) involves a conditional cash transfer program wherein the government pays families to keep their children in school and be vaccinated as a way of breaking the cycle of poverty that besets so many. In just a few years the program has become so successful that it is now being used as a model for antipoverty programs in places as diverse as New York City, Egypt, and some Eastern European countries.

Brazil's Subregions

Brazil is a federal republic consisting of 26 States and the federal district of the capital, Brasília (Fig. 5-12). As in the United States, the smallest States lie in the northeast and the larger ones farther west; their populations range from about 3 million in the huge, sparsely populated State of Amazonas to just over 40 million in burgeoning São Paulo State. Although Brazil is about as large as the 48 contiguous United States, it does not exhibit a clear physiographic regionalism. Even the Amazon Basin, which covers almost 60 percent of the country, is not entirely a plain: between the tributaries of the great river lie low but extensive tablelands. Despite this physiographic ambiguity, the Brazilian census bureau uses five formal regions, which we consider to be Brazilian subregions (Fig. 5-12):

1. **The Northeast** This subregion encompasses the States of Maranhão, Piauí, Ceará, Rio Grande do Norte, Paraíba, Pernambuco, Alagôas, Sergipe, and Bahia, and is the place of Brazil's origin, its cultural hearth. During colonial times it was the center of the sugar plantation economy, a place of maximum slave landings. This is why, today, this region is the center of Afro-Brazilian culture. The interior of the region is dry savanna (known as the *sertão*) and drought-prone (related to **17 El Niño**—periodic sea-surface-warming events off the continent's northwestern coast). This subregion contains about a third of the Brazilian population but is economically weak and has thereby become a source area of migrants to other parts of Brazil. The economy is diversifying from a reliance on sugar and cacao into other agricultural products such as grapes. There is also a petrochemical industry and a nascent technological hub.

2. **The Southeast** This subregion encompasses the States of Rio de Janeiro, Espirito Santo, São Paulo, and Minas Gerais and is the industrial core of the country. The subregion is centered on the powerhouse State of São Paulo—containing the namesake city that is one of the world's largest. This center of industrialization initially became wealthy through large coffee plantations in the State, which continue today. Brazil remains the world's largest coffee producer, but other crops are important as well such as oranges (for concentrated orange juice) and soybeans. Rio de Janeiro, Brazil's second-largest city, is the center of its cultural life and remains symbolically its heart. Gold and minerals (especially iron and gemstones) are abundant in the interior of this subregion in the State of Minas Gerais (meaning "General Mines"). These were an important part of the colonial enterprise in Brazil, but their extraction and processing continue today.

3. **The South** This subregion encompasses the States of Paraná, Santa Catarina, and Rio Grande do Sul and is the temperate zone of Brazil. Large-scale agriculture has long been practiced here and focuses principally on grains such as wheat, rice, and soybeans. One of the world's largest dams,

A woman cooking *acaraje*, a fritter of deep fried black-eyed peas, sits on an ornate balcony in Salvador, the capital of the State of Bahia in the Northeast. Salvador's Afro-infused cuisine is famous, with street vendors dressed in typical Afro-Brazilian dress. The population of the State is about 80 percent African in origin and is noted for its Afro-Brazilian culture, especially food, religion, and music. Salvador is the third-largest city in Brazil after São Paulo and Rio de Janeiro.   (© Tai Power Seeff/ The Image Bank/Getty Images)

the Itaipu, on the Paraná River, is located here and provides electricity to the entire country. Most notably, this subregion is culturally similar to the Southern Cone countries.

4. **The Central-West** This subregion encompasses the States of Goías, Mato Grosso, Mato Grosso do Sul, and Tocantins and has long been peripheral in the development of coastal-oriented Brazil. But today the subregion is more important to the economy of the country, especially since its opening up in the middle of the twentieth century with the planning and building of a new `18` **forward capital**, Brasília. The natural environment of this subregion is *cerrado*, savanna, long considered a wasteland but now believed to contain significant biodiversity and being threatened by the rapid expansion of industrial-scale soybean cultivation in the area (see photo). Significant infrastructure is still missing here, but road and rail construction are being planned to maximize the economic potential of this subregion.

5. **The North** This subregion encompasses the States of Amapá, Pará, Amazonas, Roraima, Rondônia, and Acre and is often referred to as the Amazon Basin, home of the world's largest river and largest track of remaining rainforest. Thinly populated but territorially the largest subregion in Brazil, the North has been targeted for development since the 1970s due to its bounty of natural resources (e.g., lumber, iron ore, bauxite) and perceived agricultural potential. But major tensions exist between environmentalists who wish to conserve the rainforest and those who want to develop the region economically. The city of Manaus (population of 1.8 million) has been a `19` **growth pole** since 1966, a free-trade zone assembling electronics and other durable goods in the middle of the rainforest. Due to the lack of road or rail connectivity between Manaus and the rest of Brazil, everything that is produced there needs to be exported using air freight or ships, making this industrialization rather unsustainable without subsidies. The rule of law is tenuous in this poorly connected and integrated subregion, and violence over land rights is pervasive. This is where the majority of Brazil's 200,000 indigenous people live, many on reservations that they have long fought to preserve. Threats to their existence continue as illegal incursions bring epidemic disease and violence deep into the interior of the subregion.

Brazil is the cornerstone of South America, the dominant economic force in Mercosur/l, the only dimensional counterweight to the United States in the Western Hemisphere, a maturing democracy, and an emerging global giant. The future of the entire South American realm depends on Brazil's stability, its social as well as its economic progress.

Called the *Soy Highway*, BR 163 is a 1100-kilometer-long highway that has long existed on maps as the connection between the city of Santarém on the Amazon River and the city of Cuiabá in the center of Brazil. In reality, much of the road is impassible as it was not paved and tropical rain made the road a mud track at best. However, with the rapid expansion of soybean cultivation in the center of the country and the development of a port at Santarém, a big effort is being made to pave the road to make it passable for trucks carrying soy. Here you can see a paved section that gives way to red dirt, which awaits future paving.   (© NewsCom)

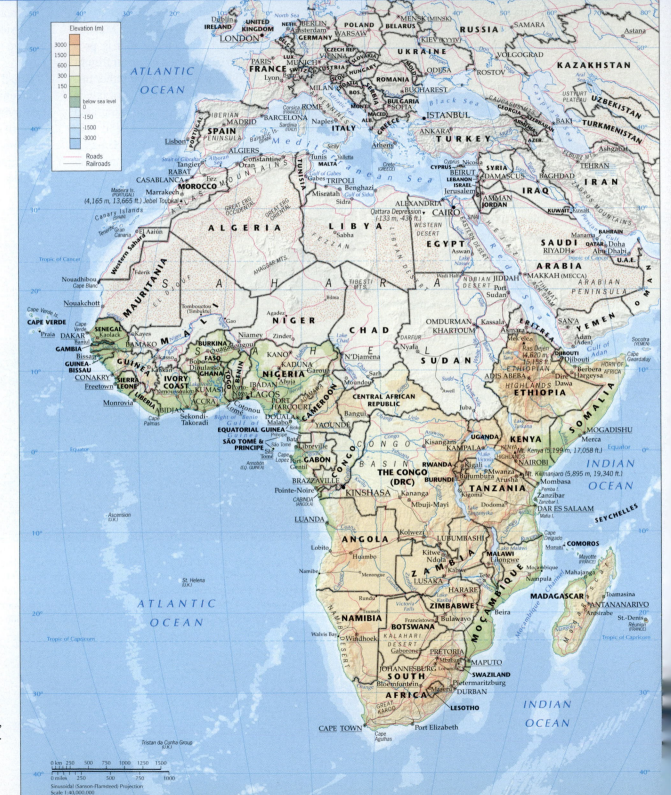

FIGURE 6-1

Map: © H. J. de Blij, P. O. Muller, and John Wiley & Sons, Inc.

6

SUBSAHARAN AFRICA

In This Chapter

- Heart of the world, cradle of humanity
- Africa's rich and varied cultures
- Why poverty remains so persistent in Africa
- South Africa: Engine of its region, beacon for the realm
- Nigeria: Cornerstone of West Africa
- The Islamic Front moving southward

CONCEPTS, IDEAS, AND TERMS

1. Human evolution
2. Rift valley
3. Continental drift
4. Land tenure
5. Colonialism
6. Land alienation
7. Green Revolution
8. Cell phone revolution
9. Medical geography
10. Endemic
11. Epidemic
12. Pandemic
13. State formation
14. Multilingualism
15. Apartheid
16. Separate development
17. Exclave
18. Landlocked state
19. Enclave
20. Periodic market
21. Islamic Front

REGIONS

SOUTHERN AFRICA
EAST AFRICA
EQUATORIAL AFRICA
WEST AFRICA
AFRICAN TRANSITION ZONE

Photos: *(upper left)* Maasai women of the Talek district, Kenya, © A. WinklerPrins; *(above)* Market in front of City Hall, Cape Town, South Africa, © Arco Images/ Age Fotostock America, Inc.

THE AFRICAN CONTINENT occupies a special place in the physical as well as the human world. You need a globe to confirm the former: the earth has a land hemisphere and a water hemisphere, and Africa lies at the center of the land hemisphere, surrounded in all directions by other landmasses. As for Africa's human significance, this is where **1 human evolution** began. In Africa we formed our first communities, spoke our first words, and created our first art. From Africa our ancestor hominids spread outward into Eurasia more than two million years ago. From Africa our own species emigrated, beginning perhaps 130,000 years ago, northward into present-day Europe and eastward via southern Asia into Australia and, much later, farther afield into the Americas. Disperse our forebears did, but we should remember that, at the source, we are all Africans.

CRADLE AND CAULDRON

For millions of years, therefore, Africa served as the cradle for humanity's emergence. For tens of thousands of years, Africa was the source of human cultures. For thousands of years, Africa led the world in countless spheres ranging from tool manufacture to plant domestication to forging trade relationships.

But in this chapter we will encounter an Africa that has been struck by a series of disasters ranging from environmental deterioration to human dislocation on a scale unmatched anywhere in the world. When we assess Africa's misfortunes, though, we should remember that they have lasted hundreds, not thousands or millions of years. Africa's catastrophic interlude will end, and Africa's time and turn will come again.

The focus in this chapter will be on Africa south of the Sahara, for which the unsatisfactory but convenient name *Subsaharan Africa* has come into use to signify not physically 'under' the great desert but directionally south of it. The African continent contains two geographic realms: the African, extending from the southern margins of the Sahara to the Cape of Good Hope, and the western flank of the realm dominated by the Muslim faith and Islamic culture whose heartland lies in the Middle East and the Arabian Peninsula. The great desert forms a formidable barrier between the two, but the powerful influences of Islam crossed it centuries before the first Europeans set foot in West Africa. By that time, the African kingdoms in what is known today as the Sahel had been converted, creating an Islamic foothold all along the northern periphery of the African realm (see Fig. G-3). As we note later, this cultural and ideological penetration had momentous consequences for Subsaharan Africa.

MAJOR GEOGRAPHIC QUALITIES OF Subsaharan Africa

1. Physiographically, Africa is a plateau continent without a major mountain range, with a set of Great Lakes, several major river basins, variable rainfall, generally low fertility soils, and mainly savanna and steppe vegetation.

2. Dozens of nations, hundreds of ethnic groups, and many smaller entities make up Subsaharan Africa's culturally rich and varied population.

3. Most of Subsaharan Africa's peoples depend on farming for their livelihood.

4. Health and nutritional conditions in Subsaharan Africa need improvement as the incidence of disease remains high and diets are often unbalanced.

The AIDS pandemic continues to be a major health crisis in this realm.

5. Africa's boundary framework is a colonial legacy; many boundaries were drawn without adequate knowledge of or regard for the human and physical geography they divided.

6. The realm is rich in raw materials vital to industrialized countries, but many economies continue to rely on primary activities—the extraction of resources—and not the better income-generating activities of assembly and manufacturing.

7. Patterns of raw-material exploitation and export routes set up in the colonial period still prevail in most of Subsaharan Africa. Interregional and international connections are poor.

8. During the Cold War, great-power competition magnified conflicts in several Subsaharan African countries, with results that will be felt for generations.

9. Severe dislocation affects many Subsaharan African countries, from Liberia to Rwanda to Zimbabwe. This realm has the largest refugee population in the world today.

10. Government mismanagement and poor leadership afflict the economies of many Subsaharan African countries.

Defining the Realm

The African continent may be partitioned into two human-geographic realms, but the landmass is indivisible. Before we investigate the human geography of Subsaharan Africa, therefore, we should take note of the entire continent's unique physical geography (Figs. 6-1, chapter opener map; and 6-2). We have already noted Africa's situation at the center of the planet's land hemisphere; moreover, no other landmass is positioned so squarely astride the equator, reaching almost as far to the south as to the north. This location has much to do with the distribution of Africa's climates, soils, vegetation, agricultural potential, and human population.

AFRICA'S PHYSIOGRAPHY

Africa accounts for about one-fifth of the earth's entire land surface. The north coast of Tunisia lies 7700 kilometers (4800 mi) from the south coast of South Africa. Coastal Senegal, in West Africa, lies 7200 kilometers (4500 mi) from the tip of the *Horn* in easternmost Somalia. These distances have major environmental implications. Much of Africa is far from maritime sources of moisture. In addition, as Figure G-8 shows, large parts of the landmass lie in latitudes where global atmospheric circulation systems produce

FIGURE 6-2   © H. J. de Blij, P. O. Muller, and John Wiley & Sons, Inc.

arid conditions. The Sahara in the north and the Kala-hari in the south form part of this globe-girdling desert zone. Water supply is one of Africa's great problems.

Rifts and Rivers

Africa's topography reveals several properties that are not replicated on other landmasses. Alone among the continents, Africa does not have a mountain backbone; neither the northern Atlas nor the southern Cape Ranges are in the same league as the Andes or Urals. Where Africa does have high mountains, as in Ethiopia and South Africa, these are really deeply eroded plateaus or, as in East Africa, high, snowcapped volcanoes. Furthermore, Africa is one of only two continents containing a cluster of Great Lakes, and the only one whose lakes result from powerful tectonic forces in the earth's crust. These lakes (with the exception of Lake Victoria) lie in deep trenches called **2 rift valleys**, which form when huge parallel cracks or faults appear in the earth's crust and the strips of crust between them sink, or are pushed down, to form great, steep-sided, linear valleys. In Figure 6-2 these rift valleys, which stretch more than 9600 kilometers (6000 mi) from the Red Sea to Swaziland, are marked by red lines.

Africa's rivers, too, are unusual: their upper courses often bear landward, seemingly unrelated to the coast toward which they eventually flow. Several rivers, such as the Nile and the Niger, have inland as well as coastal deltas. Major waterfalls, notably Victoria Falls on the Zambezi, or lengthy systems of cataracts, separate the upper from the lower courses.

Finally, Africa may be described as the plateau continent. Except for some

comparatively limited coastal plains, almost the entire continent lies above 300 meters (1000 ft) in elevation, and fully half of it lies over 800 meters (2500 ft) high. As Figure 6-2 shows, the plateau surface has sagged under the weight of accumulating sediments into a half dozen major basins (three of them in the Sahara). The margins of Africa's plateau are marked by escarpments, often steep and step-like. Most notable among these is the Great Escarpment of South Africa, marking the eastern edge of the Drakensberg Mountains.

Continental Drift and Plate Tectonics

Africa's remarkable and unusual physiography was one piece of evidence that geographer Alfred Wegener used to construct his hypothesis of **3 continental drift**. The present continents, Wegener reasoned, lay assembled as one giant landmass called *Pangaea* not very long ago in geologic time (220 million years ago). The southern part of this super-

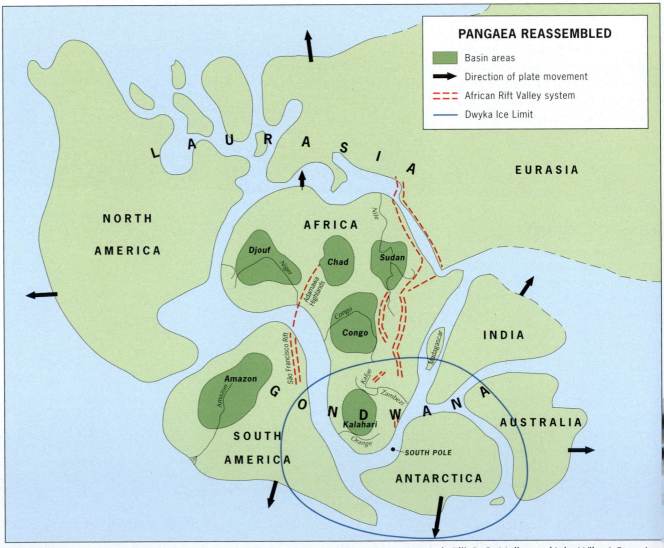

FIGURE 6-3

© H. J. de Blij, P. O. Muller, and John Wiley & Sons, Inc

continent was *Gondwana*, of which Africa formed the core (Fig. 6-3). When, about 200 million years ago, tectonic forces began to split Pangaea apart, Africa (and the other landmasses) acquired their present configurations. That process, now known as **plate tectonics**, continues, marked by earthquakes and volcanic eruptions. By the time it started, however, Africa's land surface had begun to acquire some of the features that mark it today—and make it unique. The rift valleys, for example, demarcate the zones where plate movement continues—hence the linear shape of the Red Sea, where the Arabian Plate is separating from the African Plate (Fig. G-5). And yes, the rift valleys of East Africa probably mark the further fragmentation of the African Plate (some geophysicists have already referred to a Somali Plate, which will separate, Madagascar-like, from the rest of Africa).

Africa's ring of escarpments, its rifts, its river systems, its interior basins, and its lack of significant mountains, all relate to the continent's central location in Pangaea, all pieces of the puzzle that led to the plate-tectonics solution.

NATURAL ENVIRONMENTS

Only the southernmost tip of Subsaharan Africa lies outside the tropics. Although African elevations are comparatively high, they are not high enough to ward off the heat that comes with tropical location except in especially favored locales such as the Kenya Highlands and parts of Ethiopia. And, as we have noted, Africa's bulky shape means that much of the continent lies far from maritime moisture sources. Variable weather and frequent droughts are among Africa's environmental challenges.

It is useful at this point to refer back to Figures G-8 and G-9. As that map shows, Africa's climatic regions are distributed almost symmetrically about the equator, though more so in the center of the landmass than in the east, where elevation changes the picture. The hot, rainy climate of the Congo Basin merges gradually, both northward and southward, into climates with distinct winter-dry seasons. Win-

ter, however, is marked more by drought than by cold. In parts of the area mapped *Aw* (savanna), the annual seasonal cycle produces two rainy seasons, often referred to locally as the long rains and the short rains, separated by two winter dry periods. As you go farther north and south, away from the moist Congo Basin, the dry season(s) grow longer and the rainfall diminishes and becomes less and less dependable.

Wildlife

Africa is famous for its fauna, and its rainforests and vast savannas are the home of wildlife ranging from primates to wildebeests. Gorillas and chimpanzees survive in forest habitats, while large herds of herbivores range across the savanna plains. But many species are stressed, and some are threatened with extinction as their habitats are being encroached upon. Colonialism brought with it changes in land tenure, and this event, in combination with population growth, has resulted in land enclosure to accommodate expanding agriculture, leaving less space for roaming fauna. War, civil unrest, and poaching also continue to extract a significant toll on Africa's wildlife.

Managing this wildlife has been a challenge, and many perspectives have been presented as to how to proceed with it. The biggest change in Africa's wildlife came during colonialism when human activity and animals were separated. As has been the case all over the world, Subsaharan Africa experienced a transformation from traditional land uses and land cover types to non-native livelihood systems. This was primarily due to the Europeans' colonization of peoples and landscapes from the mid-1800s to the 1960s (further discussed below). Over this time period, colonialists altered the land management practices of indigenous peoples through privatization and the creation of large protected areas that were originally created as sport hunting preserves. These protected areas led to the forced eviction of indigenous peoples from their ancestral lands. Over time, many protected areas adopted the North American conservation model in which fences and fines were consid-

ered the best practices to keep wild animals *in* and people *out*. However, this model did not take into account the migratory behavior of wild animals such as wildebeest and gorillas, nor was consideration given to the integrative nature and great variability in human use of the landscape, such as the grazing patterns of pastoralists and their cattle on savanna grasslands. As a result, an increased number of conflicts have arisen over natural resources that are critical to both people and wildlife.

Environmental Challenges to Farming

Figure 6-4 maps the population distribution of Subsaharan Africa. Many countries in Subsaharan Africa have had significant population growth. But, as data in Appendix B confirms, all the countries of this realm *combined* have a population only about half that of China alone. Much of this population continues to depend on farming for their livelihood. Although frequently thought to have only unproductive environments, Africa does have areas of good soil, ample water, and high productivity. These include the volcanic soils of Mount Kilimanjaro and those of the highlands around the Western Rift Valley, lands in the Ethiopian Highlands, those in the moister areas of higher-latitude South Africa, and parts of West Africa. The challenges to farming in Subsaharan Africa revolve around two main factors: (1) climatic variability; and (2) political and economic policies of each country.

Compare Figures G-8 and G-9 and note the steep decline in annual rainfall from more than 200 centimeters (80 in) around the equator in the Congo Basin to a mere 10 centimeters (4 in) in parts of Chad to the north and Namibia to the south. In West, Equatorial, and East Africa, rainfall is dependent on the movement of the Inter-Tropical Convergence Zone (ITCZ), a band of low-pressure air that produces rainfall. The ITCZ moves north and south of the equator at different times of the year. The degree of the ITCZ movement is a product of a number of different factors, and the reliability of rainfall is dependent on these factors.

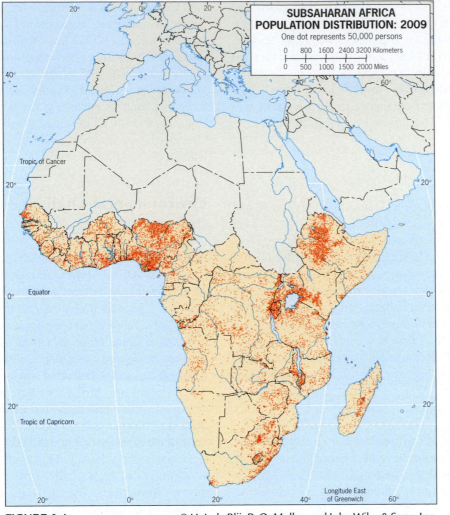

SUBSAHARAN AFRICA POPULATION DISTRIBUTION: 2009

One dot represents 50,000 persons

0   800   1600 2400 3200 Kilometers

0   500   1000 1500 2000 Miles

FIGURE 6-4   © H. J. de Blij, P. O. Muller, and John Wiley & Sons, Inc.

one night. Perhaps it is not surprising then that local populations view elephants as pests.

LAND ISSUES

While rainfall is a critical physical criterion for farming, a number of political and economic factors are also influential. Among these factors are land tenure; type of agricultural system (e.g., subsistence or commercial); type of farming system (rotational, shifting cultivation, intercropping, etc.); price for crops; government policies promoting the planting of one crop in preference to another; indigenous agricultural knowledge of particular crops; and the degree of technology adoption and mechanization.

The issue of land tenure is crucial in Africa because most Africans are still farmers. **4** **Land tenure** refers to the way people own, occupy, and use land. African traditions of land tenure are different from those of Europe or the Americas today. In most of Subsaharan Africa, communities, not individuals, customarily hold land. Occupants of the land have temporary, custodial rights to it and cannot sell it. Land may be held by large (extended) families, by a village community, or even by a traditional chief who holds the land in trust for the people. His subjects may house themselves on it and farm it, but in return they must follow his rules.

Stolen Lands

At the onset of **5** **colonialism**, colonial administrators sought to control the most fertile areas of the occupied colonies through eviction of indigenous peoples. In some cases this was done by physical force (mostly through military conquest), and in other areas by coercion. Prior to colonialism, many traditional African livelihood systems adhered to sustainable land management practices such as leaving the land fallow or rotational grazing. Viewing this land as unused, colonial planners began a process of **6** **land alienation** not unlike what happened in Middle and

Consequently, there are periods of below-normal rainfall, which can significantly affect where and when farmers are able to grow and harvest crops. Farming in semiarid regions of the world is always difficult as rainfall can be highly variable from year to year. Current climate change information indicates that this interannual variability will increase in the future, thereby increasing risk for farmers in these areas.

Millions of farmers make a living in this stressed *savanna* region, reducing their risk by mixing livestock herding with permanent agriculture. But wildlife in the savannas carries diseases that can infect livestock, adding a challenge to herding. Wildlife can also threaten crops when animals, whose movements have become constricted, invade fields during the night. Elephants are notorious raiders of maize (corn) harvests and can deplete a family's harvest in

Desertification in West Africa.   In the dryland Sahel region of West Africa, productive farmland is threatened by the encroaching Sahara from the north. Or is it? Persistent narratives in the media describe untold hardship that arises from cropland loss to the desert sand. In reality, the desert 'front' waxes and wanes over time, so some expansion of the desert is to be expected. Farming in this region is challenging because it receives only 10 to 60 centimeters (4 to 24 in) of rain per year. This rainfall comes during a three-month rainy season, but where and when it falls can be highly variable. People native to the Sahel region cope with this environment through diversification, including the herding of livestock. But these coping strategies have become more challenging as boundaries have been imposed by European colonists and changes in land use have been encouraged. The result has been cyclical droughts. In addition, scientific understanding of dryland ecologies has been limited in the past, resulting in some unsound policy decisions. All of these factors have combined to leave this region vulnerable to environmental, especially climate, change.   (© AP/Wide World Photos)

forms ranging from shifting cultivation to pastoralism, work best when the population is fairly stable and tenure is communal. Land must be left fallow to recover from cultivation, and pastures must be kept free of livestock so that the grasses can revive. The population explosion that Africa experienced during the mid-twentieth century set in motion a cycle of land overuse. When soils cannot rest and pastures are overgrazed, the land becomes degraded and yields decline.

Persistent Subsistence

Although there is commercial farming in parts of Africa, most African farmers remain subsistence farmers who grow grain crops (maize [corn], millet, sorghum) in drier areas and root crops (yams, manioc [cassava], sweet potatoes) where moisture is ample. Others herd livestock, mostly cattle and goats, as they counter environmental and climatic variability. Farmers and pastoralists alike have been unable to secure access to regular markets and stable prices for their products because of government policies that often promote one particular type of export-oriented crop (for example, peanuts in Gambia, tea and coffee in Kenya, or cacao—cocoa—in the Ivory Coast) over other crops that do not fetch high prices on the world market.

Many African farmers have had to adapt to these problems; nonetheless, farm yields in Africa have been modest for many years. Governments, in response to World Bank policies, paid greater attention to high-profile industrial projects and as a result have neglected agriculture. The decline of African agriculture has been disproportionally hard on Africa's women who, according to current estimates, produce 75 percent of local food in Subsaharan Africa. Development policies often pay little attention to this situation and the related household dynamics, thereby dooming many well-intended projects.

Another obstacle faced by Africa's farmers is the economic policies of developed nations, which advocate free trade as part of the virtues of globalization. Actually, they are protecting their own farmers

South America. Many of the most fertile and productive areas were placed under the control of colonial settlers and governments. Over time, these lands were bought and sold and became private legal property. At the end of colonialism, many newly independent African countries started a program whereby land would revert to traditional forms of ownership and management. However, the legacy of colonialism has been difficult to overcome, and African governments have adopted several neocolonial policies on land management that continue to marginalize farmers.

Rapid population growth, as Africa experienced early in the postcolonial period, makes access to land even more complicated. Traditional systems of land use, which involve subsistence farming in various

through tariffs against imports and through subsidies for their products. African commercial farmers could market their produce at prices below those of, for example, French farmers—but the French government prevents it. Economic geographers estimate that this costs African farmers as much as U.S.$200 billion annually, a sum that would make a huge difference in a realm of limited opportunities.

The **7** **Green Revolution**—the development of more productive, drought-tolerant, pest-resistant, higher-yielding types of grain (discussed at greater length in Chapter 8)—has had less impact in Africa than elsewhere. Where people depend mainly on rice and wheat, the Green Revolution pushed back the

The cell phone revolution has made a dramatic impact on farmers in many parts of the developing world, especially in Subsaharan Africa where distances can be far, land lines absent, and market information scarce. Now farmers and market women, such as the Kenyan woman pictured here, can be in touch and can better gauge when it is best to market crops.   (© AP/Wide World Photos)

prospect of increased hunger. But little research was done on Africa's dominant crops, mostly tubers, so the advances of the Green Revolution were minimal for Africa. The Green Revolution is also not an unqualified remedy: the poorest farmers, who need help the most, can least afford the more expensive, higher-yielding seeds or the pesticides that are often required.

In spite of the obstacles that most African farmers face, a second revolution is occurring in much of Subsaharan Africa and elsewhere in the developing world. The **8** **cell phone revolution** is allowing many farmers to send text (or SMS–short message system) messages to obtain critical information such as weather alerts or forecasts, and the prices for agricultural products such as the most up-to-date price estimate for a kilogram of maize or a sack of potatoes at the nearest market. In the past African farmers were constrained by poor access to reliable market information. The cell phone revolution allows them to negotiate higher prices for their products.

HIV/AIDS (see also p. 199) presents a serious challenge to farming and natural resource management in Subsaharan Africa. This disease has caused the rapid debilitation and death of so many farmers of working age that both productivity and the knowledge systems that contributed to agriculture and natural resource management are seriously threatened.

HIV/AIDS (see also p. 199)

ENVIRONMENT AND HEALTH

The study of human health in spatial context is the field of **9** **medical geography**, and medical geographers employ modern methods of analysis (including geographic information systems) to track disease outbreaks, identify their sources, detect their carriers, and prevent their repetition. Alliances between medical personnel and medical researchers, and geographers have already yielded significant results. Doctors know how a disease ravages the body; geographers know how climatic conditions such as wind direction or variations in river flow can affect the dis-

tribution and effectiveness of disease carriers. This collaboration helps protect vulnerable populations.

Tropical Africa is the source area of many serious illnesses, and has thereby become the focus of much research in medical geography. Researchers look at the carriers (*vectors*) of infectious diseases, the environmental conditions that give rise to them, and also the cultural and social geography of disease diffusion and transmission. Comparing medical, environmental, and social/cultural maps can lead to crucial evidence that helps combat the scourge.

In Africa today, hundreds of millions of people carry one or more maladies, often without knowing exactly what ails them. A disease that infects many people (the *hosts*) in a kind of equilibrium, without causing rapid and widespread deaths, is said to be **10** **endemic** to that population. People affected may not die suddenly or dramatically, but their quality of life and productive capacity are hindered as their overall health is weakened and can deteriorate quickly when an acute illness strikes. In tropical Africa, hepatitis, venereal diseases, and hookworm are among many public health threats in this category.

Epidemics and Pandemics

When a disease outbreak has local or regional dimensions, it is called **11** **epidemic**. It may claim thousands, even tens of thousands, of lives, but it remains confined to a certain area, perhaps one defined by the range of its vector. In tropical Africa, trypanosomiasis, the disease known as sleeping sickness and vectored by the tsetse fly, has regional proportions. The great herds of savanna wildlife form the *reservoir* of this disease, and the tsetse fly transmits it to livestock and people. It is endemic to wildlife, but it also kills cattle, so Africa's herders try to keep their animals in tsetse-free zones. African sleeping sickness appears to have originated in a West African source area during the fifteenth century, and it spread throughout much of tropical Africa. Its epidemic range was limited by that of the tsetse fly: where there are no tsetse flies, there is no sleeping sickness. More than anything else, the tsetse fly has kept significant parts of Subsaharan Africa's savannas free of livestock and open to wildlife.

When a disease spreads worldwide, it is described as **12 pandemic**. Africa's and the world's most deadly vectored disease is malaria, transmitted by a mosquito and killer of as many as 1 million children each year. Malaria is an ancient affliction. Hippocrates, the Greek physician of the fifth century BC, mentions it in his writings. Apes, monkeys, and several other species also suffer from it. Fever attacks, anemia, and enlargement of the spleen are its symptoms. Malaria has diffused around the world and prevails not only in tropical but also in temperate areas. Eradication campaigns against the mosquito vector have had some success, but always the carrier has come back with renewed vigor. At present, as many as 515 million people are affected by malaria globally. Current efforts at fighting the disease through an increased use of mosquito nets appears to be having good results, yet malaria causes an estimated one to three million deaths each year. The short life expectancies for tropical Africa reported in Appendix B in part reflect infant and child mortality from malarial infection.

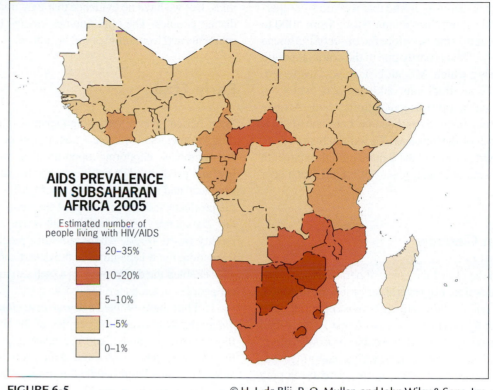

FIGURE 6-5   © H. J. de Blij, P. O. Muller, and John Wiley & Sons, Inc.

Africa's Latest Scourge

Malaria remains Africa's great affliction today. But in many ways it is overshadowed by the realm's latest scourge, HIV/AIDS, which had its start as an African epidemic and became a pandemic in just a few decades.

In late 2007, the Joint United Nations Programme on HIV/AIDS (UNAIDS) estimated that the number of HIV cases around the world stands at approximately 33 million. Of this figure, 22.5 million live in Subsaharan Africa. In addition, 61 percent of women are affected by HIV/AIDS. The highest prevalence rates can be found in Southern Africa, where in eight countries 15 percent of the total population are affected by the disease. South Africa has the highest number of people living with AIDS. Prevalence rates

are also high in western Equatorial African countries such as the Central African Republic and Gabon. In East Africa prevalence rates remain stable or are declining, as in Uganda, largely as the result of aggressive public health campaigns.

Why are more than two out of every three people who suffer from HIV/AIDS-related illnesses found in Subsaharan Africa? The answer to this question revolves around several key factors. The first is that the disease originated in Subsaharan Africa and quickly spread to all segments of society. The second is that there is social stigma associated with the disease. This makes acknowledging and treating the disease difficult, particularly in rural areas. Also HIV/AIDS is not separate from the social, political, and economic context of the region or country. This means that while a person can be infected with the disease, their likelihood of survival with the disease

is lower if they are unable to access medical centers and afford antiretroviral drugs (ARVs).

Overall the picture is less grim than it was a few years ago (Fig. 6-5). A new attitude toward the issue by politicians and societies in general, well-conducted public health campaigns, and increased availability of low-cost generic AIDS drugs are helping to contain this pandemic, although there is still a long way to go.

AFRICA'S HISTORICAL GEOGRAPHY

Africa is the cradle of humanity. Archaeological research has chronicled 7 million years of transition from Australopithecenes to Hominids to *Homo sapiens*. It is therefore ironic that we know comparatively little about Subsaharan Africa from 5000 to 500 years ago—that is, before the onset of European colonialism. This is partly due to the colonial period itself, during which African history was neglected, many African traditions and artifacts were destroyed, and many misconceptions about African cultures and institutions became entrenched. It is also a result of the absence of a written history over most of Africa south of the Sahara until the sixteenth century—and over a large part of it until much later than that.

African Genesis

Africa on the eve of the colonial period was a continent in transition. For several centuries, the habitat in and near one of the continent's most culturally and economically productive areas—West Africa—had been changing. For 2000 years, probably more, Africa had been innovating as well as adopting ideas from outside. In West Africa, cities were developing on an impressive scale; in central and Southern Africa, peoples were moving, readjusting, sometimes struggling with each other for territorial supremacy. The Romans had penetrated to southern Sudan, North African peoples were trading with West Africans, and Arab *dhows* (wooden boats with triangular sails) were sailing the waters along the eastern coasts, bringing Asian goods in exchange for gold, copper, and a comparatively small number of slaves. These same dhows today are increasingly used in tourism.

It is known that African cultures had been established in all the environmental settings shown in Figure G-9 for thousands of years and thus long before Islamic or European contact. One of these, the Nok culture, endured for over eight centuries on the Benue Plateau (north of the Niger-Benue confluence in modern Nigeria) from about 500 BC to the third century AD. The Nok people made stone as well as iron tools, and they left behind a treasure of art in the form of clay figurines representing humans and ani-

mals. But we have no evidence that they traded with distant peoples. The opportunities created by environments and technologies still lay ahead.

Early Trade

West Africa, over a north-south span of a few hundred kilometers, displayed an enormous contrast in environments, economic opportunities, modes of life, and products. The peoples of the tropical forest produced and needed goods that were different from the products and requirements of the peoples of the dry, distant north. For example, salt is a prized commodity in the forest, where humidity precludes its formation, but it is plentiful in the desert and steppe. This enabled the desert peoples to sell salt to the forest peoples in exchange for ivory, spices, and dried foods. Thus there evolved a degree of *regional complementarity* between the peoples of the forest and those of the drylands. And the savanna peoples—those located in between—found themselves in a position to channel and handle the trade (which is always economically profitable).

The markets in which these goods were exchanged prospered and grew, and urban centers arose in the savanna belt of West Africa. One of these old cities, now an epitome of isolation, was once a thriving center of commerce and learning and one of the leading urban places in the world—Timbuktu (now located in Mali). In fact, its university is one of the oldest in the world, with a library that holds valuable documents that are being preserved (see photo). Others, predecessors as well as successors of Timbuktu, have declined, some of them into oblivion. Still other savanna cities, such as Kano in the northern part of Nigeria, are still important today.

Early States

Strong and durable states arose inland in West Africa. The oldest state we know anything about is ancient Ghana, located to the northwest of the modern coun-

Timbuktu, Mali, long a place-name synonymous with nowhere, was at one point the intellectual heart of Africa. Renewed interest in its educational legacy has spurred the restoration of its library. Ancient manuscripts and books hidden for centuries in leather trunks in nomad camps and in the houses along Timbuktu's dusty streets are being preserved for future generations.   (© Candace Feit/ *The New York Times*/Redux Pictures)

try of Ghana. It covered parts of present-day Mali, Mauritania, and adjacent territory. Ghana lay astride the upper Niger River and included gold-rich streams flowing off the Futa Jallon highlands, where the Niger River has its origin. Between the ninth and twelfth centuries AD, and perhaps longer, old Ghana managed to weld various groups of people into a stable state. The country had a large capital city complete with markets, suburbs for foreign merchants, religious shrines, and, some distance from the city center, a fortified royal retreat. Taxes were collected from the citizens, and tribute was extracted from subjugated peoples on Ghana's periphery; tolls were levied on goods entering Ghana, and an army maintained control. Muslims from the northern drylands invaded Ghana in 1067, when it may already have been in decline. Even so, the Ghanaians managed to protect their capital for 14 years. However, the invaders had ruined the farmlands and destroyed the trade links with the north. Ancient Ghana could not survive. It finally broke into smaller units.

Eastward Shift

In the centuries that followed, the focus of politico-territorial organization in the West African *culture hearth* (the source area of culture) shifted almost continuously eastward—first to ancient Ghana's successor state of Mali, which was centered on Timbuktu and the middle Niger River Valley, and then to the state of Songhai, whose focus was Gao, a city on the Niger that still exists. This eastward movement may have been the result of the growing influence and power of Islam. Traditional religions prevailed in ancient Ghana, but Mali and its successor states sent huge, gold-laden pilgrimages to Mecca along the savanna corridor south of the Sahara, passing through present-day Khartoum and Cairo. Of the tens of thousands who participated in these pilgrimages, some remained behind. Today, many Sudanese trace their ancestry to the West African savanna kingdoms.

Beyond the West

West Africa's savanna region undoubtedly witnessed momentous cultural, technological, and economic developments, but other parts of Africa also progressed. Early states emerged in present-day Sudan, Eritrea, and Ethiopia. Influenced by innovations from the Egyptian culture hearth, these kingdoms were stable and durable: the oldest, Kush, lasted 23 centuries (Fig. 6-6). The Kushites built elaborate irrigation

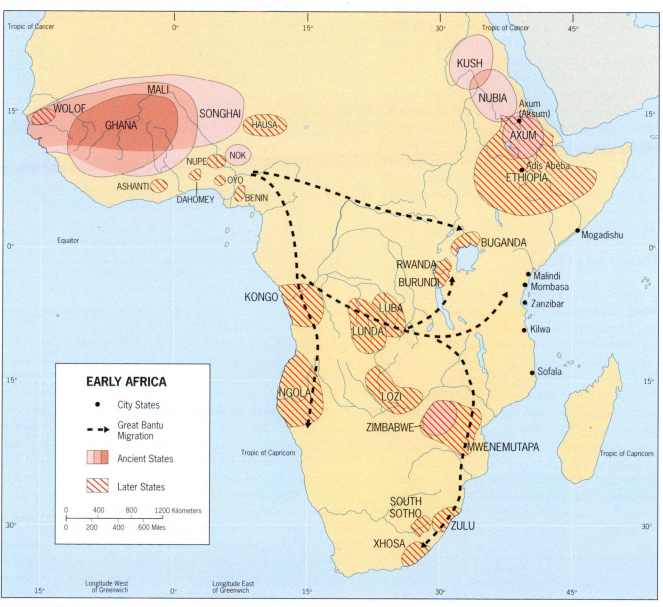

FIGURE 6-6 © H. J. de Blij, P. O. Muller, and John Wiley & Sons, Inc.

systems, forged iron tools, and built impressive structures as the ruins of their long-term capital and industrial center, Meroe, reveal. Nubia, to the southeast of Kush, was Christianized until the Muslim wave overtook it in the eighth century. And Axum was the richest market in northeastern Africa, a powerful kingdom that controlled Red Sea trade and endured for six centuries. Axum, too, was a Christian state that confronted Islam, but Axum's rulers deflected the Muslim advance and gave rise to the Christian dynasty that eventually shaped modern Ethiopia.

The process of **13 state formation** spread throughout Africa and was still in progress when the first European contacts occurred in the late fifteenth century. Large and effectively organized states developed on the equatorial west coast (notably Kongo) and on the southern plateau from the southern part of The Congo[1] to Zimbabwe. East Africa had several city-states, including Mogadishu, Kilwa, Mombasa, and Sofala.

Bantu Migration

A crucial event affected virtually all of Equatorial, West, and Southern Africa: the great Bantu migration from present-day Nigeria-Cameroon southward and eastward across the continent. This migration appears to have occurred in waves starting as long as 5000 years ago, populating the Great Lakes area and penetrating South Africa, where it resulted in the formation of the powerful Zulu Empire in the nineteenth century (Fig. 6-6).

All this reminds us that, before European colonization, Africa was a realm of rich and varied cultures, diverse lifestyles, technological progress, and external trade. It was, however, also a highly fragmented realm, its cultural mosaic (Fig. 6-7) spelling weak-

_____
[1]Two countries in Africa have the same short-form name: *Congo*. In this book, we use **The Congo** for the larger Democratic Republic of the Congo and **Congo** for the smaller Republic of Congo.

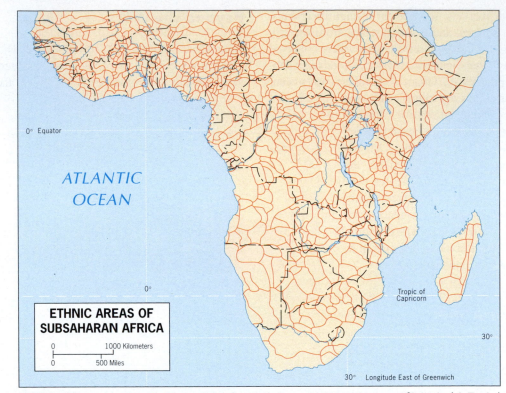

ETHNIC AREAS OF SUBSAHARAN AFRICA

0   1000 Kilometers

0   500 Miles

FIGURE 6-7   Adapted with permission from L. D. Stamp & W. T. W. Morgan, *Africa: A Study in Tropical Development*, 3rd rev. ed., after G. P. Murdock, *Africa: Its Peoples and Their Cultures*.

ness when European intervention came to change the social and political map forever.

The Colonial Transformation

European involvement in Subsaharan Africa began in the fifteenth century. It would interrupt the path of indigenous African development and irreversibly alter the entire cultural, economic, political, and social makeup of the continent. It started quietly in the late fifteenth century, with Portuguese ships groping their way along the west coast and rounding the Cape of Good Hope. Their goal was to find a sea route to the spices and riches of the Orient. Soon other European countries were sending their vessels

to African waters, and a string of coastal stations and forts sprang up. In West Africa, the nearest part of the continent to European spheres in Middle and South America, the initial impact was strongest. At their coastal control points, the Europeans traded with African intermediaries for the slaves who were destined to work New World plantations, for the gold that had been flowing northward across the desert, and for ivory and spices.

Coastward Reorientation

Suddenly, the centers of activity lay not with the inland cities of the savanna but in the foreign stations on the Atlantic coast. As the interior declined, the coastal peoples thrived. Small forest states gained unprecedented wealth, transferring and selling slaves captured in the interior to the European traders on the coast. Dahomey (now called Benin) and Benin (now part of neighboring Nigeria) were states built on the slave trade. When slavery was eventually abolished in Europe, those who had inherited the power and riches it had brought vigorously opposed abolition in both continents.

Horrors of the Slave Trade

As discussed elsewhere in the book, millions of Africans were forced to migrate from their homelands to the Americas, especially Brazil, the Caribbean region, and the United States. The slave trade was one of those African disasters alluded to above, and it was facilitated in part by what we may call the peril of proximity. The northeastern tip of Brazil, by far the largest single destination for the millions of Africans forced from their homes in bondage, lies about as far from the nearest West African coast as South Carolina lies from Venezuela. This is a short maritime intercontinental journey indeed (it is more than twice as far from West Africa to South Carolina). That proximity facilitated the forced migration of millions of West Africans to Brazil, which in turn contributed to the emergence of an African cultural diaspora in Brazil that is without equal in the New World.

Although slavery was not new to West Africa, the *kind* of slave raiding and trading the Europeans introduced certainly was. In the interior of Africa and within city-states, kings, chiefs, and prominent families traditionally took a few slaves, but the status of those slaves was unlike anything that lay in store for those who were shipped across the Atlantic. In fact, large-scale slave trading

had been introduced in East Africa long before the Europeans brought it to West Africa. African intermediaries from the coast raided the interior for able-bodied men and women and marched them in chains to the Arab markets on the coast (the island of Zanzibar was one such slave trading market). There, packed in specially built *dhows*, they were carried off to Arabia, Persia, and India. When the European slave trade took hold in West Africa, however, its volume was far greater. Europeans, Arabs, and collaborating Africans ravaged the continent, forcing perhaps as many as 30 million persons away from their homelands in captivity (Fig. 6-8). Families were destroyed, as were whole villages and cultures; those who survived their exile suffered unfathomable misery.

The European presence on the West African coast completely reoriented its trade routes, for it initiated the decline of the interior savanna states and strengthened the coastal forest states. Moreover, the Europeans' insatiable demand for slaves ravaged the population of the interior. But it did not lead to any major European thrust toward the interior or produce colonies overnight. The African intermediaries were well organized and strong, and they held off their European competitors, not just for decades but for centuries. Although the Europeans first appeared in the fifteenth century, they did not carve West Africa up until nearly 400 years later, and they did not conquer many other areas until after the beginning of the twentieth century.

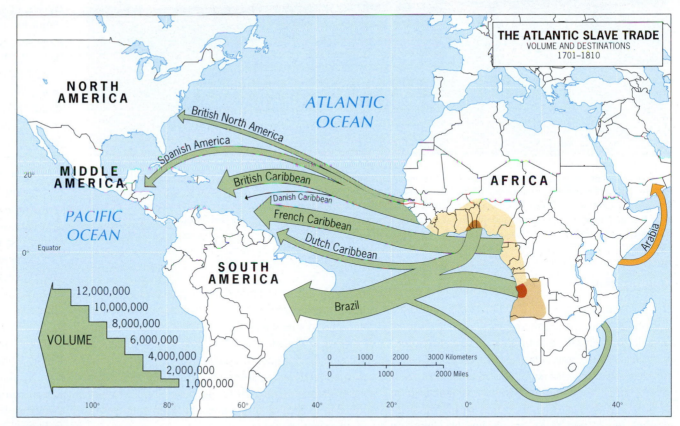

FIGURE 6-8

Adapted with permission from (1) Philip D. Curtin, *The Atlantic Slave Trade*, p. 57, University of Wisconsin Press, 1969; and (2) Donald K. Fellows, *Geography*, p. 121, John Wiley & Sons, Inc.

Elmina Castle, in Elmina, Ghana, was one of the most notorious slave ports along West Africa's Gold Coast during the peak of that dark trade. Initially constructed by the Portuguese as a trading post, it soon became a place where slaves were held before being shipped to Brazil, a Portuguese colony. Later it fell into the hands of the Dutch, who in turn lost it to the British, who bequeathed it to Ghana upon independence in 1957. The restored castle today is a tourist destination and has been designated a UNESCO World Heritage Site.   (© Robert Huberman/SUPERSTOCK)

Colonization

In the second half of the nineteenth century, whether or not they had control, the European powers finally laid claim to virtually all of Africa. Colonial competition was intense, and spheres of influence began to overlap. It was time for negotiation among the powerful, and in 1884 a conference was convened in Berlin to divide up Africa. Fourteen states participated (including the United States, which had no claims on Africa). The major colonial contestants were Britain, France, Portugal, Belgium, and Germany itself. On maps spread on a large table, representatives from these powers drew boundaries, exchanged real estate, and forged a new map that would become a liability in African decades later. As Figure 6-9 indicates, when the three-month conference was in progress most of Africa remained under traditional African rule. Not until after 1900 did the colonial powers manage to control all the areas they had marked off on their new maps.

It is important to examine Figure 6-9 carefully because the colonial powers governed their new depen-dencies in very different ways, and their contrasting legacies remain in evidence to this day in the countries their colonies spawned. Some colonial powers were democracies (Great Britain and France); others were dictatorships (Portugal and Spain). The British established a system of indirect rule over much of their domain, leaving indigenous power structures in place and making local rulers representatives of the British Crown. This was unthinkable in the Portuguese colonies, where harsh, direct control was the rule. The French sought to create culturally assimilated elites that would represent French ideals in the colonies. Unlike the British, French, or Portuguese, King Leopold II of Belgium sought to claim Congo Free State as his own personal empire. After financing the expeditions that staked Belgium's claim in Berlin, he embarked on a campaign of ruthless exploitation and murder. His enforcers mobilized almost the entire Congolese population to gather rubber (which was in high demand because of the advent of motor vehicles), kill elephants for their ivory, and build public works to improve export routes. For failing to meet production quotas, entire communities were executed and tortured. Killing and maiming became routine in a colony in which horror was the only common denominator. After the impact of the slave trade, King Leopold's reign of terror was Africa's most severe demographic disaster. By the time it ended, after growing condemnation from around the world, as many as 10 million Congolese had been murdered. In 1908, the Belgian government took over and began to mirror Belgium's own internal divisions: corporations, government administrators, and the Roman Catholic Church each pursued their sometimes competing interests. But no one thought to change the name of the colonial capital: it was Leopoldville until the Belgian Congo achieved independence in 1960.

Colonialism transformed Africa, but in its post-Berlin form it lasted less than a century. In Ghana, for example, the Ashanti (Asante) Kingdom still was fighting the British in the early years of the twentieth century; by 1957, Ghana was independent again. In a few years, much of Subsaharan Africa will have

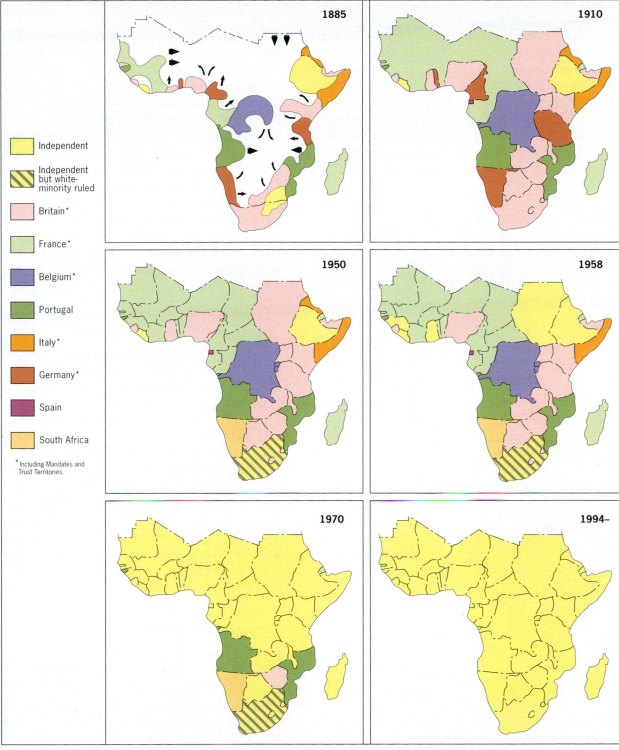

Legend:

- Independent
- Independent but white-minority ruled
- Britain*
- France*
- Belgium*
- Portugal
- Italy*
- Germany*
- Spain
- South Africa

* Including Mandates and Trust Territories.

1885

1910

1950

1958

1970

1994–

COLONIZATION AND DECOLONIZATION SINCE 1885

been independent for half a century, and the colonial period is becoming an interlude rather than a paramount chapter in modern African history, but it did leave a strong and lasting imprint on the landscape.

CULTURAL PATTERNS

We tend to think of Africa in terms of its prominent countries and famous cities, its development problems and political dilemmas, but Africans themselves have another perspective. The colonial period created states and capitals, introduced foreign languages to serve as the *linguae francae*, (common languages), and brought railways and roads. The colonizers stimulated labor movements to the mines they opened, and they disrupted other migrations that had been part of African life for many centuries. But they did not change the ways of life of most of the people. Fully 70 percent of the realm's population still live in, and work near, Africa's hundreds of thousands of villages. They speak one of more than a thousand languages in use in the realm. The villagers' concerns are local; they focus on subsistence, health, and safety. They worry that the conflicts over regional power or political ideology will engulf them, as has happened to millions in Liberia, Sierra Leone, Ethiopia, Rwanda, The Congo, and Angola since the 1970s. Africa's (numerically) largest groups of people are major nations, such as the Yoruba of Nigeria and the Zulu of South Africa. Africa's smallest groups of people number just a few thousand. As a geographic realm, Subsaharan Africa has the most complex cultural mosaic on earth.

FIGURE 6-9   © H. J. de Blij, P. O. Muller, and John Wiley & Sons, Inc.

African Languages

Africa's linguistic geography is a key component of that cultural intricacy. Most of Subsaharan Africa's more than one thousand languages do not have a written tradition, making classification and mapping difficult. Scholars have attempted to delimit an African language map, and Figure 6-10 is a composite of their efforts. One feature is common to all language maps of Africa: the geographic realm begins approximately where the Afro-Asiatic language family (mapped in yellow in Fig. 6-10) ends, although the correlation is sharper in West Africa than to the east.

In Subsaharan Africa, the dominant language family is the Niger-Congo family, of which the Kordofanian subfamily is a small, historic northeastern outlier (Fig. 6-10), and the Niger-Congo languages carry the other subfamily's name. This subfamily extends across the realm from West to East and Southern Africa. The Bantu language forms the largest branch in this subfamily, but Niger-Congo languages in West Africa, such as Yoruba and Akan, also have millions of speakers. Another important language family is the Nilo-Saharan family, extending from Maasai in Kenya northwest to Teda in Chad. No other language families are of similar extent or importance: the Khoisan family, of ancient origins, now survives among the dwindling Khoi and San peoples of the Kalahari; the small white minority in South Africa speak Indo-European languages; and Malay-Polynesian languages prevail in Madagascar, which was peopled from Southeast Asia before Africans reached it.

The Most Widely Used Languages

About 40 African languages are spoken by 1 million people or more, and a half-dozen by about 10 million or more: Hausa (50 million), Yoruba (23 million), Ibo, Swahili, Lingala, and Zulu. Although English and French have become important *linguae francae* in multilingual countries such as Nigeria and Ivory Coast (where officials even insist on spelling the name of their country—Côte d'Ivoire—in the Francophone way), African languages also serve this purpose. Hausa is a common language across the West African savanna; Swahili is widely used in East Africa. And pidgin languages, mixtures of African and European tongues, are spreading along West Africa's coast. Millions of Pidgin English (called *Wes Kos*) speakers use this medium in Nigeria and Ghana.

Language and Culture

14 **Multilingualism** can be a powerful centrifugal force in society, and African governments have tried with varying success to establish national alongside local languages. Nigeria, for example, made English its official language because none of its 250 languages, not even Hausa, had sufficient

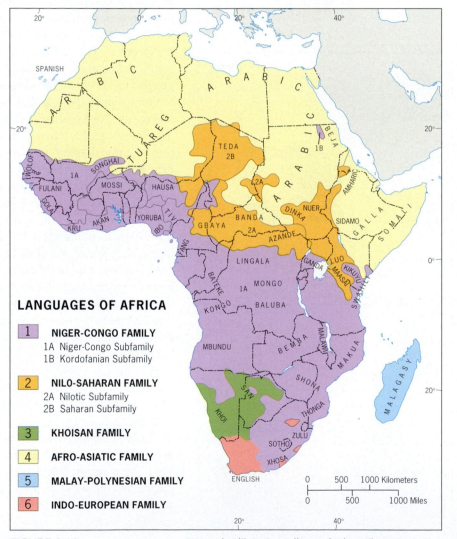

LANGUAGES OF AFRICA

1 **NIGER-CONGO FAMILY**
   1A Niger-Congo Subfamily
   1B Kordofanian Subfamily

2 **NILO-SAHARAN FAMILY**
   2A Nilotic Subfamily
   2B Saharan Subfamily

3 **KHOISAN FAMILY**

4 **AFRO-ASIATIC FAMILY**

5 **MALAY-POLYNESIAN FAMILY**

6 **INDO-EUROPEAN FAMILY**

FIGURE 6-10   © H. J. de Blij, P. O. Muller, and John Wiley & Sons, Inc.

internal interregional use. But using a European, colonial language as an official medium invites criticism, and Nigeria remains divided on the issue. On the other hand, making a dominant local language official would invite negative reactions from ethnic minorities. Language remains a potent force in Africa's cultural life.

Religion in Africa

Africans had their own belief systems long before Christians and Muslims arrived to convert them. And for all of Subsaharan Africa's cultural diversity, Africans had a consistent view of their place in nature. Spiritual forces, according to African tradition, are manifest everywhere in the natural environment, not in a supreme deity that exists in some remote place. Thus gods and spirits affect people's daily lives, witnessing every move, rewarding the virtuous, and punishing (through injury or crop failure, for example) those who misbehave. Ancestral spirits can inflict misfortune on the living. They are everywhere: in the forest, rivers, and mountains.

As with land tenure, the religious views of Africans clashed fundamentally with those of the colonists. Monotheistic *Christianity* first touched Africa in the northeast when Nubia and Axum were converted, and Ethiopia has been a Coptic Christian stronghold since the fourth century AD. But the Christian churches' real invasion did not begin until the onset of colonialism after the turn of the sixteenth century. Christianity's various denominations made inroads in different areas: Roman Catholicism in much of Equatorial Africa mainly at the behest of the Belgians, the Anglican Church in British colonies, and Presbyterians and others elsewhere. Evangelical churches are gaining adherents rapidly. Some of these churches today are more conservative than those found in Europe or North America, appealing to conservative congregations in the United States. A split in the Episcopal (Anglican) Church in the United States regarding the treatment of homosexuals has some congregations

The faithful kneel during Friday prayers at a mosque in Kano, northern Nigeria. The survival of Nigeria as a unified state is an African success story; the Nigerians have overcome strong centrifugal forces in a multi-ethnic country that is dominantly Muslim in the north, Christian in the south. In the 1990s, some Muslim clerics began calling for an Islamic Republic in Nigeria, and after the death of the dictator Abacha and the election of a non-Muslim president, the Islamic drive intensified. A number of Nigeria's northern States adopted Sharia (strict Islamic) law, which led to destructive riots between the majority Muslims and minority Christians who felt threatened by this turn of events. Can Nigeria avoid the fate of Sudan? (© M. & E. Bernheim/Woodfin Camp & Associates)

aligning themselves under the Bishop of Uganda. But almost everywhere, Christianity's penetration led to a blending of traditional and Christian beliefs, so that much of Subsaharan Africa is nominally, though not exclusively, Christian (see Fig. 7-2). Go to a church in Gabon or Uganda or Zambia, and you may hear drums instead of church bells, sing African music rather than hymns, and see African carvings alongside the usual statuary.

Islam had a different arrival and impact. By the time of the colonial invasion, Islam had advanced out of Arabia, across the desert, and part-way down the coasts of Africa. Muslim clerics converted the rulers of African states and commanded them to convert their subjects. They Islamized the savanna states and penetrated into present-day northern Nigeria, Ghana, and Ivory Coast. They encircled and isolated Ethiopia's Coptic Christians and Islamized the Somali people in Africa's Horn. They established beachheads on the Kenya coast and took over Zanzibar. Arabizing Islam and European Christianity competed for African minds, and Islam proved to be a far more pervasive force. From Senegal to Somalia, the population is virtually 100 percent Muslim, and Islam's rules dominate everyday life. The Sunni *mullahs* would never allow the kind of marriage between traditional and Christian beliefs seen in much of formerly colonial Africa. This fundamental contradiction between Islamic dogma and Christian accommodation creates a potential for conflict in countries where both religions have adherents.

MODERN MAP AND TRADITIONAL SOCIETY

The political map of Subsaharan Africa has 45 states but no nation-states (apart from some microstates and ministates in the islands and in the south). Centrifugal forces are powerful, and outside interventions during the Cold War, when communist and anticommunist foreigners took sides in local civil wars, worsened conflict within African states. Colonialism's economic legacy was not much better. In Africa, capitals, core areas, port cities, and transport systems were laid out to maximize colonial profit and facilitate exploitation of minerals and soils; the colonial mosaic inhibited interregional communications except where cooperation enhanced efficiency. Colonial Zambia and Zimbabwe, for example (then called Northern and Southern Rhodesia, respectively), were landlocked and needed outlets, so railroads were built to Portuguese-owned ports. But such routes did little to create intra-African linkages. The modern map reveals the results: in West Africa you can travel from the coast into the interior of all the coastal states along railways or adequate roads.

But no high-standard roadway was ever built to link these coastal neighbors to each other.

Supranationalism

To overcome such disadvantages, African states must cooperate internationally, continentwide as well as regionally. The Organization of African Unity (OAU) was established for this purpose in 1963 and in 2001 was superseded by the African Union (AU). In 1975 the Economic Community of West African States (ECOWAS) was founded by 15 countries to promote trade, transportation, industry, and social affairs in the region. And in the early 1990s another important step was taken when 12 countries joined in the Southern African Development Community (SADC), organized to facilitate regional commerce, intercountry transport networks, and political interaction. The AU has been an active peace-keeping force in the ongoing Darfur conflict in Sudan (discussed in Chapter 7).

WHAT'S DRIVING GEOGRAPHIC CHANGE IN THE REALM?

● Is the world finally paying attention? Africa needs appropriate **technology transfers** and consideration of **debt cancellation** more than any other realm of the world. Politicians, scholars, and philanthropists including Bill Clinton, Jeffrey Sachs, and Bill Gates have recently proposed creative ways of helping and empowering Africans.

● Will Africa see a new country evolve? In 2006 Southern **Sudan** (fully discussed in Chapter 7) signed a peace deal with the Khartoum government representing the north. Southern Sudan is not a new sovereign state yet, but is a much more autonomous political entity than before and will be part of the Subsaharan Africa realm if it becomes independent.

● **South Africa** remains the region's economic powerhouse. Recent political developments in **Zimbabwe** led to economic destabilization reinforced by massive state terrorism. As a result, many Zimbabweans were fleeing into South Africa.

● In recent years there has been good **economic growth** in Subsaharan Africa. The World Bank estimates that the rate of economic growth on the continent matches global rates—5.4 percent. However, many countries (especially the poorest ones) still rely on overseas development aid (either financially or through technical assistance) to meet the basic needs of their citizens. Indeed many of the world's poorest countries are located within Subsaharan Africa but the prospects for increased development are looking better than they have ever before.

● The arrival and adoption of cell phones has enabled farmers in Subsaharan Africa to become much more competitive locally, and even on a global scale since they can now be aware of markets. Many consider this **cell phone revolution** to be a major driver in the improved economic growth in much of the realm.

● **China** is investing in Subsaharan Africa's infrastructure by contributing to the construction of roads, railways, dams, and bridges. Technological know-how is being shared. China is of course interested in helping Africa in this way, so that its wealth of resources can more easily be extracted and used in China's booming economy.

Population and Urbanization

As can be discerned in Appendix B, Subsaharan Africa remains the least urbanized world realm, but it is urbanizing at a fast pace. By the time you read this, the percentage of urban dwellers will have passed 34. This means that more than 260 million people now live in towns and cities, of which many were founded and developed by the colonial powers. But the infrastructure of the cities has not kept pace with the number of arrivals.

African cities became centers of embryonic national core areas, and of course they served as government headquarters. This *formal sector* of the city used to be the dominant one, with government control and regulations affecting civil service, business, industry, and workers. Today, however, African cities look different. From a distance, the skyline still resembles that of a modern center. But in the streets, on the sidewalks right below the shopwindows, there are hawkers, basket weavers, jewelry sellers, garment makers, wood carvers—a second economy, most of it beyond government control. This *informal sector* now dominates many African cities. It is peopled by the rural immigrants, who also work as servants, apprentices, construction workers, and in countless other menial jobs.

Millions of urban immigrants, however, cannot find work, at least not for months or even years at a time. They live in squalid circumstances, in desperate poverty, and governments cannot assist them. As a result, the squatter rings around (and also within) many of Africa's cities are unsafe—uncomfortable, unhealthy slums without adequate shelter, water supply, or basic sanitation. Garbage-strewn (no solid-waste removal here), muddy and insect-infested during the rainy season, and stifling and smelly during the dry period, they are incubators of disease. Yet few of their residents return to their villages. Every new day brings hope.

In our regional discussion we refer to some of Subsaharan Africa's cities, all of which, to varying degrees, are stressed by the rate of population influx. Despite the plight of the urban poor and the poverty of Africa's rural areas, some of Africa's capitals remain the strongholds of privileged elites who, dominant in governments, fail to address the needs of other ethnic groups. Discriminatory policies and artificially low food prices disadvantage farmers and create even greater urban-rural disparities than the colonial period saw. But today the prospect of democracy brings hope that Africa's rural majorities will be heard and heeded in the capitals.

Regions of the Realm

On the face of it, Africa seems to be so massive, compact, and unbroken that any attempt to justify a contemporary regional breakdown is doomed to fail. No deeply penetrating bays or seas create peninsular fragments as in Europe. No major islands (other than Madagascar) provide the broad regional contrasts we see in Middle America. Nor does Africa really taper southward to the peninsular proportions of South America. And Africa is not cut by an Andean or a Himalayan mountain barrier. Given Africa's colonial fragmentation and cultural mosaic, is regionalization possible? We think it is.

Maps of environmental distributions, ethnic patterns, cultural landscapes, historic culture hearths, and other spatial data yield a four-region structure complicated by a fifth, overlapping zone as shown in Figure 6-11. Beginning in the south, we identify the following regions:

1. *Southern Africa*, extending from the southern tip of the continent to the northern borders of Angola, Zambia, Malawi, and Moçambique. Ten countries are included in this region, which includes dominant South Africa.

2. *East Africa*, where natural (equatorial) environments are moderated by elevation and where plateaus, lakes, and mountains, some still carrying permanent snow, define the countryside. Six countries, including highland Ethiopia, constitute this region. The island of Madagascar, with Southeast Asian influences, is neither Eastern nor Southern African, but is included here because of its location.

3. *Equatorial Africa*, much of it defined by the basin of the Congo River, where elevations are lower than in East Africa, where temperatures are higher and moisture more ample, and where most of Africa's surviving rainforests remain. Among the eight countries that form this region, The Congo dominates territorially and demographically.

4. *West Africa*, which includes the countries of the western coast and those on the margins of the Sahara in the interior, a populous region anchored in the southeast by Africa's demographic giant, Nigeria. Seventeen countries form this crucial African region.

5. *The African Transition Zone* is the complicating factor on the regional map of Africa. As can be seen in Figure 6-11, this zone of increasing Islamic influence completely dominates some countries (e.g., Somalia in the east and Senegal in the west) while cutting across others, creating Islamized northern areas and non-Islamic southern zones (Nigeria, Chad, Sudan), and even encompassing only a strip of others (e.g., lowland Ethiopia, Kenya, Tanzania).

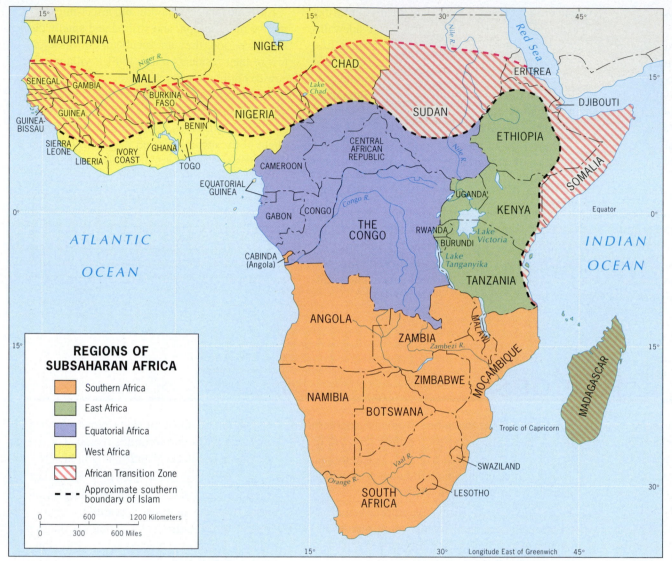

FIGURE 6-11

© H. J. de Blij, P. O. Muller, and John Wiley & Sons, Inc.

Africa's Richest Region

Southern Africa constitutes a geographic region in both physiographic and human terms. Its northern zone marks the southern limit of the Congo Basin in a broad upland that stretches across Angola and into Zambia (the tan corridor extending eastward from the Bihe Plateau in Fig. 6-2). Lake Malawi is the southernmost of the East African rift-valley 'Great Lakes'; Southern Africa has none of East Africa's volcanic and earthquake activity. Most of the region is plateau country, and the Great Escarpment is much in evidence here. There are two pivotal river systems: the Zambezi (which forms the border between Zambia and Zimbabwe) and the Orange-Vaal (South African rivers that combine to demarcate southern Namibia from South Africa).

Southern Africa is the continent's richest region materially. A great zone of mineral deposits extends through the heart of the region from Zambia's Copperbelt through Zimbabwe's Great Dyke and South Africa's Bushveld Basin and Witwatersrand to the goldfields and diamond mines of the (Orange) Free State and northern Cape Province in the heart of South Africa. Ever since these minerals began to be exploited in colonial times, many migrant laborers have come to work in the mines.

Southern Africa's agricultural diversity matches its mineral wealth. Vineyards drape the slopes of South Africa's Cape Ranges; tea plantations hug the eastern escarpment slopes of Zimbabwe. Before civil war destroyed its economy, oil-rich Angola was one of the world's leading coffee producers. South

SOUTHERN AFRICA

Southern Africa, as a geographic region, consists of all the countries and territories lying south of Equatorial Africa's The Congo and East Africa's Tanzania (Fig. 6-12). Thus defined, the region extends from Angola and Moçambique (on the Atlantic and Indian Ocean coasts, respectively) to South Africa and includes a half-dozen landlocked states. Also marking the northern limit of the region are Zambia and Malawi. Zambia is nearly cut in half by a long land extension from The Congo, and Malawi penetrates deeply into Moçambique. The colonial boundary framework, here as elsewhere, produced many liabilities.

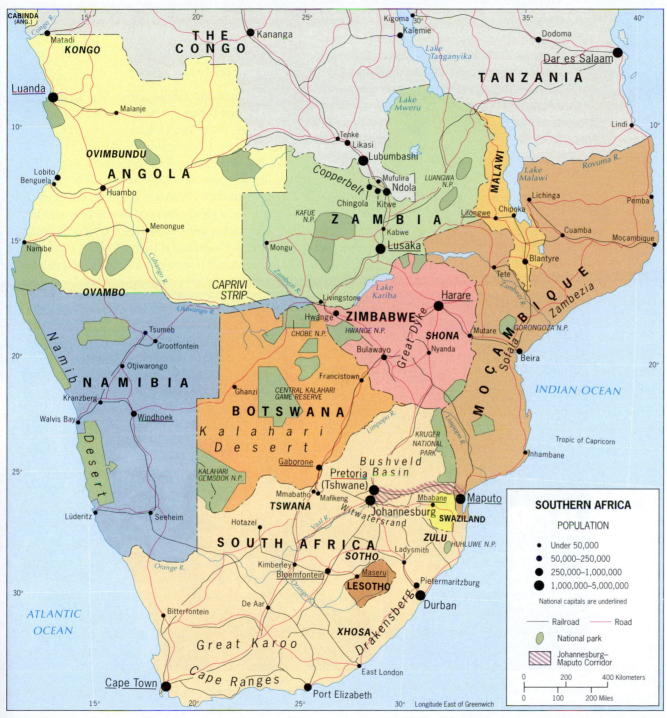

FIGURE 6-12

© H. J. de Blij, P. O. Muller, and John Wiley & Sons, Inc.

Africa's relatively high latitudes and its range of altitudes create environments for apple orchards, citrus groves, banana plantations, pineapple farms, and many other crops.

Despite this considerable wealth and potential, the countries of Southern Africa have not prospered. As Figure G-11 shows, many remain in the low-income category (Malawi, with a per-capita GNI of only U.S.$650, is one of the world's poorest states); only South Africa, Angola, Namibia, and Botswana are middle-income rank, with the latter desert state being the realm's second most sparsely populated country. A period of rapid population growth, followed by the devastating onslaught of HIV/AIDS, civil conflict, political instability, incompetent government, widespread corruption, unfair competition on foreign markets, and environmental problems have constrained economic development. Still, with the exception of Zimbabwe, Southern Africa as a region is better off than any other in Subsaharan Africa: as Figure G-11 shows, six countries here have risen above the lowest-income rank among world states. HIV/AIDS appears to be stabilizing, and agricultural production is capable of feeding increasing numbers in some places such as Malawi. Some international cooperation (e.g., IBSA–India, Brazil, South Africa), including a regional association (the African Union–AU) and a customs union, are emerging. And South Africa, the realm's most important country by many measures, gives hope for a better future.

South Africa

The Republic of South Africa is the giant of Southern Africa, an African country at the center of world attention, a bright ray of hope not only for Africa but for all humankind.

Long in the tight grip of one of the world's most notorious racial policies (**15 apartheid**, or apartness, and its derivative, **16 separate development**), South Africa today is shedding its past and is building a new future. That virtually all parties to the earlier debacle are now working cooperatively to restructure the country under a new flag, a new national anthem, and a new leadership is one of the great developments of the late twentieth century. In the twenty-first century, South Africa is poised to take its long-awaited role as the economic engine for the region—and perhaps beyond.

South Africa stretches from the warm subtropics in the north to Antarctic-chilled waters in the south. With a land area in excess of 1.2 million square kilometers (470,000 sq mi) and a heterogeneous population of 47.8 million, South Africa is the dominant state in Southern Africa. It contains the bulk of the region's minerals, most of its good farmlands, its largest cities, best ports, most productive factories, and most developed transport networks. Mineral exports from Zambia and Zimbabwe move through South African ports. Workers from as far away as Malawi and as close as Lesotho work in South Africa's mines, factories, and fields.

History and the People of South Africa

South Africa's history differs somewhat from much of the rest of Subsaharan Africa. Its lands were fought over by various African nations before Europeans arrived and the colonial 'scramble for Africa' took place. Peoples migrated southward—first the Khoisan-speakers and then the Bantu peoples—into the South African cul-de-sac. The Zulu and Xhosa nations fought over lands at about the time of the first European arrivals. Europeans arrived via the oceans and claimed the southernmost Cape. It is one of the most strategic places on earth, the gateway from the Atlantic to the Indian Ocean, a source of provisions on the route to Asia's riches. The Dutch East India Company founded Cape Town as early as 1652, and the Hollanders and their descendants, known as the Boers, have been a part of the South African cultural makeup ever since. The British took over about 150 years later, and the two colonial powers vied for power throughout South Africa's early history as a state. The British came to dominate the Cape, while the Boers trekked into the South African interior and, on the high plateau they called the *highveld*, founded their own republic. By 1910, the Boers and the British had negotiated a power-sharing arrangement, although the Boers eventually achieved hegemony between 1948 and 1994. Having long since shed their European links, they came to call themselves *Afrikaners*, their word for Africans.

The Ethnic Mosaic

In addition to the various native African nations and the Europeans that settled in South Africa are peoples from Asia. The Hollanders brought Southeast Asians to the Cape to serve as domestics and laborers. The British imported laborers from their South Asian colonies to work in sugarcane plantations, adding cultural diversity to this multiethnic state. Moreover, a substantial population of mixed ancestry comprises today's so-called Coloured sector of the country's citizenry. In the process, South Africa became Africa's most pluralistic and heterogeneous

People from a variety of ethnic backgrounds mingle at the market on Grand Parade in front of City Hall, Cape Town, South Africa. The building was completed in 1905, with limestone imported from England. It is on the steps of this building that Nelson Mandela made his first public speech upon his release from prison in 1990.   (© Ariadne Van Zandbergen/Lonely Planet Images/ Getty Images)

society. Even so, Africans now outnumber non-native Africans by about 4 to 1.

Although heterogeneity marks the spatial demography of South Africa, regionalism pervades the human mosaic. The Zulu nation still is largely concentrated in today's Kwazulu-Natal Province (Fig. 6-13). The Xhosa still cluster in the Eastern Cape, from the city of East London to the Kwazulu-Natal border and below the Great Escarpment. The Tswana still occupy ancestral lands along the Botswana border. Cape Town remains the core area of the Coloured population; Durban still has the strongest South Asian imprint. Travel through South Africa, and you will recognize the diversity of rural cultural landscapes as they change from Swazi to Ndebele to Venda.

This historic regionalism was among the factors that led the Afrikaner government to institute its *separate development* scheme, which was linked to *apartheid*, but it could not stem the tide of urbanization that intermingled the populations. Millions of workers, job-seekers, and illegal migrants converged on the cities, creating vast shantytowns on their margins. In the legal African townships such as Johannesburg's Soweto (SOuth WEstern TOwnships) and in the squatter settlements the anti-apartheid movement burgeoned, and the strength of the African National Congress (ANC) movement grew. In February 1990 Nelson Mandela, imprisoned for 28 years on Robben Island, South Africa's Alcatraz, became a free man, and following the momentous first democratic election of 1994 he became president of an ANC-dominated government in Cape Town.

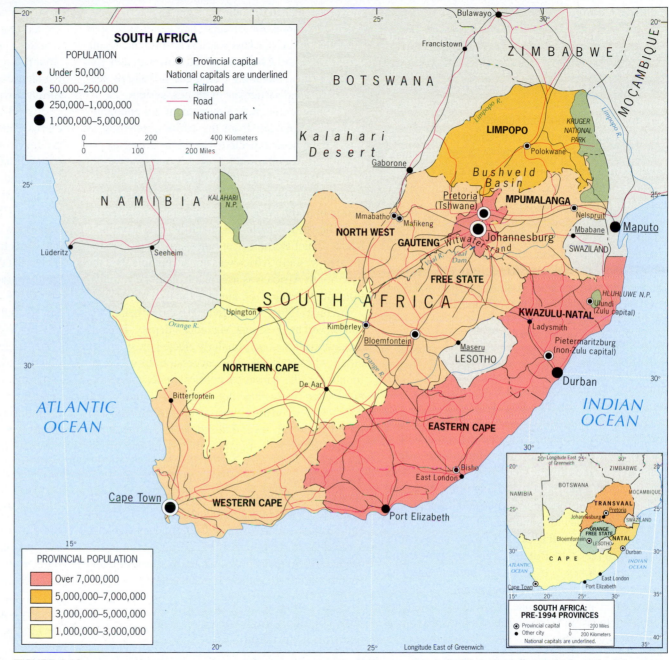

FIGURE 6-13

© H. J. de Blij, P. O. Muller, and John Wiley & Sons, Inc.

Political Change

Largely through the unmatched statesmanship of Nelson Mandela, South Africa did not undergo a massive revolution with great upheaval. Instead it is today governed by a multiracial party, the erstwhile architects of apartheid are in parliament as a legitimate opposition, and its economy is the largest and healthiest in the realm. Check Appendix B and you will see that, on the African mainland, South Africa's per-capita GNI is in a class by itself. By some calculations, South Africa produces 45 percent of all of Subsaharan Africa's GDP.

Together with political change came a change in administrative structure of the country–in fact this change occurred before the historic elections of 1994. Before 1994 South Africa was divided into four provinces: the Cape, by far the largest and centered on the legislative capital of Cape Town; Natal, anchored by the port city of Durban; the Transvaal (meaning across the Vaal River), focused on the administrative capital of Pretoria and including the great Johannesburg metropolis; and the Orange Free State, the old Boer stronghold with Bloemfontein as its headquarters (see inset map, Fig. 6-13).

This structure was replaced by nine provinces (Fig. 6-13), creating a federal arrangement in which each province has its own administration while being represented in the central government. The boundaries of Natal and the Orange Free State remained essentially unchanged, but Natal's name was changed to Kwazulu-Natal, recognizing the Zulu presence there, and the Orange Free State became simply the Free State. But the Cape Province was divided into four new provinces, one of them overlapping into the Transvaal, and the rest of the Transvaal was divided into three, one of which is Gauteng, which includes the Johannesburg-Witwatersrand-Pretoria megalopolis.

A New Era Dawns

The 1994 election produced an ANC victory in seven of these nine provinces, excepting only Kwazulu-Natal, where the Zulu vote went to a local political movement called Inkatha, and the Western Cape, where a combination of white and Coloured voters outpolled the ANC. This was a very fortunate result, proving that the ANC was not all-powerful and giving minorities, including Coloureds and whites, reason to trust and participate in the new system.

The new democratic government did encounter opposition, but not from the white opposition as might have been expected. The white extremist political parties collapsed relatively soon, and a brief terrorist episode was quickly halted. More consequential was the uncertain role of the largest single nation in South Africa, the Zulu, whose historic domain lies in Kwazulu-Natal, and whose prominent leader, Chief Buthelezi, at times hinted at secession if the new South Africa was not acceptable. But the new South Africa was acceptable, and the Zulu nation and its leaders remained in the fold.

In June 1999, ANC leader Thabo Mbeki, who had served as President Mandela's deputy, became the country's second popularly elected president. In other African states, the succession from heroic founder-of-the-nation to political inheritor-of-the-presidency often has not gone well, but South Africa's new constitution proved its worth. The new president faced several disadvantages: inevitable comparisons to the incomparable Mandela; the rising tide of AIDS, which Mbeki controversially attributed to causes other than HIV; his awkward stance toward the chaos in Zimbabwe. Nevertheless, Mbeki's economic policies, social programs, and foreign initiatives (The Congo, Rwanda) have served his country well. In 2007 Jacob Zuma succeeded Mbeki as leader of the ANC and will likely become South Africa's next president.

How the Economy Evolved: Diamonds and Gold

Ever since diamonds were discovered at Kimberley in the 1860s, South Africa has been synonymous with minerals. The Kimberley finds, made in a remote corner of the Orange Free State, set into motion a new economic geography. Rail lines were laid from the coast to the diamond capital even as fortune seekers, capitalists, and tens of thousands of African workers, many from as far afield as Lesotho, streamed to the site. One of the capitalists was Cecil Rhodes, of Rhodes Scholarship fame, who used his fortune to help Britain dominate Southern Africa.

Just 25 years after the diamond finds, prospectors discovered what was long to be the world's greatest goldfield on a ridge called the Witwatersrand (Fig. 6-13). Johannesburg became the gold capital of the world, and a new and even larger stream of foreigners arrived, along with a huge influx of African workers. Cheap labor enlarged the profits. Johannesburg grew explosively, satellite towns developed, and black townships mushroomed.

During the twentieth century, South Africa proved to be richer than had been foreseen. Additional goldfields were discovered in the Orange Free State. Coal and iron ore were found in abundance, which gave rise to a major iron and steel industry. Other metallic minerals, including chromium and platinum, yielded large revenues on world markets. Asbestos, manganese, copper, nickel, antimony, and tin were mined and sold; a thriving metallurgical industry developed in South Africa itself. Capital flowed into the country, white immigration grew, farms and ranches were laid out, and markets multiplied.

Infrastructural Gains During Apartheid

South Africa's cities grew considerably. Johannesburg was no longer just a mining town: it became an industrial complex and a financial center as well. The old Boer capital, Pretoria, just 50 kilometers (30 mi) north of the Witwatersrand, became the country's administrative center during apartheid. In the Orange Free State, major industrial growth (including oil-from-coal technology) matched the expansion of mining. While the core area developed megalopolitan characteristics, coastal cities also expanded. Durban's port served not only the Witwatersrand but a wider regional hinterland as well. Cape Town was becoming South Africa's second-largest city; its port, industries, and productive agricultural hinterland gave it primacy over a wide area.

The labor force for all this development, from mines to railroads and from farms to highways, came from the African peoples of South Africa (and, indeed, from beyond its borders as well). Even during apartheid, workers from Moçambique, Zimbabwe, Botswana, and other neighboring countries sought jobs in South Africa even as millions of South African villagers also left their homes for the towns and cities. In the process they built the best infrastructure of any African country. But apartheid ruined South Africa's prospects. The cost of the separate development program was astronomical. Social unrest during the decade preceding the end of apartheid created a vast educational gap among youngsters. International sanctions against the race-obsessed regime damaged the economy.

The Economy Today

In many respects South Africa is the most important country in Subsaharan Africa, and the entire realm's fortunes are bound up with it. No African country attracts more foreign investment or foreign workers. None has the universities, hospitals, and research facilities. No other has the military forces capable of intervening in African trouble spots. Few have the free press, effective trade unions, independent courts, or financial institutions to match South Africa's. And with a population approaching 50 million (79 percent black, just under 10 percent white, 9 percent Coloured, 2.5 percent Asian), South Africa has a large, multiracial, and growing middle class.

Despite its successes, South Africa continues to face long-range problems. The economy continues to depend far too much on the export of minerals and metals (diamonds, platinum, gold, iron, and steel) at a time when this dependence entails risks at home and abroad. At home, those union-friendly labor laws, including rising wages, are making mining less profitable. Abroad, commodity prices are unreliable. In 1970 South Africa produced nearly 70 percent of the world's gold; today it produces less than 15 per-

cent. And the manufacturing sector of the economy remains too weak. Even while the black middle class is expanding, unemployment among blacks may be as high as 50 percent—and the gap between rich and poor is growing. Conflicts between locals and immigrants are on the rise. Land reform, an urgent matter in a country where land alienation reached huge proportions, is too slow in the view of many. Add to this the scourge of AIDS and the government's failure to address this crisis effectively, and South Africa's luster is overshadowed by serious problems.

The Middle Tier

Between South Africa's northern border and the region's northern limit lie two groups of states: those with borders with South Africa and those beyond.

Five countries constitute the Middle Tier, neighboring South Africa: Zimbabwe, Namibia, Botswana, and the ministates of Lesotho and Swaziland (Fig. 6-12). As the map shows, four of these five are landlocked. Diamond-exporting (and upper-middle-income) **Botswana** occupies the heart of the Kalahari Desert and surrounding steppe; despite its lucrative diamonds, most of its 1.8 million inhabitants are subsistence farmers. In 2007 Botswana remained the most severely AIDS-afflicted country in all of tropical Africa. **Lesotho** and **Swaziland**, both traditional kingdoms, depend heavily on remittances from their workers in South African mines, fields, and factories.

The most important state in the middle tier undoubtedly is **Zimbabwe** (13.3 million), landlocked but well endowed with mineral and agricultural resources. Zimbabwe (the country is named after stone ruins in its interior) is mostly an elevated plateau between the Zambezi and Limpopo rivers, with the desert to the west and the Great Escarpment to the east. Its core area is defined by the mineral-rich Great Dyke and its environs, extending southwest from the vicinity of the capital, Harare, to the country's second city, Bulawayo. Copper, asbestos,

and chromium (of which Zimbabwe is one of the world's leading sources) are among its major mineral exports, but Zimbabwe is not just an ore-exporting country. Farms are capable of producing maize (corn), tobacco, tea, sugar, cotton, and other crops.

Two nations form most of Zimbabwe's population: the Shona (82 percent) and the Ndebele (14 percent). A tiny minority of whites owned the best farmland and organized the agricultural economy. Mounting environmental and economic problems beginning in the 1980s led to rising social and political tensions. President Mugabe and his dominant party allowed white farms to be invaded by squatters who sometimes killed the owners; corruption rose and human rights were severely curbed. Moreover, the country is unable to feed itself due to mismanagement of the agricultural sector. The economy is in ruins and the population is declining as people flee the chaos. Once-promising Zimbabwe is the tragedy of the region today; although Mugabe unleashed a reign of terror to win a sham reelection in 2008, the dictator's days finally appear to be numbered.

Southern Africa's youngest independent state, **Namibia** (2.2 million), is a former German colony with a territorial peculiarity: the so-called Caprivi Strip linking it to the right bank of the Zambezi River (Fig. 6-12), another product of colonial partitioning. Administered by South Africa from 1919 to 1990, Namibia is named after one of the world's driest deserts. This state is about as large as Texas and Oklahoma, but only its far north receives enough moisture to permit subsistence farming, which is why most of the people live near the Angolan border. Mining in the Tsumeb area and ranching in the vast steppe country of the south form the main commercial activities. The capital, Windhoek, is centrally situated opposite Walvis Bay, the main port. German influence still lingers in what used to be called South West Africa, as does an Afrikaner presence from the apartheid period. Although orderly land reform is underway, unresolved issues remain, and much of the population still lives in poverty.

The Northern Tier

In the four countries that extend across the northern tier of the region—Angola, Zambia, Malawi, and Moçambique—problems abound. **Angola** (16.7 million), formerly a Portuguese dependency, with its **17** **exclave** (outlier) of Cabinda had a thriving economy based on a wide range of mineral and agricultural exports at the time of independence in 1975. But then Angola fell victim to the Cold War, with northern peoples choosing a communist course and southerners falling under the sway of a rebel movement backed by South Africa and the United States. The results included a devastated infrastructure, idle farms, looting of diamonds, hundreds of thousands of casualties, and millions of landmines that continue to kill and maim. But Angola's oil wealth (the country ranks second in Subsaharan Africa in oil production) yields about U.S.$5 billion per year, and stability, if sustained, may attract investors to begin rebuilding this ruined country.

On the opposite coast, the other major former Portuguese colony, **Moçambique** (20.7 million), fared poorly in a different way. Without Angola's mineral base and with limited commercial agriculture, Moçambique's chief asset was its relative location. Its two major ports, Maputo and Beira, handled large volumes of exports and imports for South Africa, Zimbabwe, and Zambia. But upon independence Moçambique also chose a Marxist course with unfortunate economic and political consequences. Here, too, a rebel movement supported by South Africa caused civil conflict, created famines, and generated a stream of more than a million refugees toward Malawi. Rail and port facilities lay idle, and Moçambique at one time was ranked by the United Nations as the world's poorest country. In recent years, the port traffic has been somewhat revived, and Moçambique and South Africa are working on a joint Maputo Development Corridor (Fig. 6-12), but it will take generations for Moçambique to recover.

Landlocked **Zambia** (12.3 million), the product of British colonialism, shares the riches of the Copperbelt with The Congo's Katanga Province.

Not only have commodity prices on which Zambia depends fluctuated wildly, but Zambia's outlets—Lobito in Angola and Beira in Moçambique—and the railways leading there were made inoperative by Cold War conflicts. In recent years, China has taken an interest in Zambia's minerals, and the Chinese may invest in railroad repairs as well as mining operations. Neighboring **Malawi** (13.5 million) has an almost totally agricultural economic base and is frequently challenged by environmental degradation. But in 2007 things looked more positive as bumper harvests were produced and Malawi was able to export its surplus to Zimbabwe.

EAST AFRICA

East of the row of Great Lakes that marks the eastern border of The Congo (Lakes Albert, Edward, Kivu, and Tanganyika), the land rises from the Congo Basin to the East African Plateau. Hills and valleys, fertile soils, and copious rains mark the transition in Rwanda and Burundi. Eastward the rainforest disappears and open savanna cloaks the countryside. Great volcanoes rise above a rift-valley-dissected highland. At the heart of the region lies Lake Victoria. In the north the surface rises above 10,000 feet (3300 m), and so deep are the trenches cut by faults and rivers there that the land was called, appropriately, Abyssinia (now Ethiopia).

Five countries, in addition to the highland part of Ethiopia, form this East African region: Kenya, Tanzania, Uganda, Rwanda, and Burundi (Fig. 6-14). Here the Bantu peoples that make up most of the population met Nilotic peoples from the north.

Kenya (36.5 million) is neither the largest nor the most populous country in East Africa, but over the past half-century it has been the dominant state in the region. Its skyscrapered capital at the heart of its core area, Nairobi, home to 3.2 million, is the region's largest city and hub for many activities; its port, Mombasa, is the region's busiest.

After independence, Kenya chose a capitalist path of development, aligning itself with Western

FROM THE FIELD NOTES

Maasai Mara National Game Reserve, Kenya. "I was visiting a graduate student doing fieldwork in Kenya's Maasai Mara National Game Reserve and was able to do my own wildlife 'safari' (a Swahili word for 'journey'). Visiting a spectacular place such as the Mara is a bit like walking into a nature film with the animals (elephants, giraffes, hyena, lions, buffalo, etc.) seemingly popping out of the screen. As researchers we were able to drive in places that tourists could not go, and this was how I captured this photo of a cheetah in a tree with typical tourists viewing the wildlife. Cheetahs climb trees (there are not many in these savanna lands) to get a better overview before hunting.

The Mara, as part of the Serengeti Plains ecosystem, is famous for its annual wildebeest migration in July and draws tourists from all over the world. Tourism, particularly safari tourism, is an extremely important component of Kenya's economy, bringing in $500 million in annual revenue. This sector of the economy suffers quickly when Kenya is in the news for violence as was the case in early 2008 with post-election violence erupting in many parts of the country." (© A. WinklerPrins)

Concept Cachir

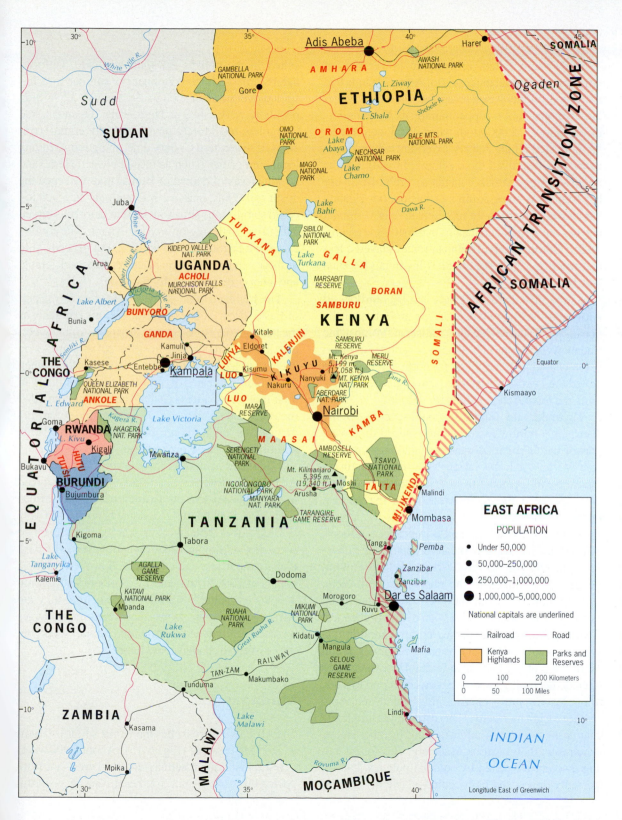

EAST AFRICA

POPULATION

- • Under 50,000
- • 50,000–250,000
- ● 250,000–1,000,000
- ⬤ 1,000,000–5,000,000

National capitals are underlined

— Railroad   — Road

Kenya Highlands   Parks and Reserves

0   100   200 Kilometers
0   50   100 Miles

Longitude East of Greenwich

interests. Without major known mineral deposits, Kenya depended on coffee and tea exports, growing agroexports of other products, and on a tourist industry based on its magnificent landscapes (see photo p. 216). Tourism became its largest single earner of foreign exchange, and Kenya prospered, apparently proving the wisdom of its economic planners.

But serious problems arose. Kenya during the 1980s had the highest rate of population growth in the world, and population pressure on farmlands and on the fringes of the wildlife reserves mounted. Poaching became widespread, and tourism declined. During the late 1990s, El Niño-induced rains affected parts of Kenya, causing landslides and washing away large segments of the crucial highway between Nairobi and Mombasa. This disaster was followed by a severe drought lasting several years, bringing famine to the interior. Meanwhile, government corruption siphoned off funds that should have been invested. Democratic principles were violated, and relationships with Western allies were strained. The AIDS epidemic brought another setback. And then Kenya sustained damage from terrorist attacks in Nairobi and Mombasa, further impacting the tourist industry.

Early in the twenty-first century Kenya was stable and relatively prosperous. However, geography, history, and politics have placed the Kikuyu (22 percent of the population) in a position of power, although there are other major peoples (see Fig. 6-14) and several smaller groups. The Luhya, Luo, Kalenjin, and Kamba together constitute about 50 percent of the population, and on the territorial margins of the country there are peoples such as the Maasai, Turkana, Boran, and Galla. Unfortunately, the contested outcome of the 2007 election—the opposition felt cheated by possibly fraudulent results keeping the Kikuyu president in office—uncorked what had been longstanding tensions, and violence erupted. Kenya's challenge is to overcome this crisis and sustain a political system that will ensure democracy and represent the interests of Kenya's disparate peoples.

FIGURE 6-14   © H. J. de Blij, P. O. Muller, and John Wiley & Sons, Inc.

Tanzania (a name derived from **Tan**ganyika plus **Zan**zibar) is the biggest and most populous East African country (39.8 million). Its total area exceeds that of the other four countries combined. Tanzania has been described as a country without a core because its clusters of population and zones of productive capacity lie dispersed—mostly on its margins on the east coast (where the capital, Dar es Salaam, is located), near the shores of Lake Victoria in the northwest, near Lake Tanganyika in the far west, and near Lake Malawi in the interior south. This is in sharp contrast to Kenya, which has a well-defined core area in the Kenya Highlands (centered on Nairobi in the heart of the country). Moreover, Tanzania is a country of many peoples, none numerous enough to dominate the state. About 100 ethnic groups coexist; one-third of the population, mainly on the coast, are Muslims.

After independence Tanzania embarked on a socialist course toward development, including a massive but poorly planned farm collectivization program. The tourist industry declined sharply, and Tanzania became one of the world's poorest countries. But Tanzania did achieve remarkable political stability and a degree of democracy that none of the other East African states attained. Since 1990, the government has changed course, but the AIDS crisis, problems with the Zanzibar merger, and involvement in the troubles of neighboring countries, including Rwanda, have been costly. Yet today, Tanzania's prospects are improving as the tourist industry has rebounded and political stability continues.

The country known today as **Uganda** (29.4 million) contained this region's most important African state when the British colonialists arrived: the Kingdom of Buganda, peopled by the Ganda, located on the northwest shore of Lake Victoria (Fig. 6-14, also see Fig. 6-6). The British established their headquarters near the Ganda capital of Kampala and used the Ganda to control Uganda through indirect rule. Thus the Ganda became the dominant nation in multicultural Uganda, and when the British left, they bequeathed Uganda a complicated federal system designed to perpetuate Ganda supremacy.

The system failed, bringing to power one of Africa's most brutal dictators, Idi Amin. Uganda had a strong economy based on coffee, cotton, and other farm exports, and on copper mining; its Asian minority of about 75,000 dominated local commerce. Amin ousted all the Asians, exterminated his opponents, and destroyed the economy. In addition, the AIDS epidemic struck Uganda severely. Recovery has been slow since Amin's expulsion, but Uganda has moved toward more representative government and has made significant gains in the struggle against AIDS. Unfortunately, Ugandan forces played a less constructive role in the conflicts involving Rwanda and The Congo.

As the map shows, Uganda is a **18** **landlocked state** and depends on Kenya for an outlet to the ocean. Its relative location adjacent to unstable Sudan, Rwanda, and The Congo constitutes a formidable challenge. And in the far north, a rebel group called the Lord's Resistance Army has been terrorizing the local Acholi people for decades.

Rwanda and **Burundi** would seem to occupy Tanzania's northwest corner, and indeed they were part of the German colonial domain conquered before World War I. But during that war Belgian forces attacked the Germans from their Congo bases and were awarded these territories when the conflict ended in 1918.

Rwanda (9.6 million) and Burundi (8.2 million), Africa's most densely populated countries, are physiographically part of East Africa, but their cultural geography is linked to the north and west. Here, Tutsi pastoralists from the north subjugated Hutu farmers, who had themselves made serfs of the local Twa (pygmy) population, setting up a conflict that was originally ethnic but became cultural. Certain Hutu were able to advance in the Tutsi-dominated society, becoming to some extent converted to Tutsi ways, leaving subsistence farming behind, and rising in the social hierarchy. These so-called moderate Hutu were—and are—often targeted by other Hutus, who resent their position in society. This longstanding discord, worsened by colonial policies, had repeatedly devastated both countries, and in 1994 re-

sulted in the horrific Rwanda genocide. Unfortunately, this conflict continues to simmer and has spilled over into The Congo. As many as 3.5 million people have perished as Hutu, Tutsi, Ugandan, and Congolese rebel forces have fought for control over areas of the eastern Congo, unleashing longstanding local animosities (such as those between Hema and Lendu around the town of Bunia) that worsened the death toll. Only massive international intervention could stabilize the situation, but the world has turned a blind eye to the region's woes—again. The conflict came to a negotiated end in 2003, but peacemaking in these circumstances is easier than peacekeeping, and local strife periodically flares up.

The highland zone of **Ethiopia** also forms part of East Africa. Adis Abeba, the historic capital, was the headquarters of a Coptic Christian, Amharic empire that held its own against the colonial intrusion except for a brief period from 1935 to 1941, when the Italians defeated it. Indeed, the Ethiopians in their mountain fortress (Adis Abeba lies about 3000 meters/10,000 ft above sea level) became colonizers themselves, taking control of much of the Islamic part of the African Horn to the east.

Ethiopia's natural outlets are toward the Gulf of Adan and the Red Sea, but its government was forced to yield independence to Eritrea (part of the African Transition Zone) and the country is now effectively landlocked. A bitter border war between the two countries, starting in 1998, cost some 100,000 lives. Physiographically and culturally, however, Ethiopia is part of East Africa, and the Amhara and Oromo peoples are neither Arabized nor Muslim: they are Africans. The map shows that functional linkages between Ethiopia and East Africa remain weak, but this is likely to change in the future.

Madagascar

Off Africa's east coast lies the world's fourth-largest island, Madagascar (Fig. 6-15). About 2000 years ago the first human settlers arrived here—not from Africa but from distant Southeast Asia. A powerful

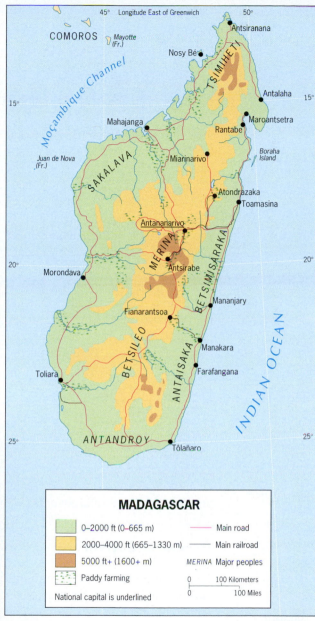

FIGURE 6-15

© H. J. de Blij, P. O. Muller, and John Wiley & Sons, Inc.

Malay kingdom of the Merina flourished in the highlands, whose language, Malagasy, became Madagascar's tongue (Fig. 6-10).

The Malay immigrants eventually brought Africans to their island, but today the Merina (5 million) and the Betsimisaraka (3 million) remain the largest of about 20 ethnic groups in the population of 18.8 million. After successfully resisting colonial invasion, the locals eventually yielded to France, and French became the *lingua franca* of the educated elite.

Madagascar is not part of either East Africa or Southern Africa. Its human as well as wildlife population is unique; its cultural landscapes still carry Southeast Asian imprints. Here the people eat rice, not corn, grown on terraced paddies. Population pressure is shrinking forests and animal refuges, the economy is weak, and poverty permeates in one of the world's most scenic outposts.

EQUATORIAL AFRICA

The term *equatorial* is not just locational but also environmental. The equator bisects Africa, but only the western part of central Africa features the conditions associated with the low-elevation tropics: intense heat, high rainfall and extreme humidity, little seasonal variation, rainforest and monsoon-forest vegetation, and enormous biodiversity. To the east, beyond the Western Rift Valley, elevations rise, and cooler, more seasonal climatic regimes prevail. As a result, we recognize two regions in these lowest latitudes: Equatorial Africa to the west and East Africa to the east.

Equatorial Africa is physiographically dominated by the gigantic Congo Basin. The Adamawa Highlands separate this region from West Africa; rising elevations and climatic change mark its southern limits (see the *Cwa* boundary in Fig. G-9). Its political geography consists of eight states, of

which The Congo (formerly Zaïre) is by far the largest in both territory and population (Fig. 6-16).

Five of the other seven states—Gabon, Cameroon, Congo, Equatorial Guinea, and São Tomé and Príncipe—all have coastlines on the Atlantic Ocean. The Central African Republic and Chad, the south of which is part of this region, are landlocked. In certain respects, the physical and human characteristics of Equatorial Africa extend even into southern Sudan. This vast and complex region is in many ways the most troubled region in the entire Subsaharan Africa realm.

The Congo

As the map shows, The Congo, long known as Zaïre and before that as the Belgian Congo, has only a tiny window, 37 kilometers (23 mi) on the Atlantic Ocean, just enough to accommodate the mouth of the Congo River. Oceangoing ships can reach the port of Matadi, inland from which falls and rapids make it necessary to move goods by road or rail to the capital, Kinshasa. Lack of navigability also necessitates transshipment between Kisangani and Ubundu, and at Kindu. Follow the railroad south from Kindu, and you reach another narrow corridor of territory at the city of Lubumbashi. That vital part of Katanga Province contains most of The Congo's major mineral resources, including copper and cobalt.

With a territory not much smaller than the United States east of the Mississippi, a population of 66.6 million, a rich and varied mineral base, and much good agricultural land, The Congo would seem to have all the ingredients needed to lead this region and, indeed, Africa. But strong centrifugal forces, arising from its physiography and cultural geography, pull The Congo apart. The immense forested heart of the basin-shaped country creates communication and transportation barriers between east and west, north and south. Many of The Congo's productive areas lie along its periphery, separated by enormous distances. These areas tend to look across the border, to one or more of The Congo's nine neighbors, for outlets, markets, and often ethnic kinship as well.

EQUATORIAL AFRICA

POPULATION
- • Under 50,000
- • 50,000–250,000
- ● 250,000–1,000,000
- ● 1,000,000–5,000,000
- ● Over 5,000,000

National capitals are underlined

 Area most affected by rebel activity, 2008

Oilfields

Potential oilfields

———— Railroad

———— Road

–○–○– Projected oil pipeline

0   200   400 Kilometers
0 100 200 Miles

Longitude East of Greenwich

FIGURE 6-16

© H. J. de Blij, P. O. Muller, and John Wiley & Sons, Inc.

Crisis in the Interior

The Congo's civil wars of the 1990s started in one such neighbor, Rwanda, and spilled over into what was then still known as Zaïre. Rwanda has for centuries been the scene of conflict between sedentary Hutu farmers and invading Tutsi pastoralists. Colonial borders and practices worsened the situation, and after independence a series of terrible crises followed. In the mid-1990s the latest of these crises generated one of the largest refugee streams ever seen in the world, and the conflict engulfed eastern (and later northern and western) Congo. The death toll will never be known, but estimates range from 3 to 4 million, a calamity that was not enough to propel the international community into concerted peacemaking action. By the beginning of 2004 a combination of power transfer in the capital, Kinshasa, negotiation among the rebel groups and the African states involved in the conflict, UN assistance, and exhaustion had produced a semblance of stability in all but some eastern areas of The Congo (Fig. 6-16). But in 2007 civil unrest erupted again and continues.

Across the River

To the west and north of The Congo and Ubangi rivers lie Equatorial Africa's other seven countries (Fig. 6-16). Two of them are landlocked.  **Chad**, straddling the African Transition Zone as well as the regional boundary with West Africa, is one of Africa's most remote countries, although recent oil discoveries in the south and assistance from China in

Civil strife in The Congo erupts periodically, especially in its eastern zone. This is a large country with a multitude of different ethnic groups, great distances, poor infrastructure, and a weak economy, a mix of circumstances that often leads to civil unrest. Here a Congolese man listens to the news on his radio as soldiers loyal to a rebel cause leave town after they captured it from government soldiers. Meanwhile, rioters in The Congo's capital city of Kinshasa burned tires and stoned police cars to protest continued instability.   (© AP/Wide World Photos)

With five neighbors, **Congo** could be a major transit hub for this region, especially for The Congo if it recovers from civil war. Its capital, Brazzaville, lies across the Congo River from Kinshasa and is linked to the port of Pointe Noire by road and railway. But devastating power struggles have negated Congo's geographic advantages.

As Figure 6-16 shows, **Equatorial Guinea** consists of a rectangle of mainland territory and the island of Bioko, where the capital of Malabo is located. A former Spanish colony that remained one of Africa's least-developed territories, Equatorial Guinea, too, has been affected by the oil business in this area. Petroleum products now dominate its exports, but, as in so many other oil-rich countries, this bounty has not significantly raised incomes for most of the people.

One other territory would seem to be a part of Equatorial Africa: **Cabinda**, wedged between the two Congos just to the north of the Congo River's mouth. But Cabinda is one of those colonial legacies on the African map—it belonged to the Portuguese and was administered as part of Angola. Today it is an *exclave* of independent Angola, and a valuable one: it contains major oil reserves.

WEST AFRICA

West Africa extends south from the margins of the Sahara to the Gulf of Guinea coast and from Lake Chad west to Senegal (Fig. 6-17). Politically, the broadest definition of this region includes all those states that lie to the south of Western Sahara, Algeria, and Libya and to the west of Chad (itself sometimes included) and Cameroon. Within West Africa, a rough division is sometimes made between the large, mostly steppe-and-desert states that extend across the southern Sahara (Chad included) and the smaller, better-watered coastal states.

Enormous cultural diversity predated the colonial era in this subregion and continues to be exhibited there. France and Britain dominated the colonial map of West Africa, and to this day the interaction among West Africa's states is limited. But West

exploiting those finds are today changing this. The **Central African Republic**, chronically unstable and poverty-stricken, never was able to convert its agricultural potential and mineral resources (diamonds, uranium) into real progress. And one country consists of two small, densely forested volcanic islands: **São Tomé and Príncipe**, a ministate with a population of only 250,000 whose economy is being transformed by recent oil discoveries.

The four coastal states present a different picture. All four possess oil reserves and share the Congo Basin's equatorial forests; oil and timber, therefore, rank prominently among their exports. In **Gabon**, this combination has produced Equatorial Africa's only upper-middle-income economy (see Fig. G-11). Of the four coastal states, Gabon also has the largest proven mineral resources, including manganese, uranium, and iron ore. Its capital, Libreville (the only coastal capital in the region), reflects all this in its high-rise downtown, bustling port, and fast-growing squatter settlements.

Cameroon, less well endowed with oil or other raw materials, has the region's strongest agricultural sector by virtue of its higher-latitude location and higher-relief topography. Western Cameroon is one of the more developed parts of Equatorial Africa and includes the capital, Yaoundé, and the port of Douala.

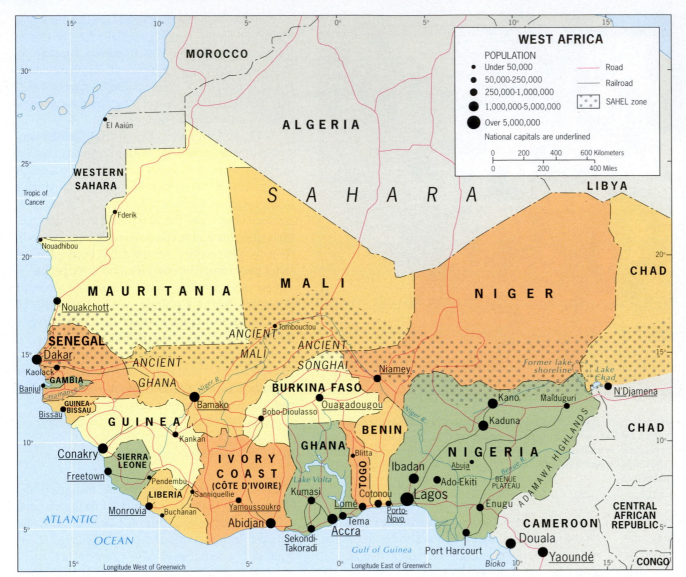

FIGURE 6-17

© H. J. de Blij, P. O. Muller, and John Wiley & Sons, Inc.

conflicts caused by outsiders, have paid a dreadful price. Some stability has returned, as exemplified by democratic elections in Sierra Leone, but strife continues in other countries.

Nigeria

Nigeria, the region's cornerstone, is home to more than 140 million people, by far the largest population total of any African country. When Nigeria achieved full independence from Britain in 1960, its new government was faced with the daunting task of administering a European political creation containing three major nations and nearly 250 other peoples ranging from several million to a few thousand in number.

For reasons obvious from the map (Fig. 6-18), Britain's colonial imprint always was stronger in the two southern regions than in the north. Christianity became the dominant faith in the south, and southerners, especially the Yoruba, took a lead role in the transition from colony to independent state. The choice of Lagos, the port of the Yoruba-dominated southwest, as the capital of a federal Nigeria (and not one of the cities in the more populous north) reflected British desires for the country's future. A three-region federation, two of which lay in the south, would ensure the primacy of the non-Islamic part of the state. But this framework did not last long. In 1967, the Ibo-dominated Eastern Region declared its independence as the Republic of Biafra, leading to a three-year civil war at a cost of 1 million lives. Since then, Nigeria's federal system has been modified repeatedly; today there are 36 States, and the capital has been moved

Africa's cultural vitality, historic legacies, populous cities, crowded countrysides, and bustling markets combine to create a regional imprint that is distinct and pervasive.

In the 1990s and into the 2000s, West Africa also was the scene of violent rebellions, collapsing governments, cultural strife, and anarchy. Sierra Leone, Liberia, and Ivory Coast (Côte d'Ivoire) were most severely afflicted, but the mayhem spilled over into Guinea and erupted in Nigeria over oil and land issues in the Niger Delta and religious issues in the north. Millions of ordinary citizens, caught up in

from Lagos to centrally located Abuja (Fig. 6-18).

Fateful Oil

Large oilfields were discovered beneath the Niger Delta during the 1950s, when Nigeria's agricultural sector produced most of its exports (peanuts, palm oil, cocoa, cotton) and farming still had priority in national and State development plans. Soon, revenues from oil production dwarfed all other sources, bringing the country a brief period of prosperity and promise. But before long Nigeria's oil wealth brought more bust than boom. Misguided development plans now focused on grand, ill-founded industrial schemes and costly luxuries such as a national airline; the continuing mainstay of the vast majority of Nigerians, agriculture, fell into neglect. Worse, poor management, corruption, outright theft of oil revenues during military misrule, and excessive borrowing against future oil income led to economic disaster. The country's infrastructure collapsed. In the cities, basic services broke down. In the rural areas, clinics, schools, water supplies, and roads to markets crumbled. In the Niger Delta area, local people beneath whose land the oil was being exploited demanded a share of the revenues and reparations for ecological damage; the military regime under General Abacha responded by arresting and executing nine of their leaders. On global indices of national well-being, Nigeria sank to the lowest rungs even as its production ranked it as high as the world's tenth-largest oil producer, with the United States its chief customer.

FIGURE 6-18

© H. J. de Blij, P. O. Muller, and John Wiley & Sons, Inc.

A woman trying to live a normal life walks past black smoke billowing from a burning oil pipeline in Nigeria. With great oil wealth that seemingly only benefits the rich, locals vandalize or tap into the oil pipes that crisscross the country. Fires often result and can burn for days as this one did in 2007. (© AP/Wide World Photos)

Islam Ascendant

In 1999, Nigeria's hopes were raised when, for the first time since 1983, a democratically elected president was sworn into office. But Nigeria's problems (now also including a deepening AIDS crisis) worsened when northern States, beginning with Zamfara, proclaimed *Sharia* (strict Islamic) law. When Kaduna State followed suit, riots between Christians and Muslims devastated the old capital city of Kaduna. There, and in 11 other northern States (Fig. 6-18), the imposition of Sharia law led to the departure of thousands of Christians, intensifying the cultural fault line that threatens the cohesion of the country. Although Kaduna State temporarily repealed its decision, the Islamic revivalism now exhibited in the north raises the prospect that Nigeria, West Africa's cornerstone and one of Africa's most important states, may succumb to devolutionary forces arising from its location in the African Transition Zone. For Africa, that would be a calamity.

Coast and Interior

Nigeria is one of 17 states (counting Chad and offshore Cape Verde) that form the region of West Africa. Four of these countries, comprising a huge territory under desert and steppe environments but containing small populations (Fig. 6-17), are landlocked: Burkina Faso, Niger, Mali, and Chad (the latter two are discussed in Chapter 7).

Burkina Faso, with the majority of its population Muslim, is fully within the Islamic world. It is a poor landlocked state with undeveloped reserves of gold; its economy relies on exports of cotton. Among the coastal states, **Ghana**, once known as the Gold Coast, was the first West African state to achieve independence, with a democratic government and a sound economy based on cocoa exports. Two grandiose postindependence schemes can be seen on the map: the port of Tema, intended to serve a vast West African hinterland, and Lake Volta, which resulted from the region's largest dam project. Mismanagement contributed to Ghana's economic collapse, but in the 1990s, a military regime was replaced by a stable and democratic government and recovery began. **Ivory Coast** (officially Côte d'Ivoire) translated three decades of autocratic but stable rule into economic progress, but excesses by the country's president-for-life cost it dearly. One of these excesses involved the construction of a Roman Catholic basilica to rival St. Peter's in Rome in the president's home village, Yamoussoukro, also designated to replace Abidjan as the country's new capital. At the turn of this century, the political succession became entangled in a north-south, Muslim-Christian schism that threatened the country's stability. Democratic **Senegal**, on the far west coast, demonstrates what stability can achieve: without oil, diamonds, or other valuable income sources and with an overwhelmingly subsistence-farming population, Senegal nevertheless managed to achieve some of the region's highest GNI levels. Over 90 percent Muslim and dominated by the Wolof, the ethnic group concentrated in the capital of Dakar, and with continuing close ties to France, Senegal has even been able to overcome both a failed effort to unify with its English-speaking **19** enclave, **Gambia**, and a secession movement in its southern Casamance District.

Other parts of West Africa have been afflicted by civil war and horror. **Liberia**, founded in 1822 by freed American slaves who returned to Africa with the help of American colonization societies, was ruled by Americo-Liberians for six generations and sold rubber and iron ore abroad. But a military coup in 1980 ended that era, and full-scale civil war beginning in 1989 embroiled virtually every ethnic group in the country. About a quarter of a million people, one-tenth of the population, perished; hundreds of thousands of others fled as refugees, many to Sierra Leone, which also was originally founded as a haven for freed slaves, in this case by the British in 1787. **Sierra Leone** went the all-too-familiar route from self-governing Commonwealth member to republic to one-party state to military dictatorship. In the 1990s worse was to follow: a rebel movement, funded by diamond sales, inflicted dreadful punishment on the local population. Early in the twenty-first century a remarkable turnabout has occurred as free elections were held and the country is stable and rebuilding.

We should remember, however, that all these conflicts we have mentioned are not representative of vast and populous West Africa. Tens of millions of farmers and herders who manage to cope with fast-changing environments in this zone between desert and ocean live remote from the newsmaking conflicts along the well-watered coast. Time-honored systems continue to serve: for example, the local village markets that drive the traditional economy. Visit the countryside, and you will find that some village markets are not open every day but, rather, every three or four days. Such a system ensures that all villages get a share in the exchange network. These **20 periodic markets** represent one of the many traditions enduring here, even as the cities beckon the farmers and burst at the seams. This region's great challenges are economic survival and nation-building through political stability under some of the most difficult circumstances in the world.

THE AFRICAN TRANSITION ZONE

As Figures 6-19 and 6-11 show, the African Transition Zone is unlike the four regions just discussed. This is where Subsaharan Africa's cultures intersect with the world of Islam, and it is Islam's gateway into the Subsaharan African realm. It encompasses some entire countries that are also part of the formal regions on the map (such as Senegal and Niger), it divides others into Muslim and non-Muslim sectors (such as Chad, Sudan, and Nigeria), and, in Africa's Horn, it comprises countries that do not form parts of other regions (Eritrea, Djibouti, Somalia). As elsewhere in the world where geographic realms meet or overlap, complications mark the African Transition Zone. In some areas, the transition from Muslim to non-Muslim society is gradual, as it remains in the northern parts of Ivory Coast and Nigeria despite the recent cultural problems erupt-

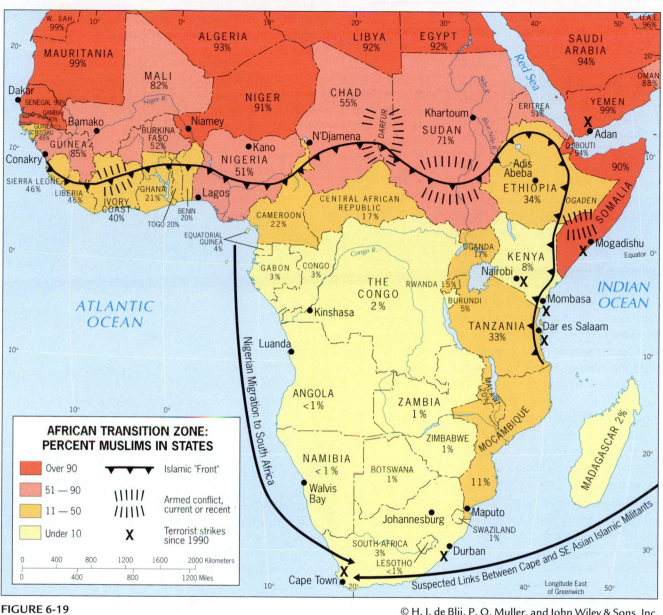

FIGURE 6-19

© H. J. de Blij, P. O. Muller, and John Wiley & Sons, Inc.

ing in these important states. Elsewhere the break is sharp, as it is in eastern Ethiopia and Sudan where a traditional border between Christian/animist-African cultures and Muslim-African communities is sudden. The southern border of the African Transition Zone, the religious frontier sometimes referred

to as Africa's **21 Islamic Front**, is therefore neither static nor everywhere the same.

Conflict marks much of the African Islamic Front (mapped in Fig. 6-19): long-term in Sudan, where a 30-year war for independence by non-Islamic African rebels in the south against the Arabized, Muslim north has cost millions of lives; intermittent in Nigeria, where Islamic revivalism is currently clouding the country's prospects; and recent in Ivory Coast, where political rivalries spurred religious strife in a once stable country. Perhaps the most conflict-prone part of the African Transition Zone, however, lies in the east, involving not only Sudan but also the historic Christian state of Ethiopia, nearly encircled by dominantly Muslim societies.

The Horn of Africa (Fig. 6-20) is an especially volatile subregion of the African Transition Zone. In landlocked **Ethiopia**, over one-third of the population of more than 78.4 million is Muslim, virtually all of it concentrated in the east. The costly conflict with neighboring **Eritrea**, which is about 50 percent Muslim, has damaged the economies of both countries (that issue was not over religion but over boundary definition). The ministate of **Djibouti**, over 90 percent Muslim, has taken on added importance in the War on Terror since it lies directly across from Yemen and overlooks the narrow entry to the Red Sea, the Bab el Mandeb Strait, a *choke point* in international commerce. But the key component in the eastern sector of the African Transition Zone is **Somalia**, where over 9 million people, virtually all Muslim, live in a harsh desert that forces cross-border migration into Ethiopia's Ogaden area in pursuit of seasonal pastures. As many as three to five million Somalis live permanently on the Ethiopian side of the border, but this is not the only division the Somali 'nation' faces. In reality, the Somali people is an assemblage of five major ethnic groups frag-

FIGURE 6-20

© H. J. de Blij, P. O. Muller, and John Wiley & Sons, Inc.

mented into hundreds of clans engaged in an endless contest for power as well as survival.

Early during this decade, Somalia's condition as a *failed state* led to the country's fragmentation into three parts (Fig. 6-20). In the north, the sector called **Somaliland** proclaimed independence in the 1990s and remains by far the most stable of the three. It functions essentially as an African state although the international community has not recognized it as such. In the east, a conclave of local chiefs declared their territory to be separate from the rest of Somalia, calling it **Puntland**, and asserted an unspecified degree of autonomy. In the south, where the official capital, Mogadishu, is located on the Indian Ocean coast, local secular warlords and Islamic militias continued their struggle for su-

premacy, the warlords supported by U.S. funding as part of the War on Terror. In June 2006, Islamic militias stormed into the capital and took control, ousting the warlords and proclaiming their determination to create an Islamic state. The combination of a failed state, Islamic fundamentalists in control, and al-Qaeda in the mix makes this part of the African Transition Zone a persistent focus of international concern.

As this chapter has underscored, Africa is a continent of infinite social diversity, and Subsaharan Africa is a realm of matchless cultural history. Relatively recent environmental change has widened the Sahara, which separates north from south. The push of Islam turned the north to Mecca and the south to

Christian proselytism. Millions of Africans were carried away in bondage; the cruelties of colonialism subjugated those who were left behind into a political straitjacket forged in Berlin. Africa's equatorial heart is an incubator for debilitating diseases second to none in the world, but Africa's health problems have, until recently, not been a top priority in the medical world. When the richer countries finally abandoned their African empires, African peoples were left a task of reconstruction in which they received little help and found their products uncompetitive on markets where rich-country producers basked in subsidies. We said at the beginning of this chapter that Africa's time and turn will come again, but it will take far more than is being done today to rectify half a millennium's malefactions.

In Eritrea, the haze-covered edge of eastern Africa's great rift valley frames a landscape littered with Ethiopian tanks left over from the 31-year-old war for independence from Ethiopia (1962–1993). Eritrea's independence has left Ethiopia a landlocked country. UN peacekeepers remain in place as border skirmishes continue between Eritrea and Ethiopia, and also between Eritrea and Djibouti. (© Robert Caputo/ Aurora/Getty Images)

FIGURE 7-1 *Map:* © H. J. de Blij, P. O. Muller, and John Wiley & Sons, Inc.

NORTH AFRICA/ SOUTHWEST ASIA

In This Chapter

- How Islam transformed a realm
- Where the oil is—and is not
- Foreign intervention, Muslim reaction
- Political theory and practical reality in Iraq
- The predicament of Israel
- Iraq, Afghanistan, and the War on Terror

CONCEPTS, IDEAS, AND TERMS

1	Cultural geography
2	Culture hearth
3	Cultural diffusion
4	Cultural landscapes
5	Hydraulic civilization theory
6	Climate change
7	Spatial diffusion (expansion; relocation; contagious; hierarchical)
8	Islamization
9	Religious revivalism
10	Wahhabism
11	Cultural revival
12	Stateless nation
13	Choke point
14	Nomadism
15	Buffer state

REGIONS

EGYPT AND THE LOWER NILE BASIN
MAGHREB AND ITS NEIGHBORS
MIDDLE EAST
ARABIAN PENINSULA
EMPIRE STATES
TURKESTAN

Photos: (upper left) Cairo, Egypt, © Sylvain Grandadam/Getty Images; (above) Saharan oasis, Libya, © Frank Krahmer/Masterfile.

229

FROM MOROCCO ON the shores of the Atlantic to the mountains of Afghanistan, and from the Horn of Africa to the steppes of inner Asia, lies a vast geographic realm of enormous cultural complexity. It stands at the crossroads where Europe, Asia, and Africa meet, and it is part of all three continents (Fig. 7-1, see chapter opener map). Throughout history, its influences have radiated to these continents and to practically every other part of the world as well. This is one of humankind's primary source areas. On the Mesopotamian Plain between the Tigris and Euphrates rivers (in modern-day Iraq) and on the banks of the Egyptian Nile arose several of the world's earliest civilizations. In its soils, plants were domesticated that are now grown from the Americas to Australia. Along its paths walked prophets whose religious teachings are still followed by hundreds of millions. And in the first decade of the twenty-first century, the heart of this realm is beset by some of the most bitter and dangerous conflicts in the world.

Defining the Realm

As you might deduce from the long name we use for this realm, it is a place of great complexity. It is therefore tempting to characterize this geographic realm in a few words and to stress one or more of its dominant features, and many do so. But we do not feel this is adequate. Let us unpack each of these shorthands.

A 'Dry World'?

This realm is, for instance, often called the 'Dry World,' containing as it does the vast Sahara as well as the Arabian Desert. But most of the realm's people live where there is water—along the Nile River, along the hilly Mediterranean coastal strip (the *tell*, meaning mound in Arabic) of northwesternmost Africa, along the Asian eastern and northeastern shores of the Mediterranean Sea, in the Tigris-Euphrates Basin, in far-flung desert oases, and along the lower mountain slopes of Iran south of the Caspian Sea and of Turkestan to the northeast. We know this world region as one where water is almost always at a premium, where peasants often struggle to make soil and moisture yield a small harvest, where nomadic peoples and their animals still circulate across dust-blown flatlands, where oases are islands of sedentary farming and trade in a sea of aridity.

But this is also the land of the Nile, the lifeline of Egypt, the crop-covered *tell* of coastal Algeria, the verdant shores of western Turkey, and the meltwater-fed valleys of Central Asia. Compare Figure 7-1 to Figure G-9, and the dominance of *B* climates becomes evident. Also consult Figure G-8, and you will see how water-dependent this realm's clustered population is.

Tinerhir in Morocco's Todra Valley, located high in the Atlas Mountains. Due to the aridity of this part of the world, the people who live here have learned to cope with limited water. In ingenious ways agricultural systems, architecture, and clothing all acknowledge that water is scarce, yet have creatively adapted to this reality.   (© Jose Fuste Raga/Corbis Images)

area who speak Arabic and related languages, but ethnologists normally restrict it to certain occupants of the Arabian Peninsula—the Arab 'source.' In any case, Turks are not Arabs, and neither are most Iranians or Israelis. Moreover, although the Arabic language prevails from Mauritania in the west across all of North Africa to the Arabian Peninsula, Syria, and Iraq in the east, it is not spoken in other parts of this realm. In Turkey, for example, Turkish is the major language, and it has Ural-Altaic rather than Arabic's Semitic or Hamitic roots. The Iranian language belongs to the Indo-European linguistic family. Other 'Arab World' languages that have separate ethnological identities are spoken by the Jews of Israel, the Tuareg people of the Sahara, the Berbers of northwestern Africa, and the peoples of the transition zone between North Africa and Subsaharan Africa to the south.

An 'Islamic World'?

Lastly, a name given to this realm is the World of Islam. The prophet Muhammad (Mohammed) was born in Arabia in AD 571, and in the centuries after his death in 632, Islam spread into Africa, Asia, and Europe. This was the age of Arab conquest and expansion. Their armies penetrated Southern Europe, their caravans crossed the deserts, and their ships plied the coasts of Asia and Africa. Along these routes they carried the Muslim (Islamic) faith, converting the ruling classes of the states of the West African savanna, threatening the Christian stronghold in the highlands of Ethiopia, penetrating the deserts of inner Asia, and pushing into India and even the island extremities of Southeast Asia. Today, the Islamic faith extends far beyond the limits of the realm under discussion (Fig. 7-2). Nor is the World of Islam entirely Muslim. Judaism, Christianity (notably in Egypt and Lebanon), and other faiths survive in the heartland of the Islamic World. So this connotation is not satisfactory either.

Is This the 'Middle East'?

This realm is commonly referred to as the Middle East. That must sound odd to someone in, say, India, who might think of a Middle West rather than a Middle East! The name, of course, reflects the biases of its source: the Western world, which saw a Near East in Turkey, a Middle East in Egypt, Arabia, and Iraq, and a Far East in China and Japan. Still, the term has taken hold, and it can be seen and heard in everyday usage by scholars, journalists, and members of the United Nations. Even so, we feel that it should only be applied to one of the regions of this vast realm, not to the realm as a whole.

An 'Arab World'?

Another shorthand often used for North Africa/ Southwest Asia is the Arab World. This term implies a uniformity that does not actually exist. First, the name *Arab* is applied loosely to the peoples of this

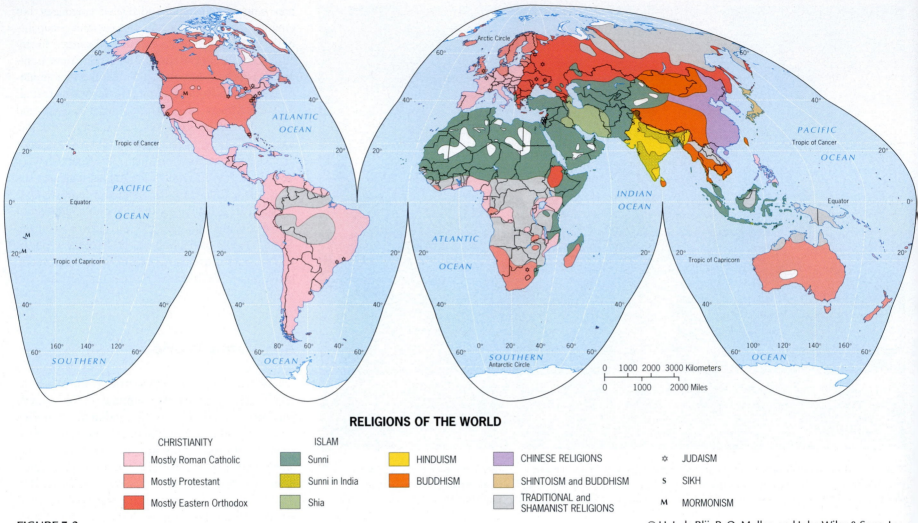

RELIGIONS OF THE WORLD

CHRISTIANITY
- Mostly Roman Catholic
- Mostly Protestant
- Mostly Eastern Orthodox

ISLAM
- Sunni
- Sunni in India
- Shia

HINDUISM

BUDDHISM

CHINESE RELIGIONS

SHINTOISM and BUDDHISM

TRADITIONAL and SHAMANIST RELIGIONS

✡ JUDAISM

S SIKH

M MORMONISM

FIGURE 7-2

© H. J. de Blij, P. O. Muller, and John Wiley & Sons, Inc.

HEARTHS OF CULTURE

This geographic realm occupies a pivotal part of the world: here Africa, the source of humanity, meets Eurasia, crucible of human cultures. Two million years ago, the ancestors of our species walked from East Africa into North Africa and Arabia and spread all across Asia. Less than one hundred thousand years ago, *Homo sapiens* crossed these lands on their way to Europe, Australia, and, eventually, the Americas. Ten thousand years ago, human communities in what we now call the Middle East began to domesticate plants and animals, learned to irrigate their fields, enlarged their settlements into towns, and formed the earliest states. The world's dominant monotheistic religions originated here, first Judaism, then Christianity, and most recently the heart of the realm was stirred and mobilized by the teachings of Muhammad and the Quran (Koran), and Islam began. Today this realm is a cauldron of religious and political activity and turmoil, weakened by conflict but empowered and enriched by oil in some places, plagued by poverty in many others.

Dimensions of Culture

In the Introduction (p. 17), we discussed the concept of culture and its regional expression in the cultural landscape. **1 Cultural geography**, we noted, is a wide-ranging and comprehensive field that studies spatial aspects of human cultures, focusing not only on cultural landscapes but also on **2 culture hearths**—the crucibles of civilization, the sources of ideas, innovations, and ideologies that changed regions and realms. Those ideas and innovations spread far and wide through a set of processes that we study under the rubric of **3 cultural diffusion**. Because we understand these processes better today, we can reconstruct many of the ancient routes by which the knowledge and achievements of culture hearths spread (that is, *diffused*) to other areas.

Another topic of cultural geography, also relevant in the context of the North Africa/Southwest Asia realm, is the study of **4 cultural landscapes** that a dominant culture creates. Human cultures exist in long-term accommodation with (and adaptation to) their natural environments, exploiting opportunities that these environments present and coping with the extremes they can impose. In the process, they fuse their physical landscape and cultural landscape into an interacting unity.

Rivers and Communities

In the basins of the major rivers of this realm (the Tigris-Euphrates system of modern-day Turkey, Syria, and Iraq, and the Nile of Egypt) lay two of the world's earliest culture hearths. Mesopotamia (land amidst the rivers) had fertile alluvial soils, abundant sunshine, ample water, and animals and plants that could be domesticated. Here, in the Tigris-Euphrates lowland between the Persian Gulf and the uplands of present-day Turkey, arose one of humanity's first culture hearths, a cluster of communities that grew into larger societies and, eventually, into the world's

first states. (Early state development probably was going on simultaneously in East Asia's river basins as well.) Mesopotamians were innovative farmers who domesticated crops such as wheat and knew when to sow and harvest various crops, water their fields, and store their surplus. Their knowledge diffused to villages near and far, and a *Fertile Crescent*, a region of significant agricultural production, evolved, extending from Mesopotamia across southeastern Turkey into Syria and the eastern Mediterranean coast beyond.

Mesopotamia

Irrigation was the key to prosperity and power in Mesopotamia, and urbanization was its reward. Among many settlements in the Fertile Crescent, some thrived, grew, enlarged their hinterlands, and diversified socially and occupationally; others failed. What determined success? One theory, the **5 hydraulic civilization theory**, holds that cities that could control irrigated farming over large hinterlands held power over others, used food as a weapon, and thrived. One such city, Babylon on the Euphrates River, endured for nearly 4000 years (from 4100 BC). A busy port, its walled and fortified center endowed with temples, towers, and palaces, Babylon for a time was the world's largest city.

Egypt and the Nile

Egypt's cultural evolution may have started even earlier than Mesopotamia's, and its focus lay upstream from (south of) the Nile Delta and downstream from (north of) the first of the Nile's series of rapids, or cataracts. This part of the Nile Valley is surrounded by inhospitable desert, and unlike Mesopotamia (which lay open to all comers), the Nile provided a natural fortress here. The ancient Egyptians converted their security into progress. The Nile was their highway of trade and interaction; it also supported agriculture through irrigation. The Nile's cyclical ebb and flow was much

more predictable than that of the Tigris-Euphrates river system. By the time Egypt fell victim to outside invaders (about 1700 BC), a full-scale urban civilization had emerged. Ancient Egypt's artist-engineers left a magnificent legacy of massive stone monuments, some of them containing treasure-filled crypts of god-kings called Pharaohs. These tombs have enabled archaeologists to reconstruct the ancient history of this culture hearth.

Today, the world continues to benefit from the accomplishments of the ancient Mesopotamians and Egyptians. They domesticated cereals (wheat, rye, barley), vegetables (peas, beans), fruits (grapes, apples, peaches), and many animals (horses, pigs, sheep). They also advanced the study of the calendar, mathematics, astronomy, government, engineering, metallurgy, and a host of other skills and technologies. In time, many of their innovations were adopted and then modified by other cultures in the Old World and eventually in the New World as well. Europe was the greatest beneficiary of these legacies of Mesopotamia and ancient Egypt, whose achievements constituted the foundations of Western civilization.

Decline and Decay

As Figure G-9 reminds us, many of the early cities of this culture realm lay in what is today desert territory. Assuming that there were no good reasons to build large settlements in the middle of deserts, we may hypothesize that **6 climate change** sweeping over this region, not a monopoly over irrigation techniques, gave certain cities in the ancient Fertile Crescent an advantage over others. Climate change, associated with shifting environmental zones after the last Pleistocene glacial retreat, may have destroyed the last of the old civilizations. Perhaps overpopulation and human destruction of the natural vegetation contributed to the process. Indeed, some cultural geographers suggest that the momentous innovations in agricultural planning and irrigation technology were not 'taught' by the seasonal flooding of the rivers but were forced on the inhabitants as they tried to survive changing environmental conditions.

The scenario is not difficult to imagine. As outlying areas began to fall dry and farmlands were destroyed, people congregated in the already crowded river valleys—and made every effort to increase the productivity of the land that could still be watered. Eventually overpopulation, destruction of the watershed, and perhaps reduced rainfall in the rivers' headwater areas dealt the final blow. Towns were abandoned to the encroaching desert; irrigation canals filled with drifting sand; croplands dried up. Those who could migrated to areas that were still reputed to be productive. Others stayed, their numbers dwindling, increasingly reduced to subsistence.

As old societies disintegrated, power emerged elsewhere. First the Persians, then the Greeks, and later the Romans imposed their imperial designs on the tenuous lands and disconnected peoples of North Africa/Southwest Asia. Roman technicians converted North Africa's farmlands into irrigated plantations whose products went by the boatload to Roman Mediterranean shores. Thousands of people were carried off as slaves to the cities of the new conquerors. Egypt was quickly colonized, as was the area we now call the Middle East. One region that lay distant, and therefore remote from these invasions, was the Arabian Peninsula, where no major culture hearth or large cities had emerged and where the turmoil had not affected Arab settlements and nomadic routes.

STAGE FOR ISLAM

In a remote place on the Arabian Peninsula, where the foreign invasions of the Middle East had had little effect on the Arab communities, an event occurred early in the seventh century that was to change history and affect the destinies of people in many parts of the world. In a town called Mecca (Makkah), about 70 kilometers (45 mi) from the Red Sea coast in the Jabal Mountains, a man named Muhammad in the year AD 611 began to receive revelations from Allah (God). Muhammad (571–632) was then in his early forties. Convinced after some initial self-doubt that he was indeed chosen to be a prophet, Muhammad committed his life to fulfilling the divine commands he believed he had received. Arab society was in social and cultural disarray, but Muhammad forcefully taught Allah's lessons and began to transform his culture. His personal power soon attracted enemies, and in 622 he fled from Mecca to the safer haven of Medina (Al Madinah), where he continued his work until his death. This moment, the *hejira* (meaning migration), marks the starting date of the Muslim era, Year 1 on Islam's calendar. Mecca, of course, later became Islam's holiest place.

The Faith

The precepts of Islam in many ways constituted a revision and embellishment of Judaic and Christian beliefs and traditions. All of these faiths have but one god, who occasionally communicates with humankind through prophets; Islam acknowledges that Moses and Jesus were such prophets but considers Muhammad to be the final and greatest prophet. What is earthly and worldly is profane; only Allah is pure. Allah's will is absolute; Allah is omnipotent and omniscient. All humans live in a world that Allah created for their use but only to await a final judgment day.

Islam brought to the Arab World not only the unifying religious faith it had lacked but also a new set of values, a new way of life, a new individual and collective dignity. Islam dictated observance of the Five Pillars: (1) repeated expressions of the basic creed, (2) daily prayer, (3) a month each year of daytime fasting (Ramadan), (4) the giving of alms, and (5) at least one pilgrimage to Mecca in each Muslim's lifetime (the *hajj*). Islam prescribed and proscribed in other spheres of life as well. It forbade alcohol, smoking, and gambling. It tolerated polygamy, although it acknowledged the virtues of monogamy. Mosques appeared in Arab settlements, not only for the (Friday) sabbath prayer, but also as social gathering places to knit communities closer together. Mecca became the spiritual center for a divided, widely dispersed people for whom a collective focus was something new.

The Arab-Islamic Empire

Muhammad provided such a powerful stimulus that Arab society was mobilized almost overnight. The prophet died in 632, but his faith and fame spread like wildfire. Arab armies carrying the banner of Islam formed, invaded, conquered, and converted wherever they went. As Figure 7-3 shows, by AD 700 Islam had reached far into North Africa, into Transcaucasia, and into most of Southwest Asia. In the centuries that followed, it penetrated Southern and Eastern Europe, Central Asia's Turkestan, West Africa, East Africa, and South and Southeast Asia, even reaching China by AD 1000.

Routes of Diffusion

The spread of Islam provides a good illustration of **cultural diffusion**. This concept focuses on the process in which ideas, inventions, and cultural practices expand geographically through a population in space and time. In 1952, the Swedish geographer named Torsten Hägerstrand published his fundamental study of **7** **spatial diffusion** entitled *The Propagation of Diffusion Waves*. He reported that diffusion takes place in two forms: **expansion diffusion**, when propagation waves originate in a strong and durable source area and spread outward, affecting an ever larger region and population; and **relocation diffusion**, in which migrants carry an innovation, idea, or (for example) a virus from the source to distant locations and it diffuses from there. Islam spread mostly through expansion diffusion (the global spread of AIDS is a case of relocation diffusion).

Islam on the March

Both expansion and relocation diffusion include several types of processes. The spread of Islam, as Figure 7-3 shows, initially proceeded by a form of *expansion diffusion* called **contagious diffusion** as the faith moved from village to village across the Arabian Peninsula, North Africa, and Southwest Asia. But Islam got powerful boosts when kings, chiefs, and other high officials were converted, who in turn propagated the faith downward through their bureaucracies to far-flung subjects. This form of expansion diffusion, called **hierarchical diffusion**, served Islam well.

But the map leaves no doubt that Islam later spread by *relocation diffusion* as well, notably to the Ganges Delta in South Asia, to present-day Indonesia, and to East Africa. That process continues to this day, but the heart of Islam remains in Southwest Asia. There, Islam became the cornerstone of an Arab Empire with Medina as its first capital. As the empire grew by expansion diffusion, its headquarters were relocated from Medina to Damascus (in present-day Syria) and later to Baghdad on the Tigris River (Iraq). And it prospered. In architecture, mathematics, and science, the Arabs overshadowed their European contemporaries. The Arabs established institutions of higher learning in many cities including Baghdad, Cairo, and Toledo

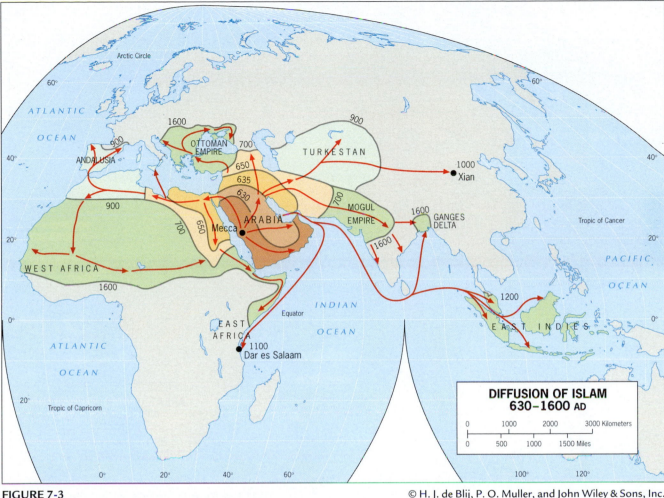

FIGURE 7-3

© H. J. de Blij, P. O. Muller, and John Wiley & Sons, Inc.

(Spain), and their distinctive cultural landscapes united their vast domain. Non-Arab societies in the path of the Muslim drive were not only Islamized, but also Arabized, adopting other Arab traditions as well. Islam had spawned a culture; it still lies at the heart of that culture today.

As we noted, Islam's expansion eventually was checked in Europe, Russia, and elsewhere. But a map showing the total area under Muslim sway in Eurasia

and Africa reveals the enormous dimensions of the domain affected by **8** **Islamization** at one time or another (Fig. 7-4). Islam continues to expand, now mainly by relocation diffusion. There are Islamic communities in cities as widely scattered as Vienna, Singapore, and Cape Town, South Africa; Islam is also growing rapidly in the United States. With more than 1.4 billion adherents today, Islam is a vigorous and burgeoning cultural force around the world.

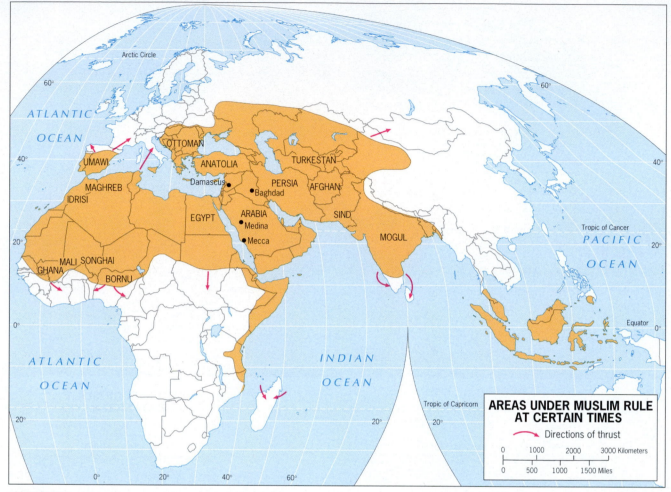

FIGURE 7-4

© H. J. de Blij, P. O. Muller, and John Wiley & Sons, Inc.

but the Shia–Sunni schism is often at the root of many conflicts in this realm.

The Strength of Shi'ism

But the Shi'ites vigorously promoted their version of the faith. In the early sixteenth century their work paid off: the royal house of Persia (modern-day Iran) made Shi'ism the only legal religion throughout its vast empire. That domain extended from Persia into lower Mesopotamia (modern Iraq), into Azerbaijan, and into western Afghanistan and Pakistan. As the map of religions (Fig. 7-2) shows, this created for Shi'ism a large culture region and gave the faith unprecedented strength. Iran remains the bastion of Shi'ism in the realm today, and the appeal of Shi'ism continues to radiate into neighboring countries and even farther afield.

During the late twentieth century, Shi'ism gained unprecedented influence in the realm. In its heartland, Iran, a *shah* (king) tried to secularize the country and to limit the power of the *imams* (mosque officials); he provoked a revolution that cost him the throne and made Iran a Shi'ite Islamic republic. Before long, Iran was at war with neighboring, Sunni-ruled Iraq, and Shi'ite parties and communities elsewhere were invigorated by the newfound power of Shi'ism. From Arabia to Africa's northwestern corner, Sunni-ruled countries warily watched their Shi'ite minorities, newly imbued with religious fervor. Mecca, the holy place for both Sunnis and Shi'ites, became a battleground during the week of the annual pilgrimage, and for a time the (Sunni) Saudi Arabian government denied entry to Shi'ite pilgrims. That schism has healed somewhat, but intra-Islamic sectarian differences run deep, as the current war in Iraq demonstrates.

ISLAM DIVIDED

For all its vigor and success, Islam still fragmented into sects. The earliest and most consequential division arose after Muhammad's death. Who should be his legitimate successor? Some believed that only a blood relative should follow the prophet as leader of Islam. Others, a majority, felt that any devout follower of Muhammad was qualified (the Sunnis). The first chosen successor was the father of Muhammad's wife (and thus not a blood relative). But this did not satisfy those who wanted to see a man named Ali, a cousin of Muhammad, made *caliph* (successor). When Ali's turn came, his followers, the Shi'ites, proclaimed that Muhammad finally had a legitimate successor. This offended the Sunnis, those who did not see a blood relationship as necessary for the succession. From the beginning of this disagreement, the numbers of Muslims who took the Sunni side far exceeded those who regarded themselves as Shia (followers) of Ali. The great expansion of Islam was largely propelled by Sunnis; the Shi'ites survived as minorities scattered throughout the realm. Today, about 85 percent of all Muslims are Sunnis,

Smaller Islamic sects further diversify this realm's religious landscape. Some of these sects play a disproportionately large role in the societies and countries of which they are a part, as we will see in our regional discussion.

Religious Revivalism in the Realm

Another cause of intra-Islamic conflict lies in the resurgence of religious fundamentalism, or as Muslims refer to it, **9** **religious revivalism**. In the 1970s, the imams in Shi'ite Iran wanted to reverse the shah's moves toward liberalization and secularization: they wanted to (and did) recast society in traditional, revivalist Islamic molds. An *ayatollah* (leader under Allah) replaced the shah in 1979; Islamic rules and punishments were instituted. Urban women, many of whom had been considerably liberated and educated during the shah's regime, were forced to return to more traditional Islamic roles. Vestiges of Westernization, encouraged by the shah, disappeared. Under these new conditions, even the war against Iraq (1980–1990), which began as a conflict over territory, became a holy war that cost more than a million lives.

Islamic revivalist fundamentalism did not rise in Iran alone, nor was it confined to Shi'ite communities. Many Muslims—Sunnis as well as Shi'ites—in all parts of the realm disapproved of the erosion of traditional Islamic values, the corruption of society by European colonialists and later by Western modernizers, and the declining power of the faith in the secular state. As long as economic times were good, such dissatisfaction remained submerged. But when jobs were lost and incomes declined, a return to fundamental Islamic ways became more appealing.

Muslim Against Muslim

This set Muslim against Muslim in all the regions of the realm. Revivalists fired the faith with a new militancy, challenging the status quo from Afghanistan to Algeria. The militants forced their governments to ban 'blasphemous' books, to resegregate the sexes in schools, to enforce traditional dress codes, to legitimize religious-political parties, and to heed the wishes of the *mullahs* (teachers of Islamic ways). Militant Muslims proclaimed that democracy inherited from colonialists and adopted by Arab nationalists was incompatible with the rules of the Quran (Koran).

The rift between moderate Muslims and militant revivalists began to spill over into other parts of the world decades ago, from Pakistan to the African Transition Zone and from the Caucasus to the Philippines. Even while Islamic countries such as Iran and Algeria struggled to overcome the instability and damage arising from their internal conflicts, militant activists planned and carried out attacks against the allies of the moderates—oil-guzzling Western countries guilty of military intervention, propping up unrepresentative regimes, mistreatment of immigrant Muslims, and cultural corruption. Before the worst of these attacks (as of mid-2008) destroyed the World Trade Center in New York and severely damaged the Pentagon outside Washington, D.C., the French had foiled a similar assault on the Eiffel Tower in Paris, the Russians had suffered hundreds of casualties in Moscow and Transcaucasia, and U.S. targets were hit in the Middle East, the Arabian Peninsula, East Africa, and elsewhere.

Back to Basics

Although it has been argued that these attacks do not represent Islam as a religion in conflict with the West, they do have a religious context. During the 1990s, following the Cold War defeat of Soviet forces in Afghanistan and the subsequent abandonment of that country by the United States and its allies, an organization named *al-Qaeda* emerged, dedicated to punishing the perceived enemies of Islamic peoples by the most effective means at its disposal: terrorist attacks. However, the leader and chief financier of this movement, Usama bin Laden, had a bigger agenda. His ultimate goal was to overthrow the regime of rich princes ruling his home country, Saudi Arabia, accusing them of collusion with the infidel United States, of apostasy (faithlessness), and worse. Bin Laden wanted to make Saudi Arabia, birthplace of Muhammad and guarantor of holy Mecca, a true Islamic state.

Bin Laden was not the first Saudi to wish to return his country to its Islamic roots. During the eighteenth century an Islamic theologian named Muhammad ibn Abd al-Wahhab (1703–1792) founded a movement to bring Islamic society on the peninsula back to the most fundamental, traditional, and puritanical form of the faith. So strict were his teachings that he was expelled from his hometown in 1744—a fateful exile because he moved to a place that happened to be the headquarters of Ibn Saud, ruler over a sizeable swath of the Arabian Peninsula. Ibn Saud formed an alliance with al-Wahhab, who had inherited a fortune when his wife died; between them, they set out on a campaign of conquest that made the Saudi dynasty the rulers of the region and **10** **Wahhabism** the dominant form of Islam. Even the relatively liberal Ottoman sultans could not suppress either the Saudis or the Wahhabis, and in 1932, when the modern Kingdom of Saudi Arabia was established, Wahhabism became its official and dominant form of Islam.

Despite its reference to the name of the founder, Wahhabism today is a term used mostly by non-Muslims and outsiders. Followers of Wahhabi doctrines call themselves *Muwahhidun* (Unitarians) to signify the strict and fundamental nature of their beliefs. These are the doctrines Usama bin Laden accuses the Saudi princes of having violated, and these are the harsh teachings so often cited in the Western media as emanating from the pulpits of Saudi Arabia's revivalist mosques.

Islam and Other Religions

Two additional major faiths had their sources in the *Levant* (the area extending from Greece eastward around the Mediterranean coast to northern Egypt),

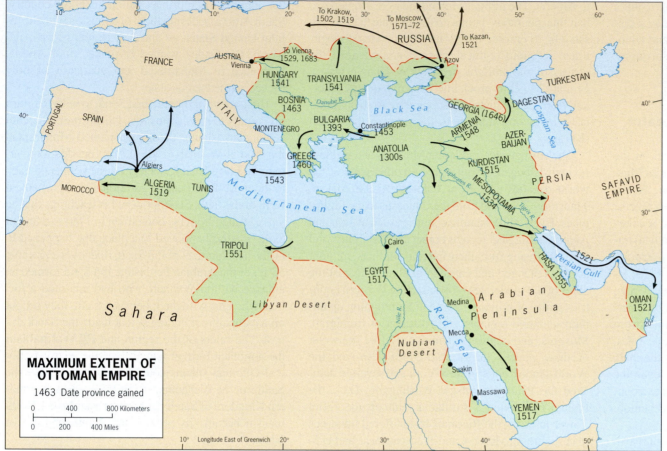

MAXIMUM EXTENT OF
OTTOMAN EMPIRE

1463 Date province gained

0   400   800 Kilometers

0   200   400 Miles

FIGURE 7-5   © H. J. de Blij, P. O. Muller, and John Wiley & Sons, Inc.

Judaism and Christianity, and both are older than Islam. Islam's rise submerged many smaller Jewish communities, but the Christians, not the Jews, waged centuries of holy war against the Muslims, seeking, through the Crusades, not only to drive Islam back but to reestablish Christian communities where they had dominated before the Islamic expansion. The aftermath of that campaign still marks the realm's cultural landscape today. A substantial Christian minority (about one-fourth of the population) remains in Lebanon, and Christian minorities also survive in Israel, Syria, Egypt, and Jordan. Strained relations between the long-dominant Christian minority in Lebanon and the Muslim majority (itself divided into five sects) contributed to the disastrous armed struggle that engulfed this country in the 1970s and 1980s. It reemerged in 2006 and Lebanon has once again fallen into turmoil.

But the most intense conflict in modern times has pitted the Jewish state, Israel, against its Islamic neighbors near and far. Israel's United Nations-sponsored creation in 1948 precipitated more than a half-century of intermittent strife and attempts at mediation; it also caused friction among Islamic states in the region.

Jerusalem—holy city for Judaism, Christianity, and Islam—lies in the crucible of this confrontation.

The Ottoman Aftermath

It is ironic that Islam's last great advance into Europe eventually led to European occupation of Islam's very heartland. The Ottomans (named after their leader, Osman I), based in what is today Turkey, conquered Constantinople (now Istanbul) in 1453 and pushed into Eastern Europe. Soon Ottoman forces were on the doorstep of Vienna; they also invaded

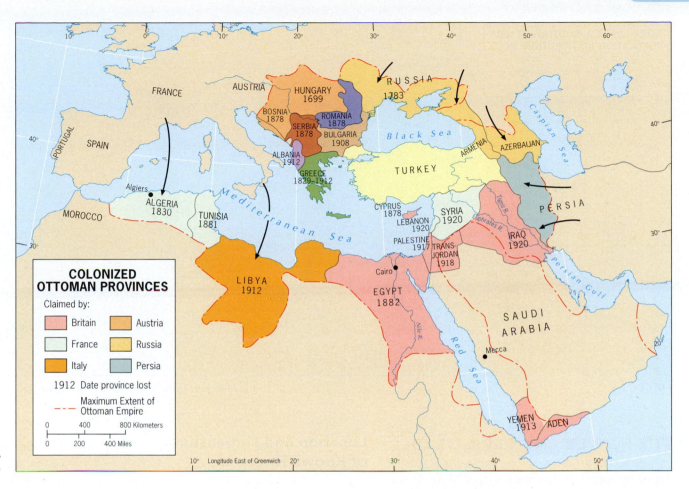

FIGURE 7-6   © H. J. de Blij, P. O. Muller, and John Wiley & Sons, Inc.

Persia, Mesopotamia, and North Africa (Fig. 7-5). At its height, the Ottoman Empire under Suleyman the Magnificent (ruled 1522–1560) was the most powerful state in western Eurasia. As Figure 7-5 shows, its armies even advanced toward Moscow, Kazan, and Krakow from a base at Azov (near present-day Rostov in Russia). The Turks also launched marine attacks on Sicily, Spain, and France.

The Ottoman Empire survived for more than four centuries (it ended in 1923), but it lost territory as time went on, first to the Hungarians, then to the Russians, and later to the Greeks and Serbs until, after

World War I, the European powers took over its provinces and made them colonies—colonies we now know by the names of Syria, Iraq, Lebanon, and Yemen (Fig. 7-6). As the map shows, the French and the British took large possessions; even the Italians annexed part of the Ottoman domain.

The boundary framework that the colonial powers created to delimit their holdings was not satisfactory. As Figure 7-7 illustrates, this realm's population of more than 580 million is clustered, fragmented, and strung out in river valleys, coastal zones, and crowded oases. The colonial powers laid out long

stretches of boundary as ruler-straight lines across uninhabited territory; they saw no need to adjust these boundaries to cultural or physical features in the landscape. Other boundaries, even some in desert zones, were poorly defined and never marked on the ground. One product of this process was Iraq, delimited by (then) British Colonial Secretary Winston Churchill in 1921 as a kingdom with Baghdad as its capital; another was Jordan, centered on Amman. Later, when the colonies had become independent states, such boundaries led to quarrels, even armed conflicts, among neighboring Muslim states.

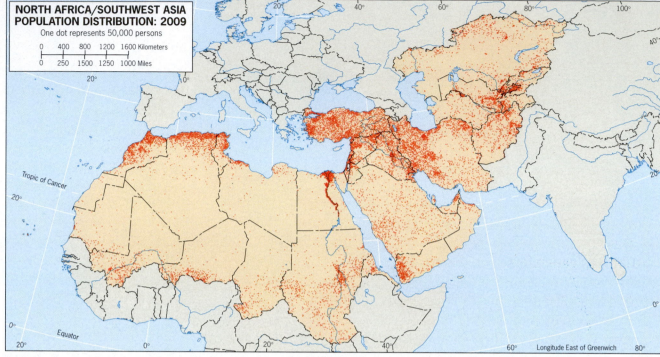

NORTH AFRICA/SOUTHWEST ASIA POPULATION DISTRIBUTION: 2009

One dot represents 50,000 persons

FIGURE 7-7

© H. J. de Blij, P. O. Muller, and John Wiley & Sons, Inc.

THE POWER AND PERIL OF OIL

Travel through the cities and towns of North Africa and Southwest Asia, talk to students, shopkeepers, taxi drivers, and migrants, and you will hear the same refrain: "Leave us in peace, let us do things our traditional way. Our problems—with each other, with the world—seem always to result from outside interference. The stronger countries of the world exploit our weaknesses and magnify our quarrels. We want to be left alone."

That wish might be closer to fulfillment were it not for two relatively recent events: the creation of the state of Israel and the discovery of some of the world's largest oil reserves. We will discuss the evolution of Israel in the regional section of this chapter. Here we focus on the realm's most valuable export product: oil.

Location and Dimensions of Known Reserves

About 30 of the world's countries have significant oil reserves. But five of these countries, all located in the North Africa/Southwest Asia realm, in combination possess larger reserves (approximately 77 percent of the world total) than the rest combined. The estimated reserves of this Big Five in 2005, averaged from several sources including the governments of the countries themselves and reported in billions of barrels, were as follows: (1) Saudi Arabia, 267; (2) Iraq, 114; (3) United Arab Emirates, 101; (4) Kuwait, 97; and (5) Iran, 91. The next-ranking country, Venezuela, had known reserves below 80 billion barrels; Russia below 60; the United States below 40; and Libya, Mexico, Nigeria, China, and Ecuador below 25. No other coun-

try's known reserves approached 20 billion barrels as of 2005. It should also be noted that Canada may possess the largest oil reserves of all—but most of this supply is contained in Alberta's *tar sands*, which requires complex and expensive recovery and refining processes. With oil prices at unprecedentedly high levels, Canada's production is likely to grow (Chapter 3, see p. 118).

In general terms, oil (and associated natural gas) exists in this realm in three discontinuous zones (Fig. 7-8). The most productive of these zones extends from the southern and southeastern part of the Arabian Peninsula northwestward around the rim of the Persian Gulf, reaching into Iran and continuing northward into Iraq, Syria, and southeastern Turkey, where it peters out. The second zone lies across North Africa and extends from north central Algeria eastward across northern Libya to Egypt's Sinai Peninsula, where it ends. The third zone begins on the margins of the realm in eastern Azerbaijan, continues eastward under the Caspian Sea into Turkmenistan and Kazakhstan, and also reaches into Uzbekistan, Tajikistan, Kyrgyzstan, and Afghanistan.

Producers and Consumers

Saudi Arabia is the world's largest oil exporter, but in recent years Russia has risen to second place. As we noted in Chapter 2, oil and natural gas are by far Russia's most valuable commodities, and the Russian state direly needs the revenues they produce. Thus Russia, with modest (though expanding) known reserves, is vigorously exporting to international markets and is now a leading factor in the global energy picture. In combination, however, production by the countries of the North Africa/Southwest Asia realm

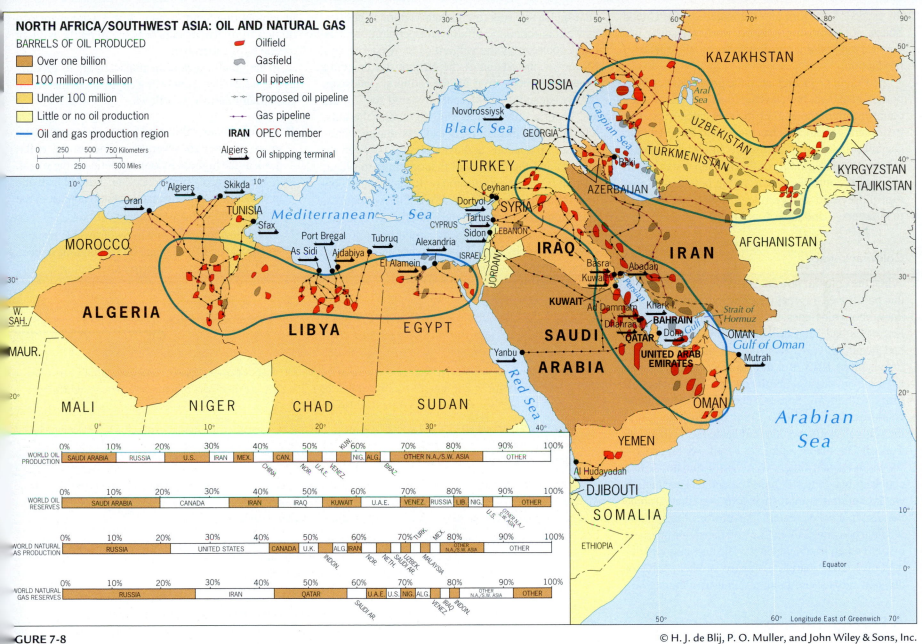

NORTH AFRICA/SOUTHWEST ASIA: OIL AND NATURAL GAS

BARRELS OF OIL PRODUCED

- Over one billion
- 100 million-one billion
- Under 100 million
- Little or no oil production
- Oil and gas production region

- Oilfield
- Gasfield
- Oil pipeline
- Proposed oil pipeline
- Gas pipeline
- **IRAN** OPEC member
- Algiers → Oil shipping terminal

FIGURE 7-8

© H. J. de Blij, P. O. Muller, and John Wiley & Sons, Inc.

far exceeds that from all other sources. The United States, not surprisingly, is the world's leading importer even while it consumes almost all of its own production.

As Figure G-11 indicates, the production and export of oil and gas has elevated several of this realm's countries into the higher-income categories. But petroleum wealth also has enmeshed these Islamic societies and their governments in global strategic affairs. When regional conflicts create instability in producing countries that have the potential to disrupt supply lines, powerful consumers are tempted to intervene—and have done so.

Colonial Legacy

When the colonial powers laid down the boundaries that partitioned this realm among themselves, no one knew about the riches that lay beneath the ground. A few wells had been drilled, and production in Iran had begun as early as 1908 and in Egypt's Sinai Peninsula in 1913. But the major discoveries came later, in some cases after the colonial powers had already withdrawn. Some of the newly independent countries, such as Libya, Iraq, and Kuwait, found themselves with wealth undreamed of when the Turkish Ottoman Empire collapsed. As Figure 7-8 shows, however, others were less fortunate. A few countries had (and still have) potential. The smaller, weaker emirates and sheikdoms on the Arabian Peninsula always feared that powerful neighbors would try to annex them (Kuwait faced this prospect in 1990 when Iraq invaded it). The unevenly distributed oil wealth, therefore, created another source of division and distrust among Islamic neighbors.

A Foreign Invasion

The oil-rich countries of the realm found themselves with a coveted energy source but not the skills, capital, or equipment to exploit it. These came from the Western world and entailed what many tradition-bound Muslims feared most: a strong foreign pres-

ence on Islamic soil, foreign intervention in political as well as economic affairs, and penetration of Islamic societies by the vulgarities of Western ways. Moreover, the advent of the oil industry created a veneer of modernity in the cultural landscape (where gleaming skyscrapers towered over ornate and historic minarets) as well as among the elites who benefited most directly from the energy riches. This set up tensions and even clashes between the modern and the traditional that continue to plague oil-rich Islamic countries today. It also created contrasts between countries that took different paths as their cultures changed. In the United Arab Emirates (UAE), it is no surprise to see a woman emerge from a supermarket by herself, get into her car, and drive home with the groceries. That would be unthinkable in neighboring Saudi Arabia, where women are prohibited from driving cars or being in public without a male family member.

In geographic terms, the impact of oil, its production and sale, in the exporting countries of this realm can be summarized as follows:

1. *High Incomes.* When oil prices on international markets are high, several countries in this realm rank as the highest-income societies in the world. Even when oil prices decline, virtually all the petroleum-exporting states remain at least in the upper-middle-income category (Fig. G-11).

2. *Modernization.* Notably in Saudi Arabia but also in Kuwait and UAE, huge oil revenues have been used to modernize infrastructure. Futuristic city

WHAT'S DRIVING GEOGRAPHIC CHANGE IN THE REALM?

- The war in **Iraq** will forever change the map of the Middle East, whatever its outcome. But what Iraq needs badly is a census: many minority Sunnis still believe that they are members of a majority whose rightful rule of the country was ended by the American-led invasion.

- The safety of **Israel** is being enhanced by a much-debated security barrier, in places a wall, elsewhere a fence, which controls the cross-border movement of Palestinians. The location and appearance of this barrier, still under construction, are contentious issues between Israelis and Palestinians (see photo, p. 255).

- Israel in 2005 made a dramatic withdrawal of all Jewish settlers from the **Gaza Strip**, but Israel continued to exercise influence in the area. In late 2007 a major peace summit was held in Annapolis, Maryland. Will these events mark the beginning of a two-state solution for this turbulent region?

- **Iran** in 2005 elected a president who declared that the Holocaust was a European myth, who stated that Israel should be 'wiped off the map,' and whose representatives defied international efforts to monitor the country's nuclear buildup. Iran is flexing its muscles and will be of concern for years.

- A deadly conflict in western Sudan's **Darfur** Province, unrelated to the north-south war that has ended, continues to cost thousands of lives as the Khartoum regime fails to protect its African citizens against Arabized militias.

- Elections in **Afghanistan** in 2005 ushered in a new era of representative government, but concern is rising over the revival of Taliban activity in many parts of the country and over the government's failure to combat the widespread cultivation of the opium poppy, of which the country produces more than 90 percent of the world's harvest.

skyscrapers are only one manifestation of this; others include modern ports, airports, superhighways, water desalinization plants, even skating rinks and ski-slopes.

3. *Industrialization.* Farsighted governments among those with oil wealth, realizing that petroleum reserves will not last forever, have invested some of their income in industrial plants that will outlast the era of oil. Petrochemical industries, plastics fabrication, and desalinization plants are among these facilities.

4. *Intra-Realm Migration.* The oil wealth has attracted millions of workers from less well-off parts of the realm to work in the oilfields, in the ports, construction, and in many other capacities. This has brought many Shi'ites to the countries of eastern Arabia; hundreds of thousands of Palestinians also work there as laborers. Saudi Arabia, with a population of 25.4 million, has more than 5 million foreign workers.

5. *Inter-Realm Migration.* There are also workers from countries outside the realm, both Muslims and non-Muslims, from countries such as Pakistan, India, and Sri Lanka, who come to work for low wages, but wages that are better than they can earn at home. These workers are often exploited as they toil as domestics, gardeners, refuse collectors, and the like.

6. *Regional Disparities.* Oil wealth and its manifestations in the cultural landscape create strong contrasts with areas not directly affected. The ultramodern east coast of Saudi Arabia is a world apart from large areas of its interior, where it becomes a land of desert, oasis, and camel, of vast distances, slow change, and isolated settlements. This phenomenon affects all oil-rich countries.

7. *Foreign Investment.* To some degree, governments and Arab businesspeople have invested oil-generated wealth in foreign countries. These investments have created a network of international involvement that links many of this realm's countries not only to the economies of foreign states, but also to growing Arab (and thus Islamic) communities in those states.

Oil brought this realm into contact with the outside world in ways unforeseen just a century ago. Oil has strengthened and empowered some of its peoples; it has dislocated and imperiled many others. It has truly been a double-edged sword.

Regions of the Realm

Identifying and delimiting regions in this vast geographic realm is quite a challenge. Population clusters are widely scattered in some areas, highly concentrated in others. Cultural transitions and landscapes—internal as well as peripheral—make it difficult to discern a regional framework. This, furthermore, is a highly changeable realm and always has been. Several centuries ago it extended into Eastern Europe; now it reaches into Central Asia, where an Islamic **11** **cultural revival** (the regeneration of a long-dormant culture through internal renewal and external infusion) is well under way.

The following are the regional components of this far-flung realm today (Fig. 7-9):

1. *Egypt and the Lower Nile Basin.* This region in many ways constitutes the heart of the realm as a whole. Egypt (together with Iran and Turkey) is one of the realm's three most populous countries. It is the historic focus of this part of the world and a major political and cultural force. It shares with its southern neighbor, Sudan, the waters of the lower Nile River.

2. *The Maghreb and its Neighbors.* Western North Africa (the Maghreb) and the areas that border it also form a region, consisting of Algeria, Tunisia, and Morocco at the center, and Libya, Chad, Niger, Mali, Burkina Faso, and Mauritania along the broad periphery. The last five of these countries lie astride or adjacent to the broad African Transition Zone where the Arab-Islamic realm of North Africa merges into Subsaharan Africa (see pp. 225–227).

3. *The Middle East.* This region includes Israel, Jordan, Lebanon, Syria, and Iraq. In effect, it is the crescent-like zone of countries that extends from the eastern Mediterranean coast to the head of the Persian Gulf.

4. *The Arabian Peninsula.* Dominated by the large territory of Saudi Arabia, the Arabian Peninsula also includes the United Arab Emirates (UAE), Kuwait, Bahrain, Qatar, Oman, and Yemen. Here lies the source and focus of Islam, the holy city of Mecca; here, too, lie many of the world's greatest oil deposits.

5. *The Empire States.* We refer to this region as the Empire States because two of the realm's giants, both the centers of historic empires, dominate its geography. Turkey, heart of the Ottoman Empire, now is the realm's most secular state and aspires to joining the European Union. Shi'ite Iran, the core of the former Persian Empire, has become an Islamic republic. To the north, Azerbaijan, once a part of the Persian Empire, lies in the turbulent, Muslim-infused Transcaucasian Transition Zone. To the south, the island of Cyprus is divided between Turkish and Greek spheres, the Turkish sector being another remnant of imperial times.

6. *Turkestan.* Turkish influence ranged far and wide in southwestern Asia, and following the Soviet collapse that influence proved to be durable and

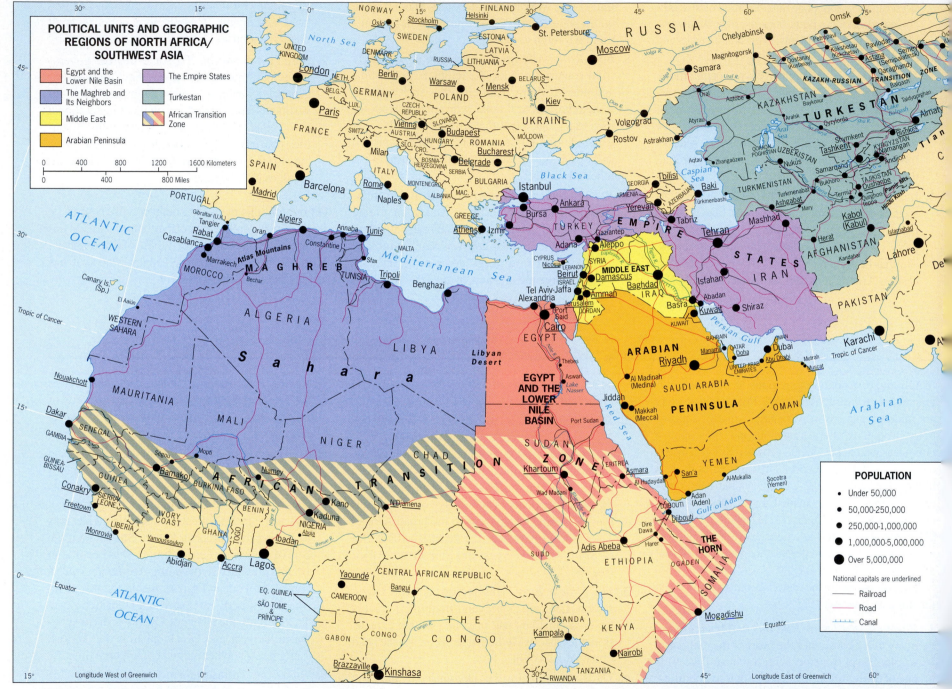

POLITICAL UNITS AND GEOGRAPHIC REGIONS OF NORTH AFRICA/SOUTHWEST ASIA

- Egypt and the Lower Nile Basin
- The Maghreb and Its Neighbors
- Middle East
- Arabian Peninsula
- The Empire States
- Turkestan
- African Transition Zone

0   400   800   1200   1600 Kilometers

0   400   800 Miles

POPULATION

- ● Under 50,000
- ● 50,000–250,000
- ● 250,000–1,000,000
- ● 1,000,000–5,000,000
- ● Over 5,000,000

National capitals are underlined

—— Railroad
—— Road
++++ Canal

FIGURE 7-9

© H. J. de Blij, P. O. Muller, and John Wiley & Sons, Inc.

strong. In the five former Soviet Central Asian republics the strength and potency of Islam vary, and the new governments deal warily (sometimes forcefully) with Islamic revivalists. Russian influence continues, democratic institutions remain weak, and Western (chiefly American) involvement has increased since the War on Terror began and Afghanistan—which also forms part of this region—became a major target of that campaign.

EGYPT AND THE LOWER NILE BASIN

Egypt occupies a pivotal location in the heart of a realm that extends over 9600 kilometers (6000 mi) longitudinally and some 6400 kilometers (4000 mi) latitudinally. At the northern end of the Nile and of the Red Sea, at the eastern end of the Mediterranean Sea, in the northeastern corner of Africa across from Turkey to the north and Saudi Arabia to the east, adjacent to Israel, to Islamic Sudan, and to Libya, Egypt lies in the crucible of this realm. Because it owns the Sinai Peninsula, Egypt, alone among states on the African continent, has a foothold in Asia, a foothold that gives it a coast overlooking the strategic Gulf of Aqaba (the northeasternmost arm of the Red Sea). Egypt also controls the Suez Canal, vital link between the Indian and Atlantic oceans and lifeline of Europe. The capital, Cairo (Al Qahira), is the realm's largest city and a leading center of Islamic civilization. We hardly need to further justify Egypt's designation (together with northern Sudan) as a discrete region.

Gift of the Nile

Egypt's Nile is the aggregate of two great branches upstream: the White Nile, which originates in the streams that feed Lake Victoria in East Africa, and the Blue Nile, whose source lies in Lake Tana in Ethiopia's highlands. The two Niles converge at Khartoum in modern-day Sudan. About 95 percent of Egypt's 78.6 million people live within 20 kilometers (12 mi) of the great river's banks or in its delta (Fig. 7-10).

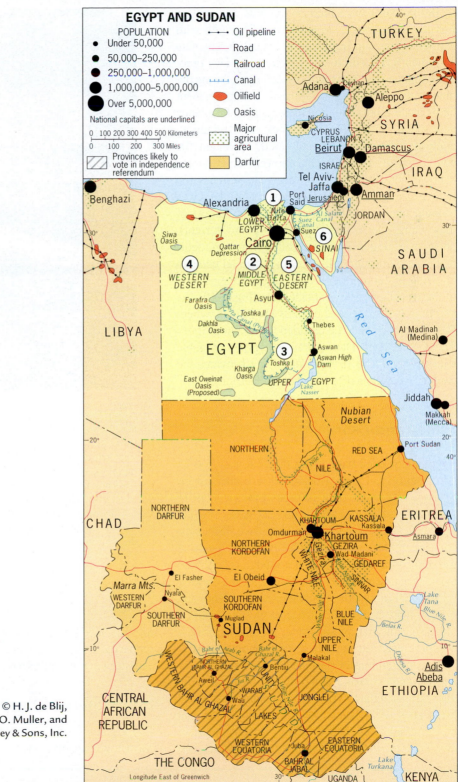

FIGURE 7-10 © H. J. de Blij, P. O. Muller, and John Wiley & Sons, Inc.

Before dams were built on the Nile, the ancient Egyptians used *basin irrigation*, building fields with earthen ridges and trapping the annual floodwaters with their fertile silt, to grow their crops. That practice continued for thousands of years until, during the nineteenth century, the construction of permanent dams made it possible to irrigate Egypt's farmlands year round. These dams, with locks for navigation, controlled the Nile's annual flood, expanded the country's cultivable area, and allowed the farmers to harvest more than one crop per year on the same field. In a single century, all of Egypt's farmland was brought under *perennial irrigation*. The largest of these dams, the Aswan High Dam (photo at right), began operating in 1968, controlling water and sediments, and bringing reliability to what had been an uncertain flood regime. It created Lake Nasser, one of the world's largest artificial lakes (Fig. 7-10). As the map shows, the lake extends into Sudan, where 50,000 people had to be resettled to make way for it. The Aswan High Dam increased Egypt's irrigable land by nearly 50 percent and today provides the country with about 40 percent of its electricity.

Valley and Delta

Egypt's elongated oasis along the Nile, just 5 to 25 kilometers (3 to 15 mi) wide, broadens north of Cairo across a delta anchored in the west by the great city of Alexandria and in the east by Port Said, gateway to the Suez Canal. The delta contains extensive farmlands, but it is a troubled area today. The ever more intensive use of the Nile's water and silt upstream is depriving the delta of much needed replenishment. And the low-lying delta is at risk due to geological subsidence and sea-level rise as a consequence of global warming, increasing the risk of salt-water intrusion from the Mediterranean Sea that can severely damage soils.

Egypt's millions of subsistence farmers, the *fella-heen*, struggle to make their living off the land, as did the peasants of the Egypt of five millennia past. Rural landscapes seem barely to have changed; ancient tools are still used, and dwellings remain rudimentary. Poverty, disease, high infant mortality rates, and

ENVIRONMENTALLY SPEAKING . . .

Aswan High Dam in Egypt.   Dams are often constructed to help propel economic development, and the Aswan High Dam is a classic example of this, bringing electricity and perennial irrigation to millions. However, megaprojects of this kind often bring environmental hardships with them, and this dam is no exception. A variety of serious problems, some anticipated and some not, are the result of the existence of the now 40-year-old dam. Built in part to control the Nile's annual flood, the dam unfortunately also blocked the critical river sediment that naturally (and without cost) fertilized the Nile's floodplain. Thus farmers today have to use artificial fertilizers to continue to be able to produce sufficiently. These are expensive and damaging to the natural environment. The costs have marginalized many smallholders, and the fertilizers and pesticides have polluted many ecosystems. The Nile no longer supports rich fish resources offshore, reducing the catch and depriving coastal populations of a badly needed resource. Recently the fish have started to return, but are not yet at the volume they were before 1968. Another environmental consequence of the Aswan High Dam was that snail-carried schistosomiasis and mosquito-transmitted malaria thrived in the standing water of the dam's reservoir, affecting the health of hundreds of thousands of people living nearby. (© Joe Arnold Images Ltd/Alamy)

low incomes prevail. The Egyptian government is embarked on grandiose plans to expand its irrigated acreage, but little of this has come to pass.

Subregions of Egypt

Egypt has six subregions, mapped in Figure 7-10. Most Egyptians live and work in Lower (i.e., north-ern) and Middle Egypt, regions ① and ②, the country's core area anchored by Cairo and flanked by the leading port and second industrial center, Alexandria. The economy has benefited from further oil discoveries in the Sinai, region ⑥, and in the Western Desert, region ④, so that Egypt now is self-sufficient and even exports some petroleum. Cotton and textiles are the other major source of external income, but the important tourist industry has repeatedly been hurt by Islamic extremists. As the population continues to grow, the gap between food supply and demand widens, and Egypt must import grain. Since the late 1970s, Egypt has been a major recipient of U.S. foreign aid.

Egypt today is at a crossroads in more ways than one. Its planners know that reducing the rather high birth rate would improve the demographic situation, but revivalist Muslims object to any programs that limit family size. Its accommodation with Israel helps ensure foreign aid but divides the people. Its government faces a fundamentalist challenge as well as a rising demand for more democracy. Egypt's future, in this crucial corner of the realm, is clouded.

Divided Sudan

As Figure 7-10 shows, Egypt is flanked by two countries that have posed challenges to its leadership: Sudan to the south and Libya to the west. Sudan, more than twice as large as Egypt and with 43.5 million people, lies centered on the confluence of the White Nile (from Uganda) and the Blue Nile (from Ethiopia). Here the twin capital, Khartoum-Omdurman, anchors a large agricultural area where cotton was planted during colonial times. The British administration combined northern Sudan, which was Arabized and Islamized, with a large area to the south, which was African and where many villagers had been Christianized. After the British left, the regime in Khartoum sought to impose its Islamic rule on this portion of the African Transition Zone that

formed southern Sudan, and a bitter civil war ensued. The cost in human lives and dislocation over the better part of the past three decades is incalculable.

Sudan has a 500-kilometer (300-mi) coastline on the Red Sea, where Port Sudan lies almost directly across from Jiddah and Makkah (Mecca) in Saudi Arabia. The country's economy for many years was typical of the energy-poor periphery, exchanging sheep, cotton, and sugar for oil, with Saudi Arabia the main trading partner. The civil war in the African Transition Zone impoverished the Islamic regime in Khartoum, and per-capita income in Sudan was one of the world's lowest. All that changed in the 1990s.

Oil in the Desert

During the 1980s it became clear that significant oil reserves lie in Sudan, and during the 1990s discoveries were made in Kordofan Province and elsewhere, including major deposits in the embattled south (Fig. 7-10). This complicated the political as well as the economic situation. In short order Sudan became an exporter instead of an importer of oil. Foreign companies, especially from China, which was willing to accept human rights abuses

in the Khartoum regime, arrived to explore and exploit. The government saw its income rise, allowing it to purchase more weapons. Local peoples living on top of or near the newly found deposits were removed. Ultramodern buildings in Khartoum reflected the newfound wealth for the few.

Leaders in the South saw oil as a ticket to self-sufficiency. The 2006 peace agreement that had set a timetable for possible independence had stipulated that the South would receive half the revenues, but the North kept the records and was not trustworthy. Southerners were divided over the independence issue—would it be better to be part of prosperous Sudan or to go it alone and become a landlocked state? The issue is not resolved. Southerners talk of a pipeline to the Indian Ocean through Kenya to avoid dependence on Sudan, but nothing has come of it yet. Sudan's oil bonanza will have political-geographical consequences either way.

Tragedy in Darfur

The current crisis in the western province of Darfur (Fig. 7-10) did not begin as a Muslim-Christian conflict even though it reflects the tension of Africa's

Another village has been torched by *janjaweed* militias empowered by Khartoum's Islamic regime, and survivors are left to flee. Relying on an understaffed, poorly equipped, and inadequately mandated African Union force to stop the violence, the international community has been slow in its response to the crisis in Darfur, arguing over the question of whether the assault by Arabized Muslim militias on sedentary African farmers in this western province of Sudan technically constitutes genocide. In mid-2008, a significant step was made by the international community when the International Criminal Court in The Hague in the Netherlands accused Sudan's President Omar al-Bashir of genocide, war crimes, and crimes against humanity. (© AP/Wide World Photos)

Islamic Front. In Darfur (which means Land of the Fur), the overwhelming majority of all the people, the Fur, are Muslims. However, northerners have Arab lifeways with pastoralism and mobility prevailing, while southerners tend to be settled farmers living in villages and creating a Subsaharan African cultural landscape. It is not clear what precipitated the conflict that engulfed Darfur; some observers reported that members of the Khartoum regime suspected the southerners of sympathizing with the rebels fighting government forces in the Nile region. In any case, in 2003 the Northern pastoralists rode into the villages and fields of the southern Fur, killing thousands, burning houses, and destroying crops. Despite worldwide condemnation and the presence of African Union peacekeeping forces, brutal attacks have continued, with *janjaweed* militia (Afro–Arab nomadic gunmen, usually on horseback) ethnically cleansing areas of people it does not want (see photo on p. 247). The Khartoum government not only failed to intervene but supports the Arabized northerners. It is estimated that well over 300,000 people have been killed and over 2.5 million driven from their homes, causing a refugee flow of immense proportions, mostly into neighboring Chad.

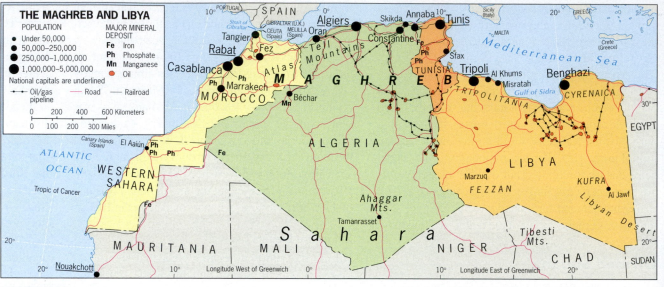

FIGURE 7-11

© H. J. de Blij, P. O. Muller, and John Wiley & Sons, Inc.

THE MAGHREB AND ITS NEIGHBORS

The countries of northwestern Africa are collectively called the *Maghreb*, but the Arab name for them is more elaborate than that: **Jezira-al-Maghreb**, or "Isle of the West," in recognition of the great Atlas Mountain range rising like a huge island from the Mediterranean Sea to the north and the sandy flatlands of the immense Sahara to the south.

The countries of the Maghreb (sometimes spelled *Maghrib*) are Morocco, last of the North African kingdoms; Algeria, a secular republic beset by the religious-political problems we noted earlier; and Tunisia, smallest and most Westernized of the three (Fig. 7-11). Neighboring Libya, facing the Mediterranean between the Maghreb and Egypt, is unlike any other North African country: an oil-rich desert state whose population is almost entirely clustered in settlements along the coast.

Atlas Mountains

Whereas Egypt is the gift of the Nile, the Atlas Mountains facilitate the settled Maghreb. These high ranges wrest from the rising air enough orographic rainfall to sustain life in the intervening valleys, where good soils support productive farming. From the vicinity of Algiers eastward along the coast into Tunisia, annual rainfall averages more than 75 centimeters (30 in), a total more than three times as high as that recorded for Alexandria in Egypt's delta.

Even 240 kilometers (150 mi) inland, the slopes of the Atlas still receive over 25 centimeters (10 in) of rainfall. The effect of the topography can be read on the world map of precipitation (Fig. G-8): where the highlands of the Atlas terminate, desert conditions immediately begin.

The Atlas Mountains trend southwest-northeast and begin in Morocco as the High Atlas, with elevations close to 4000 meters (13,000 ft). Eastward, two major ranges dominate the landscapes of Algeria proper: the Tell Atlas to the north, facing the Mediterranean, and the Saharan Atlas to the south, overlooking the great desert. Between these two mountain chains, each consisting of several parallel ranges and foothills, lies a series of intermontane basins (analogous to South America's Andean *altiplanos* but at lower elevations), markedly drier than the northward-facing slopes of the Tell Atlas. In these valleys, the rain shadow effect of the Tell Atlas is reflected not only in the steppe-like natural vegetation but also in land-use patterns: pastoralism replaces cultivation, and stands of short grass and bushes blanket the countryside.

Colonial Impact

During the colonial era, which began in Algeria in 1830 and lasted until the early 1960s, well over a million Europeans came to settle in North Africa—most of them French, and a large majority bound for Algeria—and these immigrants soon dominated commercial life. They stimulated the renewed growth of the region's towns; Casablanca, Algiers, Oran, and Tunis became the urban foci of the colonized territories. Although the Europeans dominated trade and commerce and integrated the North African countries with France and the European Mediterranean world, they did not confine themselves to the cities and towns. They recognized the agricultural possibilities of the favored parts of the Tell (the lower Tell Atlas slopes and narrow coastal plains that face the Mediterranean) and established thriving farms. Not surprisingly, agriculture here is Mediterranean. Algeria soon became known for its vineyards and wines, citrus groves, and dates; Tunisia has long been one of the world's leading exporters of olive oil; and Moroccan citrus fruits went to many European and North American markets.

The Maghreb Countries

Between desert and sea, the Maghreb states display considerable geographic diversity. **Morocco**, a conservative kingdom in a revolutionary region, is tradition-bound and weak economically. Its core area lies in the north, anchored by four major cities; but the Moroccans' attention is focused on the south, where the government is seeking to absorb its neighbor, Western Sahara (a former Spanish dependency with about 300,000 inhabitants, many of them immigrants from Morocco). Even if this campaign is successful, it will do little to improve the lives of most of Morocco's 32.7 million people. Hundreds of thousands have migrated to Europe, many of them via Spain's two tiny exclaves on the Moroccan coast, Ceuta and Melilla, or via rickety boats to the Canary Islands (Fig. 7-11).

Algeria, rich in oil and natural gas, fought a bitter war of liberation (1954–1962) against the French colonizers and has been in turmoil virtually ever since, with devastating economic and social consequences. Algiers, Algeria's primate city, is centrally situated along its Mediterranean coast and is home to 3.5 million of the country's 34.6 million citizens—but there are more Algerians in France than in the capital today. Conflict between those wanting to make Algeria an Islamic republic and the military-backed regime cost an estimated 100,000 lives; in 2005 a referendum on the matter was approved. In recent years, periodic violent flare-ups have occurred, likely instigated by al-Qaeda in the Islamic Maghreb, a terrorist group with roots in Algeria's civil war.

The smallest of the Maghreb states, **Tunisia**, lies at the eastern end of the region. As Appendix B shows, Tunisia in many ways outranks surrounding countries. Stability, national cohesion, and the maximization of limited natural resources (no oil reserves here comparable to Algeria or Libya) have yielded a higher GNI, a higher urbanization level, higher social indicators, and a lower growth rate (among its population of 10.3 million) than elsewhere in the Maghreb. Most of the country's productive capacity lies in the north, in the hinterland of its large, historic capital, Tunis. As the authoritarian national government slowly loosens its grip, Tunisia's relations with Europe—already the strongest in all of North Africa—continue to improve.

Libya has been known to sponsor terrorist groups, but after the settlement of the Lockerbie bombing claims, it is no longer a pariah toward the West. Almost rectangular in shape, Libya (population 6.2 million), lies on the Mediterranean Sea between the Maghreb states and Egypt (Fig. 7-11). What limited agricultural possibilities exist lie in the district known as Tripolitania in the northwest, centered on the capital, Tripoli, and in the northeast in Cyrenaica, where Benghazi is the urban focus. But it is oil, not farming, that drives the economy of this highly urbanized country. The oil-fields lie well inland from the Gulf of Sidra, linked by pipelines to coastal terminals. Libya's two interior corners, the desert Fezzan district in the southwest and the Kufra oasis in the southeast, are connected to the coast by two-lane roads subject to sandstorms.

Adjoining Saharan Africa

As Figure 7-11 shows, the Maghreb countries and Libya adjoin several desert-dominated states in the African Transition Zone. These are African countries with strong Islamic imprints, but as Figure 6-19 shows, they lie north of the Islamic Front with only one exception, Chad. Vast and sparsely populated **Mauritania** is the most Islamized of these countries, with about half its population of 3.4 million concentrated in and near the capital, Nouakchott, base of a small fishing fleet. Neighboring **Mali** depends on the waters of the upper Niger River, on which its capital of Bamako lies. A multiparty, democratic state, Mali is also multicultural, with a sizeable minority adhering to traditional beliefs and a small Christian minority. To the east of Mali lies **Niger**, which shares boundaries with Algeria and Libya; like Mali, this is one of the world's least urbanized countries (21 percent). Unlike Mali, however, Niger has only a short stretch of the middle course of the Niger River, where its capital, Niamey, is situated. And **Chad**, directly south of Libya, has the lowest percentage of Muslim inhabitants but the strongest division between Islamized north and Christian/animist south. As Figure 6-19 reminds us, the capital, N'Djamena, lies on the southern edge of the African Transition Zone, directly on the Islamic Front. Beginning in 2004, even as Chad was experiencing an oil boom in the south, the cultural conflict in Sudan's neighboring Darfur Province sent more than 250,000 refugees across its eastern border. By 2008 civil unrest erupted in Chad, with insurgents tied to the conflict in Darfur attempting to topple the Chadian government.

THE MIDDLE EAST: CRUCIBLE OF CONFLICT

The regional term *Middle East*, we noted earlier, is not satisfactory, but it is so common and generally used that avoiding it creates more problems than it solves. It originated when Europe was the world's dominant realm and when places were *near*, *middle*, and *far* from Europe: hence a Near East (Turkey), a Far East (China, Japan, Korea, and other countries of East Asia), and a Middle East (Egypt, Arabia, Iraq). If you check definitions used in the past, you will see that the terms were applied inconsistently: Syria, Lebanon, Palestine, and even Jordan sometimes were included in the Near East, and Persia and Afghanistan in the Middle East.

Today, the geographic designation *Middle East* has a more specific meaning. And at least half of it has merit: this region, more than any other, lies at the middle of the vast Islamic realm (Fig. 7-9). To the north and east of it, respectively, lie Turkey and Iran, with Muslim Turkestan beyond the latter. To the south lies the Arabian Peninsula. And to the west lie the Mediterranean Sea and Egypt, and the rest of North Africa. This, then, is the pivotal region of the realm, its very heart.

Five countries form the Middle East (Fig. 7-12): Iraq, largest in population and territorial size, facing the Persian Gulf; Syria, next in both categories and fronting the Mediterranean; Jordan, linked by the narrow Gulf of Aqaba to the Red Sea; Lebanon, whose survival as a unified state has come into question; and Israel, Jewish nation in the crucible of the Muslim world. Because of the extraordinary importance of this region in world affairs, we focus in some detail on issues of cultural, economic, and political geography in the discussion that follows.

War in Iraq

In the immediate aftermath of the 9/11 terrorist attacks in the United States in 2001, American forces overthrew the ruling revivalist Taliban regime in Afghanistan that had given refuge to Usama bin Laden, who had plotted the assault. Soon, however, Afghanistan lost priority among U.S. concerns as **Iraq** was designated a security risk based on intelligence reports of its terrorist links, its alleged arsenal of weapons of mass destruction (WMD), its murderous regime, and the regional threat it posed. In March 2003 the United States attacked and invaded Iraq and following a brief struggle ended the rule of the Sunni-Muslim clique (the Baath Party) led by Saddam Hussein. While the search for WMD continued, the American government now faced the task of stabilizing Iraq, reconstructing its damaged infrastructure, and reconstituting its government, a task that had been underestimated.

It is therefore important to understand key aspects of the regional geography of California-sized Iraq and its official population of approximately 31.2 million (exact numbers are difficult to determine as there is much unregistered emigration). The country constitutes nearly 60 percent of the total area of the Middle East as we define it and has more than 40 percent of its population. With its world-class oilfields, major gas reserves, and large areas of irrigable farmland, Iraq also is best endowed of all the region's countries with natural resources. Iraq is heir to the early Mesopotamian states and empires that emerged in the Tigris-Euphrates Basin, and the country is studded with matchless archaeological sites and museum collections, to which disastrous damage was done during and after the 2003 invasion from combat and looting.

The War's Wider Regional Impact

Please take a careful look at Figure 7-12, showing the location of Iraq in the region we have defined as the Middle East, which reveals that Iraq has six immediate neighbors. Perhaps the most obvious feature of this map is the lavender-striped area that not only covers northeastern Iraq but also large parts of Turkey and Iran as well as smaller parts of adjacent countries. This is the area where the majority of the inhabitants are Kurds, a people who have been discriminated against by all the countries in which they live. About 14 million Kurds are in Turkey, some 8 million in Iran, and about 4 million in Iraq, and together they form one of the world's largest **12** **stateless nations**, divided and often exploited by those who rule over them.

The Kurdish population not only in Iraq but also in several of Iraq's neighbors is only one reason it is important to study Figure 7-12 closely. Turkey is just one of the countries bordering war-torn Iraq that are directly concerned over the outcome of the struggle that started in 2003. In 2007 this concern escalated as Turkey launched attacks on Kurdish insurgents in Iraq. Also anxious is Iran to the east, which fought a bitter war with Iraq in the 1980s but which shares Shia Islam with a majority of Iraqis. To the west, Syria was Iraq's ally during Saddam Hussein's despotic rule, and like Iraq under Saddam, Syria itself still is ruled by a strongman and his minority clique. To the south, Iraq is bordered by Kuwait, the country it invaded in 1990, triggering the first Gulf War, and by Saudi Arabia, separated from populated Iraq by a wide swath of desert but troubled by the impact any resolution of Iraq's problems may have at home. And to the southwest lies Jordan, which gave Iraq some help during the first Gulf War but later turned its back not only on Saddam but also on the Sunni insurgents who waged war on American and allied forces following Saddam's ouster.

With so many land neighbors, it is no surprise that Iraq is nearly landlocked. Figure 7-12 shows how short and congested its single outlet to the Persian Gulf is; it is for this reason that Saddam tried to conquer and annex Kuwait in 1990. As a result, a network of pipelines across Iraq's neighbors must carry much of Iraq's oil export volume to coastal terminals in other countries—another reason these countries have an interest in what happens to Iraq in the years ahead.

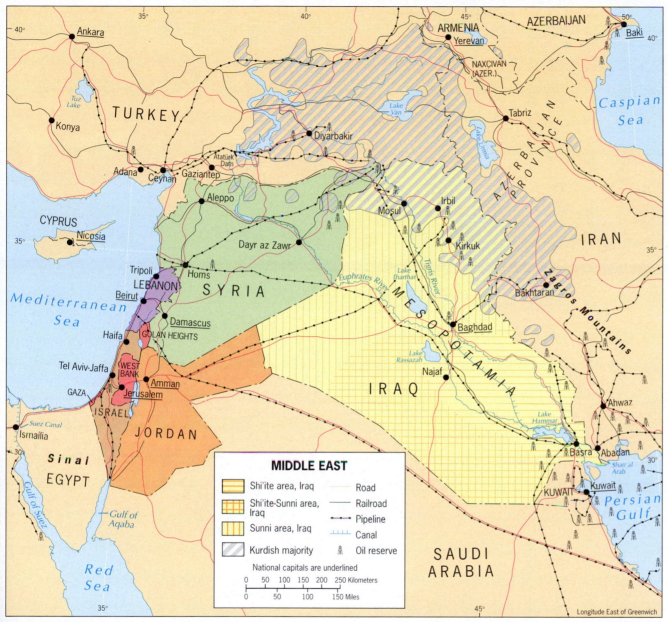

FIGURE 7-12

MIDDLE EAST

- Shi'ite area, Iraq
- Shi'ite-Sunni area, Iraq
- Sunni area, Iraq
- Kurdish majority
- Road
- Railroad
- Pipeline
- Canal
- Oil reserve

National capitals are underlined

0 50 100 150 200 250 Kilometers
0 50 100 150 Miles

Longitude East of Greenwich

© H. J. de Blij, P. O. Muller, and John Wiley & Sons, Inc.

at the heart of the country's core area. The Baghdad urbanized area has a population today of more than 6 million, a vast and often violent multicultural agglomeration of almost one quarter of the entire country's population. Only the Shia south, which extends from the eastern suburbs of Baghdad all the way south to Basra and the shores of the Persian Gulf, has more people than Baghdad. The Sunni northwest, including the Sunni Triangle where Saddam's clan had its base plus the turbulent province of Anbar (Fig. 7-13 inset map), has between 5 and 6 million, and the Kurdish sector about 4 million. Of these three cultural domains, the Kurdish sector, the most remote of the three from the worst of Iraq's violent postwar transition, shows what might have been in the rest of the country: comparative stability and security have produced progress and prosperity.

Figure 7-13 also shows the location of Iraq's major oilfields, most of which lie in the Shia domain and in the Kurdish sector and its borderland, creating an obstacle to any plan to create a three-part federation out of Iraq's political geography. The Sunni sector would be the loser; some acceptable way to allocate oil income to the Sunni provinces would have to be designed.

Recall that Iraq was a colonial creation and not a nation-state. Before the U.S.-led invasion of 2003, Iraq was held together by the brutal regime headed by Saddam Hussein, a Sunni

Cultural and Political Geography in Discord

Let us now carefully examine the regional map of Iraq (Fig. 7-13). Note that Iraq's two vital rivers, the Tigris and Euphrates, rise in Turkey. The map shows

that Iraq is divided into three cultural domains: that of the Shia majority in the south, the Sunni minority in the northwest, and the Kurdish minority in the hilly northeast. Not far from the geographic center of Iraq lies the capital, Baghdad, astride the Tigris River

(the minority Islamic sector in Iraq), which ruled through terror and mass murder (chemical weapons killed thousands of Kurds; mass graves contained tens of thousands of Shi'ites). The U.S. intervention and subsequent termination of that regime unleashed

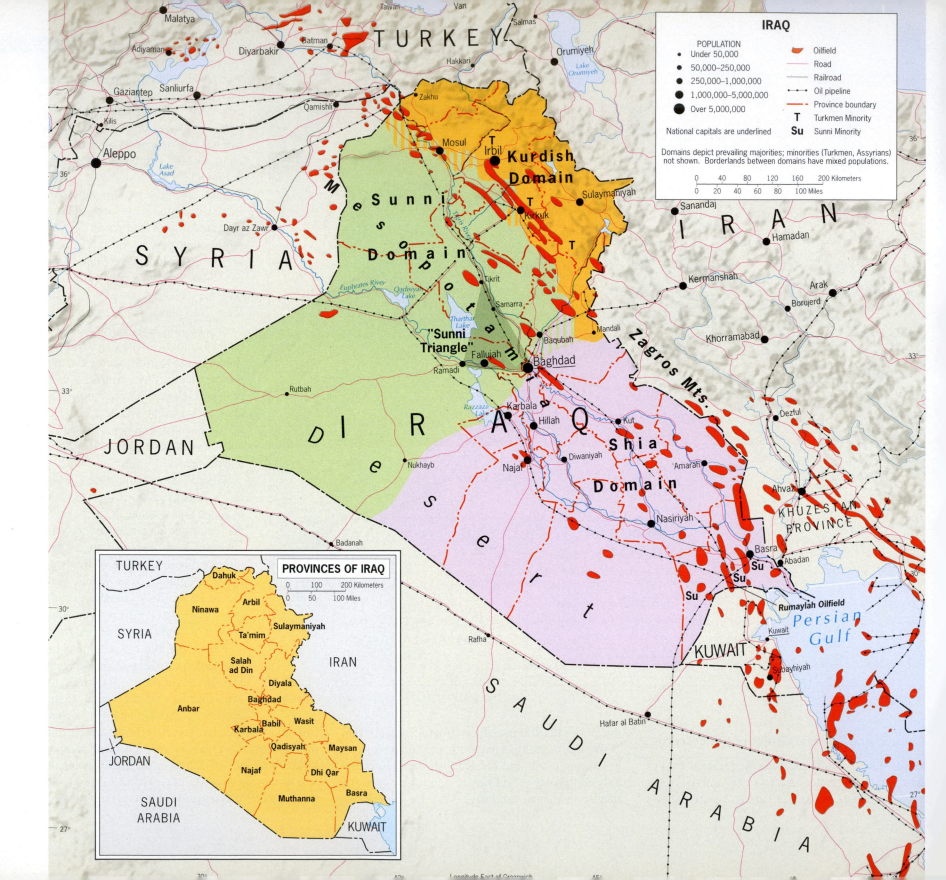

IRAQ

POPULATION
- • Under 50,000
- • 50,000–250,000
- ● 250,000–1,000,000
- ● 1,000,000–5,000,000
- ● Over 5,000,000

National capitals are underlined

Oilfield
Road
Railroad
Oil pipeline
Province boundary
T Turkmen Minority
Su Sunni Minority

Domains depict prevailing majorities; minorities (Turkmen, Assyrians) not shown. Borderlands between domains have mixed populations.

0   40   80   120   160   200 Kilometers
0   20   40   60   80   100 Miles

PROVINCES OF IRAQ

0   100   200 Kilometers
0   50   100 Miles

TURKEY
SYRIA
Dahuk
Arbil
Ninawa
Ta'mim
Sulaymaniyah
Salah ad Din
IRAN
Diyala
Baghdad
Anbar
Babil   Wasit
Karbala
Qadisyah   Maysan
Najaf   Dhi Qar
JORDAN
Muthanna   Basra
SAUDI ARABIA
KUWAIT

Malatya
Adiyaman
Diyarbakir   Batman
Gaziantep   Sanliurfa
Kilis   Qamishli
Aleppo   Lake Asad
Dayr az Zawr
SYRIA

TURKEY   Van   Taivan
Hakkari   Salmas   Orumiyeh
Zakhu   Lake Orumiyeh
Mosul   **T**   Irbil   **Kurdish Domain**
T   Kirkuk   **T**   Sulaymaniyah
Me**Sunni**   **T**   **T**
s   **Domain**   IRAN
o   Tikrit   Sanandaj
p   Samarra   Hamadan
o   Tharthar Lake   Kermanshah
t   Baqubah   Mandali   Arak
a   "Sunni Triangle"   Zagros Mts.   Borujerd
m   Fallujah   Baghdad   Khorramabad
Ramadi
Rutbah   Razzaza Lake   Karbala   Dezful
JORDAN   Hillah   Kut
D   **I**   **R**   **A**   **Q**   **Shia**
e   Nukhayb   Najaf   Diwaniyah   **Domain**   Amarah   Ahvaz
s   Nasiriyah   **KHUZESTAN PROVINCE**
e   Basra   **Su**
Badanah   **r**   **Su**   Abadan
t   **Su**
Rumaylah Oilfield
Rafha   Kuwait   **Persian Gulf**
KUWAIT
Hafar al Batin   Subayhiyah
S A U D I   A R A B I A

Longitude East of Greenwich

centrifugal forces ranging from sectarian strife and ethnic cleansing to insurgency and terrorism, even as Iraq's population repeatedly voted in elections that produced a representative government. Despite a troop surge to bring greater security to the country and a return of many middle-class Iraqis who had left to escape the unbridled violence, the elected government continues to be dysfunctional, and it remains unclear whether Iraq can achieve a transition to stability and democracy.

The Region's Other Countries

Syria, too, is ruled by a minority. Although Syria's population of 20.5 million is about 75 percent Sunni Muslim, the ruling elite comes from a smaller Islamic sect based there, the Alawites. Leaders of this powerful minority have retained control over the country for decades, at times by ruthless suppression of dissent. In 2000, president-for-life Hafez al-Assad died and was succeeded by his son, Bashar, signaling a continuation of the political status quo. For 25 years, part of this status quo was the occupation and control of neighboring Lebanon, but in 2005 this came to an end as the Syrians withdrew.

Like Lebanon and Israel, Syria has a Mediterranean coastline where crops can be raised without irrigation. Behind this densely populated coastal zone, Syria has a much larger interior than its neighbors, but its areas of productive capacity are widely dispersed. Damascus, in the southwest corner of the country, was built on an oasis and is considered to be the world's oldest continuously inhabited city. It is now the capital of Syria, with a population of 2.5 million.

The far northwest is anchored by Aleppo, the focus of cotton- and wheat-growing areas in the shadow of the Turkish border. Here the Orontes River is the chief source of irrigation water, but in the

eastern part of the country the Euphrates Valley is the crucial lifeline. It is in Syria's interest to develop its eastern provinces, and recent discoveries of oil there will contribute to this development. But the war in neighboring Iraq has caused problems along Syria's eastern border, which will delay its plans for this area.

Jordan, Syria's southern neighbor, is a classic case (and victim) of changing relative location. This desert kingdom was a product of the Ottoman collapse, but when Israel was created it lost its window on the Mediterranean Sea, the (now Israeli) port of Haifa, previously in the British-administered Mandate of Palestine. Following its independence in 1946 with about 400,000 inhabitants, newly created Israel bequeathed Jordan some 500,000 West Bank Palestinians and, later, a huge inflow of refugees. Today Palestinians outnumber original residents by more than two to one in the population of 5.9 million. It may be said that the 47-year rule by King Hussein, ending in 1999, was the key centripetal force that held the country together.

With an impoverished capital, Amman, without oil reserves, and possessing only a small and remote outlet to the Gulf of Aqaba, Jordan has survived with U.S., British, and other aid. It lost its West Bank territory in the 1967 war with Israel, including its sector of Jerusalem (then the kingdom's second-largest city). No third country has a greater stake in a settlement between Israel and the Palestinians than does Jordan.

The map suggests that Lebanon has significant geographic advantages in this region: a lengthy coastline on the Mediterranean Sea; a well-situated capital, Beirut, on its shoreline; oil terminals on its shores; and a major capital (Syria's Damascus) in its hinterland. The map at the scale of Figure 7-12 cannot reveal yet another asset: the fertile, agriculturally productive Bekaa Valley in the eastern interior.

French colonialism in the post-Ottoman era created a territory, which in 1930 was about equally Muslim and Christian. Beirut became known as the Paris of the Middle East. After independence in 1946, Lebanon functioned as a democracy and did

well economically as the region's leading banking and commercial center. But the Muslim sector of the population grew much faster than the Christian one, and in the late 1950s the Arabs launched their first rebellion against the established order. Following the first of several waves of influx by Palestinian refugees, Lebanon fell apart in 1975 in a civil war that wrecked Beirut, devastated the economy, and left the country at the mercy of Syrian forces, which entered to take control.

For so small a country (4.0 million, including more than 400,000 naturalized and noncitizen Palestinians), Lebanon is permeated with religious and ethnic factions and is prone to sectarian breakdown. The Lebanese were unhappy with what they felt to be a Syrian occupation, yet the occupation continued for several decades. In the meantime, an Iran-sponsored terrorist movement, Hizbollah, came into being, and even became a political force. The Syrians were ousted in 2005 under United Nations auspices, and a stable Lebanese government coalesced around a power-sharing agreement. However, the peace lasted only a few years. In 2006 a Hizbollah kidnapping of Israeli soldiers provoked an Israeli attack, shattering the reconstructed physical and political infrastructure. The country has been without a functioning government from late 2007 until at least mid-2008 and the situation remains precarious.

Israel, the Jewish state, lies in the heart of the Arab World (Fig. 7-9). Since 1948, when Israel was created as a homeland for the Jewish people on the recommendation of a United Nations commission, the Arab-Israeli conflict has overshadowed all else in the region.

Figure 7-14 helps us understand the complex issues involved here. In 1946 the British, who had administered this area in post-Ottoman times, granted independence to what was then called Transjordan, the kingdom east of the Jordan River. In 1948, the orange-colored area became the UN-sponsored state of Israel—including, of course, land that had long belonged to Arabs in this territory called Palestine.

As soon as Israel proclaimed its independence, neighboring Arab states attacked it. Israel, however,

FIGURE 7-13 (*left*)   © H. J. de Blij, P. O. Muller, and John Wiley & Sons, Inc.

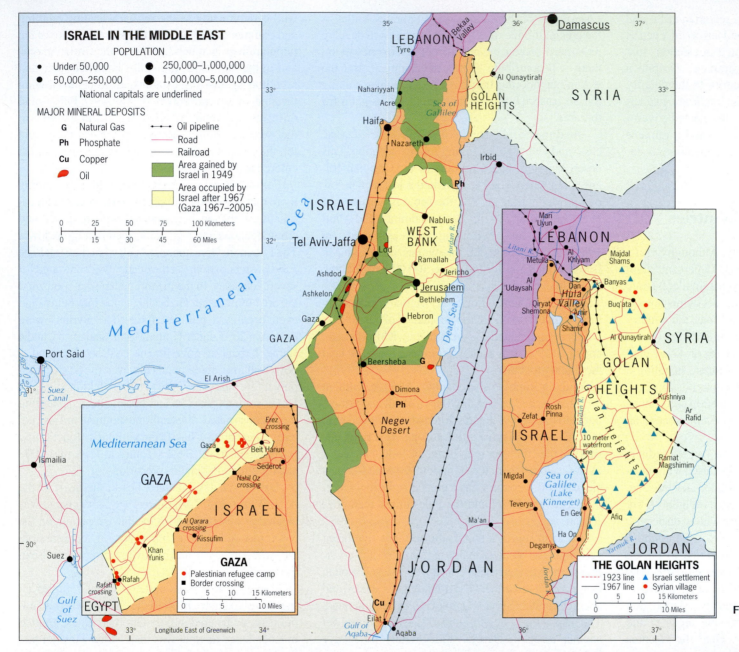

ISRAEL IN THE MIDDLE EAST

POPULATION

● Under 50,000   ● 250,000–1,000,000
● 50,000–250,000   ● 1,000,000–5,000,000

National capitals are underlined

MAJOR MINERAL DEPOSITS

G   Natural Gas
Ph   Phosphate
Cu   Copper
⬤   Oil

·•·•· Oil pipeline
—— Road
—— Railroad
Area gained by Israel in 1949
Area occupied by Israel after 1967 (Gaza 1967–2005)

0   25   50   75   100 Kilometers
0   15   30   45   60 Miles

GAZA
● Palestinian refugee camp
■ Border crossing

THE GOLAN HEIGHTS
--- 1923 line   ▲ Israeli settlement
—— 1967 line   ● Syrian village

FIGURE 7-14

© H. J. de Blij,
P. O. Muller, and
John Wiley & Sons, Inc.

not only held its own but pushed the Arab forces back beyond its borders, gaining the green areas shown in Figure 7-14. Meanwhile, Transjordanian armies crossed the Jordan River and annexed the yellow-colored area named the West Bank (of the Jordan River), including part of the city of Jerusalem. The king called his newly enlarged country Jordan.

More conflict followed. In 1967 a week-long war produced a major Israeli victory: Israel took the Golan Heights from Syria, the West Bank from Jordan, and the Gaza Strip from Egypt, and conquered the entire Sinai Peninsula all the way to the Suez Canal. In later peace agreements, Israel returned the Sinai but not the Gaza Strip to Egypt. In 2005, however, the Israeli government decided to remove all Jewish settlements from the Gaza Strip, yielding the area to the Palestinian Authority. Border skirmishes continue.

All this strife produced a huge outflow of Palestinian Arab refugees and displaced persons. The Palestinian Arabs constitute another of the region's *stateless nations*; about 1.2 million continue to live as Israeli citizens within the borders of Israel, but more than 2.5 million are in the West Bank and another 1.5 million in the Gaza Strip (the Golan Heights population is comparatively insignificant). Some Palestinians live in neighboring and nearby countries, including Jordan (approximately 3 million), Syria (460,000), Lebanon (430,000), and Saudi Arabia (320,000); another quarter million reside in Iraq, Egypt, Kuwait, and Libya; and some live in other countries around the world, including the United States. Many have been assimilated into the local societies, but tens of thousands continue to live in refugee camps. The stateless Palestinian population is estimated to total almost 10 million.

Israel is about the size of Massachusetts and has a population of 7.4 million (including its 1.2 million Arab citizens), but because of its location amid adversaries and its strong international links, these data do not reflect Israel's importance. Israel has built a powerful military even as its regional Arab neighbors have grown stronger and its policies arouse Arab and Muslim passions. As a democracy with strong Western ties, Israel has been a recipient of massive U.S. financial aid, and U.S. foreign policy has been to seek an accommodation between Jews and Palestinians, as well as between Israel and its Arab neighbors. Recently, Washington made a push toward a two-state solution that would create a Palestinian state alongside Israel, but as always there are thorny issues to be dealt with and a need for political will and accommodation on both sides that appears not to be forthcoming.

The geographic obstacles to such an accommodation include the following:

1. *The West Bank.* Even after its capture by Israel in 1967, the West Bank might have become a Palestinian homeland (and possibly a state), but Jewish immigration to the area made such a future difficult. In 1977 only 5000 Jews lived in the West Bank; by 2007 there were over 400,000, making up more than 10 percent of the population and creating a seemingly inextricable jigsaw of Jewish and Arab settlements (Fig. 7-15), making the two-state solution difficult.

2. *The Golan Heights.* The inset map on the right in Figure 7-14 suggests how difficult the Golan Heights issue is. The Heights overlook a large area of northern Israel, and they flank the Jordan River and crucial Lake Kinneret (the Sea of Galilee), the water reservoir for Israel. Relations with Syria are not likely to become normalized until the Golan Heights are returned, but in democratic Israel the political climate may make ceding this territory impossible.

3. *Jerusalem.* The United Nations intended Tel Aviv to be Israel's capital, and Jerusalem an international city. But the Arab attack and the 1948–1949 war allowed Israel to drive toward Jerusalem (see Fig. 7-14, the green wedge into

The Israeli Security Fence (officially known as the Security Barrier) against Palestinian suicide bombers forms an ugly reminder of the depth of division between two peoples. Palestinians in Ar Ram, near Jerusalem, walked on a Sunday near part of the West Bank barrier built by Israel. As Figure 7-15 shows, the Security Fence does not follow the West Bank boundary but puts sizeable parts of the West Bank on the Israeli side. Palestinians find themselves cut off from schools and farmlands, and their daily lives are inconvenienced.   (© Goran Tomasevic/Reuters)

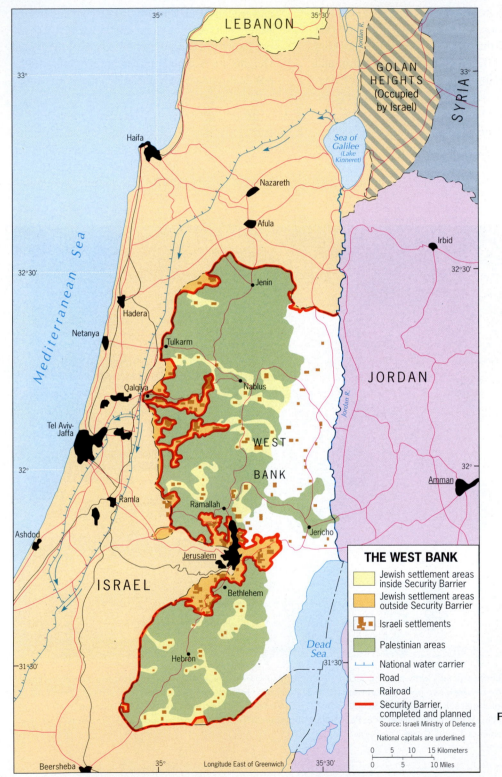

THE WEST BANK

- Jewish settlement areas inside Security Barrier
- Jewish settlement areas outside Security Barrier
- Israeli settlements
- Palestinian areas
- National water carrier
- Road
- Railroad
- Security Barrier, completed and planned

Source: Israeli Ministry of Defence

National capitals are underlined

0   5   10   15 Kilometers
0   5   10 Miles

FIGURE 7-15

the West Bank). By the time a cease-fire was arranged, Israel held the western part of the city, and Arab forces the eastern sector. But in this eastern sector lay major Jewish historic sites, including the Western Wall. Still, in 1950 Israel declared the western sector of Jerusalem its capital, making this, in effect, a *forward capital*. Figure 7-16 shows the position of the (black) armistice line, leaving most of the Old City in Jordanian hands. But then, in the 1967 war, Israel conquered all of the West Bank, including East Jerusalem; in 1980 the Jewish state reaffirmed Jerusalem's status as capital, calling on all nations to move their embassies from Tel Aviv. Meanwhile, the government redrew the map of the ancient city, building Jewish settlements in a ring around East Jerusalem that would end the old distinction between Jewish west and Arab east. This enraged Palestinian leaders, who still view Jerusalem as the eventual headquarters of a hoped-for Palestinian state.

4. *The Security Fence.* In response to the infiltration of suicide terrorists, Israel has now walled off most of the West Bank along the Security Barrier border shown in Figure 7-15 (see photo p. 255). The new wall does not conform to the de facto postwar boundary of the West Bank, cutting off about 10 percent of its territory and in effect annexing it (and its inhabitants) to Israel. This reinforcement of the West Bank border imposes a hardship on Palestinians, for some of whom it runs between their homes and their farmlands. Palestinians demand that the wall be taken down; Israelis reply that the Palestinian Authority must control the terrorists.

5. *The Gaza Strip.* This densely populated strip of land (Fig. 7-14, left inset) is a cauldron of Palestinian unrest and unhappiness. Infighting among the Palestinians here has made it all the more difficult to find a solution to the Israeli-Palestinian conflict.

© H. J. de Blij, P. O. Muller, and John Wiley & Sons, Inc.

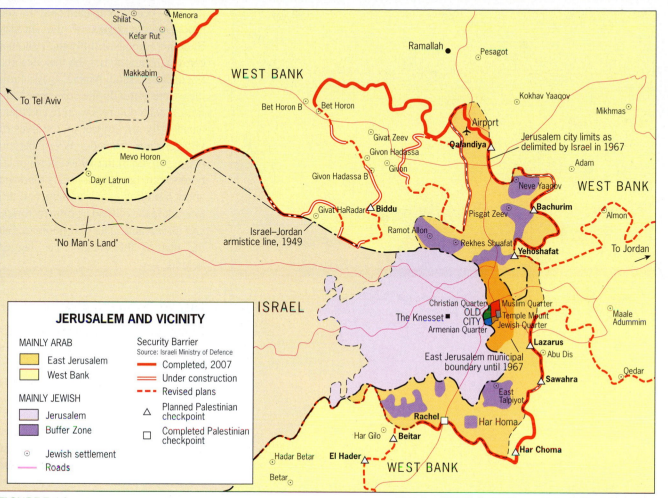

JERUSALEM AND VICINITY

MAINLY ARAB

🟨 East Jerusalem
🟡 West Bank

MAINLY JEWISH

🟪 Jerusalem
🟪 Buffer Zone

⊙ Jewish settlement
Roads

Security Barrier
Source: Israeli Ministry of Defence

━━ Completed, 2007
══ Under construction
╍╍ Revised plans
△ Planned Palestinian checkpoint
☐ Completed Palestinian checkpoint

FIGURE 7-16

© H. J. de Blij, P. O. Muller, and John Wiley & Sons, Inc.

Saudi Arabia

Saudi Arabia itself has only 25 million inhabitants (including 5 million expatriate workers) in its vast territory, but we can see the kingdom's importance in Figure 7-8: the Arabian Peninsula contains the world's largest concentration of known petroleum reserves. Saudi Arabia occupies most of this area and by some estimates may possess as much as one quarter of the world's liquid oil deposits. As Figure 7-17 shows, these reserves lie in the eastern part of the country, particularly along the Persian Gulf coast and in the Rub al Khali (Empty Quarter) to the south.

As a region, the Arabian Peninsula is environmentally dominated by a desert habitat and politically dominated by the Kingdom of Saudi Arabia (Fig. 7-17). With 2,150,000 square kilometers (830,000 sq mi), Saudi Arabia is the realm's fourth biggest state; only Kazakhstan, Algeria, and Sudan are larger. On the peninsula, Saudi Arabia's neighbors (moving clockwise from the head of the Persian Gulf) are Kuwait, Bahrain, Qatar, the United Arab Emirates, the Sultanate of Oman, and the Republic of Yemen. Together, these countries on the eastern and southern fringes of the peninsula contain almost 35 million inhabitants; the largest by far is Yemen, with 23 million.

Figure 7-17 reveals that most economic activities in Saudi Arabia are concentrated in a wide belt across the 'waist' of the peninsula, from the oil boomtown of Dhahran on the Persian Gulf through the national capital of Riyadh in the interior to the Mecca–Medina area near the Red Sea. A modern transportation and communications network has

Israel lies at the center of a fast-moving geopolitical storm; the issues raised above are only part of the overall problem (others include water rights, compensation for land expropriation, and the Palestinians' 'right to return' to pre-refugee abodes). Now, in the age of nuclear, chemical, and biological weapons and longer-range missiles, Israel's search for an accommodation with its Arab neighbors is a race against time.

ARABIAN PENINSULA

The regional identity of the Arabian Peninsula is clear: south of Jordan and Iraq, the entire peninsula is encircled by water. This is a region of old-style sheikdoms made wealthy by oil, modern-looking emirates, and the site where Islam originated.

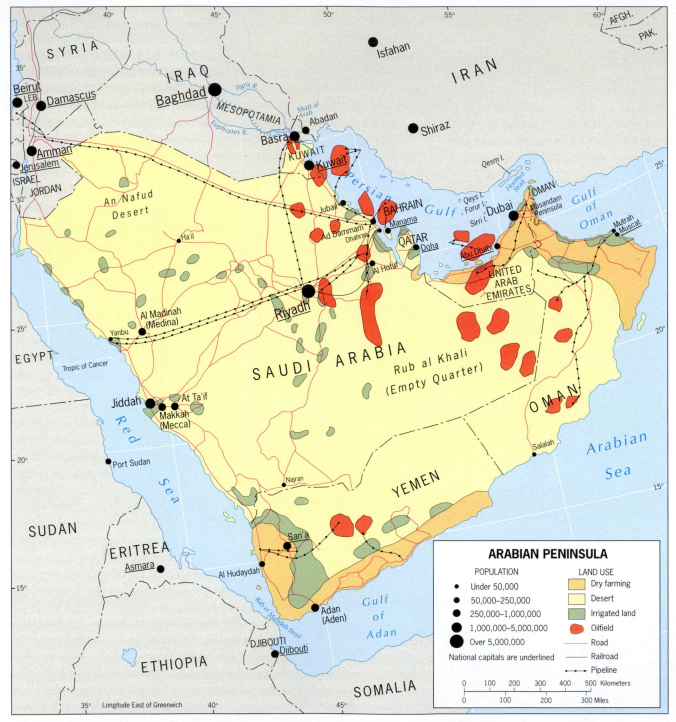

FIGURE 7-17

© H. J. de Blij, P. O. Muller, and John Wiley & Sons, Inc.

recently been completed, but in the more remote zones of the interior Bedouin nomads still ply their ancient caravan routes across the vast deserts. For decades, Saudi Arabia's royal families were virtually the sole beneficiaries of their country's wealth, and there was hardly any impact on the lives of villagers and nomads. When the oil boom arrived in the 1950s, foreign laborers were brought in (today there are about 5 million, many of them Shi'ite) to work in the oilfields, ports, factories, and as servants. The east boomed, but the rest of the country has lagged behind.

Disparities and Uncertainties

Efforts to reduce Saudi Arabia's regional economic disparities have been impeded by the enormous cost of bringing water from deep-seated sources to the desert surface to stimulate agriculture, by the sheer size of the country, and by the high rate of population growth (2.7 percent). Still, housing, health care, and education have seen major improvement, and industrialization also has been stimulated in such places as Jubail on the Persian Gulf and Yanbu, near Medina, on the Red Sea.

Saudi Arabia's conservative monarchism, official friendship with the West, and social contrasts resulting from its economic growth have raised political opposition, for which no adequate channels exist. Relations with the United States were affected by the role of Saudi suicide terrorists in 9/11, Saudi financial support for Wahhabist extremists, and other issues arising in the wake of the events of 2001. More recently, the Saudi's have resisted increasing oil production despite record prices, testing their friendship with the United States. Reliance on oil income, although currently quite profitable, is economically risky in the long run because the country's social benefits become dependent on these revenues.

On the Periphery

Five of Saudi Arabia's six neighbors on the Arabian Peninsula face the Persian and Oman gulfs (Fig. 7-17) and are monarchies in the Islamic tradition. All five also derive substantial revenues from oil. Their populations range from 0.7 to 5 million in addition to hundreds of thousands of foreign workers; these are not strong or influential states. They do, however, display considerable geographic diversity. **Kuwait**, at the head of the Persian Gulf, almost cuts Iraq off from the open sea, an issue the Gulf War did not settle. **Bahrain** is an island-state, a tiny territory with dwindling oil reserves. It has become an important banking center for the region and has some of the most progressive social policies in this part of the world. Its approximately 700,000 people are evenly split between Sunnis and Shi'ites; nearly two-thirds of its labor force is foreign. Neighboring **Qatar** consists of a peninsula jutting out into the Persian Gulf. Its oil reserves are also dwindling, but it has capitalized on its increasing natural gas reserves. The **United Arab Emirates (UAE)**, a federation of seven emirates, faces the Persian Gulf between Qatar and Oman. A reigning sheik is the absolute monarch in each of the emirates, and the seven sheiks together form the Supreme Council of Rulers.

In terms of oil revenues, however, there is no equality: two emirates—Abu Dhabi and Dubai—have most of the reserves. Dubai (Dubayy) has converted its favorable geographic location together with its oil revenues into a booming economy in which tourism, international transit functions, international higher education, banking, and trade are symbolized by ultramodern architecture, engineering feats (see photo below), and state-of-the-art entertainment complexes.

The eastern corner of the Arabian Peninsula is occupied by the Sultanate of **Oman**, another absolute monarchy, centered on the capital, Muscat. Figure 7-17 shows that Oman consists of two parts: the large eastern corner of the peninsula and a small but critical cape to the north, the Musandam Peninsula, that protrudes into the Persian Gulf to form a narrow channel or **13 choke point**—the Hormuz Strait (Iran lies on the opposite shore). Tankers that leave the other Gulf states must negotiate this narrow channel at slow speed, and during politically tense times warships have had to protect them. Iran's claim to several small islands near the Strait that are owned by the UAE is a potential source of dispute.

This brings us to what is, in a number of ways, Saudi Arabia's most substantial peninsular neighbor: **Yemen**. The boundary between Saudi Arabia and Yemen has only recently been satisfactorily delimited in an area where there may be substantial oil

reserves. Another boundary, between former North Yemen and South Yemen, was erased in 1990, when the two countries joined to form the present state. San'a, formerly the capital of North Yemen, retained that status; Adan (Aden), the only major port along a lengthy stretch of the peninsula, anchors the south. The country, with a population of 23 million—60 percent Sunni and 40 percent Shi'ite—has made remarkable progress toward representative government: the minister for human rights is a woman, Wahiba Fare. As Figure 7-17 shows, Yemen's southernmost tip forms one side of the narrow, reef-studded entry to the Red Sea, another choke point Arabs call the Bab-el-Mandeb (Gate of Grief). Directly across lies the ministate of **Djibouti**, adjoined by Somalia. Since shortly after September 11, 2001, the United States has maintained a military presence in Djibouti.

THE EMPIRE STATES

Two major states, both with imperial histories, dominate the region that lies immediately to the north of the Middle East (Fig. 7-9), where Arab ethnicity gives way but Islamic culture endures: Turkey and Iran. On its periphery lie two entities that are strongly entwined with these states: Azerbaijan to the north and the northern part of Cyprus to the south (Fig. 7-18).

Dubai is just one, and by no means the largest, of the seven United Arab Emirates; nor does it have the largest share of the country's oil. But Dubai and its people have a history of commerce and international business that have made this the Hong Kong of Arabia, as locals like to call it. A major port, a growing airport where the country's prestigious Emirates Airlines is headquartered, world-class golf courses, race tracks, tennis facilities, and a liberal vision of Islam combine to prove what is possible on the very borders of conservative and restrictive Saudi Arabia. Dubai's success is reflected in its ultramodern architecture and state-of-the-art residential communities. Jumeira Palm Island is one of the most striking of these, built out into the Gulf in the shape of a palm tree. (© AP/Wide World Photos)

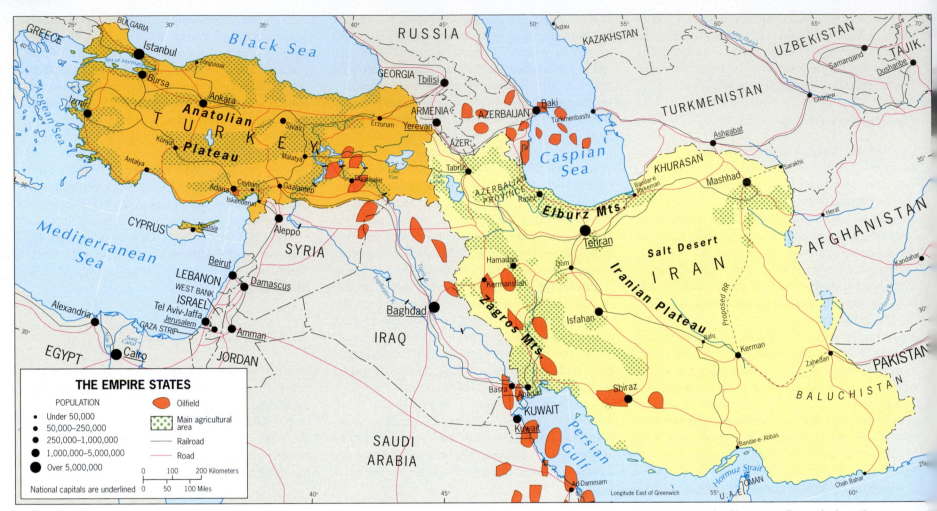

FIGURE 7-18

© H. J. de Blij, P. O. Muller, and John Wiley & Sons, In

Turkey

Turkey is a mountainous country of generally moderate relief and, as Figure G-9 indicates, considerable environmental diversity ranging from steppe to highland. On the dry Anatolian Plateau villages are small, and subsistence farmers grow cereals and raise livestock. Coastal plains are not large, but they are productive and densely populated. Textiles (from home-grown cotton) and farm products dominate the export economy, but Turkey also has substantial mineral reserves, some oil in the southeast, massive dam-building pro-

jects on the Tigris and Euphrates rivers, and a small steel industry based on domestic raw materials.

Earlier in this chapter we chronicled the historical geography of the Ottoman Empire, its expansion, cultural domination, and collapse. By the beginning of the twentieth century, the country we now know as Turkey lay at the center of this decaying and corrupt state, ripe for revolution and renewal. This occurred in the 1920s and thrust into prominence a leader who became known as the father of modern Turkey: Mustafa Kemal, known after 1933 as Atatürk, meaning 'Father of the Turks.'

Capitals New and Old

The ancient capital of Turkey was Constantinople (now Istanbul), located on the Bosporus, part of the strategic straits connecting the Black and Mediterranean seas. But the struggle for Turkey's survival had been waged from the heart of the country, the Anatolian Plateau, and it was here that Atatürk decided to place his seat of government. Ankara, the new capital, possessed certain advantages: it would remind the Turks that they were (as Atatürk always said) Anatolians; it lay nearer the center of the country than Istanbul; and it could therefore act as a stronger unifier. Istanbul lies on the threshold of Europe, with the minarets and mosques of this largest and most varied Turkish city rising above a townscape that resembles one in Eastern Europe.

Although Atatürk moved the capital eastward and inward, his orientation was westward and outward. To implement his plans for Turkey's modernization, he initiated reforms in almost every sphere of life within the country. Islam, formerly the state religion, lost its official status, and Turkey became a secular state whose army ensured that the Islamists would not take over again. The state took over most of the religious schools that had controlled education. The Roman alphabet replaced the Arabic. A modified Western code supplemented Islamic law. Symbols of old—growing beards, wearing the fez—were prohibited. Monogamy was made law, and the emancipation of women was begun. The new government emphasized Turkey's separateness from the Arab World, and it has remained aloof from the affairs that engage other Islamic states.

Turkey and Its Neighbors

From before Atatürk's time, Turkey has had a history of mistreating minorities. Soon after the outbreak of World War I, the (pre-Atatürk) regime decided to expel all the Armenians, concentrated in the country's northeast. Nearly 2 million Turkish Armenians were uprooted and brutally forced out; an estimated 600,000 died in a campaign that still arouses anti-Turkish emotions among Armenians today. In modern times, Turkey has been criticized for its treatment of its large and regionally concentrated Kurdish population. About one fifth of Turkey's population of 75.6 million is Kurdish, and successive Turkish governments have mishandled relationships with this minority nation, even prohibiting the use of Kurdish speech and music in public places during one especially repressive period. The historic Kurdish homeland lies in the southeast of Turkey, centered on Diyarbakir, but millions of Kurds have moved to the shantytowns around Istanbul—and to jobs in the countries of the European Union. With

Kurdish nationalism rising across the border in embattled Iraq, Turkey is on alert and in 2007 quickly suppressed Kurdish movements on its territory and even made incursions into Iraq to fight Kurdish insurgents there. Seeking to suppress a small but violent extremist group among the Kurds, the Turks violated the human rights of an entire minority. These rights violations are having an impact on Turkey's aspirations for the EU.

Also problematic for Turkey is its role in Cyprus, the island 80 kilometers (50 mi) to its south. Following a civil war between the majority Greeks and minority Turks in 1974, the Turkish government intervened with armed forces and the island was partitioned. In 1983, the northern 40 percent of the island, now an almost exclusively Turkish domain, declared itself the independent Turkish Republic of Northern Cyprus (TRNC). Against the wishes of the international community, Turkey alone recognized the TRNC as a state. This division on the island is a division between realms.

Turkey and the EU

With its diversified economy and secular government, Turkey is the only predominantly Islamic country potentially eligible for admission into the European Union. Its soccer team already plays in the European championships, and for a long time Turkey's western orientation, membership in NATO, its many migrants, and its economic ties to that realm made this a distinct possibility. Already in a customs union with the EU since 1995, in 1999 Turkey was formally declared an accession candidate. Formal negotiations toward admission began in late 2005. But there are various major obstacles in the way of this becoming a reality. The Cyprus and Kurdish issues have both negatively affected Turkey's prospects for joining the EU. Admitting Turkey would put Europe's boundary on the doorstep of a dangerous neighborhood, which is a concern to some, though a positive aspect for others. Another consideration is Turkey's human rights record, which remains poor. Nonetheless, there has been progress. Turkey has met the political criterion as a functioning democracy, but not

enough of the other criteria for EU membership. In addition, Europeans are perhaps not ready for a non-Christian member-state, although substantial discussion and disagreement have taken place among states, leaders, and ordinary people in many EU countries. Lately, the Turks themselves have grown less interested in joining than they were a decade ago. In 2007 they elected a more conservative government that is less likely to push for the reforms needed to accede to the EU. Indeed, the ban on head scarves for women was lifted in 2008, signaling a more conservative turn in the country. Overall, Turkey has a long way to go before it can join the EU.

Iran

Long known as *Persia*, Iran, Turkey's neighbor to the east, also has a history of imperial conquest. In 1971, the then-reigning shah and his family celebrated the 2500th anniversary of Persia's first monarchy with unmatched royal splendor. But by 1979, revolution had engulfed Iran, and Shi'ite fundamentalists drove from power the shah who had ascended his throne through American intervention. The monarchy was replaced by an Islamic republic, and a frightful wave of retribution followed.

A Crucial Location and Dangerous Terrain

As Figure 7-18 shows, Iran occupies a critical area in this turbulent realm. It controls the entire corridor between the Caspian Sea and the Persian Gulf. To the west it adjoins Turkey and Iraq, both historic enemies. To the north (west of the Caspian Sea) Iran borders Azerbaijan and Armenia, where once again Islam confronts Christianity. To the east Iran meets Pakistan and Afghanistan, and east of the Caspian Sea lies volatile Turkmenistan.

Iran, as Figure 7-18 demonstrates, is a country of mountains and deserts. The heart of the country is an upland, the Iranian Plateau, that lies surrounded by even higher mountains, including the Zagros in the west, the Elburz in the north along the Caspian Sea coast, and the mountains of Khurasan to the northeast.

This mountainous topography signals danger: here the Eurasian and Arabian tectonic plates converge, causing major and often devastating earthquakes. The Iranian Plateau therefore is actually a huge highland basin marked by salt flats and wide expanses of sand and rock. The highlands wrest some moisture from the air, but elsewhere only oases break the arid monotony—oases that for countless centuries have been stops on the area's caravan routes.

In eastern Iran, neighboring Pakistan, and Afghanistan, some people still move with their camels, goats, and other livestock along routes that are almost as old as the human history of this realm. Usually they follow a seasonal and annual cycle, visiting the same pastures year after year, pitching their tents near the same stream. It is a lifestyle especially associated with this realm: **14** **nomadism**. In Iran as elsewhere, nomads are not aimless wanderers across boundless plains. They know their terrain intimately, and they carefully judge how long to linger and when to depart based on many years of experience along the route.

City and Countryside

In ancient times, Persepolis in southern Iran (located near the modern city of Shiraz) was the focus of a powerful Persian kingdom, a city dependent on *qanats*, underground tunnels carrying water from moist mountain slopes to dry flatland sites many miles away. Today, Iran's population of 72 million is 67 percent urban, and the capital, Tehran, lies far to the north, on the southern slopes of the Elburz Mountains. This mushrooming metropolis of 7.6 million, lying at the heart of modern Iran's core area, still de-

pends in part on the same kinds of qanats that sustained Persepolis more than 2000 years ago. As such, Tehran symbolizes the internal contradictions of Iran: a country in which modernization has taken hold in the cities, but little has changed in the vast countryside, where the mullahs led their peasant followers in a revolution that overthrew a monarchy and installed a theocracy.

Energy and Conflict

As Figure 7-8 shows, Iran has a large share of the oil riches in this part of the world. Petroleum and natural gas production provide about 90 percent of the country's income. The reserves lie in a zone along the southwestern periphery of Iran's territory, and Abadan became its oil capital near the head of the Persian Gulf. But Iran is a large and populous country, and the wealth oil generated could not transform it in the way the last shah intended, a transformation that had it occurred might have staved off the revolution. Modernization remained but a veneer: in the villages away from Tehran's polluted air, the holy men continued to dominate the lives of ordinary Iranians. As elsewhere in the Muslim world, urbanites, villagers, and nomads remained enmeshed in a web of produc-

tion and profiteering, serfdom, and indebtedness that always characterized traditional society here. The revolution swept this system away, but it did not improve the lot of Iran's millions. A devastating war with Iraq (1980–1990), into which Iran ruthlessly poured hundreds of thousands of its young men, sapped both the coffers and energies of the state. When it was over, Iran was left poorer, weaker, and aimless, its revolution spent on unproductive pursuits.

In the early years of the twenty-first century, evidence abounded that the people of Iran remain divided between conservatives determined to protect the power of the mullahs and reformers seeking to modernize and liberalize Iranian society. In Iran's 2005 election, many reformist candidates were disqualified by religious conservatives who hold ultimate authority, and a president who had proved himself to be mildly reformist was succeeded by a far more extreme figure, Mahmoud Ahmadinejad, who, shortly after taking office, proclaimed that the state of Israel should be 'wiped off the map.' Meanwhile, Iran was pursuing a course toward nuclear power-generating capability with implications that trouble many countries. In addition, the Iranians want to construct a natural gas pipeline across Pakistan to India, making India highly dependent on the Iranian

Just outside the city of Isfahan, 410 kilometers (255 mi) south of the Iranian capital of Tehran, is this uranium conversion facility. The facility reprocesses uranium ore concentrate, known as yellowcake, into uranium hexafluoride gas. The gas is then taken to the Natanz site and fed into the centrifuges for enrichment. Iran's president, Mahmoud Ahmadinejad, escorted reporters on an unprecedented visit to the once-secret Natanz nuclear complex in May 2008, providing a glimpse into the underground uranium enrichment plant that the United States and Europe demand be shut down.   (© AP/Wide World Photos)

supply. The deal is currently stalled with Iran's price too high for India, and the United States trying to cancel the plan.

Iran's rise as a nuclear power and its regional ambitions should not surprise anyone. Although revolutionary Iran disavowed the imperial designs of its Persian predecessors, this does not mean that its national interests now stop at its borders. Tehran has a major stake in developments in Iraq and Afghanistan, both neighbors; it has an already-nuclear and unstable Pakistan on its eastern border; it has a strong interest in the fate of Shi'ite majorities and minorities elsewhere in the realm; it has long been an avowed enemy of Israel and a strong supporter of Palestinian causes; and it has funded organizations labeled terrorist whose actions have even reached across the Atlantic. Iran's revolution ousted the shah and ended a monarchy, but it did not extinguish all ties to an imperial past.

TURKESTAN: THE SIX STATES OF CENTRAL ASIA

For centuries Turkish (Turkic) peoples diffused over a vast Central Asian domain that extended from Mongolia and Siberia to the Black Sea. Propelled by population growth and energized by Islam, they penetrated Iran, defeated the Byzantine Empire, and colonized much of Eastern Europe. Eventually, their power declined as Mongols, Chinese, and Russians invaded their strongholds. But these conquerors could not expunge them, as the names on the modern map prove (Fig. 7-19). The latest conquerors, the Chinese and especially the Russian czars and their communist successors, created Soviet Socialist Republics named after the majority peoples within their borders. Thus the Kazakhs, Turkmen, Kyrgyz, and other Turkic peoples retained some geographic identity in what was then Soviet Central Asia.

Central Asia (*Turkestan*) is a still-changing region. In some areas, the cultural landscapes of neighboring realms extend into it, for example, in northern (Russian) Kazakhstan. In other areas, Turkestan extends into adjacent realms, as in the Xinjiang Uyghur

Autonomous Region of western China. Some areas once penetrated by Turkic peoples are no longer dominated by them—for instance, Afghanistan. And certain peoples now living in Turkestan are not of Turkic ancestry, notably the Tajiks. This is a fractious region in sometimes turbulent transition.

As we define it, Turkestan includes six states: (1) *Kazakhstan*, territorially larger than the other five combined but situated astride an ethnic transition zone; (2) *Turkmenistan*, relatively closed and dictatorial, with important frontage on the Caspian Sea and bordering Iran and Afghanistan; (3) *Uzbekistan*, the most populous state and situated at the heart of the region; (4) *Kyrgyzstan*, wedged between powerful neighbors and chronically unstable; (5) *Tajikistan*, regionally and culturally divided as well as strife-torn; and (6) *Afghanistan*, engulfed in war almost continuously since it was invaded by Soviet forces in 1979.

The States of Former Soviet Central Asia

During their hegemony over Central Asia, the Soviets tried to suppress Islam and install secular regimes (this was their objective when they invaded Afghanistan as well), but today Islam's cultural revival is one of the defining qualities of this region. From Almaty to Samarqand, mosques are being repaired and revived, and Islamic dress again is part of the cultural landscape. National leaders have made high-profile visits to Mecca; most are sworn into office touching the Quran. All of Central Asia's countries now observe Islamic holidays. In other ways, too, Turkestan reflects the norms of this realm: in its dry-world environments and the clustering of its population, its mountain-fed streams irrigating farms and fields, its sectarian conflicts, its oil-based economies. It also is a region where democratic government remains an elusive goal.

Kazakhstan is the territorial giant and borders two greater giants: Russia and China (Fig. 7-19). During the Soviet period, northern Kazakhstan was heavily Russified and became, in effect, part of Russia's

Mahmud of Kashgar. "When I had a chance to travel to the extreme west of China, I was really in Turkestan. Culturally, this is a land of Turkic people, especially the Uyghurs, and Islam. I was able to spend a day in Kashgar (known as Kashi in China), the fabled town of the Silk Route. Near Kashgar, and only about 200 kilometers (125 mi) along the Karakom Highway from the Pakistan border, is the small town of Opal, which is associated with Mohammed Kashgari (also known as Mahmud al-Kashgari), a famous Turkic eleventh-century philosopher and compiler of the first Turkic-Arabic dictionary. This book helped unify the Turkic people, and Mahmud's mausoleum, located here, is an important place for all Turkic peoples. The book, written in 1072, is still in print today. His statue stands near the mausoleum."
(© A. WinklerPrins)

Eastern Frontier (see Fig. 2-13). Rail and road links crossed the area mapped as the Kazakh–Russian Transition Zone, connecting the north to Russia. The Soviets made Almaty, in the heart of the Kazakh domain, the territory's capital. Today the Kazakhs are in control, and they in turn have moved the capital to Astana, right in the heart of the Transition Zone, where some 4.5 million Russians (who constitute about 30 percent of the total national population of 15.5 million) still live. Clearly, Astana is another *forward capital*.

Figure 7-19 reveals Kazakhstan's situation as a corridor between the oil reserves of the Caspian Basin and China. Oil and gas pipelines partially completed across Kazakhstan will eliminate, or greatly reduce, China's dependence on oil carried by tankers along distant sea lanes.

Uzbekistan, the most populous state in this region (27.0 million), occupies the heart of Turkestan and borders every other state in it. Uzbeks not only make up 75 percent of this population, but also form substantial minorities in several neighboring states. The capital, Tashkent, lies in the eastern core area of the country, where most of the people live in towns and farm villages, and the crowded Fergana Valley is the focus. In the far west lies the shrinking Aral Sea, whose feeder streams were diverted into cotton fields and croplands during the Soviet occupation; heavy use of pesticides contaminated the groundwater, and countless thousands of local people continue to suffer severe medical problems as a result.

FIGURE 7-19 © H. J. de Blij, P. O. Muller, and John Wiley & Sons, Inc.

Turkmenistan, the autocratic desert republic that extends all the way from the Caspian Sea to the borders of Afghanistan, has a population of just 5.5 million, of which about three-quarters are Turkmen. During the Soviet era, the communist planners began work on a massive project: the Garagum (Kara Kum) Canal designed to bring water from Turkestan's eastern mountains into the heart of the desert. Today the canal is 1100 kilometers (700 mi) long, and it has enabled the cultivation of some 1.2 million hectares (3 million acres) of cotton, vegetables, and fruits. The plan is to extend the canal all the way to the Caspian Sea, but meanwhile Turkestan has hopes of greatly expanding its oil and gas output from Caspian coastal reserves. But look again at Figure 7-19: Turkmenistan's relative location is not advantageous for export routes.

Kyrgyzstan's topography and political geography are reminiscent of the Caucasus. Shown in yellow in Figure 7-19, Kyrgyzstan lies intertwined with Uzbekistan and Tajikistan to the point of exclaves and enclaves along its borders. The Kyrgyz, for whom the Soviets established this republic, constitute barely 50 percent of the population of 5.3 million. Uzbeks and other minorities form a complex cultural geography; mountains and valleys isolate communities and make nation-building difficult. The agricultural economy is weak, consisting of pastoralism in the mountains and farming in the valleys. About 75 percent of the people profess allegiance to Islam, and Wahhabism has gained a strong foothold here. The town of Osh is often referred to as the headquarters of the movement in Turkestan.

Tajikistan's mountainous scenery is even more spectacular than Kyrgyzstan's, and here, too, topography is a barrier to the integration of a multicultural society. The Tajiks, who constitute about 65 percent of the population of 7.3 million, are ethnically Persian (Iranian), not Turkic, and speak a language related to Persian (and thus Indo-European). Most Tajiks, despite their Persian affinities, are Sunni Muslims, not Shi'ites. Small though Tajikistan is, regionalism plagues the state: the government in Dushanbe is often at odds with the barely connected northern area (see Fig. 7-19), a hotbed not only of Islamic revivalism but also of anti-Tajik, Uzbek activism.

Fractious Afghanistan

Afghanistan, the southernmost country in this region, exists because the British and Russians, competing for hegemony in this area during the nineteenth century, agreed to tolerate it as a cushion, or **15** **buffer state**, between them. This is how Afghanistan acquired the narrow extension leading from the main territory eastward to the Chinese border—the Vakhan Corridor (Fig. 7-20). As the colonialists delimited it, Afghanistan adjoined the domains of the Turkmen, Uzbeks, and Tajiks to the north, Persia (now Iran) to the west, and the western flank of British India (now Pakistan) to the east.

Landlocked and Fractured

Geography and history seem to have conspired to divide Afghanistan. As Figure 7-20 shows, the towering Hindu Kush range dominates the center of the country, creating three broad environmental zones: the relatively well-watered and fertile northern plains and basins; the rugged, earthquake-prone central highlands; and the desert-dominated southern plateaus. Kabol (Kabul), the capital, lies on the southeastern slope of the Hindu Kush, linked by narrow passes to the northern plains and by the Khyber Pass to Pakistan.

Across this variegated landscape moved countless peoples: Greeks, Turks, Arabs, Mongols, and others. Some settled here, their descendants today speaking Persian, Turkic, and other languages. Others left archaeological remains or no trace at all. The present population of Afghanistan (32.7 million) has no ethnic majority. This is a country of minorities in which the Pushtuns (or Pathans) of the east are the most numerous but make up barely 40 percent of the total. The second-largest minority are the Tajiks, a world away across the Hindu Kush, concentrated in the zone near Afghanistan's border with Tajikistan. The Hazaras of the central highlands and the south, the Uzbeks and Turkmen in the northern border areas, the Baluchi of the southern deserts, and other, smaller groups scattered across this country create one of the world's most complex cultural mosaics. Two major languages, Pushtun and Dari (the local variant of Persian), plus several others create a veritable Tower of Babel here.

Costs of Conflict

Episodes of conflict have marked the history of Afghanistan, but none was as costly as its involvement in the Cold War. Following the Soviet intervention of 1979, the United States supported the Muslim opposition, the *Mujahideen* (meaning strugglers), with modern weapons and money, and the Soviets were forced to withdraw. Soon the factions that had been united during the anti-Soviet campaign were in conflict, delaying the return of some 4 million refugees who had fled to Pakistan and Iran. The situation resembled the pre-Soviet past: a feudal country with a weak and ineffectual government in Kabol.

In 1994, what at first seemed to be just another warring faction appeared on the scene: the so-called *Taliban* (students of religion) from religious schools in Pakistan. Their avowed aim was to end Afghanistan's chronic factionalism by instituting strict Islamic law. Popular support in the war-weary country, especially among the Pushtuns, led to a series of successes, and by 1996 the Taliban had taken Kabol.

The Taliban's imposition of Islamic law was so strict and severe that Islamic as well as non-Islamic countries objected. Restrictions on the activities of women ended their professional education, employment, and freedom of movement, and had a devastating impact on children as well. Public amputations and stonings enforced the Taliban's code. In the process, Afghanistan became a haven for groups of revolutionaries whose goals went far beyond those of the Taliban: they plotted attacks on Western interests throughout the realm and threatened Arab regimes they deemed compliant with Western priorities. Taliban-ruled, cave-riddled, remote and isolated Afghanistan was an ideal locale for these outlaws.

Already in possession of arms and ammunition (Soviet as well as American) left over from the Cold War, they also benefited from Afghanistan's huge, illicit opium trade. It is estimated that in 2007 Afghanistan produced 92 percent of the world's opium. Much of the revenue found its way into the coffers of the conspirators.

Failed State, Terrorist Base

With such resources, the militants were able to organize and launch attacks against several Western targets in the realm and elsewhere, but events took a fateful turn in 1996 that would empower them as never before. During the conflict with the Soviets, the Mujahideen cause had been supported not only by the United States but also by a devout and revivalist Saudi Muslim named Usama bin (son of) Laden. The child of a construction billionaire who had more than 50 children including 22 sons, Usama graduated from a university in Jiddah in 1979 and headed for Afghanistan with an inherited fortune estimated at about U.S.$300 million to help the anti-Soviet campaign. Well-connected in Saudi Arabia and now a pivotal figure in Afghanistan, bin Laden saw his fame as well as his war chest grow as the Soviet intervention collapsed. Following the withdrawal of the communist forces, he returned to Saudi Arabia, where he denounced his government for allowing U.S. troops on Saudi soil during the Gulf War. The Saudi regime responded by stripping him of his citizenship and expelling him, and bin Laden fled to Sudan. There he set up several legitimate businesses to facilitate his now-global financial transactions, but he also established terrorist training camps. Under international pressure, the Khartoum regime ousted him in 1996, and bin Laden returned to a country he knew would welcome him again: Afghanistan.

Bin Laden's fateful return to Afghanistan coincided with the Taliban's conquest of Kabol, and now he helped its forces push northward into the fertile and productive northern plains. Meanwhile, a terrorist organization named **al-Qaeda** took root in the country, a global network that would further the aims of the revolutionaries once loosely allied. Afghanistan became al-Qaeda's headquarters, and bin Laden its director; its exploits were funded by numerous Muslim sources and ranged from terrorist-training in local camps to lethal attacks on American targets, including a warship in Yemen's port of Adan and two U.S. embassies in East Africa.

On February 26, 1993, terrorists exploded a massive car bomb in the basement garage of the World Trade Center in New York, but their objective—to topple the 110-story tower—failed. Even as those responsible went on trial, al-Qaeda's leaders were planning the suicide attacks of September 11, 2001 that destroyed the buildings, killed thousands, caused billions of dollars in physical damage, and inflicted an untold amount of psychological damage. Several weeks later, United States and British forces, with the acquiescence of Pakistan, attacked both the Taliban regime and the al-Qaeda infrastructure in Afghanistan, the first stage of a global War on Terror that would leave no country unaffected. Proof of bin Laden's and al-Qaeda's complicity in the September 11 assault was found in videotape and documentary form.

The Taliban and al-Qaeda leaderships may not have counted on Pakistan's compliance with Western demands, but they surely knew what the consequences of 9/11 would be for Afghanistan and its people. Once again a foreign power would invade the country and set its political course, the associated upheaval enabling rapacious warlords to exploit the peoples of border provinces far from the capital. As it turns out, Afghanistan has made significant progress toward representative government as most of the warlords were defeated or co-opted, and many of the refugees returned. Infrastructure is slowly being (re)constructed. For example, now nearly complete and mostly paved is a highway that connects all the main cities, known as the Ring Road (see Fig. 7-20). The road and others in various stages of completion will help commerce and communication significantly. But a reconstituting Taliban continues to harass American and other foreign troops in Afghanistan, the leadership of al-Qaeda remains in hiding in western Pakistan and has not been apprehended, and the heroin trade is resurgent. Still, Afghanistan is in far better shape than it was under the Soviet occupation or during Taliban rule. Despite periodic unrest, sustained stability may finally permit recovery of this long-suffering buffer state.

Afghan children ask for candies from a patrolling U.S. soldier of the 82nd Airborne in Chinar, a village in the Ghorak district of Kandahar province in southern Afghanistan. Although most of America's attention has been focused on Iraq following the U.S.-led invasion there, a comparatively small allied force has been trying to keep control in Afghanistan, where the trouble that led to 9/11 started and where al-Qaeda still has strongholds. The Taliban regime was ousted, but remnants of its movement are regrouping and threatening the stability of many areas of the country. The struggle for Afghanistan is far from over.   (© AP/Wide World Photos)

FIGURE 7-20 (left)   © H. J. de Blij, P. O. Muller, and John Wiley & Sons, Inc.

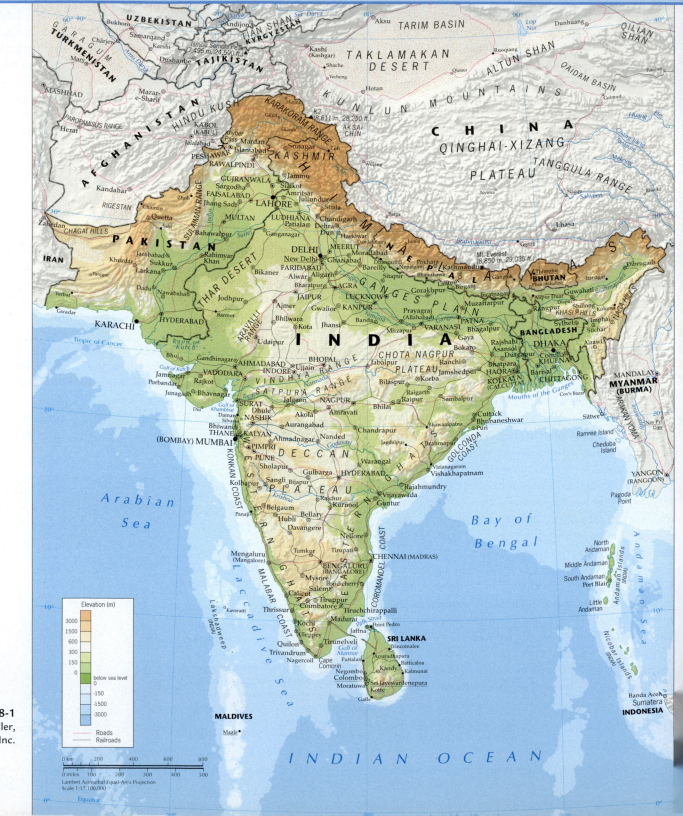

FIGURE 8-1

Map: © H. J. de Blij, P. O. Muller, and John Wiley & Sons, Inc.

8

SOUTH ASIA

In This Chapter

- The monsoon is the key to India's fortunes
- South Asia's population dilemma: Burden or bonus?
- Two nuclear powers quarrel over Kashmir
- Hindu nationalism rising in multicultural India
- India's continued economic growth: Now a global power

CONCEPTS, IDEAS, AND TERMS

1	Monsoon
2	Social stratification
3	Refugees
4	Population geography
5	Population distribution
6	Population density
7	Physiologic density
8	Rate of natural population increase
9	Demographic transition
10	Population explosion
11	Population pyramid
12	Forward capital
13	Irredentism
14	Caste system
15	Intervening opportunity
16	Natural hazards
17	Failed state
18	Buffer state
19	Insurgent state

REGIONS

PAKISTAN
INDIA
BANGLADESH
MOUNTAINOUS NORTH
ISLAND SOUTH

Photos: (*top*) Dhaka, Bangladesh © Maha Saleh El Nasser, Oxfam House; (*bottom*) Bengaluru (Bangalore), India © Pascal Sittler/REA/Redux Pictures.

269

FOR MORE THAN two decades now, the eyes of the world have been on East Asia—on the dramatic developments that transformed China from an isolated backwater to an economic power and a force in global affairs. But significant changes have also come to South Asia (Fig. 8-1, chapter opener map), and especially to the gigantic state that dominates this multicultural realm: India. South Asia may be territorially small (it is less than half the size of East Asia), but it will soon be the world's most populous realm, and India the world's most populous country. South Asia, therefore, merits close attention. Our daily lives will be affected increasingly by what happens in this crowded, restive part of the world. There is more to South Asia than India: Pakistan, Bangladesh, Nepal, and Sri Lanka also form important parts of this realm (Fig. 8-2).

Defining the Realm

Mountains, deserts, and coastlines combine to make South Asia one of the world's most vividly defined physiographic realms. To the north, the Himalaya Mountains create a natural wall between South Asia and China. To the east, mountain ranges and dense forests mark the boundary between South and Southeast Asia. To the west, rugged highlands and expansive deserts separate South Asia from its neighbors. Within these confines lies a geographic realm that is more densely populated than any other. If current population trends continue, it will soon be the most populous realm in the world.

South Asia consists of six regions (Fig. 8-2). Its keystone is India, whose population passed the 1 billion mark in 1999 and has since added another 158 million. In the west lies Pakistan. South Asia's eastern flank is centered on Bangladesh. The northern region consists of the mountainous countries of Nepal and Bhutan. And the southern region includes the islands of Sri Lanka and the Maldives. As the map shows, India divides into several subregions.

Despite South Asia's formidable physiographic barriers, it has been penetrated by peoples from afar bringing different cultures and religions. As a result, the realm today is a patchwork of religions, languages, traditions, and cultural landscapes. So complex is this mosaic that it is remarkable that its political geography numbers just seven states (only five on the mainland). Given this limited number of states despite great ethnic diversity, it is important to realize that several flashpoints flare up periodically in this realm—we will discuss these in the regional section—and there is a potential for devolution here.

Among the cultural infusions in this realm was Islam. Today, Pakistan is an Islamic republic, and Islam provides the cement for this state's 173.5 million inhabitants. Pakistan's eastern border with India is a cultural divide in more ways than one: dominantly Hindu India is a secular, not a theocratic, state. Why,

MAJOR GEOGRAPHIC QUALITIES OF South Asia

1. South Asia is clearly defined physiographically, and much of the realm's boundary is marked by mountains, deserts, and the Indian Ocean.

2. South Asia's great rivers, especially the Ganges, have for tens of thousands of years supported huge population clusters.

3. South Asia covers just over 3 percent of the earth's land area but contains nearly 23 percent of the world's human population.

4. South Asia's population continues to grow at annual rates higher than the world average. It will become the world's most populous geographic realm during the next decade.

5. Poverty remains endemic in South Asia, despite pockets of progress. Hundreds of millions of people are plagued by inadequate nutrition and poor overall health.

6. British commercial penetration and colonial domination left strong geographic imprints on this realm, ranging from its boundary framework to its cultural landscapes.

7. South Asia's annual monsoon continues to dominate life for hundreds of millions of subsistence and commercial farmers. Failure of the monsoon cycle spells economic crisis.

8. Despite its physiographic demarcation, invading armies and cultures, from ancient Greeks to modern British and from Mongols to Muslims, diversified this realm's cultural mosaic.

9. Several of the world's great religions and many minor faiths have strong bases in South Asia, including Hinduism, Islam, and Buddhism. Religion remains a powerful force in political and economic life.

10. India, the world's largest democracy and a functioning federation, is the dominant state in this realm but has contentious relations with several of its neighbors.

11. Kashmir, the mountainous territory in the northern frontier zone between India and Pakistan, is a dangerous source of friction between these two nuclear powers.

FIGURE 8-2

© H. J. de Blij, P. O. Muller, and John Wiley & Sons, Inc.

then, do we include Pakistan in the South Asia rather than the Southwest Asia/North Africa realm? One criterion is ethnic continuity, which links Pakistan to India rather than to Afghanistan or Iran. Another is historical geography. Pakistan was part of Britain's South Asian Empire, and it originated from the partition of that domain between Muslim and Hindu majorities. Although Urdu is the official national language of Pakistan, English is the *lingua franca*, as it is in India. Furthermore, the border between India and Pakistan does not signify the eastern frontier of Islam in Asia. More than 165 million of India's more than 1 billion citizens are Muslims, and in South Asia's eastern region, Bangladesh (population: 152.3 million) is more than 85 percent Muslim. Finally, Pakistan and India are locked in a struggle to control a vital mountainous area (Kashmir) in the far north, where British withdrawal left the boundary between them unresolved and a current flashpoint.

PHYSIOGRAPHIC REGIONS OF SOUTH ASIA

Before we look into South Asia's complex and fascinating cultural geography, we need to discuss the physical stage of this populous realm (Fig. 8-3). South Asia is a subcontinent of immense physiographic variety, of snow-capped peaks and forest-clad slopes, of vast deserts and broad river basins, of high plateaus and spectacular shores. The continuing collision of the Indian and Eurasian tectonic plates (Fig. G-5) has created the world's

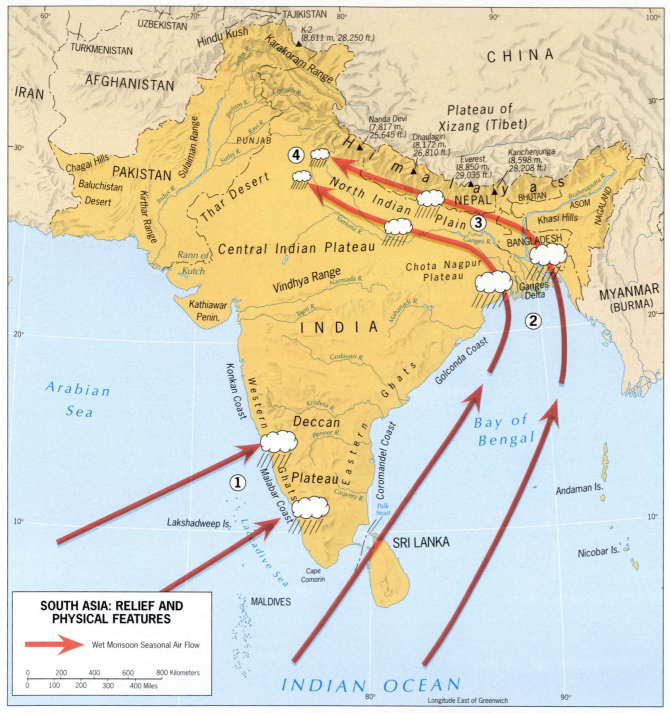

FIGURE 8-3

© H. J. de Blij, P. O. Muller, and John Wiley & Sons, Inc.

highest mountain ranges, their icy crests yielding meltwaters for great rivers below. This is a highly active tectonic zone, making it a dangerous place to live. In 2005 a series of earthquakes along the collision zone struck northern Pakistan and Kashmir, destroying countless villages, killing more than 70,000 people, and leaving over 3 million homeless.

Global circulation patterns put South Asia in the path of tropical cyclones and produce reversing seasonal windflows known as **1 monsoons**. In 2007, a powerful cyclone hit the low-lying country of Bangladesh, killing thousands of people and leaving more than a million homeless.

In general terms, we can recognize three clearly defined physiographic zones in South Asia: the northern mountains, the southern peninsular plateaus, and between them a belt of river lowlands. Superimposed on this configuration is an east-west precipitation gradient from wet (Bangladesh) to dry (western Pakistan) that is clearly visible in Figures G-8 and G-9, broken only by the strip of high moisture along peninsular India's west coast.

Northern Mountains

The northern mountains extend from the imposing Hindu Kush and Karakoram Ranges in the northwest through the Himalayas in the center (Mount Everest, the world's tallest peak, lies in Nepal) to the ranges of Bhutan and the Indian State of Arunachal Pradesh in the northeast (Fig. 8-3). Dry and barren in the west on the Afghanistan border, the ranges become green and tree-studded in Kashmir, forested in the lower sections of Nepal, and even more densely vegetated in Arunachal Pradesh. Transitional foothills, with many deeply eroded valleys cut by rushing meltwaters, lead to the river basins below.

River Lowlands

The belt of river lowlands extends eastward from Pakistan's lower Indus Valley (the area known as Sind) through the wide plain of the Ganges Valley of India and on across the great double delta of the Ganges and Brahmaputra in Bangladesh (Fig. 8-3). In the east, this physiographic region often is called the North Indian Plain. To the west lies the lowland of the Indus River, which rises in Tibet, crosses Kashmir, and then bends southward to receive its major tributaries from the Punjab (Land of Five Rivers).

Southern Plateaus

Peninsular India is mostly plateau country, dominated by the massive Deccan, a tableland built of basalt that poured out when India separated from Africa during the breakup of Gondwana (see Fig. 6-3). The Deccan (which means South) tilts to the east, so that its highest areas are in the west and the major rivers flow into the Bay of Bengal. North of the Deccan lie two other plateaus, the Central Indian Plateau to the west and the Chota-Nagpur Plateau to the east (Fig. 8-3). On the map, also note the Eastern and Western Ghats (hills) that descend from Deccan plateau elevations to the narrow coastal plains below.

The Annual Wet Monsoon

The name *South Asia* is almost synonymous with the term **monsoon**, since the annual rains that come with its onset, usually in June, are so critical to this realm. These rains are South Asia's lifeforce as they are indispensable to subsistence as well as commercial agriculture in the realm's key country, India. By June of each year, a strong low-pressure system develops over northern India, drawing moist, warm oceanic air onto the peninsula (see Fig. 8-3). As this air is pulled

The arrival of the annual rains of the wet monsoon transforms the Indian countryside. By the end of May, the paddies lie parched and brown, dust chokes the air, and it seems that nothing will revive the land. Then the rains begin, and blankets of dust turn into layers of mud. Soon the first patches of green appear on the soil, and by the time the wet monsoon ends, all is green. The photograph on the left, taken just before the onset of the monsoon in the State of Goa, shows the paddies before the rains begin; three months later the countryside is cloaked in green.   (© Steve McCurry/ Magnun Photos, Inc.)

over the Western Ghats ①, it is cooled, its moisture condenses, and rains begin which may last 60 days or more. Other onshore airflows come from the Bay of Bengal ② and get caught up in the convection over northeastern India and Bangladesh. Seemingly endless summer rains now inundate the whole North Indian Plain; the Himalayas to the north create a corridor that channels the rain westward ③ until eventually it dries out as it reaches Pakistan ④. By the end of summer the system breaks down, the wet monsoon gives way to periodic rains, and, eventually, another dry season arrives—and the anxious wait begins again for next year's life-giving monsoon.

THE HUMAN SEQUENCE

Great river basins mark the physiography of South Asia; in one of these basins, that of the Indus River in present-day Pakistan, lies evidence of the realm's oldest major civilization. It existed at the same time, and interacted with, ancient Mesopotamia, and it was centered on large, well-organized cities. From here,

influences and innovations diffused into India. In fact, India's very name is believed to derive from the ancient Sanskrit word *sindhu*. Eventually, such cities as Harappa and Mohenjo-Daro seem to have experienced the same fate as those of Mesopotamia, perhaps also because of environmental change. About 3500 years ago, Aryan (Indo-European) speaking peoples migrated into the region, penetrating India and starting the process of welding the Ganges Basin's isolated tribes and villages into an organized system. Having absorbed much of the culture of the Indus Valley, they brought a new order to this region.

Hinduism and Buddhism

In the process, a new belief system arose, and with it a new way of life. *Hinduism* was built on a complex system of **2 social stratification** in which Brahmans, powerful priests, stood at the head of a *caste system* in which soldiers, merchants, artists, peasants, and all others had their place.

Hinduism was restrictive, especially for those of the lower castes, and in the sixth century BC (more than 2500 years ago) a prince born into one of northern India's kingdoms sought a better way. Prince Siddhartha, better known as Buddha, gave up his royal position to teach religious salvation through meditation, the rejection of earthly desires, and a reverence for all forms of life. His teachings did not have a major impact on the Hindu-dominated society of his time, but what he taught was not forgotten. Centuries later, the ruler of a powerful Indian state decided to make *Buddhism* the state religion, following which the faith spread far and wide.

But first South Asia was penetrated by other foreign influences: the Persians were followed by the Greeks under Alexander the Great in the late fourth century BC. This is why, on a world map of languages, Hindi (the major domestic language of India) lies at the eastern end of the region of Indo-European languages, in the same family as Persian, Italian, and English. But the map of India's languages (Fig. 8-4) shows that southern India's languages are not Indo-European. Indeed, while northern India was a subregion of cultural infusion and turmoil, the south lay remote and isolated. This part of the peninsula had apparently been settled long before the Indus and Ganges civilizations arose, and southern India developed into a distinctive subregion with its own cultures. The *Dravidian* languages spoken here—Telugu, Tamil, Kanarese (Kannada), and Malayalam—have long literary histories.

Aśoka's Mauryan Empire

When the Greeks withdrew from the Ganges Basin and the Hindu heartland was free of foreign dominance, a powerful empire arose there—the first true empire in the realm. This, the Mauryan Empire, extended its influence over India as far west as the Indus Valley (thus incorporating the populous Punjab) and as far east as Bengal (the double delta of the Ganges and Brahmaputra); it reached as far south as the modern city of Bangalore.

This Mauryan Empire was led by a series of capable rulers who achieved stability over a vast domain. Undoubtedly the greatest of these leaders was Aśoka, who reigned for nearly 40 years during the middle of the third century BC. Aśoka was a believer in Buddhism, and it was he who elevated this religion from obscurity to regional and ultimately global importance.

In accordance with Buddha's teachings, Aśoka reordered his government's priorities from conquest and expansion to a Buddhist-inspired search for stability and peace. He sent missionaries to the outside world to carry Buddha's teachings to distant peoples, thereby also contributing to the diffusion of Indian culture. As a result, Buddhism became permanently established as the dominant religion in Sri Lanka (formerly Ceylon), and it established footholds as far afield as Southeast Asia and Mediterranean Europe. Ironically, Buddhism thrived in these remote places even as it declined in India itself. With Aśoka's death, the faith lost its strongest supporter.

The Mauryan Empire represented India's greatest political and cultural achievements in its day, and when the empire collapsed, late in the second century AD, India fragmented into a patchwork of states. Once again, India lay open to infusions from the west and northwest, and across present-day Pakistan they

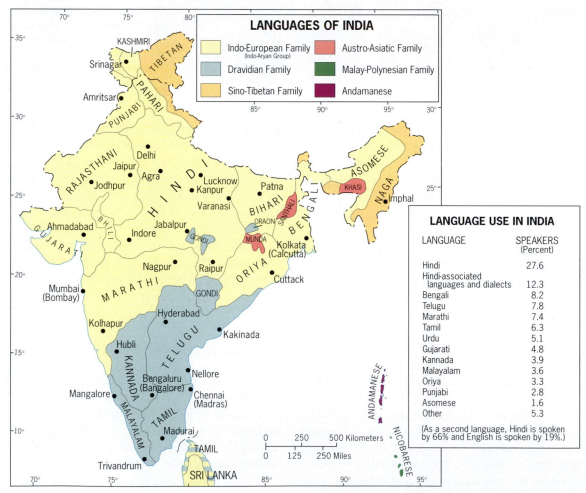

LANGUAGES OF INDIA

Legend	
Indo-European Family (Indo-Aryan Group)	Austro-Asiatic Family
Dravidian Family	Malay-Polynesian Family
Sino-Tibetan Family	Andamanese

LANGUAGE USE IN INDIA

LANGUAGE	SPEAKERS (Percent)
Hindi	27.6
Hindi-associated languages and dialects	12.3
Bengali	8.2
Telugu	7.8
Marathi	7.4
Tamil	6.3
Urdu	5.1
Gujarati	4.8
Kannada	3.9
Malayalam	3.6
Oriya	3.3
Punjabi	2.8
Asomese	1.6
Other	5.3

(As a second language, Hindi is spoken by 66% and English is spoken by 19%.)

FIGURE 8-4   © H. J. de Blij, P. O. Muller, and John Wiley & Sons, Inc.

came: Persians, Afghans, Turks, and others driven from their homelands or attracted by the lands of the Ganges.

The Power of Islam

In the late tenth century, Islam came rolling like a giant tide across the subcontinent, spreading from Persia in the west and Afghanistan in the northwest. The Indus Valley lay directly in the path of this Islamic advance, and virtually everyone was converted. Next the Muslims penetrated the Punjab, the subregion that lies astride the present Pakistan-India border, and there perhaps as many as two-thirds of the inhabitants became converts. Then Islam crossed the bottleneck where Delhi is situated and diffused east- and southeastward into the Gangetic Plain and the subregion now known as Hindustan—India's evolving core area. Here Islam's proselytizers had less success, persuading perhaps one in eight to become Muslims. In the meantime, Islam arrived at the Ganges Delta by boat, and present-day Bangladesh became overwhelmingly Islamic. Interestingly, to the south of the Ganges heartland Islam's diffusion wave lost its energy and Dravidian India never came under Muslim influence.

Why did millions of Hindus convert to Islam? The reason was a combination of compulsion and attraction. The princes of India's local states faced the choice of cooperation or annihilation, and thus the elites of the native states found it prudent to convert, just as the leaders of West Africa's states did. And among the lower castes of Hindu society, Islam was a welcome alternative to the rigid socio-religious hierarchy in which they were trapped at the bottom. Thus Islam was the faith of the rulers and of the disadvantaged, a powerful force in the heartland of Hinduism.

Just as Islam weakened in Southern Europe, so its force ultimately became spent in vast and populous India. For all the Muslims' power, they never managed to convert a majority of South Asians. They dominated the northwest corner of the realm (present-day Pakistan), where Lahore became one of Islam's greatest cities. But in all of what is today India, less than 15 percent of the population became and remained Muslim. Islam arrived like a giant tide, but Hinduism stayed afloat and outlasted the invasion.

The European Intrusion

Into this turbulent complexity of religious, political, and linguistic disunity yet another element began to intrude after 1500: European powers in search of raw materials, markets, and political influence. Because the Europeans profited from the Hindu-Muslim contest, they exploited local rivalries, jealousies, and animosities. British merchants gained control over the trade with Europe in spices, cotton, and silk goods, ousting the French, Dutch, and Portuguese. The British East India Company's ships also took over the intra-Asian sea trade between India and Southeast Asia, which had long been in the hands of Indian, Arab, and Chinese merchants. In effect, the East India Company (EIC) became India's colonial administration.

In time, however, the EIC faced problems that made it increasingly difficult to combine commerce with administration. Eventually, a mutiny among Indian troops in the service of the EIC led to the abolition of the company. The British government took over in 1857 and maintained its rule (*raj*) until 1947.

Colonial Transformation

When the British took power over South Asia, they controlled a realm with considerable industrial development (notably in metal goods and textiles) and an active trade with both Southwest and Southeast Asia. The colonialists saw this as competition, and soon the

The urban architecture of much of this realm continues to carry with it the characteristic imprint of British colonial architecture. This is Victoria Railway Station in the center of Mumbai (Bombay), India.   (© Alamy)

British made sure that India was exporting raw materials and importing manufactured goods from Europe. Local industries declined as Indian merchants lost their markets.

Unifying their realm was a tougher task for the British. In 1857, about 2 million square kilometers (750,000 sq mi) of South Asian territory still was beyond British control, including hundreds of entities that had been guaranteed autonomy by the EIC during its administration. These Native States, ranging in size from a few acres to Hyderabad's 200,000 square kilometers (80,000 sq mi), were assigned British advisors, but in fact India was a near-chaotic amalgam of modern colonial and traditional feudal systems.

Colonialism did produce assets for South Asia. The British developed one of the best transport networks in the colonial world, especially the railroad system (although the network focused on interior-seaport linkages rather than fully interconnecting the various parts of the country). British engineers laid out irrigation canals through which millions of acres of land were brought into cultivation. Settlements that had been founded by Britain developed into major cities and bustling ports, led by Bombay (now Mumbai), Calcutta (now Kolkata), and Madras (now Chennai).[1] These three cities are still three of India's largest urban centers, and their cityscapes bear the unmistakable imprint of colonialism (see photo p. 275). Modern industrialization, too, was brought to India by the British on a limited scale. In education, an effort was made to combine English and Indian traditions; the Westernization of India's elite was supported through the education of numerous Indians in Britain. Modern practices of medicine were also introduced. Moreover, the British adminis-

---

[1]Even 60 years after independence, India continues to shed colonial names. Numerous large cities have changed their names to more indigenous forms; the latest are Bengaluru instead of Bangalore and Kochi instead of Cochin. Not everyone in India is happy about this name changing as it has become a politically polarizing topic.

tration tried to eliminate or mitigate features of Indian culture that were deemed undesirable by any standards—such as the burning alive of widows on their husbands' funeral pyres, female infanticide, child marriage, and the caste system. Obviously, the task was far too great to be achieved in barely three generations of colonial rule, but independent India itself has continued these efforts.

Partition

Even before the British government decided to yield to Indian demands for independence, it was clear that British India would not survive the coming of self-rule as a single political entity. As early as the 1930s, the idea of a separate Pakistan was being promoted by Muslim activists, who circulated pamphlets arguing that British India's Muslims were a nation distinct from the Hindus and that a separate state consisting of Sind, Punjab, Baluchistan, Kashmir, and a portion of Afghanistan should be created from the British South Asian Empire in this area. The first formal demand for such partitioning was made in 1940, and, as later elections proved, the idea had almost universal support among the realm's Muslims.

As the colony moved toward independence, a political crisis developed: India's majority Congress Party would not even consider partition, and the minority Muslims refused to participate in any future unitary government. But partition would not be a simple matter. True, Muslims were in the majority in the western and eastern sectors of British India, but Islamic clusters were scattered throughout the realm (Fig. 8-5A). Any new boundaries between Hindus and Muslims to create an Islamic Pakistan and a Hindu India would have to be drawn right through areas where both sides coexisted. People by the millions would be displaced.

Nor were Hindus and Muslims the only people affected by partition. The Punjab area, for example, was home to millions of Sikhs, whose leaders were fiercely anti-Muslim. But a Hindu-Muslim border

based only on those two groups would leave the Sikhs in Pakistan. Even before independence day, and the day of partition, August 15, 1947, Sikh leaders talked of revolt, and there were some riots. But no one could have foreseen the dreadful killings and mass migrations that followed the creation of the boundary and the formation of independent Pakistan and India. Just how many people felt compelled to participate in the ensuing migrations will never be known; 15 million is the most common estimate. It was human suffering on an incomprehensible scale.

Even so large a flow of cross-border **3** refugees, however, hardly began to 'purify' India of Muslims. After the initial mass exchanges, in 1951 there still were tens of millions of Muslims in India (Fig. 8.5B). Today, the Muslim minority in Hindu-dominated India is almost as large as the entire Islamic population of Pakistan, having more than tripled since the late 1940s to over 160 million. It is the world's largest cultural minority (see Fig. 7-2), far more than a mere remnant of the days when Islam ruled the realm. This force will play a growing role in the India of the future.

SOUTH ASIA'S POPULATION DILEMMA

South Asia, as we noted earlier, is on course to become the most populous geographic realm in the world, and India the most populous country, perhaps as early as the end of the next decade (see Appendix B). That prospect has serious implications not only for this realm, but for the world as a whole. Over the past three decades, population growth in many parts of the world has been slowing, and some high-income countries are reporting decreasing populations. This means that an ever-larger proportion of the new births in this century will be occurring in low-income regions, including South Asia.

All this is part of the field of **4** population geography, in which research focuses on the dimensions, distribution, growth, and other aspects of

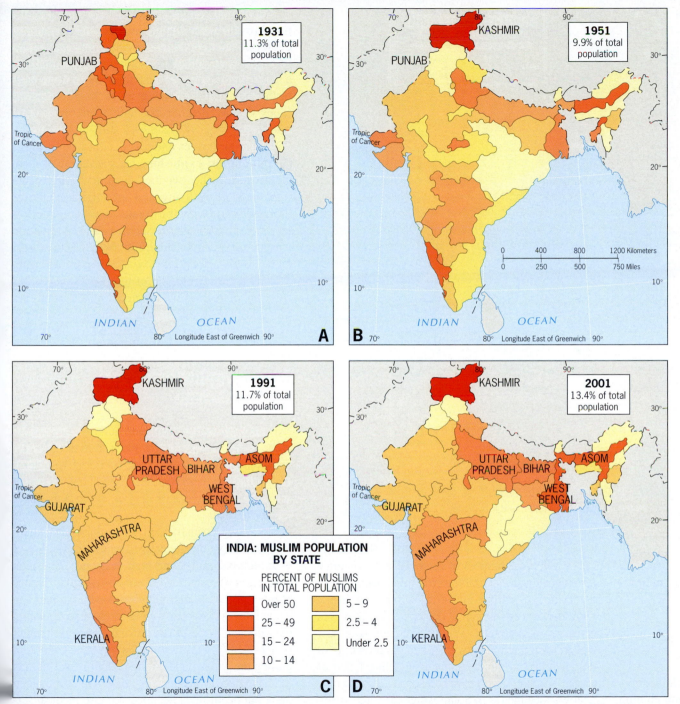

FIGURE 8-5

© H. J. de Blij, P. O. Muller, and John Wiley & Sons, Inc.

human population in a country, region, or realm as this relates to soils, climates, land ownership, social conditions, economic development, and other factors. Population geography, therefore, is the *geography of demography*. As usual, the map is a powerful ally in this geographic research.

Figure 8-6 reveals the **5 population distribution** of South Asia. You can actually discern the courses of the great Ganges, Indus, and other rivers because of the immense concentrations of humanity that inhabit their basins. But even more telling than distribution is **6 population density**, that is, the number of people per unit area (such as a square kilometer or square mile) in a country, province, or physiographic zone such as a river basin or mountain belt. Note that in Appendix B population density appears in two columns. The first of these columns shows what we call the *arithmetic* population density for a country, which is simply the number of people per square kilometer (in this case) for each country. Actually, this figure is not very meaningful despite its appearance in gazetteers and atlases, for obvious reasons. Take the case of Pakistan: tens of millions of people cluster in river basins and valleys, and almost nobody lives in deserts and high mountains, so what does 218 people per square kilometer really mean? Not much, which is why geographers prefer to use the **7 physiologic density** measure that reports the number of people in a country per unit of arable land area (suitable for agriculture). According to this measure, note that Bangladesh and Sri Lanka, not India, are the most densely

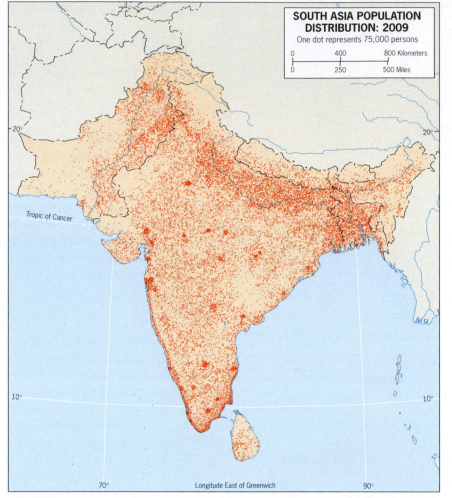

**SOUTH ASIA POPULATION
DISTRIBUTION: 2009**
One dot represents 75,000 persons

FIGURE 8-6   © H. J. de Blij, P. O. Muller, and John Wiley & Sons, Inc.

Population geographers analyze how population growth rates change as a region's economy evolves. The higher-income economies have gone through the so-called **9 demographic transition**, a four-stage sequence that took them from high birth rates and high death rates in preindustrial times to low death rates and even lower birth rates today (Fig. 8-7). Stages 2 and 3 in this model, with high birth rates and low death rates, form the **10 population explosion** that caused such great alarm during the twentieth century when death rates in industrializing and urbanizing countries dropped sharply, but birth rates took longer to follow suit. In 1900 the planet's human population was approximately 1.5 billion; by 2000 it had surpassed 6 billion, and is now at 6.7 billion. During the population explosion of the 1950s and 1960s, some scholars predicted a global calamity with as many as 10 to 12 billion people on earth by the end of the twentieth century, but this did not happen.

Demographic Prospects for South Asia

Even though world population numbers did not explode as much as predicted, some areas of the world face population (and associated economic) problems that will afflict them for decades to come, no matter what the global averages predict. South Asia might be such a place.

When the British ruled India during the nineteenth century, the country still was in the first stage, with high birth rates and high death rates; the high death rates were caused not only by a high incidence of infant and child mortality but also by famines and epidemics. As Figure 8-7 indicates, the population during Stage 1 does not grow or decline much, but it is not stable. Famines and disease outbreaks kept erasing the gains made during better times. But then India entered the second stage. Birth rates remained high, but death rates declined because medical services improved (soap came into widespread use),

populated countries of this realm (the very high Maldives number reflects the virtual absence of farmland on its tiny tropical islands).

Population Change

As we have noted, geographers are especially interested in the **8 rate of natural population change** in a country, region, or realm—that is, the number of

births minus the number of deaths, usually reported as a percentage (Natural Increase in Appendix B) or as an index per thousand of the population. These data can reveal much about the condition and prospects of a country. We have already seen how stagnant population growth produces problems of aging, how too-rapid population decline (such as Russia's) signals serious societal problems, and how a slower decline (as in South America) can be helpful in economic development.

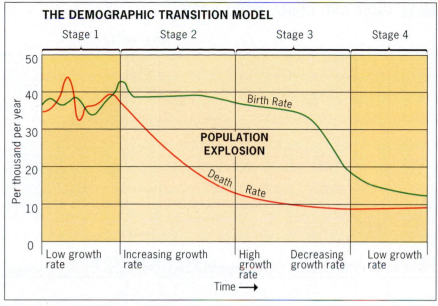

THE DEMOGRAPHIC TRANSITION MODEL

FIGURE 8-7   © H. J. de Blij, P. O. Muller, and John Wiley & Sons, Inc.

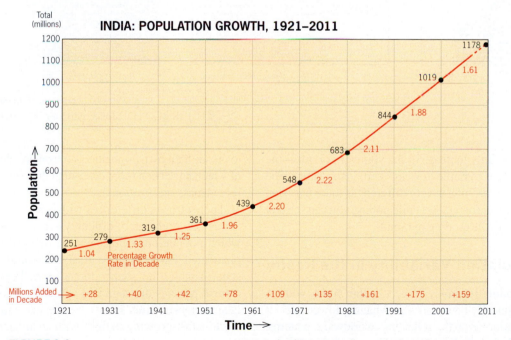

INDIA: POPULATION GROWTH, 1921–2011

FIGURE 8-8   © H. J. de Blij, P. O. Muller, and John Wiley & Sons, Inc.

food distribution networks became more effective, farm production expanded, and urbanization developed. In the 1920s, India's population still was growing at a rate of only 1.04 percent, but by the 1970s, that rate shot up to 2.22 percent per year (Fig. 8-8). Note that India gained 28 million people during the 1920s but a staggering 135 million during the 1970s.

Has India entered the third stage, when the death rate begins to level off and birth rates decline substantially, narrowing the gap and slowing the annual increase? The rate of increase suggests that this may now be the case: from 2.22 percent during the 1970s, it dropped to 2.11 percent in the 1980s and then to 1.88 percent during the 1990s (Fig. 8-8). But India also has another problem to confront. During its population explosion, its numbers grew so large that even a declining rate of natural change continues to add ever greater numbers to its total. In Figure 8-8, we see that while the decadal rate of increase dropped from 2.22 to 1.88 percent between 1971 and 2001, the millions added grew from 135 in the 1970s to 161 in the 1980s to 175 during the 1990s—taking the total past 1 billion during 1999. Even though India has entered the third stage of the demographic transition, it will not feel its effects for some time.

Some population geographers theorize that the populations of all countries will eventually stabilize at some level, just as Europe's did. Certain governments, notably China's, have instituted regulations to limit family size, but this policy is more easily implemented by dictatorships than democracies. And even if such stabilization were just one doubling of its current population away, India still would have an astronomical total of nearly two and a half billion residents. Interestingly, this population surplus is now being considered by some as a possible bonus as India will not suffer from the social consequences of population implosions such as is occurring in Japan, Russia, and several countries of Western Europe. India will have plenty of workers to fuel its economic growth.

POPULATION PYRAMIDS: INDIA AND CHINA, 2000–2025

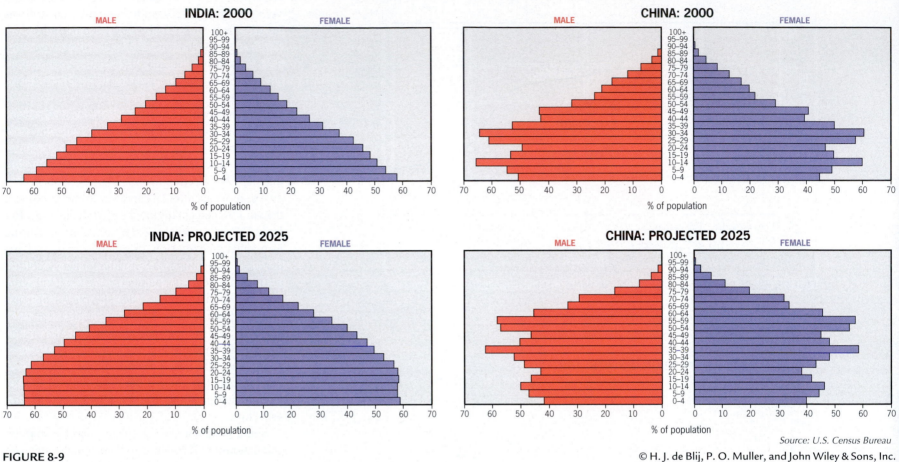

FIGURE 8-9

Source: U.S. Census Bureau

© H. J. de Blij, P. O. Muller, and John Wiley & Sons, Inc.

Population geographers and demographers often use **11** **population pyramids** to display the demographic structure of a place. Figure 8-9 displays population pyramids for India and China in 2000 and the predicted situation in 2025. These are the two most populous countries in the world and also two of the world's most dynamic economies, yet they have vastly differing cultures and political structures, as can be seen in the differences in their population profiles. India exhibits a classic population pyramid, which it is predicted to last into the future. Even so, you can see the population stabilizing (there is little increase in the younger cohorts—the sides of the pyramid are less steep) in the future. The situation in China is very different as there has been a heavily enforced one-child policy; you can see that decline in the eroding pyramid.

Population Variability in the Realm

Population dynamics do vary within India, with its economically better-off States having lower rates of population growth, reflecting worldwide patterns. Also, India's federal system allows individual States to pursue their own population-control policies, some of which have been more draconian than others (e.g., forced mass sterilizations).

There also are differences within the realm. India is a Hindu-dominated, officially secular country in which population policies can be debated and implemented. Pakistan, in contrast, is a strictly Islamic state in which no such options exist. Population-control policies are regarded as incompatible with Islamic tenets; thus Pakistan remains one of the world's fast-growing nations with an annual rate of natural increase of 2.4 percent.

SOUTH ASIA'S BURDEN OF POVERTY

In recent years, optimistic television news reports and articles in the popular press have been proclaiming a new era for South Asia, marked by rising growth rates for the realm's national economies, rewards from globalization and modernization, and increasing integration into the global economy, especially for India.

Indeed, a combination of circumstances ranging from the United States' involvement with Pakistan in the campaign against terrorism to the real estate and stock market booms in India suggest that a new era has arrived. But consider this statistic: more than two-thirds of India's nearly 1.2 billion people continue to live in poor rural areas, their villages and lives virtually untouched by what is happening for some in the cities. Many set out for cities, with rising expectations of a better life. Unfortunately, millions end up in some of the world's poorest slums. Fully a third of Pakistan's population lives in abject poverty; female literacy is below 30 percent. Half of the people of Bangladesh, and nearly half those in the realm as a whole, live on the equivalent of one U.S. dollar per day or less. It is estimated that half the children in South Asia are malnourished and underweight, a majority of them girls—this at a time when the world is able to provide adequate calories for all its inhabitants, if not adequately balanced daily meals.

Why is poverty so severe in this realm, comparable in this respect only to Subsaharan Africa? We have already noted several reasons: rapid population growth in a realm with already high physiologic densities; a continued and direct dependence, by hundreds of millions of subsistence farmers, on the vagaries of climate and weather; cultural traditions that put females at a perilous disadvantage. But there are other reasons, including skewed land ownership (in Pakistan's province of Sind, for example, 7 percent of landowners hold more than 40 percent of the land;

in India, hundreds of millions of peasants have no control over the land they cultivate) plus government inefficiency, corruption, and venality. Even as children go hungry, powerful farm lobbies pressure the government to buy and store their grain, raising prices and thus increasing the market cost for poor villagers unable to purchase enough of it.

THE LATEST INVASIONS

Although it may be premature to forecast a dramatic economic rise of the sort that has thrust China onto the world stage, there can be no doubt that South Asia today commands the world's attention. Even before the events of 9/11, the end of the Cold War had altered the geopolitics of Eurasia. Before the 1990s India had tilted toward Moscow, but the new era saw renewed linkages between India and the United States propelled by business as well as strategic motivations. Following the 2001 terrorist attacks in New York and Washington, the United States secured Pakistan's support in the War on Terror in return for massive financial assistance and support for Pakistan's authoritarian ruler (who had come to power through a coup in 1999). To get an idea of the consequences, consider this: in 2001, Pakistan's economy grew by just 2 percent (substantially below its rate of population growth); in 2006, it grew by 7 percent. Meanwhile, India became a bastion for information technology and outsourcing industries, a factor in the fortunes of many large U.S. companies. In 2005, during a presidential visit to India, the United States signed an agreement to grant India full civil nuclear energy cooperation,

WHAT'S DRIVING GEOGRAPHIC CHANGE IN THE REALM

- The United States and **India** have much in common: concern over Islamic militancy, friendly relations with Israel, military security, the rise of China, and more. But India needs energy, and Iran has it. Thus India is reluctant to join the international campaign to control Iran's nuclear programs—and a promising U.S.-India relationship suffers.

- Until its election late in 2005, **Sri Lanka** seemed headed for a peaceful resolution of its civil war, brokered by an outside team led by Norwegian negotiators. These talks have stalled, with no resolution in sight.

- A devastating earthquake in embattled **Kashmir** in 2005 seemed to bring hope that thousands had not died in vain—their plight appeared to bring India and Pakistan together in a joint effort to ease the suffering. But before long, old enmities resurfaced, and today the hope for progress has faded.

- On South Asia's eastern flank, **Bangladesh** remains vulnerable to both political and meteorological storms with a military caretaker government in place and cyclone recovery continuing.

- Keep an eye on Pakistan's southern, Iran-bordering, energy-rich province of **Baluchistan**, where many of the locals are unhappy with Islamabad's rule and where the Baluchistan Liberation Army carries on a low-level insurgency with the potential to escalate.

- **Pakistan**'s election in 2008 brought change to this key country in the War on Terror. Whether its precarious democracy will hold and keep the country together remains to be seen.

encompassing fuel supplies and technology transfers, and easing energy-poor India's way toward nuclear-energy development. Since both Pakistan and India possess nuclear weapons, the United States had withheld such cooperation because India never signed international nonproliferation agreements.

Thus South Asia today is witnessing another of its historic invasions. But this time it is an invasion of investment, commerce, technology, custom, and connectivity—in short, globalization. As we noted, hundreds of millions of people are as yet virtually unaffected by this penetration, but from the textile sweatshops of Bangladesh to the computer centers of Bengaluru (Bangalore) you can see the future. In the streets, where shops increasingly carry international goods, you see the intensifying blend of the traditional and the external. It would be impractical to draw a map of all this, because the mosaic is too intricate. India's teeming Mumbai (Bombay) has become a global metropolis, part of the international network of world cities. On the other hand, Pakistan's Lahore remains traditional and locally focused. Chaotic, dangerous Karachi is somewhere in between. The ricefields of Bangladesh and the grinding rural poverty of India's adjoining West Bengal State are a world apart from the expansive wheatfields of the Punjab.

The latest—and current—invasion of South Asia is in the process of linking this realm to the wider world as never before, not even during colonial times. As we focus on the regions of this realm, we should remember that the fortunes of the twenty-first century world will be linked directly to the fate of these constituent parts.

Regions of the Realm

PAKISTAN: SOUTH ASIA'S WESTERN FLANK

If India is the dominant entity in South Asia, why focus first on Pakistan? There are several reasons, both historic and geographic. Here lay South Asia's earliest urban civilizations, whose innovations radiated into the great peninsula. Here, too, lies South Asia's Muslim frontier, contiguous to the great Islamic realm to the west and irrevocably linked to the enormous Muslim minority to its east. Pakistan's cultural landscapes bear witness to its transitional location. Teeming, disorderly Karachi is the typical South Asian city; as in India, the largest urban center lies on the coast. Historic, architecturally Islamic Lahore is reminiscent of the scholarly centers of Muslim Southwest Asia. In Pakistan's eastern borderland, the postcolonial boundary divides a Punjab that stretches beyond the horizon on both sides of the line, a contiguous cultural landscape of villages, wheatfields, and irrigation ditches. In the northwest, Pakistan resembles Afghanistan in its huge migrant populations and its mountainous frontier. And in the far north, Pakistan and India are locked in a deadly conflict over Jammu and Kashmir. The western flank is South Asia's most critical region, especially today. Pakistan lies on the eastern periphery of the vast contiguous Islamic realm that extends from Casablanca to Kashmir, in the shadow of Afghanistan, and in the fulcrum of the War on Terror. Pakistan remains tense in the wake of the 2007 assassination of Benazir Bhutto and subsequent political developments.

Territorially, Pakistan is not large by Asian standards; its area is about the same as that of Texas plus Louisiana (Fig. 8-10). But Pakistan's population of 173.9 million makes it one of the world's ten most populous states. Among Muslim countries (officially, it is known as the Islamic Republic of Pakistan) only Southeast Asia's Indonesia is larger, but Indonesia's Islam is much less pervasive than Pakistan's.

East and West, North and South

Upon independence in 1947, Pakistan consisted not only of the territory we know as Pakistan today, which was known as West Pakistan, but also of East Pakistan—present-day Bangladesh. That union was based on their shared adherence to Islam, but it did

India and Pakistan continue their conflict over Kashmir—a dangerous flashpoint in a most volatile part of the world. Pictured here are refugees who went to India from Pakistan-occupied Kashmir, and were detained when they tried to cross back to the Pakistan side of the border. They are demonstrating that the government has continuously ignored their problems and are shouting slogans against the Round Table Conference on Kashmir that was held in New Delhi, India, in April 2007. The placards read: "Jammu and Kashmir State, have some shame," and at right, "Place all our rights here." (© AFP/Getty Images)

FIGURE 8-10

© H. J. de Blij, P. O. Muller, and John Wiley & Sons, Inc.

tled interior and by placing it on the doorstep of contested territory, Pakistan announced its intent to stake a claim to its northern frontiers. And by naming the city Islamabad, Pakistan proclaimed its Muslim foundation, here in the face of the Hindu challenge. This politico-geographical use of a national capital can be assertive, and Islamabad exemplifies the principle of the **12 forward capital**.

The Kashmir Issue

The contested territory was (and is) the northern area of Jammu and Kashmir (Kashmir for short), perhaps the most significant of the realm's flashpoints. Mountainous, remote, and unassigned when independence came in 1947, Kashmir was 75 percent Muslim but ruled by a Hindu elite. The Maharajah wanted an autonomous state that would remain outside both India and Pakistan, but a Muslim uprising put an end to that notion. Indian and Pakistani armies entered the fray, but their stalemate only postponed renewed wars in 1965 and 1971. The result was the armistice *Line of Control* mapped in Figure 8-11.

Pakistan is firmly in control of its sector of Kashmir, and the question now centers on the future of Indian-held Kashmir, part of which has been taken by China (Fig. 8-11). Even here, Muslims are in the majority, heavily concentrated in the Vale of Kashmir; the area named Ladakh on the map is strongly Buddhist; and the center and southeast are mostly minority Hindu. A referendum would undoubtedly go in Pakistan's favor, but secular India cannot abandon its minorities to the kind of Islamic dominance prevailing there. Skirmishes and acts of terrorism continue in the shadow of Islamabad; even

not last long. A political crisis in 1971 led to conflict and East Pakistan's declaration of independence as the People's Republic of Bangladesh.

When Pakistan became a sovereign state following the partition of British India in 1947, its capital was Karachi on the south coast, near the western end of the Indus Delta. As the map shows, the present capital is Islamabad, near the larger city of Rawalpindi in the north, not far from Kashmir. By moving the capital from the safe coast to the embat-

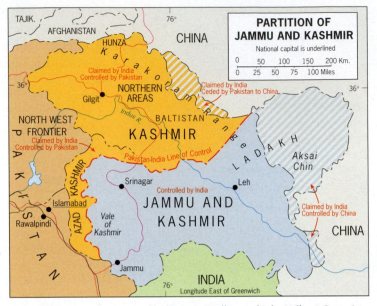

FIGURE 8-11   © H. J. de Blij, P. O. Muller, and John Wiley & Sons, Inc.

the catastrophic 2005 Kashmir earthquake—which took at least 70,000 lives and required coordinated, cross-border relief efforts—failed to reduce the political tensions. And so it is that two nuclear powers are at a stalemate over this border dispute.

Forging Centripetal Forces

At independence, Pakistan had a bounded national territory, a capital, a cultural core, and a population—but it had few centripetal forces to bind state and nation, especially while East Pakistan was still a part of it. It is easier to understand the issues confronting Pakistan if you think of it as a collection of ethnic groups and tribes rather than a nation-state. The disparate groups within Pakistan shared the Islamic faith and an aversion for Hindu India, but little else. Karachi and the coastal south, the desert of Baluchistan, the city of Lahore and Punjab, the rugged northwest along Afghanistan's border, and the mountainous far north are worlds apart, and a Pakistani nationalism to match that of India at independence did not exist. Successive Pakistani gov-

ernments, civilian as well as military, turned to Islam to provide the common bond that history and geography had denied the country. In the process, Pakistan became one of the world's most theocratic states; its common law, based on the English model, was gradually transformed into a Quranic (Koranic) system with Islamic Sharia courts and associated punishments.

But even Islam itself is not unified in restive Pakistan. About 77 percent of the people are Sunni Muslims, and the Shia minority numbers about 20 percent. Sunni fanatics intermittently attack Shi'ites, leading to retaliation and creating grounds for subsequent revenge.

Despite the Islamization of Pakistan's plural society, it remains a strongly regionalized country in which Urdu is the official language and English is still the ***lingua franca*** of the elite. Yet several other major languages prevail in diverse parts, and lifeways vary from nomadism in Baluchistan to irrigated farming in Punjab to pastoralism in the northern highlands. Centrifugal forces could overcome the centripetal forces in this region, resulting in devolution and a redrawing of the map.

The Provinces

As Figure 8-10 shows, Pakistan is administratively divided into four provinces: Punjab, Sind, North West Frontier, and Baluchistan. Pakistan was founded as a federal state and it was not intended to be either an Islamic state or a military dictatorship, both of which it has been for much of its history. The relations between these provinces were deemed to be important to the cohesion of the country, but they have been difficult.

In large part these relations are difficult because *Punjab* is disproportionately dominant in Pakistan: it is the country's core area, home to almost 55 percent of the entire population, contains the capital (Islamabad), the cultural focus (Lahore, 2000 years old, a great Muslim center with magnificent architecture), and leads the nation in almost every economic category. To the south, *Sind* lies centered on the chaotic port city of Karachi, and here the rice- and wheatfields in the lower Indus Basin form the breadbasket of Pakistan. Commercially, cotton is king, supporting major textile industries in the province's cities and towns. To the west of Sind lies desert *Baluchistan*, land of ancient caravan routes still operating and home to a sizeable Shi'ite minority, reflecting its proximity to Iran. Recent oil and gas discoveries may transform this remote province in the near future. The *North West Frontier*, the aptly named, mountainous province closest to the core area of Afghanistan anchored by Peshawar and adjoined by the unruly Tribal Areas, has repeatedly taken the brunt of events occurring across the border. When Soviet armed forces tried to install a secular regime in Kabol during the 1980s, several million Pushtun refugees streamed out of Afghanistan, settled in refugee camps in the North West Frontier, and became a major political (and Islamic-cultural) force here. Many of these refugees wanted political rights in Pakistan and campaigned to bring strict Islamic government to the North West Frontier Province, receiving cross-border support from their kinspeople still in Afghanistan. This example of **13 irredentism** posed serious problems for Pakistan, and it worsened when the repressive Taliban regime took power during the mid-1990s and many more refugees crossed the border. All this turmoil made the North West Frontier Province a haven for radicals, and today the provincial government is revivalist Islamic, challenging the authority of the federal government in Islamabad. The Tribal Areas along the border between Pakistan and Afghanistan serve as another of the realm's flashpoints. This territory, traditionally under the control of mullahs and tribal chiefs, is a hilly warren of villages and tracks where government authority fails. Long a hideout for rebels and refugees, it has taken on added significance during the War on Terror: it is believed to be the hiding place of Usama bin Laden and others in the al-Qaeda–Taliban alliance.

Pakistan's Prospects

Until September 2001, Pakistan was a typical low-income country with a troubled economy, huge debts, worrisome social indicators (see Appendix B), an unstable government, an army general having recently ousted his civilian adversary and taken control of the state. As the only country with day-to-day relations with the Taliban regime in neighboring Afghanistan, Pakistan faced another round of disputes over military versus civilian rule, Islamic versus federal courts, education policies, and other issues. But in the aftermath of the 9/11 attacks and the formal start of the War on Terror, Pakistan's ruler, General Pervez Musharraf, had to make a choice between support for the United States and the West or neutrality. It was a difficult decision: support for the United States would energize Islamic militants in his country, whereas neutrality would bring adversarial relations with its long-term ally.

When Pakistan joined the West in the War on Terror and U.S. troops, investigators, journalists, and others arrived, the benefits were financial and military and the costs were social and political. Pakistan moved to close the revivalist *madrassas* (Islamic religious schools) that had produced many of the Taliban extremists, but, as we noted, militants took over

the regional government of the North West Frontier Province. Acts of defiance and terrorism occurred from Karachi to Peshawar, and it is virtually certain that the al-Qaeda and Taliban leadership escaped to, and were given refuge in, the Tribal Areas. Political tensions in Pakistan escalated, but the quick collapse of the Taliban regime helped defuse the risk.

Still, the further radicalization of Pakistani society is likely, and the country faces long-term unrest. During the summer and fall of 2007 lawyers demonstrated on the streets of Lahore against the continuing military dictatorship of Pervez Musharraf. In the fall of 2007 he instituted a state of emergency that allowed him to remove 'uncooperating' justices from the supreme court without the usual scrutiny. Under severe international pressure he did step down from the military. Two former prime ministers returned from exile in 2007; one, Benazir Bhutto, was assassinated in December as she campaigned for her party, unleashing civil unrest throughout the country. Elections in early 2008 spurred the removal of the dictatorial and embattled President Musharraf and put into place a fractious coalition government.

An unexpected sight anywhere in the world, lawyers in business suits taking to the streets and hurling stones at riot police. Such was the situation in Lahore, Pakistan, however, in March 2007 during an anti-government rally. Lawyers boycotted court proceedings, clashed with riot police, and burned an image of President General Pervez Musharraf in a nationwide protest against the ouster of the country's top judge Iftikhar Muhammed Chaudhry. This was only the latest chapter in the political troubles of that part of the world. (© AP/Wide World Photos)

INDIA: SOUTH ASIA'S GIANT

If you have been reading the press and watching television over the past few years, you have seen the growing attention being paid to India—not just in North America but around the world. The *New York Times* in late 2005 began a series under the title "India Accelerating." *Newsweek* in March 2006 published a special feature under the banner headline "India Rising." Three months later a leading British newsmagazine, *The Economist*, produced a Survey of India introduced by an editorial asking "Can India Fly?" State visits by an Indian prime minister to Washington, an American president to New Delhi, and a senior Chinese leader to India, all within a matter of months, reflected India's growing global importance. Meanwhile, CNN kept up a drumbeat of reportage about American jobs departing for India as U.S. corporations saved money by outsourcing.

India is on the move—but will India become the next economic superpower?

Not only does it occupy three-quarters of the great land triangle of South Asia: India also is poised to overtake China to become the world's most populous country. Already, India is the planet's largest democracy, a federation of 28 States and several additional Territories with a 2008 population of 1.158 billion. For some, it is India's democratic institutions that will eventually propel it further along than still communist China. But in the meantime that democracy is messy and even chaotic, and not all components of an economic boom are occurring at once.

That India has endured as a unified country is a politico-geographical miracle. India is a cultural mosaic of immense ethnic, religious, linguistic, and economic diversity and contrast; it is a state of many nations. The period of British colonialism gave India the underpinnings of unity: a single capital, an interregional transport network, a *lingua franca*, a civil service. Upon independence in 1947, India adopted a federal system of government, giving regions and peoples some autonomy and identity, and allowing others to aspire to such status. Unlike various countries in Subsaharan Africa, where federal systems failed and where military dictatorships replaced them, India remained essentially democratic and retained a federal framework in which States have considerable local authority.

States and Peoples

The map of India's political geography shows a federation of 28 States, 6 Union Territories (UTs), and 1 National Capital Territory (NCT) (Fig. 8-12). The federal government retains direct authority over the UTs, all of which are small in both territory and population. The NCT, however, includes Delhi and the capital, New Delhi, and has more than 17 million inhabitants.

Postcolonial Restructuring

The political spatial organization shown in Figure 8-12 is mainly the product of India's restructuring following independence from Britain. Its State boundaries reflect the broad outlines of the country's cultural mosaic: overall the system recognizes languages, religions, and cultural traditions. Indians speak 14 major and numerous minor languages (see Fig. 8-4 inset),

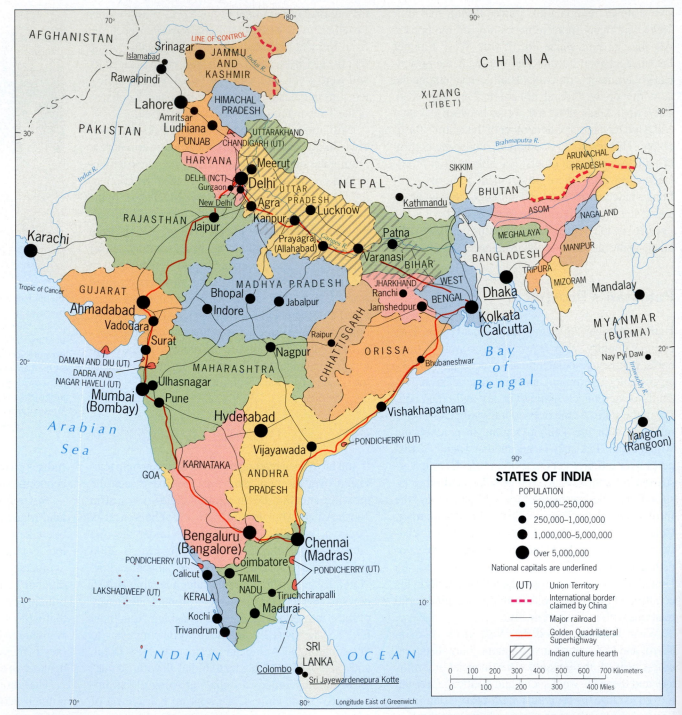

FIGURE 8-12

© H. J. de Blij, P. O. Muller, and John Wiley & Sons, Inc.

and while Hindi is the official language (and English is the *lingua franca*, especially for the educated), it is by no means universal. The map is the product of endless compromise—endless because demands for modifications of it continue to this day; as recently as late 2000, the federal government authorized the creation of three new States. In the northeast lie very small States established to protect the local traditions of small populations; minority groups in the larger States ask why they should not receive similar recognition.

With only 28 States for a national population of 1.158 billion, several of India's States contain more people than many countries of the world. As Figure 8-12 indicates, the (territorially) largest States lie in the heart of the country and in the great southward-pointing peninsula. Uttar Pradesh (190 million) and Bihar (about 94 million) constitute much of the Ganges River Basin and form the core area of modern India. Maharashtra (almost 108 million), anchored by the great coastal megacity of Mumbai and home to 19.3 million today), also has a population larger than that of most countries. West Bengal, the State that adjoins Bangladesh, has more than 87 million residents, 15 million of whom live in its urban focus, Kolkata (Calcutta).

These are staggering numbers, and they do not decline much toward the south. Southern India consists of four States linked by a discrete history and by their distinct Dravidian languages. Facing the Bay of Bengal are Andhra Pradesh (82 million) and Tamil Nadu (66 million), both part of the hinterland of the city of Chennai (home to 7.3 million today) and located on the coast near their joint border. Facing the Arabian Sea are Karnataka (58 million) and Kerala (34 million). Kerala, often at odds with the federal government in New Delhi, has long had the highest literacy rate in India and one of the lowest rates of population growth owing to strong local government and strictly enforced policies. "It's a matter of geography," explained a teacher in the Kerala city of Kochi (Cochin). "We are here about as far away as you can get from the capital, and we make our own rules."

The Northern Peripheries

As Figure 8-12 reveals, India's smaller States lie mainly in the northeast, on the far side of Bangladesh, and in the northwest, toward Jammu and Kashmir. North of Delhi, India is flanked by China and Pakistan, and physical as well as cultural landscapes change from the flatlands of the Ganges to the hills and mountains of spurs of the Himalayas. In the State of Himachal Pradesh, forests cover the hillslopes and relief reduces living space; only 6.7 million people live here, many in small, comparatively isolated clusters. Before independence and political consolidation, the colonial government called this area the Hill States.

But the map becomes even more complex in the distant northeast, beyond the narrow corridor between Bhutan and Bangladesh. The dominant State here is Asom (Assam) with more than 30 million, famed for its tea plantations and important because its oil and gas production amounts to more than 40 percent of India's total.

In the Brahmaputra Valley, Asom resembles the India of the Ganges. But in almost every direction from Asom, things change. To the north, in sparsely populated Arunachal Pradesh (1.2 million), we are in the Himalayan offshoots again. To the east, in Nagaland (2.2 million), Manipur (2.4 million), and Mizoram (1 million), lie the forested and terraced hillslopes that separate India from Myanmar (Burma) (Chapter 10). This is an area of numerous ethnic groups (more than a dozen in Nagaland alone) and of frequent rebellion against Delhi's government. And to the south, the States of Meghalaya (2.5 million) and Tripura (3.5 million), hilly and still wooded, border the teeming floodplains of Bangladesh. Here in the country's northeast, where people are restive, India faces one of its strongest regional challenges.

China's Latent Claims

Here, too, lies another of the region's flashpoints, as India faces a challenge from its powerful neighbor China. As Figure 8-12 shows, China's version of its southern border with India lies deep inside the northeast, even coinciding with the Asom boundary part of the way. China claims virtually all of Arunachal Pradesh, an Indian State, and publishes maps showing this area as part of the People's Republic. This dispute has its origins in the Simla Conference of 1913–1914, when the British negotiated a treaty with Tibet that defined their boundary in accordance with the terms laid down on the map by their chief negotiator, Sir Henry McMahon. Essentially, this *McMahon Line*, as it came to be known, ran along the most prominent crestline of the Himalayas. However, China's representatives at the conference refused to sign the agreement, arguing that Tibet was part of China and had no power to enter into treaties with foreign governments. In 1962, Chinese armed forces crossed the border and occupied Indian territory, but then withdrew. In 2003, the heads of state of India and China met on matters of mutual interest, but the Chinese leadership did not yield on this potentially explosive issue, and the matter remains unresolved today.

India's Changing Map

After independence, the Indian government began phasing out the privileged Princely States the British had protected during the colonial period. Next, the government reorganized the country on the basis of its major regional languages (see Fig. 8-4). Hindi, spoken by more than one third of the population, was designated the country's official language, but the Indian constitution gave 13 other major languages national status, including the four Dravidian languages of the south. English, it was anticipated, would become India's common language, its *lingua franca* at government, administrative, and business levels. Indeed, English not only remained the language of national administration but also became the chief medium of commerce in growing urban India. English was the key to better jobs, financial success, and personal advancement, and the language constituted a common ground in higher education. This colonial legacy has been a key to India's recent economic success.

Devolutionary Pressures

The newly devised framework based on the major regional languages, however, proved to be unsatisfactory to many communities in India. In the first place, many more languages are in use than the 14 that had been officially recognized. Demands for additional States soon arose. As early as 1960, the State of Bombay was divided into two language-based States, Gujarat and Maharashtra.

Throughout India's existence as an independent country, this devolutionary pressure has continued. In 2000, three new States were recognized: Jharkhand, carved from southern Bihar State on behalf of 18 poor districts there; Chhattisgarh, where tribal peoples had been agitating since the 1930s for separation from the State of Madhya Pradesh; and Uttarakhand, which split from India's most populous, Ganges Basin, core-area State of Uttar Pradesh on the basis of its highland character and lifeways (Fig. 8-12).

For many years India has faced quite a different set of cultural-geographic problems in its northeast, where numerous ethnic groups occupy their own niches in a varied, forest-clad topography. The Naga, a cluster of peoples whose domain had been incorporated into Asom State, rebelled soon after India's independence. A protracted war brought federal troops into the area; after a truce and lengthy negotiations, Nagaland was proclaimed a State in 1961. This led the way for other politico-geographical changes in India's problematic northeastern wing.

The Sikhs

A further dilemma involves India's Sikh population. The Sikhs (the word means disciples) adhere to a religion that was created about five centuries ago to unite warring Hindus and Muslims into a single faith. This faith's principles rejected negative aspects of Hinduism and Islam, and it gained millions of followers in Punjab and adjacent areas. During the colonial period, many Sikhs supported the British administration of India, and by doing so they won the respect and trust of the British, who employed tens of thousands of Sikhs as soldiers and policemen. By 1947, there was a large Sikh middle class in Punjab. When independence came, many left their rural homes and moved to the cities to enter urban professions. Today, they still exert a strong influence over Indian affairs, far in excess of the less than 2 percent of the population (about 20 million) they constitute.

After independence, the Sikhs demanded that the original Indian State of Panjab (Punjab) be divided into a Sikh-dominated northwest and a Hindu-majority southeast. The government agreed, so that Punjab as now constituted (Fig. 8–12) is India's Sikh stronghold, whereas neighboring Haryana State is mainly Hindu.

The Muslims

These ethnic, cultural, and regional problems are but a sample of the stresses on India's federal framework. There is no Muslim State in India, but projections based on the 2001 census showed that India today contains 165 million Muslims within its borders—the largest cultural minority in the world. As Figure 8-5D shows, the percentage of Muslims is highest in remote Jammu and Kashmir, but it also is substantial in such widely dispersed States as Kerala, Asom, and Uttar Pradesh. Moreover, the Muslim population today (14.3 percent) constitutes a larger percentage than it did after partition (9.9 percent). This Islamic minority ranks among the most rapidly growing sectors of India's population and is strongly urbanized as well—nearly one third of the population of India's largest city, Mumbai, is Muslim.

Relations between the Hindu majority and Muslim minority are complex. What makes the news is conflict—for example, in 2002 when Muslims attacked a train carrying Hindus to a contested holy site in Ayodhya, killing dozens, which was followed by retaliation in several towns (but not others) in Gujarat. What does not make the news is that a Muslim population surpassing 165 million lives and participates in the kind of democracy that is all but unknown in the Muslim world itself.

Centrifugal Forces: From India to Hindustan?

In Chapter 1 we introduced the concept of centrifugal and centripetal forces affecting the fabric of the state. No country in the world exhibits greater cultural diversity than India, and variety in India comes on a scale unmatched anywhere else on earth. Such diversity spells strong centrifugal forces, although, as we will see, India also has powerful consolidating bonds.

Class and Caste

Among the centrifugal forces, Hinduism's stratification of society into castes remains pervasive. Under Hindu dogma, *castes* are fixed layers in society whose ranks are based on ancestries, family ties, and occupations. The **14 caste system** may have its origins in the early social divisions into priests and warriors, merchants and farmers, craftspeople and servants; it may also have a racial basis, for the Sanskrit term for caste is color. Over the centuries, its complexity grew until India had thousands of castes, some with a few hundred members, others containing millions. Thus, in city as well as in village, communities were segregated according to caste, ranging from the highest (priests, princes) to the lowest (the untouchables). The term *untouchable* has such negative connotations that some scholars object to its use. Alternatives include *dalits* (oppressed), the common term in Maharashtra State but coming into general use; *harijans* (children of God), which was Gandhi's designation, still widely used in the State of Bihar; and *Scheduled Castes*, the official government label.

A person was born into a caste based on his or her actions in a previous existence. Hence, it would not be appropriate to counter such ordained caste assignments by permitting movement (or even contact) from a lower caste to a higher one. Persons of a particular caste could perform only certain jobs, wear only certain clothes, and worship only in prescribed ways at particular places. They or their children could not eat, play, or even walk with people of a higher social status. The untouchables occupying the lowest tier were the most debased, wretched members of this rigidly structured social system. Although the British ended the worst excesses of the caste system, and postcolonial Indian leaders—including Mohandas (Mahatma) Gandhi (the great spiritual leader who sparked the independence movement) and Jawaharlal Nehru (the first prime minister)—worked to modify it, a few decades cannot erase centuries of class consciousness. In traditional India, caste provided stability and continuity; in modernizing India, it constitutes an often painful and difficult legacy and can be stifling for an emerging economy.

Reforming the System

Today we can discern a geography of caste—a degree of spatial variation in its severity. Cultural geographers estimate that about 15 percent of all Indians are of lower caste, about 40 percent of backward caste (one important rank above the lower caste), and some 18 percent of upper caste, at the top of which are the Brahmans, men in the priesthood. (The caste system does not extend to the Muslims, Sikhs, and other non-Hindus in India, which is why these percentages do not total 100.) The colonial government and successive Indian governments have tried to help the lowest castes. This effort has had more effect in the urban than in the rural areas of India. In the isolated villages of the countryside, the untouchables often are made to sit on the floor of their classroom (if they go to school at all); they are not allowed to draw water from the village well because they might pollute it; and they must take off their shoes, if they wear any, when they pass higher-caste houses. But in the cities, untouchables have reserved for them places in the schools, a fixed percentage of State and federal government jobs, and a quota of seats in national and State legislatures. Gandhi, who took a special interest in the fate of the untouchables in Indian society, accomplished much of this reform.

The caste system remains a powerful centrifugal force, not only because it fragments society but also because efforts to weaken it often result in further division. Gandhi himself was killed, only a few months after independence, by a Hindu fanatic who opposed his work for the least fortunate in Indian society. But progress is being made, and while efforts to help the poorest are not always popular among the better-off, the future of India depends on it.

Hindutva

Another growing centrifugal force in India has to do with a concept known as *Hindutva* or Hinduness—a desire to remake India as a society in which Hindu principles prevail. This concept has become the guiding agenda for a political party that became a powerful component of the federal government, and it is variously expressed as Hindu nationalism, Hindu patriotism, and Hindu heritage. This naturally worries Muslims and other minorities, but it also concerns those who understand that India's secularism, its separation of religion and state, is indispensable to the survival of its democracy. *Hindutva* enthusiasts want to impose a Hindu curriculum on schools, change the flexible family law in ways that would make it unacceptable to Muslims, inhibit the activities of non-Hindu religious proselytizers, and forge an India in which non-Hindus are essentially outsiders. Moderate Hindus and non-Hindus in India oppose such notions, which are as divisive as any India has faced. They nevertheless acknowledged the appeal of the Bharatiya Janata Party (BJP), which has had mixed success in elections, reflecting internal party struggles between moderates and hardliners.

The radicalization of Hinduism and the infusion of Hindu nationalism into federal politics loom today as twin threats to India's unity, but Indian voters have not rushed to embrace these initiatives. The BJP and other Hindu-nationalist parties are potentially polarizing electorates at State as well as federal levels; the recent spate of name changes on India's map is one manifestation of this polarization. If Indian politics were to fragment along religious lines, the miracle of Indian unity would be at risk.

Centripetal Forces

In the face of all these divisive forces, what bonds have kept India unified for so long? Without question, the dominant binding force in India is the cultural strength of Hinduism, its sacred writings, holy rivers, and influence over Indian life. For most Indians, Hinduism is a way of life as much as it is a faith, and its diffusion over virtually the entire country (regardless of the Muslim, Sikh, and Christian minorities) brings with it a national cohesion that constitutes a powerful antidote to regional divisiveness. Over the long term, however, the key ingredients of this Hinduism have been its gentility and introspection, radical outbursts notwithstanding.

Democracy

Another centripetal force lies in India's democratic institutions. In a country as culturally diverse and as populous as India, reliance on democratic institutions has been a birthright ever since independence, and democracy's survival—raucous, often corrupt, always free—has been a crucial unifier.

Furthermore, communications are better in much of India than in many other countries in the global periphery, and the continuous circulation of people, ideas, and goods helps bind the disparate state together. Before independence, opposition to British rule was a shared philosophy, a strong centripetal force. After independence, the preservation of the union was a common objective, and national planning made this possible.

Accommodation

India's capacity for accommodating major changes and its flexibility in the face of regional and local demands are also a centripetal force. Boundaries have been shifted; internal political entities have been created, relocated, or otherwise modified; and secessionist demands have been handled with a mixture of federal power and cooperative negotiation. Indians in South Asia have accomplished what Europeans in former Yugoslavia could not, and India's history of success is itself a centripetal force.

Education

Still another centripetal force in India is education. Education has always been highly valued, especially science and engineering. The country takes great pride in its high literacy rates, which in urban India exceed 96 percent for both males and females. In rural areas numbers are much lower, especially for women. Overall, however, these numbers are substantially above those of most neighboring countries and reflect Indians' determination to avail themselves of every educational opportunity. Private institutes teaching English abound in the cities, so it is not surprising that when opportunities for the outsourcing of service jobs arose on the global economic scene, India had the educated and English-speaking workforce to seize them.

Leadership

Finally, no discussion of India's binding forces would be complete without mentioning the country's strong leadership. Gandhi, Nehru, and their successors did much to unify India by the strength of their compelling personalities. For many years, leadership was a family affair: Nehru's daughter, Indira Gandhi, twice took decisive control (in 1966 and 1980) after weak governments, and her son, Rajiv Gandhi (who, in 1991, like his mother seven years earlier, also was assassinated), was prime minister in the late 1980s. Since then, political leadership of India has been less dynastic—and also less cohesive.

Urbanization

India is famous for its great and teeming cities, but India is not yet an urbanized society. Only 29 percent of the population lived in towns and cities in 2008—but in terms of sheer numbers, that 29 percent amounts to over 336 million people, more than the entire population of the United States.

FROM THE FIELD NOTES

Squatters in Mumbai, India. "Searing social contrasts abound in India's overcrowded cities. Even in Mumbai (Bombay), India's most prosperous large city, hundreds of thousands of people live like this, in the shadow of modern apartment buildings. Within seconds we were surrounded by a crowd of people asking for help of any kind, their ages ranging from the very young to the very old. Somehow this scene was more troubling here in well-off Mumbai than in Kolkata (Calcutta) or Chennai (Madras), but it typified India's urban problems everywhere." (© H. J. de Blij)

www.conceptcaching.com

And India's rate of urbanization is on the upswing. People by the hundreds of thousands are arriving in the already overcrowded cities, swelling urban India by about 5 percent annually, almost three times as fast as the overall population growth. Not only do the cities attract as they do everywhere; many villagers are driven off the land by the desperate conditions in the countryside. As villagers manage to establish themselves in Mumbai or Kolkata or Chennai, they help their relatives and friends to join them in squatter settlements that often are populated by newcomers from the same area, bringing their language and customs with them and cushioning the stress of the move.

As a result, India's cities display staggering social contrasts. Squatter shacks without any amenities at all crowd against the walls of modern high-rise apartments and condominiums (photo above). Hundreds of thousands of homeless roam the streets and sleep in parks, under bridges, on sidewalks. As crowding intensifies, social stresses multiply. Disorder never seems far from the surface; sporadic rioting, often attributable to rootless urban youths unable to find employment, has become commonplace in India's cities.

Urban Tradition and Evolution

India's modern urbanization has its roots in the colonial period, when the British selected Calcutta (Kolkata), Bombay (Mumbai), and Madras (Chennai) as regional trading centers and fortified ports. Madras was fortified as early as 1640; Bombay (1664) had the situational advantage of being the closest of all Indian ports to Britain; and Calcutta (1690) lay on the margin of India's largest population cluster and had the most productive hinterland, to which the Ganges Delta's countless channels connected it. This natural transport network made Calcutta an ideal colonial headquarters, but the population of Bengal was often rebellious. In 1912 the British moved their colonial government from Calcutta to the safer interior city of New Delhi, built adjacent to the old Mughal capital of Delhi.

Figure 8-12 displays the distribution of major urban centers in India. Except for Delhi-New Delhi, the largest cities have coastal locations: Kolkata dominates the east, Mumbai the west, and Chennai the south. But urbanization also has expanded in the interior, notably in the core area. Road links among these (and other) cities remain grossly inadequate. Now, in the first decade of the twenty-first century, India is constructing a nationwide four-lane expressway that will link the four anchors of this urban system (Delhi, Mumbai, Chennai, and Kolkata) and in the process connect 15 other major cities along this national route called the *Golden Quadrilateral*.

Economic Geography

Improving India's infrastructure is not enough to overcome all serious impediments to the country's economic advancement. Take another look at Figure 8-12 and note that the Golden Quadrilateral crosses numerous State boundaries. In the United States, we are used to seeing thousands of trucks on the interstate highways, crossing from one State into another without slowing down; there are truck-weighing stations along these expressways, but in the newest ones all the truck has to do is slow down enough for electronic surveillance to record its passing.

In India, truck drivers face a very different experience. It can take as long as nine days, including more than 30 hours waiting at State-border checkpoints and tollbooths, for a loaded truck to travel from Kolkata to Mumbai via Chennai, or less than the distance from Los Angeles to New Orleans on I-10. Drivers are subject to daunting piles of paperwork and repeated demands for bribes. When mechanical problems occur, it may take days to get the truck back on the road. It will take more than expanding India's highway network to achieve the improved circulation the country so desperately needs.

Globalization

When you arrive in any Indian city, you are struck by the number of small shops everywhere—tiny businesses wedged into every available space in virtually every nonpublic building along every street. Even the upper, walkup floors are occupied by shops, their advertising signs suspended from windows and balconies. According to a study by the Indian government's Department of Consumer Affairs, India has the highest density of retail outlets of any country in the world with 15 million shops (compared to well below 1 million in the United States, where the marketplace is 13 times richer). After farming, the retail sector is India's largest provider of jobs.

What keeps all these small stores in business? Most of them earn very little and can afford to stay open only because they are part of what economic geographers call the **informal sector**: they are

Bengaluru (Bangalore) in India's Karnataka State is the epicenter of the information technology (IT) boom in this realm. That is not the whole story, however. Here a woman collects water from a common well in a slum located right next to the Infosys headquarters. Infosys is one of India's largest IT companies with about 50,000 employees worldwide. The company has benefited greatly from the country's current IT boom.   (© Frederick Remander/Redux Pictures)

essentially unregistered, pay no rent and probably no taxes, use family labor, have been handed down through generations, and survive because India's economic geography, bound by longstanding protective government regulation, has been slow to change. When you are in India you may wonder how so many shops can stay in business, but the answer is in the throngs of people on the sidewalks (spilling over into the clogged traffic in the streets). Not many of these people are wealthy, but all of them need basic goods and some can afford small luxuries, and so the shops tend to be busy all day.

Now imagine what would happen if India suddenly opened its doors to large international retailing companies like Wal-Mart. An invasion of foreign superstores would force millions of India's small shops to close, throw countless workers into unemployment, and destroy ways of life that have put most urban residents within walking distance of their daily needs. And yet this is what India faces. This is the era of globalization, and large-scale organized retailing, long held back by a combination of restraints (such as the nationwide distribution problems, but also legal and cultural obstacles), is emerging within India itself. This means that competition from foreign companies will somehow infiltrate the market, and in this respect the Indian government faces only one key question: how to keep control of the process through gradual deregulation. That issue is complicated, however, by India's chaotic democracy. Even if New Delhi approves appropriate legislation, individual States may counter it by imposing their own regulations and restrictions.

Still, the statistics show that the process is already well underway. India's middle class, now estimated to include some 300 million people (which equals the entire U.S. population), is growing rapidly, and something very new is beginning to appear in the cities: shopping malls. As recently as 2000, this vast country containing one-sixth of humankind did not have a single shopping center! By 2005, there were more than 100, and by the end of 2007 more than 350 were in operation. You will see American fast-food restaurants among the establishments in these malls as well as the brand names of numerous other foreign

companies, proving that globalization has already breached India's walls. The question remains: can India's economic transformation be achieved without severe social dislocation?

Farming's Enduring Importance

Agriculture still provides more jobs in India than any other economic sector, and India's fortunes (and misfortunes) remain strongly tied to farming. For all the emphasis on India's cities and middle-class growth, about 70 percent of the people still live on (and from) the land, traditional farming methods persist, yields per hectare remain among the world's

lowest, and hunger and malnutrition still afflict millions even as grain surpluses accrue. The relatively few areas of modernization, as in the wheatlands of the Punjab, are islands in a sea of stagnation. Thus the agricultural sector and indeed the entire country is directly vulnerable to environmental variations, with all the risks that entails. Furthermore, land reform has essentially failed; roughly one-quarter of India's entire cultivated area, including much of the best land, is still owned by less than 5 percent of the country's landholders, who have much political influence and obstruct redistribution. Perhaps half of all rural families own less than one hectare (2.5 acres) or no land at all: an estimated 175 million live and work as tenants, always uncertain of their fate.

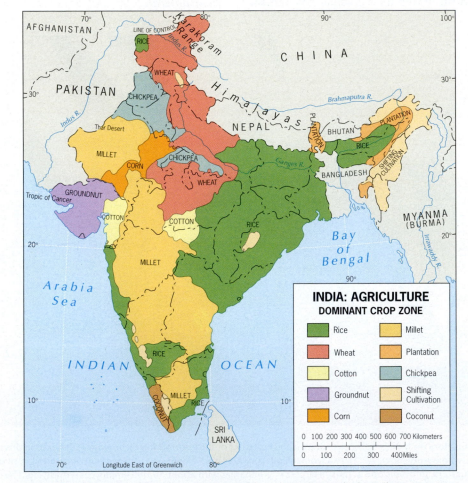

FIGURE 8-13   © H. J. de Blij, P. O. Muller, and John Wiley & Sons, Inc.

Getting produce to markets—if there is any produce to sell—is a struggle for millions of farmers. In 2007, almost half of India's 600,000 villages could not be accessed by truck or car; in this era of modern transportation, animal-drawn carts still far outnumber motor vehicles nationwide.

As you can see in Figure 8-13, India's agriculture reflects the climatic distribution shown in Figure G-9, the rainfall pattern in Figure G-8, and the monsoonal cycle depicted in Figure 8-3. Rice dominates along the Arabian Sea-facing southwestern coast and in the monsoon-drenched peninsular northeast; where drier conditions develop, wheat and other grains prevail. Although, as Appendix B reminds us, physiologic densities in India are lower than in neighboring Bangladesh to the east, the comparison is deceptive because Indian farming, especially in the rice-growing zones and despite the effects of the Green Revolution, is inefficient. For India's economic transformation to have real impact in the rural areas, nothing less than a technological revolution is required.

The Energy Problem

If you have a friend in (or from) India, you know someone who is familiar with power outages. They are a way of life in India, where electricity demand routinely exceeds available supply, where governments cannot bring themselves to require customers to pay for the actual cost of the power they consume, and where power grids, generating equipment, and other related infrastructure are in bad shape. All this while hundreds of millions of villagers still have no electricity supply at all.

Yet electrical power is key to India's modernization. Already, foreign companies doing business in India sometimes import their own generators, but others are discouraged from investing in factories and other facilities because the power supply is so unreliable. Most of India's electricity is generated in thermal plants burning coal, oil, or natural gas; about 25 percent comes from hydroelectric sources, and about 3 percent from nuclear plants. The problems are many: India has substantial coal deposits, but the railroads cannot handle the transport to power plants. So India must import coal, but the ports do not have the required capacity. And India possesses only limited oil and natural gas reserves. Add to this an increasingly inadequate national power-supply grid in a country with a still-exploding population and even faster-growing demand, and you have massive problems.

A key remedy, of course, lies in increased oil and gas imports, but here India runs into geopolitical problems. While the American government was eager to give India leeway in the nuclear arena, the United States made it clear that it does not like India's plan to buy Iranian natural gas via a pipeline across Pakistan, although this continues to be pursued. India's other options lie in interior Asia, but those sources are more distant and pipeline construction would involve further diplomatic complications. Once again, the other alternative—importing oil and gas via tankers and ports—is constrained by inadequate infrastructure.

Limitations on Manufacturing Growth

Given these problems, the geography of India's manufacturing (Fig. 8-14) is changing too slowly for the country's needs. The map is a legacy of colonial times, with coastal Mumbai, Kolkata, and Chennai

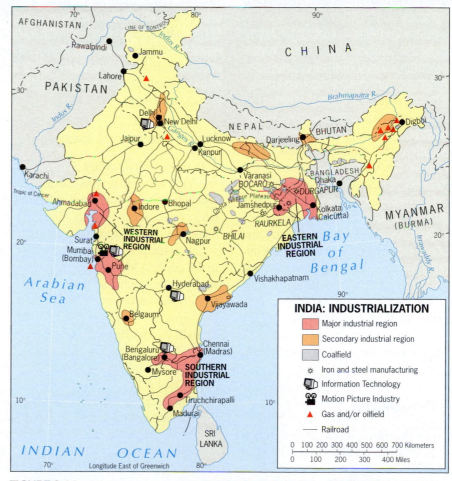

FIGURE 8-14   © H. J. de Blij, P. O. Muller, and John Wiley & Sons, Inc.

anchoring major industrial zones and textile industries—the entry-level industry of disadvantaged countries—dominating the scene. India's information technology (IT) industries, centered in and around Bengaluru (Bangalore), Hyderabad, Mumbai, and Delhi, draw much international attention, and software and IT services account for more than one quarter of merchandise exports by value; but what India needs far more are manufacturing industries competitively selling goods on world markets, putting tens of millions to work, and transforming the economy (as has happened in China, which we discuss in Chapter 9). It is almost unbelievable that as many people worked in manufacturing in 1991 as in 2001 (the two most recent census years) while China's industrial workforce more than quadrupled during the same period.

Could India follow in China's footsteps? Some economic geographers suggest that India might leapfrog China and move quickly from an underdeveloped to a postindustrial services-led economy. Certainly India has the requisite intellectual clout—it has an excellent system of universities and technical schools. Indian corporations have set up hundreds of information technology (IT) schools in China where tens of thousands of Chinese students are learning the business. But to reach India's hundreds of millions of potential wage earners, India needs a vigorous expansion of its secondary industries, those that make goods (beyond textiles) and sell them at home and abroad. Here's how it may happen: when an economy churns along the way China's has over the past three decades, labor and production costs tend to rise. That causes manufacturers to look for places where cheaper labor will reduce such costs. India, with its long history of local manufacturing, huge domestic market, and vast reservoir of capable labor, would then take its turn on the world stage.

Improving Prospects

Already, India is on the move, if not yet as dramatically as China. A growing middle class of at least 300 million people demands goods ranging from mobile phones to motor bikes. Chennai is becoming the Indian automobile center: in 2003 a South Korean manufacturer started selling cars from its local factory not only in India but was also exporting them to Europe, and BMW in 2008 was building an assembly plant nearby. India's economy today is the world's sixth largest; by 2020, it is likely to rank third. Over the past several years, the economy has grown by an average of 7 percent; while this is no match for China, it is far ahead of the growth rates in most other regions of the world. As we will see in the next chapter, China's dramatic growth resulted from decisions at the top, a transformation planned and implemented in controlled detail. In India, the economy is growing from the bottom up, with all the traditional chaos that makes India a country like no other. As with China's provinces, some of India's States will advance ahead of others, and India's incredible socioeconomic contrasts will intensify. But over time, India could achieve what China has hitherto not attained: an economic and cultural geography of consensus.

India West and East

In the map shown earlier (Fig. 8-2), we see an India that has a northern and southern divide. This is principally a cultural difference, which can be observed when you compare it with Figure 8-4, the language map. But there is another line, a north-south line, which extends from near Lucknow on the Ganges in the north to Madurai in the south near the southern tip of the peninsula. In a very general sense, this line divides an India that, to the west of it, is showing signs of the kind of economic progress that brought Pacific Rim countries a new life in recent times. To the east lies an India that has more in common, economically, with struggling Bangladesh and Myanmar.

As with other regional divides, there are exceptions to our east-west delineation. Indeed, our map seems to suggest that much of India's industrial strength lies in the east. But what the map cannot reveal is the profitability of those industries. True, the east is rich in iron and coal, but the heavy industries built by the state during the 1950s are now outdated, uncompetitive, and in decline. Thus the Eastern Industrial Region in the hinterland of Kolkata now contains India's Rustbelt. The government keeps many industries going here but at a high cost. Old industries, such as carpetmaking and cottonweaving, continue to use child labor to remain viable. The State of Bihar represents the stagnation that afflicts much of India east of our line: by several measures it ranks among the poorest of the 28 States.

Compare this to western India. The State of Maharashtra, the hinterland of Mumbai, leads India in many categories, and Mumbai leads Maharashtra. Many smaller, private industries have emerged here, manufacturing goods ranging from umbrellas to satellite dishes and from toys to textiles. Across the Arabian Sea lie the oil-rich economies of the Arabian Peninsula. Hundreds of thousands of workers from western India have found jobs there, sending money back to families from Punjab to Kerala. More importantly, many have used their foreign incomes to establish service industries back home. Outward-looking western India, in contrast to the inward-looking east, has begun to establish other ties to the outside world. Satellite and fiber-optic-cable links have enabled Bengaluru to become the center of a growing software-producing and service complex reaching world markets. The beaches of Goa, the small State immediately to the south of Maharashtra, appeal to the sun-seeking tourists of Europe. This is, in fact, a classic case of **15 intervening opportunity** because resorts have sprung up along Goa's coast, and European tourists who once traveled to the more distant Maldives and Seychelles are now coming to Goa. Maharashtra's economic success also has spilled over into Gujarat to the north, and even landlocked Rajasthan (the next State to the north) is experiencing the beginnings of what, by Indian standards, is a boom.

The boom has created political problems, however. Not only is Maharashtra State a rising economic power; it also is the base of a strong Hindu nationalist political movement whose leaders object to foreign intrusions and have blocked major development projects and other enterprises. They halted a huge industrial scheme about halfway through and closed a fast-food operation that they deemed incompatible with local culture. Such clashes between foreign interests and domestic traditions continue even as India's globalizing economy forges ahead.

VULNERABLE BANGLADESH ON SOUTH ASIA'S EASTERN FLANK

On the map of South Asia, Bangladesh looks like another State of India: the country occupies the area of the double delta of India's great Ganges and Brahmaputra rivers, and India almost completely surrounds it on its landward side (Fig. 8-15). But Bangladesh is an independent country, initially partitioned from colonial India as East Pakistan, then reborn in 1971 after a brief war for independence from Pakistan. Today it remains one of the poorest and least developed countries in the world, with a population of 152.2 million that is growing at an annual rate of 1.9 percent.

Hazard-Prone Territory

Not only is Bangladesh a poor country; it also is highly susceptible to damage from 16 **natural hazards**. During the twentieth century, several of the deadliest natural disasters in the world struck this single country. In 1970, a cyclone (as hurricanes are called in this part of the world) killed more than 500,000 people. This trend has continued in the twenty-first century, with cyclones in 2004 and 2007 each killing thousands. And every time, the storms also ruined critical crops and left hundreds of thousands homeless.

The reasons for Bangladesh's vulnerability can be deduced from Figures 8-15 and 8-1. The country is essentially floodplain as well as the double delta of the Ganges and Brahmaputra rivers, which at this massive outlet have separated into almost

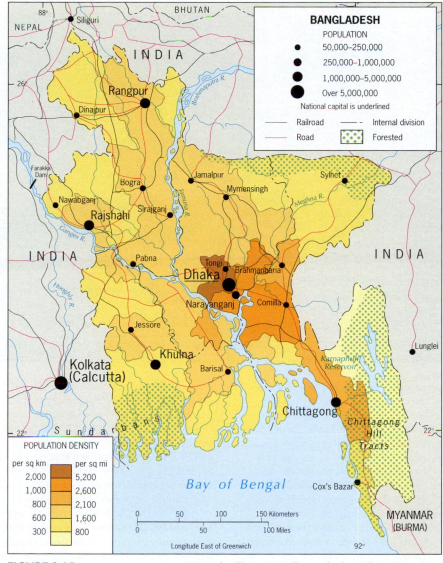

FIGURE 8-15   © H. J. de Blij, P. O. Muller, and John Wiley & Sons, Inc.

250 stream channels. The low elevations can be dangerous when there is flooding. But flooding is a natural event and provides rich sediments that create fertile alluvial soils that have attracted farmers for millennia and will keep bringing them back. Flooding is only part of the vulnerability; the shape of the Bay of Bengal forms a funnel that sends tropical cyclone-generated storm surges of wind-whipped water barreling into the delta coast. Without money to build seawalls, floodgates, elevated shelters in sufficient numbers, or adequate escape routes, hundreds of thousands of people are at continuous risk, with deadly consequences. And as if this is not enough, millions of people have been found to be exposed to excessive (natural) arsenic in the drinking water from their wells.

Flooding in Bangladesh. *The Economist* once described Bangladesh not as a country but as a lake with small islands. This is precisely how it looks during parts of the year when the rivers rise with their annual floods. The difference between fertilizing floods and a flooding calamity is very small, and Bangladesh's large population is perpetually at risk from not just one, but various potential disasters. Will there ever be improvement? Not likely. Flooding intensity has increased with upstream deforestation and overgrazing in several countries. Regional cooperation on environmental matters is in its infancy at best. Compounding the problems of the annual floods are the cyclones that can batter the coast. If these occur at the end of a heavy monsoon, then misery will ensue. The photo shows the village of Khatachira in Bangladesh that was heavily damaged by Cyclone Sidr in November 2007.

And what about global climate change? The Indian Ocean has a dearth of long-term climate data from which to develop climate models, but according to the Intergovernmental Panel on Climate Change (IPCC) there will likely be more coastal flooding due to storms as well as more flooding from the rivers. Bangladesh is at great risk from sea-level rise and does not have the resources to build barriers as do the Dutch in Europe. At minimum, houses need to be built higher and of stronger material, and flood warning systems should be improved. Drinking water sources need to be secured as this country is prone to massive outbreaks of water-borne diseases such as cholera. This will require political resolve as well as international assistance. (© Ruth Fremson/Redux Pictures)

of wheat in the crop rotation (where climate allows) have improved diets and food security. But diets remain poor and inadequate overall. The textile industry provides most of Bangladesh's foreign revenues, but the once-thriving jute industry continues its decline. The discovery of a natural gas reserve is now the subject of a national debate: home consumption or money-making export?

Bangladesh is a dominantly Muslim society, generally moderate and tolerant, but with periodic insurgent flareups. Women participate actively in politics; already there have been two female prime ministers, and 30 seats in the national legislature are reserved for women. Its relations with neighboring India have at times been strained over water resources (India controls the Ganges, which is Bangladesh's lifeline), cross-border migration (a sixth of the population is Hindu), and transit between parts of India across Bangladesh's north (refer to Fig. 8-12 to see the reason). Since early 2007 political wrangling has left the country frozen in a state of emergency with a military caretaker government running things as of mid-2008.

THE MOUNTAINOUS NORTH

As Figures 8-1 and 8-2 show, a tier of landlocked countries and territories lies across the mountainous zone that walls India off from China. **Nepal**, northeast of India's Hindu core, has a population of 27.2 million and is the size of Illinois. It has three geographic zones (Fig. 8-16): a southern, subtropical, fertile lowland called the Terai; a central belt of Himalayan foothills with swiftly flowing streams and deep valleys; and the spectacular high Himalayas themselves (topped by Mount Everest) in the north. The capital, Kathmandu, lies in the east-central part of the country in an open valley of the central hill zone.

Nepal is materially poor but culturally rich. The Nepalese are a people of many sources, including India, Tibet, and interior Asia. About 80 percent are Hindu, and Hinduism is the country's official reli-

Limits to Opportunity

Bangladesh remains a nation of subsistence farmers; urbanization is at only 23 percent, and Dhaka, the megacity capital of 13.9 million, and the southeastern port of Chittagong are the only urban centers of consequence. Moreover, Bangladesh has one of the highest physiologic densities in the world (1678 people per square kilometer [4346 per sq mi]), and only higher-yielding varieties of rice and the introduction

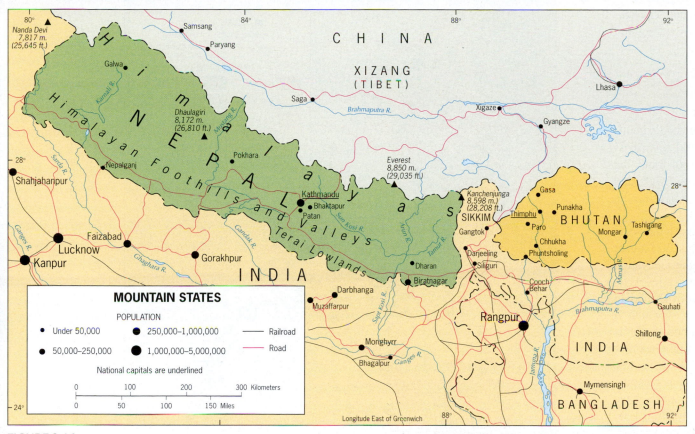

FIGURE 8-16

© H. J. de Blij, P. O. Muller, and John Wiley & Sons, Inc.

over, in 2001 the assassination of Nepal's king threatened the disintegration of the state. By 2006 Nepal was in chaos and seemed to be a **17 failed state** as the spreading Maoist insurgency raised the death toll to more than 20,000 and the new king and parliament were locked in a paralyzing struggle. In 2008, the monarchy was abolished and Nepal declared itself a republic, to be led by the Maoist Communist Party. Throughout the long period of political unrest, Nepal's famed tourist industry collapsed and its economy withered. The future remains uncertain in this fractious country.

Mountainous **Bhutan** (population 900,000), wedged between India and China's Tibet (Fig. 8-16), is a small **18 buffer state** between Asia's giants. Time seems to have stood still in this landlocked and isolated country. Bhutan is officially a constitutional monarchy, but its king rules the country with virtually absolute power; economic subsistence and political allegiance are the norms of life for most of the population of just over one million. Thimphu, the capital, has about 60,000 inhabitants. The symbols of Buddhism, the state religion, dominate its cultural landscape. Social tensions arise from the large but diminishing Nepalese minority, most of whom are Hindus and some of whom have been persecuted by the ethnically dominant Bhutia.

Forestry, hydroelectric power, and tourism all have potential here, and Bhutan has considerable mineral resources. But isolation and inaccessibility preserve traditional ways of life in this mountainous buffer state. Bhutan's king publicly rejects all forms of globalization in favor of tranquility and the pursuit of what he calls Gross National Happiness with its

gion; but Nepal's Hinduism is a unique blend of Hindu and Buddhist ideals. Thousands of temples and pagodas ranging from the simple to the ornate grace the cultural landscape, especially in the valley of Kathmandu, the country's core area. Although over a dozen languages are spoken, 90 percent of the people also speak Nepali, a language related to Indian Hindi.

As the data in Appendix B suggest, Nepal is a country suffering from severe underdevelopment reflected in its low GNI. It also faces strong centrifugal social and political forces. Environmental degradation, crowded farmlands and soil erosion, and deforestation scar the countryside. The Himalayan peaks form a world-renowned tourist attraction, but tourist spending in Nepal, always relatively modest, has

been cut back because of the recent revival of Maoist-communist terrorism, not only in the western hills where it has long been active, but also in the capital itself. Despite their name, neighboring China's government disavows Nepal's Maoists and, in fact, assists the Nepalese regime in its defense. This is another one of the realm's flashpoints and has caused political instability, stalled economic growth, and undermined social development programs.

Nepal's political geography has long been troubled. The end of absolute monarchy in 1991 and the advent of democracy did not end the country's regional divisions: the southern Terai with its tropical lowlands is a world apart from the hills of central Nepal, and the peoples of the west have origins and traditions different from those in the east. More-

four pillars of sustainable development, environmental protection, cultural preservation, and good governance.

THE ISLAND SOUTH

As Figures 8-1 and 8-2 show, South Asia's continental landmass is flanked by several sets of islands: Sri Lanka off the southern tip of India, the Maldives in the Indian Ocean to the southwest, and the Andaman Islands (part of India) marking the eastern edge of the Bay of Bengal.

The **Maldives** consists of more than a thousand tiny islands whose combined area is just 300 square kilometers (115 sq mi) and whose highest elevation is barely over 2 meters (6 ft) above sea level. Its population of just under 300,000 from Dravidian and Sri Lankan sources, one-quarter of which is concentrated on the capital island named Maale, is 100 percent Muslim. The Maldives might be unremarkable, except that, as Appendix B shows, this country has the realm's highest GNI per capita. Having hardly any arable land, the locals have translated their palm-studded, beach-fringed islands into a tourist mecca that attracts tens of thousands of mainly European visitors annually. But at the end of 2004, the Indian Ocean tsunami (Chapter 10) swept across these especially low-lying islands, severely damaging many tourist facilities that are still recovering. The very existence of this country is threatened by sea-level rise as a consequence of global climate change.

Sri Lanka: South Asian Tragedy

Sri Lanka (known as Ceylon before 1972), the compact, pear-shaped island located just 35 kilometers (22 mi) across the Palk Strait from India, became independent from Britain in 1948 (Fig. 8-17). There were good reasons to create a separate sovereignty for Sri Lanka. This is neither a Hindu nor a Muslim country: the majority of its 20 million people—about

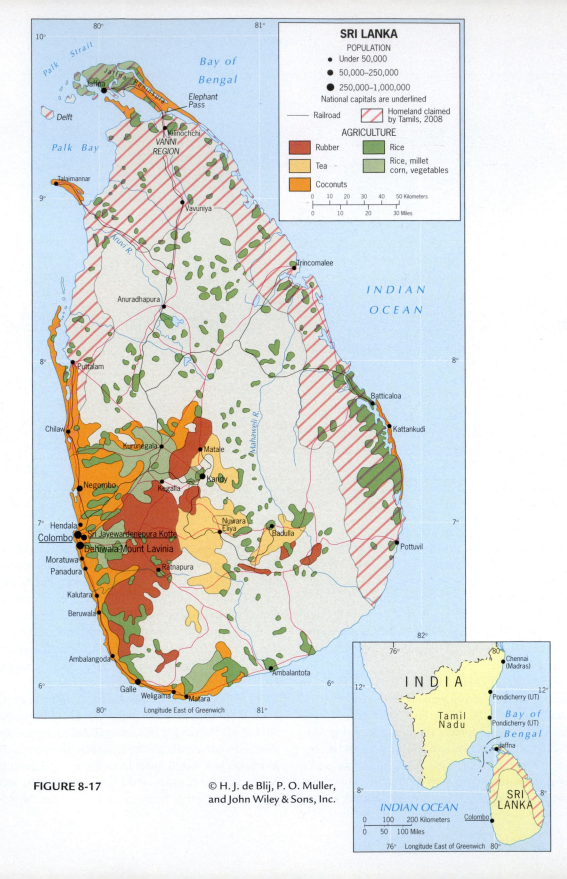

FIGURE 8-17

© H. J. de Blij, P. O. Muller, and John Wiley & Sons, Inc.

70 percent—are Buddhists. Furthermore, unlike India or Pakistan, Sri Lanka is a plantation country (tea, rubber, coconut), and commercial farming is still the mainstay of the agricultural economy.

The great majority of Sri Lanka's people are descended from migrants who came to this island from northwest India beginning about 2500 years ago. Those migrants introduced the advanced culture of their source area, building towns and irrigation systems and bringing Buddhism. Today, their descendants, known as the Sinhalese, speak a language (Sinhala) that belongs to the Indo-European language family of northern India.

The Dravidians who lived on the mainland, just across the Palk Strait, came later and in far smaller numbers—until the British colonialists intervened. During the nineteenth century the British brought hundreds of thousands of Tamils to work on their tea plantations, and soon a small minority became a substantial segment of Ceylonese society. The Tamils brought their Dravidian tongue to the island and introduced their Hindu faith. At the time of independence, they constituted more than 15 percent of the population; today they total about 18 percent. Most of the remainder are Muslims.

Hope and Disaster

When Ceylon became independent, it was one of the great hopes of the postcolonial world. The country had a sound economy and a democratic government,

and it was renowned for its tropical beauty. Its reputation soared when a massive campaign succeeded in eradicating malaria and when family-planning campaigns reduced population growth while the rest of the realm was experiencing a population explosion. Rivers from the cool, forested interior highlands fed the paddies that provided ample rice; crops from the moist southwest paid the bills, and the capital, Colombo, grew to reflect the optimism that prevailed.

In the midst of this glowing scenario, the seeds were being sown for the rise of another flashpoint in this realm. Sri Lanka's Tamil minority soon began proclaiming its sense of exclusion, demanding better treatment from the Sinhalese majority. Although the government recognized Tamil as a national language in 1978, sporadic violence marked the Tamil campaign, and in 1983 full-scale civil war began. Now many in the Tamil community demanded a separate Tamil state to encompass the north and east of the country (see Fig. 8-17), and a rebel army called the Tamil Tigers confronted Sri Lanka's national forces. Civil war has raged almost continuously in this island republic since 1983 at a cost of more than 70,000 lives and immeasurable social and economic damage.

This sequence of events fits the model of the evolution of the **19 insurgent state** discussed in Chapter 5. In Sri Lanka, the *equilibrium* stage was reached in the early 1990s, when the Tamil Tigers claimed the Jaffna Peninsula and set up their headquarters there. The *counteroffensive* stage is still in progress,

Civil unrest bubbles under the veneer of normality in the Sri Lankan capital of Colombo. Here soldiers patrol outside a school as children walk past. In early 2008, a wave of bombings blamed on Tamil Tiger rebels brought what had been a distant war to the heart of the capital city of this besieged country.   (© AP/Wide World Photos)

but the Tamil forces strike at will. The 2004 tsunami, which brought untold destruction to this country, offered a moment of peace; but it did not hold, and violence has escalated again. Even if the Tamils do not succeed in securing an independent state, they will likely force the government to make territorial concessions at some point.

9
EAST ASIA

CONCEPTS, IDEAS, AND TERMS

1. Pacific Rim
2. Confucius
3. Hanification/Sinicization
4. Extraterritoriality
5. Special Administrative Region (SAR)
6. Economic restructuring
7. Core area
8. Geography of development
9. Overseas Chinese
10. Special Economic Zone (SEZ)
11. Regional state
12. Buffer state
13. Jakota Triangle
14. Modernization
15. Areal functional organization
16. Regional complementarity
17. State capitalism
18. Economic tiger

In This Chapter

- From Xia to Shenzhen: China's ancient roots, modern transformation
- China's global economic growth
- Nationalism rising in China as well as in Japan
- Rebounding Japan: Still a regional powerhouse
- How geography could help reunite the Korean Peninsula
- Signs of progress over Taiwan

REGIONS

CHINA PROPER
XIZANG (TIBET)
XINJIANG
MONGOLIA
THE JAKOTA TRIANGLE
(JAPAN-SOUTH KOREA-TAIWAN)

Photos: *(upper left)* Bridge over the Huang He, Lanzhou, China, © A. WinklerPrins; *(above)* Three Gorges along the Chang Jiang, China, © Doug Landreth/Science Faction/Getty Images.

FIGURE 9-1 *Map:* © H. J. de Blij, P. O. Muller, and John Wiley & Sons, Inc.

EAST ASIA IS a geographic realm like no other. At its heart lies the world's most populous country. On its periphery lies one of the globe's most powerful national economies. Along its coastline, on its peninsulas, and on its islands an economic boom has transformed cities and countryside. Its interior contains the world's highest mountains as well as vast deserts. It is a storehouse of raw materials. The basins of its great rivers produce enough food to sustain more than a billion people.

Defining the Realm

The East Asian geographic realm consists of six political entities: China, Mongolia, North Korea, South Korea, Japan, and Taiwan. Note that we refer here to *political entities* rather than *states*. In East Asia, this distinction is significant. Taiwan, which its government officially calls the Republic of China, functions as a state but is regarded by mainland China (the People's Republic of China, or PRC) as a temporarily wayward province. North Korea is not a full member of the United Nations, and the division of the Korean Peninsula into North and South Korea may be temporary.

As defined here, East Asia lies between the vast expanse of Russia to the north and the populous countries of South and Southeast Asia to the south. This geographic realm extends from the deserts of Central Asia to the Pacific islands of Japan and Taiwan. Not surprisingly, environmental diversity is one of its hallmarks.

East Asia is also the hub of the evolving regional phenomenon known as the **1 Pacific Rim**. From Japan to Taiwan and from South Korea to Hong Kong (Xianggang), East Asia's Pacific Ocean frontage is undergoing a massive transformation. Skyscrapers tower over old cities whose traditional housing is being swept away. Enormous industrial complexes disgorge products that flood world markets. Millions of people are on the move, abandoning farms and villages for urban assembly lines and sweatshops. The process started in Japan and soon encompassed Taiwan, South Korea, and Hong Kong, as well as Singapore in Southeast Asia. And when China's political climate changed in the 1980s, almost all of coastal East Asia was swept up in one of the greatest regional transformations in history.

MAJOR GEOGRAPHIC QUALITIES OF East Asia

1. East Asia is encircled by snow-capped mountains, vast deserts, cold climates, and Pacific waters.

2. East Asia was one of the world's earliest culture hearths, and China is one of the world's oldest continuous civilizations.

3. East Asia is the world's most populous geographic realm, but its population is strongly concentrated in its eastern regions.

4. China, the world's largest state demographically, is the current rendition of an empire that has expanded and contracted, fragmented and unified many times during its long existence.

5. China today remains a mainly rural society, and its vast eastern river basins feed hundreds of millions of people in a historic pattern that still continues.

6. China's sparsely peopled western regions are strategically important to the state but are inhabited by ethnic minority peoples with weak linkages to the central state.

7. Along China's Pacific frontage an economic transformation is taking place, affecting all the coastal provinces and creating an emerging Pacific Rim region.

8. Increasing regional demographic disparities and fast-changing cultural landscapes are straining East Asian societies.

9. Japan, the economic giant of East Asia, has a history of colonial expansion and wartime conduct that continues to affect international relations in this realm.

10. East Asia may be home to the world's next superpower as China's economic, military, and political strength and influence grow—and if China avoids the devolutionary forces that fractured the former Soviet Union.

11. The political geography of East Asia contains a number of flashpoints that can generate conflict, including Taiwan, North Korea, and several island groups in the realm's seas.

NATURAL ENVIRONMENTS

Figure 9-1 (see chapter opener map) illustrates the complex physical geography of the East Asian realm. In the southwest lie ice-covered mountains and plateaus, a region that appears crumpled like the folds of an accordion. A gigantic collision of tectonic plates

ital, but it was felt in Beijing and Shanghai many hundreds of miles away. Approximately 70,000 people died, thousands went missing, hundreds of thousands were injured, and an estimated 5 million people were made homeless. This was the strongest earthquake in China since the 1976 Tangshan Earthquake (see Figure 9-7) and will have a significant impact for some time.

Physiography and Population

As Figure 9-2 shows, East Asia includes very densely populated areas, but its interior is among the world's most sparsely populated areas. Despite its relative lack of human inhabitants, this region is critical to the lives of hundreds of millions of people. In these high mountains rise the great rivers that flow eastward across China and southward across Southeast and South Asia. Throughout the Holocene, these rivers, fed by melting ice and snow, have been eroding the uplands and depositing sediments in the lowlands,

The May 12, 2008, earthquake in Sichuan province, China, caused enormous destruction and loss of life. Here Chinese rescue workers observe a moment of silence at the rubble of a collapsed building in Dujiangyan, Sichuan province. The Chinese government responded rapidly to the disaster, mobilizing rescue workers and troops to assist in the affected area. Unlike previous disasters and crises, the Chinese government permitted international media and rescue assistance access to the region and displayed unusual openness.   (© Vincent Yu/Associated Press)

is creating this landscape as the Indian Plate pushes northward into the underbelly of the Eurasian Plate (Fig. G-5). The result is some of the world's most spectacular scenery, and some of its most dangerous. Snow, ice, and cold are not the only dangers to human life here: earthquakes and tremors occur almost continuously, causing landslides and avalanches. As the map shows, these high mountain-studded uplands form Tibet's vast Qinghai-Xizang Plateau, which is flanked by the Himalaya range to the south. East of Tibet, the mountain ranges converge and bend southward into Southeast Asia, where they lose their high relief.

Despite low relief, the region is still prone to tectonic forces. On May 12, 2008, an earthquake measuring 7.9 on the Richter scale (according to the United States Geological Survey) occurred in Sichuan province, southeast China, devastating the region. The epicenter of the quake was 80 kilometers (50 mi) from the city of Chengdu, the provincial cap-

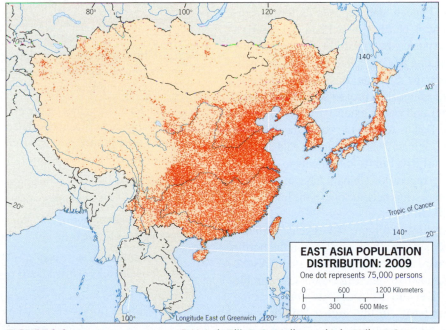

FIGURE 9-2   © H. J. de Blij, P. O. Muller, and John Wiley & Sons, Inc.

creating the alluvial basins that now sustain huge populations. Fertile alluvial soils and adequate growing seasons, combined with ample water and millions of hands to sow the wheat and plant the rice, have made possible the emergence of one of the great population concentrations on the earth.

Physiography, therefore, has much to do with East Asia's population distribution, but even the more habitable and agriculturally productive east has its limitations. The northeast suffers from severe continentality and increasing drought, with long, bitterly cold winters. High relief encircles the river basins north of the Yellow Sea and dominates much of the northern part of the Korean Peninsula, creating strong environmental contrasts. South Korea, as Figure G-9 shows, experiences relatively moderate conditions, comparable to those of the southeastern United States; North Korea has a harsh continental climate like that of North Dakota.

The Great Rivers

From the high interior come three major river systems. In the north, the Huang He (Yellow River) arises in the high mountains, crosses the Ordos Desert and the Loess Plateau, and deposits its fertile sediments in the vast North China Plain, where East Asia's earliest states emerged. In the center, the Chang Jiang (Long River), called the Yangzi downstream, crosses the Sichuan Basin and the Three Gorges (where a huge dam project will be fully operational in 2009), and waters extensive ricefields in the Lower Chang Basin. And in the south the Xi Jiang (West River) originates on the Yunnan Plateau, becoming the Pearl River in its lowest course. Its estuary, flanked by several of China's largest urban-industrial complexes, has become one of the hubs of the evolving Pacific Rim.

Further scrutiny of Figure 9-1 reveals a fourth important river system: the Liao River in the northeast and its basin, the Northeast China Plain. As the map suggests, however, the Liao is not comparable to the great rivers to its south, its course being shorter and its basin, in this higher-latitude zone, much smaller.

Interior Environments

In the interior, note the Loess Plateau, located south of the Ordos Desert, where the Huang He (Yellow River) makes a giant loop (Fig. 9-1). Glacial action pulverized rock to produce the windblown, fertile sediment known as *loess*; where water is available, loess forms a rich soil that can sustain a dense agricultural population, important for China's population distribution. To the south, deep in the interior, lies the Sichuan (Red) Basin, crossed by the Chang Jiang. This basin has supported human communities for a long time, and its current population cluster is evident in Figure 9-2. The Sichuan Basin, encircled as it is by mountains, is one of the world's most clearly defined physiographic regions, with approximately 120 million inhabitants. This is where the Sichuan earthquake of 2008 affected the lives of so many.

Still farther to the south lies the Yunnan Plateau, the source of the tributaries that feed the Xi River. Much of southeastern China is hilly and in places mountainous. This high relief has tended to limit contacts between China and Southeast Asia.

Along the Coast

East Asia's Pacific margin is a jumble of peninsulas and islands. The Korean Peninsula is a near-bridge between Asia and Japan. The Liaodong and Shandong peninsulas protrude into the Yellow Sea, which receives sediments from the Huang and Liao rivers and hence is gradually silting up. Off the mainland lie the islands that have played crucial roles in the human geography of Asia and, indeed, the world: Japan, Taiwan, and Hainan. Japan's environmental range is expressed by cold northern Hokkaido and warm southern Kyushu, but its core is its main island, Honshu. As Figure 9-1 shows, myriad smaller islands flank the mainland and dot the East and South China seas. Some of these smaller islands are of major significance in the human geography of this realm.

HISTORICAL GEOGRAPHY

East Asia has a lengthy and complex human history. Many archaeological sites in this realm have yielded

WHAT'S DRIVING GEOGRAPHIC CHANGE IN THE REALM?

- Can **China's** communist regime keep the lid on social discontent? Corrupt officials, arbitrary arrests and mistreatment, lack of legal protection, land disputes, environmental degradation, and other issues are generating unrest in rural and urban areas.

- Watch for these names on the map and in the news: Okinotori, Diaoyu (Senkaku), Chunxiao (Shirakaba), and Tianwaitian (Kashi) in the South China Sea. They are islands and oil platforms that are disputed by **China and Japan**, driving nationalist sentiment on both sides. Meanwhile, **Japan and South Korea** argue over Dokdo (Takeshima) in the East Sea (Sea of Japan).

- **Japan's** population is shrinking significantly. Deaths continue to exceed births, a trend that will result in a reduction of the total population to below 100 million by 2050.

- On the **Korean Peninsula**, North Korea periodically threatens to become nuclear, while South Korea is seeking ways to break down the barriers between itself and the communist dictatorship next door.

- It looks as though the risk of conflict over **Taiwan** is declining as more sensible approaches are producing low-level but formal discussions to resolve the standoff.

- In 2008, world attention focused on **China** because of unrest in Tibet, a major earthquake in Sichuan province, and the Summer Olympic Games in Beijing.

evidence of *Homo erectus* (a hominid), including perhaps the most famous of all: Peking Man, found in a cave not far from Beijing in the 1920s. Current anthropological theory holds that *Homo sapiens* arrived in East Asia between 40,000 and 60,000 years ago and eliminated the hominids such as Peking Man in short order.

Early Cultural Geography

Humans have inhabited the plains and river basins, foothills, and islands of this realm for a very long time. Hunting sustained both the hominids and the early human communities; fishing drew them to the coasts and onto the islands. The first crossing into Japan may have occurred as long as 10,000 to 12,000 years ago, possibly much earlier, when the Jomon people, a Caucasoid population of uncertain geographic origins, entered the islands; their modern descendants, the Ainu, spread throughout the archipelago. Today only about 20,000 people living in northernmost Hokkaido trace their ancestry to Ainu sources.

About 2300 years ago, the Yayoi people, rice farmers who had settled in Korea, appear to have crossed by boat to Kyushu, Japan's southernmost island, whence they advanced northward. The Ainu, who subsisted by fishing, trapping, and hunting, were driven back, but gene-pool studies show that much mixing of the groups took place; they also show that the Yayoi invasion was followed by other incursions from the Asian mainland. By then, powerful dynastic states had already arisen in what is today China, and early Chinese cultural traits thus found their way into Japan through the process termed **relocation diffusion** (see Chapter 7).

Mainland Cultures

People in the West take it for granted that plant and animal domestication began in what is now called the Middle East and diffused from the Fertile Crescent to other parts of Eurasia and the rest of the world. But the taming of animals and the selective farming of plants may have begun as early, or earlier, on the Asian mainland. As in Southwest Asia, the fertile alluvial soils of the great river basins and the ebb and flow of streamwater created an environment of opportunity, and millet and rice were being harvested between 7000 and 8000 years ago, during the Neolithic period.

Even during this period, East Asia was a mosaic of regional cultures. Their differences are revealed by the tools they made and the decorations on their bowls, pots, and other utensils. An especially important discovery—two 8000-year-old pots in the form of a silkworm cocoon, from China's Hebei Province—suggests a very ancient origin for one of the region's leading industries.

As noted earlier, plant and animal domestication produced surpluses and the ability to store food for future use, enabling population growth and requiring wider regional organization. Here, as elsewhere during the Neolithic period, settlements expanded, human communities grew more complex, and power became concentrated in the hands of a small group, an *elite*.

Early State Formation

The process of *state formation* is known to have occurred in only a half-dozen regions of the world, and China was one of these. Evidence about China's earliest states has long been scarce. Today, however, archaeologists are focusing on the lower Yi-Luo River Valley in the western part of Henan Province, where the first documented Chinese dynasty, the **Xia** Dynasty (2200–1770 BC), existed. The capital of this ancient state, Erlitou, has been found, and archaeologists now refer to the Xia Dynasty as the Erlitou culture. Secondary centers are being discovered, and the presence of Erlitou tools and implements in a wider area proves that the Xia Dynasty was a substantial state.

All early states were ruled by elites, but China's political history is chronicled in elite dynasties because here the succession of rulers came from the same line of descent, sometimes enduring for centuries. In the transfer of power, family ties counted for more than anything else. Particular dynasties were overthrown, but the victors did not change the system of dynastic rule, which lasted into the twentieth century (see Time Line of China's History, Fig. 9-7, pp. 310–311).

Regions of the Realm

East Asia presents us with an opportunity to illustrate the changeable nature of regional geography. Our regional delimitation is based on current circumstances and predicts ways in which the framework may change. Thus, it is anything but static. As of the beginning of the twenty-first century, we can identify five geographic regions in the East Asian realm (Fig. 9-3). These are:

1. *China Proper*. Almost any map of China's human geography—population distribution, urban centers, surface communications, agriculture, industry—emphasizes the strong concentration of Chinese culture in the country's eastern sector. This is the *real* China, where its great cities, populous farmlands, and historic sources are located. Long ago, scholars named this sector China Proper, and it is still a good regional designation. But China is a large and complex country, and a number of subregions are nested within China Proper. Some of these, such as the North China Plain and the Sichuan Basin, are old, well-established geographic units. Others are new, and notable among these is China's Pacific Rim, which is still growing, yet poorly defined. Hence it is shown as *formative* in Figure 9-3.

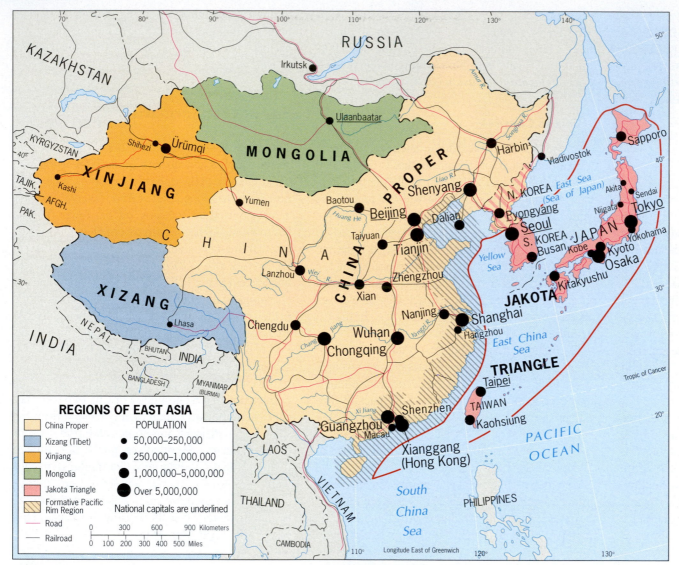

FIGURE 9-3

© H. J. de Blij, P. O. Muller, and John Wiley & Sons, Inc.

from the first two letters of each) were transformed by Pacific Rim economic development during the second half of the twentieth century. The region's recent emergence foreshadows further changes in the decades ahead as Korean unification becomes a possibility and current contrasts between the Jakota Triangle and Pacific Rim China diminish.

CHINA PROPER

When Westerners chronicle the rise of civilization, the focus tends to be on the historical geography of Southwest Asia, the Mediterranean, and Western Europe. We think of ancient Greece and Rome as the crucibles of culture and the Mediterranean and Atlantic waters as the avenues of its diffusion. From this Eurocentric perspective, China lay remote, in the 'Far East' or the 'Orient,' barely connected to the Western realm of achievement and progress. When an Italian adventurer named Marco Polo visited China during the thirteenth century and described the marvels he saw there, his account did little to change European minds. To Europeans, Europe was and always would be the center of civilization.

The Chinese, naturally, take a different view. Events on the western edge of the great Eurasian landmass were deemed irrelevant to events within China, the center of the world's most advanced and refined culture. Roman emperors were rumored to be powerful, but nothing could match the omnipotence of China's rulers. Rome was a great city, but Xian far eclipsed Rome as a center of sophistication. Chinese civilization existed long before those of ancient Greece and Rome emerged, and it was still there long after they collapsed. China, the Chinese believe, is eternal. It is, and always will be, the center of the civilized world.

2. *Xizang (Tibet)*. The high mountains and plateaus of Xizang, ruled by China but still widely known by its older name, *Tibet*, form a stark contrast to teeming China Proper. Here, next to one of the world's largest and most populous regions, lies one of its emptiest and, in terms of inhabited space, smallest regions.

3. *Xinjiang*. The vast desert basins and encircling mountains of Xinjiang form a third East Asian region. Again, physical as well as human geographic criteria come into play. Though politically part of China, culturally this region is more appropriately part of Islamic Central Asia.

4. *Mongolia*. The desert state of Mongolia forms East Asia's fourth region. Like Tibet, landlocked Mongolia, vast but sparsely peopled, stands in stark contrast to populous China Proper.

5. *The Jakota Triangle*. East Asia's fifth region is defined by its economic geography. **Ja**pan, South **Ko**rea, and **Tai**wan (the name *Jakota* is derived

We should keep this belief in mind when we study China's recent rise in global importance because 4000 years of Chinese culture and perception will not change overnight—not even in a generation. Time and again, China overcame the invasions and depredations of foreign intruders, and afterward the Chinese would close their vast country to the outside world, as they did for a long time during the Qing (Manchu) Dynasty. A mere 35 years ago, in the early 1970s, there were just a few *dozen* foreigners in the entire country with its (then) nearly 1 billion inhabitants. The communist regime required this insularity; even Soviet advisors had been thrown out. But in the early 1970s China's rulers decided that an opening to the Western world would be advantageous, and so they invited U.S. President Richard Nixon to visit Beijing. That historic occasion, in 1972, ended the latest period of Chinese isolation—as always, on China's terms (see Fig. 9-7). Since then, China has been open to tourists and businesspeople, teachers, and investors. Tens of thousands of Chinese students have been sent to study at American and other Western institutions. Long-suppressed ideas flowed into China, and a pro-democracy movement arose and climaxed in 1989 with the standoff and massacre at Tiananmen Square. China's rulers knew that their violent repression of this movement would anger the world, but that did not matter to them because they deemed foreign condemnation irrelevant at that time.

In 2008, when Tibetans demanded greater autonomy and their uprising was violently suppressed, world attention (in the form of protests during the Olympic torch relay) mattered more as Chinese officials were anxious about their presentation to the world during the Olympic year.

RELATIVE LOCATION

In contrast to another emerging power, India, periodic closure and exclusion is one of China's recurrent traditions, made possible by its relative location and Asia's physiography. In other words, China's 'splendid isolation' was made possible by geography.

Earlier we noted the role of relief and desert in encircling the culture hearth of East Asia, but equally telling is the factor of distance. Until recently, China lay far from the modern source areas of innovation and change. True, China—as the Chinese emphasize—was itself such a source, but China's contributions to the outside world remained limited, essentially, to finely made arts and crafts. China did interact with Korea, Japan, Taiwan, and parts of Southeast Asia, and eventually millions of Chinese migrated to neighboring countries. But compare these regional links to those of the Arabs, who ranged worldwide and brought their knowledge, religion, and political influence to areas extending from Mediterranean Europe to Bangladesh and from West Africa to Indonesia. Later, when Europe became the center of intellectual and material innovation, China found itself farther removed, by either land or sea, than almost any other part of the world.

Today, modern communications notwithstanding, China remains distant from almost any other place on the earth. Going by rail from Beijing to Moscow, the capital of China's Eurasian neighbor, involves a tedious journey that takes the better part of a week. Direct surface connections with India are practically nonexistent. Overland linkages with Southeast Asian countries, though improving, remain tenuous. Today, however, air travel is transforming the country and its connections to others.

For the first time in its history, China now lies near a world-class hearth of technological innovation and financial power: Japan. Although tensions between the two countries persist as a result of atrocities committed by the Japanese during World War II, proximity to an industrial and financial giant is critical for the momentous economic developments taking place in China's coastal provinces. Japanese investments and business partnerships have transformed the economic landscape of Pacific coast China. American and European trade links also are important, but Japan's role has been crucial. Japan's economic success set the Pacific Rim engine in motion, and Japan's best financial years happened to co-

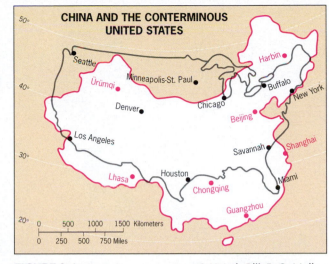

FIGURE 9-4 © H. J. de Blij, P. O. Muller, and John Wiley & Sons, Inc.

incide with China's reopening to foreigners in the 1970s. Thus, geographic and economic circumstances combined to transform the map and made *Pacific Rim* a household term around the world.

CONTINENTAL SIZE AND ENVIRONMENT

China's total area is slightly smaller than that of the United States including Alaska; both countries cover about 9.6 million square kilometers (3.7 million sq mi). As Figure 9-4 reveals, the longitudinal extent of China and the 48 contiguous U.S. States is also similar. In terms of latitude, however, China is considerably wider. Miami, near the southern limit of the United States, if located in China would lie halfway between Shanghai and Guangzhou. Thus, China's lower-latitude southern region takes on characteristics of tropical Asia. In the northeast, China incorporates the equivalent to Quebec and Ontario in North America. Westward, China's land area becomes narrower, and its physiographic similarities with North America increase—except of course, that China has no west coast.

Now look at Figure 9-5 and compare the climate maps of China and the United States (which are

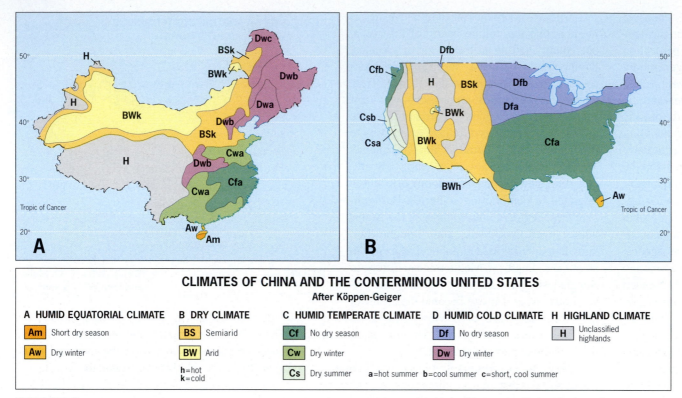

CLIMATES OF CHINA AND THE CONTERMINOUS UNITED STATES
After Köppen-Geiger

A HUMID EQUATORIAL CLIMATE	B DRY CLIMATE	C HUMID TEMPERATE CLIMATE	D HUMID COLD CLIMATE	H HIGHLAND CLIMATE
Am Short dry season	**BS** Semiarid	**Cf** No dry season	**Df** No dry season	**H** Unclassified highlands
Aw Dry winter	**BW** Arid	**Cw** Dry winter	**Dw** Dry winter	
	h=hot **k**=cold	**Cs** Dry summer **a**=hot summer **b**=cool summer **c**=short, cool summer		

FIGURE 9-5

© H. J. de Blij, P. O. Muller, and John Wiley & Sons, Inc.

enlargements of portions of the world climate map in Fig. G-9). Note that both have a large southeastern climatic region marked *Cfa* (humid, temperate, warm summer), flanked in China by a zone of *Cwa* (where winters become drier). Moving westward in both countries, the *C* climates yield to colder, drier climates. In the United States, moderate *C* climates reappear along the Pacific coast. Having no west coast, and thus no moderating effect from an ocean, China stays dry and cold, as well as high in elevation, at equivalent longitudes.

Note especially the comparative location of the U.S. and Chinese *Cfa* areas in Figure 9-5. China's *Cfa* area lies much farther to the south. In the United States, this climate extends beyond 40° North latitude, but in China, cold *D* climates with generally dry winters take over at a latitude equivalent to that of Virginia. Beijing has a warm summer but a bitterly cold and long winter. Northeast China, in the general

latitudinal range of Canada's southern Quebec and Newfoundland, has a much more severe climate than its North American equivalent. Harsh environments thus prevail over vast regions of China. From the climatic zone marked *H* (for highlands) in the west come the great rivers whose wide basins contain enormous expanses of fertile soils. Without these waters, China would likely not have such a vast population— more than four times that of the United States.

EVOLVING CHINA

Even if China is not the world's longest continuous civilization (Egypt may claim this distinction), no other state in the world can trace its cultural heritage as far back as China can. A great culture arose primarily on the North China Plain, which its citizens considered to be the 'Middle Kingdom' and the center of the

world. China's fortunes rose and fell, but over more than forty centuries its people created a society with strong traditions, values, and philosophies. The teachings of Kongfuzi (551–479 BC), known to us as **2** **Confucius**, arose during the Zhou Dynasty (see Fig. 9-7) and dominated Chinese life and thought for over twenty centuries. Kongfuzi championed the poor, but his revolutionary ideas extended to the rulers as well as the ruled. He insisted that a person's place in society should be determined by living a moral life through competence and merit. When the communists took power in 1949, they attacked Kongfuzi thought on all fronts, substituting indoctrination for education. But they were unable to eradicate two millennia of cultural conditioning in a few decades, and even today the spirit of Kongfuzi haunts China's physical and mental landscapes.

As we try to gain a better understanding of China's present political, economic, and social geography, we should keep its continuities in mind. Yes, China is changing radically today, but it has undergone convulsive change in the past and has endured and regained its coherence. Larger and more populous than Europe, China, too, had its divisive feudal periods, but unlike Europe, China always came together again under a single flag. In the process, often-dictatorial China incorporated minorities ranging from Koreans and Mongols to Uyghurs and Tibetans; in the far south, it now includes several peoples with Southeast Asian affinities. Like the Chinese citizenry itself, these minorities have experienced both benevolent government and brutal subjugation. But it has always been China's wish to *hanify* them—that is, to endow them with the elements of Chinese culture (a process known as **3** **Hanification** or **Sinicization**).

As Figure 9-6 suggests, the growth and expansion of dynastic China was the dominant process in this

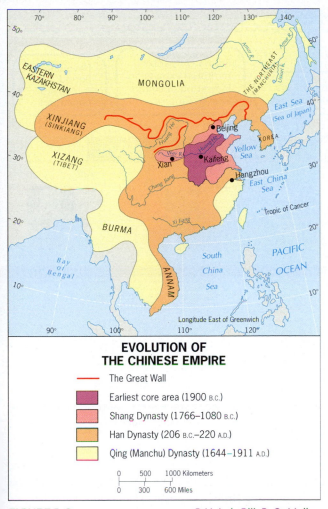

FIGURE 9-6

© H. J. de Blij, P. O. Muller, and John Wiley & Sons, Inc.

realm. The extent of the empire expanded and shrank. For example, the Chinese Empire was as large as the Roman Empire during the formative Han period but shrank to less than what is China Proper today at other times. Many of the territorial acquisitions made during the Qing Dynasty form part of the Chinese state today; others are historical justifications for potential claims on 'lost' areas such as the Russian Far East and Mongolia.

Another form of expansion has occurred more recently. Large Chinese minorities now reside in Southeast Asian countries, an emigration that began during the Qing Dynasty and changed the social and political landscape of colonial as well as modern Southeast Asia. Only Japan, while influenced by Chinese tenets, escaped incorporation into China's East Asian sphere throughout its history. Figure 9-7 presents a time line of China's history.

A Century of Convulsion

When the European colonialists appeared in East Asia, China long withstood them with a self-assured superiority based on the strength of its culture and the reassuring continuity of the state. There was no market for the British East India Company's rough textiles in a country long used to finely fabricated silks and cottons. There was little interest in the toys and trinkets the Europeans offered in exchange for Chinese tea and porcelain. Even key European inventions, such as the mechanical clock, though considered amusing and entertaining, were ignored and even deprecated as irrelevant to Chinese culture.

The Ming emperors were particularly dismissive of European manufactures, but the Ming rulers' confidence in their homemade products was sometimes misplaced. When the Manchu (Qing) forces invaded in 1644, their bows and arrows

Temple of Heaven, Beijing, China. Off-limits to the common people, this Ming Dynasty temple is an excellent expression of traditional Chinese architecture and balance. Here the emperor would make sacrifices and pray to his ancestors at the winter solstice.   (© A. WinklerPrins)

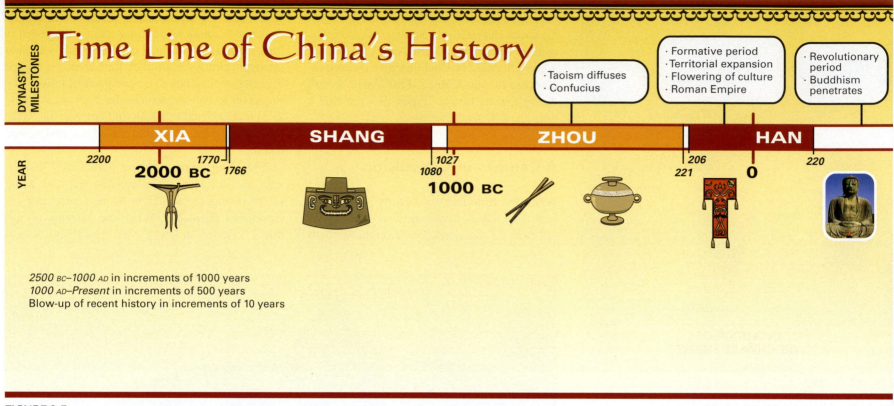

Time Line of China's History

DYNASTY MILESTONES

- Taoism diffuses
- Confucius

- Formative period
- Territorial expansion
- Flowering of culture
- Roman Empire

- Revolutionary period
- Buddhism penetrates

XIA	SHANG	ZHOU	HAN

YEAR

2200   1770   1027   206   220
2000 BC   1766   1080   221   **0**
      1000 BC

2500 *BC–1000 AD* in increments of 1000 years
1000 *AD–Present* in increments of 500 years
Blow-up of recent history in increments of 10 years

FIGURE 9-7

(Photo of Buddha: © DAJ/Getty Images)

proved superior to Chinese-manufactured muskets, which were so heavy and difficult to load that they were almost useless. Even when Europe's sailing ships gave way to steam-driven vessels and newer and better European products (including weapons) were offered in trade for China's tea and silk, China continued to reject European imports and resisted commerce in general. The Chinese kept the Europeans confined to small peninsular and island outposts such as Macau, and interacted with them only minimally. Long after India had succumbed to mercantilism and economic imperialism, China maintained its established order. This was no surprise to the Chinese. After all, they had held a position of undisputed superiority in their Celestial Kingdom as long as could be remembered, and they had dealt with foreign invaders before.

The Opium Wars

During the later period of Qing (Manchu) rule, however, the balance of power shifted decisively in favor of the colonialists. On two fronts in particular—economic and political—the European powers made inroads on China's invincibility. Economically, they succeeded in lowering the cost and improving the quality of their manufactured goods, especially textiles, and the handicraft industries of China began to collapse in the face of unbeatable competition. Polit-

ically, the demands of the British merchants and the growing British presence in China led to conflicts.

In the early part of the nineteenth century, the central issue was opium, a dangerous and addictive intoxicant that was imported from British India. Opium destroyed the fabric of Chinese culture, weakening the society and rendering China easy prey for colonial profiteers. As the Qing government moved to stamp out the opium trade in 1839, armed hostilities broke out, and soon the Chinese found themselves losing a war on their own territory. The First Opium War (1839–1842) ended in disaster: China's rulers were forced to yield to British demands, and the breakdown of Chinese sovereignty was under way.

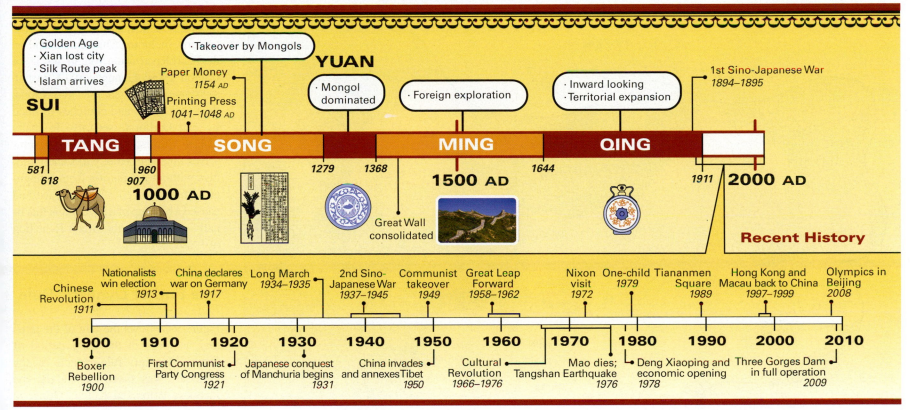

- Golden Age
- Xian lost city
- Silk Route peak
- Islam arrives

Paper Money
1154 AD

Printing Press
1041–1048 AD

· Takeover by Mongols

YUAN

· Mongol dominated

· Foreign exploration

· Inward looking
· Territorial expansion

1st Sino-Japanese War
1894–1895

SUI

TANG   SONG   MING   QING

581   618   907   960   1279   1368   1644   1911   2000 AD

1000 AD   1500 AD

Great Wall consolidated

Recent History

Chinese Revolution 1911

Nationalists win election 1913

China declares war on Germany 1917

Long March 1934–1935

2nd Sino-Japanese War 1937–1945

Communist takeover 1949

Great Leap Forward 1958–1962

Nixon visit 1972

One-child 1979

Tiananmen Square 1989

Hong Kong and Macau back to China 1997–1999

Olympics in Beijing 2008

1900   1910   1920   1930   1940   1950   1960   1970   1980   1990   2000   2010

Boxer Rebellion 1900

First Communist Party Congress 1921

Japanese conquest of Manchuria begins 1931

China invades and annexes Tibet 1950

Cultural Revolution 1966–1976

Tangshan Earthquake

Mao dies; 1976

Deng Xiaoping and economic opening 1978

Three Gorges Dam in full operation 2009

(Photo of Great Wall: © Corbis Digital Stock)

© H. J. de Blij, P. O. Muller, and John Wiley & Sons, Inc.

Drugs and Defeat

British forces penetrated up the Chang Jiang and controlled several areas south of it (Fig. 9-8); Beijing hurriedly sought a peace treaty by which it granted leases and concessions to foreign merchants. China ceded Hong Kong Island to the British and opened five ports, including Guangzhou (Canton) and Shanghai, to foreign commerce. No longer did the British have to accept a status inferior to that of the Chinese in order to do business; henceforth, negotiations would be pursued on equal terms. Opium flooded into China, and its impact on Chinese society became even more devastating. Fifteen years after the First Opium War, the Chinese again tried to stem

the disastrous narcotic tide, and the foreigners who had attached themselves to their country again defeated them. The scourge of this drug abuse was not defeated until after the revival of Chinese power in the twentieth century.

Colonial Indignities

Before China regained control of its destiny, it lost heavily to the colonial powers (Fig. 9-8). In 1898 the Germans obtained a 'lease' on the city of Qingdao on the Shandong Peninsula. The French acquired a sphere of influence in the far south at Zhanjiang. The Portuguese confirmed their hold on Macau. The Russians took control over Liaodong in the north-

east. Even Japan got into the act by annexing the Ryukyu Islands (1879) and, more important, Formosa (Taiwan) in 1895.

One of the most humiliating practices forced upon the Chinese by the Europeans was the doctrine known as **4** **extraterritoriality**. Under this doctrine, foreign states and their representatives are immune from the jurisdiction of the country in which they are based. This doctrine has traditionally been applied to embassies and diplomatic personnel. But in Qing China, it went far beyond that. The European and Japanese invaders established as many as 90 so-called *treaty ports*, which were, in effect, extraterritorial enclaves where traders as well as diplomats were exempt from Chinese law. The best residential areas

FIGURE 9-8

CHINA:
COLONIAL SPHERES,
TERRITORIAL LOSSES

Qing (Manchu) Dynasty at its greatest extent
The Long March, 1934–1935
Administered by Japan, 1937–1945
Territorial losses 19th and 20th Century
People's Republic of China (1949–)

19TH CENTURY COLONIAL INFLUENCE
Russian   British   French   German

0   300   600   900 Kilometers
0 100 200 300 400 500 Miles

© H. J. de Blij, P. O. Muller, and John Wiley & Sons, Inc.

in major cities were declared extraterritorial and made inaccessible to Chinese citizens, including Sha Mian Island on the Pearl River waterfront in the city of Canton (now Guangzhou). Christian missionaries fanned out into China, their residences and churches fortified by extraterritorial security. In many places, Chinese citizens found themselves unable to enter parks and buildings without permission from foreigners. This involved a loss of face and caused a buildup of a resentment that finally exploded in the Boxer Rebellion of 1900.

The End of Dynastic Rule

Figure 9-8 shows the fate of the territory acquired in the early period of Qing (Manchu) rule. Much of the northeast was taken by Russia; the Japanese colonized Korea; Mongolia became independent under foreign auspices; large parts of Turkestan were lost, also to Russia; the French took over in Indochina and the British in Burma. But in the early twentieth century the Chinese had had enough. Bands of revolutionaries roamed city and countryside, attacking the hated foreigners as well as Chinese who had adopted Western ways. The Boxer Rebellion (a loose translation of the Chinese name for these revolutionaries) was put down with much bloodshed by what today would be called a multinational force: British, French, German, Italian, Russian, Japanese, and American soldiers participated. But at the same time a domestic Chinese revolutionary movement, aimed against the

Qing Dynasty itself, was gaining support. In 1911 the emperor's garrisons throughout China were attacked, and in a few months the 267-year-old dynasty was overthrown.

The so-called Nationalist forces that ousted the last emperor proclaimed a republican China and negotiated an end to the extraterritorial treaties, but China remained a badly divided country. Even while the famous Chinese revolutionary leader Sun Yat-sen tried to unify the country from his base in Canton (Guangzhou) in the south, another government tried to rule from Beijing, the old imperial headquarters in the north. Meanwhile, a group of intellectuals in Shanghai founded the Chinese Communist Party. A prominent member of this group was Mao Zedong.

Nationalists and Communists

During the chaotic 1920s, the Nationalists and the Communist Party at first cooperated, making the remaining foreign presence their joint target. After Sun Yat-sen's death in 1925, Chiang Kai-shek became the Nationalists' leader, and by 1927 the foreigners were on the run, escaping by boat and train or falling victim to rampaging Nationalist forces. But soon the Nationalists began purging communists even as they pursued foreigners, and in 1928, when Chiang established his Nationalist capital in the city of Nanjing, it appeared that the Nationalists would emerge victorious. They had driven the communists ever deeper into the interior, and by 1933 Nationalist forces were on the verge of encircling the last communist stronghold, located in the area of Ruijin in Jiangxi Province.

This led to a momentous event in Chinese history: the *Long March*. Nearly 100,000 people—soldiers, peasants, leaders—marched westward from Ruijin in 1934, a communist column that included Mao Zedong and Zhou Enlai. The Nationalist forces continually attacked the marchers, and about three-quarters of them were killed. But new sympathizers joined along the way (see the route marked on Fig. 9-8), and the approximately 20,000 survivors found a refuge in the mountainous interior of Shaanxi Province, 10,000 kilometers (6000 mi) away. There they prepared for a renewed campaign that would bring them to power.

Japan in China

Although many foreigners fled China during the 1920s and 1930s, others seized the opportunity presented by the contest between the Nationalists and the communists. The Japanese took control over the Northeast (formerly referred to as Manchuria), and when the Nationalists proved unable to dislodge them, they set up a puppet state there, appointed a Manchu ruler to represent them, and called their possession Manchukuo.

Full-scale war between the Chinese and the Japanese broke out in 1937, with the Nationalists bearing the brunt of it (thus giving the communists an opportunity to regroup). The gray boundary in Figure 9-8 shows how much of China the Japanese conquered. The Nationalists moved their capital to Chongqing, and the communists controlled the area centered on Yanan. China had been broken into three pieces.

The Japanese committed unspeakable atrocities during their campaign in China. Millions of Chinese citizens were shot, burned, drowned, subjected to gruesome chemical and biological experiments, and otherwise victimized. During the 1980s and 1990s, when China's economic reforms led to a renewed Japanese presence in China, the Chinese public and its leaders called for Japan to acknowledge and apologize for these wartime abuses. In Japan, this pitted apologists against strident nationalists, causing a political crisis. In 1992 Japan's Emperor Akihito visited China and referred to the war but stopped short of a formal apology. And during his term in office Japanese Prime Minister Junichiro Koizumi annually visited the Yasukuni Shrine, where Japanese soldiers—including war criminals—are buried. These visits inflamed public opinion in China and continue to be a contentious issue between the countries.

After the U.S.-led Western powers defeated Japan in 1945, the civil war in China quickly resumed. The United States, hoping for a stable and friendly government in China, sought to mediate the conflict, but at the same time it recognized the Nationalists as China's legitimate government. The United States also aided the Nationalists militarily, destroying any chance of genuine and impartial mediation. By 1948 it was clear that Mao Zedong's well-organized militias would defeat Chiang Kai-shek. Chiang kept moving his capital—back to Guangzhou, the seat of Sun Yat-sen's first Nationalist government, then to Chongqing. Late in 1949, after a series of disastrous defeats in which hundreds of thousands of Nationalist forces were killed, the remnants of Chiang's faction gathered Chinese treasures and valuables and fled to the island of Taiwan. There they took control of the government and proclaimed their own Republic of China.

Meanwhile, on October 1, 1949, standing in front of the assembled masses at the Gate of Heavenly Peace in Beijing's Tiananmen Square, Mao Zedong proclaimed the birth of the People's Republic of China.

China under Communist Rule

After more than a half-century of communist rule, China has been transformed. It has been said that 1949 actually marked the beginning of a new dynasty not so different from the old, an autocratic system that dictated from the top. In that view, Mao Zedong simply bore the mantle of his dynastic predecessors. Only the family lineage had fallen away; now communist 'comrades' would succeed each other.

Certainly, some of China's old traditions continued during the communist era, but in many other ways Chinese society was totally overhauled. Benevolent or otherwise, the dynastic rulers of old China headed a country in which—for all its splendor, strength, and cultural richness—the fate of landless people and serfs was indescribably miserable; in which floods, famines, and diseases could decimate the populations of entire regions without any help from the state; in which local lords could (and often did) repress the people with impunity; in which children were sold and brides were bought. The European intrusion made things even worse, bringing slums, starvation, and deprivation to millions of people who had moved to the cities in hopes of improving their condition.

The communist regime, dictatorial though it was, attacked China's weaknesses on many fronts, mobilizing virtually every able-bodied citizen in the process. Land was taken from the wealthy; farms were collectivized; dams and levees were built; the threat of hunger receded; health conditions improved; child labor was reduced. But China's communist planners also made terrible mistakes. Perhaps the worst of these was the *Great Leap Forward*, in which the peasantry was organized into communal brigades to speed industrialization and make farming more productive. It had the opposite effect, so disrupting agriculture that as many as 30 million people died of starvation between 1958, when the program was implemented, and 1962, when it was abandoned.

Mao ruled China from 1949 to 1976, long enough to leave a lasting mark on the state. Like the Soviets, he refused to impose or even recommend any population policy, arguing that such a policy would represent a capitalist plot to constrain China's human resources. As a result, China's population grew explosively during his rule.

Yet another costly episode of Mao's rule was the so-called *Great Proletarian Cultural Revolution*, launched during his last decade in power (1966–1976). Fearful that Maoist communism was being contaminated by Soviet 'deviationism' and worried about his own stature, Mao unleashed a campaign against what he viewed as emerging elitism in Chinese society. He mobilized young people living in cities and towns into cadres known as Red Guards and ordered them to attack *bourgeois* elements throughout China, criticize Communist Party officials, and root out suspected opponents of the system. He shut down all of China's schools, persecuted untrustworthy intellectuals, and encouraged the Red Guards to engage in what he called a renewed revolutionary experience. The results were disastrous: Red Guard factions took to fighting among themselves, and anarchy, terror, and economic paralysis followed. Thousands of China's leading intellectuals died; moderate leaders were purged; and teachers, elderly citizens, and older revolutionaries were tortured to make them confess to crimes they had not committed. As the economy suffered, agricultural and industrial production declined. Violence and famine killed as many as 30 million people as the Cultural Revolution spun out of control. One of those who survived was a Communist Party leader who had himself been purged and then reinstated—Deng Xiaoping. Deng was destined to lead the country in the post-Mao period of economic transformation, discussed later in this section.

Political and Administrative Divisions

Different types of government often use their power to impose certain administrative units on their territories. So, before we investigate the human geography of contemporary China, we should acquaint ourselves with the country's current political and administrative framework (Fig. 9-9). For administrative purposes, China is divided into the following units:

- Four central-government-controlled municipalities (*shi's*)
- Five autonomous regions
- Twenty-two provinces
- Two special administrative regions

The four central-government-controlled *municipalities* are the capital, Beijing; the nearby port city of Tianjin; China's largest metropolis, Shanghai; and the Chang River port of Chongqing in the interior.

These *shi's* form the cores of China's most populous and important subregions, and the capital's direct control over them entrenches the central government's power.

We should note that the administrative map of China continues to change—and to pose problems for geographers. The city of Chongqing was made a *shi* in 1996, and its 'municipal' area was enlarged to incorporate not only the central urban area but a huge hinterland covering all of eastern Sichuan Province. As a result, the 'urban' population of Chongqing is officially 30 million, making this the world's largest metropolis—but actually, the central urban area has no more than about 5 million inhabitants. Because Chongqing's population is officially *not* part of Sichuan, the province that borders it to the west (Fig. 9-9), the official population of Sichuan declined by 30 million when the ***Chongqing Shi*** was created.

The five *Autonomous Regions* (ARs) were established to recognize the non-Han minorities living there. Some laws that apply to Han Chinese do not apply to these populations. However, Han Chinese immigrants now outnumber several of these minorities in their own regions, making their administration more complex today than in the past. The five are: (1) Nei Mongol AR (Inner Mongolia); (2) Ningxia Hui AR (adjacent to Inner Mongolia); (3) Xinjiang Uyghur AR (China's northwest corner); (4) Guangxi Zhuang AR (far south, bordering Vietnam); and (5) Xizang AR (Tibet).

China's 22 *provinces*, like U.S. States, tend to be smaller in the east and larger toward the west. The territorially smallest are the three easternmost provinces on China's coastal bulge: Zhejiang, Jiangsu, and Fujian. The two largest are Qinghai, flanked by Xizang AR, and Sichuan, China's 'Midwest.'

As with all large countries, some provinces are more important than others. The Province of Hebei nearly surrounds Beijing and occupies much of the core of the country. The Province of Shaanxi is centered on the great ancient city of Xian. In the southeast, momentous economic developments are occurring in the Province of Guangdong, whose urban focus is Guangzhou. In the remainder of the chapter,

FIGURE 9-9

© H. J. de Blij, P. O. Muller, and John Wiley & Sons, Inc.

Province has 14 million more inhabitants than does France.

The One-Child Policy

With a population of 1.325 billion, China is the largest country in the world demographically, and it inherited from the communist period a high rate of natural increase. Aware of the economic costs of rapid population growth, after Mao's death in 1976 the Chinese government embarked on a vigorous population-control program. Families were ordered to have only one child, and those who violated the policy were penalized by the loss of tax advantages, educational opportunities, and even housing privileges. In the early 1970s, the annual rate of natural increase was about 3 percent; by the mid-1980s, it was down to 1.2 percent. Today, China officially reports a moderate growth rate of 0.6 percent. If you refer back to Figure 8-9 and compare the shapes of the population pyramids, you can see that population control was successfully implemented in China.

A serious ramification of the one-child policy in this patriarchal society, where male children are preferred, is that rates of female fetus abortion, infanticide, and abandonment have been high. The result is a gender imbalance that is entering a critical period as children born since 1979 (when the policy was instituted) reach marriageable age. The Chinese government estimates that by 2020 it will be short some 30 million brides, an astonishing number. The official gender imbalance is 118 boys for every 100 girls, but in some villages it is much higher for a variety of reasons. This has resulted in the trafficking of women, both within China and from neighboring countries

when we refer to a particular province or other administrative unit, Figure 9-9 can serve as a useful locational guide.

In 1997, the British dependency of Hong Kong (Xianggang) was taken over by China and became the country's first **5 Special Administrative Region (SAR)**. In 1999, Portugal similarly transferred Macau, opposite Hong Kong on the Pearl River estuary, to Chinese control, creating the second SAR under Beijing's administration.

POPULATION ISSUES

Many Chinese provinces, like many of India's States, have populations that are larger than those of most of the world's countries. If Chongqing Shi is included, Sichuan province has almost 120 million inhabitants. Approaching 100 million population are Henan and Shandong provinces (see Fig. 9-9). There are almost as many people in Guangdong Province as there are in Mexico. Jiangsu

such as North Korea, to be married to the millions of single Chinese men. This situation will only become worse, and government authorities predict that it will affect 'social stability.'

The one-child policy also had a devastating impact in the aftermath of the earthquake that occurred in Sichuan in May 2008. Since the earthquake hit in the middle of the day, many children were at school. Poorly built schools collapsed while other structures stood on either side of them. These schools entombed thousands of children. Grieving parents were openly angry with communist party officials who may have permitted shoddy school construction that resulted in the deaths of about 10,000 children. Many families

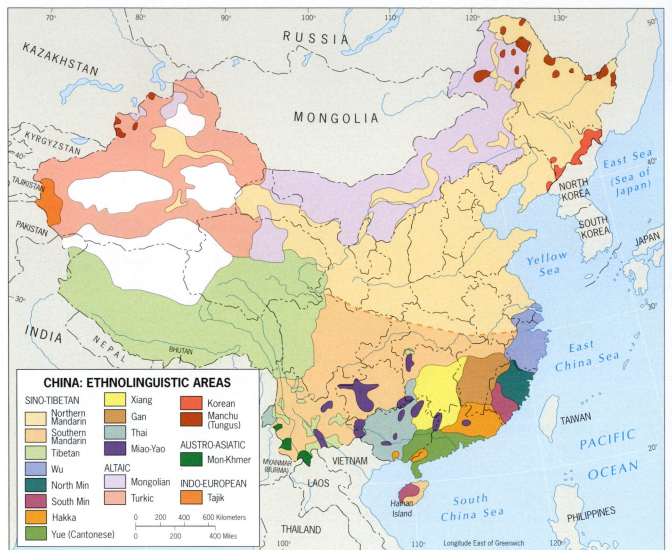

FIGURE 9-10

© H. J. de Blij, P. O. Muller, and John Wiley & Sons, Inc.

CHINA: ETHNOLINGUISTIC AREAS

SINO-TIBETAN
- Northern Mandarin
- Southern Mandarin
- Tibetan
- Wu
- North Min
- South Min
- Hakka
- Yue (Cantonese)
- Xiang
- Gan
- Thai
- Miao-Yao

ALTAIC
- Mongolian
- Turkic

Korean
Manchu (Tungus)

AUSTRO-ASIATIC
- Mon-Khmer

INDO-EUROPEAN
- Tajik

0 200 400 600 Kilometers
0 200 400 Miles

lost their only child, leaving gaping holes in their lives and in their prospects for old-age. Although the government permits couples to have another child if the first one dies, many couples will be unable to do so as they are too old or have been sterilized.

The May 12, 2008, earthquake in Sichuan province, China, had a particularly harsh impact on parents. Since Chinese couples are limited to having only one child, many parents lost their only child in the earthquake. Here a grieving mother holds the photo of her daughter at the site of one of the collapsed schools. (© AP/Wide World Photos)

The Minorities

When the Soviet Union collapsed, some observers described that event as the *end of empire*—the breakdown of the last empire on earth. But they forgot about China. Though diminished from its Qing Dynasty dimensions, the Beijing government nonetheless controls territories and peoples that are, in effect, colonized, as can be seen in the ethnolinguistic map (Fig. 9-10).

The map gives added meaning to the term ***China Proper*** as the home of the People of Han, the ethnic (Mandarin-speaking) Chinese depicted in tan and light orange in Figure 9-10. From the upper northeast to the border with Vietnam, and from the Pacific coast to the margins of Xinjiang, this Chinese majority dominates. When this map is compared to that of population distribution (Fig. 9-2), it is clear that the minorities constitute only a small percentage of the country's total population. The Han Chinese form the largest and densest clusters.

China also controls non-Chinese areas that are vast, if not populous. The Buddhist Tibetan group numbers less than 3 million, but it extends over all of settled Xizang. Muslim Turkic peoples remain dominant in Xinjiang. Thai, Vietnamese, and Korean minorities also occupy areas on the margins of Han China. As we will see later, the Southeast Asian minorities in China have participated strongly in the Pacific Rim developments on their doorsteps. Hundreds of thousands have migrated from their autonomous regions in search of greater economic opportunity in the coastal regions.

Numerically, Chinese dominate in China to a far greater degree than Russians dominated their Soviet Empire. But territorially, China's minorities extend over a proportionately larger area. Truly, the Ming and Qing (Manchu) rulers bequeathed an empire to the People of Han.

PEOPLE AND PLACES OF CHINA PROPER

A map of East Asia's population distribution (Fig. 9-2) reveals the continuing relationship between the physical world and its human occupants. The pattern in the developed world has been that people typically shake off their dependence on what the land can provide and move to cities and other areas where economic opportunities are greater. This depopulates rural areas that may once have been densely inhabited. In China, that stage has not yet been reached. Whereas China has large cities, a sizable majority (63 percent) of its people still live in rural areas. Thus, the map of population distribution reflects the high productivity of China's basins, lowlands, and plains. If one compares Figures 9-2 and 9-5A, China's continuing dependence on soil, water, and warmth is evident.

The population map also suggests that in certain areas environmental limitations are being overcome. Industrialization in the Northeast, irrigation in the Inner Mongolia Autonomous Region, and oil-well drilling in Xinjiang enabled millions of Chinese to migrate from China Proper into these frontier zones, where they now collectively outnumber the indigenous minorities.

Nevertheless, physiography and demography remain closely linked in China. To grasp this relationship, it is useful to compare Figures 9-1 (physiography) and 9-2 (population). On the population map, the darker the color, the denser the population; in some places, using this means we can follow the courses of major rivers. Look, for example, at China's Northeast. A ribbon of population follows the lower Liao and Songhua rivers. Also note the dense nearly circular population concentration to the west of central China's main population cluster; this is the Sichuan Basin, in the upper course of the Chang Jiang. In fact, more than three-quarters of China's 1.3 billion people are contained within four major river basins:

1. The Liao-Songhua Basin or Northeast China Plain
2. The Lower Huang He (Yellow River) Basin, known as the North China Plain
3. The Upper and Lower Basins of the Chang Jiang (Yangzi River)
4. The Basins of the Xi (West) and Pearl Rivers

Not all the people living in these river plains are farmers, of course. China's great cities are also found in these populous areas, from the aging industrial centers of the Northeast to the Pacific Rim upstarts of the south. Both Beijing and Shanghai lie in major river basins. And, as the map shows, the hilly areas between the river basins are not sparsely populated either. Note that the hill country south of the Chang Basin (opposite Taiwan) still has clear, dense population clusters.

The Northeast China Plain

The Northeast China Plain is the heartland of China's Northeast, the ancestral home of the Manchus who founded China's last dynasty; it used to be called Manchuria. As the administrative map (Fig. 9-9) shows, there are three provinces here: Liaoning in the south, facing the Yellow Sea; Jilin in the center; and Heilongjiang, by far the largest, in the north.

The Northeast has experienced many ups and downs. Although Japanese colonialism was ruthless and exploitive, the Japanese did build railroads, roads, bridges, factories, and other components of the regional infrastructure. At one time, half of China's total railroad mileage was in the Northeast. After the Japanese were ousted, the Soviets looted the area of machinery, equipment, and other assets. During the late 1940s, the Northeast was a ravaged frontier. But then the communists took power and made the industrial development of the Northeast a priority. From the 1950s until the 1970s, the Northeast led the nation in manufacturing growth. Its population, just a few million in the 1940s, mushroomed to over 100 million. Towns and cities grew exponentially. All this growth was based on the Northeast's considerable mineral wealth, and for a time it worked. During the 1970s, the Northeast contributed fully one-quarter of the country's entire industrial output.

But then came the **6 economic restructuring** of the post-Mao era, resulting from a new, market-driven economic order. The huge, inefficient, state-supported industries could not adapt, and in short order the Northeast became a rustbelt, its contribution to China's industrial output plummeting below 10 percent. As factories closed, hundreds of thousands of workers found themselves destitute, their factory jobs and pensions eliminated and their families impoverished. They staged protests, blocking roads, railroads, and even airport runways to demand government help, a rarity in tightly controlled China.

Recently, the government has begun to address the Northeast's problems, in part by awarding laid-off workers a small stipend but also by seeking ways to revive the region's fortunes. One initiative involves Russia, whose sliver of coastal territory blocks the

Northeast's Jilin Province from access to the sea. The Chinese want Russia to lease them a coastal enclave where China would build a small Special Economic Zone complete with port facilities, linking Jilin, and thus the Northeast, to the outside world. The Russians, fearing a Chinese takeover of the territory, have resisted, arguing that their own small ports, Zarubino and Posyet, can handle any transit trade generated by China's Northeast. China proposed building a facility where oil and natural gas can be imported from Russia; the Russians responded by suggesting a joint venture with China instead of a Chinese lease on any part of their coast. Under these conditions, energy-hungry China has few options in its attempts to bring the geographically disadvantaged Northeast into the Pacific Rim sphere.

The North China Plain

The North China Plain is one of the world's most heavily populated agricultural areas. This density is visible in Figure 9-2. Most of this part of China has a population density of more than 400 inhabitants per square kilometer (1000 per sq mi), and in some parts the density is twice as high. Here the ultimate hope of the Beijing government lay less in land redistribution than in raising yields through improved fertilization, expanded irrigation facilities, and more intensive use of labor. A series of dams on the Huang River, including the Xiaolangdi Dam upstream in Henan Province, now reduce the danger of flooding, but outside the irrigated areas the ever-present problem of rainfall variability and drought per-

FIGURE 9-11

© H. J. de Blij, P. O. Muller, and John Wiley & Sons, Inc.

sists. The North China Plain has not produced any substantial food surplus even under normal circumstances; thus, when the weather turns unfavorable the situation soon becomes precarious. The specter of famine may have receded, but the food situation is still uncertain in this critical part of China Proper.

The North China Plain is an excellent example of the concept of a national **7** **core area**. Not only is it a densely populated, highly productive agricultural zone, but it is also the site of the capital and other major cities, a substantial industrial complex, and several ports, among which Tianjin ranks as one of China's largest (Fig. 9-11). Tianjin, on the Bohai Gulf, is linked by rail and highway (less than a two-hour drive) to Beijing, the capital. Like many of China's harbors, that of Tianjin's river port is not particularly good; but Tianjin is well situated to serve not just the northern sector of the North China Plain and the capital but the Upper Huang Basin and Inner Mongolia beyond as well. Tianjin was formerly a treaty port, but the city's major growth occurred under communist rule. For decades it was a center for light industry and a flood-prone harbor, but after 1949 the communists constructed a new port and flood canals. They also chose Tianjin as a site for major industrial development and made large investments in the chemical industry (in which Tianjin still leads the nation), iron and steel production, heavy-machine manufacturing, and textiles. Today, with a population of 7.4 million, Tianjin is China's fifth-largest metropolis and the center of one of its leading industrial complexes.

Beijing, unlike Tianjin, is China's political, cultural, and educational center. Its industrial development has not matched Tianjin's. The communist administration did, however, greatly expand the municipal area of Beijing, which, as a *shi*, is not controlled by the Province of Hebei but is directly under the authority of the central government. In one direction, Beijing was enlarged all the way to the Great Wall—50 kilometers (30 mi) to the north—so that the urban area includes hundreds of thousands of farmers. In central Beijing, old, traditional neighborhoods called *hutongs* are being razed to make way

ENVIRONMENTALLY SPEAKING . . .

Air Pollution in China.   The new China Central Television headquarters building in Beijing is shrouded in fog and pollution. Rapid economic development through industrialization and urbanization like that currently taking place in China has numerous negative impacts, including environmental degradation. China faces a number of serious environmental problems, air pollution being perhaps one of the most visible. Chinese authorities tolerate particulate matter at levels much higher than those accepted elsewhere, with serious health consequences. Air pollution is not new in China—throughout much of the nation's history, high-sulphur coal has been used for heat (directly) and energy (indirectly)—but it is reaching critical levels due to the staggering pace of China's economic growth. China has vast stocks of coal and meets two-thirds of its energy demands by burning coal, consuming more coal per year than the United States, Japan, and Europe combined.

Only 1 percent of China's 500 million urban residents breathe air that would meet European Union standards; three out of four Chinese breathe air that does not even meet China's own air-quality standards. Blue skies are virtually unknown to millions of city dwellers, and hundreds of thousands of people die prematurely each year due to indoor or outdoor air pollution. In Beijing alone, the addition of approximately 1000 new private cars per day is a major contributor to air pollution. (© AP/Wide World Photos)

for eight-lane highways, skyscrapers, and new housing. The ongoing transformation of the capital's cityscape is almost unimaginable. Beijing, with its 11.3 million inhabitants, now ranks among the world's most dynamic megacities.

Inner Mongolia

Northwest of the core area as defined by the North China Plain, along the border with the state of Mongolia, lies Inner Mongolia, administratively defined as the Nei Mongol Autonomous Region (see Fig. 9-9). Originally established to protect the rights of the approximately 5 million Mongols who live outside the Mongolian state, Inner Mongolia has experienced massive immigration by Han Chinese. Today, Chinese outnumber Mongols here by nearly four to one. Irrigation and industry have created an essentially Chinese landscape, so different from independent Mongolia that the location of the border can be observed from an aircraft. Fences and eroded land mark the Chinese side, whereas grassy steppe characterize the Mongolian side. The AR's capital, Hohhot, has been eclipsed by Baotou on the Huang River, which supports a corridor of farm settlements as it crosses the dry land here on the margins of the Gobi and Ordos deserts. At about 24 million, Inner Mongolia's population cannot compare to the huge numbers that crowd the North China Plain, but recent mineral discoveries have boosted industry in Baotou, and livestock herding is expanding. Nei Mongol may retain its special administrative status, but it functions as part of Han China in all but name.

FIGURE 9-12

© H. J. de Blij, P. O. Muller, and John Wiley & Sons, Inc.

The Basins of the Chang/Yangzi

In contrast to the contiguous, flat agricultural-urban-industrial North China Plain defined by the sediment-laden Huang River, the basins and valleys of the Chang Jiang (Long River) display variation in elevation and relief. The Chang River, whose name becomes the Yangzi along its lower course, is an artery like no other in China. Near its mouth lies the country's largest city, Shanghai. Part of its middle course has been transformed by a gigantic engineering project. Farther upstream, the Chang crosses the popu-

lous, productive Sichuan Basin. And unlike the Huang, the Chang Jiang is navigable by oceangoing ships for over 1000 kilometers (600 mi), from the coast all the way to Wuhan (Fig. 9-11). Smaller ships can reach Chongqing. Several of the Chang River's tributaries also are navigable, with the result that its basin is served by 30,000 kilometers (18,500 mi) of water transport routes. Thus, the Chang Jiang constitutes one of China's leading transit corridors. With its tributaries, it handles the trade of a vast area, including nearly all of central China and sizable parts of the north and south. The North China Plain may be the core area of China, but in many ways the Lower Chang Basin is its heart.

A Pivotal Corridor

As Figure 9-12 shows, the Lower Chang Basin is an area of both rice and wheat farming, offering further proof of its pivotal situation between the south and the north. Shanghai (population: 15.3 million) lies at the coastal gateway to this productive region, on a small tributary of the Chang (here Yangzi) River, the Huangpu. The city has an immediate hinterland of some 50,000 square kilometers (20,000 sq mi)—an area half the size of Ohio—containing more than 50 million people. About two-thirds of this population consists of farmers who produce food, silk filaments, and cotton for the city's industries.

Travel upriver along the Yangzi/Chang Jiang, and you meet an unending stream of vessels large and small, including numerous barge trains—as many as six, sometimes more, barges pulled by a single tug in an effort to save fuel (they are slow, but time is not the primary concern). The traffic between Shanghai and Wuhan—Wuhan is short for Wuchang, Hanyang, and Hankou, a conurbation of three cities—makes the Yangzi one of the world's busiest waterways. Smoke-belching boat engines create a persistent plume of pollution here, worsening the regional smog created by the factories of Nanjing and others perched on the waterfront. Here China's Industrial Revolution is in full gear, with all its environmental consequences.

The Three Gorges Dam

Above Wuhan, river traffic dwindles because the shallower depth of the Chang reduces the size of vessels that can reach Yichang. Riverboats carry coal, rice, building materials, barrels of fuel, and many other items of trade. But the middle course of the Chang River is becoming more than a trade route. In the northward bend of the great river, between Yichang and Chongqing, where the Chang has cut deep troughs, a massive engineering project was completed in 2006 and will become fully operational by 2009. It has several names: the Sanxia Project, the Chang Jiang Water Transfer Project, the Three Gorges Dam, the New China Dam. It is most commonly called the Three Gorges Dam because it is here that the great river flows through a 240-kilometer (150-mi) series of steep-walled valleys less than 110 meters (360 ft) wide. Near the lower end of this natural trough, a concrete dam has risen to a height of over 180 meters (600 ft) above the valley floor to a width of 2.1 kilometers (1.3 mi), creating a reservoir that inundates the Three Gorges and extends over 600 kilometers (385 mi) upstream. The rising water required the relocation of close to 2 million people and has had a significant impact on the surrounding ecosystems. But China's pro-development officials feel that the benefits of the dam—the end of the river's rampaging flood cycle, enhanced navigation,

increased development along the new lake perimeter, and the provision of at least one tenth, and perhaps as much as one eighth, of China's electrical power supply—far outweigh its environmental and social impacts. Because it will transform the heart of China, many Chinese like to compare this gigantic project to the Great Wall and the Grand Canal, and they call it the New China Dam.

Chongqing and Sichuan

The Three Gorges Dam will strongly affect the fortunes of the city of Chongqing, upstream from the reservoir. A diversion channel around the dam will enhance Chongqing's river-port functions, allowing larger vessels to reach it. Although Chengdu is Sichuan's capital, Chongqing, whose status was recently elevated to that of *shi*, is the province's key outlet. With 85 million people, immense energy resources, and farms that produce crops ranging from grains to tea, sugarcane, fruits, and vegetables, Sichuan is a Chinese breadbasket, and Chongqing may become the country's number-one growth area.

The Basins of the Xi (West) and Pearl Rivers

As Figure 9-11 shows, the Xi River and its basins are no match for the Chang or Huang, or even for the Liao. This southernmost river even seems to have the wrong name: Xi means West! It reaches the coast in a complex area where it forms a delta immediately adjacent to the estuary of the Pearl River. In this subtropical part of China, local relief is higher than in the lowlands and basins of the center and north, so farmlands are more limited in their extent. Especially in the interior areas, water supply is a recurrent problem. On the other hand, the warm climate permits double-cropping of rice. The region's food production, however, has never approached that of the North China Plain or the Lower Chang Basin. And, as Figure 9-2 shows, the population of this southern part of China is smaller than that of the more northerly heartlands.

Again unlike the Huang and Chang rivers, which rise among snow-covered interior mountains, the Xi Jiang is a shorter stream whose source is on the Yunnan Plateau (Fig. 9-1). Just up the coast from its delta, however, lies one of the most important areas of modern China. As we will note later, the mouth of the Pearl River is flanked by some of China's fastest-growing economic complexes, including Guangzhou, the capital of Guangdong Province; Hong Kong, the former British dependency; and adjoining Shenzhen, one of the world's fastest-growing cities.

This area's economic growth is all the more remarkable because, as the maps emphasize, South China is not especially well endowed with geographic advantages, locational or otherwise. The

The massive Three Gorges Dam on the Chang River above Yichang in Hubei province became operational in 2006. This is the world's largest-ever hydroelectric project and began providing power in 2003. Capacity since then has steadily expanded, and the facility is on target to reach full production in 2009.   (© Zuma Press/Newscom)

Xinjiang. Here, in this remote western periphery adjoining Turkestan, China meets Muslim Central Asia.

XIZANG (TIBET)

As the climate and physiographic maps demonstrate, harsh physical environments dominate Xizang—pronounced "sheedz-AHNG." This vast, sparsely peopled region has been designated as an autonomous region but in fact is an occupied society.

In considering Xizang's human geography, we should take note of two subregions. The first is the core area of Tibetan Buddhist culture, located on the southeastern edge of the Qinghai-Xizang Plateau just north of the Himalayas. In this area, some valleys lie below 2130 meters (7000 ft); here the climate is comparatively mild and some cultivation is possible. Here also lies Tibet's main population cluster, including the crossroads capital of Lhasa (Fig. 9-14). In fact, China's determination to strengthen its presence in Tibet led to the construction of the world's highest railroad, linking Lhasa to Beijing, which opened in July 2006. The second area of interest is the Qaidam Basin to the north. This basin lies thousands of feet below the surrounding Kunlun and Altun Mountains and has always contained a concentration of nomadic pastoralists. Recently, however, exploration has revealed the presence of oilfields and coal reserves below the surface of the Qaidam Basin, and these resources are now being developed.

Conquest by China

As Figure 9-6 reminds us, Xizang (population around 3 million) came under Chinese domination

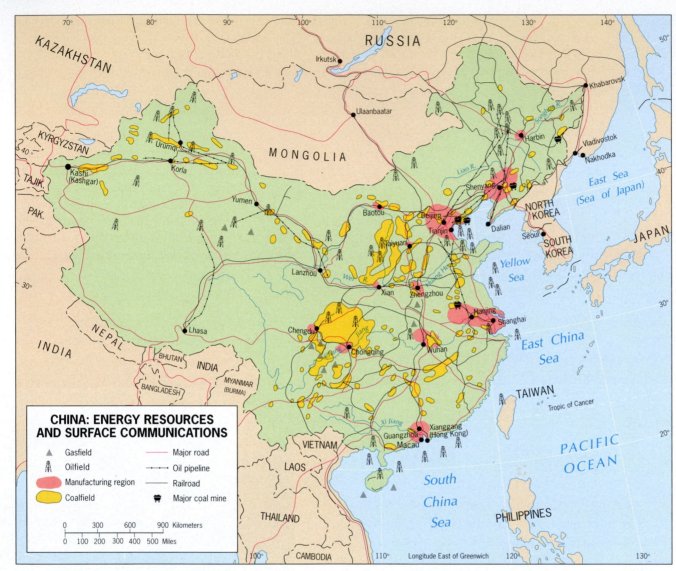

FIGURE 9-13

© H. J. de Blij, P. O. Muller, and John Wiley & Sons, Inc.

CHINA: ENERGY RESOURCES AND SURFACE COMMUNICATIONS

- ▲ Gasfield
- ⚒ Oilfield
- 🔴 Manufacturing region
- 🟡 Coalfield
- — Major road
- ••• Oil pipeline
- — Railroad
- 🛒 Major coal mine

Pearl River is navigable by larger ships only to Guangzhou, and the Xi by small riverboats no farther than Wuzhou. Guangdong Province, the key province of southern China, lies far from China's heartlands. Nor is South China blessed with the resources that propelled the industrial development of the Northeast and other parts of China. Indeed, one of the principal objectives of the Three Gorges Dam project is to provide South China with ample elec-

tricity. As Figure 9-13 shows, northern and western China, not the south, have most of the coal, oil, and natural gas resources. Still, the south is where China's economic transformation has been focused.

China Proper, with its populous and powerful subregions, is in many ways the dominant region in the East Asian geographic realm, but it is only one part of China. To the west lie two vast regions with relatively sparse populations: Xizang (Tibet) and

CHINA'S WESTERN FLANK

- ● Under 50,000
- ● 50,000–250,000
- ● 250,000–1,000,000
- ● 1,000,000–5,000,000
- ● Over 5,000,000

■ Oil/gas reserve
■ Irrigated area
—○— Existing pipeline
–○– Proposed pipeline

National capitals are underlined

0 200 400 600 800 Kilometers
0 100 200 300 400 500 Miles

FIGURE 9-14   © H. J. de Blij, P. O. Muller, and John Wiley & Sons, Inc.

during the Manchu (Qing) Dynasty in 1720 but regained its separate status in the late nineteenth century. China's communist regime took control of the region in 1950; in 1959, it crushed an uprising by Tibetan villagers. Tibetan society had been organized around the fortress-like monasteries of Buddhist monks who paid allegiance to their supreme leader, the Dalai Lama. The Chinese wanted to modernize this feudal system, but the Tibetans clung to their traditions. They proved no match for the Chinese armed forces; the Dalai Lama was forced into exile and the monasteries were emptied. The Chinese also destroyed much of Xizang's cultural heritage, looting its religious treasures and works of art. Their harsh rule devastated Tibetan society, but after Mao's death in 1976 the Chinese relaxed their control and some amends were made. Since its formal annexation in 1965, Xizang has been administered as an autonomous region but remains a contentious region. The central Chinese government continues its process of *Hanification* by encouraging Han Chinese to immigrate to the region. Tibetan discontent because of cultural repression bubbled over in 2008 with demonstrations that were quickly repressed. In an era of the Internet and with the increased visibility of the Olympics, the situation will likely remain unsettled for some time.

XINJIANG

China's northwestern corner consists of the giant Xinjiang Uyghur Autonomous Region, constituting over one-sixth of the country's land area and situated in a strategic part of Central Asia (Fig. 9-14). Linked to China Proper via the vital Hexi Corridor in Gansu Province to the east, Xinjiang (meaning *New Territory* and pronounced shin-jee-AHNG) is China's largest single administrative area. Culturally, however, it is closer to the Central Asian republics and is sometimes referred to as Chinese Turkestan as its people are Turkic and practice Islam.

Even today, after more than a half-century of vigorous Hanification (Sinicization), Han Chinese remain a minority here. About 40 percent of the region's population of over 20 million is Chinese (up from 5 percent in 1949), yet the Chinese control virtually all aspects of life, including the clock: although Xinjiang lies 2500 kilometers (1500 mi) west of the capital, it must maintain Beijing time! The majority of the region's people are Muslim Uyghurs, who make up approximately half of the total, and Kazakhs, Kyrgyz, and others with cultural affinities to the republics of former Soviet Central Asia. The Uyghurs like to call this area East Turkestan, and some defiantly set their watches two hours behind Beijing's, calling this Ürümqi Time.

Physiography

The physical geography of Xinjiang is dominated by high mountain ranges and vast basins (Fig. 9-1). The southern periphery is defined by the Kunlun Shan (Mountains), beyond which lies Xizang (Tibet). Across the north-center of the region lies the mountain wall of the Tian Shan. Between the Kunlun and Tian Shan extends the vast, dry Tarim Basin, within which lies the Taklamakan Desert, and north of the Tian Shan lies another depression, the Junggar Basin (Fig. 9-1). Rivers rising on the mountain slopes disappear beneath the sediments of the basins, sustaining a ring of oases where the water approaches the surface or where wells and *qanats* (an ancient irrigation system) make agriculture possible.

Energy and Technology

While this would not seem to be the most favorable geography, Xinjiang has other assets that China exploits. The Qing (Manchu) armies that fought the nomadic Muslim warlords here in the 1870s saw the region's agricultural possibilities and planted its first crops. After 1949, China's communist rulers exiled their political opponents, as well as criminals, to notorious prison camps in this remote frontier. Then Xinjiang turned out to have extensive reserves of oil and natural gas (Figs. 9-14). Here, too, China could experiment with space technology, rocketry, and atomic weaponry far from densely populated areas (and shielded from foreign spies). And now, as Figure 9-14 reminds us, China's westernmost borders lie relatively close to the great energy reservoirs of the Caspian Sea Basin—and only one neighbor lies in the way. Construction of a pipeline across Kazakhstan to Xinjiang is already nearing completion.

Most of the Han Chinese who live in Xinjiang are concentrated in the north-central cluster of cities and towns between the capital, Ürümqi (population 2.3 million), and the oil center of Karamay. Also centered on the Ürümqi-Karamay corridor is Xinjiang's largest agricultural zone, where cotton and fruits, especially grapes and melons, yield export revenues. A ring of oases and clusters of irrigated agriculture encircles both the Tarim and Junggar basins (see Fig. 9-18).

Beijing has made extensive efforts to integrate Xinjiang into the Chinese state. A railroad and a highway link Ürümqi and Lanzhou via the Hexi Corridor. Westward from Ürümqi, a railroad crosses the border with Kazakhstan and connects to the Öskemen-Almaty line; another railroad, opened in 2000, connects the capital to Kashi (Kashgar) in the distant southwest (Fig. 9-14). But vast spaces in Xinjiang remain distant and inaccessible. Just one road, recently upgraded, loops around the vast southern and eastern flank of the Tarim Basin, from Kashi (Kashgar) via Hotan to Korla.

China's western frontier regions in some ways resemble the Soviet Union's former Central Asian republics. Here Chinese modernizers meet traditional Buddhism and Islam; here locals argue that they are colonized by Beijing and at times demand independence. But the map leaves no doubt as to these regions' importance and value to China. Soviet-style devolution is unlikely to happen here.

Xinjiang Uyghur Autonomous Region is home to the Uyghur people whose language and religion (Islam) are very different from the Han Chinese culture of eastern China. Picture here on the central square, the International Grand Bazaar of Ürümqi, are several Uyghurs looking at books for sale. (© A. WinklerPrins)

CHINA'S PACIFIC RIM

From Dalian on the doorstep of the Northeast China Plain to Hainan Island off the country's southernmost point, China is booming. In the cities, old neighborhoods are being bulldozed to make way for skyscrapers. In the countryside of the coastal provinces, thousands of factories employing millions of workers who used to farm the land are changing life in the towns and villages. New roads, airports, power plants, and dams are being built. This process of modernization is an uneven, sometimes chaotic one that is affecting certain stretches of China's coast more than others. It is penetrating parts of China's interior, but meanwhile social gaps are widening: between advantaged and disadvantaged,

well-off and poor, urbanite and villager. Modernization is creating regional economic disparities that China has never seen before. And it is producing what may become a discrete geographic region along China's Pacific littoral, a region that has more in common with Japan, South Korea, and Taiwan than with Yunnan or Gansu.

In the Introduction, we discussed issues related to the geography of development, emphasizing that in this complex, globalizing world, the concept of developed versus underdeveloped countries (DCs and UDCs) is no longer tenable. By almost any measure, China remains what used to be called a UDC. As Figure G-11 shows, it modestly ranks as a lower-middle-income economy. But China now has its own core-periphery duality: in its Pacific coast provinces, per-capita incomes today are consider-

ably above the national average GNI of U.S. $6600, while hundreds of millions of villagers in China's remote interior earn much less. In its burgeoning Pacific Rim, China resembles a modernizing economy with *developed* characteristics. In much of its interior, it does not. This contrast is one reason that a discrete Pacific Rim region appears to be forming in China Proper.

The Geography of Development

Many yardsticks are used to gauge a country's development. These include GNI per capita; percentage of workers in farms, factories, or other kinds of employment; amount of energy consumed; efficiency of transport and communications; use of manufactured metals like aluminum and steel in the economy; productivity of the labor force; and social measures such as literacy, nutrition, medical services, and savings rates. All these data are calculated per person, yielding an overall measure that ranks the country among all those providing statistics. These data do not, however, tell us *why* some countries, or parts of countries, exhibit the level of development that they do.

The **8** **geography of development** connects raw-material distributions, environmental conditions, cultural traditions, historical factors (such as the lingering effects of colonialism), and forces of location. In the emergence of China's Pacific Rim, relative location played a major role, as we will see shortly.

A Development Model

Geographers tend to view the development process spatially, whereas other scholars focus on structural aspects. Among the latter is economist Walt Rostow, who in 1960 formulated a global model of the development process that is still much discussed today. This model suggested that all developing countries

FROM THE FIELD NOTES

Globalization in Xinjiang. "The capital of Xinjiang, Ürümqi, has long been known for its central bazaar. More Central Asian than East Asian, the bazaar sold merchandise for the region's dominant Uyghur population. Uyghurs are Turkic people whose cultural roots and language link them to Central and Southwest Asia. They are Muslims and have a different cuisine than do the Han Chinese from the East. The central bazaar was a labyrinth of passageways and alleys, fascinating to tourists but disturbing to officials who felt that these alleys could harbor terrorists. In 2002 the old bazaar was razed and a new one, the International Grand Bazaar, was built with an organized layout. Although the new bazaar is modern, it was fascinating to walk around, with sights and smells that were very non-Western. Yet, rounding a corner I encountered the scene pictured here, the entrance to a mall, with underground parking and a modern Carrefour supermarket. Carrefour, the French Wal-Mart, had found a home in this International Bazaar. On the photo you can see Chinese characters as well as the Arabic script of the written Uyghur language and a little bit of English."

(© A. WinklerPrins)

Concept Caching

www.conceptcaching.com

follow an essentially similar path through five inter-related growth stages, as follows:

Stage	Stage Name	Societal Activity
1	**Traditional society**	The society engages mainly in subsistence farming, is locked in a rigid social structure, and resists technological change.
2	**Preconditions for takeoff**	Old ways are abandoned, workers move from farming into manufacturing, and transport improves.
3	**Takeoff**	The region experiences an industrial revolution; sustained economic growth takes hold, urbanization proceeds, and technological and mass-production breakthroughs occur.
4	**Drive to maturity**	Sophisticated industrial specialization and increasing international trade take place.
5	**High mass consumption**	This stage is marked by high incomes, widespread production of consumer goods and services, and most workers employed in the tertiary and quaternary economic sectors.

Although at first glance this is a useful model for explaining the development process, it does not explain the whole story. The model treats countries as homogeneous entities and takes little account of core-periphery contrasts such as those found in China. It also assumes that all countries develop along the same pathway, which we know does not occur. (Compare India with China, for example.) In China, takeoff conditions (stage 3) and evidence of maturation (stage 4) exist in much of the Pacific Rim, but other major areas of this vast country remain in stage 1. China has passed through the transition from tradition-bound to progressive leadership, but the effect of that leadership's reformist policies has not been the same in all parts of the country. The most fertile ground for these conditions lies along the Pacific; here China has a history of contact with foreign enterprises and is most open to the world. It is from this region that many Chinese departed for Southeast Asian (and other) countries. These **9** **Overseas Chinese** were positioned to play a crucial role in China's Pacific Rim development: many had left kin and community there, and those links facilitated the investment of hundreds of millions of dollars when the opportunity arose. Many of the Pacific Rim's factories were built by means of this repatriated money.

Transforming the Economic Map

China's communist era has witnessed the metamorphosis of the world's most populous nation. Despite the dislocations associated with communist rule, as a communist state China has managed to stave off famine and hunger, control its population growth rate, strengthen its military capacity, improve its infrastructure, and enhance its world position. It took over Hong Kong from the British and Macau from the Portuguese, ending the era of colonialism. Its trade with neighbors near and far continues to grow. Foreign investment in China has mushroomed. How could the government manage the coexistence of communist politics and market economics? That was the key question confronting Deng Xiaoping and his comrades when they took power in 1979. In terms of ideology, this objective would seem unattainable; an 'open-door' economic policy would surely lead to rising pressures for political democracy. But Deng thought otherwise. If China's economic experiments could be spatially separated from the bulk of the country, their political impact would be minimized. At the outset, the new economic policies would apply mainly to China's bridgehead on the Pacific Rim, leaving most of the vast country comparatively unaffected. Accordingly, the government introduced a complicated system of **10** **Special Economic Zones (SEZs)**, so-called *Open Cities*, and *Open Coastal Areas*, which would attract technologies and investments from abroad and transform the economic geography of eastern China (Fig. 9-15).

In these economic zones, investors are offered many incentives. Taxes are low. Import and export regulations are eased. Land leases are simplified. The hiring of labor under contract is allowed. Products made in the economic zones may be sold on foreign markets and, under some restrictions, in China as well. Even Taiwanese enterprises may operate here; profits made may be sent back to the investors' home countries.

When Deng's government made the decisions that would reorient China's economic geography, location was a prime consideration. Beijing wanted China to participate in the global market economy, but it also wanted that participation to have as little impact on interior China as possible—at least in the early stages. The obvious answer was to position the Special Economic Zones along the coast. Initially, the government established four SEZs, all of which had been, or were located near, colonial treaty ports (Fig. 9-15):

- *Shenzhen*, adjacent to Hong Kong on the Pearl River Estuary in Guangdong Province.
- *Zhuhai*, across from Macau, also on the Pearl River Estuary in Guangdong Province.
- *Shantou*, opposite southern Taiwan, also in Guangdong Province and the source of many Chinese now living in Thailand.
- *Xiamen*, on the Taiwan Strait in Fujian Province, the source of many Chinese now based in Singapore, Indonesia, and Malaysia.

In 1988 and 1990, respectively, two additional SEZs were proclaimed:

- *Hainan Island*, declared an SEZ in its entirety, its potential success linked to its location closest to Southeast Asia.

- **Pudong**, across the river from Shanghai, China's largest city. Pudong differed from other SEZs because it was a giant state-financed project designed to attract large multinational companies.

In 2006, still another SEZ was added:

- **Binhai New Area**, the coastal zone of the northern port city of Tianjin, a long-established Open City with considerable foreign investment, now elevated to SEZ status, projected to outperform even Shanghai-Pudong and Shenzhen itself.

China also opened fourteen other coastal cities for preferential treatment of foreign investors (Fig. 9-15). Again, most of these cities had been treaty ports in the colonial period. They were chosen for their size, overseas trading histories, links to Overseas Chinese, level of industrialization, and pool of local talent and labor. These Open Cities are, from north to south:

Dalian (Liaoning Province)	Shanghai Shi
Qinhuangdao (Hebei)	Ningbo (Zhejiang)
Tianjin Shi	Wenzhou (Zhejiang)
Yantai (Shandong)	Fuzhou (Fujian)
Qingdao (Shandong)	Guangzhou (Guangdong)
Lianyungang (Jiangsu)	Zhanjiang (Guangdong)
Nantong (Jiangsu)	Beihai (Guangxi Zhuang AR)

When all these Pacific Rim initiatives were implemented, China's market-conscious communist leaders had reason to expect an economic boom in the coastal provinces. (Note in Fig. 9-15 that the new economic policy affected every coastal province, as well as the single coastal AR.) Japan already was a highly developed economy, having reached the final stage of the Rostow model. South Korea and Taiwan were well beyond the takeoff stage. Hong Kong had demonstrated what could be done on this side of the Pacific. And Singapore, Southeast Asia's most successful state, is at least three-quarters (77 percent) Chinese. In truth, China's pragmatic planners were unsure of just what forces they might unleash, which is why foreign investment rules were stricter in the Open Cities and Open Coastal Areas than in the freewheeling SEZs.

CHINA'S PACIFIC RIM TODAY

When China's planners laid out the SEZs, they held the highest hopes for Shenzhen, immediately adjacent to the thriving British dependency of Hong Kong. They knew that the kinds of attractions the SEZs offered to foreign investors would entice many factory owners paying Hong Kong wages to move

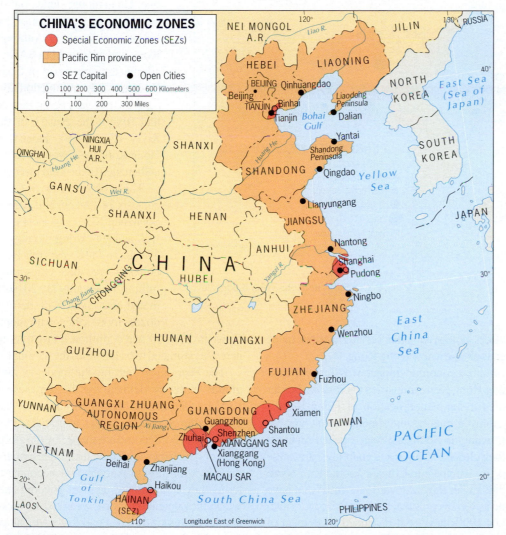

FIGURE 9-15   © H. J. de Blij, P. O. Muller, and John Wiley & Sons, Inc.

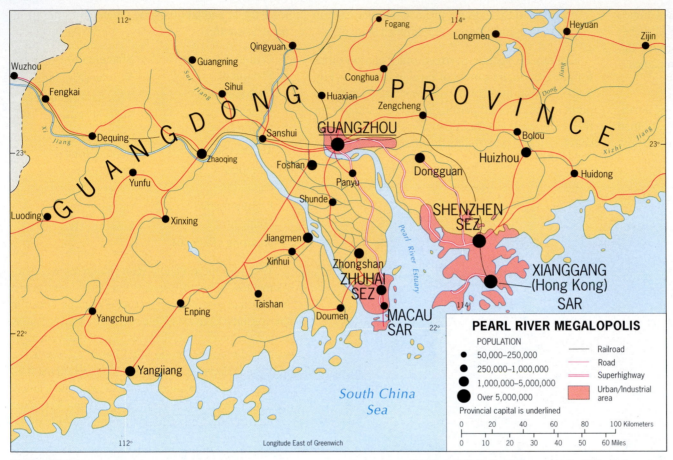

FIGURE 9-16

© H. J. de Blij, P. O. Muller, and John Wiley & Sons, Inc.

Guangdong Province and the Pearl River Estuary

As Figure 9-15 shows, eight provinces (including Hainan Island) and one autonomous region face the Pacific and form the basis for recognizing an emerging Pacific Rim region in eastern China. But among these provinces one, Guangdong, stands out by making the largest contribution by far to the national economy. Although Hong Kong and Macau, both SARs, are not administratively a part of Guangdong Province, they are functionally crucial parts of the vast economic complex that fringes the estuary of the Pearl River (Fig. 9-16). Indeed, without Hong Kong there would not have been a Shenzhen of the kind just described.

The economic core of Guangdong consists of Guangzhou (the capital), Shenzhen SEZ, and Zhuhai SEZ, as well as Hong Kong (Xianggang) and Macau. In a few years this Pearl River hub will have three cities with over 5 million inhabitants. By 2010, it is estimated that there will be at least a dozen more cities with populations between 500,000 and 5 million. Per-capita incomes here are estimated to be ten times the national average.

The Pearl River hub fits the model of the **11 regional state** as defined by the Japanese scholar Kenichi Ohmae and discussed earlier in the book. These *natural economic zones*, as he called them, defy old political borders and are shaped by the global economy of which they are a part; their representatives deal directly with foreign partners and negotiate the best terms they can with the national governments under which they operate (Europe, we noted, has several examples of this kind). The Pearl River Megalopolis has moved so far ahead of China as a whole that it is a political as well as an economic force in the country.

across the border to Shenzhen. Successful Chinese living in Southeast Asia could use Hong Kong as a base for investment in Shenzhen; despite the political standoff between Beijing and Taipei, Taiwanese could channel investments into this favored SEZ.

Those hopes were not disappointed. In the late 1970s, Shenzhen was a sleepy fishing and duck-farming village of about 20,000. By 2008, its population neared 8 million—the fastest growth of any urban center in human history. Thousands of factories, large and small, moved into the SEZ from Hong Kong. Workers by the hundreds of thousands came from Guangdong Province, Guangxi Zhuang AR, and beyond. High-rise buildings housing banks, cor-

porate offices, hotels, and other service facilities made Shenzhen look like a new Hong Kong.

Shenzhen's economic impact was far-reaching. Persuading Hong Kong industries to cross the border precipitated a sharp modification of Hong Kong's own economy; attracting investments from Overseas Chinese, mainly in Southeast Asia, opened financial coffers long closed to China; and enabling Taiwanese investors to participate in China's Pacific Rim boom created links with the 'wayward province' that have economic as well as political benefit. By its very success (in its 1990s heyday it produced 15 percent of China's exports by value), Shenzhen signaled the way for the other SEZs and Open Cities.

FIGURE 9-17   © H. J. de Blij, P. O. Muller, and John Wiley & Sons, Inc.

kets around the globe) and then in electrical equipment, appliances, and countless other consumer goods. Meanwhile, Hong Kong quietly served as a back door to China, which returned the favor by supplying fresh water and staple foods. The once-desultory British colony had become an *economic tiger*, its GNI larger than that of a hundred countries.

Hong Kong also became one of the world's leading financial centers, its skyline rivaling those of the most impressive cities on earth. When China changed course at the end of the 1970s and embarked on its open-door policy, Hong Kong was ready. Not only did Hong Kong factory owners move their plants to Shenzhen, but Hong Kong banks also financed Shenzhen's industries (and later those of other SEZs).

Meanwhile, the Chinese and the British agreed on a transfer of authority from London to Beijing, effective in mid-1997. The Chinese promised to allow Hong Kong's freewheeling economy to continue, and not to reverse Hong Kong's belated progress toward democratic government. China clearly hoped that this One Country/Two Systems policy might persuade democratic Taiwan to accept Chinese authority as well. True to the letter (if not always the spirit) of this commitment, the name Hong Kong has continued in general use, although the territory's official name is now the Xianggang SAR. Occasional quarrels notwithstanding, the Hong Kong Chinese have lived with their new political circumstances relatively satisfactorily. They do enjoy greater freedom and a more secure human rights environment than do citizens elsewhere in the People's Republic—for example, every year they freely and loudly commemorate the 1989 Tiananmen Square massacre that suppressed student-led demonstrations for greater democracy—but their votes indicate that they are more concerned with economics than with politics.

In 1999 China took control over the last remaining European colony, Portuguese Macau, and made it a SAR as well, guaranteeing, as in Hong Kong, the continuation of existing social and economic systems for 50 years.

Hong Kong

As noted earlier, Hong Kong played a key role in the early success of China's new economic policies. As a British colony, Hong Kong had become an economic powerhouse on a mere 1000 square kilometers (400 sq mi) of fragmented, hilly territory crowded with more than 7 million inhabitants and possessing no resource base—not even an adequate water supply—except a key one: its geographic location on China's doorstep.

Hong Kong (the name means Fragrant Harbor) consists of a peninsula flanked by numerous islands (Fig. 9-17). When China's role in the Korean War during the 1950s led to a United Nations embargo and closure of the Hong Kong-China border, the colony lost its hinterland and needed a means of survival. This came in the form of a textile industry based on imported raw materials, low factory wages, and international sales. The money made by the textile exporters was invested in light industries (plastic toys made in Hong Kong appeared on mar-

Qingdao in Shandong province, often called 'Little Shanghai,' is part of China's burgeoning Pacific rim, reflected in its modernistic skyline. Well known for its parks and beaches, it has China's best research institutes related to marine studies and is the site of the sailing events in the 2008 Summer Olympic Games. The city was once the piece of Chinese territory leased to the Germans (Fig. 9-8) and an enduring legacy of that era is the German tradition of brewing beer—Qingdao is the home of the now internationally known Tsingtao beer.   (© A. WinklerPrins)

ize on this mushrooming market, and before long China's domestic industry will undoubtedly challenge Japanese, American, and European brands on their home turf. Meanwhile, people by the millions are streaming toward the cities in one of the greatest rural-to-urban migrations in the history of humanity; in mid-2007 this migration required the *monthly* addition of a metropolis the size of Orlando (2 million) just to keep pace with the influx! Not surprisingly, many of urban China's traditional neighborhoods are being destroyed in the process as a new cultural landscape evolves with astonishing speed.

Coast and Interior

We noted earlier that these developments are reshaping China's regional geography and that a new region is forming on its Pacific Rim. Inevitably, the concentration of industrial and commercial expansion in the coastal provinces intensifies the regional disparities that have always marked this vast country. In the past, villagers in the rural areas had little recourse in the face of depredations by representatives of emperors or, later, by deputies of the Communist Party, and their protests could be ruthlessly put down with ease and in secret. But this is the era of the Internet and the cell phone, and news of mass protests reaches far and wide. It is no longer as easy as it once was to conceal such protests—or the reasons that they occur. Poverty, restrictions on movement, exploitation, administrative corruption, environmental degradation, forced relocation, and other grievances create a rising tide of anger.

China's planners are now addressing this potentially dangerous problem by channeling economic development inland, thereby attempting to reverse the growth in the gap between coast and interior. The key initiative involves the Yangzi-Chang corridor and includes the Three Gorges Dam, the city of Chongqing, the Sichuan Basin, and Yunnan Province, where SEZ-type privileges are being extended to corporations. As Figures 9-9 and 9-15

The Burgeoning Pacific Rim

Not only did China grant economic advantages to corporations doing business in the coastal SEZs; it also made enormous investments in infrastructure to enhance the efficiency of these zones. Although the old SEZ system still prevails officially, the entire coastal zone is being transformed at a staggering pace. New airports, bridges, superhighways, and port facilities are springing up; cities are being transformed by office towers, residential skyscrapers, ultramodern hotels, and gated communities. Beijing's metamorphosis befits its role as capital; Shanghai's Pudong district, across the Huangpu River from the preserved colonial riverfront, rivals Shenzhen in the speed of its construction and makes Shanghai look like the most futuristic city in the world. Roads that just two decades ago were clogged with bicycles are now jammed with automobiles. Carmakers from all over the world have operations in China to capital-

China's rapid rate of urbanization and economic growth are radically changing the look of many of its older cities. Old neighborhoods are being swept away to be replaced by modern apartment towers. Social networks and communities are being destroyed in the process. Numerous neighborhoods were torn down in Beijing to make room for the Olympic facilities. Shown here is a traditional housing complex in Shanghai awaiting the wrecker's ball to clear a site that will become another modern skyscraper. This photo is taken from Renmin street looking east. In the background is the Oriental Pearl Tower, already a famous icon of the Shanghai skyline. (© A. WinklerPrins)

show, however, this project involves mainly central and southern China, yet the greatest disparities lie in the north, in Shanxi, Shaanxi, and Gansu provinces. Channeling development to these provinces will prove far more difficult. Nevertheless, the need to improve the living conditions of farmers and others in the interior provinces is growing. China must restrain the influx of migrants from countryside to city, but at best this can only be a temporary remedy. A more lasting solution will require closing the prosperity gap between the coast and the interior.

CHINA: GLOBAL SUPERPOWER?

China in the early twenty-first century appears likely to become more than an economic force of world proportions: it also appears to be on course to achieve global superpower stature. As long as it retains its autocratic form of government (which allowed it to impose draconian population policies, engage in comprehensive economic experiments, and ignore environmental issues without having to consult the electorate), China will be able to practice the kind of state capitalism that—in another guise—made South Korea an economic power. Moreover, China's military is the largest standing army in the world, with nearly 3 million soldiers and some 1.2 million reserves. Its equipment is being updated, and the government raised its armed forces budget enormously in 2001. And unlike Japan or South Korea, China has no constraints on its military power; its army served as security forces during the pro-democracy turbulence of 1989. China is already a nuclear power and has a growing arsenal of medium-range and intercontinental ballistic missiles. It has also placed a manned spacecraft in orbit and declared its intention to reach the Moon.

During the twentieth century, the United States and the Soviet Union were locked in a 45-year Cold War that repeatedly risked nuclear conflict. That fatal exchange never happened, in part because it was a struggle between superpowers that understood each other comparatively well. While the politicians and military strategists were plotting, the cultural doors were never closed: American audiences listened to Russian composers and watched Russian ballet; they read Russian literature as the Soviets cheered American musicians and authors and lionized American dissidents. In short, it was an *intra-cultural* Cold War, a characteristic that served to reduce the threat of mutual destruction.

The twenty-first century, in contrast, may witness a far more dangerous geopolitical struggle in which the adversaries could well be the United States and China. U.S. power and influence still prevail in the western Pacific, but it is easy to discern areas where Chinese and American interests will diverge (Taiwan is only one example). American bases in Japan, thousands of American troops in South Korea, and American warships in the East and South China seas are potential grounds for dispute. All this might generate the world's first *intercultural* Cold War, in which the risk of fatal misunderstanding is incalculably greater than it was during the last.

How can such a Cold War be averted? Trade, scientific and educational links, and cultural exchanges are obvious remedies: stronger ties between nations make conflict between those nations less likely. Westerners should learn as much about China as possible in order to understand it better, to appreciate its cultural characteristics, and to recognize the historico-geographical factors underlying China's views of the West and its approach to development today.

For more than forty centuries, China has known authoritarianism of both the brutal and the benevolent kind, has been fractured by regionalism only to be unified once again, and has depended on communalism to survive environmental and despotic depredations. For nearly twenty-five centuries, it has been guided by Confucianism. Time and again, China has been opened to, and ruled by, foreigners, and time and again it has retreated into isolationism when things went wrong. Such a reversal may no longer be possible, given what we have seen in this chapter. But now a renewed force, nationalism, is rising in China. A spate of recent books, including one by Song Qiang et al., titled *China Can Say No*, reflects the frustrations of many Chinese over what they view as American arrogance and insensitivity to China's traditions and interests. The 'inadvertent' bombing of the Chinese embassy in Belgrade by American aircraft operating under NATO command in 1999, and the midair collision between what the Chinese regarded as a U.S. spy plane and one of its own aircraft near the island of Hainan in 2001, are the kinds of incidents that contribute to this view and energize Chinese nationalism. American scorn for China's human rights failings is another source of irritation.

China today is on the world stage: literally, through its hosting of the Summer Olympics in 2008, and figuratively, through its global economic and increasingly political engagement. Constraining the forces that tend to lead to world-power competition between China and the United States will be the geopolitical challenge of the twenty-first century.

MONGOLIA

Between China's Inner Mongolia and Xinjiang to the south and Russia's Eastern Frontier region to the north lies a vast, landlocked, isolated country called Mongolia that constitutes a discrete region of East Asia (Fig. 9-3). With only 2.7 million inhabitants in an area larger than Alaska, Mongolia is a steppe-and-desert-dominated vacuum between two of the world's most powerful countries.

From the late 1600s until the revolutionary days of 1911, Mongolia was part of the Chinese Empire. With Soviet support, the Mongols held off Chinese attempts to regain control, and in the 1920s the country became a People's Republic on the Soviet model. Free elections in 1990 ushered in a new political era, but Mongolia's economy, still based largely on animal products (chiefly cashmere wool), is troubled by mismanagement and environmental problems.

The map reflects Mongolia's difficulties. Despite its historical associations, ethnic affinities, and cultural involvement with China, Mongolia's incipient core area, including the capital of Ulaanbaatar, adjoins Russia's Eastern Frontier. The country's 800,000 herders and their millions of sheep follow nomadic tracks along the fenceless fringes of the vast Gobi Desert. When the extreme cold of a harsh Siberian winter descends on the steppes, human and livestock losses can be severe and the economy breaks down.

This is a weak country in a vulnerable part of inner Asia, a **12** **buffer state** wedged, Tibet-like, between populous and powerful neighbors. Will it escape the fate of other Asian buffers?

THE JAKOTA TRIANGLE

We turn now to a region that exemplifies the economic success of East Asia. Along the Pacific Rim, from Japan (through South Korea, Taiwan, and China's Guangdong) to Singapore in Southeast Asia,

rapid economic development has transformed community and society. In China, as noted earlier, a Pacific Rim region is being created, still discontinuous at present but likely to extend all along the coast in the future. In Japan, South Korea, and Taiwan we can observe that future today. That is why we recognize an East Asian region that we call the **13** **Jakota Triangle**, consisting of Japan, Korea (South Korea at present but probably a reunited Korea someday), and Taiwan (Fig. 9-18). This is a region of great cities, huge consumption of raw materials from all over the world, voluminous exports, and global financial linkages. It also is a region of increasing social problems, political uncertainties, and economic vulnerabilities.

JAPAN

When we assess China's prospects of becoming a superpower, we should remember what happened in Japan in the nineteenth century. In 1868, a group of reform-minded modernizers seized power from an old guard, and by the end of the century Japan was a military and economic force. From the factories in and around Tokyo and from urban-industrial complexes elsewhere poured a stream of weapons and equipment that the Japanese used to embark on colonial expansion. By the mid-1930s, Japan lay at the center of an empire that included all of the Korean Peninsula, the whole of China's Northeast (which the Japanese called Manchukuo), the Ryukyu Islands, and Taiwan, as well as the southern half of Russia's Sakhalin Island (called Karafuto). Not even a disastrous earthquake, which destroyed much of Tokyo in 1923 and killed 143,000 people, could slow the Japanese drive.

Colonial Wars and Recovery

During World War II Japan expanded its domain farther than the architects of the 1868 modernization could have anticipated. By early December 1941,

FIGURE 9-18

© H. J. de Blij, P. O. Muller, and John Wiley & Sons, Inc.

economic victories in a new global arena. Japan became an industrial giant, a technological pacesetter, a fully urbanized society, a political power, and an affluent nation. Cities everywhere have reliable Japanese cars on their streets; tourists across the globe snap pictures with Japanese cameras; laboratories the world over use Japanese optical equipment. From microwave ovens to DVDs, from oceangoing ships to digital cameras, Japanese-designed goods flood world markets.

Japan's brief colonial adventure helped lay the groundwork for other economic successes along the western Pacific Rim. The Japanese ruthlessly exploited the natural and human resources of Korea and Taiwan, but they also installed a new economic order there. After World War II, this infrastructure facilitated an economic transition—and soon made both Taiwan and South Korea competitors in world markets.

British Tutelage

After the modernizers took control of Japan in 1868—an event known as the *Meiji Restoration* (the return of 'enlightened rule' centered on the Emperor Meiji)—they turned to Britain for guidance in reforming their nation and its economy. In the decades that followed, the British advised the Japanese on the layout of cities and the construction of a railroad network, the location of industrial plants, and the organization of education. The British influence still is visible in the Japanese cultural landscape: The Japanese, like the British, drive on the left side of the road. Consider how this affects the effort to open the Japanese market to U.S.-made automobiles!

The Japanese reformers of the late nineteenth century undoubtedly saw many geographic similarities between Britain and Japan. At that time, most of what mattered in Japan was concentrated on the country's largest island, Honshu (literally, *mainland*). The ancient capital, Kyoto, lay in the interior, but the modernizers wanted a coastal,

Japan had conquered large parts of China Proper, all of French Indochina to the south, and most of the small islands in the western Pacific. Then, on December 7, 1941, Japanese-built aircraft carriers moved Tokyo's warplanes within striking range of Hawai'i, and the surprise attack on Pearl Harbor underscored Japan's confidence in its war machine. Soon the Japanese overran the Philippines, the (then) Netherlands East Indies, Thailand, and British Burma and Malaya, and drove a wide corridor through the heart of China to the border with Vietnam.

A few years later, Japan's expansionist era was over. Its armies had been driven from virtually all its possessions, and when American nuclear bombs devastated two Japanese cities in 1945, the country lay in ruins. But once again, aided this time by an enlightened U.S. postwar administration, Japan surmounted disaster.

Japan's economic recovery and its rise to the status of world economic superpower was the success story of the second half of the twentieth century. Japan lost a war and an empire, but it scored many

FIGURE 9-19

outward-looking headquarters. So they chose the town of Edo, situated on a large bay where Honshu's eastern coastline turns sharply (Fig. 9-19). They renamed the place *Tokyo* (meaning eastern capital), and little more than a century later it was the largest urban agglomeration in the world. Honshu's coasts were near mainland Asia, where raw materials and potential markets for Japanese products could be found. The notion of a greater Japanese empire followed naturally from the British example.

Spatial Contrasts and Constraints

In other ways, the British and Japanese archipelagoes, at opposite ends of the Eurasian landmass, differ considerably. In total area, Japan is larger. In addition to Honshu, Japan has three other large islands—Hokkaido to the north and Shikoku and Kyushu to the south—as well as numerous small islands and islets, for a total land area of about 377,000 square kilometers (146,000 sq mi). Much of this area is mountainous and steep-sloped, geologically young, earthquake prone, and studded with volcanoes. Britain has lower relief, is older geologically, does not suffer from severe earthquakes, and has no active volcanoes. And in terms of the raw materials for industry, Britain was much better endowed than Japan. Self-sufficiency in iron ore and high-quality coal gave Britain a head start that lasted a century.

Japan's high-relief topography also constrained its economic development. Except for the ancient capital of Kyoto,

JAPAN: MANUFACTURING, LAND, AND LIVELIHOODS

- Primary region
- Secondary region
- Agriculture
- Major railroad
- Core area

0   50   100   150 Kilometers
0   25   50   75 Miles

all of Japan's major cities are perched along the coast, and virtually all lie partly on artificial land claimed from the sea. Sailing into Kobe harbor, one passes artificial islands designed for high-volume shipping and connected to the mainland by automatic space-age trains. Enter Tokyo Bay, and the refineries and factories to the east and west stand on huge expanses of landfill that have pushed the bay's shoreline outward. With 128.1 million people, the vast majority (79 percent) living in towns and cities, Japan uses its habit-able living space intensively—and ex-pands it wherever possible.

As Figure 9-20 shows, farmland in Japan is both limited and regionally fragmented. Urban sprawl has in-vaded much of the cultivable land. In the hinterland of Tokyo lies the Kanto Plain; around Nagoya, the Nobi Plain; around Osaka, the Kansai District—each a major farming zone under re-lentless urban pressure. All three of these plains lie within Japan's frag-mented but well-defined core area (delimited by the red line on the map), the heart of Japan's prodigious manu-facturing complex.

Modernization, Japanese Style

The reformers who set Japan on a new course in 1868 set in motion a process of **14 modernization**, but in so doing they managed to build on, not replace, Japan-ese cultural traditions. We in the West-ern world tend to equate modernization with Westernization: urbanization, the spread of transport and communica-tions facilities, the establishment of a

FIGURE 9-20

© H. J. de Blij, P. O. Muller, and John Wiley & Sons, Inc.

market (money) economy, the breakdown of local traditional communities, the proliferation of formal schooling, and the acceptance and adoption of foreign innovations. In the non-Western world, the process is often viewed differently. There, modernization is seen as an outgrowth of colonialism, the perpetuation of a system of wealth accumulation introduced by foreigners driven by greed. In this view, the local elites who replaced the colonizers in the newly independent states merely continue the disruption of traditional societies, rather than truly modernizing them. Traditional societies, they argue, can be modernized without being Westernized.

In this context, Japan's modernization is unique. Having long resisted foreign intrusion, the Japanese did not achieve the transformation of their society by importing a Trojan horse; it was done by Japanese planners, building on the existing Japanese infrastructure, to fulfill Japanese objectives. Certainly, Japan imported foreign technologies and adopted innovations from the British and others, but the Japan that was built, a unique combination of modern and traditional elements, was basically an indigenous achievement.

The Role of Relative Location

Japan's changing fortunes over the past century reveal the influence of *relative location* in the country's development. When the Meiji Restoration took place, Britain, located on the other side of the Eurasian landmass, lay at the center of a global empire. The colonization and Europeanization of the world were in full swing. The United States was still a developing country, and the Pacific Ocean was an avenue for European imperial competition. Japan, even while it was conquering and consolidating its first East Asian colonies (the Ryukyus, Taiwan, Korea), lay remote from the mainstream of global change.

Then Japan became embroiled in World War II and dealt severe blows to the European colonial armies in Asia. The Europeans never recovered: the French lost Indochina, and the Dutch were forced to abandon their East Indies (now Indonesia). When the war ended, Japan was defeated and devastated, but at the same time the Japanese had done much to diminish the European presence in the Pacific Basin. Moreover, the global situation had changed dramatically. The United States, Japan's trans-Pacific neighbor, had become the world's wealthiest and most powerful, whereas Britain and its global empire were fading. Suddenly Japan was no longer remote from the mainstream of global action: now the Pacific was becoming an avenue to the world's richest markets. Japan's relative location—its situation relative to the economic and political foci of the world—had changed. Therein lay much of the opportunity that the Japanese seized after the postwar rebuilding of their country.

Spatial Organization

As important as relative location is spatial organization. Imagine 128 million people crowded into a territory the size of Montana (population: 975,000), most of it mountainous, subject to frequent earthquakes and volcanism, with no domestic oilfields, little coal, few raw materials for industry, and not much level land for farming. If Japan today were an underdeveloped country in need of food relief and foreign aid, there would be abundant explanations for its condition, including overpopulation, inefficient farming, and energy shortages.

True, only an estimated 18 percent of Japan's national territory is designated as habitable. And Japan's large population is crowded into some very big cities. Moreover, Japan's agriculture is not especially efficient. But Japan defeated these odds by calling on old Japanese virtues: organizational efficacy, massive productivity, dedication to quality, and adherence to common goals. Even before the Meiji Restoration, Japan was a tightly organized country of some 30 million citizens.

All this proved invaluable to the modernizers when they set Japan on its new course. The country's industrial growth could be based on the urban and manufacturing development that was already taking place. As noted earlier, Japan does not possess major domestic raw-material sources, so no substantial internal reorganization was necessary. However, some cities were better situated than others relative to those limited local resources and, more important, external sources of raw materials. As Japan's regional organization took shape, a hierarchy of cities developed; Tokyo took and kept the lead, but other cities rapidly developed into industrial centers.

Areal Functional Organization

The evolution of Japan's regional organization is governed by a geographic principle known as *areal functional organization*. Human activity has a *spatial focus*. It is concentrated in some locale, whether a farm or a factory or a store. Every one of these establishments occupies a particular *location*; no two of them can occupy exactly the same spot on the earth's surface (even in highrises there is a vertical form of absolute location). Nor can any human activity proceed in total isolation, so *interconnections* develop among these various establishments. This system of interconnections grows more complex as human capacities and demands expand. Each system (for example, farmers sending crops to market and buying equipment at service centers) forms a unit of **15 areal functional organization**.

In the Introduction, we referred to *functional regions* as systems of spatial organization; we can map units of areal functional organization as regions. These regions evolve because of so-called creative imaginations, in which people apply their cultural experience and technological know-how to organize and rearrange their living space. Finally, we can recognize **levels of development** in areal functional organization, a ranking of places and regions based on the type, extent, and intensity of exchange. Those levels are *subsistence, transitional*, and *exchange*.

Coastal Development

Japan's level of development, exchange, is the highest of the three categories. It is reflected in the organizational map (Fig. 9-20). The map also tells us much about the nature of Japan's exchange economy, its external orientation, and its dependence on foreign trade. All of the country's primary and secondary regions lie on the coast.

Dominant among these regions is the **Kanto Plain** (Fig. 9-20), the heart of Japan's core area, which is focused on the Tokyo urban area and contains about one-third of the country's population. Among its advantages are an unusually extensive area of low relief, a fine natural harbor at Yokohama, a relatively mild and moist climate, and a central location with respect to the country as a whole. Its major disadvantage lies in its vulnerability to earthquakes. The Kanto Plain and its Tokyo-centered megalopolis (population 26.7 million) lie at the convergence of three tectonic plates, and Toyko has a centuries-long history of devastating earthquakes that have struck the region, on average, about every 70 years since 1633 (Figs. G-5, G-6).

The second-ranking economic region in Japan is the **Kansai District**, containing the Osaka-Kobe-Kyoto triangle and located at the eastern end of the Seto Inland Sea. Osaka and Kobe are major industrial centers and busy ports, but the Kansai District also yields large harvests of rice, Japan's staple food. Between the Kanto Plain and the Kansai District lies the **Nobi Plain** (Fig. 9-20), where Nagoya is the key city. And, as the map shows, the Japanese core area is anchored in the west by the conurbation centered on **Kitakyushu**, situated not on Honshu Island but on northern Kyushu. This five-city conurbation is growing rapidly, favored by its location relative to Japan's Pacific Rim partners.

Japan's Pacific Rim Prospects

During the late 1990s and the first years of the twenty-first century, Japan's economy failed to sustain the growth of previous decades. A large part of the prob-

lem involved mismanagement and government inefficiency, and some of it had to do with a downturn in Southeast Asian economies, with which Japan's was closely linked. But Japan also faced tougher competition from other economies on the Pacific Rim, including South Korea (which has cut into Japan's lead in automobile sales) and Taiwan (which dominates the computer field). By 2006, however, and confounding the predictions of many economists, Japan's economy was steadily reviving and continues to do so late in the decade. Government policies, including privatization and more flexible labor regulations, were the main stimulus.

Coupled with these economic issues are issues of international relations. Japan has unfinished World War II business that hurts its status. It never signed a peace treaty with the Soviet Union in the aftermath of the war because the Russians had occupied and retained four small islands in the Kurile chain northeast of Hokkaido (Fig. 9-19). Negotiations for the return of these 'Northern Territories' failed despite Japan's offer of a massive aid program to develop the Russian Far East, including Vladivostok. As a result, Japan lost the opportunity to play a role in the economic development of a hinterland from which it might have gained vast energy as well as mineral supplies.

Elsewhere, relations with South Korea have been eroded by Korean memories of Japanese misconduct during World War II and Japan's refusal to acknowledge those actions. Similar issues damage Japan's carefully nurtured links with China.

Population Change

All the issues just discussed must be considered in the context of Japan's changing society. Japan's population of 128 million is aging rapidly; population growth stopped around 2007 and is expected to decline to about 100 million in 2050 and only 67 million in 2100. Over the long term, therefore, Japan faces serious challenges in maintaining its present level of prosperity with a shrinking and aging population. Immigration could help alleviate this problem, but Japan has a long history of cultural insularity. It is ethnically very homogeneous and has historically resisted immigration. Only recently has it recruited 'return' migrants from the Japanese-Brazilian and Japanese-Peruvian communities in an effort to alleviate labor shortages.

Thus, Japan's future place in the geopolitical and economic frameworks of the western Pacific is uncertain. More than six decades after the end of World War II, American forces are still based on Japanese

Tokyo is the largest city in the world and bustles with people. Pictured here is the Shibya, a part of Tokyo known for its brightly lit billboards that make night appear as bright as day. Despite its dense population, Tokyo functions relatively efficiently with an excellent public transit system. At this pedestrian crossing, as at so many in the city, all traffic is stopped at the same time so that all pedestrians can safely cross together. (© A. WinklerPrins)

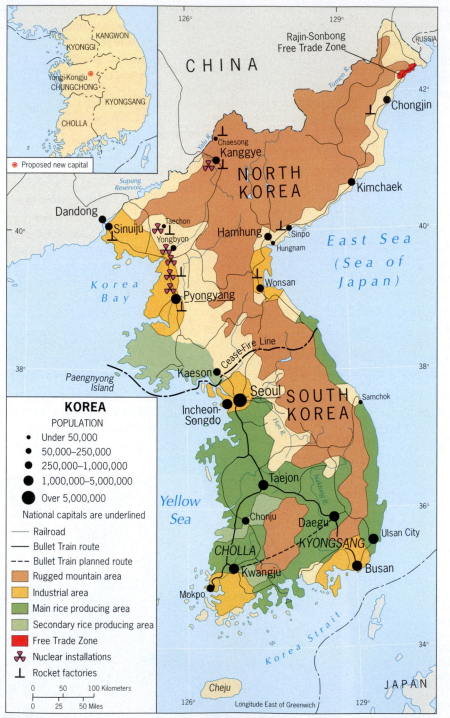

KOREA
POPULATION
● Under 50,000
● 50,000–250,000
● 250,000–1,000,000
● 1,000,000–5,000,000
● Over 5,000,000

National capitals are underlined
— Railroad
▬ Bullet Train route
--- Bullet Train planned route
Rugged mountain area
Industrial area
Main rice producing area
Secondary rice producing area
Free Trade Zone
☢ Nuclear installations
⊥ Rocket factories

0   50   100 Kilometers
0   25   50 Miles

FIGURE 9-21   © H. J. de Blij, P. O. Muller, and John Wiley & Sons, Inc.

territory. China and Korea accept this arrangement, which constrains Japan from rearming. But in Japan, where nationalism is rising and concern over a nuclear-armed North Korea is deepening, domestic military preparedness is a growing political issue. This situation is fraught with potential problems. Japan therefore finds itself at an economic, social, and geopolitical crossroads. Its future will profoundly influence not only East Asia and the Pacific, but the world as a whole.

KOREA

On the Asian mainland, directly across the Sea of Japan (East Sea), lies the peninsula of Korea (Fig. 9-21), a territory about the size of the State of Idaho, much of it mountainous and rugged, with a total population of about 73 million. Unlike Japan, however, Korea has long been a divided nation. For centuries Korea has been a pawn in the struggles of its more powerful neighbors. It has been a dependency of China and a colony of Japan. When it was freed from Japan's oppressive rule at the end of World War II, the victorious Allied powers divided Korea for administrative purposes. That division gave North Korea (north of the 38th parallel) to the forces of the Soviet Union and South Korea to those of the United States. In effect, Korea traded one master for two new ones. For the rest of the century the nation remained divided because North Korea immediately entered the communist ideological sphere and became a dictatorship. South Korea, with massive American aid, became part of East Asia's capitalist perimeter. Here again, the will of external powers prevailed over the desires of the Korean people.

War and Aftermath

The Korean War (1950–1953), which began when communist forces from North Korea invaded the south in a forced-unification drive, devastated much of the peninsula before a cease-fire line became the de facto boundary between the present-day countries of South and North Korea (Fig. 9-21). Tens of thousands of U.S. troops continue to guard this border between ideological adversaries. The postwar fates of those adversaries have differed greatly. In the decades since the war ended, South Korea, with 45 percent of the peninsula's land but two-thirds of its population, has become an economic powerhouse, while North Korea has languished under strict authoritarian rule. Although the two Koreas are in a potential situation of **16 regional complementarity**, in which North Korea has raw materials that South Korea needs and South Korea produces food that North Korea needs, the political status of the two sides prevents them from taking advantage of it.

South Korea

As shown in Figure 9-21, South Korea (population 48.9 million) is a key component of the Jakota Triangle. From the ravages of war it emerged as the world's largest shipbuilding nation, a major automobile manufacturer, and a producer of iron and steel as well as chemicals and electronics. Despite corrupt dictatorial rule, political instability, and social unrest, successive South Korean regimes managed to keep the country's economy growing, its **17 state capitalism** propelled by powerful industrial conglomerates in cahoots with the politicians. Eventually, however, democratic government took hold in South Korea, and the transition to democratic rule, coupled with the economic downturn in other parts of Asia in the 1990s, slowed the country's growth somewhat.

By global standards, South Korea is a prosperous country; its economic prowess is evident from the map (Fig. 9-21). The capital, Seoul, with about 10 million inhabitants, ranks among the world's megacities and is the anchor of a huge industrial complex at the waist of the Korean Peninsula facing the Yellow Sea. Hundreds of thousands of farm families migrated to the Seoul area after the end of the Korean War, with the result that today only 20 percent of South Koreans live in rural areas.

Migrants also streamed into Busan, the nucleus of the area called **Kyongsang**, the country's second-largest manufacturing zone, which is located on the Korea Strait opposite the western tip of Honshu Island (Japan). And the government-supported, urban-industrial drive here continues. Just 30 years ago, Ulsan City, 60 kilometers (40 mi) north of Busan along the coast, was a fishing center with perhaps 50,000 inhabitants; today its population exceeds one million, nearly half of them the families of workers in the Hyundai automobile factories and the local shipyards and docks.

The third industrial area shown in Figure 9-21, anchored by the city of Kwangju, has advantages of relative location that will spur its development. This area is called **Cholla**, and although it has traditionally been marginalized by Seoul, its relative location will help it overcome this problem.

North Korea

North Korea, after six decades of communist rule, has become one of the poorest, hungriest, and most regimented nations on earth. The capital, Pyongyang, has a population about one third that of Seoul; in the industrial zone facing Korea Bay, industries ranging from mining (coal) to manufacturing (textiles) are outdated and inefficient. Collectivized farming produces far less than North Korea's 23.5 million people need, but Pyongyang issues no information about such matters. Refugees' and escapees' stories tell of poverty and misery. But North Korea has progressed on one significant front: nuclear capability and associated weaponry (Fig. 9-21). The country's nuclear capability became an international concern in 2003, when North Korea itself proclaimed its nuclear aspirations and achievements. Since then, a multinational effort by the United States, Russia, China, Japan, and South Korea continues, and in 2007 significant progress was made in de-nuclearizing North Korea.

Many people on both sides of the ideological divide hope for the eventual reunification of the two

North Korea's communist regime has impoverished and starved its citizens and stifled freedoms, but in recent years its policies in the nuclear and missile arenas have become matters of international concern. By its own admission, North Korea has a (still-small) arsenal of nuclear weapons; the country also has tested missiles, one of which overflew Japan and plunged into the Pacific beyond. Diplomatic efforts to curb Pyongyang's nuclear ambitions, involved six nations most directly concerned since 2006. This is one of the country's older installations, the Yongbyon Nuclear Center north of the capital. As Figure 9-21 shows, North Korea's nuclear and missile network extends across much of the country. (© AP/Wide World Photos)

Koreas, which over the long term could create one of East Asia's most productive and prosperous societies. In 2005, South Koreans were allowed to cross the cease-fire line by train for brief visits, although no northerners were permitted to visit the South. This breakthrough notwithstanding, the prospect for reunification in the foreseeable future remains dim,

FIGURE 9-22

© H. J. de Blij, P. O. Muller, and John Wiley & Sons, Inc.

and the map of Korea continues to illustrate the high cost of the Cold War. In 2007 North Korea, under pressure from the international community and in dire need of humanitarian assistance, agreed to stop refining uranium. In 2008, concerts by the New York Philharmonic and Eric Clapton created a small cultural opening. Only time will tell if the North will join the global community or remain a separate world.

TAIWAN

The third component of the Jakota Triangle is another economic success story on the Asian Pacific Rim: Taiwan (Fig. 9-22). Along with a small archipelago in the Taiwan Strait and two small islands on the doorstep of mainland China, Taiwan became the last refuge of China's Nationalists under Chiang Kai-shek, who fled there in 1949 when the communists triumphed. Arriving in the capital, Taipei, they took control and proclaimed the Republic of China (ROC), the 'legitimate' government of the country now ruled from Beijing by Mao Zedong and his comrades.

Peoples and Rulers

Taiwan's earliest inhabitants were Malay–Polynesians; Taiwan became a formal part of China only during the last Qing (Manchu) Dynasty, following a hunger-driven Chinese emigration from Fujian Province. It was known as Formosa during the colonial era. Colonial powers established *treaty ports* here as they had on the mainland. In 1895 the Manchus were forced to cede all of Taiwan to the Japanese, and during half a century of rule the Japanese exploited Taiwan but also endowed it with major infrastructure, including railways, roads, hydroelectric schemes, irrigation projects, factories, and mines. Though damaged during the war, this infrastructure was of inestimable value to the Nationalists when they took control, enabling them to speed a massive reconstruction project, with American help, that quickly revived Taiwan's economy.

In the process, Taiwan became one of the **18 economic tigers** on the Pacific Rim. As Figure 9-22 shows, this is not a large island. In fact, it is smaller than Switzerland, though it has a much larger population (22.9 million), most of which is concentrated in an arc lining the western and northern coasts. The

Chungyang Mountains, an area of high elevations (some over 3000 meters [10,000 ft]), steep slopes, and dense forests, dominate the eastern half of the island. Westward, these mountains yield to a zone of hilly topography and, facing the Taiwan Strait, a substantial coastal plain. Streams flowing down from the mountains irrigate the paddyfields, and farm production has more than doubled since 1950 even as hundreds of thousands of farmers have left the fields for work in Taiwan's expanding industries.

A Strong Economy

Today the lowland urban-industrial corridor of western Taiwan is anchored by the capital, Taipei (Taibei), at the island's northern end and rapidly growing Kaohsiung (Gaoxiong) in the far south.* The Japanese developed Chilung (Jilong), Taipei's port, to export nearby coal, but now raw materials flow the other way: Taiwan imports raw cotton for its textile industry, bauxite (for aluminum) from Indonesia, oil from Brunei, and iron ore from Subsaharan Africa. Taiwan has a developing iron and steel industry, nuclear power plants, shipyards, a large chemical industry, and modern transport networks. Increasingly, however, it is exporting the products of its budding high-technology industries: personal computers, telecommunications equipment, and precision electronic instruments.

Taiwan's economy was boosted by the creation of China's SEZs, where the rules permitted Taiwanese companies to set up factories, and by the SAR status of Hong Kong, which allows Taiwanese business-people to enter China via this 'back door' since direct travel is prohibited. The Taiwanese have skillfully exploited these opportunities, and the result is one of the world's strongest economies. Today, annual per-capita income in Taiwan exceeds U.S. $16,000, which is higher than that of many European countries.

---

*The Taiwanese have retained the old Wade-Giles spelling of place names; China now uses the **Pinyin** system for romanizing Chinese ideograms. Names in parentheses are written according to the Pinyin system.

Geopolitical Risks

But Taiwan faces serious political difficulties. The communist regime in Beijing regards Taiwan as a wayward province that must be reunited with the motherland, possibly by force. In 1971, the ROC was ousted from the United Nations, its seat taken by the PRC; later, the United States, Taiwan's only ally, publicly subscribed to a One-China policy. Taiwan's ROC leaders never proclaimed an independent state on their island, and now the Taiwan question has devolved into a complicated, uneasy standoff. Unlike communist mainland China, Taiwan has progressed from authoritarianism to democracy; it is the ROC, not the PRC, which has managed to combine economic success with democratization.

There are signs of hope. A majority of Taiwan's voters have recently signaled that they are not in favor of risky moves toward sovereignty, and Beijing's rulers have shown signs of flexibility by agreeing to talk directly with Taiwanese representatives. All sides seem to be aware that an armed conflict over Taiwan would produce no winners, although some Taiwanese politicians like to test the waters. After September 2003, for example, Taiwanese passport holders found the name *Taiwan* beneath the (routine) Republic of China on the cover, identification not previously used. And some prominent politicians raise the prospect of referendums to test the people's views on matters that might be camouflaged as a vote on independence. This enrages leaders in mainland China, who do not wish such views to be tested.

In sum, the East Asian realm is as fraught with risk as it is filled with promise. As China's power rises and Japan's, comparatively, wanes, the role of the United States in this realm will change, a transition that will substantially define the route to the New World Order to which we alluded at the beginning of this book.

10

SOUTHEAST ASIA

In This Chapter

- Natural hazards abound: A dangerous part of the world
- An intricate ethnic and cultural mosaic
- Infusion of religions from near and far
- Overseas Chinese and their economic power
- Indonesia's struggle to unify its vast archipelago
- Troubles continue for East Timor

CONCEPTS, IDEAS, AND TERMS

1 **Buffer zone**
2 **Shatter belt**
3 **Wallace's Line**
4 **Overseas Chinese**
5 **Genetic boundary classification (antecedent; subsequent; superimposed; relict)**
6 **Territorial morphology (compact; protruded; elongated; fragmented; perforated)**
7 **Domino theory**
8 **Choke point**
9 *Entrepôt*
10 **Archipelago**
11 **Transmigration**

REGIONS

MAINLAND SOUTHEAST ASIA
INSULAR SOUTHEAST ASIA

Photos: (*top*) Pyapon, Irrawaddy Delta, Myanmar © AP/Wide World Photos; (*bottom*) Klong canal, Bangkok, Thailand © A. WinklerPrins.

SOUTHEAST ASIA IS a realm of peninsulas and islands, a corner of Asia bounded by India on the northwest and China on the northeast (see chapter opener map, Fig. 10-1). Its western coasts are washed by the Indian Ocean, and to the east stretches the vast Pacific. From all these directions, Southeast Asia has been penetrated by outside forces. From India came traders; from China, settlers; from across the Indian Ocean, Arabs to engage in commerce and Europeans to build empires; and from across the Pacific, Americans. Southeast Asia has been the scene of countless contests for power and primacy—the competitors have come from near and far.

Southeast Asia's geography in some ways resembles that of Eastern Europe. It is a mosaic of smaller countries on the periphery of two of the world's largest states. It has been a **1** **buffer zone** between powerful adversaries. It is a **2** **shatter belt** in which stresses and pressures from without and within have fractured the political geography. Like Eastern Europe, Southeast Asia exhibits great cultural diversity. It is a realm of hundreds of cultures, numerous languages and dialects, and several major religions.

Defining the Realm

Figure 10-2 shows the dimensions of the Southeast Asian geographic realm, but note the disconformity between the eastern boundary of the realm and the eastern limits of its most populous state, Indonesia. Papua, the easternmost part of Indonesia, constitutes the western half of the island of New Guinea, where indigenous cultures are not Southeast Asian but Pacific. Here lie three of Indonesia's administrative provinces, but in effect Papua is an Indonesian colony. We discuss Papua in the context of Indonesia, but all of New Guinea in Chapter 12.

Because the politico-geographical map (Fig. 10-2) is so complicated, it should be studied attentively. One good way to strengthen your mental map of this realm is to follow the mainland coastline from west to east. The westernmost state in the realm is Myanmar (called Burma before 1989 and still referred to by that name in antiregime circles), the only country in Southeast Asia that borders both India and China. Myanmar shares the neck of the Malay Peninsula with Thailand, heart of the *Mainland Region*. The south of the peninsula is part of Malaysia—except for Singapore, at the very tip of it. Facing the Gulf of Thailand is Cambodia. Still moving generally eastward, we reach Vietnam, a strip of land that extends all the way to the Chinese border. And surrounded by its neighbors is landlocked Laos, remote and isolated.

MAJOR GEOGRAPHIC QUALITIES OF Southeast Asia

1. Southeast Asia extends from the peninsular mainland to the archipelagos offshore. Because Indonesia controls part of New Guinea, its functional region reaches into the neighboring Pacific geographic realm.

2. Southeast Asia, like Eastern Europe, has been a political shatter belt between powerful adversaries and has a fractured cultural and political geography shaped by foreign intervention.

3. Southeast Asia's physiography is dominated by high relief, with crustal instability marked by volcanic activity and earthquakes. The predominantly tropical environment has produced considerable natural resources that continue to be relentlessly exploited by both outside interests and insiders.

4. A majority of Southeast Asia's more than half-billion people live on the islands of just two countries: Indonesia, with the world's fourth-largest population, and the Philippines. The rate of population increase in the insular region of Southeast Asia exceeds that of the mainland.

5. Although the overwhelming majority of Southeast Asians have the same ancestry, cultural divisions and local traditions abound, which the realm's divisive physiography sustains.

6. The legacies of powerful foreign influences, historical (colonialism) and present-day, Asian as well as non-Asian, continue to affect the cultural landscapes of Southeast Asia.

7. Southeast Asia's political geography exhibits a variety of boundary types and several categories of state territorial morphology.

8. The Mekong River, Southeast Asia's Danube, has its source in China and borders or crosses five Southeast Asian countries, sustaining tens of millions of farmers, fishing people, and boat owners.

9. The realm's giant in terms of territory as well as population, Indonesia, has not asserted itself as the dominant state because of mismanagement and corruption; but Indonesia has enormous potential and is making progress.

FIGURE 10-2

© H. J. de Blij, P. O. Muller, and John Wiley & Sons, Inc.

1976, released to United Nations supervision in 1999, and granted independence in 2002.

COLONIALISM'S HERITAGE

These are countries of a geographic realm that has no dominant state—no China, no India, no Brazil—although one country, Indonesia, contains 40 percent of its total population and has the potential to emerge as a commanding force. Neither did any single, dominant core of indigenous culture develop here as it did in East Asia. In the river basins and on the plains of the mainland, as well as on the islands offshore, a flowering of cultures produced a diversity of societies whose languages, religions, arts, music, foods, and other achievements formed an almost infinitely varied mosaic—but none of those cultures rose to imperial power. The European colonizers forged empires here, often by playing one state off against another; the Europeans divided and ruled. Out of this foreign intervention came the modern map of Southeast Asia, as only Thailand (formerly Siam) survived the colonial era as an independent entity. Thailand was useful to two competing powers, the French to the east and the British to the west. It was a convenient buffer, and while the colonists carved pieces off Thailand's domain, the kingdom endured.

Indeed, the Europeans accomplished what local powers could not: the formation of comparatively large, multicultural states that encompassed diverse peoples and societies and welded them together. Were it not for the colonial intervention, it is unlikely that the 17,000 islands of far-flung

This leaves the islands that constitute Southeast Asia's *Insular Region*: the Philippines in the north and Indonesia in the south, and between them the offshore portion of Malaysia, situated on the largely Indonesian island of Borneo. Also on Borneo lies the ministate of Brunei, small, but as we shall see, important in the regional picture. And finally, a new state has recently appeared on the map in the realm's southeastern corner: East Timor (Timor-Leste), a former Portuguese colony annexed by Indonesia in

Located in the Pacific Ring of Fire, Indonesia is particularly susceptible to tectonic activity. Here residents of Minahasa, on the island of Sulawesi, observe volcanic Mount Soputan belching more ash just days after it erupted in October 2007, sending ash raining down on at least one village and spewing smoke about 1500 meters (5000 ft) into the air. Residents were kept on high alert around the same time for a second volcano on the main island of Jawa.   (© Sonny Tembelaka/AFP/Getty Images)

Indonesia would today constitute the world's fourth-largest country in terms of population. Nor would the nine sultanates of Malaysia have been united, let alone with the peoples of northern Borneo across the South China Sea. For good or ill, the colonial intrusion consolidated a realm of few culture cores and numerous ministates into less than a dozen countries.

PHYSICAL GEOGRAPHY

As Figure 10-1 shows, Southeast Asia is a realm in which high relief dominates the physiography. From the Arakan Mountains in western Myanmar (Burma) to the glaciers (yes, glaciers!) of the Indonesian part of New Guinea, elevations rise above 3300 meters (10,000 ft) in many locales.

The relief map reminds us that this is not only the Pacific Rim but also the Pacific Ring of Fire, where the crust is unstable, earthquakes are common, and volcanoes are active (Figures G-5 and G-6). Eruptions of Indonesian volcanoes have darkened skies worldwide and even changed human history, for example, Krakatau in 1883. More recently a seafloor earthquake off westernmost Indonesia at the end of 2004 generated an Indian Ocean *tsunami* that killed some 300,000 people from Sumatera[1] to East Africa. It is rare for a year to pass without a newsworthy eruption in Southeast Asia. Since most of the islands of this realm are volcanic creations, it is important to note that their soils are highly fertile, often resulting in high population densities, which are highly vulnerable to natural disasters.

Among the islands, Borneo is the sole exception. Borneo is a nonvolcanic slab of ancient crust, pushed high above sea level by tectonic forces and eroded

[1]As in Africa and South Asia, names and spellings have changed with independence. In this chapter, we will use the contemporary spellings, except when we refer to the colonial period. Thus Indonesia's four major islands are Jawa, Sumatera, Kalimantan (the Indonesian part of Borneo), and Sulawesi. The Dutch called them Java, Sumatra, Borneo, and Celebes, respectively.

into its present mountainous topography. Unfortunately, this means that its soils are not nearly as fertile as the excellent soils found on Jawa and other volcanic islands.

Four Rivers

As Figure 10-1 underscores, rivers rise in the highland backbones of the islands and peninsulas, and deposit their sediments as they wind their way toward the coast; the physiography of Sumatera demonstrates this unmistakably. The volcanic hills, plateaus, and better-drained lowlands are fertile and, in the warmth of tropical climates, can yield multiple crops of rice.

On the peninsular mainland we see a pattern that is already familiar: rivers rising in the Asian interior that create alluvial plains and deltas. The Mekong River is the Chang/Yangzi of Southeast Asia: you can trace it all the way from China via Laos, Thailand, and Cambodia into southern Vietnam, where it forms a massive and populous delta. In the west, Myanmar's key river is the Irrawaddy; Thailand's is the Chao Phraya. In the north, the Red River Basin is the breadbasket of northern Vietnam.

No survey of the physical geography of Southeast Asia would be complete without reference to the realm's seas, gulfs, straits, and bays. One of these bays, the Bay of Bengal, is a funnel for tropical cyclones. A large storm devastated coastal Myanmar in May 2008. Irregular and indented coastlines such as these, with thousands of islands near and far, create difficult problems when it comes to drawing **maritime boundaries** in the waters offshore. Southeast Asia has one of the most complex maritime boundary frameworks in the world.

Tropical Rainforests

Southeast Asia contains one of the world's three main areas of tropical rainforest. The global distribution of tropical rainforests is closely linked to climate, partic-

Deforestation in Southeast Asia. Since colonial times, Southeast Asia has been exploited for its tropical hardwood species, especially teak. This hardwood is prized for its ability to withstand rot when wet. It is, therefore, a key wood type for boat construction, outdoor furniture, and other uses. Teak is grown in uniform stands, making it quite easy to extract. It has been estimated that a prime teak tree can fetch a price of U.S.$40,000. Mainland Southeast Asia has traditionally been the source of teak, with extensive stands in Thailand, Myanmar, and Laos. But insular Southeast Asia has also become an important source of other tropical hardwoods.

In the last few decades, logging has accelerated, and some scientists worry that in the absence of drastic measures, the biodiversity of the realm will be seriously reduced in the next decade. Part of this deforestation is due to the strong demand for lumber initiated by the Japanese. Because Japan's own forests are no longer available for harvesting (most of them have been cleared, and the remainder are now protected), the Japanese have turned to Southeast Asia for their timber needs. Malaysia, the major exporter of tropical roundwood, ships all its lumber exports to the Japanese and Chinese markets.

Recently, in response to extensive overlogging of teak and other valuable hardwoods, the Thai government has placed a moratorium on its logging and has begun several large reforestation projects, many of them through the king's direct involvement. Nonetheless, logging of teak in Myanmar and Laos continues. (© Wayne G. Lawler/Photo Researchers, Inc.)

ularly the tropical humid climate types (*Af, Am*) where consistent rainfall and warm temperatures are predominant in equatorial/tropical regions of the world (see Fig. G-8 for precipitation and Fig. G-9 for climates). Tropical rainforests are perhaps the most complex of terrestrial (land) biomes on earth. Generally, they feature a staggering variety of trees and other vegetation growing in very close proximity. Interestingly, in tropical rainforests the greatest concentration of nutrients is in the vegetation and not in the soil. This means that tropical rainforest soils, in general, are infertile despite the lush vegetation and the appearance of fertility. Traditional users of these areas have developed strategies for using these soils despite their infertility; but when farming methods are introduced that assume soil fertility, crops often fail.

Biodiversity in Southeast Asia

Southeast Asia is perhaps not as well known as South America for its biodiversity, but in fact Indonesia ranks second in the world (Brazil is first) in the biological diversity of its plant and animal species. It is estimated that 10 percent of the world's forests, plants, birds, and mammal species live here. This biodiversity is in part due to sea-level changes and tectonic plate movements in the realm over millions of years. About 16,000 years ago, Southeast Asia's land level dropped significantly during a glacial period. As a result, the seafloor between Southeast Asia's mainland and the islands that now form Indonesia was exposed, creating a land bridge between the two. During this time, many species migrated from the mainland to Southeast Asia's insular region. Animals also migrated to the insular region from the Australian Plate to the south. Not all Southeast Asian seabeds were exposed: the deep trench between Bali and Lombok (eastern Indonesian islands) remained filled with water, which prevented any further migration of land-based species. Thus the species on either side of this line could not mix and remained

separate from one another. This ecological division is called **3** **Wallace's Line**, named after the naturalist, Alfred Russel Wallace, who first realized the contrast between the species of these two areas (for further discussion, see p. 376).

The biodiversity of the realm has attracted outsiders for centuries, beginning with foreigners who sought the plants that we know today as spices. Most of the well-known spices we commonly use today (e.g., black pepper, ginger, cinnamon, turmeric, cloves, and nutmeg) come from plants that were domesticated in Southeast Asia and grown on the so-called *Spice Islands* in today's Indonesian archipelago. The Spice Islands usually refer to the Maluku Islands (or Moluccas), located between the Indonesian islands of Sulawesi and western New Guinea. Indian, Arab, and Chinese traders took an interest in these spices long before Europeans fought over the islands in order to gain direct access to these valuable species. The spice trade of course was the beginning of colonization in Southeast Asia, a topic we return to later in the chapter.

POPULATION GEOGRAPHY

Compared to the huge population numbers and densities in the habitable regions of South Asia and China, demographic totals for the countries of Southeast Asia, with the exception of Indonesia, seem modest (Fig. 10-3). Again, comparisons with Europe come to mind. Three countries—Thailand, the Philippines, and Vietnam—have populations between 60 and 90 million. Laos, quite a large country territorially (comparable to the United Kingdom), had just 6 million inhabitants in 2008. Cambodia, half the size of Germany, had only 14.7 million. Of Southeast Asia's 582 million inhabitants, just over 55 percent live on the islands of Indonesia and the Philippines, leaving the realm's mainland countries with only 39 percent of the population.

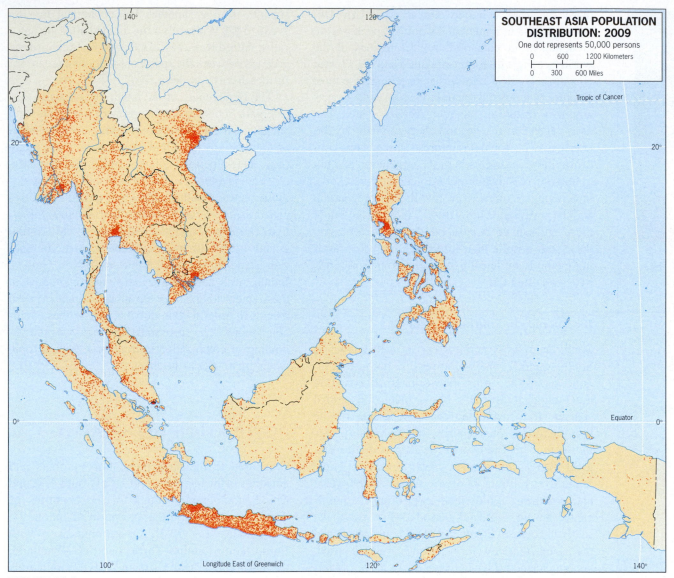

FIGURE 10-3

© H. J. de Blij, P. O. Muller, and John Wiley & Sons, Inc.

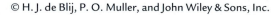

The Ethnic Mosaic

Southeast Asia's peoples come from a common stock just as (Caucasian) Europeans do, but this has not prevented the emergence of regionally or locally discrete ethnic or cultural groups. Figure 10-4 displays the broad distribution of ethnolinguistic groups in the realm, but be aware that this is a generalization. At the scale of this map, numerous small groups cannot be depicted.

The map shows the rough spatial coincidence, on the mainland, between major ethnic group and modern political state. The Burman dominate in the country once known as Burma (Myanmar); the Thai occupy the state once known as Siam (now Thailand); the Khmer form the nation of Cambodia and extend northward into Laos; and the Vietnamese inhabit the long strip of territory facing the South China Sea.

Territorially, by far the largest population is classified in Figure 10-4 as Indonesian, the inhabitants of the great archipelago that extends from Sumatera west of the Malay Peninsula to the Malukus (Moluccas) in the east and from the Lesser Sunda Islands in the south to the Philippines in the north. Collectively, all these peoples—the Filipinos, Malays, and Indonesians—shown in Figure 10-4 are known as Indonesians, but they have been divided by history and politics. Note, on the map, that the Indonesians in Indonesia itself include Javanese, Madurese, Sundanese, Balinese, and other large groups; hundreds of smaller ones are not shown. In the Philippines, too, island isolation and contrasting ways of life are reflected in the cultural mosaic. Also part of this Indonesian ethnic-cultural complex are the Malays, whose heartland lies on the Malay Peninsula but who form minorities in other areas as well. Like most Indonesians, the Malays are Muslims, but Islam is a more powerful force in Malay society than, in general, in Indonesian culture.

In the northern part of the mainland region, numerous minorities inhabit remote parts of the countries in which the Burman (Burmese), Thai, and Vietnamese dominate. Those minorities, as a comparison of Figures 10-2 and 10-4 indicates, tend to occupy areas on the peripheries of their countries, away from the core areas, where the terrain is mountainous and the forest is dense, and where the governments of the national states do not

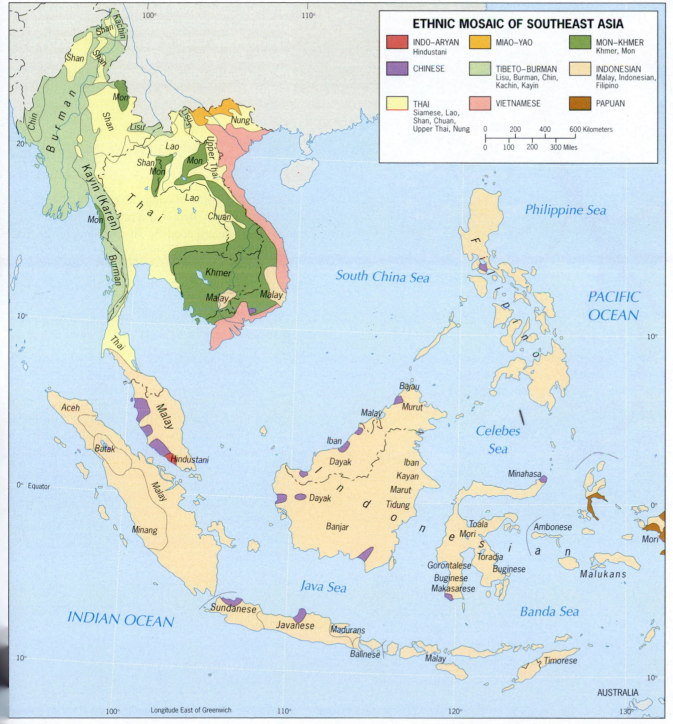

ETHNIC MOSAIC OF SOUTHEAST ASIA

- **INDO–ARYAN** Hindustani
- **MIAO–YAO**
- **MON–KHMER** Khmer, Mon
- **CHINESE**
- **TIBETO–BURMAN** Lisu, Burman, Chin, Kachin, Kayin
- **INDONESIAN** Malay, Indonesian, Filipino
- **THAI** Siamese, Lao, Shan, Chuan, Upper Thai, Nung
- **VIETNAMESE**
- **PAPUAN**

0   200   400   600 Kilometers

0   100   200   300 Miles

FIGURE 10-4   © H. J. de Blij, P. O. Muller, and John Wiley & Sons, Inc.

have complete control. This remoteness and sense of detachment give rise to notions of secession, or at least resistance to governmental efforts to establish authority, often resulting in bitter ethnic conflict.

Immigrants

Figure 10-4 also reminds us that, again like Eastern Europe, Southeast Asia is home to major ethnic minorities from outside the realm. On the Malay Peninsula, note the South Asian (Hindustani) cluster. Hindu communities with Indian ancestries exist in many parts of the peninsula, but in the southwest they form the majority in a small area. In Singapore, too, South Asians form a significant minority. These communities arose during the colonial period, but South Asians had arrived in this realm many centuries earlier, propagating Buddhism and leaving architectural and cultural imprints on places as far away as Jawa and Bali.

The Chinese

By far the largest immigrant minority in Southeast Asia, however, is Chinese. The Chinese began arriving here during the Ming and early Qing (Manchu) dynasties, but the largest exodus occurred during the late colonial period (1870–1940), when as many as 20 million immigrated. The European powers at first encouraged this influx, using the Chinese in administration and trade. But soon these **4** **Overseas Chinese** began to move into the major cities, where they established Chinatowns and gained control over much of the

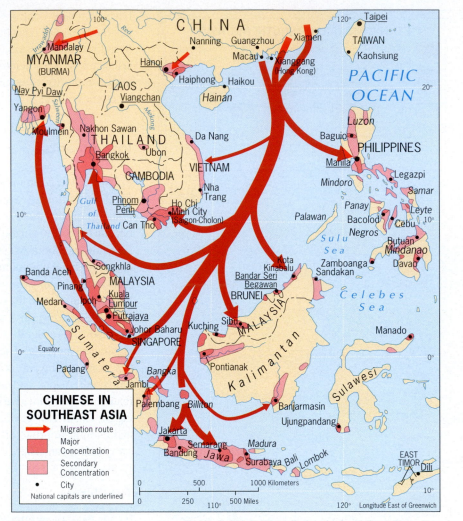

FIGURE 10-5

© H. J. de Blij, P. O. Muller, and John Wiley & Sons, Inc.

Most originated in China's Fujian and Guangdong provinces, and many invested much of their wealth back in China when it opened up to foreign businesses. The Overseas Chinese of Southeast Asia played a major role in the economic miracle of the Pacific Rim.

HOW THE POLITICAL MAP EVOLVED

The leading colonial competitors in Southeast Asia were the Dutch, British, French, and Spanish (with the Spanish replaced by the Americans in their stronghold, the Philippines). The Japanese had colonial objectives here as well, but these aspirations came and went during the course of World War II.

The Dutch acquired the greatest prize: control over the vast archipelago now called Indonesia (formerly the Netherlands East Indies). France established itself on the eastern flank of the mainland, controlling all territory east of Thailand and south of China. The British conquered the Malay Peninsula, gained power over the northern part of the island of Borneo, and established themselves in Burma as well. Other colonial powers also gained footholds but not for long. The exception was Portugal, which held on to its eastern half of the island of Timor (Indonesia) well after the Dutch had been ousted from their East Indies.

Figure 10-6 shows the colonial framework in the late nineteenth century, before the United States assumed control over the Philippines in 1898. Note that while Thailand survived as an independent state, it lost territory to the British in Malaya and Burma and to the French in Cambodia and Laos.

The Colonial Imprint

The colonial powers divided their possessions into administrative units as they did in Africa and elsewhere. Some of these political entities became

commerce. By the time the Europeans tried to reduce Chinese immigration, World War II was about to start and the colonial era would soon end.

Today, Southeast Asia is home to as many as 30 million Overseas Chinese, more than half the world total. Their lives have often been difficult. The Japanese relentlessly persecuted those Chinese who lived in Malaya during World War II. Later, during the 1960s, Chinese in Indonesia were accused of communist sympathies, and hundreds of thousands were killed. More recently, in the late 1990s, Indonesian mobs again attacked Chinese and their property, this time because of their relative wealth and because many Chinese became Christians during the colonial period, now targeted by Islamic throngs. Resentment continues today with episodic flare-ups against Chinese in various parts of Southeast Asia.

Figure 10-5 shows the migration routes and current concentrations of Chinese in Southeast Asia.

independent states when the colonial powers withdrew or were removed (Fig. 10-6).

French Indochina

France, one of the mainland's leading colonial powers, divided its Southeast Asian empire into five units. Three of these units lay along the east coast: Tonkin in the north next to China, centered on the basin of the Red River; Cochin China in the south, with the Mekong Delta as its focus; and between these two, Annam. The other two French territories were Cambodia, facing the Gulf of Thailand, and Laos, landlocked in the interior. Out of these five French dependencies there emerged the three states of Indochina. The three east coast territories ultimately became one state, Vietnam; the other two (Cambodia and Laos) each achieved separate independence.

The French had a name for their empire: *Indochina*. The *Indo* part of *Indochina* refers to the cultural imprints from South Asia: the Hindu presence, the importance of Buddhism (which came to Southeast Asia via Sri Lanka [Ceylon] and its seafaring merchants), the influences of Indian architecture and art (especially sculpture), writing and literature, and social structures and patterns. The *China* in the name *Indochina* signifies the role of the Chinese here. Chinese emperors coveted Southeast Asian lands, and China's power reached deep into the realm. Social and political upheavals in China, combined with the opportunities created by the European colonists, sent millions of Sinicized

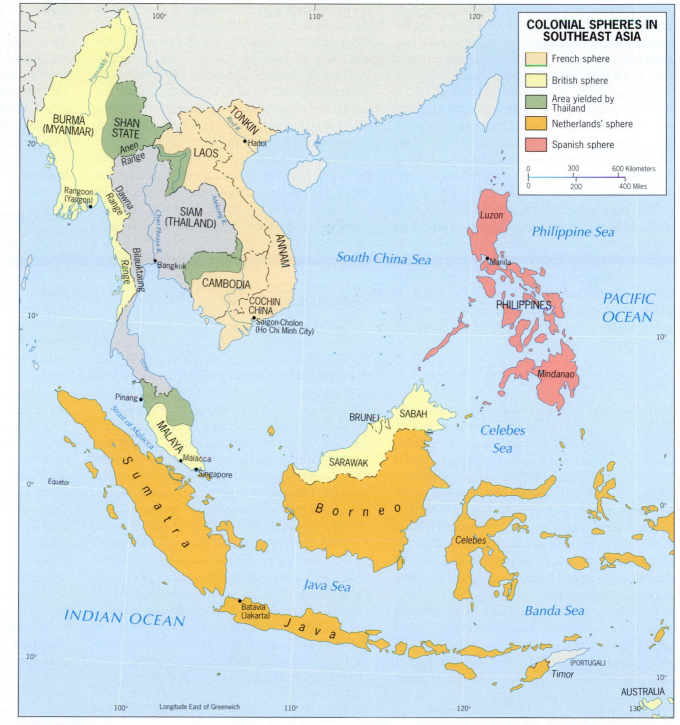

FIGURE 10-6

© H. J. de Blij, P. O. Muller, and John Wiley & Sons, Inc.

people southward. Chinese traders, pilgrims, sea-farers, fishermen, and others sailed from south-eastern China to the coasts of Southeast Asia and established settlements there. Over time, those settlements attracted more Chinese immigrants, and Chinese influence in the realm grew (Fig. 10-5). Not surprisingly, relations between the Chinese settlers and the earlier inhabitants of Southeast Asia have at times been strained, even violent. The Chinese presence in Southeast Asia is long term, but the invasion has continued into modern times.

The name *Indochina* can only refer to a part of Mainland Southeast Asia, however; it cannot be used to refer to the realm as a whole. Although the *Indo* segment of this regional name can be taken to also refer to the Buddhist influences that dominate here, it makes no reference to the momentous arrival of Islam, introduced by Arab seafarers in the twelfth and thirteenth centuries and destined to change the cultural geography of the realm.

British Imperialism

The British ruled two major entities in Southeast Asia (Burma and Malaya) in addition to a large part of northern Borneo and many small islands in the South China Sea. Burma was attached to Britain's Indian Empire; from 1886 until 1937 it was governed from distant New Delhi. But when British India became independent in 1947 and split into several countries, Burma was not part of the grand design that created West and East Pakistan (the latter now Bangladesh), Ceylon (now Sri Lanka), and India. Instead, in 1948 Burma (today called Myanmar) was given the status of a sovereign republic.

In Malaya, the British developed a complicated system of colonies and protectorates that eventually gave rise to the equally complex, far-flung Malaysian Federation. Included were the former Straits Settlements (Singapore was one of these colonies), the nine protectorates on the Malay Peninsula (former sultanates of the Muslim era),

the British dependencies of Sarawak and Sabah on the island of Borneo, and numerous islands in the Strait of Malacca and the South China Sea. The original Federation of Malaysia was created in 1963 by the political unification of recently inde-pendent mainland Malaya, Singapore, and the for-mer British dependencies on the largely Indonesian island of Borneo. Singapore, however, left the Fed-eration in 1965 to become a sovereign city-state, and the remaining units were later restructured into peninsular Malaysia and, on Borneo, Sarawak and Sabah. Thus the term *Malaya* properly refers to the geographic area of the Malay Peninsula, including Singapore and other nearby islands; the term *Malaysia* identifies the politico-geographical entity of which Kuala Lumpur is the capital city.

Netherlands 'East Indies'

Following in the footsteps of the first colonizers in the realm, the Portuguese, the Dutch took control of the Spice Islands through their Dutch East India Company, and the wealth that they extracted from what is today Indonesia brought the Netherlands its Golden Age. It may seem odd in today's world that spices could be so important for commerce, but at that time, in the pre-refrigeration era, spices con-served food and added taste to otherwise bland food. From the mid-seventeenth to the late eighteenth cen-tury, the Dutch could develop their East Indies sphere of influence almost without challenge, for the British and French were preoccupied with the Indian subcontinent. By playing the princes of Indonesia's states against one another in the search for economic concessions and political influence, by placing the Chinese in positions of responsibility, by imposing systems of forced labor in areas directly under its control, and by rearranging land ownership and power structures, the Company had a disastrous ef-fect on the Indonesian societies it subjugated.

Java (Jawa), the most populous and productive is-land, became the focus of Dutch administration; from its capital at Batavia (now Jakarta), the Dutch East

India Company extended its sphere of influence into Sumatra (Sumatera), Celebes (Sulawesi), much of Borneo (Kalimantan), and the smaller islands of the East Indies. This was not accomplished overnight, and the struggle for territorial control was carried on long after the Company had yielded its administra-tion to the Netherlands government. Dutch colonial-ism thus threw a girdle around Indonesia's more than 17,000 islands, paving the way for the creation of the realm's largest and most populous state (231.9 mil-lion today).

From Spain to the United States

In the colonial tutelage of Southeast Asia, the Philippines, long under Spanish domination, had a unique experience. As early as 1571, the islands north of Indonesia were under Spain's control (they were named for Spain's King Philip II). Spanish rule began when Islam was reaching the southern Philippines via northern Borneo. The Spaniards spread their Roman Catholic faith with great zeal, and between them the soldiers and priests consoli-dated Hispanic dominance over the mostly Malay population. Manila, founded in 1571, became a profitable waystation on the route between south-ern China and western Mexico (Acapulco was the main trans-Pacific destination for the galleons leav-ing Manila's port). There was much profit to be made, but the indigenous people shared little in it. Great landholdings were awarded to loyal Spanish civil servants and to men of the church. Oppression eventually yielded revolution, and Spain was con-fronted with a major uprising in the Philippines when the Spanish-American War broke out else-where in 1898.

As part of the settlement of that war, the United States replaced Spain in Manila as colonial propri-etor. That was not the end of the revolution, how-ever. The Filipinos now took up arms against their new foreign ruler, and not until 1905, after terrible losses of life, did American forces manage to pacify their new dominion. Subsequently, U.S. administration in the Philippines was more progressive than Spain's had been. In 1934, Congress passed the Philippine Independence Law, providing for a ten-year transition to sovereignty. But before independence could be arranged, World War II intervened. In 1941, Japan conquered the islands, temporarily ousting the Americans; U.S. forces returned in 1944 and, with strong Filipino support, defeated the Japanese in 1945. The agenda for independence was resumed, and in 1946 the sovereign Republic of the Philippines was proclaimed.

Today, all of Southeast Asia's states are independent, but centuries of colonial rule have left strong cultural imprints. In their urban landscapes, their education systems, and countless other ways, this realm still carries the marks of its colonial past.

SOUTHEAST ASIA'S POLITICAL GEOGRAPHY

Political geographers study the rise and decline of nations and states, as well as their interactions. They also study state morphology, specifically a territory's shape and the boundaries between states. Southeast Asia's states display these boundaries and shapes in great variety. We focus first on national boundaries on land (deferring maritime boundaries until Chapter 12) and then on the territorial morphology, or shape, of this realm's states.

Boundaries

Boundaries are sensitive parts of a state's anatomy: just as people are territorial about their individual properties, so nations and states are sensitive about their territories and borders. The saying that "good fences make good neighbors" certainly applies to states, but, as we know, the boundaries between states are not always good fences. Boundaries are also part of the cultural-geographic legacies that

exist in regions, signatures on the map and landscapes of events and peoples of the past.

Boundaries, in effect, are contracts between states. Such a contract takes the form of a treaty that contains the *definition* of the boundary in the form of elaborate description. Next, cartographers perform the *delimitation* of the treaty language, drawing the boundary on official, large-scale maps. Throughout human history, states have used those maps to build fences, walls, or other barriers in a process called *demarcation*.

Classifying Boundaries

Once established, we can classify boundaries geographically. Some are sinuous, conforming to rivers or mountain crests (*physiographic*) or coinciding with breaks or transitions in the cultural landscape (*anthropogeographic*). As any world political map shows, many boundaries are simply straight lines, delimited without reference to physical or cultural features. These *geometric* boundaries can lead to problems when the cultural landscape changes where they exist.

In general, the boundaries of Southeast Asia were better defined than those of several other postcolonial areas of the world, notably Africa, the Arabian Peninsula, and Turkestan. The colonial powers that established the original treaties tried to define boundaries to lie in remote and/or sparsely peopled areas: for example, across interior Borneo. Nonetheless, certain Southeast Asian boundaries have created problems, among them the geometric boundary between Papua, the portion of New Guinea ruled by Indonesia, and the country of Papua New Guinea, which occupies the eastern part of the island. The artificiality of this boundary is resulting in increasing secessionist feelings among the population of Indonesian Papua.

Even on a small-scale map of the kind we use in this chapter, we can categorize the boundaries of this realm. A comparison between Figures 10-2 and 10-4 reveals that the boundary between Thailand and Myanmar over long segments is anthropogeographic

(ethnic-cultural), notably where the name *Kayin* (*Karen*), the Myanmar minority, appears in Figure 10-4. Figure 10-1 shows that a large segment of the Vietnam-Laos boundary is physiographic-political, coinciding with the Annamite Cordillera (Highlands).

Genetic Classification System

Boundaries also can be classified genetically, that is, as their evolution relates to the cultural landscapes they traverse. A leading political geographer, Richard Hartshorne (1899–1992), proposed a four-level **5** **genetic boundary classification**. All four of these boundary types can be observed in Southeast Asia.

Certain boundaries, Hartshorne reasoned, were defined and delimited before the present-day human landscape developed. In Figure 10-7 (upper-left map), the boundary between Malaysia and Indonesia on the island of Borneo is an example of the first boundary type, an **antecedent boundary**. Most of this border passes through sparsely inhabited tropical rainforest, and the break in settlement can even be detected on the small-scale world population map (Fig. G-10).

A second category of boundaries evolved as the cultural landscape of an area took shape and became part of the ongoing process of accommodation between several states. These **subsequent boundaries** are represented in Southeast Asia by the map in the upper right of Figure 10-7, which shows in some detail the border between Vietnam and China. This border is the result of a long process of adjustment and modification, the end of which may not yet have come.

The third category involves boundaries drawn forcibly across a unified or at least homogeneous cultural landscape. The colonial powers did this when they divided the island of New Guinea by delimiting a boundary in a nearly straight line (curved in only one place to accommodate a bend in the Fly River), as shown in the lower-left map of Figure 10-7. The **superimposed boundary** they delimited gave the

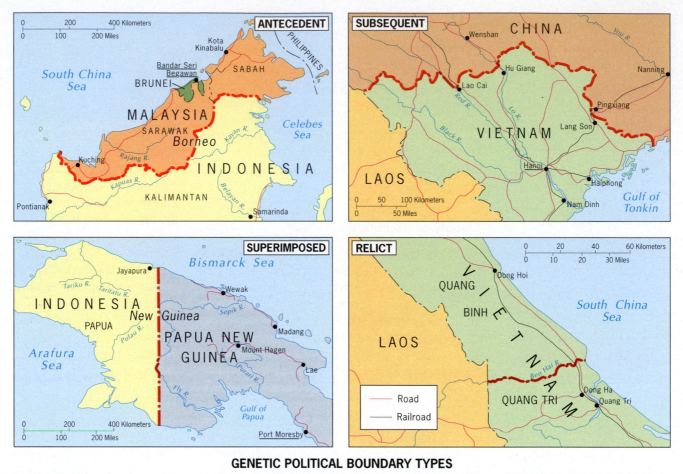

GENETIC POLITICAL BOUNDARY TYPES

FIGURE 10-7

© H. J. de Blij, P. O. Muller, and John Wiley & Sons, Inc.

Netherlands the western half of New Guinea. When Indonesia became independent in 1949, the Dutch did not yield their part of New Guinea, which is peopled mostly by ethnic Papuans, not Indonesians. In 1962, the Indonesians invaded the territory by force of arms, and in 1969 the United Nations recognized its authority there. This made the colonial, superimposed boundary the eastern border of Indonesia and had the effect of projecting Indonesia from Southeast Asia into the Pacific Realm. Geographically, all of New Guinea forms part of the Pacific Realm.

The fourth genetic boundary type is the so-called **relict boundary**—a border that has ceased to func-

tion but whose imprints (and sometimes influence) are still evident in the cultural landscape. The boundary between the former North and South Vietnam (Fig. 10-7, lower-right map) is a classic example: once demarcated militarily, it has had relict status since 1976 following the reunification of Vietnam in the aftermath of the Indochina War (1964–1975).

Southeast Asia's boundaries have colonial origins, but they have continued to influence the course of events in postcolonial times. Take one instance: the physiographic boundary that separates the main island of Singapore from the rest of the Malay Peninsula, the Johor Strait (see Fig. 10-13). That physio-

graphic-political boundary facilitated, perhaps crucially, Singapore's secession from the state of Malaysia in 1965. Without it, Malaysia might have been persuaded to stop the separation process; at the very least, territorial issues would have arisen to slow the sequence of events. As it was, no land boundary needed to be defined. The Johor Strait demarcated Singapore and left no question as to its limits.[2]

---

[2]Except one: a tiny island at the eastern entrance to the Strait named Pedra Blanca (as Singapore calls it) or Pulau Batu Putih (the Malaysian version), still disputed today.

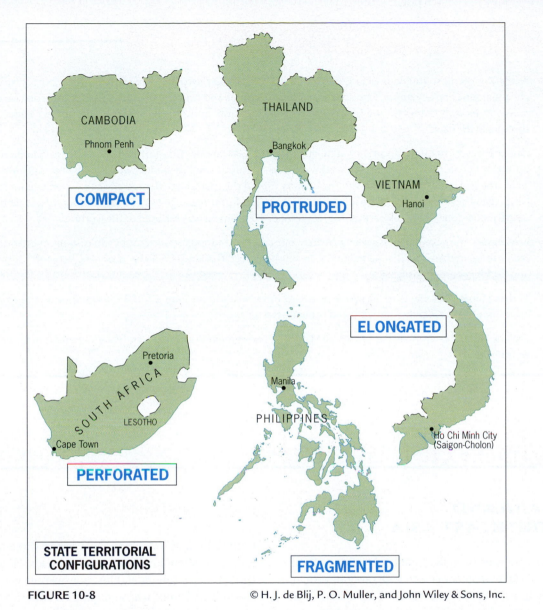

CAMBODIA
Phnom Penh

COMPACT

THAILAND
Bangkok

PROTRUDED

VIETNAM
Hanoi

ELONGATED

Pretoria

SOUTH AFRICA
LESOTHO
Cape Town

PERFORATED

Manila

PHILIPPINES

Ho Chi Minh City
(Saigon-Cholon)

**STATE TERRITORIAL
CONFIGURATIONS**

FRAGMENTED

FIGURE 10-8   © H. J. de Blij, P. O. Muller, and John Wiley & Sons, Inc.

Jawanese from the most populous island to many of the others.

Political geographers identify five dominant state territorial configurations, all of which we have encountered in our world regional survey but which we have not categorized until now. All but one of these shapes is represented in Southeast Asia, and Figure 10-8 provides the terminology and examples:

- *Compact states* have territories shaped somewhere between round and rectangular, without major indentations. This encloses a maximum amount of territory within a minimum length of boundary. Southeast Asian example: Cambodia.

- *Protruded states* (sometimes called **extended**) have a substantial, usually compact territory from which extends a peninsular corridor that may be landlocked or coastal. Southeast Asian examples: Thailand and Myanmar.

- *Elongated states* (also called **attenuated**) have territorial dimensions in which the length is at least six times the average width, creating a state that lies astride environmental or cultural transitions. Southeast Asian example: Vietnam.

- *Fragmented states* consist of two or more territorial units separated by foreign territory or by water. Subtypes are mainland-mainland, mainland-island, and island-island. Southeast Asian examples: Malaysia, Indonesia, the Philippines, and East Timor.

- *Perforated states* completely surround the territory of other states, so that they have a 'hole' in them. No Southeast Asian example; the most illustrative current case is South Africa, perforated by Maryland-sized Lesotho.

State Territorial Morphology

Boundaries define and delimit states; they also create the mosaic of often interlocking territories that give individual countries their shape. This shape or **6 territorial morphology** can affect a state's condition, even its survival. Vietnam's extreme elongation has influenced its existence since time immemorial. As we will see, Indonesia has tried to redress its fragmented nature (thousands of islands) by promoting unity through the 'transmigration' of

In the discussion that follows, we will have frequent occasion to refer to this geographic property of Southeast Asia's states. For so comparatively small a realm with so few states, Southeast Asia displays a considerable variety of state morphologies.

One point of caution: states' territorial morphologies do not determine their viability, cohesion, unity, or lack thereof; they can, however, influence these qualities. Cambodia's compactness has not ameliorated its divisive political geography, for example. But as we will find in the pages that follow, shape plays a key role in the still-unfolding political and economic geography of Southeast Asia.

WHAT'S DRIVING GEOGRAPHIC CHANGE IN THE REALM

● As the world's most populous Islamic country, **Indonesia** is a battleground between moderates, who form the great majority, and extremists, who seek to radicalize Muslims and who carry out terrorist attacks. Look for the names of Jemaah Islamiyah, the Islamic terrorist organization, and Abu Bakar Bashir, its alleged leader, to show up in the news.

● One of the world's most important malfunctioning states, **Myanmar**, may be headed for change. Monks rose up and led demonstrations for a few weeks in September 2007 until the junta put down the uprising. Meanwhile, the paranoid generals who run the country moved the capital city to a poorly connected interior site, Naypyidaw, because they fear invasion—but by whom and from where? In May 2008, the country was hit by a disastrous cyclone; in its lack of response, Myanmar's government was exposed to the world.

● In **Malaysia**, the government is seeking greater unity among its various ethnic groups while Islamic moderation seems to be preferred to Islamic militancy. In southernmost **Thailand**, however, problems between Muslims and Buddhists are intensifying.

● Thriving **Singapore** wants to lease one or two of Indonesia's more than 17,000 islands in order to expand its commercial, research, financial, and other operations.

● Seeking resources from nearby places, the **Chinese** government is offering significant aid and is involved in various Southeast Asian countries.

Regions of the Realm

Southeast Asia's first-order regionalization must be based on its mainland-island fragmentation. But as we have noted there are physiographic, historical, and cultural reasons to include the Malaysian (southern) part of the Malay Peninsula in the insular region, as shown in Figure 10-2. Using the political framework as our grid, we see that the regions of Southeast Asia are constituted as follows:

Mainland Region: Vietnam, Cambodia, Laos, Thailand, Myanmar (Burma)

Insular Region: Malaysia, Singapore, Indonesia, East Timor, Brunei, the Philippines

Note, however, that the realm boundary excludes the Indonesian zone of New Guinea (Papua), which is part of the Pacific geographic realm.

MAINLAND SOUTHEAST ASIA

Five countries form the Mainland region of Southeast Asia: two of them protruded, one compact, one elongated, and one landlocked. Two colonial powers, buffered by Thailand, shaped its modern historical geography. One religion, Buddhism, dominates cultural landscapes, but this is a multicultural, multiethnic region. Although one of the least urbanized regions in the world, it contains several major cities. And as Figure 10-2 shows, two countries (Vietnam and Myanmar) possess more than one core area each. We approach the region from the east.

INDOCHINA

Former French Indochina gave rise to three modern states: Vietnam, Cambodia, and Laos. Here the United States fought and lost a disastrous war that ended in 1975 but whose impact on America continues to be felt today.

After the Indochina War started formally in 1964 (U.S. involvement actually began earlier), some scholars warned that the conflict might spill over from Vietnam into Laos and Cambodia, and hence into Thailand, Malaysia, and even Myanmar (Burma). This view was based on the **7 domino theory**, which holds that destabilization and conflict from any cause in one country can result in the

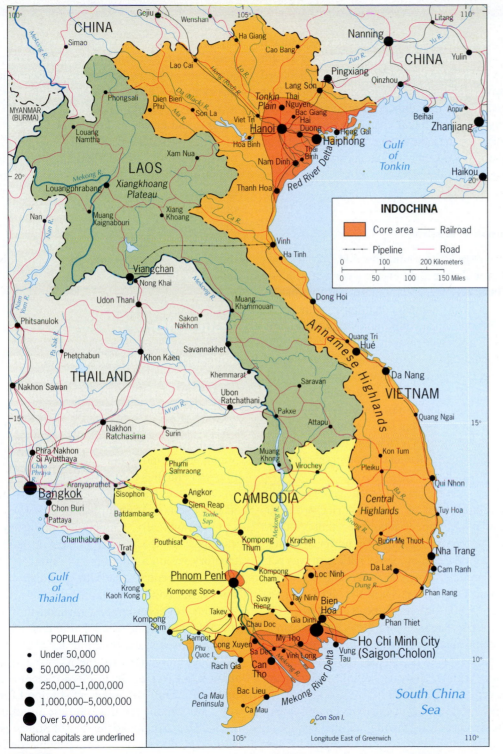

FIGURE 10-9   © H. J. de Blij, P. O. Muller, and John Wiley & Sons, Inc.

collapse of order in one or more neighboring countries, triggering a chain of events that can affect a series of contiguous states in a region.

History proved these scholars wrong; Cambodia and Laos were affected but not the other states. This seemed to invalidate the domino theory, which was grounded in the capitalist-communist struggle of the twentieth century. But communist insurgency was (and is) only one way a country may be destabilized. Ethnic conflict such as found in east-central Africa and cultural strife (e.g., in the former Yugoslavia)—even environmental and economic causes—can set the domino effect in motion.

Vietnam

Vietnam (population: 86.6 million) still carries the scars of war, although the vast majority of Vietnamese have no personal memory of that terrible conflict. The more immediate concerns in Vietnam today are to reconnect the country with the outside world and to integrate its 2000-kilometer (1200-mi) strip of attenuated territory through better infrastructure (Fig. 10-9). But it is an emerging economy and is making considerable economic progress. Poverty has dropped dramatically in the last decade as Vietnam has become a player in the global economy and is a new 'hot' location for assembly plants, attracted by low wages (lower than China) and economic development zones.

French Legacy

The French colonizers recognized that Vietnam, whose average width is under 240 kilometers (150 mi), was not a homogeneous colony, so they divided it into three units: (1) Tonkin, land of the Red River Delta and centered on Hanoi in the north; (2) Cochin China, region of the Mekong Delta and centered on Saigon in the south; and (3) Annam, focused on the ancient city of Hué, in the middle (Fig. 10-6). Today, the Vietnamese prefer to use *Bac Bo*, *Nam Bo*, and *Trung Bo*, respectively, to designate these areas.

The Vietnamese (or Annamese, also Annamites, after their cultural heartland) speak the same language, although the northerners can easily be distinguished from southerners by their accent. As elsewhere in their colonial empire, the French made their language the *lingua franca* of Indochina, but their tenure was cut short by the Japanese, who invaded Vietnam in 1940. During the Japanese occupation, Vietnamese nationalism became a powerful force, and after the Japanese defeat in 1945, the French could not regain control. In 1954, the French suffered a disastrous final trouncing on the battlefield at Dien Bien Phu in the far northwest and were ousted from the country.

North and South

But even after its forces routed the colonizers, Vietnam did not become a unified state. Separate regimes took control: a communist one in Hanoi and a noncommunist counterpart in Saigon. Vietnam's pronounced elongation had made things difficult for the French; now it played its role during the postcolonial period. Note, in Figure 10-9, that Vietnam is widest in the north and south, with a narrow 'waist' in its middle zone. North and South Vietnam were worlds apart, and those worlds were represented in Hanoi by communism and in Saigon by anticommunism. For more than a decade the United States tried to prop up the Saigon regime that controlled the south, but the communists prevailed, and like China, Vietnam still has a communist government today. As many as 2 million Vietnamese refugees set out on often-flimsy boats onto the South China Sea; of those who survived, a majority settled in the United States.

Today, contrasts between north and south continue, although they are diminishing. The capital, Hanoi, carries the imprints of its Soviet-era tutelage in the form of Ho Chi Minh's mausoleum and rows of faceless apartment buildings representative of the Soviet socialist city. With 4.5 million residents, Hanoi anchors the northern (Tonkin Plain) core area of Viet-

nam, the lower basin of the Red River (its agricultural hinterland). In the paddies, irrigation water is still raised by bucket. On the roads, goods are still moved by human- or animal-drawn cart, although the roads themselves are being improved. The south, however, is undergoing a profound change, with significant economic growth as the government slowly, and on its own terms, is opening the country up to the global economy, and in the process is lifting many out of poverty.

Vietnam in Transition

In the first decades of communist rule, private enterprise was abolished and farmers were compelled to join collectives on the communist Chinese model, resulting in near-famine conditions in the early 1980s. This coincided with China's economic opening, and Hanoi's leaders were quick to follow suit, though more timidly. By the mid-1990s, free enterprise was encouraged; farmers were allowed to cultivate for profit, and foreign investment was welcomed, though under far more restrictive terms than in China. The government has continued to provide adequate social services with health indicators better than those in China and with secondary school attendance rising rapidly.

One focus of Vietnam in recent years has been on agricultural production. Long a rice producer, in 2002 the government announced that Vietnam had become the world's largest exporter of coffee, a lucrative global commodity. That claim was perhaps a bit premature, as Vietnam is still far behind number-one Brazil, but it is now the second-largest producer of bulk coffee in the world. However, since mass-produced bulk coffee is grown plantation-style in the sun, the expansion of this commodity in the interior highlands has resulted in deforestation. This activity is having a significant environmental impact, leading to worsening annual floods affecting the once-stable countryside.

Vietnam has a fast-growing national economy, averaging annual growth of about 7.5 percent dur-

ing the early twenty-first century. Ho Chi Minh City (Saigon's official name since 1976) is booming again and now has a Special Economic Zone based on the Chinese model of such zones, as well as a New Saigon business and residential district. Saigon and environs now contribute 25 percent of Vietnam's industrial output and pay one third of its taxes. In 2007, Vietnam was admitted to the World Trade Organization, and more such changes are yet to come. Still, Hanoi's communist planners have kept tighter economic control over the country than their Chinese counterparts and are determined not to allow runaway capitalism of the Chinese variety—nor the wealth disparities that come with it.

Cambodia

Compact Cambodia is heir to the ancient Khmer Empire whose capital was Angkor and whose legacy is a vast landscape of imposing monuments including Buddhist-inspired Angkor Wat. Today, 90 percent of Cambodia's 14.7 million inhabitants are Khmers, with the remainder divided between Vietnamese and Chinese. The present capital, Phnom Penh, lies on the Mekong River (Fig. 10-9), which crosses Cambodia before it enters and forms its great delta in Vietnam.

Geographically, Cambodia enjoys several advantages; compact states enclose a maximum of territory within a minimum of boundary, and cultural homogeneity tends to diminish centrifugal forces. But neither spatial morphology nor homogeneous ethnicity could withstand the impact of the Indochina War, which led to communist revolution and the systematic murder of as many as 2 million Cambodians by the Maoist terror group, the *Khmer Rouge*. Once self-sufficient and a food exporter, Cambodia today must import food. Corrupt and violent politics, chronic instability, and rural dislocation make this one of Southeast Asia's poorest countries. Its postwar trauma continues.

Laos

Landlocked Laos has no fewer than five neighbors, one of which is East Asia's giant, China (Fig. 10-9). The Mekong River forms a long stretch of its western boundary, and the important sensitive border with Vietnam to the east lies in mountainous terrain. With 6.4 million people (about half of them ethnic Lao, related to the Thai of Thailand), Laos lies surrounded by comparatively powerful states. The country has no railroads, just a few miles of paved roads, and very little industry; it is only 21 percent urbanized (the capital, Viangchan, lies on the Mekong and has an oil pipeline to Vietnam's coast).

Laos has long included one small corner of the Golden Triangle of opium-poppy cultivation, but under international (especially U.S. and European) pressure, the communist regime has forced the mainly hill-tribe people who produce most of it to abandon their crops. Since this was the only way they could make a living, these farmers were sent to resettlement villages in the lowlands, where they contracted malaria and other diseases not prevalent in their upland domain and were subject to cultural disintegration. The reward for the authorities was the continuation of foreign aid; the cost to the powerless hill people (especially the women) is incalculable. Here is an outstanding example of the power of the globalizing core reaching into the weakest of peripheries.

The Mighty Mekong

No discussion of mainland Southeast Asia would be complete without a look at its greatest river. From its source among the snowy peaks of China's Qinghai and Xizang, the Mekong River rushes and flows some 4200 kilometers (2600 mi) to its delta in southernmost Vietnam (Fig. 10-9). This Danube of Southeast Asia crosses or borders five of the realm's countries, supporting rice farmers and fishing people, forming a transport route where roads are few and providing electricity from dams upstream. Tens of millions of people depend on the waters of the Mekong, ranging from subsistence farmers in Laos to apartment dwellers in China. The Mekong Delta in southern Vietnam is one of the realm's most densely populated areas and produces enormous harvests of rice.

But problems loom. China is building a series of dams across the Lancang (as the Mekong is called there) to supply Yunnan Province with electricity. Although such hydroelectric dams should not interfere with water flow, countries downstream worry that a severe dry spell in the interior would impel the Chinese to slow the river's flow to keep the reservoirs full. Cambodia is concerned over the future of the Tonle Sap, a large natural lake filled by the Mekong. In Vietnam, farmers worry about salt water invading the paddies should the Mekong's level drop. All along the river, fish catches are already falling, and rare species, from the Irrawaddy dolphin to the Siamese crocodile, face extinction. And the Chinese may not be the only dam builders in the future: Thailand has expressed an interest in building a dam on the Thai-Laotian border where it is demarcated by the Mekong.

In such situations, the upstream states have an advantage over those downstream. Several international organizations have been formed to coordinate development in the Mekong Basin, including the Mekong River Commission (MRC) founded half a century ago (China and Myanmar have so far refused to join). China has offered to sell electricity from its dams to Thailand, Laos, and Myanmar. Coordinated efforts to reduce logging in the Mekong's drainage basin have had some effect. After consultations with the MRC, Australia built a bridge linking Laos and Thailand. There is even a plan to make the Mekong navigable from Yunnan to the coast, creating an alternative outlet for interior China.

Sail the Mekong today, however, and you are struck by the slowness of development along this artery. Wooden boats, thatch-roofed villages, and teeming paddies mark a river still crossed by antiquated ferries and flanked by few towns. Of modern infrastructure, one sees little. Yet the Mekong and its basin form the lifeline of mainland Southeast Asia's dominantly rural societies.

THAILAND

In virtually every way, Thailand is the leading state of the Mainland region. In contrast to its neighbors, Thailand has been a strong participant in the Pacific Rim's economic development. Its capital, Bangkok (population: 6.8 million), is the largest urban center in the Mainland region and one of the world's most prominent primate cities. The country's population, 66.1 million in 2008, is growing at the second slowest rate in this geographic realm (only fully urbanized Singapore grows more slowly). Over the past few decades, only political instability and uncertainty have inhibited economic progress. Thailand is a constitutional monarchy with an elected parliament; its progress toward a stable democracy has experienced a history of setbacks. The latest of these occurred in 2006, when the armed forces ousted a controversial prime minister and took control amid the usual promises of a return to representative government.

Thailand has a compact heartland in which lie the core area, capital, and major areas of productive capacity, while a 1000-kilometer (600-mi) protrusion or corridor of land, in places less than 32 kilometers (20 mi) wide, extends southward to the border with Malaysia (Fig. 10-10). The boundary that defines this protrusion runs down the length of the upper Malay Peninsula to the Kra Isthmus, where neighboring Myanmar peters out and Thailand fronts the Andaman Sea (an arm of the Indian Ocean) as well as the Gulf of Thailand. The popular tourist resorts that lay along the Andaman Sea side of the peninsula were severely impacted by the 2004 tsunami. Many were on vacation in and around the Phuket resort when the tsunami hit.

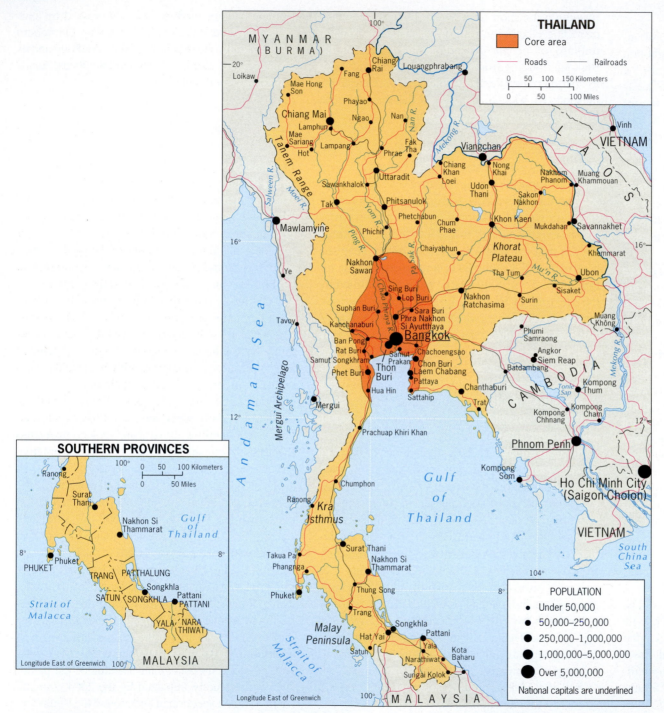

FIGURE 10-10

© H. J. de Blij, P. O. Muller, and John Wiley & Sons, Inc.

The Restive Peninsular South

In the entire country, no place lies farther from the capital than the southern end of this tenuous protrusion. Consider this spatial situation in the context of Figure 10-4. Note that the Malay ethnic population group extends from Malaysia more than 300 kilometers (200 mi) into Thai territory. In Thailand's five southernmost provinces, 85 percent of the inhabitants are Muslims (the figure for Thailand as a whole is less than 4 percent). The political border between Thailand and Malaysia is porous; in fact, you can cross it at will almost anywhere along the Kolok River by canoe, and inland it is a matter of walking a forest trail. For more than a century, the southern provinces have had closer ties with Malaysia across the border than with Bangkok 1000 kilometers (600 mi) away, and the Thai government has not tried to restrict movement or impose onerous rules on the Muslim population here.

But in the new era of the War on Terror, and in view of rising violence in this southern frontier, the south is coming to national attention. The Pattani United Liberation Organization, named after the functional capital of the Muslim south, was active in the 1960s and 1970s but had been dormant since. Then a series of bombings at a Pattani hotel, several schools, a Chinese shrine, and a Buddhist temple in 2001 and 2002 raised fears that Muslim-inspired violence was on the rise again. Thailand's pronounced protrusion, its porous border, its dependence on tourism, and the vulnerable location of its Andaman coast tourist facilities combine to raise concern over this distant part of its national territory.

This view of Bangkok from the Chao Phraya River shows a modern skyline. Built on the floodplain near the river's mouth, the city suffers from serious congestion but is unable to build a subway system due to the swampy ground. Instead, the river serves as a transportation artery to help relieve legendary street-level traffic jams. A new, above-ground 'skytrain' has been built, but it has a limited route network and rides are expensive.   (© A. WinklerPrins)

this soggy terrain has compounded its notorious traffic problems as a desperately needed underground transit system cannot be constructed.

During the 1990s Thailand seemed destined to become a Pacific Rim tiger producing an array of goods including toys and cars for the world market. But while workers toiled, government and financial mismanagement undercut the economic gains made. Then in 1997 the country's currency, the *baht*, began a slide that undermined the economy and led to bank failures. Exports and worker productivity declined, and the real estate market stalled. Partly finished buildings in Bangkok are a reminder of this tailspin. Thailand became the first of various economic setbacks the region suffered.

But Thailand's infrastructure had been much improved during the boom years, not just in the core but throughout much of the country. Today it has a diversified economy, producing goods ranging from specialized rice to automobiles. It also has a significant tourist industry. Though heavily impacted by the 2004 tsunami in the southwest, it has recovered sufficiently and continues to draw tourists from throughout the world. Providing Thailand can control its political turmoil and keep its restive south from erupting, it will indeed become an economic tiger, offering a much improved quality of life for its citizens.

Bangkok on the Chao Phraya

As Figure 10-2 shows, Thailand occupies the heart of the Mainland region of Southeast Asia. While Thailand has no Red, Mekong, or Irrawaddy Delta, its central lowland is watered by a set of streams that flow off the northern highlands and the Khorat Plateau in the east. One of these streams, the Chao Phraya, is the Rhine of Thailand. From the head of the Gulf of Thailand to Nakhon Sawan, this river is a highway of traffic. Barge trains loaded with rice head for the coast, ferry boats head upstream, freighters transport tin and tungsten (of which

Thailand is among the world's leading producers). Bangkok sprawls on both sides of the lower Chao Phraya floodplain, here flanked by skyscrapers, pagodas, factories, boatsheds, ferry landings, luxury hotels, and modest dwellings all built in crowded confusion on very swampy ground. On the right bank of the Chao Phraya, Bangkok's west side, lie the city's remaining *klong* (canal) neighborhoods, where waterways and boats form the transport system (see photo p. 343). Bangkok still is known as the Venice of Asia, although many *klongs* have been filled in and paved over to serve as roadways. Unfortunately Bangkok's location on

MYANMAR

Thailand's neighbor, Myanmar (referred to as Burma in antiregime circles), is one of the world's poorest and most isolated countries where, it seems, time has stood still for centuries. Long languishing under one of the world's most corrupt and brutal military dictatorships, it has potential as an oil producer, but its promise awaits regime change that would make multinationals more welcome. Myanmar shares with Thailand the neck of the Malay Peninsula—but look again at Figure 10-10 and note the contrast in surface communications.

FROM THE FIELD NOTES

Yangon, Myanmar. "Little did I realize, as I walked the central business district of Yangon (Myanmar) in mid-April 2008, that these streets would be turned into chaos by Cyclone Nargis just days later. This is Sule Pagoda Road, a few blocks north of the famed Octagonal Pagoda, the golden landmark in the center of town. Many signs in English remind you that this was once a British colony."

(© H. J. de Blij)

Concept Caching

www.conceptcaching.com

Morphology and Structure

Myanmar's issues are complicated by a shift in the Burmese core area that took place during colonial times. Prior to the colonial period, the core of embryonic Burma lay in the so-called dry zone between the Arakan Mountains and the Shan Plateau, which covers the country's triangular eastern extension toward the Laotian border (Fig. 10-11). The urban focus of the state was Mandalay, which had a central situation and relative proximity to the non-Burmese highlands all around. Then the British developed the agricultural potential of the Irrawaddy Delta, and Rangoon (now called Yangon) became the hub of the colony. The Irrawaddy waterway links the old and the new core areas, but the center of gravity has shifted to the south.

FIGURE 10-11

© H. J. de Blij, P. O. Muller, and John Wiley & Sons, Inc.

In September 2007, Buddhist monks initiated and led several street marches in protest against the repressive military government in Yangon, Myanmar. Relying on the reverence toward monks that Burmese have, the protestors hoped to make further progress in improving the plight of the country's people. However, this did not happen and the regime clamped down on all the demonstrators after several marches.   (© AP/Wide World Photos)

As Figure 10-4 indicates, the peripheral peoples (11 minorities) of Myanmar occupy a significant part of the state. A closer look at Figure 10-11 shows that their domains have the status of State, of which there are seven; the Burman-dominated areas (60 percent of a population of 52.1 million) are designated as Divisions. The Shan of the northeast and far north, who are related to the neighboring Thai, account for about 8 percent of the population, or about 4 million. The Kayin (Karen), who constitute just over 8 percent (4.3 million), live in the neck of Myanmar's protrusion and have proclaimed their desire to create an autonomous territory within a federal Myanmar. The Mon (2.3 percent, or 1.2 million) were in what is today Myanmar long before the Burmans and introduced Buddhism to the area; they want the return of ancestral lands from which they were ousted. Although the powerful military has dealt such aspirations a series of setbacks, centrifugal forces continue to bedevil the central regime. The military government's response has been to exert power by all available means rather than to accommodate these forces, even to the point of stifling political discourse—let alone opposition—among the Burman themselves.

In 2007 the generals moved the capital city from Yangon to a newly constructed governmental center, ostensibly a city but it can barely be called that as it is poorly connected to anywhere else and has few city amenities. The center is located next to the interior town of Pyinmana and is known as Naypyidaw. The government asserts that its more central location will yield greater administrative efficiency as well as better protection against (unspecified) enemies.

Myanmar today ranks among the world's least developed countries not because of its intrinsic indigence but because a rapacious military regime, tolerated and even commercially engaged by the international community, represses its peoples and destroys their aspirations.

Two recent events have really shaken up Myanmar and exposed its failings in governance to the world in a way its secretive junta does not like (see photos). In September 2007, a sizeable uprising by the people was led by Buddhist monks, and the world held its breath to see if this most repressive of states would finally open up. This was not to be: the junta once again clamped down, proclaiming that dissent is not to be tolerated. Then in May of 2008, a large cyclone hit the Irrawaddy Delta, killing at least 130,000 and displacing 2 to 3 million people. The world watched as the junta initially did not acknowledge the severity of the situation and only slowly mobilized relief for those affected. At first, foreign relief workers were prevented from bringing assistance that the government itself could not provide, and only weeks later was limited entry permitted of workers and supplies.

In early May 2008, a major cyclone named Nargis devastated much of the Irrawaddy Delta in Myanmar. Here a man walks through his damaged house on the outskirts of Yangon. Due to the repressive regime's refusal to help the victims and its reluctance to let in outside help, relief was slow to arrive to help millions of survivors, many of whom were left to suffer without food or shelter. (© AP/Wide World Photos)

INSULAR SOUTHEAST ASIA

On the peninsulas and islands of Southeast Asia's southern and eastern periphery lie six of the realm's 11 states (Fig. 10-2). Few regions in the world contain so diverse a set of countries. Malaysia, the former British colony, consists of two major areas separated by hundreds of miles of South China Sea. The realm's southernmost state, Indonesia, sprawls across thousands of islands from Sumatera in the west to New Guinea in the east. North of the Indonesian archipelago lies the Philippines, a country that once was a U.S. colony. These are three of the most severely fragmented states on earth, and each has faced the challenges that such politico-spatial division brings. This Insular region of Southeast Asia also contains two small but important sovereign entities: a city-state and a sultanate. The city-state is Singapore, once a part of Malaysia (and one instance in which internal centrifugal forces were too great to be overcome). The sultanate is Brunei, an oil-rich Muslim territory on the island of Borneo that seems transplanted from the Persian Gulf. In addition, a third small entity, East Timor, achieved independence in 2002. Few parts of the world are more varied or interesting geographically.

MAINLAND-ISLAND MALAYSIA

The state of Malaysia represents one of the three types of fragmented states discussed earlier: the mainland-island type, in which one part of the national territory lies on a continent and the other on an island. Malaysia is a colonial political artifice that combines two quite disparate components into a single state: the southern end of the Malay Peninsula and the northern part of the island of Borneo. These are known, respectively, as West Malaysia and East Malaysia (Fig. 10-2). The name *Malaysia* came into use in 1963, when the original Federation of Malaya, on the Malay Peninsula, was expanded to incorporate the areas of Sarawak and Sabah in Borneo. When the name Malaya is used, it refers to the peninsular part of the Federation, whereas Malaysia refers to the total entity.

Ethnic Components

The Malays of the peninsula, traditionally a rural people, displaced older aboriginal communities there and today make up just over 58 percent of the country's population of 27.8 million. They possess a strong cultural identity expressed in adherence to the Muslim faith, a common language, and a sense of territoriality that arises from their perceived Malayan origins and their collective view of Chinese, Indian, European, and other foreign intruders.

The Chinese came to the Malay Peninsula and to northern Borneo in substantial numbers during the colonial period, and today they constitute about one fourth of Malaysia's population (they are the largest single group in Sarawak).

Hindu South Asians were in this area long before the Europeans, and for that matter before the Arabs and Islam arrived on these shores. Today they still form a substantial minority of over 8 percent of the population, clustered, like the Chinese, on the western side of the peninsula (Fig. 10-4).

The Dominant Peninsula

The populous peninsular part of Malaysia remains the country's dominant sector with 11 of its 13 States and fully 80 percent of its population. Here the Malay-dominated government has strictly controlled economic and social policies while pushing the country's modernization. During the Asian economic boom of the 1990s, Malaysia's planners embraced the notion of symbols: the capital, Kuala Lumpur, was endowed with the world's tallest building; a space-age airport outpaced Malaysia's needs; a high-tech administrative capital was built at Putrajaya; and a nearby development was called Cyberjaya—all part of a so-called *Multimedia Supercorridor* to anchor Malaysia's core area (Fig. 10-12).

The chief architect of this program was Malaysia's long-term and autocratic head of state, Mahathir bin Mohamad, leader of the Malay-dominated party that forms the majority in government. Mahathir had the support not only of the great majority of the country's Malays, but also the important and influential ethnic Chinese minority, which saw him as the only acceptable alternative to the more fundamentalist Islamic party challenging his rule. But Malaysia's headlong rush to modernize caused a backlash among more conservative Muslims, which, in 2001, led to Islamist victories in two States, tin-producing Kelantan and energy-rich but socially poor Terengganu. As the fundamentalist governments in those two States imposed strict religious laws, Malaysians talked of two corridors marking their country: the Multimedia Supercorridor in the west and the Mecca Corridor in the east (Fig. 10-12). But after Mahathir's resignation in 2003 and the appointment of Abdullah Badawi as his more moderate successor, Islamist fervor in the Mecca Corridor, which had featured calls for a *jihad* in Malaysia, began to wane.

The Malay Peninsula's primacy began long ago, during colonial times, when the British created a substantial economy based on rubber plantations, palm-oil extraction, and mining (tin, bauxite, copper, iron). Singapore, at the southern end of it, was a prized possession until Singapore seceded from the Malaysian Federation in 1965. The Strait of Malacca (Melaka) to the west continues to be one of the world's busiest and most strategic waterways, a **8 choke point** (a narrow waterway that constrains navigation) in the flow of resources and goods between world realms. Piracy, long a significant concern in the strait, has been significantly curtailed through increased patrolling that began with the aid given to the region after the 2004 tsunami.

Malaysia, despite the loss of Singapore and notwithstanding its recurrent ethnic troubles, has become a major player on the burgeoning Pacific Rim.

The strong skills and modest wages of the local work-force have attracted many companies, and the government has capitalized on its opportunities, for example, by encouraging the creation of a high-technology manufacturing complex on the island of Pinang, where Chinese outnumber Malays by two to one.

Malaysian Borneo

The decision to combine the 11 Sultanates of Malaya with the States of Sabah and Sarawak on Borneo, creating the country now called Malaysia, had far-reaching consequences. These two States make up 60 percent of Malaysia's territory (although they represent only 19 percent of the population). They endowed Malaysia with major energy resources and huge stands of timber. They also complicated Malaysia's ethnic makeup because each State is home to more than two dozen indigenous groups (in fact, the immigrant Chinese form the largest single group in Sarawak). These locals complain that the federal government in Kuala Lumpur treats East Malaysia as a colony, and politics here are contentious and fractious. It is likely that Malaysia will eventually confront devolutionary forces in East Malaysia.

BRUNEI

Also located on Borneo—where Sarawak and Sabah meet—is Brunei, a rich, oil-exporting Islamic sultanate far from the Persian Gulf (Fig. 10-7, upper-left map). Brunei, the remnant of a former Islamic kingdom that once controlled all of Borneo and areas beyond, came under British control and was granted independence in 1984. Just slightly larger than Delaware and with a population of 430,000, the sultanate is a mere ministate—except for the discovery of oil in 1929 and natural gas in 1965, which made this Southeast Asia's richest country after Singapore. And there are indications that further discoveries will be

WEST MALAYSIA

'Multimedia Supercorridor'

States with greatest Islamic strength

States with largest Chinese and Hindu presence

—— Highway

—— Main road

—— Railroad

✈ Airport

POPULATION
● 50,000–250,000
● 250,000–1,000,000
● 1,000,000–5,000,000

National capital is underlined

FIGURE 10-12

© H. J. de Blij, P. O. Muller, and John Wiley & Sons, Inc.

made in the offshore zone owned by Brunei. The Sultan of Brunei rules as an absolute monarch; his palace in the capital, Bandar Seri Begawan, is reputed to be the world's largest. He will have no difficulty finding customers for his oil in energy-poor eastern Asia.

SINGAPORE

In 1965, a fateful event occurred in Southeast Asia. Singapore, crown jewel of British colonialism in this realm, seceded from the recently independent (1963) Malaysian Federation and became a sovereign state, albeit a ministate (Fig. 10-13). With its magnificent relative location, its multiethnic and well-educated population, and its firm government, Singapore then overcame the limitations of space and the absence of raw materials to become one of the economic tigers on the Pacific Rim.

With a mere 619 square kilometers (239 sq mi) of territory, space is at a premium in Singapore, and this is a constant worry for the government. Singapore's only local spatial advantage over Hong Kong is that its small territory is less fragmented (there are just a few small islands in addition to the compact main island). With a population of 4.6 million and an expanding economy, Singapore must develop space-conserving high-tech and service industries.

The Port of Singapore is one of the world's largest container ports and is strategically located near the entrance to the Strait of Malacca—through which virtually all commercial shipping between East Asia and Europe must go. Singapore has taken excellent advantage of its relative location and built a prosperous city-state.   (© R. Ian Lloyd/Masterfile)

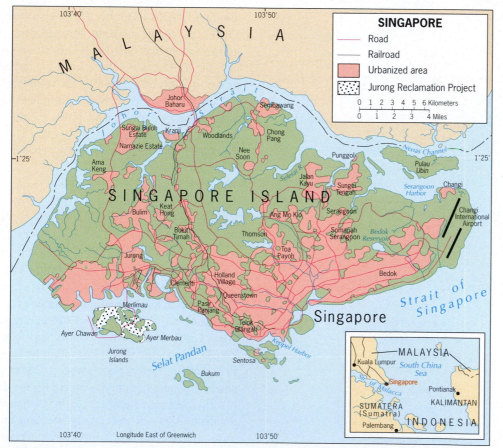

FIGURE 10-13   © H. J. de Blij, P. O. Muller, and John Wiley & Sons, Inc.

Benefiting from its relative location, the old port of Singapore had become one of the world's busiest (by numbers of ships served) even before independence. It thrived as an **9 entrepôt** between the Malay Peninsula, Southeast Asia, Japan, and other emerging economic powers on the Pacific Rim and beyond. Crude oil from Southeast Asia still is unloaded and refined at Singapore, then shipped to Asian destinations. Raw rubber from the adjacent peninsula and from Indonesia's island of Sumatera is shipped to Japan, the United States, China, and other countries. Timber from Malaysia, rice, spices, and other foodstuffs are processed and forwarded via Singapore. In return, automobiles, machinery, and equipment are imported into Southeast Asia through Singapore.

But that is the old pattern. Singapore's leaders are redirecting the city-state's economy and moving toward high-tech industries for the future. In Singapore the essentially one-party government tightly controls business as well as other aspects of life and borders on authoritarian. Some newspapers and magazines have been banned for criticizing the regime, and there are even fines for chewing gum, eating on the subway, failing to flush a public toilet, and wearing hair deemed longer than appropriate. However, its overall success after secession has tended to keep the critics quiet: while GNI per capita from 1965 to 2003 multiplied by a factor of more than 15—to U.S. $29,780—that of neighboring Malaysia reached just U.S. $10,320. Among other things, Singapore became (and for many years remained) the world's largest producer of disk drives for small computers.

To accomplish its revival, Singapore has moved in several directions. First, it is concentrating on three growth areas: information technology, automa-tion, and biotechnology. Second, it espouses notions of a Growth Triangle involving Singapore's devel-oping neighbors, Malaysia and Indonesia; those two countries would supply the raw materials and cheap labor, and Singapore the capital and technical know-how. Third, Singapore opened its doors to capitalists of Chinese ancestry who left Hong Kong when China took it over and who wanted to relocate their enterprises in Singapore. The population in Singa-pore is 77 percent Chinese, 14 percent Malay, and 8 percent South Asian. The government is Chinese-dominated, and its policies have served to sustain Chinese control. Indeed, China itself often cites Sin-gapore's combination of authoritarianism and eco-nomic success as proving that communism and mar-ket economies can coexist.

INDONESIA

The fourth-largest country in the world in terms of human numbers is also the globe's most expansive **10 archipelago**. Spread across a chain of more than 17,000 mostly volcanic islands, Indonesia's more than 230 million people live both separated and clustered—separated by water and clustered on is-lands large and small.

The complicated map of Indonesia requires close attention (Fig. 10-14). Five large islands dominate the archipelago territorially, but one of these, New Guinea in the east, is not part of the Indonesian cul-ture sphere, although its western half is under In-donesian control. The other four major islands are

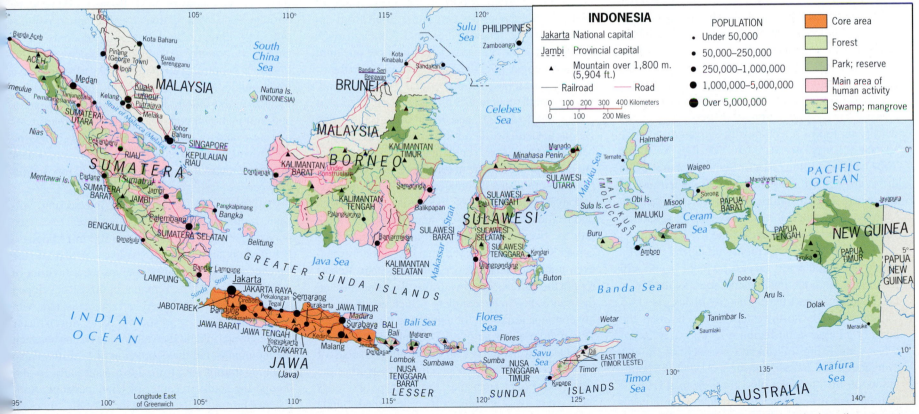

© H. J. de Blij, P. O. Muller, and John Wiley & Sons, Inc.

collectively known as the Greater Sunda islands: *Jawa* (Java), smallest but by far the most populous and important; *Sumatera* (Sumatra) in the west, directly across the Strait of Malacca from Malaysia; *Kalimantan*, the Indonesian sector of large, compact, minicontinent Borneo; and wishbone-shaped, *Sulawesi* (Celebes) to the east. Extending eastward from Jawa are the Lesser Sunda Islands, including Bali and, near the eastern end, Timor. Another important island chain within Indonesia is the Maluku (Molucca) Islands, between Sulawesi and New Guinea. The central water body of Indonesia is the Java Sea.

The Major Islands

Indonesia is a Dutch colonial creation, and Jawa was chosen as their colonial headquarters with Batavia (now Jakarta) as the capital. Today, Jawa remains the core of Indonesia. With about 130 million inhabitants, Jawa is one of the world's most densely peopled places (see Figs. G-10 and 10-3) and one of the most agriculturally productive with rice paddies rising up on the flanks of active and soil-fertilizing volcanoes. This combination of extreme population density on top of a tectonically active crustal zone has the potential for disaster. You may recall the 2004 tsunami off the coast of Sumatera, but in 2006 there was a significant earthquake centered just 25 kilometers (15 mi) from the city of Yogyakarta. Moreover, Mount Merapi, also near Yogyakarta, was threatening to erupt at the time.

Jawa also is the most highly urbanized part of a country in which nearly 60 percent of the people still live on the land; the Pacific Rim boom of the 1990s had a strong impact here. The city of Jakarta (14.5 million), on the northwestern coast, became the heart of a larger conurbation now known as *Jabotabek*, consisting of *Ja*karta as well as *Bo*gor, *Ta*ngerang, and *Bek*asi. During the 1990s the population of this megalopolis grew from 15 to 20 million, and it is predicted to reach 30 million by

2010. Already, Jabotabek is home to over 10 percent of Indonesia's entire population and 25 percent of its urban population. Thousands of factories, their owners taking advantage of low prevailing wages, were built in this area, straining its infrastructure and overburdening the port of Jakarta. On an average day, hundreds of ships lie at anchor, awaiting docking space to offload raw materials and take on finished products.

As has always been the case in Indonesia, Jawa is where the power lies. As a cultural group (though itself heterogeneous), the Jawanese constitute about 45 percent of the country's population. In the middle of the island, a politically powerful sultanate centers on Yogyakarta; Jawa also is the main base for the two national Islamic movements, one comparatively moderate and the other increasingly revivalist.

Sumatera, Indonesia's westernmost island, forms the western shore of the busy Strait of Malacca; Singapore lies across the Strait from approximately the middle of the island. Although much larger than Jawa, Sumatera has only about one third as many people (49 million). In colonial times the island became a base for rubber and palm-oil plantations; its high relief makes possible the cultivation of a wide range of crops, and neighboring Bangka and Belitung yield petroleum and natural gas. Palembang is the key urban center in the south, but current attention focuses on the north. There, the Batak people accommodated colonialism and Westernization and made Medan one of Indonesia's Pacific Rim boom cities. Farther north the Aceh fought the Dutch into the twentieth century and, after Indonesia became a sovereign country, demanded autonomy and even outright independence for their State. Rebels fought the Indonesian army to a costly stalemate, and thousands died in the conflict—which would probably continue today but for a dramatic turn of events. The seafloor epicenter of the December 26, 2004 Indian Ocean tsunami lay near the far northern coast of Sumatera, and Aceh was directly in the path of the most powerful ocean waves. Entire towns and roads were swept away and tens of thousands died; Banda

Aceh, the State capital, was devastated (Fig. 10-14). The international relief effort opened Aceh to foreigners in ways the Indonesians had long prevented, and the Indonesian army as well as the rebels were engaged in rescue missions rather than warfare. This combination of circumstances facilitated a truce and negotiations resulting in an agreement under which the rebels agreed to drop their demand for independence and the Indonesian army withdrew.

Kalimantan is the Indonesian part of the island of Borneo, a slab of the earth's crystalline crust whose backbone of tall mountains is of erosional, not volcanic, origin. Larger than Texas, Borneo has a deep, densely rainforested interior that is a last refuge for some 35,000 orangutans as well as dwindling numbers of Asian elephants, rhinoceroses, and tigers. These and a number of indigenous peoples survive even as loggers undermine their habitat (photo p. 347). Borneo's Pleistocene heritage sustains a relatively small human population (14 million on the Indonesian side, only 6 percent of the country's population) on poor soils. Indigenous peoples, principally the Dayak clans, have traditionally had less impact on the natural environment than the Indonesian and Chinese immigrants and multinational corporations who log the forests and clear land for farms. As Figure 10-14 shows, the only towns of any size in Kalimantan lie on or near the coast; routes into the interior are still few and far between.

Sulawesi consists of a set of intersecting, volcanic mountain ranges rising above sea level (see photo p. 346); the 800-kilometer (500-mi) Minahasa Peninsula is still growing by volcanic action into the Celebes Sea. This northern peninsula, a favorite of the Dutch colonizers, remains the most developed part of an otherwise rugged and remote island, with Manado its relatively prosperous focus. Seven major ethnic groups inhabit the valleys and basins between the mountains, but the population of 18 million also includes immigrants from Jawa, especially in and around the southern center of Ujungpandang. Subsistence farming is the leading mode of life, although gging, some mining, and fishing augment the economy. Clashes between Muslims and Christians ccur intermittently in remote areas.

Papua, the Indonesian name for the western part f the island of New Guinea, has become an issue in ndonesian politics. Bounded on the east by a clasic superimposed geometric boundary (Figs. 10-7, 0-14), it was taken over by Indonesia from the Dutch in 1969. Papua constitutes about 22 percent of ndonesia's territory, but its population is barely 2.5 million—just over 1 percent of the national total. The indigenous inhabitants of this territory, which is n effect a colony, are Papuan, most living in remote eaches of this mountainous and densely forested isand. Papua is economically important to Indonesia, or it contains what is reputed to be the world's richst gold mine and its second-largest open-pit copper nine, but political consciousness has reached the Papuans. The Free Papua Movement has become inreasingly active, holding small rallies in the capital, ayapura, displaying a Papuan flag, and demanding ecognition.

Diversity in Unity

ndonesia's survival as a unified state is as remarkble as India's and Nigeria's. With more than 300 iscrete ethnic clusters, over 250 languages, and just bout every religion practiced on earth (although slam dominates), actual and potential centrifugal orces are powerful here. Wide waters and high nountains perpetuate cultural distinctions and diferences. Indonesia's national motto is *bhinneka unggal ika*: diversity in unity.

What Indonesia has achieved is etched against the country's continuing cultural complexity. There are dozens of distinct aboriginal cultures; virtually every coastal community has its own roots and traditions. And the majority, the rice-growing Indonesians, include not only the numerous Jawanese—who have heir own cultural identity—but also the Sundanese who constitute 14 percent of Indonesia's popula-

tion), the Madurese (8 percent), and others. Perhaps the best impression of the cultural mosaic comes from the string of islands that extends eastward from Jawa to Timor (Fig. 10-14). The rice-growers of Bali adhere to a modified version of Hinduism, giving the island a unique cultural atmosphere; the population of Lombok is mainly Muslim, with some Balinese Hinduism; Sumbawa is a Muslim community; Flores is mostly Roman Catholic. In western Timor, Protestant groups dominate; in the now-independent east, where the Portuguese ruled, Roman Catholicism prevails. Nevertheless, Indonesia nominally is the world's largest Muslim country: overall, 88 percent of the people adhere to Islam, and in the cities the silver domes of neighborhood mosques rise above the townscape. Although until recently Indonesian Islam has been relatively moderate, lately more overt Islamization has been on the rise, with new laws banning public displays of affection and limiting the types of clothes women may wear in public.

Indonesia has not been immune from terrorism. Serious terrorist attacks have occurred in the tourist areas of Bali and in central Jakarta, mounted by a terrorist organization named *Jemaah Islamiyah (JI)*, apparently headquartered in eastern Jawa. In addition to Indonesians, many Australians, for whom Indonesia, and especially Bali, is a popular tourist destination, have lost their lives in these attacks. In late 2005, the Australian government announced the capture of a JI cell active in Australia itself.

Transmigration and the Outer Islands

As noted earlier, Indonesia's population of 231.9 million makes it the world's fourth most populous country, but Jawa, we also noted, contains approximately 55 percent of it. With about 130 million people on an island the size of Louisiana, population pressure is enormous here. Moreover, Indonesia's annual rate of population growth is at the realm's average at 1.4 percent. To deal with this problem, and

at the same time to strengthen the core's power over outlying areas, the Indonesian government long pursued a policy known as **11** **transmigration** (known as *transmigrasi* in Indonesian), inducing many from the densely populated inner islands (e.g., Jawa, Bali, Madura) to relocate to the sparsely inhabited outer islands (e.g. Sumatera, Sulawesi, Kalimantan). During the last few decades of the twentieth century, millions moved to peripheral locales as part of the government-sponsored program.

In fact, the Dutch colonialists started the Transmigration Program, but President Sukarno abandoned it and it was not revived until 1974, when President Suharto saw the economic and political opportunities it presented. Over the next 25 years, nearly 8 million people were relocated; the World Bank gave Indonesia nearly U.S.$1 billion to facilitate the program before it pulled out amid reports of forced migration.

The results were mixed. World Bank project reviews suggested that the vast majority of the transmigrants were happy with the land, schools, medical services, and other amenities they were provided, but independent surveys painted a different picture. About half of the migrants never managed to rise above a life of subsistence as environmental and social conditions were quite different from what they left behind. Many never received even the most basic needs for survival in their new environment. Land was simply taken away from indigenous inhabitants, notably in Kalimantan, but the soils were poor and not suitable for the cultivation techniques used by people from agriculturally superior Jawa. In addition, the newcomers inadvertently and deliberately did much damage to the traditional cultures they invaded; the program also led to massive forest and wetland destruction. Moreover, thousands of migrants died in clashes with local inhabitants.

In 2001, the Transmigration Program was canceled as one of the final acts of the short-lived government of President Abdurrahman Wahid, but the damage it did will trouble Indonesia for generations. From northern Sumatera to western Kalimantan to central Sulawesi to the islands of the Malukus to Papua, cultural strife and its polarizing effect will challenge future Indonesian governments.

EAST TIMOR

The easternmost of the Lesser Sunda Islands is Timor, the eastern part of which was a Portuguese colony, overrun by Indonesia in 1975 and annexed in 1976, which became the scene of a bitter struggle for independence. When the people of East Timor in 1999 were allowed to express their views on independence, this Connecticut-sized territory with only about 800,000 inhabitants took the first steps toward statehood.

East Timor (officially known as *Timor-Leste*) became an independent state in May 2002. But as Figure 10-14 shows, its sovereignty creates certain politico-geographical complications. The political entity of East Timor consists of a main territory, where the capital named Dili is located, and a small *exclave* to the west on the north coast of (Indonesian) western Timor called Ocussi (spellings vary, and it is sometimes mapped as Ocussi-Ambeno or Ambeno Province). Although there is a road from the exclave to the main territory, relations between East Timor and Indonesia may not make this link viable, so that the two parts of the new state have to be connected by boat traffic.

Indeed, the sea may come to mean more to East Timor than mere connectivity. Offshore to the southeast, beneath the waters of the Timor Sea between East Timor and Australia, lie valuable oil and gas reserves whose ownership will depend on how the new state's maritime boundaries (a concept discussed in Chapter 12) are to be drawn. After some difficult negotiations, East Timor and Australia agreed in 2005 to defer the maritime-boundary issue for 50 years while East Timor receives a satisfactory share of the revenues from the known oil-yielding reserves. If its continued political turmoil can be calmed, this will go a long way toward providing East Timor with the resources it needs to sustain itself.

THE PHILIPPINES

North of Indonesia, across the South China Sea from Vietnam, and south of Taiwan lies an archipelago of more than 7000 islands (only about 460 of them larger than 1 square kilometer [0.4 sq. mi] in area) inhabited by 90.1 million people. The islands of the Philippines can be viewed as three groups: (1) Luzon, largest of all, and Mindoro in the north, (2) the Visayan group in the center, and (3) Mindanao, second largest, in the south (Fig. 10-15). Southwest of Mindanao lies a small group of islands, the Sulu Archipelago, nearest to Malaysian Borneo, where Muslim-based insurgencies have kept the area in turmoil.

Few of the generalizations we have been able to make for Southeast Asia could apply in the Philippines without qualification. The country's location relative to the mainstream of change in this part of the world has had much to do with this situation. The islands, inhabited by peoples of Malay ancestry with Indonesian strains, shared with much of the rest of Southeast Asia an early period of Hindu cultural influence, which was strongest in the south and southwest and diminished northward. Next came a Chinese invasion, felt more strongly on the largest island of Luzon in the northern part of the Philippine archipelago. Islam's arrival was delayed somewhat by the position of the Philippines well to the east of the mainland and to the north of the Indonesian islands. The few southern Muslim beachheads were soon overwhelmed by the Spanish invasion during the sixteenth century. Today the Philippines, adjacent to the world's largest Muslim state (Indonesia), is 83 percent Roman Catholic, 9 percent Protestant, and only 5 percent Muslim.

Muslim Insurgency

The Philippines' small Muslim population, concentrated in the southeastern flank of the archipelago (Fig. 10-15), has long decried its marginalization in this dominantly Christian country. Over the past 30 years a half-dozen Muslim organizations, including the Moro (Moor) National Liberation Front and the Moro Islamic Liberation Front, have promoted the Muslim cause with tactics ranging from peaceful negotiation with the government to violent insurgency. Densely forested Basilan Island became the base for an especially extreme group, Abu Sayyaf, which received support through the al-Qaeda network and brought the War on Terror to the Philippines when American troops joined Filipino forces in pursuit. The Muslim challenge is pirating a disproportionate share of Manila's operating budget.

People and Culture

Out of the Philippines melting pot, where Malay, Arab, Chinese, Japanese, Spanish, and American elements have met and mixed, has emerged the distinctive Filipino culture. It is not a homogeneous or unified culture, as is reflected by the nearly 90 Malay languages in use in the islands, but it is in many ways unique. At independence in 1946, the largest of the Malay languages, Tagalog (also called Pilipino), became the country's official language. But English is widely learned as a second language, and a Tagalog-English hybrid, Taglish, is increasingly heard today. The Chinese component of the population is small (less than 2 percent) but dominant in local business.

The Philippines population, concentrated where the good farmlands lie, is densest in three general areas (Fig. 10-3): (1) the northwestern and south-central part of Luzon, (2) the southeastern extension of Luzon, and (3) the islands of the Visayan Sea between Luzon and Mindanao. Luzon is the site of the capital, Manila-Quezon City (11.3 million—one eighth of the entire national population), a megacity facing the South China Sea. Alluvial as well as volcanic soils, together with ample moisture in this tropical environment, produce self-sufficiency in rice and other staples and make the Philippines a net exporter of farm products despite a rather high population growth rate of 2.1 percent.

Perhaps more than any other people, Filipinos take jobs in foreign countries in massive numbers, proving their capacity to succeed in jobs they cannot find

FIGURE 10-15   © H. J. de Blij, P. O. Muller, and John Wiley & Sons, Inc.

Pacific Rim's economic growth. It has participated in offshore manufacturing in its numerous Export Processing Zones (which are similar to the maquiladoras discussed in Chapter 4), but the necessary trade linkages have not been created and sustained. Also, governmental mismanagement and political instability have slowed the country's participation, but during the 1990s the situation improved. Despite a series of jarring events—the ouster of U.S. military bases, the damaging eruption of a volcano near the capital, the violence of Muslim insurgents, and a dispute over the nearby Spratly Islands in the South China Sea—the Philippines made substantial economic progress during the decade. Its electronics and textile industries (mostly in the Manila hinterland) expanded continuously, and more foreign investment arrived. But agriculture continues to dominate the Philippines' economy, unemployment remains high, further land reform is badly needed, and social restructuring (reducing the controlling influence of a comparatively small group of families over national affairs) must occur. However, progress is being made. The country now is a lower-middle-income economy, and given a longer period of stability and success in reducing the population growth rate, it will rise to the next level and finally take its place among Pacific Rim growth poles.

at home. The merchant marine would not exist without Filipino sailors, and Filipina nurses and domestic workers can be found from Dubai to Dubuque. Remittances from the emigrants make the Philippines a world leader in monetary inflow.

Prospects

The Philippines seems to get little mention in discussions of developments on the Asian Pacific Rim, and yet it would seem to be well positioned to share in the

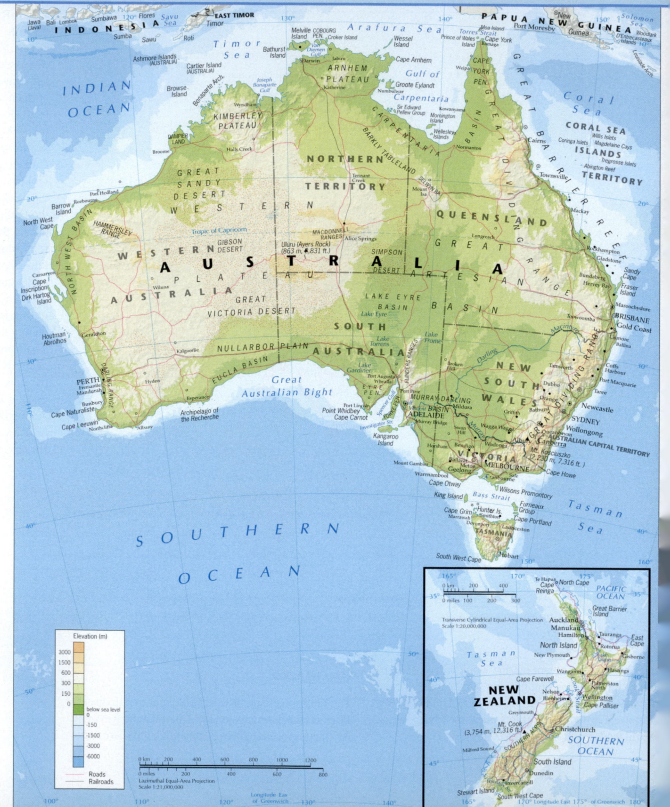

FIGURE 11-1
Map: © H. J. de Blij, P. O. Muller,
and John Wiley & Sons, Inc.

Elevation (m)

3000
1500
600
300
150
0
below sea level
0
-150
-1500
-3000
-6000

Roads
Railroads

11

AUSTRAL REALM

In This Chapter

- Australia's amazing biogeography
- Australia's changing population
- Aboriginal claims to land and resources
- China covets Australian commodities
- Foreign policy dilemmas downunder
- New Zealand's matchless physiography

CONCEPTS, IDEAS, AND TERMS

1 Austral
2 Southern Ocean
3 Subtropical Convergence
4 West Wind Drift
5 Biogeography
6 Wallace's Line
7 Aboriginal population
8 Outback
9 Federation
10 Unitary state
11 Import-substitution industries
12 Aboriginal land issue
13 Environmental degradation
14 Peripheral development

REGIONS

AUSTRALIA
NEW ZEALAND

Photos: (*top*) Sydney Harbor, Australia © Rob Crandall; (*bottom*) Akaroa, New Zealand © A. WinklerPrins.

THE AUSTRAL REALM is geographically unique (Fig. 11-1; see chapter opener map). It is the only geographic realm that lies entirely in the Southern Hemisphere. It is also the only realm that has no land link of any kind to a neighboring realm and is thus completely surrounded by ocean and sea. It is second only to the Pacific as the world's least populous realm. Appropriately, its name refers to its location (1 **Austral** means south)—a location far from the sources of its dominant cultural heritage but close to its newfound economic partners on the western Pacific Rim.

Defining the Realm

Two countries constitute the Austral Realm: Australia, in every way the dominant one, and New Zealand, physiographically more varied but demographically much smaller than its giant partner (Fig. 11-2). Between them lies the Tasman Sea. To the west lies the Indian Ocean, to the east the Pacific, and to the south the frigid Southern Ocean.

This southern realm is at a crossroads. On the doorstep of populous Asia, its Anglo-European legacies are now infused by many other cultural strains. Polynesian Maori in New Zealand and Aboriginal communities in Australia are demanding greater rights and more acknowledgment of their cultural heritage. Pacific Rim markets are buying large quantities of raw materials. Japanese and other Asian tourists fill hotels and resorts. Queensland's tropical Gold Coast resembles Honolulu's Waikiki. The streets of Sydney and Melbourne display a multicultural panorama unimagined just two generations ago. All these changes have stirred political debate. Issues ranging from immigration quotas to indigenous land rights dominate, exposing social fault lines (city versus Outback in Australia; North versus South Island in New Zealand). Aborigines and Maori were first in this realm, then the Europeans arrived, and now Asians are a significant economic and cultural element.

LAND AND ENVIRONMENT

Physiographic contrasts between massive, compact Australia and elongated, fragmented New Zealand are related to their locations with respect to the earth's tectonic plates (consult Fig. G-5). Australia, with some of the geologically most ancient rocks on the planet, lies at the center of its own plate, the Australian Plate. New Zealand, younger and less stable, lies at the convulsive convergence of the Australian and Pacific plates. Earthquakes are rare in Australia and volcanic eruptions are unknown; New Zealand has plenty of both. This locational contrast is also reflected by differences in relief (Fig. 11-1). Australia's highest relief occurs in what Australians call the Great Dividing Range, the mountains that line the east coast from the Cape York Peninsula to southern Victoria, with an outlier in Tasmania. The highest point along these old, now eroding mountains is Mount Kosciusko, 2230 meters (7316 ft) tall. In New Zealand, entire ranges are higher than this— Mount Cook, for example, reaches 3754 meters (12,316 ft).

West of Australia's Great Dividing Range, the physical landscape generally has low relief, with some local exceptions such as the Macdonnell Ranges near the center; plateaus and plains dominate (Fig. 11-1). The Great Artesian Basin is a key physiographic region, providing underground water sources in what is otherwise desert country; to the

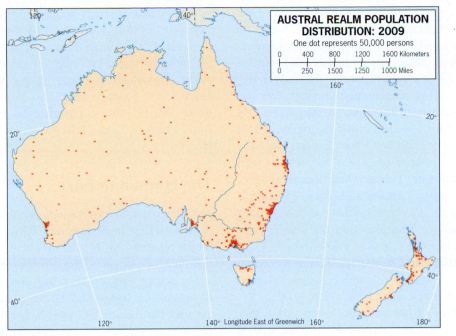

FIGURE 11-2   © H. J. de Blij, P. O. Muller, and John Wiley & Sons, Inc.

south lies the continent's predominant river system, the Murray-Darling. The area mapped as *Western Plateau and Margins* in Figure 11-3 contains much of Australia's mineral wealth.

Climates

Figure G-9 reveals the effects of latitudinal position and interior isolation on Australia's climatology. In this respect, Australia is far more varied than New Zealand, its climates ranging from tropical in the far north, where rainforests flourish, to Mediterranean in parts of the south. The interior is dominated by desert and steppe conditions, the steppes providing the grasslands that sustain tens of millions of livestock. Only in the east does Australia have an area of humid temperate climate, and here lies most of the country's economic core area. New Zealand, by contrast, is totally under the influence of the Southern and Pa-

cific oceans, creating moderate, moist conditions, temperate in the north and colder in the south.

The Southern Ocean

Twice now we have referred to the **2** **Southern Ocean**, but try to find this ocean on maps and globes published by famous cartographic organizations such as the National Geographic Society and Rand McNally. From their maps you would conclude that the Atlantic, Pacific, and Indian oceans reach all the way to the shores of Antarctica. Australians and New Zealanders know better. They experience the frigid waters and persistent winds of this great weather-maker on a daily basis.

For us geographers, it is a good exercise to turn the globe upside down now and then. After all, the usual orientation is quite arbitrary. Modern mapmaking started in the Northern Hemisphere, and the cartogra-

phers put their hemisphere on top and the other at the bottom. That is now the norm, and it can distort our view of the world. In bookstores in the Southern Hemisphere, you see upside down maps showing Australia and Argentina at the top, and Europe and Canada at the bottom. But this matter has a serious side. A reverse view of the globe shows us how vast the ocean encircling Antarctica is. The Southern Ocean may be remote, but its existence is real.

Where do the northward limits of the Southern Ocean lie? This ocean is bounded not by land but by a marine transition called the **3** **Subtropical Convergence**. Here the cold, extremely dense waters of the Southern Ocean meet the warmer waters of the Atlantic, Pacific, and Indian oceans. It is quite sharply defined by changes in temperature, chemistry, salinity, and marine fauna. Flying over it, you

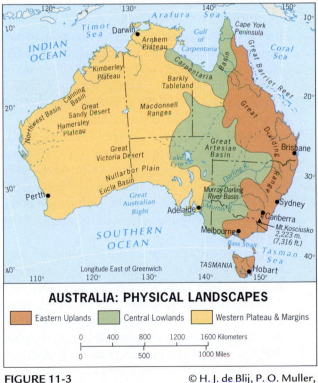

AUSTRALIA: PHYSICAL LANDSCAPES

Eastern Uplands   Central Lowlands   Western Plateau & Margins

FIGURE 11-3   © H. J. de Blij, P. O. Muller, and John Wiley & Sons, Inc.

can actually observe it in the changing colors of the water: the Antarctic side is a deep gray, the northern side a greenish blue.

Although the Subtropical Convergence moves seasonally, its position does not vary far from latitude 40° South, which also is the approximate northern limit of Antarctic icebergs. Defined this way, the great Southern Ocean is a huge body of water that moves clockwise (from west to east) around Antarctica, which is why we also call it the **4** **West Wind Drift**.

Biogeography

One of this realm's defining characteristics is its wildlife. Australia is the land of kangaroos and koalas, wallabies and wombats, possums and platypuses. These and numerous other *marsupials* (animals whose young are born very early in their development and then are carried in an abdominal pouch) owe their survival to Australia's early isolation during the breakup of Gondwana (see Fig. 6-3). Before more advanced mammals could enter Australia and replace the marsupials, as happened in other parts of the world, the landmass was separated from Antarctica and India, and today it contains the world's largest assemblage of marsupial fauna.

Australia's vegetation also has distinctive qualities, notably the hundreds of species of eucalyptus trees native to this geographic realm. Many other plants form part of Australia's unique flora, some with unusual adaptation to the high temperatures and low humidity that characterize much of the continent.

The study of fauna and flora in spatial perspective combines the disciplines of biology and geography in a field known as **5** **biogeography**, and Australia is a giant laboratory for biogeographers. In the Introduction, we noted that several of the world's climatic zones are named after the vegetation that marks them: tropical savanna, steppe, tundra. When climate, soil, vegetation, and animal life reach a long-term, stable adjustment, vegetation forms the most visible element of this ecosystem.

Biogeographers are especially interested in the distribution of plant and animal species, as well as in the relationships between plant and animal communities and their natural environments. (The study of plant life is called *phytogeography*; the study of animal life is called *zoogeography*.) These scholars seek to explain the distributions the map reveals. In 1876 one of the founders of biogeography, Alfred Russel Wallace, published a book entitled *The Geographical Distribution of Animals* in which he fired the first shot in a long debate: where does the zoogeographic boundary of Australia's fauna lie? Wallace's fieldwork in the area revealed that Australian forms exist not only in Australia itself but also in New Guinea and in some islands to the northwest. As already introduced in Chapter 10, Wallace proposed that the faunal boundary should lie between Borneo and Sulawesi, and just east of Bali (Fig. 11-4).

6 **Wallace's Line** soon was challenged by other researchers, who found species Wallace had missed and who visited islands Wallace had not. There was no question that Australia's zoogeographic realm ended somewhere in the Indonesian archipelago, but where? Western Indonesia was the habitat of non-marsupial animals such as tigers, rhinoceroses, and elephants, as well as primates; New Guinea clearly was part of the realm of the marsupials. How far had the more advanced mammals progressed eastward along the island stepping stones toward New Guinea? The zoogeographer Max Weber found evidence that led him to postulate his own *Weber's Line*, which, as Figure 11-4 shows, lay very close to New Guinea.

The Human Impact

Not all research in zoogeography or phytogeography deals with such large questions. Much of it focuses on the relationships between particular species and their habitats—that is, the environment they normally occupy and of which they form a part. Such environments change, and the changes can spell disaster for the species. Climate change forces species to advance or retreat and is a topic often investigated by biogeographers. But it is not always clear whether environmental change or human intervention caused the problem. In Australia, the arrival of the **7** **Aboriginal population** (about 50,000 years ago) appears

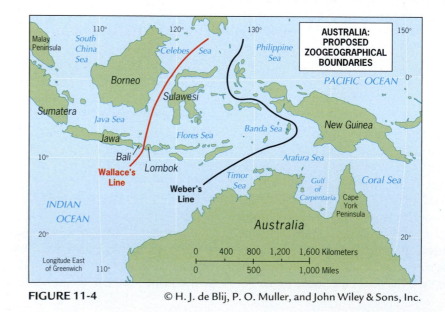

FIGURE 11-4

© H. J. de Blij, P. O. Muller, and John Wiley & Sons, Inc.

to have caused an ecosystem collapse because there is no evidence of significant climate change at the time. In an article in *Science*, Gifford H. Miller and his colleagues report that widespread burning of the existing forest, shrub, and grassland vegetation across Australia appears to have led to the spread of desert scrub and to the rapid extinction of most of the continent's large mammals soon after the human invasion occurred. The species that survived faced a second crisis when European colonizers introduced their livestock, leading to the further destruction of remaining wildlife habitats. Survivors include marsupials such as the koala bear and the wombat, but the list of extinctions is far longer.

WHAT'S DRIVING GEOGRAPHIC CHANGE IN THE REALM

- **Aboriginal** and **Maori** activism and demands for compensation including land claims are challenging the status quo and require creative solutions.

- Proximity to the world's largest Islamic state (Indonesia) and changing demographics as Asian and Pacific-islander immigrants arrive in the realm are changing the **cultural fabric** of the Austral Realm.

- Australia will continue to dominate this realm as it has a **strong and resilient economy** in a favorable location on the Pacific Rim. Its new prime minister is a China expert and fluent in Mandarin, and will strengthen ties to Asia rather than elsewhere.

- **Drought** continues to be an issue in Australia, likely made worse by global climate change, and is challenging the productive capacity of the agricultural sector.

Regions of the Realm

Australia is the dominant component of the Austral Realm, a continent-scale country in a size category that also includes China, Canada, the United States, and Brazil. For two reasons, however, Australia has fewer regional divisions than do the aforementioned countries: the relatively uncomplicated physiography of Australia and its diminutive human numbers. Our discussion, therefore, uses the core-periphery concept as a basis for investigating Australia and focuses on New Zealand as a region by itself.

AUSTRALIA

On January 1, 2001, Australia celebrated its 100th birthday as a state, the Commonwealth of Australia, recognizing (still) the British monarch as the head of state and entering its second century with a strong economy, stable political framework, high standard of living for most of its people, and favorable prospects ahead. Positioned on the Pacific Rim, nine-tenths as large as the 48 contiguous U.S. States, well endowed with farmlands and vast pastures, rivers, underground water, minerals, and energy resources,

served by good natural harbors, and populated by 20.9 million mostly-well-educated people, Australia is one of the most fortunate countries in the world.

Not everyone in Australia shares adequately in all this good fortune, however, and the less advantaged made their voices heard during the celebrations. The country's indigenous (Aboriginal) population, though a small minority today of about 450,000, remains disproportionately disadvantaged in almost every way, from lower life expectancies to higher unemployment than average, from lower high school graduation rates to higher imprisonment ratios. But the nation is now embarked on a campaign to address these ills, with a formal apology issued by the government in early 2008; its conciliatory actions range from enhanced social services for Aboriginals to favorable court decisions involving Aboriginal land claims.

When Australia was born as a federal state, its per-capita GNI was the highest in the world. Australia's bounty fueled a huge flow of exports to Europe, and Australians prospered. That golden age could not last forever, and eventually the country's share of world trade declined. Still, Australia today

ranks among the top 15 countries in the world in terms of GNI, and for the vast majority of Australians life is comfortable. In terms of the indicators of development discussed in Chapter 9, Australia is far ahead of all its western Pacific Rim competitors except Japan. As Australians celebrated their first century, they were, on average, earning far more than Thais, Malaysians, or Koreans. In terms of consumption of energy per person, number of automobiles and kilometers of roads, levels of health, and literacy, Australia is a high-income developed country.

Distance

Australians often talk about distance. One of their leading historians, Geoffrey Blainey, labeled it a tyranny—an imposed remoteness from without and a divisive part of life within. Even today, Australia is far from nearly everywhere on earth. A jet flight from Los Angeles to Sydney takes 14 hours nonstop and is correspondingly expensive. Freighters carrying products to European markets take ten days to two weeks to get there. Inside Australia, distances also are of continental proportions, and Australians pay the price—literally. Until some upstart private airlines started a price war, Australians paid more per mile for their domestic flights than air passengers anywhere else in the world.

But distance also was an ally, permitting Australians to ignore the obvious. Australia was a British progeny, a European outpost. Once you had arrived as an immigrant from Britain or Ireland, there were a wide range of environments, magnificent scenery, vast open spaces, and seemingly limitless opportunities. When the Japanese Empire expanded, Australia's remoteness saved the day. When immigration became an issue, Australia in its comfortable isolation could adopt an all-white admission policy that was not officially terminated until 1976. When boat people by the hundreds of thousands fled Vietnam in the mid-1970s aftermath of the Indochina War, almost none reached Australian shores.

Immigrants

Today Australia is changing and rapidly so. Immigration policy now focuses on the would-be immigrants' qualifications, skills, financial status, age, and facility with the English language. With regard to skills, high-technology specialists, financial experts, and medical personnel are especially welcome. Relatives of earlier immigrants, as well as a quota of genuine asylum-seekers, also are admitted. In recent years, total immigration has been about 120,000 annually, which keeps the country's population growing. According to the latest data, Australia's declining natural rate of increase today stands at 0.6 percent. The country is fast becoming a truly multicultural society. In Sydney, for instance, one in five residents is now of Asian ancestry. Overall, a quarter of Australia's 2008 population is foreign-born, and another quarter consists of first-generation Australians.

Core and Periphery

As Figure 11-2 shows, Australia is a large landmass, but its population is heavily concentrated in a core

area that lies in the east and southeast, most of which faces the Pacific Ocean (here named the Tasman Sea between Australia and New Zealand). Figure 11-5 shows that this crescent-like Australian heartland extends from north of the city of Brisbane to the vicinity of Adelaide and includes the largest city, Sydney, the capital, Canberra, and the second-largest city, Melbourne. A secondary core area has developed in the far southwest, centered on Perth and its outport, Fremantle. Beyond lies the vast periphery, which the Australians call the **8** **Outback**.

To better understand the evolution of this spatial arrangement, it helps to refer again to the map of world climates (Fig. G-9). Environmentally, Australia's most favored strips face the Pacific and Southern oceans, and they are not large. We can describe the country as a coastal rimland with cities, towns, farms, and forested slopes giving way to the vast, arid, interior Outback. On the western flanks of the Great Dividing Range lie the extensive grassland pastures that catapulted Australia into its first commercial age—and on which still graze one of the

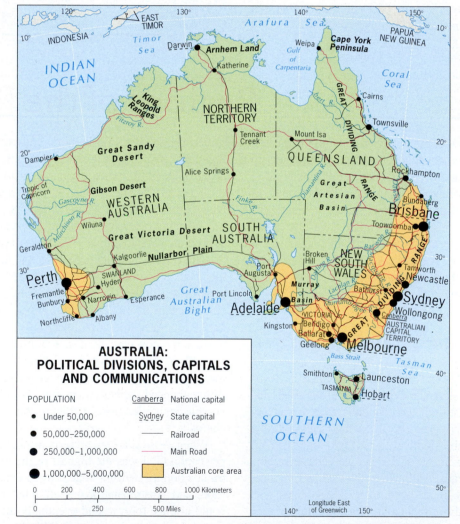

FIGURE 11-5

© H. J. de Blij, P. O. Muller, and John Wiley & Sons, Inc.

largest sheep herds in the world (over 160 million sheep, producing more than one fifth of all the wool sold in the world). Where it is moister, to the north and east, cattle by the millions graze on ranchlands. This is frontier Australia, over which livestock have ranged for nearly two centuries.

Aborigines and the British

Aboriginal Australians reached this landmass as long as 50,000 years ago, crossed the Bass Strait into Tasmania, and had developed a patchwork of indigenous cultures when Captain Arthur Phillip sailed into what is today Sydney Harbor (1788) to establish the beginnings of modern Australia. The Europeanization of Australia doomed the continent's Aboriginal societies. The first to suffer were those situated in the path of British settlement on the coasts, where penal colonies and free towns were founded. Distance protected the Aboriginal communities of the northern interior longer than elsewhere; but in Tasmania, the indigenous Australians died off in just decades after having lived there for perhaps 45,000 years.

The Seven Colonies

Eventually, the major coastal settlements became the centers of seven different colonies, each with its own hinterland; by 1861, Australia was delimited by its now-familiar pattern of straight-line boundaries (Fig. 11-5). Sydney was the focus for New South Wales; Melbourne, Sydney's rival, anchored Victoria. Adelaide was the heart of South Australia, and Perth lay at the core of Western Australia. Brisbane was the nucleus of Queensland, and Hobart was the seat of government in Tasmania. The largest clusters of surviving Aboriginal people were in the so-called Northern Territory, with Darwin, on Australia's tropical north coast, its colonial city. Notwithstanding their shared cultural heritage, the Australian colonies were at odds not only with London over colonial policies but also with each other over economic and political issues. The building of an Aus-

tralian nation during the late nineteenth century was a slow and difficult process.

A Federal State

On January 1, 1901, following years of difficult negotiations, the Australia we know today finally emerged: the Commonwealth of Australia, consisting of six States and two Federal Territories (Fig. 11-5). The two Federal Territories are the *Northern Territory*, assigned to protect the interests of the large Aboriginal population concentrated there and agitating for statehood, and the *Australian Capital Territory*, carved from southern New South Wales to accommodate the federal capital of Canberra, inaugurated in 1927.

Australia's six States are New South Wales (capital Sydney), at 7.1 million the most populous and politically powerful; Queensland (Brisbane), with the Great Barrier Reef offshore and tropical rainforests in its north; Victoria (Melbourne), small but populous by Australian standards with 5.3 million inhabitants; South Australia (Adelaide), where the Murray-Darling river system reaches the sea; Western Australia (Perth) with barely 2 million people in an area of nearly 2.5 million square kilometers (1 million sq mi); and Tasmania (Hobart), the island across the Bass Strait from the mainland and in the path of the storms of the Southern Ocean.

Successful Federation

In earlier chapters, we have referred to the concept of federalism, an idea with ancient Greek and Roman roots familiar to Americans, Canadians, South Asians, and, more recently, Russians. It is a notion of communal association whose name comes from the Latin *foederis*, implying alliance and coexistence, a union of consensus and common interest—a **9 federation**. It stands in contrast to the idea that states should be centralized, or unitary. For this, too, the ancient Romans had a term: *unitas*, meaning unity. Most European countries are **10 unitary states**, in-

cluding the United Kingdom of Great Britain and Northern Ireland. Although the majority of Australians came from that tradition (a kingdom, no less), they managed to overcome their differences and establish a Commonwealth that was, in effect, a federation of States with different viewpoints, economies, and objectives, separated by vast distances along the rim of an island continent.

An Urban Culture

Despite the vast open spaces and romantic notions of frontier and Outback, Australia is an urban country, with 91 percent of all Australians living in cities and towns. As noted earlier, the core of Australia lies in the southeast of the continent (Fig. 11-5). Yet for all its vastness and youth, Australia has developed a remarkable cultural identity, a sameness of urban and rural landscapes that persists from one end of the continent to the other. Sydney, often called the New York of Australia, lies on a spectacular estuarine site, its compact, high-rise central business district overlooking a port bustling with ferry and freighter traffic (photo p. 373). Sydney is a vast, sprawling metropolis of 4.5 million, with multiple outlying centers studding its far-flung suburbs; brash modernity and reserved British ways blend here. Melbourne (3.7 million), sometimes regarded as the Boston of Australia, prides itself on its more interesting architecture and more cultured ways. Brisbane, the capital of Queensland, which also anchors Australia's Gold Coast and adjoins the Great Barrier Reef, is the Miami of Australia; unlike Miami, however, its residents can find nearby relief from the summer heat in the mountains of its immediate hinterland (as well as at its beaches). Perth, Australia's San Diego, is one of the world's most isolated cities, separated from its nearest Australian neighbor by two-thirds of a continent and from Southeast Asia and Africa by thousands of miles of ocean.

And yet, each of these cities—as well as the capitals of South Australia (Adelaide), Tasmania (Hobart), and, to a lesser extent, the Northern

Territory (Darwin)—exhibits an Australian character of unmistakable quality. Life is orderly and unhurried. Streets are clean, slums are few, graffiti rarely seen. By American and even European standards, violent crime (though rising) is uncommon. Standards of public transport, city schools, and health-care provision are high. Spacious parks, pleasing waterfronts, and plentiful sunshine make Australia's urban life more acceptable than that almost anywhere else in the world. Critics of Australia's way of life say that this very pleasant state of affairs has persuaded Australians that hard work is not really necessary. But a

Both Australia and New Zealand are experiencing a large increase in the number of immigrants from Asia, and this is changing the cultural make-up of both societies. This street scene from Federation Square in Melbourne, Australia, underscores some of this diversity.   (© Alamy)

few days' experience in the commercial centers of the major cities contradicts that assertion: the pace of life is quickening. The country's cultural geography evolved as that of a European outpost, prosperous and secure in its isolation. Now Australia is reinventing itself as a major link in an Australo-Asian chain, a Pacific partner in a transformed regional economic geography.

Economic Geography

Because of its early history as a treasure trove of raw materials, Australia to this day is seen worldwide as a country whose economy depends primarily on its exportable natural resources. But in fact, Australia's economy depends overwhelmingly on services, not commodity exports. Tourism alone contributes around 5 percent—about the same as the value of mineral exports. The action is in those bustling coastal cities far more than in the Outback.

When Australia became established as a state, however, it needed goods from overseas, and here the tyranny of distance played a role. Imports from Britain (and later the United States) were expensive mainly because of transport costs. This encouraged local entrepreneurs to set up their own industries to produce such goods more cheaply. Economic geographers call such industries **11** **import-substitution industries**, and this is how local industrialization got its start.

When the prices of foreign goods became lower because transportation was more efficient and therefore cheaper, local businesses demanded protection from the colonial governments, and high tariffs were erected against imported goods. Local products now could continue to be made inefficiently because their market was guaranteed. If Japan could not afford this, how could Australia? We can see the answer on the map (Fig. 11-6). Even before federation in 1901, all the colonies could export valuable minerals whose earnings shored up those inefficient, uncompetitive local in-

dustries. By the time the colonies unified, pastoral industries were also contributing income. So the miners and the farmers paid for those imports Australians could not produce themselves, plus the products made in the cities. No wonder the cities grew: here were secure manufacturing jobs, jobs in state-run service enterprises, and jobs in the growing government bureaucracy. When we noted earlier that Australians once had the highest per-capita GNI in the world, this was initially achieved in the mines and on the farms, not in the cities.

Agricultural Abundance

Australia has material assets of which other countries on the Pacific Rim can only dream. In agriculture, sheep-raising was the earliest commercial venture, but it was the technology of refrigeration that brought world markets within reach of Australian beef producers. Wool, meat, and wheat have long been the country's big three income earners; Figure 11-6 displays the vast pastures in the east, north, and west that constitute the ranges of Australia's huge herds. The zone of commercial grain farming forms a broad crescent extending from northeastern New South Wales through Victoria into South Australia, and covers much of the hinterland of Perth. Keep in mind the scale of this map: Australia is only slightly smaller than the 48 contiguous States of the United States! Commercial grain farming in Australia is big business. As the climatic map (Fig. G-9) would suggest, sugarcane grows along most of the warm, humid coastal strip of Queensland, and Mediterranean crops (including grapes for Australia's highly successful and expanding wine industry) cluster in the hinterlands of Adelaide and Perth. Mixed horticulture concentrates in the basin of the Murray River system, including rice, grapes, and citrus fruits, all under irrigation. And, as elsewhere in the world, dairying has developed near the large urban areas. With its considerable range of environments, Australia yields a diversity of crops.

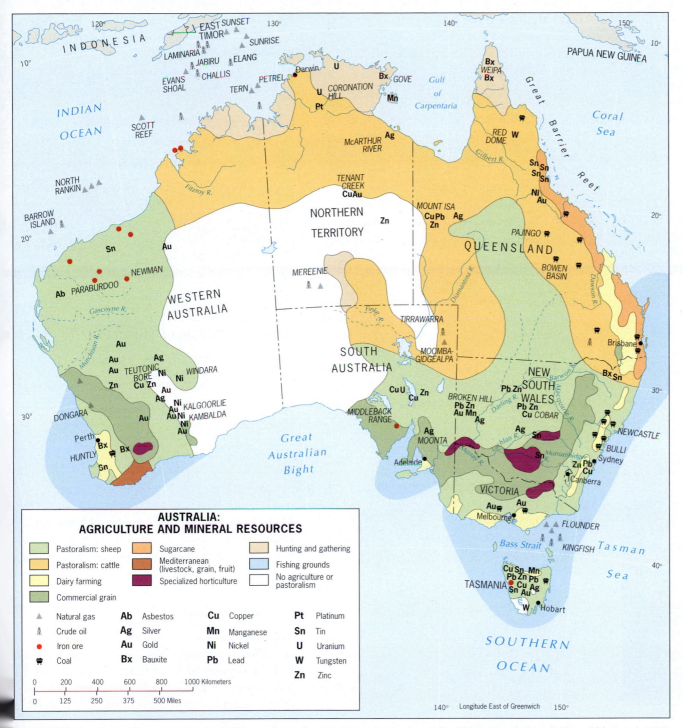

**AUSTRALIA:
AGRICULTURE AND MINERAL RESOURCES**

- Pastoralism: sheep
- Pastoralism: cattle
- Dairy farming
- Commercial grain
- Sugarcane
- Mediterranean (livestock, grain, fruit)
- Specialized horticulture
- Hunting and gathering
- Fishing grounds
- No agriculture or pastoralism

Symbol	Mineral		Symbol	Mineral		Symbol	Mineral
▲ Natural gas		**Ab**	Asbestos	**Cu**	Copper	**Pt**	Platinum
Crude oil		**Ag**	Silver	**Mn**	Manganese	**Sn**	Tin
● Iron ore		**Au**	Gold	**Ni**	Nickel	**U**	Uranium
Coal		**Bx**	Bauxite	**Pb**	Lead	**W**	Tungsten
						Zn	Zinc

0   200   400   600   800   1000 Kilometers
0   125   250   375   500 Miles

140°   Longitude East of Greenwich   150°

FIGURE 11-6   Adapted (in part) and updated with permission from *The Jacaranda Atlas*, 4 rev. ed., p. 29. Jacaranda Press (Milton, Qld., Aust.), 1992.

Mineral Wealth

Australia's mineral resources, as Figure 11-6 shows, also are diverse. Major gold discoveries in Victoria and New South Wales produced a ten-year gold rush starting in 1851 and ushered in a new economic era. By the middle of that decade, Australia was producing 40 percent of the world's gold. Subsequently, the search for more gold led to the discoveries of other minerals. New finds are still being made today, and even oil and natural gas have been found both inland and offshore (see the symbols in Fig. 11-6 in the Bass Strait between Tasmania and the mainland, and off the northwestern coast of Western Australia). Coal is mined at many locations, notably in the east near Sydney and Brisbane but also in Western Australia and even in Tasmania; before coal prices fell, this was a valuable export. Major deposits of metallic and non-metallic minerals abound—from the complex at Broken Hill and the mix of minerals at Mount Isa to the huge nickel deposits at Kalgoorlie and Kambalda, the copper of Tasmania, the tungsten and bauxite of northern Queensland, and the asbestos of Western Australia. A closer look at the map reveals the wide distribution of iron ore (the red dots), and for this raw material as for many others, Japan was Australia's best customer for many years. Today, China has become Australia's leading customer.

Manufacturing's Limits

Australian manufacturing, as we noted earlier, remains oriented to domestic markets. Australian automobiles, electronic equipment, and cameras are not challenging the Pacific Rim's economic tigers for a place on world markets—not yet, at any rate. Australian manufacturing is diversified, producing some machinery and equipment made of locally produced steel as well as textiles, chemicals, paper, and many other items. These industries cluster in and near the major urban areas where the markets are. The domestic market in Australia is not large, but it remains relatively affluent. This makes it attractive to foreign producers, and Australia's shops are full of high-priced goods from Japan, South Korea, Taiwan, and Hong Kong. Indeed, despite its long-term protectionist practices, Australia still does not produce many goods that could be manufactured at home. Overall, the economy continues to display symptoms of a still-developing rather than a fully developed country.

In early 2008, newly elected Prime Minister Kevin Rudd offered a formal apology to the Aboriginal people of Australia for the mistreatment they suffered in the past. This is considered to be an act of enormous symbolic importance; however, the apology will unleash requests for reparations, which will be much harder to provide.
(© Mick Tsikas/Reuters/Zuma Press)

Australia's Future

The Commonwealth of Australia is changing, but its neighbors in Southeast and South Asia are changing even faster. Australia's European bonds are weakening as its Asian ties are strengthening and it plays a bigger role in the Pacific Rim. Australia does face some challenges at home as well. These include: (1) Aboriginal claims, (2) concerns over immigration, (3) environmental degradation, and (4) issues involving Australia's status and regional role.

Aboriginal Issues

For several decades the Aboriginal issues focused on two questions: the formal admission by the government and majority of mistreatment of the Aboriginal minority with official apologies and reparations, and land ownership. The first was resolved in 2008 when newly elected Prime Minister Kevin Rudd offered a formal apology for the historic mistreatment of the Aboriginals. The second question has major geographic implications. Although making up only 2 percent of the total population, the Aboriginal population of 450,000 (including many of mixed ancestry) has been gaining influence in national affairs, and in the 1980s Aboriginal leaders began a campaign to obstruct exploration on what they designated as ancestral and sacred lands. When the Europeans arrived as colonists they had invoked the doctrine of *terra nullis* (land owned by no one) and occupied land that had been used communally by the indigenous people. The situation is similar to what happened in the Americas when Amerindian ideas of land ownership conflicted with the private ownership ideas of the Europeans. Until 1992, Australians had taken it for granted that Aboriginals had no right to land ownership, but in that year the Australian High Court made the first of a series of rulings in favor of Aboriginal claimants. A subsequent court decision implied that vast areas (probably as

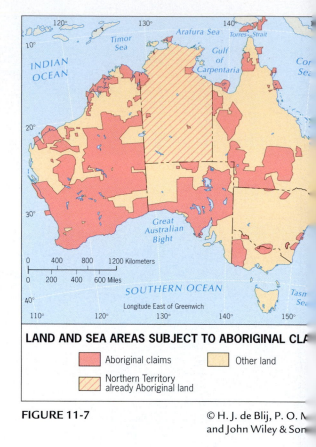

LAND AND SEA AREAS SUBJECT TO ABORIGINAL CLA

- Aboriginal claims
- Other land
- Northern Territory already Aboriginal land

FIGURE 11-7 © H. J. de Blij, P. O. M and John Wiley & Son

much as 78 percent of the whole continent) could potentially be subject to Aboriginal claims (Fig. 11-7). Today the **12 Aboriginal land issue** remains mainly (though not exclusively) an Outback issue, but it has the potential to overwhelm Australia's court system and to inhibit economic growth.

Immigration Issues

The immigration issue is older than Australia itself. Fifty years ago, when Australia had less than half the population it has today, 95 percent of the people were of European ancestry, and more than three-quarters of them came from the British Isles.

Eugenic (race-specific) immigration policies maintained this situation until the 1970s. Today, the picture is dramatically different: of 20.9 million Australians, only about one-third have British-Irish origins, and Asian immigrants outnumber both European immigrants and the natural increase each year. During the early 1990s, nearly 150,000 legal immigrants arrived in Australia annually, most from Hong Kong, Vietnam, China, the Philippines, India, and Sri Lanka. Immigration quotas have since been reduced, most recently to 80,000, but Asian immigrants continue to outnumber those from Western sources. This situation upsets some in the long-dominant Anglo-Australian society, and restrictions on foreign ownership of Australian real estate reflect fears of what extremist groups call the Asianization of the country. But the success of many Asian settlers in Australia evinces the opportunities still available in this free and open society. Sydney, the main recipient of the Asian influx, has become a mosaic of ethnic districts, some of which have gone through periods of gang violence and drug dealing but have stabilized and even prospered over time. Still, multiculturalism will remain a long-term challenge for Australia in the twenty-first century.

Environmental Issues

13 **Environmental degradation** is almost synonymous with Australia. Although the Aboriginals had an impact on the land as well, the European arrival triggered great shocks to the unique Australian environment. Great stands of magnificent forest were destroyed. In Western Australia, centuries-old trees were simply ringed and left to die so that the sun could penetrate through their leafless crowns to nurture the grass below for pasture for introduced livestock. In Tasmania, where Australia's native eucalyptus tree reaches its greatest dimensions (comparable to North American redwood stands), tens of thousands of hectares of this irreplaceable treasure have been lost to chain saws and pulp mills.

ENVIRONMENTALLY SPEAKING . . .

Drought in Australia. As already pointed out, Australia suffers from wide and long-term climate variability. Yet it is a dominantly dry continent, and droughts are to be expected. Indeed, Australia's history is replete with devastating dry spells. Australia is vulnerable to El Niño events, but recent global warming may also be playing a role in the process. Predictions are that drought will become more normal and wet years rare. One such serious drought, regarded as the worst in 1000 years, has been ongoing for the past decade and is devastating Australia's agricultural sector. Even though some rain fell in late 2007, there has been so little overall that the Murray-Darling river system (in New South Wales and South Australia; see Figs. 11-1 and 11-3), Australia's breadbasket, is threatened as water for irrigation is depleted. By some estimates, an area that accounts for 40 percent of Australia's agricultural output and contains 85 percent of its irrigated land is on the verge of ruin. Here a rancher surveys his dwindling water supply near Goulburn, New South Wales.

Part of this situation is caused by nature, but other factors involve the increasing demand for water in Australia's burgeoning urban centers and the excessive damming, well-drilling, and water diversion in the Murray-Darling Basin's upstream tributaries. The Australian government wants to institute a coordinated drainage-basin control program for the Murray-Darling watershed, but this effort requires cooperation from above (the State governments involved) and below (local farmers who survive by diverting water for their crops). So far the plan has not been adopted, but if the drought continues to deepen, Australia's environmental crisis will leave no alternative—and even then it may be too little too late. (© Torsten Blackwood/AFP/Getty Images)

Many of Australia's unique marsupial species have been driven to extinction, and many more are endangered or threatened. "Never have so few people wreaked so much havoc on the ecology of so large an area in so short a time," observed a geographer in Australia recently. But awareness of this environmental degradation is growing. In Tasmania, the Green environmentalist political party has become a force in State affairs, and its activism has slowed deforestation, dam-building, and other development projects. Still, many Australians fear the environmentalist movement as an obstacle to economic growth at a time when the economy needs stimulation. This, too, is an issue for the future.

Issues of Status and Role

Several issues involving Australia's status at home, relations with neighbors, and position in the world are also stirring up national debate. A persistent domestic question is whether Australia should become a republic, ending the status of the British monarch as the head of state, or whether it should continue its status quo in the British Commonwealth.

Relations with neighboring *Indonesia* and *East Timor* have been complicated. For many years, Australia had what may be called a special relationship with Indonesia, whose help it needs in curbing illegal seaborne immigration. It was also profitable for Australia to counter international (UN) opinion and recognize Indonesia's 1976 annexation of Portuguese East Timor, for in doing so Australia could deal directly with Jakarta for the oil reserves under the Timor Sea (see Fig. 10-14). Thus Australia gave neither recognition nor support to the rebel movement that fought for independence in East Timor. But this story had a relatively happy ending. When the East Timorese campaign for independence succeeded in 1999 and Indonesian troops began an orgy of murder and destruction, Australia sent an effective peacekeeping force and spearheaded the United Nations effort to stabilize the situation. Today, a new chapter has

opened in Australia's relations with these northern neighbors.

Australia also has a long-term relationship with *Papua New Guinea (PNG)*, as is noted in Chapter 12. This association, too, has gone through difficult times. In recent years, the inhabitants of Papua New Guinea have strongly resisted privatization, World Bank involvement, and globalization generally. Australia assists PNG in several spheres, but its motives are sometimes questioned. In 2001, the construction of a projected gas pipeline from PNG to the Australian State of Queensland precipitated fighting among tribespeople over land rights, resulting in dozens of casualties, and Australian public opinion reflected doubts over the appropriateness of this venture.

When political violence and chaos overtook the *Solomon Islands* east of Papua New Guinea in 2003, Australian forces intervened: a failing state in Australia's neighborhood could become a base for terrorist activity.

Immediately after the 9/11 attack on U.S. targets in New York and Washington, Australian leaders expressed strong support for the American campaign against terrorism, but they made it a point also to assure the Indonesian government that the War on Terror was not a war on Islam. Although the great majority of Australians supported this stance and troops were sent to Iraq, their will was severely tested in 2002 when a terrorist attack on a nightclub on the Indonesian island of Bali killed 88 vacationing Australians—and again in 2005 when another attack in Bali was followed by the apprehension of a terrorist cell preparing for an assault within Australia itself. The elections of 2007 brought in a new Labor government with a different orientation, especially toward the United States, and Australia's troops have been pulled out of Iraq.

Territorial dimensions, relative location, and raw-material wealth have helped determine Australia's place in the world and, more specifically, on the Pacific Rim. Australia's population, still

barely 20 million in the first decade of the twenty-first century, is smaller than Malaysia's and only slightly larger than that of the Caribbean island of Hispaniola. But Australia's importance in the international community far exceeds its human numbers.

NEW ZEALAND

Fifteen hundred miles east-southeast of Australia, in the Pacific Ocean across the Tasman Sea, lies New Zealand, also known as *Aotearoa* in Maori (meaning 'land of the long white cloud'). In an earlier age, New Zealand would have been part of the Pacific geographic realm because its population was all Maori, a people with Polynesian roots. But New Zealand, like Australia, was invaded and occupied by Europeans. Today, its population of 4.2 million is almost 75 percent European, and the Maori form a substantial minority of about 650,000, with many of mixed Euro-Polynesian ancestry (including Pacific Islanders).

New Zealand consists of two large mountainous islands and many scattered smaller islands (Fig. 11-8). The two large islands, with the South Island somewhat larger than the North Island, look diminutive in the great Pacific Ocean, but together they are larger than Britain. In contrast to Australia, the two main islands are mostly mountainous or hilly, with several peaks rising far higher than any on the Australian landmass. The South Island has a spectacular snowcapped range appropriately called the Southern Alps, with numerous peaks reaching beyond 3300 meters (10,000 ft). The smaller North Island has proportionately more land under low relief, but it also has an area of central highlands along whose lower slopes lie the pastures of New Zealand's chief dairying district. Hence, while Australia's land lies relatively low in elevation and exhibits much low relief, New Zealand's is on the average high and is dominated by rugged terrain.

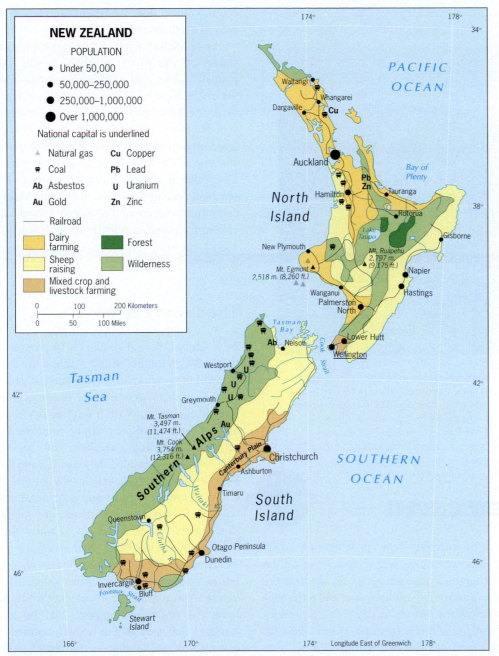

Human Spatial Organization

Thus the most promising areas for habitation are the lower-lying slopes and lowland fringes on both islands. On the North Island, the largest urban area, Auckland, occupies a comparatively low-lying peninsula. On the South Island, the largest lowland is the agricultural Canterbury Plain, centered on Christchurch. What makes these lower areas so attractive, apart from their availability as cropland, is their magnificent pastures. The range of soils and pasture plants allows both summer and winter grazing. Moreover, the Canterbury Plain, the chief farming region, also produces a wide variety of vegetables, cereals, and fruits. About half of all New Zealand is pasture land, and much of the farming provides fodder for the pastoral industry. Sixty million sheep and eight million cattle dominate these livestock-raising activities, with wool, meat, and dairy products providing about two-thirds of the islands' export revenues.

Despite their contrasts in size, shape, physiography, and history, New Zealand and Australia share a number of characteristics. Apart from their joint British heritage, they share a sizeable pastoral economy with growth in specialty goods such as wines, a small local market, the problem of great distances to world markets, and a desire to stimulate (through protection) domestic manufacturing. The high degree of urbanization in New Zealand (86 percent of the total population) again resembles Australia: substantial employment in city-based industries, mostly the processing and packing of livestock and farm products, as well as government jobs.

Spatially, New Zealand also shares with Australia its pattern of **14** **peripheral development** (Fig. 11-2), imposed not by desert but by high rugged mountains and the fragmented nature of the country. The country's major cities—Auckland and the capital of Wellington on the North Island; Christchurch and Dunedin on the South Island—are all located on the coast, and the entire rail and road system is thereby peripheral in its configuration. Moreover,

Tourism in New Zealand. "Tourism has recently overtaken agricultural products as the main earner of foreign exchange for New Zealand. This is not surprising as New Zealand is a wonderful vacation destination, its only drawback being the distance from anywhere else. The climate is mild (the photo above was taken in winter), people are friendly, outdoor activities abound, it is easy to get around. Landscapes such as these dominate; sheep are everywhere, usually framed by beautiful mountains in the background. This photo was taken on the Canterbury Plain in the South Island, one of the most productive agricultural regions in the country, with the magnificent Southern Alps in the background. We had just visited one of the many wineries in the region specializing in white wines, and we were headed through the mountains to the west coast of the South Island." (© A. WinklerPrins)

www.conceptcaching.com

Concept Caching

the two main islands are separated by Cook Strait, a wind-swept waterway that can only be crossed by ferry or air (Fig. 11-8). On the South Island, the Southern Alps are New Zealand's most formidable barrier to surface communications.

The Maori Factor and New Zealand's Future

Like Australia, New Zealand has had a history of difficult relations with its indigenous population.

The Maori, who account for 15 percent of the country's population today, appear to have reached the islands during the tenth century AD. By the time the European colonists arrived, they had had a tremendous impact on the islands' ecosystems, especially on the North Island where most Maori lived. In 1840, the Maori and the British signed a treaty at Waitangi that granted the colonists sovereignty over New Zealand but guaranteed Maori rights over established tribal lands. Although the British abrogated parts of the treaty in 1862, the Maori had rea-

son to believe that vast reaches of New Zealand, as well as offshore waters, were theirs.

As in Australia, judicial rulings during the 1990s supported the Maori position, which led to expanded claims and growing demands. Culturally, the declaration of Maori as an official language in New Zealand and its teaching throughout the school system are seen as significant progress toward an acceptance of the Maori cultural heritage. But one of the Maori's persistent complaints is the slow pace of integration of the minority into modern New Zealand society. Although Maori claims encompass much of the South Island, they also cover prominent sites in the major cities. Today, the Maori question is the leading national issue in the country.

The Green Factor

New Zealand is well known for its progressive politics and high quality of life. Among the factors that contribute to superior environmental progressiveness is its status as one of the leading 'green' societies in the world, with a long-active Green Party and an established program of environmental conservation. Although the Maori and then the European colonists degraded the New Zealand landscape, their descendants have been exemplary in observing environmentally friendly and sustainable policies.

Environmental scientists recently ranked New Zealand as number one (the United States was twenty-eighth) in a report that examined a range of environmental indices such as clean water, air pollution, renewable energy, and biodiversity conservation. Approximately 30 percent of its land area is protected from development, and in 2007 New Zealand declared that it would become the first carbon-neutral country in the world. Already 70 percent of its energy comes from renewable sources (hydro and geothermal), but it aspires to a goal of 90 percent by 2025. This is far beyond what any country has set out to achieve. It is also a nuclear-free country that does not permit even visiting naval vessels with nuclear capacities to stop at its ports. The country has also established Environmental Courts that hear cases involving environmental management decisions. With its green initiatives, New Zealand demonstrates that a country, albeit one with a small population, can work to improve the environment if it possesses the appropriate political will at the highest level.

Dominant cultural heritage and prevailing cultural landscape form two criteria on which the delimitation of the Austral Realm is based. But in both Australia and New Zealand, the cultural mosaic is changing, and the convergence with neighboring realms is proceeding.

New Zealand's landscapes are varied and spectacular. This view shows Akaroa Harbor, on the Banks Peninsula just south of Christchurch on the South Island. The harbor is the crater of an ancient volcano filled in by seawater. This area was settled by the French before it was taken over by the British.   (© A. WinklerPrins)

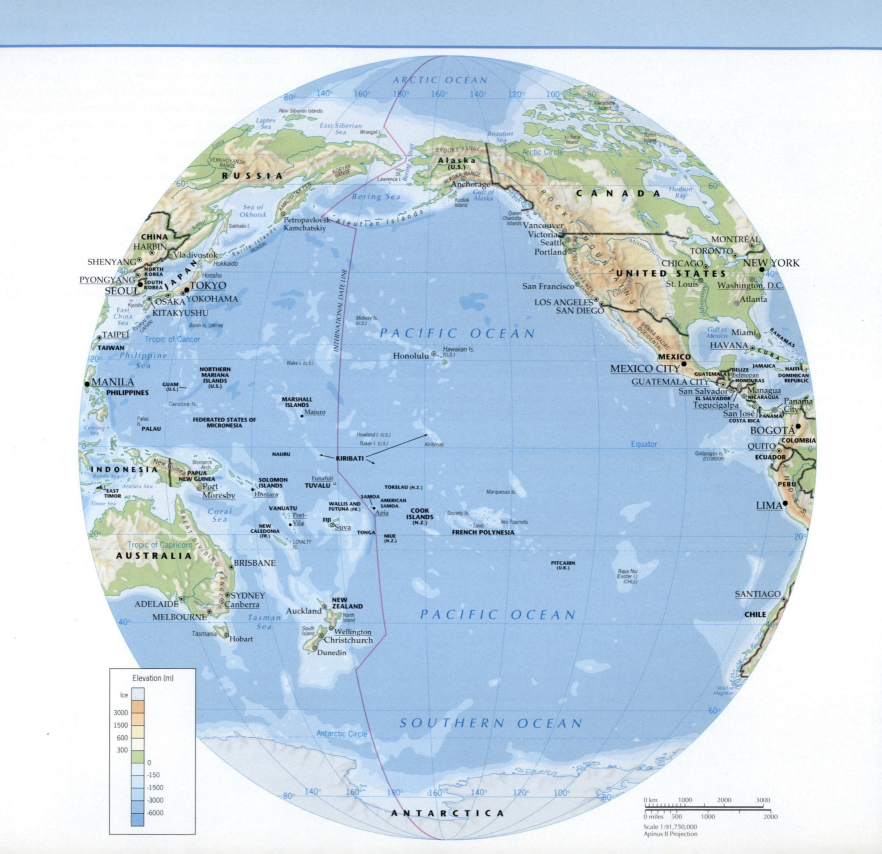

12

PACIFIC REALM

In This Chapter

- Water, water everywhere—but who owns it?
- Islands high and low
- The conundrum of divided Papua
- Colonists and foreigners: Still there, still dominant
- The troubled legacy of Fiji
- Native Hawaiians hope for a sliver of sovereignty

CONCEPTS, IDEAS, AND TERMS

1	Marine geography
2	Territorial sea
3	High seas
4	Continental shelf
5	Exclusive Economic Zone (EEZ)
6	Maritime boundary
7	Median-line boundary
8	High-island cultures
9	Low-island cultures
10	Antarctic Treaty

REGIONS

MELANESIA

MICRONESIA

POLYNESIA

Photos: (*upper left*) Florida Island in the Solomon Islands © National Geographic/Getty Images; (*above*) Bora Bora, French Polynesia © Sylvain Grandadam/Getty Images.

BETWEEN THE AMERICAS to the east and the western Pacific Rim to the west lies the vast Pacific Ocean, larger than all the world's land areas combined. In this greatest of all oceans lie tens of thousands of islands, some large (New Guinea is by far the largest), most small (many are uninhabited). Together, the land area of these islands is a mere 974,000 square kilometers (376,000 sq mi), about the size of Texas plus New Mexico, and over 90 percent of this lies in New Guinea.[1]

Defining the Realm

The Pacific geographic realm—land and water—covers nearly an entire hemisphere of this world, the one commonly called the Sea Hemisphere (Fig. 12-1, see chapter opener map). This Sea Hemisphere meets the Russian and North American realms in the far north and merges into the Southern Ocean in the south. Despite the preponderance of water, this fragmented, culturally complex realm does possess regional identities. It includes the Hawaiian Islands, Tahiti, Tonga, and Samoa—fabled names in a world apart.

In terms of modern cultural and political geography, Indonesia and the Philippines are not part of the Pacific Realm, although Indonesia's political system reaches into it; nor are Australia and New Zealand part of it. Before the European invasion and colonization, Australia would have been included because of its Aboriginal population and New Zealand because of its Maori population's Polynesian affinities. But the Europeanization of their countries has engulfed indigenous Australians and Maori New Zealanders, and the regional geography of Australia and New Zealand today is decidedly not Pacific. In New Guinea, on the other hand, Pacific peoples remain numerically and culturally the dominant element.

COLONIZATION AND INDEPENDENCE

The Pacific islands were colonized by the French, British, and Americans; an indigenous Polynesian kingdom in the Hawaiian Islands was annexed by the United States and is now the fiftieth State. Still, today, the map is an assemblage of independent and colonial territories (Fig. 12-1). Paris controls New Caledonia and French Polynesia. The United States administers Guam and American Samoa, the Line Islands, Wake Island, Midway Islands, and several smaller islands; the United States also has special relationships with other territories, former dependencies that are now nominally independent. The British, through New Zealand, have responsibility for the Pitcairn group of islands, and New Zealand administers and supports the Cook, Tokelau, and Niue Islands. Easter Island, the storied speck of land in the south-

MAJOR GEOGRAPHIC QUALITIES OF The Pacific Realm

1. The Pacific Realm 's total area is the largest of all geographic realms. Its land area, however, is the smallest, as is its population.

2. The island of New Guinea, with 8.8 million people, alone contains over 80 percent of the Pacific Realm's population.

3. The Pacific Realm, with its wide expanses of water and numerous islands, has been strongly affected by United Nations Law of the Sea provisions regarding states' rights over economic assets in their adjacent waters.

4. The highly fragmented Pacific Realm consists of three regions: Melanesia (including New Guinea), Micronesia, and Polynesia.

5. Melanesia forms the link between Papuan and Melanesian cultures in the Pacific.

6. The Pacific Realm's islands and cultures may be divided into volcanic high-island cultures and coral-based low-island cultures.

7. In Micronesia, U.S. influence has been particularly strong and continues to affect local societies.

8. In Polynesia indigenous culture exhibits remarkable consistency and uniformity throughout the region, its enormous dimensions and dispersal notwithstanding. Yet at the same time, local cultures are nearly everywhere severely strained by external influences. In Hawai'i, as in New Zealand, indigenous culture has been largely submerged by Westernization.

[1]The figures in Appendix B do not match these totals because only the political entity of Papua New Guinea is listed, not the Indonesian province of Papua that occupies the western part of the island. Here, as Figure 10-2 shows, the political and the realm boundaries do not coincide.

astern Pacific, is part of Chile. Indonesia rules apua, the western half of the island of New Guinea.

Other island groups have become independent ates. The largest are Fiji, once a British dependency, the Solomon Islands (also formerly British), and Vanuatu (until 1980 ruled jointly by rance and Britain). Also on the current map, however, are such microstates as Tuvalu, Kiribati, Nauru, and Palau. Foreign aid is crucial to the survival of most of these countries. Tuvalu, for example, has a total area of around 25 square kilometers 10 sq mi), a population of some 10,000, and a per-apita GNI of about U.S. $1200, derived from fishing, copra sales (coconut meat used to make oil), nd some tourism. But what really keeps Tuvalu oing is an international trust fund set up by Australia, New Zealand, the United Kingdom, Japan, nd South Korea. Annual grants from that fund, as well as money sent back to families by workers who have left for New Zealand and elsewhere, llow Tuvalu to survive.

THE PACIFIC REALM AND ITS MARINE GEOGRAPHY

Certain land areas may not be part of the Pacific Realm (we cited the Philippines and New Zealand), but the Pacific Ocean extends from the shores of North and South America to mainland East and Southeast Asia and from the Bering Sea to the Subtropical Convergence. This means that several seas, ncluding the Sea of Japan (East Sea), the East China Sea, and the South China Sea, are part of the Pacific Ocean. As we will see below, this relationship matters. Pacific coastal countries, large and small, mainland and island, compete for jurisdiction over the waters that bound them.

The Pacific Realm and its ocean, therefore, form an ideal place to focus on **1 marine geography**. This field encompasses a variety of approaches to the study of oceans and seas; some marine geographers focus on the biogeography of coral reefs, others on the geomorphology of beaches, and still others on the

movement of currents and drifts. A particularly interesting branch of marine geography has to do with the definition and delimitation of political boundaries at sea. Here geography meets political science and maritime law.

The State at Sea

Littoral (coastal) states do not end where atlas maps suggest they do. States have claimed various forms of jurisdiction over coastal waters for centuries, closing off bays and estuaries and ordering foreign fishing fleets to stay away from nearby fishing grounds. Thus arose the notion of the **2 territorial sea**, where all the rights of a coastal state would prevail. Beyond lay the **3 high seas**, free, open, and unfettered by national interests.

It was in the interest of colonizing, mercantile states to keep territorial seas narrow and high seas wide, thus interfering as little as possible with their commercial fleets. In the seventeenth and eighteenth centuries the territorial sea was 3, 4, or at most 6 nautical miles wide, and the colonizing powers claimed the same widths for their colonies (1 nautical mile = 1.85 kilometers or 1.15 statute miles).[2]

Nautical Dimensions	Nautical Miles	English Units	Metric Units
What is a nautical mile?	1	1.15 (statute) mi	1.85 km
Territorial Sea	12	13.8 (statute) mi	19.31 km
Exclusive Economic Zone (EEZ)	200	230 (statute) mi	370 km

[2]Here and in the rest of this section we will give all distances involving maritime boundaries in nautical miles only. Throughout the modern history of maritime law, the latter unit has been the only one used for this type of boundary-making. Any distance stated in nautical miles can be converted to its metric equivalent by multiplying it by 1.85. The table on p. A-1 will aid in this understanding.

In the twentieth century these constraints weakened. States without trading fleets saw no reason to limit their territorial seas. States with nearby fishing grounds traditionally exploited by their own fleets wanted to keep the increasing number of foreign trawlers away. States with shallow **4 continental shelves**, offshore continuations of coastal plains, wished to control the resources on and below the seafloor, made more accessible by improved technology. States disagreed on the methods by which offshore boundaries, whatever their width, should be defined. Early efforts by the League of Nations (the precursor to the UN) in the 1920s to resolve these issues had only partial success, mainly in the technical area of boundary delimitation.

Scramble for the Oceans

In 1945, the United States helped precipitate what has become known as the scramble for the oceans. President Harry S. Truman issued a proclamation that claimed U.S. jurisdiction and control over all the resources "in and on" the continental shelf down to its margin, around 100 fathoms (600 ft or 183 m) deep. In some areas, the shallow continental shelf of the United States extends more than 300 kilometers (200 mi) offshore, and Washington did not want foreign countries drilling for oil just beyond the 3-mile territorial sea.

Few observers foresaw the impact the Truman Proclamation would have, not only on U.S. waters but on the oceans everywhere, including the Pacific. It set off a rush of other claims. In 1952, a group of South American countries, some with little continental shelf to claim, issued the Declaration of Santiago, claiming exclusive fishing rights up to a distance of 200 nautical miles off their coasts. Meanwhile, as part of the Cold War competition, the Soviet Union urged its allies to claim a 12-mile territorial sea.

UNCLOS Intervention

At this point the United Nations intervened, and a series of UNCLOS (United Nations Conference on the Law of the Sea) meetings began. These meetings addressed issues ranging from the closure of bays to the width and delimitation of the territorial sea, and after three decades of negotiations they achieved a convention that changed the political and economic geography of the oceans forever. Among its key provisions were the authorization of a 12-mile territorial sea for all countries and the establishment of a 200-mile (230-statute-mi/370 km)-wide **5 Exclusive Economic Zone (EEZ)** over which a coastal state would have total economic rights. Resources in and under this EEZ (fish, oil, minerals) belong to the coastal state, which could either exploit them or lease, sell, or share them as it saw fit. We already mentioned these zones in Chapter 2 in regard to the

new competition for the Arctic Ocean, but as you will see EEZ's take on even greater importance in this Pacific Realm.

These provisions had a far-reaching impact on the world's oceans and seas (Fig. 12-2), especially the Pacific. Unlike the Atlantic Ocean, the Pacific is studded with islands large and small, and a microstate consisting of one small island suddenly acquired an EEZ covering 166,000 square nautical miles. European colonial powers still holding minor Pacific possessions (notably France) saw their maritime jurisdictions vastly expanded. Small low-income archipelagos could now bargain with large, rich fishing nations over fishing rights in their EEZs. And for all the UNCLOS Convention's provisions

for the "right of innocent passage" of shipping through EEZs and via narrow straits, the world's high seas have obviously been diminished.

Maritime Boundaries

The extension of the territorial sea to 12 nautical miles and the EEZ to an additional 188 nautical miles created new **6 maritime boundary** problems. Waters less than 24 nautical miles wide separate many countries all over the world, so that **7 median lines**, equidistant from opposite shores, have been delimited to establish their territorial seas. And even more countries lie closer than 400 nautical miles apart, requiring further maritime-boundary de-

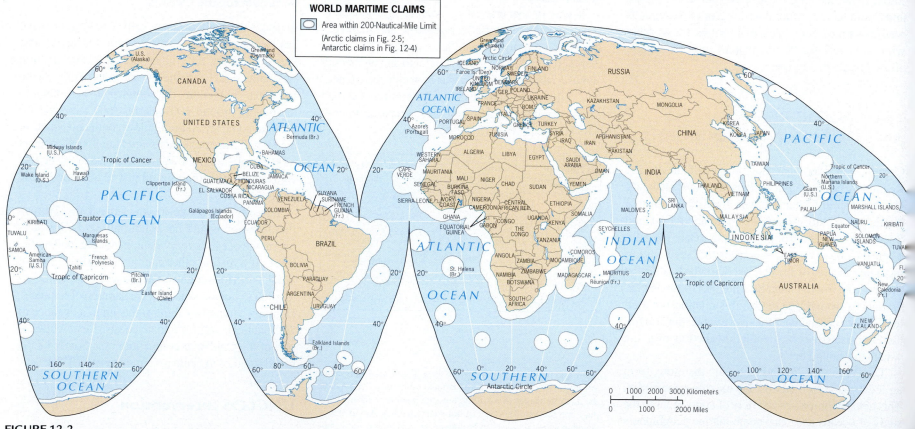

FIGURE 12-2

© H. J. de Blij, P. O. Muller, and John Wiley & Sons, In

limitation to determine their EEZs. In such maritime regions as the North Sea, the Caribbean Sea, and the Japan, East China, and South China seas (and recently the Arctic Ocean), a maze of maritime boundaries emerged, some of them subject to dispute. Political changes on land can lead to significant modifications at sea.

A case in point involves newly independent East Timor and its neighbor across the Timor Sea, Australia. Australia had divided the waters and seafloor of the Timor Sea with Indonesia while recognizing Indonesia's 1976 annexation of East Timor. When East Timor achieved independence in 2002, the so-called *Timor Gap* became an issue: where was the median line that would divide Timorese and Australian claims to the oil and gas reserves in this zone? The Australians argued that since the line defined with Indonesia predated East Timor's independence, it should continue in effect, giving Australia the bulk of the energy resources. But UNCLOS regulations, to which Australia subscribed, required a redelimitation, giving East Timor a much larger share. This led Australia to withdraw from UNCLOS in 2002 so that it would not be bound by its rules. After difficult negotiations, the Australians in 2005 offered East Timor a half share of the energy revenues from the natural gas reserve that was in dispute, in addition to income already flowing from another major gasfield in the so-called Joint Petroleum Development Area. In return, East Timor's government agreed to defer the maritime-boundary issue for 50 years, leaving a future generation to solve this question. It was estimated that, over the next 30 years, East Timor might receive U.S. $13 billion from these sources, and, depending on the life and capacity of the reserves, perhaps even more. This will go a long way toward meeting the young state's most urgent needs.

EEZ Implications

The UNCLOS provisions created opportunities for some states to expand their spheres of influence. Wider territorial-sea and EEZ allocations raised the stakes: claiming an island now entailed potential control over a huge maritime area. In Chapter 2 we highlighted the need to carve up the Arctic as the ice melts and resources beckon. In Chapters 9 and 10, we referred to several island disputes off mainland East Asia, which involve Japan and Russia, Japan and South Korea, Japan and China, China and Vietnam, and China and the Philippines. Ownership of many islands there is uncertain, and small specks of island territory have become large stakes in the scramble for the oceans. For example, in the case of the Spratly Islands (Fig. 10-2), six countries claim ownership, including both Taiwan and China. China's island claims in the South China Sea support Beijing's contention that this body of water is part and parcel of the Chinese state—a position that worries other states with coasts facing it.

Figure 12-2 reveals what EEZ regulations have meant to Pacific Realm countries such as Tuvalu, Kiribati, and Fiji, with nearly circular EEZs now surrounding the clusters of islands in this vast ocean space. Japan, Taiwan, and other fishing nations have purchased fishing rights in these EEZs from the island governments. Nonetheless, violations of EEZ rights do occur; recently, Vanuatu and the Philippines were at odds over unauthorized Filipino fishing in Vanuatu's EEZ.

The process of boundary delimitation continues. In an earlier edition of this book, we included a map of the South China Sea and nearby waters off East and Southeast Asia showing the median-line boundaries delimited according to UNCLOS specifications. But that map changed when coastal states engaged in bilateral and multilateral negotiations—and argued over island ownership with major boundary implications. Indeed, the Pacific (and world) map of maritime boundaries remains a work in progress.

This point is illustrated by a development that occurred late in 1999, when the implications of provisions in Article 76 of the 1982 UNCLOS Convention were reexamined. Where a continental shelf extends beyond the 200-mile limit of the EEZ, the Convention apparently allows a coastal state to claim that extension as a natural prolongation of its landmass. Although there is as yet no regulation that would also allow the state to extend its EEZ to the edge of the continental shelf, it is not difficult to foresee such an amendment to the Convention. As it stands, states are now delimiting their proposed natural prolongations under a 2009 deadline, which puts poorer states at a disadvantage since the required marine surveys are very expensive. Lately, this issue has become quite heated, especially in the Arctic Ocean as countries scramble to claim prolongations. A recent article in *Science* reported that the big winners are likely to include the United States, Canada, Australia, New Zealand, Russia, and India—although a total of 60 coastal countries could ultimately benefit to some extent from the provision.

Thus the scramble for the oceans continues, and with it the constriction of the world's high seas and open waters.

WHAT'S DRIVING GEOGRAPHIC CHANGE IN THE REALM

- Are you following developments in **Hawai'i**? Native Hawaiians, led by Senator Daniel Akaka, are seeking federal recognition of a 'native Hawaiian governing entity' that would not, as some native Hawaiians demand, lead to independence for part of the archipelago but would give them their own office within the federal government, greater authority over their remaining lands, and protection against legal interference.

- **Sea-level rise** as a consequence of global climate change could have a major impact on many of the islands in this realm, especially the low-lying islands of Micronesia, possibly even eliminating or displacing nations.

Regions of the Realm

Sail across the Pacific Ocean, and one spectacular vista follows another. Dormant and extinct volcanoes, sculpted by erosion into basalt spires draped by luxuriant tropical vegetation and encircled by reefs and lagoons, tower over azure waters. Low atolls with nearly snow-white beaches, crowned by stands of palm trees, seem to float in the water. Pacific islanders, where foreign influences have not overtaken them, appear to take life with enviable ease.

More intensive investigations reveal that such Pacific sameness is more apparent than real. Even the Pacific Realm, with its long sailing traditions, its still-diffusing populations, and its historic migrations, has a durable regional framework. Figure 12-3 outlines the three regions that constitute the Pacific Realm:

Melanesia Papua (Indonesia), Papua New Guinea, Solomon Islands, Vanuatu, New Caledonia (France), Fiji

Micronesia Palau, Federated States of Micronesia, Northern Mariana Islands, Republic of the Marshall Islands, Nauru, western Kiribati, Guam (United States)

Polynesia Hawaiian Islands (United States), Samoa, American Samoa, Tuvalu, Tonga, eastern Kiribati, Cook and other New Zealand-administered islands, French Polynesia, Easter Island (Chile)

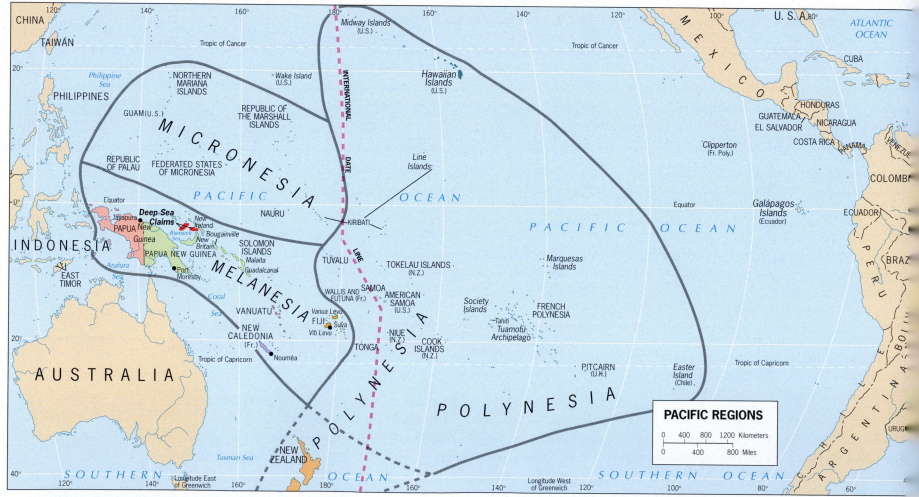

FIGURE 12-3

© H. J. de Blij, P. O. Muller, and John Wiley & Sons, I

Ethnic, linguistic, and physiographic criteria are among the bases for this regionalization of the Pacific realm, but we should not lose sight of the dimensions. Not only is the land area small, but so is the population. In 2008 the population of this entire realm (including Indonesia's Papua Province) was only about 11.5 million—9 million without Papua—about the same as one very large city. Even fewer people live in this realm than in another vast area of far-flung settlements—the oases of North Africa's Sahara.

MELANESIA

The large island of New Guinea lies at the western end of a Pacific region that extends eastward to Fiji and includes the Solomon Islands, Vanuatu, and New Caledonia (Fig. 12-3). The human mosaic here is complex, both ethnically and culturally. Most of the 7.8 million people of New Guinea (including the Indonesian part—the province of Papua—and the independent state of Papua New Guinea) are Papuans, and a large minority is Melanesian. Altogether there are as many as 700 communities speaking different languages; the Papuans are most numerous in the densely forested highland interior and in the south, while the Melanesians inhabit the north and east. The region as a whole has more than 7 million inhabitants, making this the most populous Pacific region by far.

With 6.3 million people today, **Papua New Guinea (PNG)** became a sovereign state in 1975 after nearly a century of British and Australian administration. Almost all of PNG's limited development is taking place along the coasts, while most of the interior remains hardly touched by the changes that transformed neighboring Australia. Perhaps four-fifths of the population lives in the self-sufficient subsistence economy, growing root crops and hunting wildlife, raising pigs, and gathering forest products. Old traditions of the kind lost in Australia persist here, protected by remoteness and the rugged terrain.

Welding this disparate population into a nation is a task hardly begun, and PNG faces numerous obstacles in addition to its cultural complexity. Not only are hundreds of languages in use, but more than half the population is illiterate. English, the official language, is used by the educated minority but is of little use beyond the coastal zone and its towns. The capital, Port Moresby, has just under 350,000 inhabitants, reflecting the low level of urbanization (13 percent) in this developing economy.

Yet Papua New Guinea is not without economic opportunities. Oil was discovered in the 1980s, and by the late 1990s crude oil was PNG's largest export by value. Gold now ranks second, followed by copper, silver, timber, and several agricultural products including coffee and cocoa, reflecting the country's diversity of environments and resources. Pacific Rim developments have affected even PNG: most exports go to the nearest neighbor, Australia, but Japan ranks close behind.

Turning eastward, it is a measure of Melanesia's cultural fragmentation that as many as 120 languages are spoken in the approximately 1000 islands that make up the **Solomon Islands** (about 80 of these islands support almost all the people, numbering about 500,000). Inter-island, historic animosities among the islanders were worsened by the events of World War II, when thousands of Malaitans were moved by U.S. forces to Guadalcanal. This started a postwar cycle of violence that led Australia to intervene in 2003.

New Caledonia, still under French rule, is in a very different situation. Only around 43 percent of the population of about 240,000 are Melanesian; 37 percent are of French ancestry, many of them descended from the inhabitants of the penal colony France established here in the nineteenth century. Nickel mines, based on reserves that rank among the world's largest, dominate New Caledonia's export economy. The mining industry attracted additional French settlers, and social problems arose. Most of the French population lives in or near the capital city of Nouméa, steeped in French cultural landscapes, in the southeastern quadrant of the island. Melanesian demands for an end to colonial rule have led to violence, and the two communities are still in the process of coming to terms.

On its eastern margins, Melanesia includes one of the Pacific Realm's most interesting countries, **Fiji**. On two larger and over 100 smaller islands live nearly 800,000 Fijians, of whom 51 percent are Melanesians and 44 percent South Asians, the latter brought to Fiji from India during the British colonial occupation to work on the sugar plantations. When Fiji achieved independence in 1970, the native Fijians owned most of the land and held political control, while the Indians were concentrated in the towns (chiefly Suva, the capital) and dominated commercial life. It was a recipe for trouble, and it was not long in coming when, in a later election, the politically active Indians outvoted the Fijians for seats in the parliament. A coup by the Fijian military was followed by a revision of the constitution, which awarded a majority of seats to ethnic Fijians. Before long, however, the Fijian majority splintered, but a coalition government for some time proved the constitution workable—until 2000. In 1999, Fiji's first prime minister of Indian ancestry had taken office, angering some ethnic Fijians to the point of staging another coup. The following year the prime minister and members of the government were taken hostage at the parliament building in Suva, and the perpetrators demanded that Fijians would henceforth govern the country.

This action, and Fiji's inability to counter it, had a devastating impact on the country. Foreign trading partners stopped buying Fijian products. The tourist industry suffered severely. In 2006, another coup took place, this time led by the armed forces, and a military government remains in place. The sugar industry was damaged by the nonrenewal of Indian-held leases by Fijian landowners, which also resulted in a substantial movement of Indians from the countryside to the towns, where unemployment is already high. Fiji's future remains clouded.

Melanesia, the most populous region in the Pacific Realm, is bedeviled by centrifugal forces of many kinds. No two countries present the same form of multiculturalism; each has its own challenges to confront, and some of these challenges spill over into neighboring (or more distant foreign) islands.

Sea-Level Rise in the Pacific. The impact of global climate change is multifaceted in this realm. The most obvious impact is seen in the Pacific's low-lying islands (as in Bora Bora above). Especially in Micronesia, islands are at high risk of being flooded by rising sea level as a consequence of global warming. Entire countries might even disappear, which will cause enormous social and economic upheavals as people are displaced, become environmental refugees, and have to be resettled in other countries.

But the residents of this realm face other environmental challenges apart from flooding. Global climate change is already impacting the coral reefs that surround many islands. Coral reefs have been dying, and in the process they are eroding, losing their abilities to buffer coasts from storm surges. Rising sea-surface temperatures are also having an impact on offshore fisheries, with resulting changes in an industry vital to the economies and food security of many nations. Various governments of the microstates of Micronesia have declared the Micronesia Challenge—a conservation effort aimed at confronting environmental threats amplified by climate change. Although conserving marine resources and related activities may not mitigate the major environmental changes that appear inevitable, it is at least a committed local effort to face the reality of the challanges that lie ahead. (© Art Wolfe/Stone/Getty Images)

MICRONESIA

North of Melanesia and east of the Philippines lie the islands that constitute the region known as Micronesia (Fig. 12-3). The name (*micro* means small) refers to the size of the islands: the 2000-plus islands of Micronesia are not only tiny (many of them no larger than 1 square kilometer (half a square mile), but they are also much lower-lying, on average, than those of Melanesia. Some are volcanic islands (**8** high islands, as the people call them), but they are outnumbered by islands composed of coral, the **9** low islands that barely lie above sea level. Guam, with 550 square kilometers (210 sq mi), is Micronesia's largest island, and no island elevation anywhere in Micronesia reaches 1000 meters (3300 ft).

The high-island/low-island dichotomy is useful not only in Micronesia, but also throughout the realm. Both the physiographies of these islands and the economies they support differ in crucial ways. High islands wrest substantial moisture from the ocean air; they tend to be well watered and have good volcanic soils. As a result, agricultural products show some diversity, and life is reasonably secure. Populations tend to be larger on these high islands than on the low islands, where drought is the rule and fishing and the coconut palm are the mainstays of life. Small communities cluster on the low islands, and over time many of these have died out. The major migrations, which sent fleets to populate islands from Hawai'i to New Zealand, tended to originate in the high islands.

Until the mid-1980s, Micronesia was largely a United States Trust Territory (the last of the post-World War II trusteeships supervised by the United Nations), but that status has now changed. As Figure 12-3 shows, today Micronesia is divided into countries bearing the names of independent states. The **Marshall Islands**, where the United States tested nuclear weapons (giving prominence to the name *Bikini*), now is a republic in 'free association' with the United States, having the same status as the Federated States of Micronesia and (since 1994) Palau. The **Northern Mariana Islands** are a commonwealth

'in political union' with the United States. In effect, the United States provides billions of dollars in assistance to these countries, in return for which they commit themselves to avoid foreign policy actions that are contrary to U.S. interests. There are other conditions: **Palau**, for example, granted the United States rights to existing military bases for 50 years following independence.

Also part of Micronesia is the U.S. territory of **Guam**, where independence is not in sight and where U.S. military installations and tourism provide the bulk of income, and the remarkable Republic of **Nauru**. With a population of barely 12,000 and only 20 square kilometers (8 sq mi) of land, the latter got rich by selling its phosphate deposits to Australia and New Zealand, where they are used as fertilizer. Per-capita incomes rose to U.S.$11,500, making Nauru one of the Pacific's high-income societies. But the phosphate deposits have run out, and an island scraped bare now faces an economic crisis.

In this region of tiny islands, most people subsist on farming or fishing, and virtually all the countries need infusions of foreign aid to survive. The natural economic complementarity between the high-island farming cultures and the low-island fishing communities all too often is negated by distance, spatial as well as cultural. Life here may seem idyllic to the casual visitor, but for the Micronesians it often is a daily challenge.

POLYNESIA

To the east of Micronesia and Melanesia lies the heart of the Pacific, enclosed by a great triangle stretching from the Hawaiian Islands to Chile's Easter Island to New Zealand. This is Polynesia (Fig. 12-3), a region of numerous islands (*poly* means many), ranging from volcanic mountains rising above the Pacific's waters (Mauna Kea on Hawai'i reaches over 4200 meters [nearly 13,800 ft]), clothed by luxuriant tropical forests and drenched by well over 250 centimeters (100 in) of rainfall each year, to low coral atolls where a few palm trees form the only vegeta-

ancestries. In the U.S. State of Hawai'i—actually an archipelago of more than 130 islands—Polynesian culture has been both Europeanized and Orientalized.

Its vastness and the diversity of its natural environments notwithstanding, Polynesia clearly constitutes a geographic region within the Pacific Realm. Polynesian culture, though spatially fragmented, exhibits a remarkable consistency and uniformity from one island to the next, from one end of this widely dispersed region to the other. This consistency is particularly expressed in vocabularies, technologies, housing, and art forms. The Polynesians are uniquely adapted to their maritime environment, and long before European sailing ships began to arrive in their waters, Polynesian seafarers had learned to navigate their wide expanses of ocean in huge double canoes as long as 45 meters (150 ft). They traveled hundreds of miles to favorite fishing zones and engaged in inter-island barter trade, using maps constructed from bamboo sticks and cowrie shells and navigating by the stars. However, modern descriptions of a Pacific Polynesian paradise of emerald seas, lush landscapes,

Easter Island has long been known for its large carvings and unique Polynesian culture. Its environment has been much degraded as successive peoples exploited its resources   (© H. J. de Blij)

tion and where drought is a persistent problem. Anthropologists differentiate between these original Polynesians and a second group, the Neo-Hawaiians, who are a blend of Polynesian, European, and Asian

FROM THE FIELD NOTES

Pre-European Hawai'i.   "Hawai'i is a fascinating place to visit with its complex layered cultural geography and its varied physical environment. Here at Honaunau Historical Park on the western side of the big island of Hawai'i, a traditional Hawaiian village has been preserved. Part of Honaunau was a place of refuge as well as a place of absolution for lawbreakers in the past when Hawai'i was a Polynesian kingdom. That is, if they could reach it—the only way to get to the sanctuary was by swimming through strong ocean currents. Today you can reach the site via a road and walk around the *pahoehoe* (ropy) lava tidepools, visit houses and other traditional buildings, and see wood carvings known as *ki'i* made the way they were before 1819 when Europeans arrived on the island and forever changed the cultures of Hawai'i."   (© A. WinklerPrins)

Concept Caching

and gentle people distort harsh realities. Polynesian society was forced to get used to much loss of life at sea when storms claimed their boats; families were ripped apart by accident as well as migration; hunger and starvation afflicted the inhabitants of smaller islands; and the island communities were often embroiled in violent conflicts and cruel retributions.

The political geography of Polynesia is complex. In 1959, the Hawaiian Islands became the fiftieth State to join the United States. The State's population is now 1.4 million, with over 80 percent living on the island of Oahu. There, the superimposition of cultures is symbolized by the panorama of Honolulu's skyscrapers against the famous extinct volcano at nearby Diamond Head. The Kingdom of **Tonga** became an independent country in 1970 after seven decades as a British protectorate; the British-administered Ellice Islands were renamed **Tuvalu**, and along with the Gilbert Islands to the north (now renamed **Kiribati**), they received independence from Britain in 1978. Other islands continue under French control (including the Marquesas Islands and Tahiti), under New Zealand's administration (Rarotonga), and under British, U.S., and Chilean flags.

In the process of politico-geographical fragmentation, Polynesian culture has suffered severe blows. Land developers, hotel builders, and tourist dollars have set **Tahiti** on a course along which Hawai'i has already traveled far. The Americanization of eastern **Samoa** has created a new society different from the old. Polynesia has lost much of its ancient cultural consistency; today, the region is a patchwork of new and old—the new often bleak and barren, with the old under intensifying pressure.

The countries and cultures of the Pacific Realm lie in an ocean on whose rim a great drama of economic and political transformation will play itself out during the twenty-first century. Already, the realm's own former margins—in Hawai'i in the north and in New Zealand in the south—have been so recast by foreign intervention that little remains of the kingdoms and cultures that once prevailed. Now the Pacific Basin faces changes far greater even than those brought here by European colonizers. Once upon a time the shores and waters of the Mediterranean Sea formed an arena of regional transformation that changed the world. Then it was the Atlantic, avenue of the Industrial Revolution and stage of fateful war. Now it seems to be the turn of the Pacific as the world's largest country and next superpower (China) faces the richest and most powerful (the United States). Giants will jostle for advantage in the Pacific; how will the microstates of the Pacific Realm fare?

A FINAL CAVEAT: PACIFIC AND ANTARCTIC

South of the Pacific Realm lies Antarctica and its encircling Southern Ocean. The combined area of these two geographic expanses constitutes 40 percent of the entire planet—two-fifths of the earth's surface containing a mere one one-thousandth of the world's population.

Is Antarctica a geographic realm? In physiographic terms, yes, but not on the basis of the criteria we use in this book. Antarctica is a continent, nearly twice as large as Australia, but virtually all of it is covered by a dome-shaped ice sheet nearly 3.2 kilometers (2 mi) thick near its center. The continent often is referred to as the white desert because, despite all of its ice and snow, annual precipitation is low—less than 15 centimeters (6 in) per year. Temperatures are frigid, with winds so strong that Antarctica also is called the home of the blizzard. For all its size, no functional regions have developed here, nor have any towns or transport networks except the supply lines of research stations. Antarctica still is a frontier, even a scientific frontier still slowly giving up its secrets. Underneath all that ice lie some 70 lakes of which Lake Vostok is the largest with over 14,000 square kilometers (5400 sq mi). It may be as much as 600 meters (2000 ft) deep. No one has yet seen a sample of its water.

Like virtually all frontiers, Antarctica has always attracted pioneers and explorers. Whale and seal hunters destroyed huge populations of Southern Ocean fauna during the eighteenth and nineteenth centuries, and explorers planted the flags of their countries on Antarctic shores. Between 1895 and 1914, the quest for the South Pole became an international obsession; Roald Amundsen, the Norwegian, reached it first in 1911. All this led to national claims in Antarctica during the interwar period (1918–1939).

The geographic effect was the partitioning of Antarctica into pie-shaped sectors centered on the South Pole (Fig. 12-4). In the least frigid area of the continent, the Antarctic Peninsula, British, Argentinian, and Chilean claims overlapped—and still do. One sector, Marie Byrd Land (shown in neutral beige on the main map), was never formally claimed by any country.

Why should states be interested in territorial claims in so remote and difficult an area? As we described in Chapter 2 in the section on the Arctic,

With global environmental change and many people's searches for ever more exotic travel destinations, Antarctica has been popular as a place to visit. The challenge is that its environment remains harsh, and given the international nature of the continent, there is no responsible coast guard. (© AP/Wide World Photos)

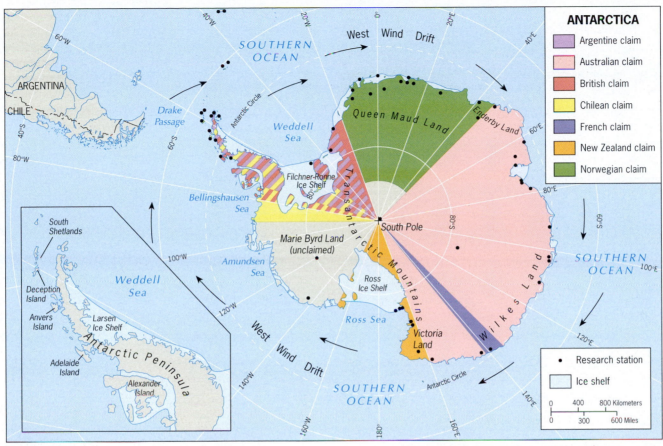

FIGURE 12-4

© H. J. de Blij, P. O. Muller, and John Wiley & Sons, Inc.

these remote areas near the poles contain substantial resources and are attractive for possible future use, despite the difficulties of access. Antarctica is no different as both land and sea contain raw materials that may some day become crucial: proteins in the waters, and fuels and minerals beneath the ice. Antarctica (14.2 million square kilometers [5.5 million sq mi]) is, as already noted, almost twice as large as Australia, and the Southern Ocean is nearly as large as the North and South Atlantic. However distant actual exploitation may be, countries want to keep their stakes here. What is different in the Antarctic from the Arctic is that Antarctica has already been 'carved up' for many decades. During the late 1950s, the various claimant states (those with territorial claims)

recognized the need for cooperation, and they joined in the International Geophysical Year (IGY) that launched major research programs and established a number of permanent research stations throughout the continent. This spirit of cooperation led to the 1961 signing of the **10** **Antarctic Treaty**, which ensures continued scientific collaboration, prohibits military activities, protects the environment, and holds national claims in abeyance. In 1991, when the treaty was extended under the terms of the Wellington Agreement, concerns were raised that it does not do enough to control future resource exploitation. What also remains murky is who patrols the waters of the Antarctic. This became an issue in 2007 when a cruise ship sank off the Antarctic Peninsula (see

photo at left). The region has experienced significant growth in tourism, but it remains a very dangerous place for ships as icebergs abound, and there are no territorial coast guards or other safety patrols.

In an age of growing national self-interest and increasing raw material consumption, the possibility exists that Antarctica and its offshore waters may yet become an arena for international rivalry. Until now, its remoteness and its forbidding environments have saved it from that fate. The entire world benefits from this because evidence is mounting that Antarctica plays a critical role in the global environmental system, so that human modifications may have worldwide (and unpredictable) consequences.

APPENDIX A: METRIC (STANDARD INTERNATIONAL [SI]) AND CUSTOMARY UNITS AND THEIR CONVERSIONS

Appendix A provides a table of units and their conversions from older (British) units to metric (SI) units for the measures of length, area, and temperature that are used in this book.

NOTE: Additional appendices may be found on the book's website at www.wiley.com/college/deblij

Length

Metric Measure

1 kilometer (km)	= 1000 meters (m)
1 meter (m)	= 100 centimeters (cm)
1 centimeter (cm)	= 10 millimeters (mm)

Nonmetric Measure

1 mile (mi)	= 5280 feet (ft)
	= 1760 yards (yd)
1 yard (yd)	= 3 feet (ft)
1 foot (ft)	= 12 inches (in)
1 fathom (fath)	= 6 feet (ft)

Conversions

1 kilometer (km)	= 0.6214 mile (mi)
1 meter (m)	= 3.281 feet (ft)
	= 1.094 yards (yd)
1 centimeter (cm)	= 0.3937 inch (in)
1 millimeter (mm)	= 0.0394 inch (in)
1 mile (mi)	= 1.609 kilometers (km)
1 foot (ft)	= 0.3048 meter (m)
1 inch (in)	= 2.54 centimeters (cm)
	= 25.4 millimeters (mm)

Area

Metric Measure

1 square kilometer (km^2)	= 1,000,000 square meters (m^2)
	= 100 hectares (ha)
1 square meter (m^2)	= 10,000 square centimeters (cm^2)
1 hectare (ha)	= 10,000 square meters (m^2)

Nonmetric Measure

1 square mile (mi^2)	= 640 acres (ac)
1 acre (ac)	= 4840 square yards (yd^2)
1 square foot (ft^2)	= 144 square inches (in^2)

Conversions

1 square kilometer (km^2)	= 0.386 square mile (mi^2)
1 hectare (ha)	= 2.471 acres (ac)
1 square meter (m^2)	= 10.764 square feet (ft^2)
	= 1.196 square yards (yd^2)
1 square centimeter (cm^2)	= 0.155 square inch (in^2)
1 square mile (mi^2)	= 2.59 square kilometers (km^2)
1 acre (ac)	= 0.4047 hectare (ha)
1 square foot (ft^2)	= 0.0929 square meter (m^2)
1 square inch (in^2)	= 6.4516 square centimeters (cm^2)

Temperature

To change from Fahrenheit (F) to Celsius (C)

$$°C = \frac{°F - 32}{1.8}$$

To change from Celsius (C) to Fahrenheit (F)

$$°F = °C \times 1.8 + 32$$

APPENDIX B: AREA AND DEMOGRAPHIC DATA FOR THE WORLD'S STATES

APPENDIX B is a valuable resource and should be consulted throughout your reading. Like all else in this book, Appendix B is subject to continuous revision and modification. In this fourth edition, we have deleted some indices, elaborated others, and introduced new ones. For example, in a world with ever-slower population growth, the so-called Doubling Time index—the number of years it will take for a population to double in size based on its current rate of natural increase—has lost most of its utility. On the other hand, when it comes to **Life Expectancy** and **Literacy**, general averages conceal significant differences between male and female rates, which in turn reflect conditions in individual countries, so we report these by gender. Also in this edition, we continue to use the **Corruption Index**, not available for all countries but an important reflection of a global problem.

The **Big Mac Price** index, a measure introduced by the journal *The Economist*, tells you much more than what a hamburger with all the trimmings would cost in real dollars in various countries of the world—it also reflects whether those countries' currencies are overvalued or undervalued. And the final column, in which we formerly used to reported the per-capita GNP (Gross National Product), now reveals the **GNI**, that is, the Gross National Income per person and what this would buy in each country. In the language of economic geographers, this is called the GNI-PPP, the per-capita Gross National Income in terms of its Purchasing Power Parity.

Indexes that may not be immediately obvious to you include **Arithmetic Population Density**, the number of people per square kilometer in each country; **Physiologic Population Density**, the number of people per square kilometer of agriculturally productive land; **Birth** and **Death Rates** per thousand in the population, resulting in the national population's rate of **Natural Increase**; a population's **Infant Mortality**, the number of deaths per thousand in the first year of life, thus reflecting largely the number of deaths at birth; **Child Mortality**, the deaths per thousand of children in their first five years; the **Corruption Index**, based on Transparency International data in which 10.0 is perfect and 0.1 is the worst; the **Big Mac Price** index, which tells you why China in 2007 was the best place to buy a hamburger in U.S. dollars; and the **Per Capita GNI ($US)**, the GNI-PPP index referred to above, which tells you how spendable income varies around the globe. For additional details on sources and data, please consult the Data Sources section of the Preface.

Area and Demographic Data for the World's States

	Land Area Sq km	Land Area Sq mi	Population 2008 (Millions)	Population 2025 (Millions)	Population Density Arithmetic	Population Density Physiologic	Birth Rate	Death Rate	Natural Increase %	Infant Mortality per 1,000	Child Mortality per 1,000	Life Expectancy Male (years)	Life Expectancy Female (years)	Percent Urban Pop.	Literacy Male %	Literacy Female %	Corruption Index	Big Mac Price ($US)	Per Capita GNI ($US)
WORLD	134,134,451	51,789,601	6707.4	7940.0	50	432	21	9	1.2	52	76	65	69	50	83.4	69.9			**$9,190**
Europe	5,919,355	2,284,950	590.1	587.9	32	254	10	12	−0.1	9	11	71	79	75	99.3	98.0			**$21,120**
Albania	28,749	11,100	3.3	3.5	113	452	14	6	0.8	16	18	72	79	45	95.5	88.0	2.6		$5,420
Austria	83,859	32,378	8.3	8.7	99	550	9	9	0.0	4	5	76	82	66	100.0	100.0	8.6		$33,140
Belarus	207,598	80,154	9.6	9.4	46	149	9	15	−0.6	10	12	63	75	72	99.7	99.2	2.1		$7,890
Belgium	30,528	11,787	10.5	10.8	345	1,379	11	10	0.1	4	5	76	82	97	100.0	100.0	7.3		$32,640
Bosnia	51,129	19,741	3.9	3.7	76	587	9	9	0.1	13	15	71	77	43	96.5	76.6	2.9		$7,790
Bulgaria	110,908	42,822	7.6	6.6	69	167	9	15	−0.5	12	15	69	76	70	99.1	98.0	4.0		$8,630
Croatia	56,539	21,830	4.4	4.3	78	298	9	11	−0.2	6	7	71	78	60	99.4	97.3	3.4		$12,750
Cyprus	9,249	3,571	1.0	1.1	109	681	11	7	0.4	4	5	75	80	66	98.7	95.0	5.6		$22,230
Czech Republic	78,860	30,448	10.3	10.2	130	303	10	11	−0.1	3	4	73	79	74	100.0	100.0	4.8	$2.51	$20,140
Denmark	43,090	16,637	5.4	5.6	126	225	12	10	0.2	4	5	76	80	85	100.0	100.0	9.5	$5.08	$33,570
Estonia	45,099	17,413	1.3	1.2	29	106	11	13	−0.2	6	7	66	78	71	99.9	99.6	6.7		$15,420
Finland	338,149	130,560	5.3	5.4	16	225	11	9	0.2	3	4	75	82	62	100.0	100.0	9.6		$31,170
France	551,497	212,934	61.7	63.4	112	320	13	9	0.4	4	5	77	84	76	98.9	98.7	7.4		$30,540
Germany	356,978	137,830	82.1	82.0	230	657	8	10	−0.2	4	5	76	82	88	100.0	100.0	8.0		$29,210
Greece	131,960	50,950	11.1	11.4	84	280	10	10	0.0	4	5	77	81	61	98.6	96.0	4.4		$23,620
Hungary	93,030	35,919	10.0	9.6	108	196	10	13	−0.3	7	8	69	77	65	99.5	99.3	5.2	$3.33	$16,940
Iceland	102,999	39,768	0.3	0.3	3	296	14	6	0.8	2	3	79	83	93	100.0	100.0	9.6	$7.44	$34,760
Ireland	70,279	27,135	4.3	4.5	61	304	15	7	0.8	5	6	75	80	60	100.0	100.0	7.4		$34,720

| | Land Area | | Population | | Population Density | | Birth Rate | Death Rate | Natural Increase % | Infant Mortality per 1,000 | Child Mortality per 1,000 | Life Expectancy | | Percent Urban Pop. | Literacy | | Corruption Index | Big Mac Price ($US) | Per Capita GNI ($US) |
	Sq km	Sq mi	2008 (Millions)	2025 (Millions)	Arithmetic	Physiologic						Male (years)	Female (years)		Male %	Female %			
Italy	301,267	116,320	59.0	58.7	196	529	10	10	0.0	4	4	78	83	90	98.9	98.1	4.9		$28,840
Latvia	64,599	24,942	2.3	2.2	35	122	9	14	−0.5	9	11	67	77	68	99.8	99.6	4.7	$2.52	$13,480
Liechtenstein	161	62	0.1	0.1	252	1,006	11	6	0.5	3	4	79	82	21	100.0	100.0	—		—
Lithuania	65,200	25,174	3.4	3.1	52	112	9	13	−0.4	7	9	66	78	67	99.7	99.4	4.8	$2.45	$14,220
Luxembourg	2,587	999	0.5	0.5	195	779	12	8	0.4	4	5	75	81	91	100.0	100.0	8.6		$65,340
Macedonia	25,711	9,927	2.0	2.1	78	300	11	9	0.2	15	17	71	76	59	94.2	83.8	2.7		$7,080
Malta	321	124	0.4	0.4	1,250	3,678	9	7	0.2	5	6	77	81	91	91.4	92.8	6.4		$18,960
Moldova	33,701	13,012	4.0	3.8	118	179	11	12	−0.2	14	16	65	72	45	99.6	98.3	3.2		$2,150
Montenegro	13,812	5,333	0.6	0.6	44	156	13	9	0.3	12	15	—	—	—	100.0	100.0	—		—
Netherlands	40,839	15,768	16.5	16.9	405	1,499	12	8	0.3	4	5	77	81	90	100.0	100.0	8.7		$32,480
Norway	323,878	125,050	4.7	5.2	15	487	12	9	0.3	3	4	78	83	78	100.0	100.0	8.8	$6.63	$40,420
Poland	323,249	124,807	38.1	36.7	118	251	10	10	0.0	6	7	71	79	62	99.8	99.8	3.7	$2.51	$13,490
Portugal	91,981	35,514	10.6	10.4	115	397	10	10	0.1	4	5	75	81	55	94.8	90.0	6.6		$19,730
Romania	238,388	92,042	21.5	18.1	90	205	10	12	−0.2	16	19	68	75	55	99.1	97.3	3.1		$8,940
Serbia	88,357	34,115	9.5	9.2	108	284	13	12	0.1	12	15	69	75	52	100.0	100.0	3.0		—
Slovakia	49,010	18,923	5.4	5.2	110	324	10	10	0.0	7	8	70	78	56	100.0	100.0	4.7	$2.13	$15,760
Slovenia	20,251	7,819	2.0	2.0	99	705	9	9	0.0	3	4	74	81	49	98.6	96.8	6.4		$22,160
Spain	505,988	195,363	45.7	46.2	90	231	11	9	0.2	4	5	77	84	78	100.0	100.0	6.8		$25,820
Sweden	449,959	173,730	9.1	9.9	20	289	11	10	0.1	3	4	78	83	84	100.0	100.0	9.2	$4.86	$31,420
Switzerland	41,290	15,942	7.5	7.4	182	1,520	10	8	0.2	4	5	79	84	68	99.5	97.4	9.1	$5.20	$37,080
Ukraine	603,698	233,089	46.1	41.7	76	129	9	17	−0.8	13	17	63	74	68	100.0	100.0	2.8		$6,720
United Kingdom	244,878	94,548	60.7	65.8	248	954	12	10	0.2	5	6	76	81	89	99.0	99.0	8.6	$4.01	$32,690
Russia	**17,075,323**	**6,592,819**	**140.6**	**130.0**	**8**	**137**	**10**	**16**	**−0.6**	**14**	**18**	**59**	**72**	**73**	**99.8**	**99.2**	**2.5**	**$2.03**	**$10,640**
Armenia	29,800	11,506	3.0	3.4	101	507	13	9	0.4	26	29	67	75	65	99.4	98.1	2.9		$5,060
Azerbaijan	86,599	33,436	8.7	9.7	100	456	17	6	1.1	74	89	70	75	52	98.9	95.9	2.4		$4,890
Georgia	69,699	26,911	4.4	3.9	63	422	12	11	0.1	41	45	69	75	52	99.7	99.4	2.8		$3,270
North America	**19,599,647**	**7,567,466**	**335.6**	**387.0**	**17**	**138**	**14**	**8**	**0.6**	**6**	**7**	**75**	**81**	**79**	**95.7**	**95.3**			**$40,980**
Canada	9,970,600	3,849,670	32.9	37.6	3	66	11	7	0.3	5	6	77	82	80	95.7	95.3	8.5	$3.68	$32,220
United States	9,629,047	3,717,796	302.7	349.4	31	157	14	8	0.6	6	7	75	80	79	95.7	95.3	7.3	$3.41	$41,950
Middle America	**2,712,660**	**1,047,364**	**194.8**	**234.2**	**72**	**444**	**23**	**6**	**1.7**	**24**	**31**	**70**	**75**	**68**	**81.9**	**83.8**			**$7,475**
Antigua and Barbuda	440	170	0.1	0.1	233	1,292	18	6	1.3	11	12	69	74	37	—	—	—		$11,700
Bahamas	13,880	5,359	0.3	0.3	22	2,205	19	9	1.0	13	15	67	73	89	95.4	96.8	—		$15,800
Barbados	430	166	0.3	0.3	706	1,811	14	8	0.6	11	12	70	74	50	98.0	96.8	6.7		—
Belize	22,960	8,865	0.3	0.4	14	341	27	5	2.3	15	17	67	74	50	76.7	77.1	3.5		$6,740
Costa Rica	51,100	19,730	4.4	5.6	86	959	17	4	1.3	11	12	77	81	61	95.5	95.7	4.1	$2.18	$9,680
Cuba	110,859	42,803	11.4	11.8	103	251	11	7	0.4	6	7	75	79	76	96.5	96.4	3.5		—
Dominica	751	290	0.1	0.1	135	676	15	7	0.8	13	15	71	77	71	—	—	4.5		$5,560
Dominican Republic	48,731	18,815	9.3	11.6	191	616	23	6	1.7	26	31	66	69	64	84.0	83.7	2.8		$7,150

(Continued)

| | Land Area | | Population | | Population Density | | Birth Rate | Death Rate | Natural Increase % | Infant Mortality per 1,000 | Child Mortality per 1,000 | Life Expectancy | | Percent Urban Pop. | Literacy | | Corruption Index | Big Mac Price ($US) | Per Capita GNI ($US) |
	Sq km	Sq mi	2008 (Millions)	2025 (Millions)	Arithmetic	Physiologic						Male (years)	Female (years)		Male %	Female %			
El Salvador	21,041	8,124	7.3	9.1	346	887	26	6	2.0	23	27	67	73	59	81.6	76.1	4.0		$5,120
Grenada	339	131	0.1	0.1	302	943	19	7	1.2	17	21	—	—	39	—	—	3.5		$7,260
Guadeloupe	1,709	660	0.5	0.5	298	1,989	16	6	1.0	—	—	75	82	100	89.7	90.5	—		—
Guatemala	108,888	42,042	13.7	20.0	126	701	34	6	2.8	32	43	63	71	44	76.2	61.1	2.6		$4,410
Haiti	27,749	10,714	8.9	13.0	321	971	36	13	2.3	84	120	51	54	36	51.0	46.5	1.8		$1,840
Honduras	112,090	43,278	7.8	10.7	69	385	31	6	2.5	31	40	67	74	47	72.5	72.0	2.5		$2,900
Jamaica	10,989	4,243	2.8	3.0	252	1,008	19	6	1.3	17	20	69	73	52	82.5	90.7	3.7		$4,110
Martinique	1,101	425	0.4	0.4	368	1,842	14	7	0.7	—	—	76	82	95	96.0	97.1	—		—
Mexico	1,958,192	756,062	112.0	129.4	57	409	22	5	1.7	22	27	73	78	75	93.1	89.1	3.3	$2.69	$10,030
Netherlands Antilles	800	309	0.2	0.2	252	2,524	13	8	0.5	—	—	72	79	69	96.6	96.6	—		—
Nicaragua	129,999	50,193	5.9	7.7	45	205	29	5	2.4	30	37	66	70	59	64.2	64.4	2.6		$3,650
Panama	75,519	29,158	3.4	4.2	45	502	22	5	1.7	19	24	73	78	62	92.6	91.3	3.1		$7,310
Puerto Rico	8,951	3,456	3.9	4.1	441	4,899	13	7	0.6	—	—	73	81	94	93.7	94.0	—		—
Saint Lucia	619	239	0.2	0.2	333	1,189	20	5	1.5	12	14	72	77	28	89.5	90.6	—		$5,980
St. Vincent & the Grenadines	391	151	0.1	0.1	261	933	18	7	1.1	17	20	68	74	45	—	—	—		$6,460
Trinidad and Tobago	5,131	1,981	1.3	1.3	256	1,068	14	8	0.6	17	19	67	73	74	99.0	97.5	3.2		$13,170
South America	**17,855,070**	**6,893,881**	**389.0**	**465.2**	**22**	**348**	**21**	**6**	**1.4**	**25**	**28**	**69**	**76**	**80**	**89.9**	**88.6**			**$8,210**
Argentina	2,780,388	1,073,514	39.8	46.4	14	143	18	8	1.1	15	18	71	78	90	96.9	96.9	2.9	$2.67	$13,920
Bolivia	1,098,575	424,162	9.5	12.1	9	433	31	8	2.2	52	65	62	66	63	92.1	79.4	2.7		$2,740
Brazil	8,547,360	3,300,154	192.4	228.9	23	322	21	6	1.4	31	33	68	76	83	85.5	85.4	3.3	$3.61	$8,230
Chile	756,626	292,135	16.8	19.1	22	738	16	5	1.0	8	10	75	81	87	95.9	95.5	7.3	$2.97	$11,470
Colombia	1,138,906	439,734	48.2	58.3	42	1,058	20	5	1.5	17	21	69	75	77	91.8	91.8	3.9	$3.06	$7,420
Ecuador	283,560	109,483	13.9	17.5	49	444	27	6	2.1	22	25	71	77	61	93.6	90.2	2.3		$4,070
French Guiana	89,999	34,749	0.2	0.3	2	234	31	4	2.7	—	—	72	79	75	83.6	82.3	—		—
Guyana	214,969	83,000	0.7	0.7	3	167	22	9	1.3	47	63	72	80	36	99.0	98.1	2.5		$4,230
Paraguay	406,747	157,046	6.5	8.6	16	267	22	5	1.7	20	23	69	73	57	94.4	92.2	2.6	$1.90	$4,970
Peru	1,285,214	496,224	29.1	34.1	23	756	19	6	1.3	23	27	67	72	74	94.7	85.4	3.3	$3.00	$5,830
Suriname	163,270	63,039	0.5	0.5	3	315	21	7	1.4	30	39	66	73	74	95.9	92.6	3.0		—
Uruguay	177,409	68,498	3.3	3.5	19	268	15	10	0.5	14	15	71	79	93	97.4	98.2	6.4	$2.17	$9,810
Venezuela	912,046	352,143	27.9	35.2	31	765	22	5	1.7	18	21	70	76	88	93.3	92.7	2.3	$3.45	$6,440
Subsaharan Africa	**21,786,509**	**8,411,818**	**761.1**	**1,090.2**	**35**	**464**	**40**	**16**	**2.4**	**98**	**164**	**47**	**49**	**34**	**72.7**	**55.2**			**$1,970**
Angola	1,246,693	481,351	16.7	25.9	13	446	49	22	2.6	154	260	39	42	35	55.6	28.5	2.2		$2,210
Benin	112,620	43,483	9.2	14.3	82	511	41	12	2.9	89	150	53	55	40	47.8	23.6	2.5		$1,110
Botswana	581,727	224,606	1.8	1.7	3	309	26	27	−0.1	87	120	35	33	54	74.4	79.8	5.6		$10,250
Burkina Faso	274,000	105,792	14.3	23.2	52	435	44	19	2.5	96	191	48	49	18	31.2	13.1	3.2		$1,220
Burundi	27,829	10,745	8.2	14.0	296	689	46	18	2.7	114	190	44	45	10	56.3	40.5	2.4		$640

| | Land Area | | Population | | Population Density | | Birth Rate | Death Rate | Natural Increase % | Infant Mortality per 1,000 | Child Mortality per 1,000 | Life Expectancy | | Percent Urban Pop. | Literacy | | Corruption Index | Big Mac Price ($US) | Per Capita GNI ($US) |
	Sq km	Sq mi	2008 (Millions)	2025 (Millions)	Arithmetic	Physiologic						Male (years)	Female (years)		Male %	Female %			
Cameroon	475,439	183,568	18.1	24.3	38	238	37	14	2.3	87	149	50	52	52	81.8	69.2	2.3		$2,150
Cape Verde Islands	4,030	1,556	0.5	0.7	130	1,185	30	5	2.5	26	35	68	74	55	84.3	65.3	—		$6,000
Central African Republic	622,978	240,533	4.5	5.5	7	238	37	19	1.7	115	193	43	44	43	59.6	34.5	2.4		$1,140
Chad	1,283,994	495,753	10.6	17.2	8	274	48	20	2.8	124	208	43	45	25	66.9	40.8	2.0		$1,470
Comoros	2,230	861	0.7	1.0	333	628	37	7	2.9	53	71	62	66	33	63.5	49.1	—		$2,000
Congo	341,998	132,046	3.9	5.9	11	1,139	40	14	2.6	81	108	50	52	51	87.5	74.4	2.2		$810
Congo, The	2,344,848	905,351	66.6	108.0	28	711	45	14	3.1	129	205	49	52	35	86.6	67.7	2.0		$720
Djibouti	23,201	8,958	0.8	1.1	36	3,580	31	12	1.9	88	133	52	54	82	65.0	38.4	—		$2,240
Equatorial Guinea	28,050	10,830	0.5	0.8	19	207	43	20	2.3	123	205	43	44	39	92.5	74.5	2.1		$7,580
Eritrea	117,598	45,405	4.9	7.4	41	1,033	39	11	2.8	50	78	53	57	19	43.9	33.4	2.9		$1,010
Ethiopia	1,104,296	426,371	78.4	107.8	71	646	39	15	2.4	109	164	48	50	15	83.7	70.0	2.4		$1,000
Gabon	267,668	103,347	1.5	1.8	5	272	33	13	2.0	60	91	53	55	81	79.8	62.2	3.0		$5,890
Gambia	11,300	4,363	1.6	2.4	140	665	38	12	2.7	97	137	52	55	50	43.8	29.6	2.5		$1,920
Ghana	238,538	92,100	23.7	32.7	99	431	33	10	2.3	68	112	57	58	45	79.5	61.2	3.3		$2,370
Guinea	245,860	94,927	10.4	15.2	42	702	41	13	2.8	98	150	54	54	30	55.1	27.0	1.9		$2,240
Guinea–Bissau	36,120	13,946	1.5	2.4	41	316	50	20	3.0	124	200	44	46	48	53.0	21.4	—		$700
Ivory Coast	322,459	124,502	20.7	27.1	64	279	39	14	2.5	118	195	49	53	48	54.6	38.5	2.1		$1,490
Kenya	580,367	224,081	36.5	49.4	63	785	40	15	2.5	79	120	49	47	40	89.0	76.0	2.2		$1,170
Lesotho	30,349	11,718	1.8	1.7	60	542	28	25	0.3	102	132	35	36	13	73.6	93.6	3.2		$3,410
Liberia	111,369	43,000	3.6	5.8	32	808	50	21	2.9	157	235	41	44	45	69.9	36.8	—		$130
Madagascar	587,036	226,656	18.8	28.2	32	641	40	12	2.7	74	119	53	57	30	87.7	72.9	3.1		$880
Malawi	118,479	45,745	13.5	23.8	114	542	44	18	2.6	79	125	44	47	17	74.5	46.7	2.7		$650
Mali	1,240,185	478,838	14.8	24.0	12	298	50	18	3.2	120	218	48	49	27	47.9	33.2	2.8		$1,000
Mauritania	1,025,516	395,954	3.4	5.0	3	330	42	14	2.8	78	125	53	55	51	50.6	29.5	3.1		$2,150
Mauritius	2,041	788	1.3	1.4	647	1,245	15	7	0.8	13	15	69	76	42	87.7	81.0	5.1		$12,450
Moçambique	801,586	309,494	20.7	27.6	26	647	41	20	2.0	100	145	41	42	34	59.9	28.4	2.8		$1,170
Namibia	824,287	318,259	2.2	2.5	3	262	29	15	1.4	46	62	47	47	33	82.9	81.2	4.1		$7,910
Niger	1,266,994	489,189	15.4	26.4	12	304	55	21	3.4	150	256	44	44	21	23.5	8.3	2.3		$800
Nigeria	923,766	356,668	141.0	199.5	153	449	43	19	2.4	100	194	43	44	43	72.3	56.2	2.2		$1,040
Réunion	2,510	969	0.8	1.0	328	2,185	19	5	1.4	—	—	72	80	89	84.8	89.2	—		—
Rwanda	26,340	10,170	9.6	13.8	364	866	43	17	2.7	118	203	46	48	17	73.7	60.6	2.5		$1,320
São Tomé and Príncipe	961	371	0.2	0.2	219	509	34	9	2.5	75	118	62	64	38	70.2	39.1	—		—
Senegal	196,720	75,954	12.6	17.3	64	534	39	10	2.9	77	136	55	58	50	47.2	27.6	3.3		$1,770
Seychelles	451	174	0.1	0.1	226	1,509	18	8	1.0	12	13	66	76	50	82.9	85.7	3.6		$15,940
Sierra Leone	71,740	27,699	6.0	8.7	83	1,039	46	23	2.3	165	282	39	42	36	50.7	22.6	2.2		$780
Somalia	637,658	246,201	9.4	14.9	15	739	46	17	2.9	133	225	46	50	34	85.8	84.5	—		—
South Africa	1,221,034	471,444	47.8	48.0	39	301	23	18	0.5	55	68	45	49	58	68.3	46.0	4.6	$2.22	$12,120
Swaziland	17,361	6,703	1.1	1.0	63	577	29	28	0.1	110	160	33	35	23	80.9	78.7	2.5		$5,190

(Continued)

	Land Area		Population		Population Density		Birth Rate	Death Rate	Natural Increase %	Infant Mortality per 1,000	Child Mortality per 1,000	Life Expectancy		Percent Urban Pop.	Literacy		Corruption Index	Big Mac Price ($US)	Per Capita GNI ($US)
	Sq km	Sq mi	2008 (Millions)	2025 (Millions)	Arithmetic	Physiologic						Male (years)	Female (years)		Male %	Female %			
Tanzania	945,087	364,900	39.8	53.6	42	843	42	17	2.5	76	122	44	45	32	84.1	66.6	2.9		$730
Togo	56,791	21,927	6.6	9.6	117	272	38	12	2.6	78	139	53	57	33	72.2	42.6	2.4		$1,550
Uganda	241,040	93,066	29.4	55.5	122	359	47	16	3.1	79	136	47	47	12	77.7	57.1	2.7		$1,500
Zambia	752,607	290,583	12.3	16.4	16	234	41	23	1.9	102	182	38	37	35	85.2	71.2	2.6		$950
Zimbabwe	390,759	150,873	13.3	14.4	34	378	30	23	0.7	81	132	37	34	34	95.5	89.9	2.4		$1,940
North Africa/ Southwest Asia	**19,318,887**	**7,459,064**	**581.0**	**754.1**	**30**	**328**	**26**	**7**	**1.8**	**47**	**51**	**66**	**70**	**55**	**76.2**	**56.9**			**$5,674**
Afghanistan	652,086	251,772	32.7	50.3	50	418	48	22	2.6	165	257	41	42	23	51.0	20.8	—		—
Algeria	2,381,730	919,591	34.6	43.1	15	485	21	4	1.7	34	39	74	76	81	75.1	51.3	3.1		$6,770
Bahrain	689	266	0.7	1.0	1,053	11,700	21	3	1.8	9	11	73	75	71	91.0	82.7	5.7		$21,290
Egypt	1,001,445	386,660	78.6	101.1	78	2,616	27	6	2.1	28	33	67	72	43	66.6	43.7	3.3	$1.68	$4,440
Iran	1,633,182	630,575	72.0	89.0	44	401	18	6	1.2	31	36	69	72	67	83.5	70.4	2.7		$8,050
Iraq	438,319	169,236	31.2	44.7	71	547	36	10	2.6	102	125	57	60	68	70.7	45.0	1.9		—
Israel	21,059	8,131	7.4	9.3	353	1,681	21	5	1.5	5	6	78	82	91	97.9	94.3	5.9		$25,280
Jordan	89,210	34,444	5.9	7.9	66	1,316	29	5	2.4	22	26	71	72	82	94.9	84.4	5.3		$5,280
Kazakhstan	2,717,289	1,049,151	15.5	16.0	6	48	18	10	0.8	63	73	61	72	57	99.1	96.1	2.6		$7,730
Kuwait	17,819	6,880	2.8	3.9	157	15,672	19	2	1.7	9	11	77	79	97	84.3	79.9	4.8		$24,010
Kyrgyzstan	198,499	76,641	5.3	6.6	27	385	21	7	1.4	58	67	64	72	35	98.6	95.5	2.2		$1,870
Lebanon	10,399	4,015	4.0	4.6	386	1,244	19	5	1.5	27	30	70	74	87	92.3	80.4	3.6		$5,740
Libya	1,759,532	679,359	6.2	8.3	4	351	27	4	2.4	18	19	74	78	86	90.9	67.6	2.7		$5,530
Morocco	446,548	172,413	32.7	38.8	73	332	21	6	1.6	36	40	68	72	55	61.9	36.0	3.2		$4,360
Oman	212,459	82,031	2.7	3.1	13	1,273	24	4	2.0	10	12	73	75	71	80.4	61.7	5.4		$14,680
Palestinian Territories	6,260	2,417	4.2	7.1	665	33,240	37	4	3.3	21	23	71	74	57	—	—	—		—
Qatar	11,000	4,247	0.8	1.2	75	7,508	18	2	1.6	18	21	71	76	100	80.5	83.2	6.0		—
Saudi Arabia	2,149,680	829,996	25.4	35.6	12	591	30	3	2.7	21	26	70	74	86	84.1	67.2	3.3	$2.40	$14,740
Sudan	2,505,798	967,494	43.5	61.3	17	248	36	9	2.6	62	90	57	59	37	36.0	14.0	2.0		$2,000
Syria	185,179	71,498	20.5	28.1	111	369	29	4	2.5	14	15	71	75	50	88.3	60.4	2.9		$3,740
Tajikistan	143,099	55,251	7.3	9.3	51	730	30	8	2.2	59	71	61	66	26	99.6	98.9	2.2		$1,260
Tunisia	163,610	63,170	10.3	11.6	63	197	17	6	1.1	20	24	71	75	65	81.4	60.1	4.6		$7,900
Turkey	774,816	299,158	75.6	86.0	98	257	19	5	1.3	26	29	69	74	62	93.6	76.7	3.8	$3.66	$8,420
Turkmenistan	488,099	188,456	5.5	6.6	11	281	25	8	1.6	81	104	58	67	47	98.8	96.6	2.2		—
United Arab Emirates	83,600	32,278	5.0	7.1	60	6,027	15	1	1.3	8	9	75	80	74	75.5	79.5	6.2	$2.72	$24,090
Uzbekistan	447,397	172,741	27.0	33.0	60	504	23	7	1.6	57	68	63	70	36	98.5	96.0	2.1		$2,020
Western Sahara	252,120	97,344	0.4	0.7	2	4	28	8	2.0	—	—	62	66	93	—	—	—		—
Yemen	527,966	203,849	23.0	38.8	44	1,452	41	9	3.2	76	102	59	62	26	67.4	25.0	2.6		$920
South Asia	**4,487,760**	**1,732,734**	**1,532.9**	**1,841.9**	**342**	**682**	**25**	**8**	**1.7**	**58**	**76**	**62**	**64**	**30**	**65.9**	**39.5**			**$3,330**
Bangladesh	143,998	55,598	152.2	190.0	1,057	1,678	27	8	1.9	54	73	61	62	23	51.7	29.5	2.0		$2,090

	Land Area Sq km	Land Area Sq mi	Population 2008 (Millions)	Population 2025 (Millions)	Population Density Arithmetic	Population Density Physiologic	Birth Rate	Death Rate	Natural Increase %	Infant Mortality per 1,000	Child Mortality per 1,000	Life Expectancy Male (years)	Life Expectancy Female (years)	Percent Urban Pop.	Literacy Male %	Literacy Female %	Corruption Index	Big Mac Price ($US)	Per Capita GNI ($US)
Bhutan	47,001	18,147	0.9	1.3	20	655	20	7	1.3	65	75	62	64	31	61.1	33.6	6.0		—
India	3,287,576	1,269,340	1,158.0	1,363.0	352	618	24	8	1.7	56	74	62	63	29	68.6	42.1	3.3		$3,460
Maldives	300	116	0.3	0.4	1,029	10,287	18	3	1.5	33	42	70	70	27	96.3	96.4	—		$4,635
Nepal	147,179	56,826	27.2	36.2	185	879	31	9	2.2	56	74	62	63	14	59.1	21.8	2.5		$1,530
Pakistan	796,098	307,375	173.9	228.8	218	753	33	9	2.4	79	99	61	63	38	57.6	27.8	2.2	$2.31	$2,350
Sri Lanka	65,610	25,332	20.4	22.2	311	1,073	19	6	1.3	12	14	71	77	20	94.5	88.9	3.1	$1.75	$4,520
East Asia	**11,773,125**	**4,545,629**	**1,550.6**	**1,699.4**	**132**	**1,068**	**12**	**7**	**0.5**	**21**	**24**	**71**	**75**	**43**	**93.3**	**80.4**			**$9,050**
China	9,572,855	3,696,100	1,324.5	1,476.0	138	988	12	7	0.6	23	27	70	74	37	92.3	77.4	3.3	$1.45	$6,600
Japan	377,799	145,869	128.1	121.1	339	2,607	9	8	0.0	3	4	79	86	79	100.0	100.0	7.6	$2.29	$31,410
Korea, North	120,541	46,541	23.5	25.8	195	1,219	16	7	0.9	42	55	68	73	60	99.0	99.0	—		—
Korea, South	99,259	38,324	48.9	49.8	493	2,592	9	5	0.4	5	5	74	81	82	99.2	96.4	5.1	$3.14	$21,850
Mongolia	1,566,492	604,826	2.7	3.1	2	170	18	6	1.2	39	49	64	68	57	99.2	99.3	2.8		$2,190
Taiwan	36,180	13,969	22.9	23.6	634	2,536	9	6	0.3	—	—	73	79	78	97.6	90.2	5.9	$2.28	$16,250
Southeast Asia	**4,494,792**	**1,735,449**	**581.7**	**682.0**	**129**	**621**	**21**	**6**	**1.4**	**30**	**39**	**66**	**71**	**39**	**92.8**	**85.7**			**$4,530**
Brunei	5,770	2,228	0.4	0.5	72	3,585	20	3	1.7	8	9	72	77	72	94.7	88.2	—		—
Cambodia	181,040	69,900	14.7	19.6	81	369	30	9	2.1	98	143	57	63	19	79.7	53.4	2.1		$2,490
East Timor	14,869	5,741	1.1	1.9	71	591	42	15	2.7	52	61	54	57	22	65.0	52.0	2.6		$600
Indonesia	1,904,561	735,355	231.9	263.7	122	716	20	6	1.4	28	36	67	72	42	91.9	82.1	2.4	$1.75	$3,720
Laos	236,800	91,429	6.4	8.7	27	899	36	13	2.3	62	79	53	56	21	73.6	50.5	2.6		$2,020
Malaysia	329,750	127,317	27.8	34.6	84	351	20	4	1.6	10	12	72	76	64	91.5	83.6	5.0	$1.60	$10,320
Myanmar/Burma	676,577	261,228	52.1	59.0	77	482	21	10	1.1	75	105	57	63	29	89.0	80.6	1.9		—
Philippines	299,998	115,830	90.1	115.7	300	910	27	5	2.1	25	33	67	72	59	95.5	95.2	2.5	$1.85	$5,300
Singapore	619	239	4.6	5.2	7,357	367,860	10	4	0.6	3	3	78	82	100	96.4	88.5	9.4	$2.59	$29,780
Thailand	513,118	198,116	66.1	70.2	129	322	14	7	0.7	18	21	68	75	34	97.2	94.0	3.6	$1.78	$8,440
Vietnam	331,689	128,066	86.6	102.9	261	1,186	19	5	1.3	16	19	70	73	27	95.7	91.0	2.6		$3,010
Austral Realm	**8,011,714**	**3,093,340**	**25.0**	**29.2**	**3**	**44**	**13**	**6**	**0.7**	**5**	**6**	**78**	**83**	**91**	**100.0**	**100.0**			**$29,352**
Australia	7,741,184	2,988,888	20.9	24.6	3	39	13	6	0.6	5	6	78	83	91	100.0	100.0	8.7	$2.95	$30,610
New Zealand	270,529	104,452	4.2	4.6	15	128	14	7	0.7	5	6	77	81	86	100.0	100.0	9.6	$5.89	$23,030
Pacific Realm	**975,341**	**376,804**	**9.0**	**11.4**	**24.5**	**757**	**30**	**9**	**2.0**	**44**	**58**	**59**	**61**	**22**	**69.1**	**59.0**			**$2,540**
Federated States of Micronesia	699	270	0.1	0.1	149	286	26	6	2.0	34	42	67	67	22	67.0	87.2	—		—
Fiji	18,270	7,054	0.8	0.9	45	282	21	6	1.4	16	18	66	71	46	95.0	90.9	—		$5,960
French Polynesia	3,999	1,544	0.3	0.3	77	962	18	5	1.3	—	—	72	77	53	94.9	95.0	—		—
Guam	549	212	0.2	0.2	377	1,712	21	4	1.6	—	—	75	81	93	99.0	99.0	—		—
Marshall Islands	179	69	0.1	0.1	597	3,512	38	5	3.3	51	58	—	—	68	92.4	90.0	—		—
New Caledonia	18,581	7,174	0.2	0.3	11	1,102	17	5	1.2	—	—	71	77	71	57.4	58.3	—		—
Papua New Guinea	462,839	178,703	6.3	8.2	14	1,351	32	11	2.1	55	74	55	56	13	63.4	50.9	2.4		$2,370
Samoa	2,841	1,097	0.2	0.2	74	171	29	6	2.4	24	29	72	74	22	100.0	100.0	—		$6,480
Solomon Islands	28,899	11,158	0.5	0.7	18	607	34	8	2.6	24	29	62	63	16	62.4	44.9	—		$1,880
Vanuatu	12,191	4,707	0.2	0.4	17	172	31	6	2.5	31	38	66	69	21	57.3	47.8	—		$3,170

GLOSSARY

(*Note:* Words in **boldface** type within an entry are defined elsewhere in this Glossary.)

Aboriginal land issue The legal campaign in which Australia's **indigenous peoples** have claimed title to traditional land in several parts of that country. The courts have upheld certain claims, fueling Aboriginal activism that has raised broader issues of indigenous rights.

Aboriginal population See **indigenous peoples**.

Absolute location The position or place of a certain item on the surface of the earth as expressed in degrees, minutes, and seconds of **latitude**, 0° to 90° north or south of the equator, and **longitude**, 0° to 180° east or west of the *prime meridian* passing through Greenwich, England (a suburb of London).

Acculturation Cultural modification resulting from intercultural borrowing. In **cultural geography**, the term refers to the change that occurs in the **culture** of **indigenous peoples** when contact is made with a society that is technologically superior.

Advantage The most meaningful distinction that can now be made to classify a country's level of economic **development**. Takes into account geographic **location**, **natural resources**, government, political stability, productive skills, and much more.

Agglomeration Process involving the clustering or concentrating of people or activities.

Agrarian Relating to the use of land in rural communities, or to agricultural societies in general.

Alluvial Refers to the mud, silt, and sand (collectively *alluvium*) deposited by rivers and streams. *Alluvial plains* adjoin many larger rivers; they consist of these renewable deposits that are laid down during floods, creating fertile and productive soils. Alluvial **deltas** mark the mouths of rivers such as the Nile and the Ganges.

Altiplano High-elevation plateau, basin, or valley between even higher mountain ranges, especially in the Andes of South America.

Altitudinal zones The five vertical regions defined by physical-environmental zones at various elevations (see Fig. 4-13 and discussion on page 150), particularly in the highlands of South and Middle America. The five zones (going from the lowest to the highest zone) *Tierra caliente* (hot land), *Tierra templada* (temperate land) *Tierra fría* (cold land), *Tierra helada* (icy land), and *Tierra Nevada* (snowy land).

American Manufacturing Belt North America's near-rectangular Core Region, whose corners are Boston, Milwaukee, St. Louis, and Baltimore. Dominated the industrial geography of the U.S. and Canada during the industrial age; still a formidable economic powerhouse that remains the realm's geographic heart.

Amerindian Person ethnically linked to the people who inhabited the Americas before Europeans arrived.

Animistic religion The belief that inanimate objects, such as hills, trees, rocks, rivers, and other elements of the natural landscape, possess souls and can help or hinder human efforts on Earth.

Antarctic Treaty International cooperative agreement on the use of Antarctic territory (see p. 399).

Antecedent boundary See **Boundaries.**

Apartheid Literally, *apartness*. The Afrikaans term for South Africa's pre-1994 policies of racial separation, a system that produced highly segregated socio-geographical patterns.

Aquifer An underground reservoir of water contained within a porous, water-bearing rock layer.

Arable Land fit for cultivation by one farming method or another. See **physiologic density**.

Archipelago A set of islands grouped closely together, usually elongated into a *chain*.

Area A term that refers to a part of the earth's surface with less specificity than **region**. For example, *urban area* alludes generally to a place where urban development has occurred, whereas *urban region* requires certain specific criteria upon which such a designation is based (e.g., the spatial extent of commuting or the built townscape).

Areal functional organization A geographic principle for understanding the evolution of regional organization, whose five interrelated tenets are applied to the spatial development of Japan on p. 334.

Areal interdependence A term related to **functional specialization**. When one area produces certain goods or has certain raw materials and another area has a different set of raw materials and produces different goods, their needs may be *complementary*; by exchanging raw materials and products, they can satisfy each other's requirements.

Arithmetic density A country's population, expressed as an average per unit area, without regard for its **distribution** or the limits of **arable** land—see also **physiologic density**.

Aryan From the Sanskrit *Arya* ('noble'), a name applied to an ancient people who spoke an **Indo-European language** and who moved into northern India from the northwest.

Atmosphere The earth's envelope of gases that rests on the oceans and land surface and penetrates open spaces within soils. This layer of nitrogen (78 percent), oxygen (21 percent), and traces of other gases is densest at the Earth's surface and thins with altitude.

Austral South.

Autocratic A government that holds absolute power, often ruled by one person or a small group of persons who control the country by despotic means.

Balkanization The fragmentation of a **region** into smaller, often hostile political units.

Barrio Term meaning 'neighborhood' in Spanish. Usually refers to an urban community in a Middle or South American city; also applied to low-income, inner-city concentrations of Hispanics in such western U.S. cities as Los Angeles.

Bauxite Aluminum ore; usually deposited at shallow depths in the wet tropics.

Biogeography The study of *flora* (plant life) and *fauna* (animal life) in spatial perspective.

Biome Broadest possible subdivision of the plant and animal world; an assemblage and association of plants and animals which forms a regional ecological unit of subcontinental dimensions.

Birth rate The *crude birth rate* is expressed as the annual number of births per 1000 individuals within a given population.

Boundaries The divisions between states. There are various types of boundaries. An *antecedent* boundary is a political boundary that existed before the **cultural landscape** emerged and stayed in place while people moved in to occupy the surrounding area. *Geometric* boundaries are political boundaries **defined** and **delimited** (and occasionally **demarcated**) as straight lines or arcs. A *relict* boundary is a political boundary that has ceased to function, but the imprint of which can still be detected on the **cultural landscape**. A *subsequent* boundary developed contemporaneously with the evolution of the major elements of the **cultural landscape** through which it passes. A *superimposed* boundary is emplaced by powerful outsiders on a developed human landscape. Usually ignores preexisting cultural-spatial patterns, such as the border that still divides North and South Korea.

Break-of-bulk point A location along a transport route where goods must be transferred from one carrier to another. In a port, the cargoes of ocean-going ships are unloaded and put on trains, trucks, or perhaps smaller river boats for inland distribution.

Buffer state See **buffer zone**.

Buffer zone A country or set of countries separating ideological or political adversaries. In southern Asia, Afghanistan, Nepal, and Bhutan were parts of a buffer zone between British and Russian-Chinese imperial spheres. Thailand was a *buffer state* between British and French colonial domains in mainland Southeast Asia.

CAFTA-DR The Central American free-trade agreement between the United States, all the Central American countries, and the Dominican Republic.

Campesinos Spanish term used in Middle and South America to denote **peasants**.

Cartogram A specially transformed map not based on traditional representations of **scale** or area.

Cartography The art and science of making maps, including data compilation, layout, and design. Also concerned with the interpretation of mapped patterns.

Caste system The strict **social stratification** and segregation of people—specifically in India's Hindu society—on the basis of ancestry and occupation.

Cay A low-lying small island usually composed of coral and sand. Pronounced *kee* and often spelled 'key.'

Cell phone revolution The changes that are occurring throughout the world, but especially in lower income countries, related to the availability of information via cellular phones (see p. 198).

Central business district (CBD) The downtown heart of a central city, the CBD is marked by high land values, a concentration of business and commerce, and the clustering of the tallest buildings.

Centrality The strength of an urban center in its capacity to attract producers and consumers to its facilities; a city's 'reach' into the surrounding region.

Centrifugal forces A term employed to designate forces that tend to divide a country—such as internal religious, linguistic, ethnic, or ideological differences.

Centripetal forces Forces that unite and bind a country together—such as a strong national culture, shared ideological objectives, and a common faith.

Cerrado The physiographic region southeast of the Amazon Basin that has greater seasonality in its precipitation than the rainforest to the north.

Chaebol Giant corporation controlling numerous companies and benefiting from government connections and favors, dominant in South Korea's **economic geography**; key to the country's **development** into an **economic tiger**, but more recently a barrier to freemarket growth.

Charismatic Personal qualities of certain leaders that enable them to capture and hold the popular imagination, to secure the allegiance and even the devotion of the masses. Gandhi, Mao Zedong, and Franklin D. Roosevelt were good examples in the twentieth-century.

China Proper The eastern and northeastern portions of China that contain most of the country's huge population; mapped in Figure 9-11.

Choke point A narrowing of an international waterway causing marine traffic congestion, requiring reduced speeds and/or sharp turns, and increasing the risk of collision as well as vulnerability to attack. When the waterway narrows to a distance of less than 38 kilometers (24 mi), this necessitates the drawing of a **median line (maritime) boundary**. Examples are the Hormuz Strait between Oman and Iran at the entrance to the Persian Gulf, and the Strait of Malacca between Malaysia and Indonesia.

City-state An independent political entity consisting of a single city with (and sometimes without) an immediate **hinterland**. The ancient city-states of Greece have their modern equivalent in Singapore.

Climate The long-term conditions (over at least 30 years) of aggregate **weather** over a region, summarized by averages and measures of variability; a synthesis of the succession of weather events we have learned to expect at any given location.

Climate change theory An alternative to the **hydraulic civilization theory**; holds that changing **climate** (rather than a monopoly over **irrigation** methods) could have provided certain cities in the ancient Fertile Crescent with advantages over others.

Climate region A **formal region** characterized by the uniformity of the **climate** type within it. Figure G-9 maps the global distribution of such regions.

Climatology The geographic study of **climates**. Includes not only the classification of climates and the analysis of their regional distribution, but also broader environmental questions that concern climate change, interrelationships with soil and vegetation, and human-climate interaction.

Coal A fossil fuel used to generate energy. When heated and converted to coking coal or *coke*, it is used to make steel. *Anthracite* coal is the hardest and highest carbon-content coal, and therefore of the highest quality. *Bituminous* coal is softer coal of lesser quality than anthracite, but of higher grade than lignite. *Lignite* is the lowest-grade (brown-colored) variety of coal.

Collectivization The reorganization of a country's **agriculture** under communism that involves the expropriation of private holdings and their incorporation into relatively large-scale units, which are farmed and administered cooperatively by those who live there.

Colonialism Rule by an autonomous power over a subordinate and alien people and place. Although often established and maintained through political structures, colonialism also creates unequal cultural and economic relations. Because of the magnitude and impact of the European colonial thrust of the last few centuries, the term is generally understood to refer to that particular colonial endeavor. Also see **imperialism**.

Command economy The tightly controlled economic system of the former Soviet Union, whereby central planners in Moscow assigned the production of particular goods to particular places, often guided more by socialist ideology than the principles of **economic geography**.

Commercial agriculture For-profit **agriculture**.

Common market A **free-trade area** that not only has created a **customs union** (a set of common tariffs on all imports from outside the area) but also has eliminated restrictions on the movement of capital, labor, and enterprise among its member countries.

Compact state See **State territorial morphology.**

Complementarity Exists when two regions, through an exchange of raw materials and/or finished products, can specifically satisfy each other's demands.

Confucianism A philosophy of ethics, education, and public service based on the writings of Confucius (*Kongfuzi*); traditionally regarded as one of the cornerstones of Chinese **culture**.

Congo See the footnote on p. 202, which makes the distinction between the two Equatorial African countries, Congo and The Congo.

Coniferous forest A forest of cone-bearing needleleaf evergreen trees with straight trunks and short branches, including spruce, fir, and pine. Also see **taiga**.

Contagious diffusion See diffusion.

Conterminous United States The 48 contiguous or adjacent States that occupy the southern half of the North American realm. Alaska is not contiguous to these States because western Canada lies in between; neither is Hawai'i, separated from the mainland by over 2000 miles of ocean.

Continental drift The slow movement of continents controlled by the processes associated with **plate tectonics**.

Continental shelf Beyond the coastlines of many landmasses, the ocean floor declines very gently until the depth of about 200 meters (660 ft). Beyond the 200-meter line the sea bottom usually drops off sharply, along the *continental slope*, toward the much deeper mid-oceanic basin. The submerged continental margin is called the continental shelf, and it extends from the shoreline to the upper edge of the continental slope.

Continentality The variation of the continental effect on air temperatures in the interior portions of the world's landmasses. The greater the distance from the moderating influence of an ocean, the greater the extreme in summer and winter temperatures. Continental interiors also tend to be dry when the distance from oceanic moisture sources becomes considerable.

Conurbation General term used to identify a large multi-metropolitan complex formed by the coalescence of two or more major **urban areas**. The Atlantic Seaboard **Megalopolis**, extending along the northeastern U.S. coast from southern Maine to Virginia, is a classic example.

Cordillera Mountain chain consisting of sets of parallel ranges, especially the Andes in northwestern South America.

Core See **core area**; **core-periphery relationships**.

Core area In geography, a term with several connotations. *Core* refers to the center, heart, or focus. The core area of a **nation-state** is constituted by the national heartland, the largest population cluster, the most productive region, and the part of the country with the greatest **centrality** and **accessibility**—probably containing the capital city as well.

Core-periphery relationships The contrasting spatial characteristics of, and linkages between, the *have* (core) and *have-not* (periphery) components of a national or regional **system**.

Corridor In general, refers to a spatial entity in which human activity is organized in a linear manner, as along a major transport route or in a valley confined by highlands. More specifically, the politico-geographical term for a land extension that connects an otherwise **landlocked state** to the sea.

Cultural diffusion The **process** of spreading and adoption of a cultural element, from its place of origin across a wider area.

Cultural ecology The multiple interactions and relationships between a **culture** and its **natural environment**.

Cultural geography The wide-ranging and comprehensive field of geography that studies spatial aspects of human **cultures**.

Cultural landscape The forms and artifacts sequentially placed on the **natural landscape** by the activities of various human occupants. By this progressive imprinting of the human presence, the physical (natural) landscape is modified into the cultural landscape, forming an interacting unity between the two.

Cultural pluralism See **plural(istic) society**.

Cultural revival The regeneration of a long-dormant **culture** through internal renewal and external infusion.

Culture The sum total of the knowledge, attitudes, and habitual behavior patterns shared and transmitted by the members of a society. This is anthropologist Ralph Linton's definition; hundreds of others exist.

Culture area See **culture region**.

Culture hearth Heartland, source area, innovation center; place of origin of a major **culture**.

Culture region A distinct, culturally discrete spatial unit; a **region** within which certain cultural norms prevail.

Customs union A **free-trade area** in which member countries set common tariff rates on imports from outside the area.

Death rate The *crude death rate* is expressed as the annual number of deaths per 1000 individuals within a given population.

Deciduous A deciduous tree loses its leaves at the beginning of winter or the onset of the dry season.

Definition In **political geography**, the written legal description (in a treaty-like document) of a boundary between two countries or territories—see **delimitation**.

Deforestation The clearing and destruction of forests (especially tropical rainforests) to make way for expanding settlement frontiers and the exploitation of new economic opportunities.

Deglomeration Deconcentration.

Delimitation In **political geography**, the translation of the written terms of a boundary treaty (the **definition**) into an official cartographic representation (map).

Delta Alluvial lowland at the mouth of a river, formed when the river deposits its alluvial load on reaching the sea. Often triangular in shape, hence the use of the Greek letter whose symbol is Δ.

Demarcation In **political geography**, the actual placing of a political boundary on the **cultural landscape** by means of barriers, fences, walls, or other markers.

Demographic transition model Multi-stage **model**, based on Western Europe's experience, of changes in population growth exhibited by countries undergoing industrialization. High **birth rates** and **death rates** are followed by plunging death rates, producing a huge net population gain; this is followed by the convergence of birth and death rates at a low overall level. See Figure 8-7.

Demography The interdisciplinary study of population—especially **birth rates** and **death rates**, growth patterns, longevity, **migration**, and related characteristics.

Desert An arid area supporting sparse vegetation, receiving less than 10 inches (25 cm) of precipitation per year. Usually exhibits extremes of heat and cold because the moderating influence of moisture is absent.

Desertification The **process** of **desert** expansion into neighboring **steppelands** as a result of human degradation of fragile semiarid environments.

Development The economic, social, and institutional growth of national **states**.

Devolution The **process** whereby regions within a **state** demand and gain political strength and growing autonomy at the expense of the central government.

Dialect Regional or local variation in the use of a major language, such as the distinctive accents of many residents of the U.S. South or New England.

Diffusion The spatial spreading or dissemination of a **culture** element (such as a technological innovation) or some other phenomenon (e.g., a disease outbreak). There are various channels of outward geographic spread from a source area: **contagious** (via person-to-person contact), **expansion** (through a growing population), **hierarchical** (trickling down from larger to smaller adoption units), and **relocation** (elements are moved with carrier units) **diffusion**. See pp. 234-235 for discussion.

Distance decay The various degenerative effects of distance on human spatial structures and interactions.

Diurnal Daily.

Divided capital In **political geography**, a country whose central administrative functions are carried on in more than one city is said to have divided capitals. The Netherlands and South Africa are examples.

Domestication The transformation of a wild animal or wild plant into a domesticated animal or a cultivated crop to gain control over food production. A necessary evolutionary step in the development of humankind: the invention of **agriculture**.

Domino theory The belief that political destabilization in one **state** can result in the collapse of order in a neighboring state, triggering a chain of events that, in turn, can affect a series of **contiguous** states.

Double cropping The planting, cultivation, and harvesting of two crops successively within a single year on the same plot of farmland.

Dry canal An overland rail and/or road **corridor** across an isthmus dedicated to performing the transit functions of a canalized waterway. Best adapted to the movement of containerized cargo, there must be a port at each end to handle the necessary **break-of-bulk** unloading and reloading.

Ecology The study of the many interrelationships between all forms of life and the natural environments in which they have evolved and continue to develop. The study of *ecosystems* focuses on the interactions between specific organisms and their environments. See also **cultural ecology**.

Economic geography The field of geography that focuses on the diverse ways in which people earn a living, and how the goods and services they produce are expressed and organized spatially.

Economic restructuring The transformation of China into a market-driven economy in the post-Mao era, beginning in the late 1970s.

Economic tiger One of the burgeoning beehive countries of the western **Pacific Rim**. Following Japan's route since 1945, these countries have experienced significant modernization, industrialization, and Western-style economic growth since 1980. Three leading economic tigers today are South Korea, Taiwan, and Singapore. Term is increasingly used more generally to describe any fast-developing economy.

Economies of scale The savings that accrue from large-scale production wherein the unit cost of manufacturing decreases as the level of operation enlarges. Supermarkets operate on this principle and are able to charge lower prices than small grocery stores.

Ecosystem See **ecology**.

Ecumene The habitable portions of the earth's surface where permanent human settlements have arisen.

Edge City See **suburban downtown**.

Elite A small but influential upper-echelon social class whose power and privilege give it control over a country's political, economic, and cultural life.

El Niño-Southern Oscillation (ENSO) A periodic, large-scale, abnormal warming of the sea surface in the low latitudes of the eastern Pacific Ocean that has global implications, disturbing normal **weather** patterns in many parts of the world, especially South America.

Elongated state See **State territorial morphology**.

Emigrant A person **migrating** away from a country or area; an out-migrant.

Empirical Relating to the real world, as opposed to theoretical abstraction.

Enclave A piece of territory that is surrounded by another political unit of which it is not a part.

Endemic A species or disease that is high localized in it spatial extent.

Endemism Refers to a disease in a host population that affects many people in a kind of equilibrium without causing rapid and widespread deaths.

Entrepôt A place, usually a port city, where goods are imported, stored, and transshipped; a **break-of-bulk point**.

Environmental degradation The accumulated human abuse of a region's **natural landscape** that, among other things, can involve air and water pollution, threats to plant and animal ecosystems, misuse of **natural resources**, and generally upsetting the balance between people and their habitat.

Epidemic A local or regional outbreak of a disease.

Escarpment A cliff or very steep slope; frequently marks the edge of a plateau.

Estuary The widening mouth of a river as it reaches the sea; land subsidence or a rise in sea level has overcome the tendency to form a **delta**.

Ethnic cleansing The slaughter and/or forced removal of one **ethnic** group from its homes and lands by another, more powerful ethnic group bent on taking that territory.

Ethnicity The combination of a people's **culture** (traditions, customs, language, and religion) and racial ancestry.

European state model A **state** consisting of a legally defined territory inhabited by a population governed from a capital city by a representative government.

European Union (EU) **Supranational** organization constituted by 27 European countries to further their common economic interests. See Figure 1-10 on p. 44.

Exclave A bounded (non-island) piece of territory that is part of a particular **state** but lies separated from it by the territory of another state. Alaska is an exclave of the United States.

Exclusive Economic Zone (EEZ) An oceanic zone extending up to 200 **nautical miles** from a shoreline, within which the coastal **state** can control fishing, mineral exploitation, and additional activities by all other countries.

Expansion diffusion See **diffusion**.

Extraterritoriality Politico-geographical concept suggesting that the property of one **state** lying within the boundaries of another actually forms an extension of the first state.

Failed state A country whose institutions have collapsed and in which anarchy prevails. Afghanistan during the late-1990s rule of the Taliban is a recent example. Today, Somalia is the textbook case.

Fatwa Literally, a legal opinion or proclamation issued by an Islamic cleric, based on the holy texts of Islam, long applicable only in the *Umma*, the realm ruled by the laws of Islam. In 1989 the Iranian ayatollah Khomeini extended the reach of the *fatwa* by condemning to death a British citizen and author living in the United Kingdom.

Favela Shantytown on the outskirts or even well within an urban area in Brazil.

Federal state A political framework wherein a central government represents the various subnational entities within a **nation-state** where they have common interests—defense, foreign affairs, and the like—yet allows these various entities to retain their own identities and to have their own laws, policies, and customs in certain spheres.

Federation See **federal state**.

Fertile Crescent Crescent-shaped zone of productive lands extending from near the southeastern Mediterranean coast through Lebanon and Syria to the **alluvial** lowlands of Mesopotamia (in Iraq). Once more fertile than today, this is one of the world's great source areas of **agricultural** and other innovations.

First Nations Canada's **indigenous peoples** of American descent, whose U.S. counterparts are called Native Americans.

Fjord Narrow, steep-sided, elongated, and inundated coastal valley deepened by glacier ice that has since melted away, leaving the sea to penetrate.

Floodplain Low-lying area adjacent to a mature river, often covered by **alluvial** deposits and subject to the river's floods.

Forced migration Human **migration** flows in which the movers have no choice but to relocate.

Formal region A type of **region** marked by a certain degree of homogeneity in one or more phenomena; also called *uniform region* or *homogeneous region*.

Formal sector The regulated (licenses, taxes) portion of the economy, in contrast to the **informal sector,** which is unregulated.

Forward capital Capital city positioned in actually or potentially contested territory, usually near an international border; it confirms the **state's** determination to maintain its presence in the region in contention.

Fossil fuels The energy resources of **coal**, natural gas, and petroleum (oil), so named collectively because they were formed by the geologic compression and transformation of tiny plant and animal organisms.

Four Motors of Europe Rhône-Alpes (France), Baden-Württemberg (Germany), Catalonia (Spain), and Lombardy (Italy). Each is a high-technology-driven region marked by exceptional industrial vitality and economic success not only within Europe but on the global scene as well.

Fragmented state See **state territorial morphology.**

Francophone French-speaking. Quebec constitutes the heart of Francophone Canada.

Free-trade area A form of economic integration, usually consisting of two or more **states**, in which members agree to remove tariffs on trade among themselves. Usually accompanied by a **customs union** that establishes common tariffs on imports from outside the trade area, and sometimes by a **common market** that also removes internal restrictions on the movement of capital, labor, and enterprise.

Free Trade Area of the Americas (FTAA) The ultimate goal of **supranational** economic integration in North, Middle, and South America: the creation of a single-market trading bloc that would involve every country in the Western Hemisphere between the Arctic shore of Canada and Cape Horn at the southern tip of Chile.

Frontier Zone of advance penetration, usually of contention; an area not yet fully integrated into a national **state.**

FTAA See **Free Trade Area of the Americas.**

Functional region A **region** marked less by its sameness than its dynamic internal structure; because it usually focuses on a central node, also called *nodal region* or *focal region.*

Functional specialization The production of particular goods or services as a dominant activity in a particular location. See also **local functional specialization.**

Fundamentalism See **revivalism (religious).**

Geographic change Evolution of **spatial** patterns over time.

Geographic realm The basic spatial unit in our world regionalization scheme. Each realm is defined in terms of a synthesis of its total human geography—a composite of its leading cultural, economic, historical, political, and appropriate environmental features.

Geography of development The subfield of economic geography concerned with spatial aspects and regional expressions of **development.**

Geometric boundaries See **boundaries.**

Geomorphology The geographic study of the configuration of the earth's solid surface—the world's landscapes and their constituent landforms.

Ghetto An intraurban region marked by a particular **ethnic** character. Often an inner-city poverty zone, such as the black ghetto in U.S. central cities. Ghetto residents are involuntarily segregated from other income and racial groups.

Glaciation See **Pleistocene Epoch.**

Globalization The increase in economic and cultural integration around the world. This has resulted in a reduction of regional contrasts at the world scale.

Global climate change The changes to climate that occur naturally, often in cycles, although in more recent uses of the term it refers to the human-induced changes (such as the increase of carbon dioxide in the atmosphere) that are accelerating natural climate change.

Green Revolution The successful recent development of higher-yield, fast-growing varieties of rice and other cereals in certain developing countries.

Gross domestic product (GDP) The total value of all goods and services produced in a country by that state's economy during a given year.

Gross national income (GNI) The total income earned from all goods and services produced by the citizens of a country, within or outside of its borders, during a calendar year.

Gross national product (GNP) The total value of all goods and services produced in a country by that state's economy during a given year, plus all citizens' income from foreign investment and other external sources.

Growth pole An urban center with certain attributes that, if augmented by a measure of investment support, will stimulate regional economic development in its **hinterland.**

Hacienda Literally, a large estate in a Spanish-speaking country. Sometimes equated with the **plantation**, but there are important differences between these two types of agricultural enterprise (see p. 140).

Hanification See **Sinicization.**

Hegemony The political dominance of a country (or even a region) by another country. The former Soviet Union's postwar grip on Eastern Europe, which lasted from 1945 to 1990, was a classic example.

Hierarchical diffusion See diffusion.

Hierarchy An order or gradation of phenomena, with each level or rank subordinate to the one above it and superior to the one below. The levels in a national urban hierarchy are constituted by hamlets, villages, towns, cities, and (frequently) the **primate city.**

High island Volcanic islands of the Pacific Realm that are high enough in elevation to wrest substantial moisture from the tropical ocean air (see **orographic precipitation**). They tend to be well watered, their volcanic soils enable productive agriculture, and they support larger populations than **low islands**—which possess none of these advantages and must rely on fishing and the coconut palm for survival.

High seas Areas of the oceans away from land, beyond national jurisdiction, open and free for all to use.

Hinterland Literally, 'country behind,' a term that applies to a surrounding area served by an urban center. That center is the focus of goods and services produced for its hinterland and is its dominant urban influence as well. In the case of a port city, the hinterland also includes the inland area whose trade flows through that port.

Historical inertia A term from manufacturing geography that refers to the need to continue using the factories, machinery, and equipment of heavy industries for their full, multiple-decade lifetimes to cover major initial investments—even though these facilities may be increasingly obsolete.

Holocene The current *interglaciation* epoch (the warm period of glacial contraction between the glacial expansions of an **ice age**); extends from 10,000 years ago to the present. Also known as the *Recent Epoch.*

Human evolution The long-term biological maturation of the human species. Geographically, all evidence points toward East Africa as the source of humankind. Our species, *Homo sapiens*, emigrated from this hearth to eventually populate the rest of the **ecumene.**

Hydraulic civilization theory The theory that cities able to control **irrigated** farming over large **hinterlands** held political power over other cities. Particularly applies to early Asian civilizations based in such river valleys as the Chang (Yangzi), the Indus, and those of Mesopotamia.

Hydrologic cycle The **system** of exchange involving water in its various forms as it continually circulates between the **atmosphere**, the oceans, and above and below the land surface.

Ice age A stretch of geologic time during which the Earth's average atmospheric temperature is lowered; causes the equatorward expansion of continental ice sheets in the higher latitudes and the growth of mountain glaciers in and around the highlands of the lower latitudes.

Immigrant A person **migrating** into a particular country or area; an in-migrant.

Imperialism The drive toward the creation and expansion of a colonial empire and, once established, its perpetuation.

Import-substitution industries The industries local entrepreneurs establish to serve populations of remote areas when transport costs from distant sources make these goods too expensive to import.

Inaccessibility See **accessibility.**

Indentured workers Contract laborers who sell their services for a stipulated period of time.

Indigenous peoples Native or *Aboriginal* peoples; often used to designate the inhabitants of areas that were conquered and subsequently colonized by the **imperial** powers of Europe.

Indo-European languages The major world language family that dominates the European **geographic realm** (Fig. 1-7). This language family is also the most widely dispersed globally, and about half of humankind speaks one of its languages.

Industrial Revolution The term applied to the social and economic changes in agriculture, commerce, and especially manufacturing and urbanization that resulted from technological innovations and specialization in late-eighteenth-century Europe.

Informal sector Dominated by unlicensed sellers of homemade goods and services, the primitive form of capitalism found in many developing countries that takes place beyond the control of government. Compare with **Formal sector.**

Infrastructure The foundations of a society: urban centers, transport networks, communications, energy distribution systems, farms, factories, mines, and such facilities as schools, hospitals, postal services, and police and armed forces.

Inner City The non-suburban residential areas within older cities located beyond the **central business district.**

Insular Having the qualities and properties of an island. Real islands are not alone in possessing such properties of **isolation**: an **oasis** in the middle of a **desert** also has qualities of insularity.

Insurgent state Territorial embodiment of a successful guerrilla movement. The establishment by antigovernment insurgents of a territorial base in which they exercise full control; thus a state within a state.

Intercropping The planting of several types of crops in the same field; commonly used by **shifting cultivators.**

Interglaciation See **Pleistocene Epoch.**

Intermontane Literally, between mountains. The location can bestow certain qualities of natural protection or **isolation** to a community.

Internal migration Migration flow within a country, such as ongoing westward and southward movements toward the **Sunbelt** in the United States.

International migration Migration flow involving movement across an international boundary.

Intervening opportunity In trade or **migration** flows, the presence of a nearer opportunity that greatly diminishes the attractiveness of sites farther away.

Inuit Indigenous peoples of North America's Arctic zone, formerly known as Eskimos.

Irredentism A policy of cultural extension and potential political expansion by a **state** aimed at a community of its nationals living in a neighboring state.

Irrigation The artificial watering of croplands.

Islamic Front The southern border of the African Transition Zone that marks the religious **frontier** of the **Muslim** faith in its southward penetration of Subsaharan Africa (see Fig. 6-19).

Islamization Introduction and establishment of the **Muslim** religion (see Fig. 7-2). A **process** still under way, most notably along the **Islamic Front,** that marks the southern border of the African Transition Zone.

Isohyet A line connecting points of equal rainfall total.

Isolated state See **von Thünen's Isolated State model.**

Isolation The condition of being geographically cut off or far removed from mainstreams of thought and action. It also denotes a lack of receptivity to outside influences, caused at least partially by poor **accessibility.**

Isotherm A line connecting points of equal temperature.

Isthmus A **land bridge**; a comparatively narrow link between larger bodies of land. Central America forms such a link between Mexico and South America.

Jakota Triangle The easternmost region of the East Asian realm, consisting of *Ja*pan, (South) *Ko*rea, and *Ta*iwan.

Juxtaposition Contrasting places in close proximity to one another.

Karst The distinctive natural landscape associated with the chemical erosion of soluble limestone rock.

Ladinos In Guatemala this term is used to referred to **mestizos.**

Land alienation One society or culture group taking land from another. In Subsaharan Africa, for example, European **colonialists** took land from **indigenous** Africans and put it to new uses.

Land bridge A narrow **isthmian** link between two large landmasses. They are temporary features—at least in terms of geologic time—subject to appearance and disappearance as the land or sea level rises and falls.

Land reform The spatial reorganization of **agriculture** through the allocation of farmland (often expropriated from landlords) to **peasants** and tenants who never owned land.

Land tenure The way people own, occupy, and use land.

Landlocked An interior **state** surrounded by land. Without coasts, such a country is disadvantaged in terms of **accessibility** to international trade routes, and in the scramble for possession of areas of the **continental shelf** and control of the **exclusive economic zone** beyond.

'Latin' American city model The Griffin-Ford model of intraurban spatial structure in the Middle American and South American realms, diagrammed in Figure 5-8.

Latitude Lines of latitude are **parallels** that are aligned east-west across the globe, from 0° latitude at the equator to 90° North and South latitude at the poles.

Leached soil Infertile, reddish-appearing, tropical soil whose surface consists of oxides of iron and aluminum; all other soil nutrients have been dissolved and transported downward into the subsoil by percolating water associated with heavy rainfall.

Leeward The protected or downwind side of a **topographic** barrier with respect to the winds that flow across it.

Levant The area extending from Greece eastward around the Mediterranean coast to northern Egypt.

Lingua franca A "common language" prevalent in a given area; a second language that can be spoken and understood by many peoples, although they speak other languages at home.

Littoral Coastal or coastland.

Llanos The interspersed **savanna** grasslands and scrub woodlands of the Orinoco River's wide basin that covers much of interior Colombia and Venezuela.

Local functional specialization A hallmark of Europe's **economic geography** that later spread to many other parts of the world, whereby particular people in particular places concentrate on the production of particular goods and services.

Location Position on the Earth's surface; see **absolute location** and **relative location.**

Location theory A logical attempt to explain the locational pattern of an economic activity and the manner in which its producing areas are interrelated. The agricultural location theory that underlies the **von Thünen model** is a leading example.

Loess Deposit of very fine silt or dust that is laid down after having been windborne for a considerable distance. As part of a developed soil it is notable for its fertility under **irrigation** and its ability to stand in steep vertical walls.

Longitude Angular distance (0° to 180°) east or west as measured from the *prime meridian* (0°) that passes through the Greenwich Observatory in suburban London, England. For much of its length across the mid-Pacific Ocean, the 180th meridian functions as the *international date line.*

Low island Low-lying coral islands of the Pacific Realm that—unlike **high islands**—cannot wrest sufficient moisture from the tropical ocean air to avoid chronic drought. Thus productive agriculture is impossible, and their modest populations must rely on fishing and the coconut palm for survival.

Madrassa Revivalist (**fundamentalist**) religious school where the curriculum focuses on Islamic religion and law and requires rote memorization of the Qu'ran (Koran), Islam's holy book. Founded in former British India, these schools were most numerous in present-day Pakistan but have **diffused** to Turkey in the west and Indonesia in the east.

Maghreb The region occupying the northwestern corner of Africa, consisting of Morocco, Algeria, and Tunisia.

Main Street Canada's dominant **conurbation** that is home to nearly two-thirds of the country's inhabitants; extends southwestward from Quebec City in the mid-St. Lawrence Valley to Windsor on the Detroit River.

Mainland-Rimland framework Twofold regionalization of the Middle American realm based on its modern cultural history. The Euro-Amerindian *Mainland*, stretching from Mexico to Panama (minus the Caribbean coastal strip), was a self-sufficient zone dominated by **hacienda land tenure.** The Euro-African *Rimland*, consisting of that entire Caribbean coastal strip plus all of the Caribbean islands to the east, was the zone of the **plantation** that heavily relied on trade with Europe. See Figure 4-7.

Maquiladora The term given to assembly plants in Mexico. First constructed in the U.S. border zone, these assembly plants are now also found in other parts of Mexico. These foreign-owned factories assemble imported components and/or raw materials, and then export finished manufactures, mainly to the United States.

Marine geography The geographic study of oceans and seas. Its practitioners investigate both the physical (e.g., coral-reef **biogeography**, ocean-atmosphere interactions, coastal **geomorphology**) and human (e.g., **maritime boundary**-making, fisheries, beachside development) aspects of oceanic environments.

Maritime boundary An international boundary that lies in the ocean. Like all boundaries, it is a vertical plane, extending from the seafloor to the upper limit of the air space in the atmosphere above the water.

Median-line boundary An international **maritime boundary** drawn where the width of a sea is less than 400 **nautical miles.** Because the states on either side of that sea claim **exclusive economic zones** of 200 nautical miles, it is necessary to reduce those claims to a (median) distance equidistant from each shoreline. **Delimitation** on the map almost always appears as a set of straight-line segments that reflect the configurations of the coastlines involved.

Medical geography The study of health and disease within a geographic context and from a spatial perspective. Among other things, this field of geography examines the sources, **diffusion** routes, and distributions of diseases.

Megacity Informal term referring to the world's most heavily populated cities; in this book, the term refers to a **metropolis** containing a population of greater than 10 million.

Megalopolis When spelled with a lower-case *m*, a synonym for **conurbation,** one of the large coalescing supercities forming in diverse parts of the world.

When capitalized, refers specifically to the multi-metropolitan corridor that extends along the north-eastern U.S. seaboard from north of Boston to south of Washington, D.C. (Fig. 3-7).

Mercantilism Protectionist policy of European **states** during the sixteenth to the eighteenth centuries that promoted a **state's** economic position in the contest with rival powers. Acquiring gold and silver and maintaining a favorable trade balance (more exports than imports) were central to the policy.

Mesoamerica A term used by anthropologists to refer to the Middle American **culture hearth** which extended southeast from the vicinity of present-day Mexico City to central Nicaragua.

Meridian Line of **longitude**, aligned north-south across the globe, which together with **parallels** of **latitude** forms the global grid system. All meridians converge at both poles and are at their maximum distances from each other at the equator.

Mestizo Derived from the Latin word for *mixed*, refers to a person of mixed European (white) and Amerindian ancestry.

Metropolis Urban **agglomeration** consisting of a (central) city and its suburban ring. See **urban (metropolitan) area**.

Metropolitan area See **urban (metropolitan) area**.

Migration A change in residence intended to be permanent. See also **forced, internal, international**, and **voluntary migration**.

Model An idealized representation of reality built to demonstrate its most important properties. A **spatial** model focuses on a geographical dimension of the real world, such as the **von Thünen model** that explains agricultural location patterns in a commercial economy.

Modernization In the eyes of the Western world, the Westernization **process** that involves the establishment of **urbanization**, a market (money) economy, improved circulation, formal schooling, adoption of foreign innovations, and the breakdown of traditional society. Non-Westerners mostly see 'modernization' as an outgrowth of **colonialism** and often argue that traditional societies can be modernized without being Westernized.

Monsoon Refers to the seasonal reversal of wind and moisture flows in certain parts of the subtropics and lower-middle latitudes. The *dry monsoon* occurs during the cool season when dry offshore winds prevail. The *wet monsoon* occurs in the hot summer months, which produce onshore winds that bring large amounts of rainfall. The air-pressure differential over land and sea is the triggering mecha-

nism, with windflows always moving from areas of relatively higher pressure toward areas of relatively lower pressure. Monsoons make their greatest regional impact in the coastal and near-coastal zones of South Asia, Southeast Asia, and East Asia.

Mosaic culture The emerging cultural-geographic framework of the United States, dominated by the fragmentation of specialized social groups into homogeneous communities of interest marked not only by income, race, and **ethnicity** but also by age, occupational status, and lifestyle. The result is an increasingly heterogeneous socio-spatial complex, which resembles an intricate mosaic composed of myriad uniform—but separate—tiles.

Mulatto A person of mixed African (black) and European (white) ancestry.

Multilingualism A society marked by a mosaic of many local languages. Constitutes a **centrifugal force** because it impedes communication within the larger population. Often a *lingua franca* is used as a "common language," as in many countries of Sub-saharan Africa.

Multinationals Internationally active corporations capable of strongly influencing the economic and political affairs of many countries they operate in.

Muslim An adherent of the Islamic faith.

Muslim Front See **Islamic Front**.

NAFTA (North American Free Trade Agreement) The **free trade area** launched in 1994 involving the United States, Canada, and Mexico.

Nation Legally a term encompassing all the citizens of a **state**, it also has other connotations. Most definitions now tend to refer to a group of tightly-knit people possessing bonds of language, **ethnicity**, religion, and other shared **cultural** attributes. Such homogeneity actually prevails within very few states.

Nation-state A country whose population possesses a substantial degree of **cultural** homogeneity and unity. The ideal form to which most **nations** and **states** aspire—a political unit wherein the territorial state coincides with the area settled by a certain national group or people.

NATO (North Atlantic Treaty Organization) Established in 1950 at the height of the Cold War as a U.S.-led **supranational** defense pact to shield postwar Europe against the Soviet military threat. NATO is now in transition, expanding its membership while modifying its objectives in the post-Soviet era.

Natural hazard A natural event that endangers human life and/or the contents of a **cultural landscape**.

Natural increase rate Population growth measured as the excess of live births over deaths per 1000 individuals per year. Natural increase of a population does not reflect either **emigrant** or **immigrant** movements.

Natural landscape The array of landforms that constitutes the earth's surface (mountains, hills, plains, and plateaus) and the physical features that mark them (such as water bodies, soils, and vegetation). Each **geographic realm** has its distinctive combination of natural landscapes.

Natural resource Any valued element of (or means to an end using) the environment; includes minerals, water, vegetation, and soil.

Nautical mile By international agreement, the nautical mile—the standard measure at sea—is 6076.12 feet in length, equivalent to approximately (1.85 km) (1.15 mi).

Near Abroad The 14 former Soviet republics that, together with the dominant Russian Republic, constituted the U.S.S.R. Since the 1991 breakup of the Soviet Union, Russia has asserted a sphere of influence in these now-independent countries, based on its proclaimed right to protect the interests of ethnic Russians who were settled there in substantial numbers during Soviet times.

Neocolonialism The term used by developing countries to underscore that the entrenched **colonial** system of international exchange and capital flow has not changed in the postcolonial era—thereby perpetuating the huge economic advantages of the developed world.

Nomadism Cyclical movement among a definite set of places. Nomadic peoples mostly are **pastoralists**.

North American Free Trade Agreement See **NAFTA**.

North Atlantic Drift The warm water ocean current that flows across the north Atlantic from North America to northern Europe.

Nucleation Cluster; agglomeration.

Oasis An area, small or large, where the supply of water (from an **aquifer** or a major river such as the Nile) permits the transformation of the immediately surrounding **desert** into productive cropland.

Occidental Western. Also see **Oriental**.

Offshore banking Term referring to financial havens for foreign companies and individuals, who channel their earnings to accounts in such a country (usually an 'offshore' island-state) to avoid paying taxes in their home countries.

Oligarchs Opportunists in post-Soviet Russia who used their ties to government to enrich themselves.

OPEC (Organization of Petroleum Exporting Countries) The international oil *cartel* or syndicate formed by a number of producing countries to promote their common economic interests through the formulation of joint pricing policies and the limitation of market options for consumers. The 11 member-states (as of late 2008) are: Algeria, Indonesia, Iran, Iraq, Kuwait, Libya, Nigeria, Qatar, Saudi Arabia, United Arab Emirates (UAE), and Venezuela.

Oriental The root of the word *oriental* is from the Latin for *rise*. Thus it has to do with the direction in which one sees the sun 'rise'—the east; *oriental* therefore means Eastern. **Occidental** originates from the Latin for fall, or the "setting" of the sun in the west; occidental therefore means Western.

Orographic precipitation Mountain-induced precipitation, especially where air masses are forced to cross **topographic** barriers. Downwind areas beyond such a mountain range experience the relative dryness known as the **rain shadow effect**.

Outback The name given by Australians to the vast, peripheral, sparsely-settled interior of their country.

Outer city The non-central-city portion of the American **metropolis**; no longer 'sub' to the 'urb,' this outer ring was transformed into a full-fledged city during the late twentieth century.

Overseas Chinese The more than 50 million ethnic Chinese who live outside China. Over half live in Southeast Asia (see Fig. 10-5), and many have become quite successful. A large number maintain links to China, and as investors played a major economic role in stimulating the growth of **SEZs** and Open Cities in China's **Pacific Rim**.

Pacific Rim A far-flung group of countries and parts of countries (extending clockwise on the map from New Zealand to Chile) sharing the following criteria: they face the Pacific Ocean; they evince relatively high levels of economic development, industrialization, and urbanization; their imports and exports mainly move across Pacific waters.

Pacific Ring of Fire Zone of crustal instability along **tectonic plate** boundaries, marked by earthquakes and volcanic activity, that ring the Pacific Ocean basin.

Paddies (paddyfields) Rice fields.

Pandemic An outbreak of a disease that spreads worldwide.

Pangaea A vast, singular landmass consisting of most of the areas of the present-day continents. This supercontinent began to break up more than 200 million years ago when still-ongoing **plate** divergence and **continental drift** became dominant **processes** (see Fig. 6-3).

Parallel An east-west line of **latitude** that is intersected at right angles by **meridians** of **longitude**.

Pastoralism A form of **agricultural** activity that involves the raising of livestock.

Peasants In a **stratified** society, peasants are the lowest class of people who depend on agriculture for a living. But they often own no land at all and must survive as tenants or day workers. In parts of Middle and South America they are sometimes referred to as campesinos.

Peninsula A comparatively narrow, finger-like stretch of land extending from the main landmass into the sea. Florida and Korea are examples.

Peon (*peone*) Term used in Middle and South America to identify people who often live in serfdom to a wealthy landowner; landless **peasants** in continuous indebtedness.

Per capita Capita means *individual*. Income, production, or some other measure is often given per individual.

Perforated state See **state territorial morphology**.

Periodic market Village market that opens every third day or at some other regular interval. Part of a regional network of similar markets in a preindustrial, rural setting where goods are brought to market on foot and barter remains a major mode of exchange.

Peripheral development Spatial pattern in which a country's or region's development (and population) is most heavily concentrated along its outer edges rather than in its interior. Australia, with its peripheral population distribution, is a classic example (note its **core area** in Fig. 11-5), and nearby New Zealand exhibits a similar configuration of people and activities.

Periphery See **core-periphery relationships**.

Permafrost Permanently frozen water in the near-surface soil and bedrock of cold environments, producing the effect of completely frozen ground. Surface can thaw during brief warm season.

Physical geography The study of the geography of the physical (natural) world. Its subfields include **climatology, geomorphology, biogeography, soil geography, marine geography,** and **water resources**.

Physical landscape Synonym for **natural landscape**.

Physiographic political boundaries Political boundaries that coincide with prominent physical features in the **natural landscape**—such as rivers or the crest ridges of mountain ranges.

Physiographic region (province) A **region** within which there prevails substantial **natural-landscape** homogeneity, expressed by a certain degree of uniformity in surface **relief, climate,** vegetation, and soils.

Physiography Literally means *landscape description*, but commonly refers to the total **physical geography** of a place; includes all of the natural features on the earth's surface, including landforms, **climate,** soils, vegetation, and water bodies.

Physiologic density The number of people per unit area of **arable** land.

Pilgrimage A journey to a place of great religious significance by an individual or by a group of people (such as a pilgrimage to Mecca for **Muslims**).

Plantation A large estate owned by an individual, family, or corporation and organized to produce a cash crop. Almost all plantations were established within the tropics; in recent decades, many have been divided into smaller holdings or reorganized as cooperatives.

Plate tectonics Plates are bonded portions of the earth's mantle and crust, averaging 100 kilometers (60 mi) in thickness. More than a dozen such plates exist (see Fig. G-5), most of continental proportions, and they are in motion. Where they meet one slides under the other, crumpling the surface crust and producing significant volcanic and earthquake activity; a major mountain-building force.

Pleistocene Epoch Recent period of geologic time that spans the rise of humankind, beginning about 2 million years ago. Marked by *glaciations* (repeated advances of continental ice sheets) and milder *interglaciations* (ice sheet contractions). Although the last 10,000 years are known as the **Holocene** Epoch, Pleistocene-like conditions seem to be continuing and we are most probably now living through another Pleistocene interglaciation; thus the glaciers likely will return.

Plural(istic) society A society in which two or more population groups, each practicing its own **culture,** live adjacent to one another without mixing inside a single **state**.

Polder Land reclaimed from the sea adjacent to the shore of the Netherlands by constructing dikes and then pumping out the water trapped behind them.

Political geography The study of the interaction of geographic space and political **process**; the spatial analysis of political phenomena and processes.

Pollution The release of a substance through human activity that chemically, physically, or biologically alters the air or water it is discharged into. Such a discharge negatively impacts the environment, with possible harmful effects on living organisms—including humans.

Population decline A decreasing national population. Russia, which now loses about 800,000 people per year, is the best example. Also see **population implosion**.

Population density The number of people per unit area. Also see **arithmetic density** and **physiologic density** measures.

Population distribution The way people have arranged themselves in geographic space. One of human geography's most essential expressions because it represents the sum total of the adjustments that a population has made to its natural, cultural, and economic environments.

Population explosion The rapid growth of the world's human population during the past century, attended by accelerating *rates* of increase.

Population geography The field of geography that focuses on the spatial aspects of **demography** and the influences of demographic change on particular places.

Population implosion The opposite of the **population explosion**; refers to the declining populations of many European countries and Russia in which the **death rate** exceeds the **birth rate** and **immigration** rate.

Population movement See **migration; migratory movement**.

Population projection The future population total that demographers forecast for a particular country. For example, in Appendix B such projections are given for all the world's countries for 2025.

Population (age-sex) structure Graphic representation (*profile*) of a population according to age and gender.

Postindustrial economy Emerging economy, in the United States and a handful of other highly advanced countries, as traditional industry is increasingly eclipsed by a higher-technology productive complex dominated by services, information-related, and managerial activities.

Primary economic activity Activities engaged in the direct extraction of **natural resources** from the environment such as mining, fishing, lumbering, and especially **agriculture**.

Primate city A country's largest city—ranking atop the urban **hierarchy**—most expressive of the national culture and usually (but not always) the capital city as well.

Productive activities The major components of the spatial economy. For individual components see: **primary economic activity, secondary eco-** nomic activity, tertiary economic activity, quaternary economic activity, and quinary economic activity.

Protruded state See **State territorial morphology**.

Push-pull concept The idea that **migration** flows are simultaneously stimulated by conditions in the source area, which tend to drive people away, and by the perceived attractiveness of the destination.

Qanat In **desert** zones, particularly in Iran and western China, an underground tunnel built to carry **irrigation** water by gravity flow from nearby mountains (where **orographic precipitation** occurs) to the arid flatlands below.

Quaternary economic activity Activities engaged in the collection, processing, and manipulation of *information*.

Quinary economic activity Managerial or control-function activity associated with decision making in large organizations.

Rain shadow effect The relative dryness in areas downwind of mountain ranges caused by **orographic precipitation,** wherein moist air masses are forced to deposit most of their water content as they cross the highlands.

Rate of natural population increase See **natural increase rate**.

Realm See **geographic realm**.

Refugees People who have been dislocated involuntarily from their original place of settlement.

Region A commonly used term and a geographic concept of central importance. An **area** on the earth's surface marked by specific criteria, which are discussed on pp. 7-8.

Regional boundary In theory, the line that circumscribes a **region**. But razor-sharp lines are seldom encountered, even in nature (e.g., a coastline constantly changes depending upon the tide). In the **cultural landscape,** not only are regional boundaries rarely self-evident, but when they are ascertained by geographers they most often turn out to be **transitional** borderlands.

Regional character The personality or 'atmosphere' of a **region** that makes it distinct from all other regions.

Regional complementarity See **complementarity**.

Regional concept The geographic study of **regions** and regional distinctions, as discussed on pp. 7-8.

Regional disparity The spatial unevenness in standard of living that occurs within a country, whose "average," overall income statistics invariably mask the differences that exist between the extremes of the wealthy **core** and the poorer **periphery**.

Regional geography Approach to geographic study based on the spatial unit of the **region**. Allows for an all-encompassing view of the world, because it utilizes and integrates information from geography's topical (**systematic**) fields; diagrammed in Figure G-13.

Regional state A 'natural economic zone' that defies political boundaries, and is shaped by the global economy of which it is a part; its leaders deal directly with foreign partners and negotiate the best terms they can with the national governments under which they operate.

Regionalism The consciousness and loyalty to a **region** considered distinct and different from the **state** as a whole by those who occupy it.

Relative location The regional position or **situation** of a place relative to the position of other places. Distance, **accessibility**, and connectivity affect relative location.

Relict boundary See **boundaries**.

Relief Vertical difference between the highest and lowest elevations within a particular area.

Religious revivalism See **revivalism (religious)**.

Relocation diffusion See **diffusion**.

Restrictive population policies Government policy designed to reduce the **rate of natural population increase**. China's one-child policy, instituted in 1979 after Mao's death, is a classic example.

Revivalism (religious) Religious movement whose objectives are to return to the foundations of that faith and to influence state policy. Often called *religious fundamentalism*; but in the case of Islam, **Muslims** prefer the term revivalism.

Rift valley The trough or trench that forms when a strip of the earth's crust sinks between two parallel faults (surface fractures).

Rural-to-urban migration The dominant **migration** flow from countryside to city that continues to transform the world's population, most notably in the less advantaged geographic realms.

Russification Demographic resettlement policies pursued by the central planners of the Soviet Empire, whereby ethnic Russians were encouraged to emigrate from the Russian Republic to the 14 non-Russian republics of the U.S.S.R.

Sahel Semiarid **steppeland** zone extending across most of Africa between the southern margins of the arid Sahara and the moister tropical **savanna** and forest zone to the south. Chronic drought, **desertification**, and overgrazing have contributed to severe famines in this area since 1970.

Savanna Tropical grassland containing widely spaced trees; also the name given to the tropical wet-and-dry climate type (*Aw*).

Scale Representation of a real-world phenomenon at a certain level of reduction or generalization. In **cartography**, the ratio of map distance to ground distance; indicated on a map as a bar graph, representative fraction, and/or verbal statement. *Macroscale* refers to a large area of national proportions; *microscale* refers to a local area no bigger than a county.

Scale economies See **economies of scale**.

Secondary economic activity Activities that process raw materials and transform them into finished industrial products. The *manufacturing* sector.

Sedentary Permanently attached to a particular area; a population fixed in its location. The opposite of **nomadic**.

Separate development The spatial expression of South Africa's 'grand' **apartheid** scheme, whereby nonwhite groups were required to settle in segregated "homelands." The policy was dismantled when white-minority rule collapsed in the early 1990s.

Sequent occupance The notion that successive societies leave their cultural imprints on a place, each contributing to the cumulative **cultural landscape**.

Shantytown Unplanned slum development on the margins of cities in disadvantaged countries, dominated by crude dwellings and shelters mostly made of scrap wood and iron, and even pieces of cardboard. These are called *favelas* in Brazil.

Sharecropping Relationship between a large landowner and farmers on the land whereby the farmers pay rent for the land they farm by giving the landlord a share of the annual harvest.

Sharia The criminal code based in Islamic law that prescribes corporal punishment, amputations, stonings, and lashing for both major and minor offenses. Its occurrence today is associated with the spread of **religious revivalism** in **Muslim** societies.

Shatter belt Region caught between stronger, colliding external cultural-political forces, under persistent stress, and often fragmented by aggressive rivals. Eastern Europe and Southeast Asia are classic examples.

Shifting agriculture Cultivation of crops in recently cut and burned tropical-forest clearings, soon to be abandoned in favor of newly cleared nearby forest land. Also known as *slash-and-burn agriculture*.

Sinicization Giving a Chinese cultural imprint; Chinese **acculturation**. This process is sometimes called **Hanification**.

Site The internal locational attributes of an urban center, including its local spatial organization and physical setting.

Situation The external locational attributes of an urban center; its **relative location** or regional position with reference to other non-local places.

Social stratification See **stratification (social)**.

Southern Ocean The ocean that surrounds Antarctica (see discussion on p. 375).

Spatial Pertaining to space on the earth's surface. Synonym for *geographic(al)*.

Spatial diffusion See **diffusion**.

Spatial interaction See **complementarity, intervening opportunity**, and **transferability**.

Spatial model See **model**.

Spatial process See **process**.

Spatial system The components and interactions of a **functional region**, which is defined by the areal extent of those interactions. Also see **system**.

Special Administrative Region (SAR) Status accorded the former dependencies of Hong Kong and Macau that were taken over by China, respectively, from the United Kingdom in 1997 and Portugal in 1999. Both SARs received guarantees that their existing social and economic systems could continue unchanged for 50 years following their return to China.

Special Economic Zone (SEZ) Manufacturing and export center within China, created in the 1980s to attract foreign investment and technology transfers. Seven SEZs currently operate: Shenzhen, adjacent to Hong Kong; Zhuhai; Shantou; Xiamen; Hainan Island, in the far south; Pudong, across the river from Shanghai; and Binhai New Area, next to the port of Tianjin.

Squatter settlement See **shantytown**.

State A politically organized territory that is administered by a sovereign government and is recognized by a significant portion of the international community. A state must also contain a permanent resident population, an organized economy, and a functioning internal circulation system.

State boundaries The borders that surround **states** which, in effect, are derived through contracts with neighboring states negotiated by treaty. See **definition, delimitation**, and **demarcation**.

State capitalism Government-controlled corporations competing under free-market conditions, usually in a tightly regimented society. South Korea is a leading example. Also see *chaebol*.

State formation The creation of a **state** based on traditions of human **territoriality** that go back thousands of years.

State planning Involves highly centralized control of the national planning process, a hallmark of communist economic systems. Soviet central planners mainly pursued a grand political design in assigning production to particular places; their frequent disregard of the principles of economic geography contributed to the eventual collapse of the U.S.S.R.

State territorial morphology The shape of a territory or state. There are five shapes: a *compact* **state** that is a roughly circular, oval, or rectangular and the distance from the geometric center to any point on the boundary exhibits little variance. Poland and Cambodia are examples of this shape category. An *elongated* **state** is decidedly long and narrow and its length is at least six times greater than its average width. Chile and Vietnam are two classic examples. A *fragmented* **state** consists of several separated parts, not a contiguous whole. The individual parts may be isolated from each other by the land area of other states or by international waters. The United States and Indonesia are examples. A *perforated* **state** is a state whose territory completely surrounds that of another state. South Africa, which encloses Lesotho and is perforated by it, is a classic example. A *protruded* **state** exhibits a narrow, elongated land extension (or *protrusion*) leading away from the main body of territory. Thailand is a leading example.

Stateless nation A national group that aspires to become a **nationstate** but lacks the territorial means to do so; the Palestinians and Kurds of Southwest Asia are classic examples.

Steppe Semiarid grassland; short-grass prairie. Also the name given to the semiarid climate type (*BS*).

Stratification (social) In a layered or stratified society, the population is divided into a **hierarchy** of social classes. In an industrialized society, the working class is at the lower end; **elites** that possess capital and control the means of production are at the upper level. In the traditional **caste system** of Hindu India, the "untouchables" form the lowest class or caste, whereas the still-wealthy remnants of the princely class are at the top.

Subduction In **plate tectonics**, the **process** that occurs when an oceanic plate converges head-on with a plate carrying a continental landmass at its leading edge. The lighter continental plate overrides the denser oceanic plate and pushes it downward.

Subsequent boundary See **boundaries**.

Subsistence Existing on the minimum necessities to sustain life; spending most of one's time in pursuit of survival.

Subsistence agriculture Agriculture practiced primarily to sustain only a household or community with little to no commercial purpose.

Subtropical Convergence A narrow marine **transition zone**, girdling the globe at approximately latitude 40°S, that marks the equatorward limit of the frigid **Southern Ocean** and the poleward limits of the warmer Atlantic, Pacific, and Indian oceans to the north.

Suburban downtown In the United States (and increasingly in other advantaged countries), a significant concentration of major urban activities around a highly accessible suburban location, including retailing, light industry, and a variety of leading corporate and commercial operations. The largest are now coequal to the American central city's **central business district (CBD)**. Also called an **edge city.**

Sunbelt The popular name given to the southern tier of the United States, which is anchored by the mega-States of California, Texas, and Florida. Its warmer climate, superior recreational opportunities, and other amenities have been attracting large numbers of relocating people and activities since the 1960s; broader definitions of the Sunbelt also include much of the western U.S., particularly Colorado and the coastal Pacific Northwest.

Superimposed boundary See **boundaries**.

Supranational A venture involving three or more **states**—political, economic, and/or cultural cooperation to promote shared objectives. The **European Union** is one such organization.

System Any group of objects or institutions and their mutual interactions. Geography treats systems that are expressed spatially, such as in **functional regions**.

Systematic geography Topical geography: **cultural, political, economic geography**, and the like.

Taiga The subarctic, mostly **coniferous** snowforest that blankets northern Russia and Canada south of the **tundra** that lines the Arctic shore.

Taxonomy A **system** of scientific classification.

Technopole A planned techno-industrial complex (such as California's Silicon Valley) that innovates, promotes, and manufactures the products of the **postindustrial** informational economy.

Terracing The transformation of a hillside or mountain slope into a step-like sequence of horizontal fields for intensive cultivation.

Territoriality A country's or more local community's sense of property and attachment toward its territory, as expressed by its determination to keep it inviolable and strongly defended.

Territorial sea Zone of seawater adjacent to a country's coast, held to be part of the national territory and treated as a segment of the sovereign **state**.

Tertiary economic activity Activities that engage in *services*—such as transportation, banking, retailing, education, and routine office-based jobs.

Tierra caliente See **Altitudinal zones**.

Tierra fría See **Altitudinal zones**.

Tierra helada See **Altitudinal zones**.

Tierra nevada See **Altitudinal zones**.

Tierra templada See **Altitudinal zones**.

Topography The surface configuration of any segment of **natural landscape**.

Toponym Place name.

Transculturation Cultural borrowing and two-way exchanges that occur when different **cultures** of approximately equal complexity and technological level come into close contact.

Transferability The capacity to move a good from one place to another at a bearable cost; the ease with which a commodity may be transported.

Transhumance Seasonal movement of people and their livestock in search of pastures. Movement may be vertical (into highlands during the summer and back to lower elevations in winter) or horizontal, in pursuit of seasonal rainfall.

Transition zone An area of spatial change where the **peripheries** of two adjacent realms or regions join; marked by a gradual shift (rather than a sharp break) in the characteristics that distinguish these neighboring geographic entities from one another.

Transmigration The now-ended policy of the Indonesian government to induce residents of the overcrowded, **core-area** island of Jawa to move to the country's other islands. Sometimes referred to using the Indonesian *transmigrasi*.

Treaty ports **Extraterritorial enclaves** in China's coastal cities, established by European colonial invaders under unequal treaties enforced by gunboat diplomacy.

Tropical deforestation See **deforestation**.

Tropical savanna See **savanna**.

Tsunami A seismic (earthquake-generated) sea wave that can attain gigantic proportions and cause coastal devastation. The tsunami of December 26, 2004, centered in the Indian Ocean near the Indonesian island of Sumatera, produced the first great natural disaster of the twenty-first century.

Tundra The treeless plain that lies along the Arctic shore in northernmost Russia and Canada, whose vegetation consists of mosses, lichens, and certain hardy grasses.

Turkestan Northeasternmost region of the North Africa/Southwest Asia realm. Known as Soviet Central Asia before 1992, its five (dominantly Islamic) former S.S.R.'s have become the independent countries of Kazakhstan, Uzbekistan, Turkmenistan, Kyrgyzstan, and Tajikistan. Today Turkestan has expanded to include a sixth state, Afghanistan.

Unitary state A **nation-state** that has a centralized government and administration that exercises power equally over all parts of the state.

Urbanization A term with several connotations. The proportion of a country's population living in urban places is its level of urbanization. The **process** of urbanization involves the movement to, and the clustering of, people in towns and cities—a major force in every geographic realm today. Another kind of urbanization occurs when an expanding city absorbs rural countryside and transforms it into suburbs; in the case of cities in disadvantaged countries, this also generates peripheral **shantytowns**.

Urban (metropolitan) area The entire built-up, nonrural area and its population, including the most recently constructed suburban appendages. Provides a better picture of the dimensions and population of such an area than the delimited municipality (central city) that forms its heart.

Urban realms model A spatial generalization of the contemporary large American city. It is shown to be a widely dispersed, multicentered metropolis consisting of increasingly independent zones or *urban realms*, each focused on its own **suburban downtown**; the only exception is the shrunken central realm, which is focused on the central city's **central business district**.

Voluntary migration Population movement in which people relocate in response to perceived opportunity, not because they are forced to **migrate**.

Von Thünen's Isolated State model Explains the location of **agricultural** activities in a commercial economy. A **process** of spatial competition allocates various farming activities into concentric rings around a central market city, with profit-earning capability the determining force in how far a crop locates from the market. The original (1826) Isolated State model now applies to the continental scale (see Fig. 1-5).

Wahhabism A particularly virulent form of (Sunni) **Muslim revivalism** that was made the official faith when the modern **state** of Saudi Arabia was founded in 1932. Adherents call themselves 'Unitarians' to signify the strict fundamentalist nature of their beliefs.

Wallace's Line As shown in Figure 11-4, the zoogeographical boundary proposed by Alfred Russel Wallace that separates the marsupial fauna of Australia and New Guinea from the nonmarsupial fauna of Indonesia.

Weather The immediate and short-term conditions of the **atmosphere** that impinge on daily human activities.

West Wind Drift The clockwise movement of water around Antarctica in the **Southern Ocean**.

Wet monsoon See **monsoon**.

Windward The exposed, upwind side of a **topographic** barrier that faces the winds that flow across it.

INDEX

A Reference with an italicized number refers to figures, tables, boxes and captions. An italic n *refers to a footnote.*

Abadan, Iran, 262
A-B-C islands, 142
Aberdeen, Scotland, 57
Aboriginal land issue, 377, 382
Aboriginal population, 376–377
Aborigines, 374, *377*, 379, 382
Abrupt climate change, *6*
Absolute location, 8
Abu Dhabi, 259
Acculturation, 144
Activity space, 2
Adamawa Highlands, 219
Adelaide, Australia, 379
Adis Abeba, Ethiopia, 218
Adygeya, 97
Afghanistan, *242*, 245, 265–267
Africa, 6. *See also* North Africa/Southwest Asia; Subsaharan Africa
  early states, *201*
  land issues, 196–198
  languages, 206–207
  physiography, *192*, 193–195
  wildlife, 195
African Plate, 195
African Transition Zone, *6*, 225–227, 247, 249
  Djibouti, 226
  Eritrea, 226
  Ethiopia, 226
  Somalia, 226, 227
Afro-Asiatic languages, 206
Agglomerative forces, 33
Agrarian revolution, Europe, 33
Agriculture:
  agrarian revolution, 33
  ancient Rome, 32
  animals in, 8–9
  Austral Realm, 370, 380, *381*, *383*
  East Asia, 304, 305, 320, 321, 324, 335, *338*, 339
  ethanol crops, 126–127
  Europe, 33, 44, 45, 55, 57, 62, 63, 65, 67
  and globalization, 22
  Green Revolution, 198
  Mesopotamia, 233
  Middle America, 136–140, *141*, 147, 150–155, *156*, *157*, *158*, *159*, *160*
  North Africa/Southwest Asia, 234, *245*, 246, 249, 253, 260, 265

North America, 57, 109, 118–119, 122, 126, 127
Pacific Realm, 385, 395, 396
Russia, 79, 99, 101
South America, 169–170, *172*, 179, 182–184, 186, 188
South Asia, *283*, 284, 292–293, 296, *298*, 299
Southeast Asia, 358
Subsaharan Africa, *192*, 195–198, 210–211, 216, 217, 221, 223
  von Thünen rings, 33
AIDS, *192*, 198, 199, 211, 217, 218
Akaroa, New Zealand, 373
Akaroa Harbor, New Zealand, *387*
Alabama, 117
Alaska, 107, 129
Albania, 45, 66, 70, 72–73
Alberta, *118*, 121, 124
Algeria, 247, 249
Algiers, Algeria, 249
Alpine Mountains (Europe), 31, *32*
Alpine states (Europe), 54–55
  Austria, 54, 55
  Liechtenstein, 54
  Switzerland, 54, 55
Alps, 31
al-Qaeda, 237, 249, 267, 285
Altaic languages, *38*, *316*
Altiplano, 165–166, 180
Altitudinal zones, Middle America, 150
Amazonas State (Brazil), 188
Amazon Basin, 164, 166, 167, *169*, 172, 179, 188, 189
Amazonians, 166–167
Amazon River, 164, 179
American Manufacturing Belt, 114, 119, 125
American Samoa, 390
Amerindians, 165
  Middle America, 135–139, 144–146, 150–153, 155
  South America, 165–167, 170, *173*, 177, 180, 184
Amsterdam, Netherlands, *37*
Anatolian Plateau, 260
Ancient Greece, 33, 34, 64
Ancient Rome, 33, 34
Ancona Line, 62
Andalusia, 63
Andamanese, *274*

Andaman Islands, 298
Andaman Sea, 359
Andean Community, 170
Andean South America, *see* West (South America)
Andean zone (Ecuador), 179, 180
Andes Mountains, 150, 164, 165, 168, 176, 179, 180
Andes region, Argentina, 182
Andorra, 47
Angara River, 101
Angola, 210, 211, 216
Ankara, Turkey, 260–261
Annam, 357
Antarctica, 6, 398–399
Antarctic Icesheet, 11
Antarctic Peninsula, 398
Antarctic Treaty, 399
Antecedent boundaries, 353, *354*
Anthropogeographic boundaries, 353
Antigua, Guatemala, *133*
Apartheid, 212, 213
Appalachian Highlands, 107, 117
Appennine Mountains, 31
Arabian desert, *230*
Arabian Peninsula, 242, 257–259
  Bahrain, 259
  Djibouti, 259
  Kuwait, 259
  Oman, 259
  Qatar, 259
  Saudi Arabia, 257–259
  United Arab Emirates, 259
  Yemen, 259
Arabian Plate, 195
"Arab World," 231
Arakan Mountains, 362
Aral Sea, 264
Arauca, Colombia, 175
Archipelagos, 136, 161, 370
Arctic, 80–83
Arctic icecap, 12
Arctic National Wildlife Refuge, 129
Arctic Ocean, 81, *82*, *83*, 392, 393
Area, 3, 8
Areal functional organization, 336
Argentina, 181–182
  economy, 164

independence, 168
megacities, 172
population, 171, 181
soybeans, *169*
Spanish in, 167
Arizona, 128
Armenia, 76, 97, 99
Aruba, 142
Asom, India, 287
Astana, Kazakhstan, 264
Asturias, 63
Aswan High Dam, Egypt, 246
Atacama Desert, 168, 183
Athens, Greece, 40, 65
Atlanta, Georgia, 127
Atlantic Multi-Decadal Oscillation, *158*
Atlantic Provinces (Canada), 121, 125
Atlantic Seaboard Megalopolis, 114
Atlas Mountains, 193, 247, 248
Attenuated states, 355
Auckland, New Zealand, 385
Austral (as term), 374
Australia, 377–384
  Aborigines, 376–377, 382
  climate, 375, *377*, *383*
  core and periphery, 378–379
  cultural geography, 379–380
  distance in, 377–378
  economic geography, 374, *377*, 380–382
  as federal state, 379
  immigration, 378
  maritime boundaries, 393
  Pacific Rim trade, 129
  physical geography, *375*
  population, *374*, 378, 384
  zoogeographical boundaries, *376*
Australian Capital Territory, 379
Australian Plate, 347, 374
Austral Realm, 372–387
  Australia, 377–384
  biogeography, 376
  climates, 375
  human impact, 376–377
  New Zealand, 384–387
  physical geography, *372*, 374–377
  population distribution, *375*
  regions, 377
  Southern Ocean, 375–376

Austria, 54, 55
Austro-Asiatic languages, 274, *316*
Autocracies, *94*, 224, 331
Autonomous Communities (ACs), Europe, 62, 63
Autonomous Regions (ARs), China, 314
Autonomous Soviet Socialist Republics
    (ASSRs), 87, 90
Axum, 202
Aymara, 178
Azerbaijan, 76, 97, 99–100
Aztec civilization, 136, 137, 167

Bab-el-Mandeb Strait, 259
Babylon, 233
Baden-Württemberg, Germany, 47
Baghdad, Iraq, 251
Bahamas, 6
Bahia, Brazil, 185, 186
Bahrain, 259
Baja California Peninsula, 142
Baki (Baku), Azerbaijan, 100
Bali, 368
Balkanization, 65
Balkan Mountains, 31, 70
Balkan Peninsula (Balkans), 65
Balsas Lowland, 146
Baluchistan province (Pakistan), *281*, 284
BAM (Baykal-Amur Mainline) Railroad, 102
Banana republic, 142
Bangka, 368
Bangkok, Thailand, 359, 361, *361*
Bangladesh, 275, 277, *281*, 295–296
Bantu migration, 202
Barbados, 142
Barcelona, Spain, 62
Barents Sea, 80
Barrios, 173
Basilan Island, 370
Basin-and-Range, 108
Basque Country, 46, 62
Bass Strait, 380
Baykal, Lake, 80. *See also* Lake Baykal area
    (*Baykaliya*)
Baykal-Amur Mainline (BAM) Railroad, 102
Beijing, China, 314, 319, 330
Belarus, 6, 45, 66–68, 77, 93
Belém, Brazil, *163*
Belfast, Northern Ireland, 57
Belgium, 36, 46, 54, 204
Belitung, 368
Belize, 149–152
Benelux, 48, 54
    Belgium, 54
    Luxembourg, 54
    Netherlands, 54

Be-Ne-Lux Agreement, 42
Bengal, Bay of, 295, 346
Bengaluru (Bangalore), India, *269*
Benin, 203
Bering land bridge, 136
Berlin, Germany, 41
Beslan, North Ossetia, Russia, *100*
Bhutan, 297–298
Biafra, 222
Bihar State, India, 294
Binhai New Area SEZ, 327
Biodiversity, 347, 376
Biofuels, 126–127, 186
Biogeography, 376
Bioko island, 221
Birmingham, England, 57
Birth rates:
    Brazil, 184–185
    China, 315
    Russia, 93, *94*
    South Asia, 276, 278, 279
Blue Nile, 245, 247
Bogotá, Colombia, 173, 175
Bohemia, 68
Bolivarian Revolution, 177
Bolivia, 167, 177, 180
Bonaire, 142
Bora Bora, French Polynesia, *389*, *396*
Border Industrialization Program (Mexico), 147
Borneo, 346, 364, 365, 368
Bosnia, 36n. 1, 71
Botswana, 211, 215
Boundaries, 3, 8
    Brazil, 184
    classifying, 353
    East Timor, 370
    Guatemala, 151
    maritime, 346, 391–393
    North Africa/Southwest Asia, *230*, 239
    political vs. economic, 23
    of and within realms, 18
    South America, 170
    Southeast Asia, 353–355
    Subsaharan Africa, *192*
    and transition zones, 6
    U.S.-Mexico, 142
Boxer Rebellion, 312–313
Brahmaputra Valley, 287
Brasília, Brazil, 189
Brazil, *164*, *173*, 174, 184–189
    African heritage, 185, 186
    change campaign, 164
    climates, 184
    development prospects, 186–188
    economy, 184, 187, 188

in IBSA, 20
immigration, 170
income distribution, *186*
inequality in, 186
megacities, 172
periphery, *19*
population, 171, 184–185
soybeans, *169*
states, *185*
subregions, *185*, 188–189
Brazilian Highlands, 164
Break-of-bulk function, 59
Brisbane, Australia, 379
Britain (island), 55
Britain (nation), *see* United Kingdom
British Columbia, 121, 123, 124, 131
British Guiana, 177
British Honduras, 151
British Isles, *29*, *48*, 55–58
    Republic of Ireland, 57–58
    United Kingdom, 57
Brunei, 364–366
Brussels, Belgium, *39*, 54
Bucharest, Romania, 70
Buda, Hungary (Budapest), 68
Buddhism, *232*, 274
    East Asia, 317, 322
    South Asia, *270*, 284, 297, 299
    Southeast Asia, 351, 356
Buenos Aires, Argentina, 173, 182
Buffalo Commons, 127
Buffer states, 265, 297, 332
Buffer zones, 344
Buganda, Kingdom of, 218
Bulgaria, 43, 65, 70
Burkina Faso, 224
Burma, *see* Myanmar
Burundi, 218
Buryat Republic, 101
Busan, South Korea, 339

Cabinda, 216, 221
Cairo, Egypt, *229*, 245
Calabria, 61
California, 130
Callao, Peru, 178
Cambodia, 347, 351, 356–358
Cameroon, 219, 221
Camisea reserve (Peru), 179
Canada, 120–124
    and Arctic melting, 82, 83
    contrasts between United States and, 107
    cultural/political geography, 122–124
    economic geography, 124
    indigenous peoples, 110

oil, 240
Pacific Rim trade, 129
population, 107, *120*, 121–122
Quebec, *106*
and United States, 106
Canadian Pacific Railway, 122
Canadian Shield, 107, 117, 124, 129
Canals. *See also specific canals*
    dry, 149, 153
    Europe, 30, *32*, *35*, *49*, *56*
    Middle America, 137, 149, 153, *154*
    North Africa/Southwest Asia, *245*, *251*
    Russia, 96
    South Asia, 276
Canal Zone (Panama), 153
Candomblé, 185
Cantabria, 63
Canterbury Plain, New Zealand, 385, *386*
Canton, China, 312
Cape Province (South Africa), 213
Cape Ranges, 193, 210
Cape Town, South Africa, *191*, 212, 214–215
Caprivi Strip, 215
Caracas, Venezuela, 176
Cardiff, Wales, 57
Caribbean Basin, *134*, *139*, 154–161
    economic and social patterns, 155–156
    ethnicity, 155–156
    Greater Antilles, 156–161
    independence movements, 142
    islands of, 136
    Lesser Antilles, 161
    Spanish in, 138, 139
Caribbean Plate, 136
Caribbean Sea, 393
Caribbean South America, *see* North (South America)
Carpathian Mountains, 31
Carreterra Transoceanica, 179
Cartograms, 171
Cartography, 2, 375. *See also* Maps
Cascades mountains, 108, 128
Caspian Sea, 100
Caste system, 273–274, 288–289
Castile-La Mancha, 63
Castile-Leon, 63
Catalonia, 46, 47, 62–63
Catholicism, *see* Roman Catholicism
Caucasus Mountains, 79, 80
Cauca Valley, 175
Cayman Trench, 158
Cell phone revolution, 198, *208*
Celtic Tiger, 57
Central African Republic, 219, 221
Central America, 136. *See also* Middle America

Central American Common Market, 150
Central American Free Trade Agreement-Dominican Republic (CAFTA-DR), *141*, 150
Central American republics, *134*, 149–154
    altitudinal zones, 150
    Belize, 151–152
    climates, 150
    Costa Rica, 153
    economic geography, 150
    El Salvador, 152
    Guatemala, 151
    Honduras, 152
    independence movement, 142
    Nicaragua, 152–153
    Panama, 153–154
    physiography, 150
    population, 149, 150
Central Asia, *see* Turkestan
Central Asian Ranges, 79, 80
Central Business Districts (CBDs), 115, 173, *187*
Central Indian Plateau, 273
Central Industrial Region (Russia), 95
Central Siberian Plateau, 80
Central Uplands (Europe), 30, *32*
Central Valley, 108
Central-West (Brazil), 189
Centrifugal forces, 36
    Europe, 36, 45–46
    Pacific Realm, 395
    South America, 180, 182
    South Asia, 284, 288–289
    Southeast Asia, 364
    Subsaharan Africa, 208, 219
Centripetal forces, 36, 284, 289–291
Cerrado, 169, 189
Cerro de Pasco, Peru, 179
Ceuta, 249
Chaco, Argentina, 182
Chad, 219–221, 249
Chang Jiang (Long) River, 304
Chang Jiang River Basin, 317, 320–321
Chang River Basin, 320–321
Chao Phraya River, 346, 361
Chaparral, *130*
Chechnya, 90, 97, 99
Chennai (Madras), India, 291
Chesapeake Bay colony, 113
Chiapas State (Mexico), 145
Chicago, Illinois, 115
Chile, 182–184
    Amerindians, 167
    climate, 183
    economy, 164, 183
    independence, 168

Pacific Rim trade, 129
    subregions, 182–183
    urban population, 171
China, *302, 304. See also* Taiwan
    air pollution, *319*
    boundary, 353
    climates, 307–308
    under communism, 313–315
    dams, 359
    economic zones, 326–328
    and globalization, *22*
    history of, 308–313
    independence movement, 312–313
    and India, 287
    Little Ice Age in, 11–12
    move of manufacturing to, 147
    oil, 240
    one-child policy, 315–316
    Pacific Rim, 324–331
    Pacific Rim trade, 129
    and Panama Canal, 154
    political divisions, *315*
    population, 11, 16, *280*, 315–317, *323, 328*
    relative location, 307
    size and environment, 307–308
    in South America, *173*
    in Southeast Asia, 351, 352, *356*
    state formation, 305
    in Subsaharan Africa, *208*
    as superpower, 331–332
    and Taiwan, 340, 341
    Xinjiang, 323–324
    Xizang (Tibet), 322–323
    in Zambia, 216
China Proper, 305–322
    Chang/Yangzi Basins, 320–321
    Inner Mongolia, 319–320
    North China Plain, 318, 319
    Northeast China Plain, 317–318
    population, 317, *318*
    Xi (West) and Pearl River Basins, 321–322
Chinese religions, *232*
Chittagong, Bangladesh, 296
Choke points, 226, 259, 364
Cholla, 339
Chongqing, China, 314, 321
Chota-Nagpur Plateau, 273
Christianity, *232*
    Europe, 37, 66
    North Africa/Southwest Asia, *230*, 231, 238
    North America, 116
    Russia, 92, 93, 99
    Southeast Asia, 370
    Subsaharan Africa, 202, 207, 222, 224, 225
Chungyang Mountains, 340

Cities:
    ancient Rome, 32
    Austral Realm, 378–380
    on cartograms, 171
    East Asia, 317, 319, 327, 328, 335
    Europe, 32, 36, 49–50, 53, 54, 56, 57, 59, 65
    Middle America, 137, *138*, 143–144, 155
    North Africa/Southwest Asia, 233, 262
    North America, 107
    primate, 40
    site and situation of, 51
    South America, 170–173, 186
    South Asia, 273, 276, 290
    Subsaharan Africa, 209
City Hall, Cape Town, South Africa, *212*
City-states, 136, 202
Ciudad del Este, Paraguay, 181
Ciudad Juarez, Mexico, 144
Climate(s), 10–15
    Austral Realm, 375, *383*
    climate change, 10–12
    East Asia, 304, 307–308
    Europe, 29–30, *31*, 60
    ice ages, 10–12
    Middle America, 135, 143, 150, 153
    North Africa/Southwest Asia, *230*
    North America, 108
    Pacific Realm, 396–397
    regions, climatic, 12–15
    Russia, 78–79, 102
    South America, 183, 184
    South Asia, 272, 273
    Subsaharan Africa, 195–196
    as technical term, 12
    weather vs., 79
Climate change, *6*, 10–12, 233, 376. *See also* Global climate change
Climatology, 78
Coahuila State (Mexico), 147
Coal:
    Austral Realm, 380, *385*
    East Asia, 322, *334*
    Europe, 29, 34, *35, 56*, 57
    Middle America, *143*
    North America, 117, 118
    South America, *174, 178*
    South Asia, 293
Coastal zone (Ecuador), 179
Cochin China, 357
Cocos Plate, 136
Cold polar climate, *13*
Cold polar climates, 14
Cold War, 21, 28, 37, 76, *192*, 216, 265, 331
Colombia, 153, 167, *174*–177
Colombo, Sri Lanka, *299*

Colón, Panama, 154
Colón Free Zone, 154
Colonialism, 312–313
    Austral Realm, 379, 386
    core-periphery dichotomy in, 19, 20
    East Asia, 310–311, *312*, 332–333, 336, 338, 340
    Europe, 28, 32, 36, 56–57, 62, 65–66
    as globalization, 23
    Middle America, 137–139, 142, 151, 155–157, 159
    North Africa/Southwest Asia, 234, 238–239, 242, 247, 249, 251, 253, 263, 265
    North America, 113, 121–122
    Pacific Realm, 390–391, 398
    Russia, 83
    South America, *166*, 167–170, 178
    South Asia, *270*, 275–276, 291
    Southeast Asia, *344*, 345–346, 350–353, 357–358, 366
    Subsaharan Africa, 195–197, 202–205, 208, 221–222
Colorado Plateau, 108, 128
Colorado River, 109
Columbia Plateau, 108
Columbia River, 108, 109, 131
Command economies, 89
Commercial agriculture, 169
Common Agricultural Policy (CAP), 44
Common Markets:
    Central America, 150
    Europe, 43, 50, 61
    South America, 170
Communism:
    China, 307, 313–314, 326
    East Germany, 50–51
    North Korea, *339*
    Russia, 83–87
    Vietnam, 358
    Yugoslavia, 70–71
Compact states, 355
Complementarity, 39, 339
Confucius, 308
Congo (Republic of the Congo), 202n. 1, 219, 221
The Congo (Democratic Republic of the Congo), 202n. 1, 219–220
Congo Basin, 195, 219
Connecticut, 126
Constantinople, Turkey, 260
Contagious diffusion, 235
Continental climates, 14
Continental drift, 9, 194–195
Continental Interior (North America), 126–127
Continentality, 79, 109

Continental shelves, 391, 393
Contras, 152
Conurbations, 54, 114–115, 127, 130, 321, 337, 368
Cook, Mount, 374
Cook Islands, 390
Cook Strait, 386
Copenhagen, Denmark, 59
Córdoba, Argentina, 182
Core areas, 19, 80, 94
    Australia, 378–379
    China, 319
    India, 275, 287
    Indochina, *357*
    Japan, 335
    Mexico, 146
    Myanmar, 362
    Nepal, 297
    Nicaragua, 152
    Nigeria, *223*
    Pakistan, 284
    Philippines, *361*
    Southeast Asia, *345*
    Subsaharan Africa, 209
    Taiwan, *340*
    Thailand, *360*
Core-periphery relations, 19, 24
Core Region (North America), 114, 125. *See also* American Manufacturing Belt
Corporations, power of, 20–21
Corsica, 46
Costa Rica, 150, 153
Council of Europe, 42
Cowlitz-Puget Sound Lowland, 108
Cree nation, 123
Crete, 63
Crime and corruption:
    Austral Realm, 380
    Europe, 41, 65, 70
    Middle America, 146, 149, 154, 160
    Russia, 76, 92
    South America, 164, 182, 186, 187
    South Asia, 281
    Southeast Asia, *344*, 358, 364
Crimea Peninsula, 69
Croatia, 36*n*. 1, 45, 71
Cuba, 142, 155–158, 177
Cultural diffusion, 233, 234
Cultural geography, 16–17, 25, 233
Cultural landscapes, 16, 233
Cultural pluralism, 107
    Middle America, 135
    North America, *106*, 107, 121
    South America, *164*, 169
Cultural revival, 243

Culture, *6*
Culture hearths, 113, 233
    East Asia, *302*
    Middle America, 136
    North Africa/Southwest Asia, *230*, 232–234
    North America, 113
    South Asia, *286*
    Subsaharan Africa, 201
Curaçao, 142
Currency:
    Brazil, 187
    Dominican Republic, 160
    European Union, 44
    Thailand, 361
    United Kingdom, 57
Cuyo, Argentina, 182
Cuzco, Peru, 165, 167
Cyberjaya, Malaysia, 364
Cyclades, 63
Cyclones, 272, 295, 363. *See also* Hurricanes
Cyprus, 43, 63, 65, 261
Czechoslovakia, 45, 66, 68
Czech Republic, 43, 68

Dagestan, 97
Dahomey, 203
Dams, 224, *246*, *301*, 318, 359
Damascus, Syria, 253
Danube River Basin, 68
Darfur, Sudan, *242*, *245*, 247–248
Darien Province (Panama), 153
Darwin, Australia, 379
Days of the Dead, *145*
Death rates, 93, 278, 279
Debt, 21, *208*
Deccan, 273
Deforestation:
    Middle America, 137
    Southeast Asia, *347*
    soybeans, *169*
    tropical, 153
Deglomerative forces, 33
Demarcation, 353
Democracy:
    East Asia, 307
    Europe, 36, 62, 65
    Middle America, 146, 153, 158
    North Africa/Southwest Asia, 255
    Russia, 76, 91–93
    South America, 164, 165, 184
    South Asia, 285, 289, 297
    Southeast Asia, 359
    Subsaharan Africa, 214, 224
Democratic Republic of the Congo, *see* The Congo
Demographic transition, 278

Denmark, 43, 44, 59
Deserts, 9, 14
    Austral Realm, 375
    formation of, 233–234
    North Africa/Southwest Asia, 230, 249, *258*
    Russia, 79
    Subsaharan Africa, 193–194
Desertification, 11, *197*
Developing countries, growth model for, 325–327
Devolution, 45
    Europe, 45–46, 54, 55, 57, 59, 66, 71
    North America, *106*, 124
    South Asia, 288
Dhaka, Bangladesh, *269*, 296
Dialects, 115
Disamenity, zone of, 173
Distance decay, 92
Distribution of features, 3–4
Districts, 5
Djibouti, 226, 259
Dnieper River, 69
Domains, 5
Dominican Republic, 142, 150, 155, 156, 159–160
Domino theory, 356, 357
Donets Basin, 69
Drakensberg Mountains, 193
Dravidian languages, 274, *274*
Drought, 195, *377*, *383*
Drugs:
    East Asia, 310–311
    Middle America, 149, 154
    North Africa/Southwest Asia, *242*, *266*, 267
    Russia, 93
    South America, 175–176
    Southeast Asia, 359
Dry canals, 149, 153
Dry climates, *13*, 14
"Dry World," 230
Dubai, 259
Durban, South Africa, 214
Dutch Guiana, 177

Earthquakes, 9–10
    Austral Realm, 374
    East Asia, 303, 304, 316, 332, 334, 337
    Middle America, 136, 152, 161
    South America, *179*
    South Asia, 272, 284
    Southeast Asia, 346, 368
    Subsaharan Africa, 195, 210
East Africa, 202, 216–219
    Burundi, 218
    Ethiopia, 218
    Kenya, 216, 217

    population, *217*
    Rwanda, 218
    Tanzania, 218
    Uganda, 218
East African Plateau, 216
East Asia, 300–341
    China Proper, 306–322
    China's Pacific Rim, 324–331
    cultural geography, 305
    geographic change drivers, *304*
    historical geography, 304–305
    Jakota Triangle, 332
    Japan, 332–338
    Korea, 338–340
    Mongolia, 332
    physical geography, *300–301*, 302–304
    political geography, *302*
    population, 11, 16, *303–304*, *306*
    regions, 305–306
    Taiwan, 340–341
    Xinjiang, 323–324
    Xizang (Tibet), 322–323
East China Sea, 391, 393
Easter Island, 390–391, *397*
Eastern Europe, *29*, *48*, 65–73
    Albania, 72–73
    Belarus, 68
    Bosnia, 71
    Bulgaria, 70
    Croatia, 71
    cultural legacies, 65–66
    Czech Republic, 68
    devolution, 46
    and EMU, 44
    ethnicity, 65, *69*, *71*
    geographic framework, 66, 67
    Hungary, 68
    Kosovo, 72
    Latvia, 68
    Lithuania, 68
    Macedonia, 71, 72
    Moldova, 70
    Montenegro, 72
    Ottoman invasion of, 32
    Poland, 67–68
    population, *66*, *67*
    religion, 37
    Romania, 70
    Serbia, 72
    Slovakia, 68
    Slovenia, 71
    Ukraine, 69–70
Eastern Frontier (Russia), 100–101
    Kuznetsk Basin (*Kuzbas*), 100
    Lake Baykal area (*Baykaliya*), 101

Eastern Ghats, 273
Eastern Highlands (Russia), 79–80
Eastern Industrial Region, 294
East Germany, 18, 49–51
East Indies, 352
East Malaysia, 364, 365
East Timor, 345, 364, 370, 384, 393
Economic geography, 18–25
    causes of contrast in, 19–20
    classification of development, 18–19
    and corporate power, 20–21
    and debt, 21
    and globalization, 21–24
Economic growth, *208*
Economic integration, Europe, *47*
Economic restructuring, 317
Economic tigers, 329, 340, 361
Economies, world, *17*
Economies of scale, 119
Ecotourism, 152, 153
Ecuador, 167, 177, 179–180, 240
Ecumene, 121
Edge Cities, 115
Edinburgh, Scotland, 57
Education:
    Chile, 184
    India, 290, 294
    and Islamization, 235
    Mexico, 147
Egypt, 233, 245–247
Eire, 55. *See also* Republic of Ireland
Ejidos, 145, 147
Elba (island), 32
Elbe River, 68
Elburz Mountains, 261
Elite residential sectors, 173
Ellice Islands, 398
Elmina Castle, Elmina, Ghana, *204*
El Niño, 188
Elongated states, 182, 355
El Salvador, 150, 152
Emigration:
    East Asia, 308, 328
    Europe, 34, 57
    Middle America, 152, 155, 160
    Southeast Asia, 349–350, 352, 358, 364,
        370–371
Empire States, 259–263
    Iran, 261–263
    Turkey, 260–261
Enclaves, 175, 224, 311, 312
Endemic diseases, 198
Energy resources. *See also specific types, e.g.:* Oil
    Austral Realm, 386
    East Asia, *322*, 324

Middle America, 147, 149
    North America, 117–118
    South America, 179, 180, 186
    South Asia, 293
England, 45, 46, 55, 113. *See also* United Kingdom
English Channel, 56
English language, 23, 37, 115
Entrepôts, 59, 154, 366
Entre Rios, Argentina, 182
Environmental degradation, 297, 383, 384
Environmental issues. *See also* Climate change
    Africa, 195–196
    Australia, 383–384
    California, 130
    and Colombia flower industry, 176
    Green Revolution, 198
    New Zealand, 386–387
    Pacific Realm, *396*
    from tourism, 156
Epidemics, 198–199
Equatorial Africa, 219–221
    Cabinda, 221
    Cameroon, 221
    Central African Republic, 221
    Chad, 220–221
    Congo, 221
    The Congo, 219–220
    Equatorial Guinea, 221
    Gabon, 221
    population, *220*
    São Tomé and Príncipe, 221
Equatorial Guinea, 219, 221
Eritrea, 218, 226, *227*
Estonia, 43, 59, 88
Ethanol, 126–127
Ethiopia, 218, 225, 226
Ethnic cleansing:
    Darfur, Sudan, 247
    Eastern Europe, 65
    Serbia, 72
    Subsaharan Africa, 218, 220
    Turkey, 261
Ethnicity:
    Austral Realm, 382–384
    East Asia, 305, 308, 316–317, 320, 324, *325*,
        340
    Europe, 30, 40, 57, 59, 60, 65, 68–73
    Middle America, 135, 139, *140*, 144, 145,
        151–153, 155–156, 158–159
    North Africa/Southwest Asia, 231, 247, *251*,
        252, 253, 255, *256*, *257*, 258, 261,
        263–265, *266*
    North America, 107, 116, 131
    Pacific Realm, 395, 397
    Russia, 77, 78, 84–87, 97, 99, 102

South America, 165–170, 174, 175, 177–180,
        184, *188*
    South Asia, 270, 271, 280, 288, 299
    Southeast Asia, 348, 349, *350*, *356*, 358, 360,
        363, 364, 368–370
    Subsaharan Africa, *202*, 204, 212–213, 215,
        217, 218, 226, 227
Euphrates River, 233, 251
Euphrates Valley, 253
Eurasian Plate, 271, 303
Euro-African, 139
Euro-Amerindian, 139
Europe, 26–73
    British Isles, 55–58
    centrifugal forces, 45–46
    climates, 29–30, *31*
    contemporary, 36–42
    cultural geography, 39
    Eastern, 65–73
    eastern border of, 28–29
    economic geography, 46–47
    economies, *28*, 37, 42
    ethnicity, 30
    expansions, 45, 64–65
    geographic change drivers, *47*
    glaciation in, 11
    historical geography, 31–32
    Little Ice Age in, 11–12
    locational advantages, 30
    Mediterranean, 60–65
    modern transformation of, 42–47
    natural resources, 29
    Northern (Nordic), 58–60
    physical geography, *26*, 30–31
    population, 16, *28*, 33, 40–42, *49*, 61
    regions, 47–48
    revolutions in modernization of, 33–36
    and Russia, 28–29, 93, 94
    subsidies in, 19
    unification of, 42–44
    von Thünen rings, *34*
    Western, 48–55
European Community (EC), 43, 48
European Economic Community (EEC), 43
European Monetary Union (EMU), 44
European Recovery Program, 42
European state model, 18
European Union (EU), 36, 43–45
    Alpine states in, 55
    and crime, 41
    Eastern European countries in, 67–68, 70–72
    economic geography, 46–47
    Greenland's withdrawal from, 59
    Mediterranean European countries in, 61,
        63–65

and Turkey, 251
    United Kingdom in, 57
Everest, Mount, 272, 296
Exclaves, 67, 99, 216, 221, 224, 249, 370
Exclusive Economic Zones (EEZs), 83, 392–393
Expansion diffusion, 234, 235
Extended states, 355
Extraterritoriality, 311–312
Extremadura, 63

Failed states, 180
    Afghanistan, 267
    Bolivia, 164, 165, 180
    Nepal, 297
    Somalia, 227
Falkland Islands, 182
Fall Line cities, 109
Far East (Russia), *91*, 102–103
Farm resource regions, *118*, 119
Faroe Islands, 59
Favelas, 173, 186, *187*
Federalism:
    Austral Realm, 379
    Europe, 45, 57
    North America, *106*, 120–121
    Russia, 88, 90–92
    South America, 188
    South Asia, 284, 286
Federal systems, 91
Federations, 88, 379
Fertile Crescent, 233
Fiji, 391, 395
Finland, 37, 59
Finnmark (Norway), 59
First Nations (Canada), 110, 123, 129
First World countries, 18–19
Fishing:
    Europe, 29, 59, 60
    maritime boundaries, 391–393
    North America, 126
    Pacific Realm, *396*
    Russia, 102
    South America, 178
Flemish region, Belgium, 54
Flood-control technology, 36, 54
Florence, Italy, 61
Florida Island (Solomon Islands), *389*
Food shortages, China, 319
Forests:
    Austral Realm, 383, *385*
    Europe, 29
    in interglacials, 11
    Middle America, 137, 153
    North Africa/Southwest Asia, *266*
    Russia, *101*

Forests (continued)
South America, 166–167, 183
South Asia, 287, 295, 297
Formal regions, 8
Formal sectors, 209
Forward capitals, 83, 189, 256, 264, 283
Four Motors of Europe, 47, 62
Fragmented states, 355
France, 36, 49, 51–52, 63
devolution, 46
in East Asia, 311, 312
economy, 54
as European Union force, 48
and Germany, 51, 54
and globalization, 22
growth center in, 47
in Middle America, 139, 142, 159
in North Africa/Southwest Asia, 249
in North America, 113, 121–122
in Pacific Realm, 390, 395
religion, 41
in South America, 173
in Southeast Asia, 350–352, 357–358
in Subsaharan Africa, 204, 219, 221
Francophones, 121
Frankfurt, Germany, 50
Fraser Valley, 108
Free Trade Area of the Americas (FTAA), 165, 170, 173
French Canada, 122, 126
French Guiana, 177
French Polynesia, 390
French Revolution, 36
Frontiers, 3, 129, 317, 398
Functional regions, 8, 336

Gabon, 219, 221
Galicia, 46, 63
Gambia, 224
Ganges River Basin, 11, 16, 273, 277
Ganges Valley, 273
Garagum (Kara Kum) Canal, 265
Gaza Strip, 242, 254–256
Genetically modified crops, 137
Genetic boundary classification, 353–354
Geneva, Switzerland, 55
Geographic change drivers, 6. See also under specific realms
Geographic information systems (GISs), 2
Geographic realms, 4–7
criteria for, 6–7
cultural landscapes of, 17
and population distributions, 15–16
regions of, 24
states in, 18

Geographic regions, 5, 18. See also under specific realms
criteria for, 7–8
and population distributions, 15–16
in realms, 24
Geography:
classification systems in, 4–5
cultural, 16–17, 25
of demography, 277
of development, 3, 325
economic, 18–25
human, 3, 12
maps in, 2–3
marine, 391
medical, 198
physical, 8–15
political, 18, 25
population, 276–277
regional, 5, 25
systematic, 25
urban, 25
Geometric boundaries, 353
Georgia (Transcaucasia), 76, 93, 97, 99
Geothermal power, 386
Germany, 49–51
in Africa, 204
in China, 311
emergence of, 36
as European Union force, 48
and France, 51, 54
growth center in, 47
religion, 37, 41–42
Ghana, 200–201, 204, 224
Ghettos, 115
Gibraltar, Strait of, 62
Gilbert Islands, 398
Glaciations, 10–11
Glasgow, Scotland, 57
Global climate change, 3, 11–12, 78
Arctic melting, 80–83, 94
Canada, 124
and coastal flooding, 296
Europe, 31
and forest expansion, 101
and hurricanes, 158
and ice ages, 10–12
North Africa/Southwest Asia, 233–234
in Pleistocene, 31
Russia, 94
sea-level rise, 393, 396
and wildfires, 130
Globalization, 21
and European colonialism, 28
and free trade, 197–198
India, 291–292

Mexico, 148
and population growth, 16
Venezuela, 177
Xinjiang, 325
Goa, India, 294
Golan Heights, 254, 255
Golden Quadrilateral, 291
Gondwana, 195, 376
Grand Canyon, 108
Great Artesian Basin, 374
Great Basin, 108
Great Dividing Range, 374, 378
Greater Antilles, 134, 136, 154, 156–161
Cuba, 157, 158
Dominican Republic, 159–160
Haiti, 158–159
Jamaica, 158
Puerto Rico, 160–161
Greater Sunda islands, 368
Great Escarpment, South Africa, 193, 210
Great European Plain, 31
Great Lakes:
North America, 109
Subsaharan Africa, 193, 210
Great Lakes regional state, 124
Great Leap Forward, 314
Great Plains (North America), 107, 109, 110, 126, 127
Great Proletarian Cultural Revolution, 314
Great Salt Lake, 108
Greece, 31, 63–65. See also Ancient Greece
Greenhouse effect, 12, 78, 81
Greenland, 59
Green Line (Cyprus), 65
Green Revolution, 198
Grenada, 155
Gross domestic (national) product (GDP), 21n. 2
Gross National Income (GNI), 21
Growth pole, 189
Groznyy, Chechnya, 90, 97, 99
Guadalajara, Mexico, 144
Guadeloupe, 142, 156, 161
Guam, 390, 396
Guangdong Province (China), 314, 315, 321, 328
Guatemala, 149–151
Guayaquil, Ecuador, 179
Guayas River, 179
Guiana Highlands, 164, 176
Guianas, 174. See also French Guiana; Guyana; Suriname
Guinea, 222
Gulf-Atlantic Coastal Plain, 107
Gulf Coast, Mexico, 146
Gulf of Mexico, 109

Guyana, 177
Gypsies, 68

Haciendas, 140, 144, 146, 167, 179, 180, 182, 183
The Hague, 54
Haifa, Israel, 253
Hainan Island SEZ, 304, 326
Haiti, 156, 158–159
Hamburg, Germany, 50
Hanification, 308, 323
Hanoi, Vietnam, 358
Havana, Cuba, 155, 157
Hawai'i, 107, 390, 393, 397, 398
Hawaiian Islands, 390, 398
Health issues. See also AIDS
Amerindians, 166, 167
and Aswan High Dam, 246
Brazil, 189
Egypt, 246
Haiti, 159
Russia, 93
South Asia, 281
Subsaharan Africa, 192, 198–199, 209
Uzbekistan, 264
Hebei Province (China), 314
Helsinki, Finland, 59
Henan Province (China), 315
Hidrovia, 170
Hierarchical diffusion, 235
High islands, 390, 396
Highland climates, 13, 14–15
High seas, 391, 392
Highveld, 212
Himalaya Mountains, 270, 272, 296, 303
Hinduism, 232
South Asia, 270, 273–276, 284, 288, 296–297, 299
Southeast Asia, 351, 369, 370
Hindu Kush Mountains, 265, 272
Hindustan, 275
Hindutva, 289
Hinterland, 8
Hispaniola, 136, 156, 159. See also Dominican Republic; Haiti
Historical perspective, 3
Hobart, Australia, 379
Ho Chi Minh City, Vietnam, 358
Hohhot, Inner Mongolia, 320
Hokkaido island (Japan), 304, 305, 334
Holocene, 10, 11
Homogeneity, 8
Honduras, 150, 152
Hong Kong (Xianggang), 315, 326–329, 340
Honshu island (Japan), 304, 333, 334

Hormuz Strait, 259
Horn of Africa, 226
Huang He Basin, 317
Huang He (Yellow) River, *301*, 304, 318
Human evolution, 192
Human geography, 3, 12
Humid cold climates, *13*, 14
Humid equatorial climates, *13*, 14
Humid temperate climates, *13*, 14
Hungary, 36, 43, 55, 68
Hurricanes, 136, 152, 153, 155, *158*, 161. *See also* Cyclones
Hutongs, 319
Hydraulic civilization theory, 233
Hydroelectric power:
    Austral Realm, 386
    East Asia, 321
    Europe, 55, 62
    North Africa/Southwest Asia, *246*
    North America, 123, 126, 131
    Russia, 102
    South America, *186*, 189
    South Asia, 293, 297
    Southeast Asia, 359
Hydrography (surface water), 109, 324. *See also* Rivers

Iberian Peninsula, 32, 62–63
    Portugal, 63
    Spain, 62–63
Ice ages, 10–12
Iceland, 36, 45, 59–60
Idaho, 128
Île de la Cité, 51
Immigration:
    Austral Realm, 378, *380*, 382–384
    East Asia, 314, 320, 337
    Europe, 40–42, *47*, 49, 55, 62, 68
    Middle America, 142, 148, 152, 156
    North Africa/Southwest Asia, 249
    North America, *106*, 111, 112, 116, 122, *124*, 127–128
    Russia, 93
    South America, 170, 182, 184, 185
    Southeast Asia, 349–350, 352, 364
    Subsaharan Africa, 209, 215
Imperialism, 83, 352
Import-substitution industries, 380
Incas, 165–166
India, *270*, 285–295
    agriculture, 292–293
    centripetal forces, 289–291
    economic geography, 281, 291–292
    energy problem, 293
    in IBSA, 20

independence movement, 276
language-based reorganization of, 287–289
manufacturing, 293–294
north-south divide in, 294–295
political geography, 286–287
population, 270, *279*, *280*, 285, *286*, 287
and United States, *281*
Indian Plate, 271, 303
Indigenous peoples:
    Austral Realm, 374, 376–377, 379, 382, 386
    East Asia, 317
    Europe, 40, 59
    Middle America, 136–137, 158, 159
    North America, 110, *116*, 123, 129
    Pacific Realm, *390*
    Russia, 81
    South America, 165–167, 170, 180, 184, 189
    Southeast Asia, 344, 368, 369
    Subsaharan Africa, 195–197
Indochina, 312, 351–352, 356–357
    Cambodia, 358
    Laos, 359
    Mekong River, 359
    population, *357*
    Vietnam, 357–358
Indochina War, 358
Indo-European languages, 37, *38*, 206, *274*, *316*
Indonesia, *344*, *356*, 368–370
    and Australia, 384
    biodiversity, 347
    Jawa, 368
    Kalimantan, 368
    Papua, 369
    population, *344*, 345, 347, 369
    Sulawesi, 368–369
    Sumatera, 368
    tectonic activity, *346*
    territorial configuration, 355
Indus River, 273, *277*
Industrial intensification, 33
Industrialization:
    Europe, 33, 34, *35*
    North Africa/Southwest Asia, 243
    North America, *106*, 119–120, 126, 131
    South America, 187, 188
    Russia, 95, 96
    South Asia, 276, 293
Industrial Revolution, 33
    Argentina, 182
    Canada, 122
    Europe, 33, 34, *35*
    Russia, 95, 96
    United Kingdom, 57
    United States, 113
Industry. *See also specific industries, e.g.:* Tourism
    Austral Realm, 380, *382*, 385

East Asia, 317, 319, 321, 329, 330, 335, *338*, 339, 340
    Europe, 54, 55, 57, 59, 67, 69
    Middle America, 147, 152, 160, 161
    North Africa/Southwest Asia, 260
    North America, 119–120, 122, 125, 127, 129
    Russia, 95, 96, 100, 103
    South America, 176, 182, 183
    South Asia, 294, 296
    Southeast Asia, 366, 368
    technopoles, 119–120
Indus Valley, 273, 275
Informal sectors, 173, 209, 291–292
Infrastructure, 32
Ingushetiya, 90
Inner cities, 115
Inner Mongolia, 319–320
In situ accretion, zone of, 173
Insurgent states, 175, 299
Interconnections, 8
Interglacials, 10, 11
Interior Lowlands (North America), 107, 109
Intermontane, 108, 248
Intermontane Basins and Plateaus, 108
Internal Southern Periphery (Russia), 96–99
Internet, *119*, 120
Inter-Tropical Convergence Zone (ITCZ), 195
Intervening opportunity, 39, 294
Iran, 240, *242*, 261–263
Iranian Plateau, 261–262
Iraq, 240, *242*, 250–253
Ireland, 43, 55. *See also* Northern Ireland; Republic of Ireland
Irkutsk, Russia, 101
Iron Curtain, 28
Irrawaddy Delta, Myanmar, 363
Irrawaddy River, *346*, *362*
Irredentism, 68, 284–285
Irtysh River, 80
Isfahan, Iran, *262*
Islam and Muslims, *232*
    Arab-Islamic empire, 234–235
    areas under rule of, *236*
    diffusion of, 231, 234–235
    divisions within, 236–237
    East Asia, 317
    Europe, 37, 39–42, 66, 70, 72
    and European Union, 45
    North Africa/Southwest Asia, *230*, 234–239, 249, 251, 253, 261, 263, 265
    and other religions, 237–238
    and Ottoman Empire, 238–239
    principles of faith, 234
    Russia, 90, 97, 99

South America, 184
South Asia, 270, 271, 275, 276, *277*, 282, 284, 288, 296, 298, 299
South Asia, *270*, *271*, 275, 276, *277*, 282, 284, 288, 296, 298, 299
Southeast Asia, 352, *356*, 364, 365, 369, 370
Subsaharan Africa, 192, 201, 202, 207, 224, 225, 227
Islamic Front, 225–226
"Islamic World," 231
Islamization, 235
Isohyets, 109
The Isolated State, 33
Israel, *230*, *242*, 253–257
Israeli Security Fence, *255*, 256
Istanbul, Turkey, 260–261
Isthmus, 136
Italy, 36, 46, 47, 60–62
Ivanovo, Russia, 95
Ivory Coast, 222, 224, 225

Jabotabek, 368
Jaffna Peninsula, 299
Jakota Triangle, 306, 332, *333*, 339, 340
Jalisco State (Mexico), 147
Jamaica, 142, 155, 156, 158
Jammu and Kashmir, *see* Kashmir
Japan, *302*, *304*, 332–338
    Britain in, 333, 334
    in China, 311–313, 317
    coastal development, 337
    colonial wars and recovery, 332–333
    cultural geography, 305
    economy, 19, 20, 327, 337
    Jakota Triangle, 332
    Kurile Islands, 93
    modernization, 335–336
    Pacific Rim trade, 129, 337–338
    population, *334*, 336–338
    relative location, 336
    spatial issues, 334–336
    subsidies in, 19
    timber, *347*
    trade with Chile, 183
    in Vietnam, 358
Japan, Sea of (East Sea), 393
Java Sea, 368
Jawa, 352, 368, 369
Jerusalem, Israel, 238, 253, 255, 256, *257*
Jezira-al-Maghreb, 248. *See also* Maghreb
Jiangsu Province (China), 315
Johannesburg, South Africa, 214
Johor Strait, 354
Jordan, 253, 254
Judaism, 116, *230*, 231, *232*, 238
Junggar Basin, 324
Jutland Peninsula, 59

Kabol (Kabul), Afghanistan, 265
Kalaallit Nunaat, 59
Kalahari Desert, 194
Kalimantan, 368
Kaliningrad, 67, 68
Kamchatka Peninsula, 80
Kano, Nigeria, *207*
Kansai District, 335, 337
Kanto Plain, 335, 337
Kaohsiung, Taiwan, 340
Karachi, Pakistan, 282, 283
Karakoram Range, 272
Karamay, Xinjiang, 324
Kashmir, *270,* 271, 272, *281, 282,*
    283–284
Kathmandu, Nepal, 296
Katowice, Poland, 67
Kazakhstan, 6, 77, 263, 264
Kazan, Russia, 96
Kenya, 216, 217
Khabarovsk, Russia, 102, 103
Khatachira, Bangladesh, *296*
Khoisan languages, 206
Khorat Plateau, 361
Khyber Pass, 265
Kiev (Kyyiv), Ukraine, 69
Kilwa, 202
Kingston, Jamaica, 155, 158
Kiribati, 391, 398
Kitakyushu, 337
Klong canal, Bangkok, Thailand, *343*
Klyuchevskaya, Mount, 80
Kobe, Japan, 337
Kola Peninsula, 80
Kolkata (Calcutta), India, 291
Kolok River, 360
Komsomolsk, Russia, 102, 103
Kongo, 202
Korea, 332, 338–340. *See also* North Korea;
    South Korea
Korean Peninsula, 302, 304
Korean War, 338–339
Kosciusko, Mount, 374
Kosovo, 18, 36*n.* 1, 66, 72
Kra Isthmus, 359
Krakatau, 346
Krakow, Poland, 67
Kuala Lumpur, Malaysia, 364
Kunda, Kenya, *22*
Kunlun Shan, 324
Kurds, 250, 251, 261
Kurile Islands, 93
Kush, 201, 202
Kuwait, 240, 242, 250, 259
Kuznetsk Basin (*Kuzbas*), 100

Kwangju, South Korea, 339
Kyongsang, 339
Kyoto, Japan, 333
Kyrgyzstan, 265
Kyushu island (Japan), 304, 334

Ladinos, 151
Lagos, Nigeria, 222
Lahore, Pakistan, 282, 285
Lake Baykal area (*Baykaliya*), 101
Land alienation, 167, 196–197
Land bridges, 136, 150
Land hemisphere, 30, 192
Land issues:
    in Arctic, 82–83
    Australia, 382
    South Asia, 281
    Subsaharan Africa, 196–198
Landlocked locations, 55, 218, 220, 224, 296
Landmasses, 5
Land reform:
    Brazil, 187
    Ecuador, 180
    El Salvador, 152
    India, 292
    in Mexico, 144–145
    South America, 172
Landscapes:
    cultural, 16
    natural, 9
Land tenure, 196
Land use:
    Japan, *335*
    Middle America, 137, *141*
    Northern Frontier (North America), 129
    population, *258*
    South America, 169–170
    Subsaharan Africa, 195
Languages. *See also Lingua franca*
    ancient Rome, 32
    as centrifugal force, 36
    East Asia, *316*
    Europe, 37, *38,* 54, *56,* 59, 60
    and globalization, 23
    Middle America, 135, 136, 138, 144, 152, 153,
        156, 158
    North Africa/Southwest Asia, 231, 265
    North America, 115, 121–123
    Pacific Realm, 395
    South America, 178, 180, 184
    South Asia, 271, 274, 280, 281, 284, 287–289,
        299
    Southeast Asia, 358, 369, 370
    Subsaharan Africa, 204–207, 219
Laos, 347, 351, 356, 357, 359

La Plata, 167
Latin America, 135. *See also* Middle America;
    South America
'Latin' American city model, 172–173
Latvia, 43, 68, 88, 93
Lebanon, 253
Leeward Islands, 161
Leith, Scotland, 57
Lena River, 101–102
Leningrad, Soviet Union, 84
León, Mexico, 144
Leopoldville, Belgian Congo, 204
Lesotho, 215
Lesser Antilles, *134,* 136, 155, 161
Lesser Sunda Islands, 368
Levant, 237
Lhasa, Xizang, 322
Liaodong Peninsula, 304, 311
Liao River, 304, 317
Liao-Songhua Basin, 317
Liberia, 222, 224
Libya, 240, 247, *248,* 249
Liddonia, Missouri, *127*
Liechtenstein, 54
Lima, Peru, 167, 168, 178
Line Islands, 390
Lingua franca, 37
    Middle America, 152
    South Asia, 271, 284, 286, 287
    Southeast Asia, 358
    Subsaharan Africa, 205, 206, 219
Lisbon, Portugal, *62,* 63
Lithuania, 43, 68, 88, 93
Little Ice Age, 11–12
Littoral (coastal) states, 391, 392
Liverpool, England, 57
Llanos, 176
Local functional specialization, 32
Localization, 24
Location, 3, 8
Location theory, 3, 33
Loess, 304
Loess Plateau, 304
Lombardy, 46, 47, 61
London, England, 34, 40, 56, 57
Londonderry, Northern Ireland, 57
Long March, 313
Los Angeles, California, 130
Lower Canada, 122
Lower Egypt, 246
Lower Nile Basin, 247–248
Low islands, *390,* 396
Luxembourg, 54
Luzon, 370
Lyon, France, 53

Maasai Mara National Game Reserve, Kenya,
    216
Macau, 311, 315, 326, 328, 329
Macdonnell Ranges, 374
Macedonia, 36*n.* 1, 65, 71, 72
Machu Picchu, 166
McMahon Line, 287
Madagascar, 218–219
Madrassas, 285
Madrid, Spain, 63
Magdalena Valley, 175
Maghreb:
    Algeria, 249
    Libya, 249
    Morocco, 249
    population, *248*
    Tunisia, 249
Maharashtra State, India, 294
Maine, 125
Mainland, *135,* 136, 139–140
Mainland-Rimland framework, 139
Main Street, 121
Malacca, Strait of, 364
Malaria, 199
Malawi, 210, 211, 216
Malawi, Lake, 210
Malaya, 364, 365
Malay Peninsula, 349, 356, 361, 364
Malay-Polynesian languages, 206, *274*
Malaysia, *356,* 364–365
    boundary, 354
    Pacific Rim trade, 129
    population, *365*
    and Thailand, 360
    timber exports, *347*
Maldives, 278, 298
Mali, *1,* 201, 249
Malta, 43, 44, 65
Maluku Islands, 347, 368
Malvinas, 182
Manado, Sulawesi, 368
Managua, Lake, 154
Managua, Nicaragua, 152
Manaus, Brazil, 189
Manchester, England, 57
Mandalay, Myanmar, 362
Manila, Philippines, 352
Manila-Quezon City, Philippines, 370
Manitoba, 121, 124
Manufacturing:
    Austral Realm, 380–382
    East Asia, *322, 335*
    Europe, 34, 56, 59, 62, 65
    Middle America, 147, 152, 160
    North America, *106,* 114, 119, 122, 126, 130

Russia, 96
  South America, 182–184, 187
  South Asia, 293–294
Maori, 374, *377*, 384
Maps, 2–4, 7, 375
Maquiladoras, 147, 149
Maracaibo, Lake, 176
Maracaibo Lowlands, Venezuela, 176
Marie Byrd Land, 398
Mariinsk canals, 96
Marine geography, 391
Marine West Coast climate, 14
Maritime boundaries, 346, 391–393
Maritime Northeast (North America), 125–126
Marquesas Islands, 398
Marshall Islands, 396
Marshall Plan, 42, 49
Martinique, 142, 161
Massachusetts, 126
Maturity, zone of, 173
Mauritania, 249
Mauryan Empire, 274–275
Maya civilization, 136–137, 151, 167
Mecca (Makkah), 234
Medan, Sumatera, 368
Median lines, 83, 392
Medical geography, 198
Medina (Al Madinah), 234, 235
Mediterranean climate, 14
Mediterranean Europe, 29, 48, 60–65
  climate, 60
  Cyprus, 65
  great civilizations in, 31
  Greece, 63–65
  Iberian Peninsula, 62–63
  Italy, 61–62
  Malta, 65
  population, 60, 64
Mediterranean Sea, 31, 32
Megacities, 143–144, 172, *187*
Megalopolises, 114
  East Asia, 328, 337
  North America, 118, 121
  Southeast Asia, 368
Megalopolitan growth, 114, 115
Meiji Restoration, 333, 336
Mekong Delta, 359
Mekong River, *344*, 346, 359
Melanesia, *390*, 395
Melbourne, Australia, 379, *380*
Melilla, 249
Mendoza, Argentina, 182
Mensk (Minsk), Belarus, 68
Mental maps, 2
Mercantilism, 32, 36, 56

Mercosur/l, 170, 181, 183
Mérida, Mexico, 146
Meroe, 202
Mesoamerica, 136–137
Mesopotamia, 230, 233, 273
Mestizos, 139, 144, 151, 152, 179
Mexico (country), *134, 135*, 142–149
  boundary of, 6
  climates, 143
  cultural mix in, 144
  economic geography, *141*, 146–148
  independence movement, 142
  oil, 240
  physiography, 142–144
  population, 135, 140, *143, 144*
  regions of, 145–147
  revolutions in, 144–145
  states of, 143, *144*
Mexico (Mexican State), 143
Mexico City, Mexico, 135, 143–144
Mezzogiorno, 61, 62
Michigan, 19
Micronesia, 396, *396*
Microstates, 47, 54
Mid-Atlantic Ridge, 60
Middle America, 132–161
  Caribbean Basin, 154–161
  Central American republics, 149–154
  climate, 135
  cultural collisions, 137–138
  cultural geography, *135*
  geographic drivers in, *141*
  Greater Antilles, 156–161
  Lesser Antilles, 161
  mainland and rimland, 138–140, *141*
  Mesoamerican legacy, 136–137
  Mexico, 142–149
  physical geography, *132–133*, 136
  political geography, *135*, 140–142
  population, *134, 135*
  pre-Columbian, 136–137
  regions, 18, *134*, 142
Middle Atlantic region, 113
Middle Chile, 182
Middle East, *230*, 250–257
  Iraq, 250–253
  Israel, 253–257
  Jordan, 253
  Lebanon, 253
  Syria, 253
  as term for North Africa/Southwest Asia,
    250–257
Middle Egypt, 246
Midlands (England), 57
Midway Islands, 390

Migration:
  East Asia, 317, 330, 339
  Middle America, 144, 148
  North Africa/Southwest Asia, 243, 249, 262
  North America, *106*, 110–112
  Pacific Realm, 396
  rural-to-urban, 171, 172
  South America, 172, 177, 184
  South Asia, 276, 296
  Southeast Asia, 350, 369–370
  Subsaharan Africa, 202, 212, 215
Milan, Italy, 61, 62
Minahasa, Sulawesi, *346*
Minahasa Peninsula, 368
Minas Gerais State (Brazil), 188
Mindanao, 370
Mindoro, 370
Mineral resources:
  ancient Rome, 32
  Austral Realm, 375, 380, 381, *385*
  East Asia, 320, *334*
  Europe, 29, 34, *35*
  Middle America, *143*, 147, 151, *156*, 157, 158,
    159, 160
  North Africa/Southwest Asia, *248, 254*, 260
  North America, 117, 124, 129
  Pacific Realm, 395
  Russia, 101, 102
  South America, 176, 179, 180, 183, *185*,
    186–188
  Southeast Asia, 361
  Subsaharan Africa, 210–211, 214–216, 219,
    221
Mining:
  Austral Realm, 380
  Brazil, *19*
  China, *322*
  Europe, 32
  Middle America, 152
  North America, 129
  Pacific Realm, 395
  Russia, 101, 103
  South America, 179, 183
  Subsaharan Africa, 215
Ministates, 54, 65, 215
Minnesota, 127
Mississippi-Missouri river network, 109
Moçambique, 216
Models, 33, 172
Modernization, 335
  Japan, 335–336
  Malaysia, 364
Mogadishu, 202
Moldova, 45, 70, 78, 88
Mombasa, 202

Monaco, 47
Monarchies, 33, 258, 261, 297, 359
Mongolia, 306, 312, 332
Monroe Doctrine, 142
Monsoons, 14, *270*, 272, 273
Montenegro, 18, 36*n*. 1, 66, 72
Monterrey, Mexico, 144, 147
Montevideo, Uruguay, 184
Monument to the Discoveries, 62
Morelia State, Mexico, *133*
Mormonism, *232*
Morocco, 247, 249
Mosaic culture, United States as, 116
Moscow, Russia, *75*, 92, 95
Moscow Canal, 96
Moscow Region, *see* Central Industrial Region
    (Russia)
Mountains, 9. *See also specific ranges*
  Austral Realm, 374, 384
  East Asia, 302–303
  Europe, 55, 68
  Middle America, 136, 142, 143, 150, 152, 153,
    157
  North Africa/Southwest Asia, 261–262
  North America, 107, 108
  Russia, 79, 80
  South America, 164, 183
  South Asia, 272–273
  Southeast Asia, 346
  Subsaharan Africa, 193
Mulattos, 156
Multilingualism, 206–207
Multinational corporations, 20–21
Mumbai (Bombay), India, 282, *290*, 291, 294
Murmansk, Russia, 80
Murray-Darling Basin, 375, *383*
Murray River system, 380
Musandam Peninsula, 259
Muslims, *see* Islam and Muslims
Myanmar, 344, 352, *356*, 361–363

Nagorno-Karabakh, 99
Nakhodka, Russia, 102
Namibia, 211, 215
Nasser, Lake, 246
Natal province (South Africa), 213
Nations, 36
Nationalism, 36, 284, 289, 294–295, 313, 332,
    338
Nation-states, 28, 36
Native Americans, 110, *111*, 128, 129
NATO, 93
Natural gas:
  Austral Realm, 380, 384, *385*
  East Asia, *322, 323*, 324

Natural gas (*continued*)
   Europe, 29, *56*, 57, *58*, 59
   and maritime boundaries, 393
   Middle America, 147
   North Africa/Southwest Asia, *230*, *241*, *245*, *248*, 262–263
   North America, 117, 124, 129
   Russia, *94*, 96, 97, 100–102, *103*
   South America, 174, *178*, 179, 180, *185*
   South Asia, 284, 293
   Southeast Asia, 365–366, 368, 370
   Subsaharan Africa, *220*
Natural hazards, 9–10, 136, *141*, 295. *See also specific hazards, e.g.:* Earthquakes
   Bangladesh, *296*
   Canada, *124*
   Caribbean Basin, 155
   East Asia, 303
   Honduras, 152
   Kenya, 217
   Lesser Antilles, 161
   Southeast Asia, 346
Natural landscapes, 9
Natural resources:
   Antarctica, 399
   East Asia, 336, 339, 340
   Europe, 29, 59, 70
   and maritime boundaries, 391–393
   North America, *106*, 129
   Russia, 81, 101, 102
   South America, 175, *185*, 189
   Subsaharan Africa, *192*, 210, 211, 221
Nauru, 391, 396
N'Djamena, Chad, 249
Negative population growth, 40. *See also* Population decline
Nei Mongol Autonomous Region, 320
Neocolonialism, 20
Nepal, 296–297
The Netherlands, 54, 139, 142, 173, 212, 350, 352
Netherlands Antilles, *141*, 142, 161
Nevada, 128
New Brunswick, 121, 122, 126
New Caledonia, 390, 395
New England, 113, 125–126
Newfoundland and Labrador, 121, 126
Newfoundland Island, 126
New France, 122
New Granada, 167, 353–354
New Guinea, 344, 376, 390, 395
New Hampshire, 125
New Mexico, 128
New South Wales, Australia, 379, 380
New world order, 2

New York City, New York, *105*, *113*, 120, 125
New Zealand, *374*, 375, 384–387
Nicaragua, 150, 152–153
Niger, 249
Niger-Congo languages, 206
Nigeria, 206–207, 222–225, 240
Niger River, 193, 249
Nile Basin, 233
Nile River, 193, 230, 233, 245–246
Nile Valley, 233
Nilo-Saharan languages, 206
Niue, 390
Nizhniy Novgorod, Russia, 95
Nobi Plain, 335, 337
Nok culture, 200
Nomadism, 262
North (Argentina), 182
North (Brazil), 189
North (South America), 173, 174–177
   Colombia, 174–176
   French Guiana, 177
   Guyana, 177
   population, *174*
   Suriname, 177
   Venezuela, 176–177
North Africa/Southwest Asia, 6, 228–267
   Arabian Peninsula, 257–259
   Chad, 249
   climate, *230*
   and climate change, 233–234
   cultural geography, 233
   culture hearths, 232–234
   desertification in, 11
   deserts in, 230
   Egypt and Lower Nile Basin, 245–248
   Empire States, 259–263
   Islam in, 234–239
   Maghreb, 248–249
   Mali, 249
   Mauritania, 249
   Middle East, 250–257
   names given to, 231
   Niger, 249
   oil issues in, 240–243
   physical geography, *228*
   political units, *244*
   population, *230*, *240*
   regions, 243, *244*, 245
   Turkestan, 263–267
North America, 104–131
   Canada, 120–124
   climate, 108
   Continental Interior, 126–127
   as core area, 19
   Core Region, 125

   energy resources, *117*
   French Canada, 126
   geographic and social contrasts, 107
   glaciation in, 10, 11
   indigenous peoples, 110
   Maritime Northeast, 125–126
   megalopolitan growth, *114*
   Northern Frontier, 129
   Pacific Hinge, 129–131
   physical geography, *104*, 107–109
   population, *106*, *113*, 127
   regions, 7, 125
   South, 127–128
   Southwest, 128
   United States, 111–120
   Western Frontier, 128–129
North American Free Trade Agreement (NAFTA), *106*, *124*, *135*, *141*, 146–148
North American Plate, 136
North Atlantic Current, 30
North China Plain, 317–319
Northeast (Brazil), 188
Northeast China Plain, 317–318
Northern California, 130
Northern England, 57
Northern (Nordic) Europe, *29*, 48, 58–60
   Denmark, 59
   Estonia, 59
   Finland, 59
   Iceland, 59–60
   Norway, 59
   population, 58, *58*
   Sweden, 59
Northern European Lowland (Europe), 31
Northern Frontier (North America), 129
Northern Ireland, 37, 46, 55, *56*, 57. *See also* United Kingdom (Britain)
Northern Mariana Islands, 396
Northern Territory (Australia), 379
North European Lowland, 67
North Indian Plain, 273
North Island (New Zealand), 384, 385
North Korea, 302, 304, 338–340
North Pole, *83*
North Sea, 57, 393
North West Frontier, Pakistan, 284
Northwest Passage, 82, 83
Northwest Territories, 121
Norway, 45, 58, 59
Nouakchott, Mauritania, 249
Nova Scotia, 121, 122, 126
Novokuznetsk, Russia, 100
Novorossiysk, Russia, 100
Novosibirsk, Russia, 100
Nubia, 202

Nuclear power:
   China, 331
   France, 36
   India, 293
   Iran, 262–263
   Korea, *338*
   North Korea, 339, *339*
   Sweden, 59
   Taiwan, 340
Nuevo León State (Mexico), 147
Nunavut, 121, 123, 129

Oasis, Libya, *229*
Oaxaca State (Mexico), 145
Ob River, 80, 101–102
Oceans, rights to, 5, 391–393
Ocussi, East Timor, 370
Offshore banking, 152, 161
Oil, 240, 242
   Austral Realm, 380
   East Asia, 322, 324
   Europe, 29, 36, *56*, 57, *58*, 59, *64*
   and maritime boundaries, 393
   Middle America, *143*, 146, 147, *149*
   North Africa/Southwest Asia, *230*, 240–243, 245, 247, *248*, 249, *251*, *252*, 253, *254*, 257–260, 262–265
   North America, 117, 124, 126, 129
   Pacific Realm, 395
   Russia, 81, *94*, 96, 97, *98*, 100–102, *103*
   sources of, *118*
   South America, 165, 174, 176–177, *178*, 179, 180, 182, *185*, 186
   South Asia, 293
   Southeast Asia, 365–366, 368, 370
   Subsaharan Africa, *220*, 221, 223
Oligarchs, 92, 181
Oman, 259
Ontario, 57, 121, 122
Open Cities, 326, 327
Open Coastal Areas, 326, 327
Opium Wars, 310–311
Orange Free State (South Africa), 213
Orange Revolution (Ukraine), 70
Orange-Vaal River, 210
Ordos Desert, 304
Øresund bridge-tunnel, 59
Organization for European Economic Cooperation (OEEC), 42, 50
Oriente (Ecuador), 179
Oriente (Peru), 179
Orinoco Basin, 164, 176
Orontes River, 253
Osaka, Japan, 337
Osh, Kyrgyzstan, 265

Ottoman Empire, 238–239, 260
Outback, 378
Overfishing, 60
Overseas Chinese, 326, 328, 349–350

Pacific Hinge (North America), 129–131
Pacific Northwest, 130–131
Pacific Northwest regional state, 124
Pacific Ocean, 390–392
Pacific Plate, 374
Pacific Realm, 388–398
    colonization and independence, 390–391
    geographic change drivers, 393
    marine geography, 391–393
    Melanesia, 395
    Micronesia, 396
    physical geography, 388
    Polynesia, 396–398
    population, 395
    regions, 394–395
Pacific Rim, 129, 302, 307
    Australia, 377, 382
    Austral Realm, 374
    Chile, 183
    China, 324–331
    East Asia, 302
    Jawa, 368
    Malaysia, 364, 365
    North America, 129
    Philippines, 371
    Russia, 102, 103
    Thailand, 359, 361
Pacific Rim region (China), 324–331
    current trends in, 330
    economy, 326–327
    geography of development, 325–326
    Guangdong Province, 328
    Hong Kong, 329
    Pearl River estuary, 328
    regional disparities, 330–331
Pacific Ring of Fire, 10, 346
Pakistan, 267, 276, 280–285
Palau, 391, 396
Palembang, Sumatera, 368
Pampa, Argentina, 181, 182
Panama, 136, 138, 142, 150, 153–154
Panama Canal, 141, 150, 153, 154
Panama City, Panama, 154
Pan American Highway, 153
Pandemics, 199
Pangaea, 9, 10, 194–195
Papua, 344, 353, 369, 391, 395
Papua New Guinea (PNG), 353, 384, 395
Paraguay, 169, 177, 180–181
Paraguay River, 180

Paraná-Paraguay Basin, 164, 170, 180
Paraná River, 180, 189
Paris, France, 27, 31, 40, 42, 51, 53
Patagonia, Argentina, 182
Patagonia Plateau, 164
Pattern (use of term), 3
Peace of Westphalia, 36
Pearl River Basin, 304, 317, 321–322
Pearl River estuary, 304, 328
Pearl River megalopolis, 328
Pennines mountains, 56
Pension system, Russia, 92
Peones, 144
People's Republic of China, see China
Perforated states, 355
Periodic markets, 225
Peripheral development, 19–20, 385
Peripheral squatter settlements, zone of, 173
Permafrost, 80
Persia, 261
Perth, Australia, 379
Peru, 167, 170, 177–179
Pest, Hungary (Budapest), 68
Petén region, 151
The Philippines, 22, 344, 347, 352–353, 370–371
Physical geography, 8–15. See also under spe-
        cific areas
    climate, 10–15
    natural hazards, 9–10
    natural landscapes, 9
Physiographic boundaries, 353
Physiographic provinces, 107
Physiography, 30
    Austral Realm, 374
    East Asia, 303–304, 324
    Europe, 30–31
    Middle America, 136, 142–144, 150
    North America, 107, 108
    regions, 77
    Russia, 77, 79–83
    South America, 175, 180, 188
    South Asia, 270, 271–273
    Southeast Asia, 344, 346
    Subsaharan Africa, 192, 193–195
Physiologic density, 277–278
Phytogeography, 376
Piedmont (Italy), 61
Piedmont (North America), 107
Pinang, Malaysia, 365
Pinyin system, 340n.
Pisco, Peru, 179
Pitcairn islands, 390
Plantations, 140
    Middle America, 139, 140, 142, 150, 152, 153,
        155, 157

South America, 167, 169, 174, 188
South Asia, 299
Southeast Asia, 368
Subsaharan Africa, 210
Plata estuary, 184
Plateau of Mexico, 143
Plateaus:
    Austral Realm, 374
    East Asia, 302–303
    Europe, 30
    Middle America, 136
    South America, 164
    South Asia, 273
    Subsaharan Africa, 193
Plate tectonics, 195. See also Tectonic plates
Pleistocene, 10, 11
Plural societies, 169. See also Cultural pluralism
Poland, 36, 45, 67–68
Polders, 54
Political geography, 18, 25
Polynesia, 390, 396–398
Population decline, 15, 28, 40, 76, 91, 93
Population density, 16
Population distribution, 11, 15–16. See also
        under specific areas
Population explosion, 278, 279
Population geography, 276–277
Population implosion, 6, 40, 47
Population pyramids, 280
Po River Basin, 62
Port-au-Prince, Haiti, 155, 159
Porto, Portugal, 27, 63
Portugal, 62, 63, 167, 174, 202, 204, 311
Post-industrial economy, United States as,
        119–120
Potosí, Bolivia, 180
Poverty:
    Europe, 62, 63, 70, 72–73
    and globalization, 21
    Middle America, 145, 148–149, 152–159
    North Africa/Southwest Asia, 246
    North America, 107, 127
    Russia, 92
    South America, 165, 172, 173, 177, 180, 181,
        186–188
    South Asia, 270, 281, 290, 295
    Southeast Asia, 357, 361
    Subsaharan Africa, 209, 216
Prague, Czech Republic, 68
Prairie Provinces (Canada), 57, 121, 126
Precipitation, 12–14
Pretoria, South Africa, 214
Primary activities, 117
Primate cities, 40, 55, 68, 120
Prince Edward Island, 121, 126

Productive activities, 117
Protestantism, 232
    Europe, 37, 55, 57
    North America, 116
    South America, 184
    Southeast Asia, 369, 370
    Subsaharan Africa, 207
Protruded states, 355
Pudong SEZ, 22, 23, 327
Puebla, Mexico, 144
Puerto Rico, 142, 155, 156, 160–161
Pull factors, 112, 143, 172
Punjab (India), 275, 276, 282, 284, 288
Punjab Province (Pakistan), 284
Puntland, 227
Push factors, 112, 143, 172
Putrajaya, Malaysia, 364
Pyapon, Irrawaddy Delta, Myanmar, 343
Pyongyang, North Korea, 339
Pyrenees, 32, 62

Qaidam Basin, 322
Qanats, 262, 324
Qatar, 259
Qingdao, China, 311, 330
Qinghai-Xizang Plateau, 303, 322
Quaternary activities, 117
Quaternary sectors, 124
Quebec (Canadian province), 106, 121–123, 126
Quebec City, Quebec, 123
Queensland, Australia, 379, 380
Quinary activities, 117

Rainforests, 169
    in interglacials, 11
    South America, 167, 169, 189
    Southeast Asia, 346, 347
    Subsaharan Africa, 195, 216
Rain shadow effect, 109
Randstad, Netherlands, 54
Rangoon, Myanmar, 362
Rarotonga, 398
Rate of natural population change, 278
Red Sea, 195
Refugees:
    North Africa/Southwest Asia, 249, 253, 255
    South Asia, 276, 282, 284–285
    Subsaharan Africa, 192, 224
Regions:
    climatic, 12–15
    formal, 8
    functional, 8
    geographic, see Geographic regions
Regional complementarity, 200, 339
Regional concept, 7

Regional disparity, 20
Regional geography, 5, 17–18, 25
Regionalism, Canada, *124*
Regional states, 47, 124, 328
Relative location, 8, 30
    China, 307
    and economic development, 19
    France, 53
    Japan, 336
    Moçambique, 216
    Philippines, 370
    and population distribution, 15
    Singapore, 366
    Uganda, 218
Relict boundaries, 354, *354*
Relief (elevation), 80
Religions, *232*. See also *specific religions*
    Europe, 37, 39, 55, 59, 60, 69, 70
    Middle America, 138
    North Africa/Southwest Asia, *230*, 231, 253
    North America, 116
    Russia, 92
    South America, 185
    South Asia, *270*
    Southeast Asia, *356*, 369
    Subsaharan Africa, 201, 207–208
Religious revivalism, 237
Relocation diffusion, 234–235, 305
Remittances, 151, 152, 160, 294, 391
Renaissance, 32
Republic of China, *see* Taiwan
Republic of Ireland, 55, 57–58
Republic of the Congo, *see* Congo
Republic of Transdniestria, 70
Reykjavik, Iceland, 60
Rhode Island, 126
Rhône-Alpes Region, 47
Rift valleys, 195
Rimland, 139–140, 151
Rio de Janeiro, Brazil, 188
Rio Ebro, 62
Rivers, 9
    East Asia, 303–304
    Europe, 30, 32, 53
    North Africa/Southwest Asia, 233
    South America, 164, 170
    South Asia, *270*, 273
    Southeast Asia, 346
    Subsaharan Africa, 193, 195, 210
Rocky Mountains, 107, 128
Roman Catholicism, *232*
    Europe, 37, 55, 57, 66, 70
    Middle America, 138
    North America, 116
    South America, 184, 185

Southeast Asia, 369, 370
    Subsaharan Africa, 207
Romance languages, 32
Romania, 43, 70
Rotterdam, Netherlands, 54
Ruhr industrial complex, 49, 50
Rural-to-urban migration, 171, 172
Russia, 74–103
    changing role of, 76–77
    and China, 311, 312, 317–318
    climate, 78–79, 102
    Core and Peripheries, 94–100
    culture, 83
    demographic changes, 93
    Eastern Frontier, 100–101
    economy, 76, *78*, 85, 88–90, 102
    ethnicity, 77, 78, 84–87, 97, 99, 102
    and Europe, 28–29, 93, 94
    Far East, 102–103
    historical geography, 78, 83–90, *84–86*
    human geography, 102
    and Kazakhstan, 263, 264
    oil, 240
    physical geography, *74*, 77–83
    political geography, 82–83, 85–88, 90–92, 102
    population, 79, *89*, *91*, 92, 93, 96, *98*, *103*
    regions, 95, *95*
    Siberia, 101–102
    social geographies, 92–93
    trade, 82–83
    transition zones, 77–89
Russian Plain, 80
Russification, 84, 88
Rustbelt, 119
Rwanda, 218, 220
Ryukyu Islands, 311

Sabah, 364, 365
Sahara Desert, 11, 14, 192, 194, *229*, 230
Saharan Atlas Mountains, 248
Sahel region, West Africa, 192, *197*, 222
St. Lawrence River, 126
St. Lucia, 161
St. Petersburg, Russia, 83, 95
St. Vincent, 142
Saint Dominique, 159
Saint John-Halifax crescent, 121
Sakartvelos, 99
Sakha (Yakut Republic), 102
Sakhalin, 102
Salta, Argentina, 182
Salvador, Bahia State, Brazil, *188*
Samoa, 398
San Andreas Fault, 130

San Francisco, California, 120
San Francisco Bay Area, 130
San Joaquin-Sacramento Valley, 108
San Juan, Puerto Rico, 155, *161*
San Marino, 47
Santa Clarita, California, *130*
Santa Cruz Province (Bolivia), 180
Santa Fe, Argentina, 182
Santa Fe, New Mexico, *128*
Santiago, Chile, 173, 182
Santo Domingo, Dominican Republic, 155
São Paulo, Brazil, *163*, 184, *187*, 188
São Paulo State (Brazil), 184, 188
São Tomé and Príncipe, 219, 221
Saratov, Russia, 96
Sarawak, 364, 365
Sardinia, 61
Saskatchewan, 121, 124
Satellites, information from, 2–3
Saudi Arabia, 240, 257–259
Savanna, 14
Saxony, 49, 50
Scale, maps, 3–4
Scandinavian Peninsula, 58
Schengen Agreement, 48
Scotland, 46, 55, 57. See also United Kingdom (Britain)
Scottish Lowlands, 57
"Scramble for the oceans," 391, 393
Seas, claims to, 391–393
Sea hemisphere, 390
Sea-level rise, 11, *54*, 298, *393*, *396*
Sea of Japan (East Sea), 391
Sea-surface temperatures, *396*
Secondary activities, 117
Second World countries, 19
Sectors, Arctic, 83
Segregation, 115. See also Apartheid
Senegal, 224
Seoul, South Korea, 339
Separate development, 212, 213
Serbia, 18, 66, 71, 72
Serbia-Montenegro, 36*n*. 1, 72
Shaanxi Province (China), 314
Shamanist religions, *232*
Sha Mian Island, 312
Shandong Peninsula, 304
Shandong Province (China), 315
Shanghai, China, 314, 330
Shan Plateau, 362
Shantou SEZ, 326
Sharia, *207*, *223*, 224, 284
Shatter belt, 65, 344
Shenzhen SEZ, 326–329
Shi'ite Muslims, 236, 237, 251, 259, 284

Shikoku island, Japan, 334
Shining Path movement, 179
Shintoism, *232*
Shipbuilding, 57, 59
Shipping, 154, 366, 392
Siberia, 8, 80, *91*, 101–102
Sichuan (Red) Basin, 304, 317, 346
Sichuan Province (China), 9, *10*, 303, 315, 316, 321
Sicily, 61, 62
Sierra Leone, 222, 224
Sierra Madre Occidental, 142
Sierra Madre Oriental, 142, 143
Sierra Maestra, 157
Sierra Nevada, 108, 128
Sikhs, *232*, 276, 288
Silesia, Poland, 49, 50, 67
Silicon Valley, 120
Sinai Peninsula, 136, 242, 254
Sinai region, 246
Sind Province (Pakistan), 273, 284
Singapore, 129, 327, 354, *356*, 364, 366–368
Sinicization, 308
Sino-Tibetan languages, *274*, *316*
Site, 51
Situation, 51
Sjaelland, 59
Slavery and slave trade:
    Amerindians, 167
    Caribbean Basin, 139, 156
    North Africa/Southwest Asia, 234
    North (Caribbean) South America, 174
    South America, 167–168
    Subsaharan Africa, 202–203
Slovakia, 43, 68
Slovenia, 36*n*. 1, 43–45, 71
Small-scale maps, 3
Social landholdings, Mexico, 145
Social stratification, 273
Sofala, 202
Soils:
    Austral Realm, 385
    and climatic regions, 14
    East Asia, 304, 305
    Europe, 29
    in interglacials, 11
    Middle America, 135, 145, 150, 151
    North America, 109
    Pacific Realm, 396
    South America, 186
    South Asia, 284
    Southeast Asia, 346, 347, 370
    Subsaharan Africa, 195
Solomon Islands, 384, 391, 395
Somalia, 226, 227

Somaliland, 227
Somali Plate, 195
Songhai, 201
Songhua River, 317
Soputan, Mount, *346*
South (Brazil), 188–189
South (North America), 127–128, 173
South (South America), 181–184
   Argentina, 181–182
   Chile, 182–184
   population, 171, *181*
   Uruguay, 184
South Africa, *208*, 210–215
   apartheid, 212, 213
   economy, 214, 215
   ethnicity, 212–213
   in IBSA, 20
   population, *213*, 276–280
   provinces, 213
South America, 6, 162–189
   Argentina, 181–182
   Bolivia, 180
   Brazil, 184–189
   Chile, 182–184
   Colombia, 174–176
   cultural fragmentation in, 168–170
   economic integration in, 170
   Ecuador, 179–180
   French Guiana, 177
   geographic change drivers, *173*
   Guyana, 177
   human sequence in, 165–168
   independence movements, 168
   North (Caribbean), 174–177
   Paraguay, 180–181
   Peru, 177–179
   physical geography, *162*
   political units, 168
   population, *164*, *165*, 168, *171*
   regions, 168, 173–174
   South (mid-Latitude), 181–184
   Suriname, 177
   urbanization in, 170–173
   Uruguay, 184
   Venezuela, 176–177
   West (Andean), 177–181
South American Plate, 136
South Asia, 6, 268–299
   Bangladesh, 295–296
   Bhutan, 297–298
   geographic change drivers, *281*
   historical geography, 271
   human sequence, 273–276
   India, 285–295
   Maldives, 298

modern invasions in, 281–282
   Nepal, 296–297
   Pakistan, 282–285
   physical geography, *268*, *270*, 271–273
   population, 16, *270*, 276–280
   poverty, 281
   regions and subregions, 270, *271*
   Sri Lanka, 298, 299
South Australia, 379, 380
South China Sea, 391, 393
Southeast (Brazil), 188
Southeast Asia, 342–371
   boundaries, 18
   Brunei, 365–366
   colonialism, 345–346, 350–353
   East Timor, 370
   geographic change drivers, *356*
   Indochina, 356–357
   Indonesia, 368–370
   Insular, 364–371
   Mainland, 356–363
   Malaysia, 364–365
   Myanmar, 361–363
   the Philippines, 370–371
   physical geography, *342*, *344*, 346–347
   political geography, *344*, *345*, 353–356
   population, 347–350
   regions, 356
   Singapore, 366–368
   states, 355–356
   Thailand, 359–361
Southern Africa, 200, 210–216
   Angola, 216
   Botswana, 215
   Lesotho, 215
   Malawi, 216
   Moçambique, 216
   Namibia, 215
   natural resources, 210, 211
   population, *211*
   South Africa, 212–215
   Swaziland, 215
   Zambia, 216
   Zimbabwe, 215
Southern Alps, 384, 386
Southern California, 130
Southern Cone, *see* South (South America)
Southern England, 57
Southern Europe, *see* Mediterranean Europe
Southern Highlands (Mexico), 146
Southern Ocean, 375–376, 398, 399
Southern Utah, *105*
South Island (New Zealand), 384–386
South Korea, 129, 304, 327, 332, 337–339
South Pole, 398

South Tyrol, 46
Southwest (North America), 128
Soviet Socialist Republics (SSRs), 87, 88
Soviet Union, *see* Union of Soviet Socialist
   Republics
Soybeans, *169*, 189
Spain, 36, 60, 62–63
   in Africa, 204
   in Cuba, 157
   devolution, 46
   growth center in, 47
   in Hispaniola, 158–159
   in Middle America, 137–139
   in Peru, *178*
   in South America, 166–168, 174
   in Southeast Asia, 350, 352
Spatial diffusion, 234–235
Spatial interaction, 39
Spatial issues, Japan, 334–336
Spatial perspective, 3
Spatial systems, 8
Special Administrative Region (SAR), China,
   315
Special Economic Zones (SEZs), China,
   326–327, *329*, 330, 358
Spice Islands, 347, 352
Spine ('Latin' American city model), 173
Spratly Islands, 371, 393
Sri Lanka, 274, 277, *281*, 298, 299
States, 7*n*. 1, *17*, 18
   failed, 180
   insurgent, 175
   origin of, 233
   political entities vs., 302
   territorial morphology, 355–356
   unitary, 379
State capitalism, 331, 339
State formation, 202, 305
Stateless nations, 250, 255
Steppes, 14
Subnational entities, 6
Subregions, 5
Subsaharan Africa, 6, 190–227
   African Transition Zone, 225–227
   climate, 195–196
   cultural geography, 205–208
   East Africa, 216–219
   economies, *192*
   Equatorial Africa, 219–221
   geographic change drivers, *208*
   health issues, 198–199
   historical geography, 199–205
   land issues, 196–198
   Madagascar, 218–219
   natural environments, 195–196

physical geography, *190*, 193–195
   political geography, 208
   population, 196, 209
   regions, 209, *210*
   Southern Africa, 210–216
   supranationalism, 208
   urbanization, 209
   West Africa, 221–225
Subsequent boundaries, 353, *354*
Subsistence agriculture, 169, 179, 197–198, 246,
   296
Subtropical Convergence, 375–376
Suburbs, in United States, 115, 129
Sudan, 16–17, *208*, 225, *245*, 247–248
Suez Canal, 245
Sulawesi, 368–369
Sulu Archipelago, 370
Sumatera, 346, 368
Sunbelt, 112
Sunni Muslims, 236, 237, 251, 259, 284
Superimposed boundaries, 353–354
Superpowers, *6*
   Britain, 36
   China, 331–332
   and Eastern Europe, 66
   Soviet Union, 7
   United States, 7
Supranationalism, 43
   Europe, 42–43, *44*, 47
   South America, 170
   Subsaharan Africa, 208
Suriname, 177
Swaziland, 215
Sweden, 36, 44, 59
Switzerland, 45, 54, 55
Sydney, Australia, 378, 379, 383
Sydney Harbor, Australia, 373
Syria, 250, 253
Systematic geography, 25

Tahiti, 398
Taiga, 80
Taipei, Taiwan, 340
Taiwan, 302, *304*, 340–341
   economic development, 327
   Jakota Triangle, 332
   Pacific Rim trade, 129
   and Panama, 154
Tajikistan, 265
Taklamakan Desert, 324
Talek district, Kenya, *191*
Taliban regime, 285
Tallinn, Estonia, 59
Tamil Tigers, 299
Tampa, Florida, 127

Tampere, Finland, 59
Tanzania, 218
Tarim Basin, 324
Tar sands, Alberta, *117*
Tashkent, Uzbekistan, 264
Tasmania, Australia, 379, 380, 383, 384
Tasman Sea, 378
Tatarstan, 90
Tbilisi, Georgia, 99
Technology transfers, *208*
Technopoles, 120, 125, 147, 307, 364
Tectonic plates, 9
    Africa, 194–195
    Austral Realm, 374
    East Asia, 302–303
    Mid-Atlantic Ridge, 60
    Middle America, 136
    South Asia, 271
    Southeast Asia, 347
    West (South America), *179*
Tehran, Iran, 262
Tehuantepec, Isthmus of, 142
Tell Atlas Mountains, 248, 249
Temperatures, 12–14
Temple of Heaven, Beijing, China, *309*
Temporal perspective, 3
Tenochtitlán, 137
Teotihuacán, 137
Territorial morphology, 355–356
Territorial seas, 391–393
Terrorism, *100*
    Austral Realm, 384
    East Asia, 312, 314
    North Africa/Southwest Asia, 237, 249, 250,
        253, 258, 267
    Russia, *94*, 99
    and Russian-Western relations, 94
    South America, 175, 181
    South Asia, 281, 284–285
    Southeast Asia, *356*, 358, 360, 369
    Subsaharan Africa, 204, 217, *225*
Tertiary activities, 117
Tertiary sectors, 124
Thailand, 129, 345, 347, *356*, 359–361
The Congo (Democratic Republic of the Congo),
    202*n*, 219–220
Thimphu, Bhutan, 297
Third World countries, 18–19
Three Gorges Dam, *301*, 321
Tianjin, China, 314, 319
Tian Shan mountains, 324
Tibet, *see* Xizang
Tierra caliente, 150, 179
Tierra fría, 150, 179
Tierra helada, 150, 179

Tierra nevada, 150
Tierra templada, 150, 153, 175
Tigris-Euphrates Basin, 230, 233
Tigris River, 251
Tijuana, Mexico, *142*, 144
Timbuktu, Mali, 200, *200*
Timor, 368, 370
Timor-Leste, *see* East Timor
Timor Sea, 393
Tinerhir, Morocco, *231*
Titicaca, Lake, 180
Titicaca Basin, 180
Tohono Ó'odham nation, *111*
Tokelau Islands, 390
Tokyo, Japan, *1*, 334, 337
Toluca, Mexico, *20*
Tomsk, Russia, 100
Tonga, 398
Tonkin, 357
Tonle Sap, 359
Topography, 80
Tornadoes, 109
Tourism:
    Antarctica, 399
    Austral Realm, 374, 380, *386*
    Europe, 63, 65
    Middle America, 149, 151–153, 156, 158, 160,
        161
    North Africa/Southwest Asia, 247
    North America, 126
    Pacific Realm, 395, 396
    South America, 184
    South Asia, 294, 297, 298
    Southeast Asia, 359, 361, 369
    Subsaharan Africa, *216*, 217, 218
Towns, Spanish, 138
Trade. *See also* Slavery and slave trade
    Austral Realm, 377, 380
    and core-periphery dichotomy, 20
    East Asia, 307–309, *338*, 340
    economic unions for, 23
    Europe, 30, 54, 59, 62, 63, 68, 70
    and globalization, 22–23
    Middle America, 136–137, 146–150, 152, 154,
        155, 157, 160
    North Africa/Southwest Asia, 247
    North America, 106, 124, 128, 129, 131
    Pacific Realm, 395
    Russia, 82–83, 95, 102
    South America, *169*, 170, 176, 180, 183, 184,
        186, 187
    South Asia, 275
    Southeast Asia, 347, 352, 366, 368
    Subsaharan Africa, 200, 223
Trade routes, 82

Transcaucasia, 99–100
    Armenia, 99
    Azerbaijan, 99–100
    Georgia, 99
Transcaucasus, 77
Transculturation, 144
Transdniestria, Republic of, 70
Transferability, 39
Transition zones, 6, 18
    Italy, 61–62
    Kazakhstan, 264
    Mexico, 146
    North Africa/Southwest Asia, *230*
    Russia, 77–89
    Southeast Asia, *345*
Transjordan, 253, 254
Transmigration, 369
Transnational corporations, 20–21
Transoceanic Highway (Peru), 179
Transportation:
    ancient Rome, 32
    Austral Realm, *378*
    East Asia, 307, *318*, 321, *322*, 324, *328*, *329*,
        330, *333–335*, *338*, *340*
    Europe, *29*, 36, *49*, *55*, *56*, *58*, 59, *60*, 62, *64*,
        66, *67*
    Middle America, *134*, 137, 149, 154, *156*, *159*,
        160
    North Africa/Southwest Asia, *245*, *248*, *251*,
        252, *254*, *256*, 257, *258*, 260, *264*, *266*,
        267
    North America, *112*, 114, 115, *120*
    Russia, *89*, 95, 96, *98*, 101–103
    South America, 170, 174, *178*, 179, *181*, 182,
        184, *185*, *189*
    South Asia, 276, *283*, *286*, 291, 293, *295*, *298*
    Southeast Asia, *345*, *357*, 358, 359, *360–362*,
        361, *365*, *366*, 370
    Subsaharan Africa, 208, *211*, *213*, *217*, 219,
        220, *222*, *223*, *226*
Trans-Siberian Railroad, 102, 103
Transvaal Province (South Africa), 213
Tree line, *101*
Tribal Areas, Pakistan, 284, 285
Tricultural region, 128
Trinidad, 156
Trinidad and Tobago, 142
Triple Frontier, 180, 181
Tropical deforestation, 153
Tropical rainforests, *see* Rainforests
Tsunamis, 298, 346, 359, 368
Tucumán, Argentina, 182
Tula, Russia, *76*, 95
Tundra, 14, 80
Tunisia, 247, 249

Turin, Italy, 61
Turkestan, 263–267, 312
    Afghanistan, 265–267
    Kazakhstan, 263, 264
    Kyrgyzstan, 265
    population, *264*
    Tajikistan, 265
    Turkmenistan, 265
    Uzbekistan, 264
Turkey, *47*, 260–261
    and Cyprus, 65
    and European Union, 45
    Kurds, 250
    language, 231
Turkish Republic of Northern Cyprus, 65, 261
Turkmenistan, 265
Turku, Finland, 59
Tuscany, 61
Tuvalu, 391, 398

Uganda, 218
Ukraine, 45, 69–70, 86, 96
Ulsan City, South Korea, 339
Union of South American Nations (UNASUR),
    170
Union of Soviet Socialist Republics (USSR),
    84–85, *88*
    Cold War, 331
    command economy, 89
    historical geography, *84–85*
    and Iron Curtain, 28
    republics of, 88
    as superpower, 7
Unitary states, 91, 379
United Arab Emirates (UAE), 240, 242, 259
United Kingdom (Britain), 36, 57, 334
    in Africa, 204
    areas encompassed by, 55
    and Australia, 379, 384
    in Canada, 122
    in Central America, 151
    in China, 312
    in Common Market, 43
    Cyprus, 65
    devolution, 45–46
    and EMU, 44
    Hong Kong, 329
    and industrialization, 33
    in Malaysia, 364
    in Middle America, 139, 142
    in New Zealand, 386
    Opium Wars, 310–311
    in Pacific Realm, 390, 391
    religion, 41, 42
    in South Africa, 212

in South America, 173
in South Asia, 275–276
in Southeast Asia, 350, 352
in Subsaharan Africa, 218, 221, 222
United National Conference on the Law of the
    Sea (UNCLOS), *390*, 391–393
United States, 111–120
    and Afghanistan, 265
    in Africa, 204
    and Azerbaijan's oil, 100
    boundary of, 6
    and Canada, 106
    in Caribbean Basin, 139
    and Chile, 183
    and China, 313, 331–332
    cities, 115
    climates, 307–308
    Cold War, 331
    contrasts between Canada and, 107
    cultural geography, 115–116
    and drug production, 175
    economic geography, 116–120
    and El Salvador, 152
    European reconstruction, post-WWII, 42
    and globalization, 22
    and India, *281*
    indigenous peoples, 110
    industrial urbanization, 113–115
    and Mexico, 142, 149
    in Micronesia, *390*
    in Middle America, 142
    in Nicaragua, 152
    oil, 240, 242
    in Pacific Realm, 390, 396
    Pacific Rim trade, 129
    and Pakistan, 285
    in Panama, 153
    population, 107, 111–113, 128
    and Puerto Rico, 160–161
    and Saudi Arabia, 258
    and scramble for the oceans, 391
    in Southeast Asia, 350, 352–353
    subsidies in, 19
    as superpower, 7
    and Venezuela, 177
    War in Iraq, 20
U.S. Virgin Islands, 161
Upper Canada, 122
Uralic languages, *38*
Ural Mountains, 28, 78, 80
Urals Region, *91*, 96
Urban agriculture, *172*
Urban geography, 25
Urbanization, 16, 25, 100, 170–171
    ancient Rome, 32

Austral Realm, 379–380, 385
East Asia, 314, 330, *331*
Europe, 39, 40, 53, 61
Middle America, 138, 143–144, 146, 155
North Africa/Southwest Asia, 233
North America, 106, 107, 113–115, 126
South America, *164*, 170–172, 182, 185
South Asia, 290–291
Southeast Asia, 359
Subsaharan Africa, 209, 213
Urban planning; *51*
Uruguay, 167, 171, 184
Ürümqi, Xinjiang, 324, *325*
Utah, *105*, 128
Uzbekistan, 264

Vakhan Corridor, 265
Valle Central (Costa Rica), 153
Valley of Mexico, 136–137, 143
Valparaíso, Chile, 182–183
Vancouver, British Columbia, 121, 131
Vanuatu, 391, 393
Varna, Bulgaria, 70
Vegetation, 109, *130*, 376, 377
Veneto, 61
Venezuela, 165, 167, 174, 176–177, 240
Venezuelan Highlands, 176
Venice, Italy, 61
Vermont, 125
Victoria, Australia, 379, 380
Victoria Railway Station, Mumbai, India, *275*
Vienna, Austria, 55
Vietnam, 147, 347, 351, 353–355, 357–358
Vilnius, Lithuania, 68
Visayan islands, 370
Vladivostok, Russia, 92, 102, 103
Volcanoes, 9–10
    Austral Realm, 374
    East Asia, 334
    Middle America, 135, 136, 155, 161
    Pacific Realm, 394, 396
    Russia, 80
    South America, *179*
    Southeast Asia, 346, 368
    Subsaharan Africa, 195, 210
Volga Region (*Povolzhye*), *91*, 95–96
Volga River, 95, 96
Volgograd, Russia, 96
Von Thünen rings, 33, *34*, 118
Vostok, Lake, 398

Wahhabism, 237
Wake Island, 390
Wales, 46, 55, 57. *See also* United Kingdom
    (Britain)

Wallace's Line, 347, 376
Walloon region, Belgium, 54
War in Iraq, 250–251, 253
War on Terror, 226, 227, 267, 281, 285, 360, 370,
    384
Washington, D.C., 120, 125
Water highways, 170
Water resources:
    China, 321
    East Asia, 304
    North Africa/Southwest Asia, *230, 231*
    Saudi Arabia, 258
    South Asia, *296*
Weather, 12, 14, 79, 109. *See also* Climate(s)
Weber's Line, 376
West (South America), 173, 177–181
    Bolivia, 180
    Ecuador, 179–180
    Paraguay, 180–181
    Peru, 177–179
    population, 171, *178*
West Africa, 200, 221–225
    Burkina Faso, 224
    colonialism, 203
    desertification, *197*
    Gambia, 224
    Ghana, 224
    Ivory Coast, 224
    Liberia, 224
    Nigeria, 222–224
    population, *222*
    Senegal, 224
    Sierra Leone, 224
West Bank, 253–255, *256*
Westermann Islands, 60
Western Australia, 379, 380
Western Desert (Egypt), 246
Western Europe, *29*, 48–55
    Alpine states, 54–55
    Benelux, 54
    economic development in, 19, 20
    France, 51–52
    Germany, 49–51
Western Frontier (North America), 117,
    128–129
Western Ghats, 273
Western Plateau and Margins (Australia), 375
Western Uplands (Europe), 31
West Germany, 18, 49–51
West Malaysia, 364
Westphalia, Treaty of, 36
West Siberian Plain, 80
West Wind Drift, 376
White Nile, 245, 247
Wildfires, *31, 130*

Wildlife:
    Arctic, 81
    Australia, 376, 377, 384
    Southeast Asia, 347, 359
    Subsaharan Africa, 195, 196
Willamette Valley, 108
Windward Islands, 161
Windward Passage, 158
Wisconsinan Glaciation, 11
World Bank, 19
World Trade Organization, 20–23, 358
World War II, 33, 56, 58, 64, 65, 67, 336, 353
Wroclaw, Poland, 67
Wuhan, China, 321
Wyoming, 128

Xia Dynasty, 305
Xiamen SEZ, 326
Xian, China, 306
Xiaolangdi Dam, 318
Xi Jiang (West) River, 304
Xingu River, 166
Xinjiang Uyghur Autonomous Region (China),
    263, 306, 323–324, *325*
Xi (West) River Basin, 317, 321–322
Xizang (Tibet), 303, 306, 307, 317, 322–323

Yakut Republic, 102
Yakutsk, Russia, 102
Yakutsk Basin, 80
Yangon, Myanmar, 362, *363*
Yangzi Basin, 320–321
Yangzi River, 304
Yaroslavl, Russia, 95
Yemen, 259
Yenisey River, 101–102
Yerevan, Armenia, *75*
Yi-Luo River Valley, 305
Yucatán Peninsula, 142
Yucatán State (Mexico), 146, 147
Yugoslavia, 36, 66, 70–71
Yukon territory, 121
Yunnan Plateau, 304
Yunnan Province (China), 330–331

Zagros Mountains, 261
Zambezi River, 210
Zambia, 208, 210, 216
Zanzibar, 203
Zeeland province (Netherlands), *54*
Zhuhai SEZ, 326, 328
Zimbabwe, 208, 210, 215
Zoogeography, 376
Zuider Zee, *54*
Zürich, Switzerland, 55

LIST OF MAPS AND FIGURES

Front Endpapers   States of the World, 2009

INTRODUCTION

G-1   World Regional Geography: Global Perspectives (p. 1)

G-2   Effect of Scale (p. 4)

G-3   World Geographic Realms (pp. 5)

G-4   World Map Circa 1492 (p. 7)

G-5   World Tectonic Plates (p. 9)

G-6   Recent Earthquakes and Volcanic Eruptions (p. 10)

G-7   Extent of Glaciation During the Pleistocene (p. 11)

G-8   Average Annual Precipitation of the World (p. 12)

G-9   World Climates (p. 13)

G-10 World Population Distribution (p. 15)

G-11 States and Economies of the World, 2007 (p. 17)

G-12 World Geographic Realms and Their Constituent Regions (p. 24)

G-13 Diagram: The Relationship Between Regional and Systematic Geography (p. 25)

CHAPTER 1

1-1   Europe (p. 26)

1-2   Europe (p. 29)

1-3   Europe's Climates (p. 30)

1-4   Europe's Physical Landscapes (p. 32)

1-5   Von Thünen Rings: Europe (p. 34)

1-6   Europe: Industrialization (p.35)

1-7   Languages of Europe (p. 38)

1-8   Europe Population Distribution: 2009 (p. 40)

1-9   Europe: Muslims as Percentage of National Population, 2008 (p. 41)

1-10 European Supranationalism (p. 44)

1-11 Europe: Foci of Devolutionary Pressures, 2008 (p. 46)

1-12 Regions of Europe (p. 48)

1-13 Western Europe (p. 49)

1-14 States (*Länder*) of Reunified Germany (p. 50)

1-15 Site and Situation of Paris, France (p. 52)

1-16 Regions of France (p. 53)

1-17 The British Isles (p. 56)

1-18 Northern (Nordic) Europe (p. 58)

1-19 Mediterranean Europe (p. 60)

1-20 Regions of Italy (p. 61)

1-21 Autonomous Communities of Spain (p. 63)

1-22 Northeastern Mediterranean (p. 64)

1-23 Former Eastern Europe (1919–1991) (p. 66)

1-24 Changing Eastern Europe (p. 67)

1-25 Ethnic Mosaic of Eastern Europe (p. 69)

1-26 Serbia and Its Neighbors (p. 71)

CHAPTER 2

2-1   Russia (p. 74)

2-2   Russia: Relief and Physiographic Regions (p. 77)

2-3   Climates of Russia and Neighboring States (p. 79)

2-4   The Arctic: Ice Cap Melting (p. 81)

2-5   The Arctic: Geopolitical Situation (p. 82)

2-6   Russian/Soviet Historic Time Line (pp. 84–85)

2-7   Growth of the Russian Empire (p. 86)

2-8   Peoples of Russia (p. 87)

2-9   The Former Soviet Empire (p. 88)

2-10 The Russian Realm (p. 89)

2-11 Russian Federal Districts and Population Decline (p. 91)

2-12 Russian Realm Population Distribution: 2009 (p. 92)

2-13 Regions of the Russian Realm (p. 95)

2-14 Russia's Manufacturing Regions (p. 96)

2-15 Oil and Gas Regions of Russia and Its Western and Southern Margins (p. 97)

2-16 Southern Russia: Interior and Exterior Peripheries (p. 98)

2-17 The Russian Far East (p. 103)

CHAPTER 3

3-1   North America (p. 104)

3-2   North America: Physiography (p. 108)

3-3   North America: Indigenous Domains (p. 110).

3-4   The United States of America (p. 112)

3-5   North America: Population Distribution: 2008 (p. 113)

3-6   North American Manufacturing (p. 114)

3-7   North America: Megalopolitan Growth (p. 114)

3-8   Diagram: Model of a U.S. Metropolis (p. 115)

3-9   Spatial Distribution of U.S. Ethnic Minorities (p. 116)

3-10 North America: Major Deposits of Fossil Fuels (p. 117)

3-11 Farm Resource Regions (p. 118)

3-12 Internet Backbone Cities of the United States (p. 119)

3-13 Canada (p. 120)

3-14 Percentage of Canada's Population in 1996 Whose Mother Tongue Was French (p. 121)

3-15 Regions of the North American Realm (p. 125)

CHAPTER 4

4-1   Middle America (p. 133)

4-2   Regions of Middle America (p. 134)

4-3   Middle America Population Distribution: 2009 (p. 135)

4-4   Idealized Layout and Land Uses in a Colonial Spanish Town (p. 138)

4-5   Amerindian Languages of Middle America (p. 138)

4-6   Caribbean Region: Colonial Spheres ca. 1800 (p. 139)

4-7   Middle America: Mainland and Rimland (p. 140)

4-8   Mexico (p. 143)

4-9   States of Mexico (p. 144)

4-10 Regions of Mexico (p. 146)

4-11 Mexico: Economic and Political Disparities (p. 148)

4-12 Central America (p. 149)

4-13 Diagram: Altitudinal Zones (p. 150)

4-14 Cuba (p. 156)

4-15 Hispaniola: Haiti and Dominican Republic (p. 159)

4-16 Puerto Rico (p. 160)

CHAPTER 5

5-1   South America (p. 162)

5-2   South America Population Distribution: 2009 (p. 165)

5-3   Indigenous and Colonial Domains of South America (p. 166)

5-4   South America: Political Units and Modern Regions (p. 168)

5-5   South America: Agricultural Systems (p. 169)

5-6   South America: Dominant Ethnic Groups (p. 170)

5-7   Population Cartogram of South America (p. 171)

5-8   Diagram: A Generalized Model of Latin American City Structure (p. 172)

5-9   The North: Caribbean South America (p. 174)

5-10 The West: Andean South America (p. 178)

5-11 The Southern Cone: Mid-Latitude South America (p. 181)

5-12 Brazil: Subregions, States, and Resources (p. 185)

5-13 Brazil: The Geography of Wealth (p. 186)